GLOSSARY OF BIOTECHNOLOGY AND AGROBIOTECHNOLOGY TERMS

5TH EDITION

GLOSSARY OF BIOTECHNOLOGY AND AGROBIOTECHNOLOGY TERMS

5TH EDITION

KIMBALL NILL

CRC Press
Taylor & Francis Group
Boca Raton London New York

CRC Press is an imprint of the
Taylor & Francis Group, an Informa business

CRC Press
Taylor & Francis Group
6000 Broken Sound Parkway NW, Suite 300
Boca Raton, FL 33487-2742

Printed on acid-free paper
Version Date: 20160502

International Standard Book Number-13: 978-1-4987-5820-8 (Hardback)

Library of Congress Cataloging-in-Publication Data

Names: Nill, Kimball R., author.
Title: Glossary of biotechnology and agrobiotechnology terms / Kimball Nill.
Other titles: Glossary of biotechnology terms
Description: Fifth edition. | Boca Raton : Taylor & Francis, 2016. | Includes index. | Preceded by Glossary of biotechnology terms / Kimball Nill. 3rd ed. c2002.
Identifiers: LCCN 2016007230 | ISBN 9781498758208 (alk. paper)
Subjects: LCSH: Biotechnology--Dictionaries. | Agriculture--Dictionaries. | MESH: Biotechnology | Agriculture | Dictionary
Classification: LCC TP248.16 .F54 2016 | NLM TP 248.16 | DDC 660.03--dc23
LC record available at http://lccn.loc.gov/2016007230

Visit the Taylor & Francis Web site at
http://www.taylorandfrancis.com

and the CRC Press Web site at
http://www.crcpress.com

Printed and bound in the United States of America by Publishers Graphics, LLC on sustainably sourced paper.

Preface

I began writing this book, more than a decade ago, when it became obvious to me that the various specialists working in the then-emerging field of biotechnology (e.g., geneticists chemists, molecular biologists, intellectual property attorneys, marketers) were often having difficulty simply understanding the terms utilized by their colleagues. The first three editions of my book steadily expanded and found great acceptance among those professionals.

After the third edition appeared, a number of ill-informed lay groups with interests in environmental matters and food security, especially those skeptical of genetically modified crops and large agro-business enterprises, raised objections about the use of certain nanotechnologies, including various forms of genetic modification as well as other technologies that professionals widely regard today as beneficial and easily implemented. So the fourth edition included many of the "nanotech" terms now routinely part of the biotechnology toolkit. The goal was to reduce the level of concern by providing people with clear and easy-to-understand definitions of new terms.

Therefore, the fourth edition naturally included many new terms—some misunderstood and, therefore, controversial. This last edition was also well received especially by those seeking to explain novel scientific advances to a skeptical and uninformed public. Biotechnological advances continue to accelerate, and the need to provide clear definitions of terms used in a fast-expanding field generates a continuing need for references like this.

Recently, the importation and distribution of some U.S.-origin agricultural products were delayed or prohibited because these products were created using new biotechnologies. This occurred in spite of contravening World Trade Organization rules. Once again, assertions of supposed safety concerns about the now-emerging techniques—this time concerning genome editing, synthetic biology, etc.—require the publication of a fifth edition.

The fifth edition of this book is nearly twice the length of the fourth edition and includes at least 30% of entirely new terms. It is the only publication of its kind, containing definitions of all the new techniques utilized in agricultural biotechnology. The world's nations are now considering whether to regulate, or perhaps overregulate, these new techniques just as they did for "GMO" crops. The results could be costly for a world struggling to feed itself. It is important that those considering any steps to limit the use of biotechnology in the creation of reliable food sources be well informed. I hope that this book will fill a vacuum and help agricultural marketers, scientists, regulators, patent attorneys, venture capitalists, and university tech transfer staff explain science and bring products created via these techniques to fruition.

I offer this work in good faith and in the hope that it will assist individuals who seek to gain some understanding of the terminology as it is currently used. However, the reader should be aware that all fields of biotechnology are rapidly expanding and evolving: New terms have entered the nomenclature (and thus this book) at a rapid pace. In fact, the meaning(s) of some of the newest terms will undoubtedly be expanded or contracted as the technologies further develop.

Kimball R. Nill
Director of Market Development
Minnesota Soybean Research & Promotion Council
Mankato, Minnesota

Author

Kimball Nill is the director of market development at the Minnesota Soybean Research & Promotion Council, where he works to expand domestic and overseas markets for Minnesota-origin soybeans and soy products.

Kimball has authored numerous papers and articles on various aspects of the marketing, technology, and sustainability of agricultural products for U.S. and European journals, plus he has authored four editions of this book designed to explain agricultural biotechnology to the public.

Kimball grew up on a farm in the U.S. state of North Dakota. He holds a bachelor of science degree in chemistry from North Dakota State University and an MBA from the Wharton Business School in Pennsylvania.

A

A. flavus

See *Aspergillus flavus.*

aAI-1

See *Alpha-amylase inhibitor-1.*

AARS

See *Ribosomes, Aminoacyl-transfer RNA synthetases.*

Ab Initio Gene Prediction

(*ab initio* = "from the beginning") The prediction of a gene's (exon) structure via algorithms (e.g., in a bioinformatics computer), based on the protein coded for by the gene.

See also *Gene, Protein, Exon, Algorithm (bioinformatics), Sequence (of a DNA molecule), Sequence (of a protein molecule).*

ABA

Acronym for abscisic acid. See *Abscisic acid.*

ABA Pathway

See *Abscisic acid.*

ABC

See *ABC transporters.*

ABC Transport Proteins

See *ABC transporters.*

ABC Transporters

Refer to a class of *membrane transporter proteins* that "transfer" across cell membranes

- Sugar molecules (i.e., used by cells as "fuel")
- Inorganic ions (needed to catalyze certain cellular processes)
- Polypeptides (i.e., protein molecules)
- Certain anticancer drugs (i.e., out of tumor cells, thereby making it harder to halt certain cancer tumors via use of pharmaceuticals)
- Certain antibiotics (i.e., out of some species of pathogenic bacteria, thereby conferring *antibiotic resistance* to those pathogenic bacteria)
- Certain plant metabolites (out of plant cells, thereby conferring *resistance to some fungal diseases* to the plants possessing those ABC transporters)

ABC transporter molecules are embedded in the plasma membrane (i.e., surface "skin") of cells. See also *Cell, Plasma membrane, Protein, Catalyst, Ion, Polypeptide (protein), Cancer, Chemotherapy, Antibiotic resistance, Species, Bacteria, Metabolite, Fungus.*

Abiogenesis

Spontaneous generation.

See also *Biogenesis.*

Abiotic

Refers to the absence of living organisms.

See also *Abiotic stresses.*

Abiotic Stresses

The stress caused (e.g., to crop plants) by nonliving, environmental factors such as cold, drought, flooding, salinity, ozone, toxic-to-that-organism metals (e.g., aluminum, for plants), and ultraviolet-B light. See also *Citrate synthase (CSB) gene, Abscisic acid, Ethylene, PARP, Cytochrome p450, Drought tolerance.*

Abrin

A potent natural toxin derived from the seed of the rosary pea or crab's eye vine (*Abrus precatorius*).

See also *Ricin, Phytochemicals, Toxin.*

ABS

Acronym for "Africa Biofortified Sorghum," a new type of sorghum (i.e., a crop grown in arid regions, especially in Africa) being developed that would contain more of the nutrients needed (e.g., iron, zinc, beta-carotene, vitamin A) to help reduce the prevalence of human malnutrition there. See *Nutraceuticals, Vitamin, Beta-carotene, Golden rice.*

A

Abscisic Acid

A phytohormone (plant hormone) utilized to control

- The size of *stomatal pores*—that is, the openings in leaves through which plants exchange oxygen and carbon dioxide (and water inadvertently) with the atmosphere. Abscisic acid levels increase in plants during drought stress conditions, which leads to closing of the stomatal pores to conserve water in the plant.
- Fruit ripening in some plants (i.e., in grapes).
- Abscission (e.g., shedding of flowers and fruits).
- Dormancy (i.e., causing seeds to become properly dehydrated and their metabolism to slow down so that they can wait for proper environmental conditions to germinate).
- Seed germination.
- Root tissue growth.

In addition to those functions, abscisic acid also sometimes acts as a plant *stress hormone* (i.e., a hormone that signals the plant to respond in a defensive way to a particular environmental stress). For example, the increase in abscisic acid in plant cells under drought conditions leads to initiation of the ABA pathway (i.e., a sequential series of chemical reactions), via which specific genes are "switched on" or "switched off" to cause increased water uptake and storage by the plant and reduced water loss. See also *Plant hormone, GPA1, Abscission, Signaling, Small ubiquitin-related modifier, Stress hormones, Pathway, Stress hormones, Receptor engineering.*

Abscisic Acid Pathway

See *Abscisic acid, Receptor engineering, Pathway.*

Abscission

The act of a plant shedding its flower(s), fruit(s), etc. See also *Abscisic acid, Ethylene.*

Absolute Configuration

The configuration of four different substituent groups around an asymmetric carbon atom, in relation to D- and L-glyceraldehyde.

See also *Dextrorotary* (*D*) *isomer, Levorotary* (*L*) *isomer.*

Absorbance (A)

A measure of the amount of light absorbed by a substance suspended in a matrix. The matrix may be gaseous, liquid, or solid in nature. Most biologically active compounds (e.g., proteins) absorb light in the ultraviolet or visible light portion of the spectrum. Absorbance is used to quantitate (measure) the concentration of the substance in question (e.g., substance dissolved in a liquid).

See also *Optical density* (*OD*), *Spectrophotometer.*

Absorption

From the Latin *ab*, "away," and *sorbere*, "to suck into." The taking up of nutrients, water, etc., by assimilation (e.g., transport of the products of digestion from the intestinal tract across the cell membranes that comprise the gut and into the blood).

See also *ADME tests, Digestion* (*within organisms*).

Abzymes

Catalytic antibodies that are synthetic constructs. They either stabilize the transition state of a chemical reaction or bind to a specific substrate, thereby increasing the reaction rate of that chemical reaction. See also *Catalytic antibody, Transition state, Substrate* (*chemical*).

ACC

Abbreviation/acronym for the compound 1-aminocyclopropane-1-carboxylic acid, which is produced from *S*-adenosylmethionine (SAM) in the fruit of certain plants. When the "sam-k" gene is inserted into the genome of those plants, the level of SAM is greatly reduced in their fruit, which inhibits (slows) ripening/softening of that fruit via a reduction/slowdown in the production of ethylene (hormone that causes fruit to ripen/soften). See also *ACC synthase, Ethylene, SAM-K gene, Genetic engineering, Genome, Plant hormone.*

ACC Synthase

Aminocyclopropane carboxylic acid (ACC) synthase/deaminase is one of the most critical enzymes in the metabolic pathway that creates the hormone ethylene inside fruits. Because ethylene causes certain fruit (e.g., tomatoes) to ripen (soften), it is possible to significantly delay the softening (i.e., spoilage) process by controlling creation of ACC synthase via manipulation of the ACC synthase gene. See also *ACC, Metabolism, Enzyme, Metabolite, Intermediary metabolism, Pathway, Plant hormone, Polygalacturonase* (*PG*), *Ethylene, SAM-K gene.*

ACCase

See *Acetyl-CoA carboxylase.*

Acceptor Control

The regulation of the rate of respiration by the availability of ADP as phosphate acceptor. See also *Respiration, Adenosine diphosphate* (*ADP*).

Acceptor Junction Site

The junction between the right 3' end of an intron and the left 5' end of an exon. See also *Intron, Exon, Donor junction site.*

Accession (Germplasm)

The addition of germplasm deposits to existing germplasm storage banks. See also *American type culture collection* (*ATCC*), *Germplasm.*

Accession (Sequence Data)

The addition (e.g., to major public database) of the sequence data for a newly determined gene or protein molecule. See also *Gene, Protein, Sequence* (*of a DNA molecule*), *Sequence* (*of a protein molecule*), *Algorithm* (*bioinformatics*).

Acclimatization

The biological process whereby an organism adapts to a new environment. For example, the body of a mountain climber who has spent significant time high up on Mount Everest (e.g., 20,000 ft. above sea level) produces twice as many red blood cells (to transport oxygen) as it does at sea level.

Often, this adaptation actually occurs on a molecular level. One example is when natural microorganisms adapt so that they feed on and degrade toxic chemical wastes or change from using one sugar as a fuel source to another.

Another type of acclimatization is *cold hardening* (e.g., when the approach of winter causes certain plants to produce specific proteins that protect those plants from freezing temperatures). For example, prior to cold hardening, the crop plant known as rye (*Secale cereale*) usually dies after several hours of exposure to a temperature of −5°C (23°F). If the rye plants are first exposed to gradually decreasing daily temperatures (e.g., typical autumn season weather in North America), such acclimated rye plants can survive temperatures as low as −30°C (−22°F).

In microgravity (e.g., plants growing on the international space station or in the U.S. space shuttle), some gene expression and some cell structure/function are also altered. See also *Sugar molecules, Catabolism, Red blood cells, Cold hardening, Pharmacoenvirogenetics, Cell, Gene, Gene expression.*

Ac-CoA

Abbreviation for acetyl-coenzyme A. Ac-CoA is a chemical that is synthesized in cell mitochondria by combining the thiol (molecular group) of coenzyme A with an *acetyl group* (i.e., from breakdown/digestion of fats, carbohydrates, or proteins), utilizing the pyruvate dehydrogenase enzyme.

Plants subsequently use Ac-CoA as the starting material to manufacture fatty acids. See also *Coenzyme, Fats, Acetylcholine, Gluconeogenesis, Acetyl-CoA carboxylase, Cholinesterase, Cell, Mitochondria, Fatty acid, Protein.*

ACE

Angiotensin-converting enzyme (ACE) is an enzyme that is crucial (within the human vascular system) for catalyzing the formation of angiotensin II, a hormone that causes narrowing/restriction of blood vessels, which increases the body's blood pressure as the blood is "squeezed" through those narrowed blood vessels.

Overactivity of ACE can contribute to coronary heart disease (CHD). The action of ACE can be inhibited by the pharmaceuticals known as ACE inhibitors. Research indicates that consumption of egg proteins or whey protein can also result in inhibition of ACE. See also *Enzyme*, Hormone, ACE inhibitors, CHD.

ACE Inhibitors

Refer to "family" of naturally occurring proteins or to a group of chemically similar pharmaceuticals utilized to lower blood pressure in humans, by blocking the formation of a hormone (angiotensin II) that narrows/restricts blood vessels. In 2009, research by Jianping Wu and Kaustav Majumder showed that certain proteins in boiled or fried eggs can be converted via enzymes in the human stomach and small intestine into peptides with angiotensin-converting enzyme (ACE)-inhibitory activity. See also *ACE, Protein, Enzyme, Hormone.*

Acentric Chromosome

Refers to a chromosome fragment that lacks a centromere. See also *Chromosomes, Centromere.*

Acetobacter aceti

A bacterium that can "spoil" alcohol-containing beverages by turning the ethanol into vinegar (acetic acid). Discovered by Louis Pasteur in the 1800s. See also *Bacteria.*

Acetolactate Synthase

See *ALS.*

Acetyl Carnitine

One of the metabolites of mitochondria; it is a substrate (i.e., substance that is acted upon) for acylcarnitine transferase (which converts the acetyl carnitine to carnitine).

Research indicates that consumption of acetyl carnitine helps to increase the levels of acetylcholine and nerve growth factor in the brain. See also *Metabolite, Mitochondria, Acylcarnitine transferase, Substrate* (*chemical*), *Carnitine, Acetylcholine, Nerve growth factor* (*NGF*).

Acetyl Coenzyme A

See *Ac-CoA.*

Acetylation

Refers to addition of an acetyl molecular group to a protein molecule.

See *Histones, Posttranslational modification of protein.*

Acetylcholine

A neurotransmitter (i.e., one of several relatively small, diffusible molecules utilized by the human body to "transmit" nerve impulses) that is synthesized (i.e., manufactured) near the ends of axons (i.e., one type of neuron). That synthesis is accomplished by the "transfer" of an acetyl group (portion of molecule) from Ac-CoA to a choline molecule (i.e., available in the body via consumption of soybean lecithin or certain other foods), in a chemical reaction catalyzed by cholinesterase.

Research indicates that consumption of a chemical compound known as "huperzine A," which is extracted from the Chinese club moss (*Huperzia serrata*), inhibits the enzyme within the human body that breaks down acetylcholine molecules.

Increased amount of acetylcholine in the (human) brain has been shown to reduce the symptoms of Alzheimer's disease. See also *Neurotransmitter, Neuron, Choline, Ac-CoA, Lecithin, Alzheimer's disease, Thymus, Enzyme, Cholinesterase, Endothelial nitric oxide synthase* (*eNOS*).

A

Acetylcholinesterase

An enzyme that hydrolyzes (i.e., cuts into smaller pieces) molecules of the neurotransmitter acetylcholine, after the acetylcholine molecules have accomplished "transmission" of a nerve impulse. That hydrolysis (cutting into pieces) of acetylcholine molecules serves to thus prepare the neurons (cells of the body's nervous system) to be able to transmit other, later nerve impulses. See also *Enzyme, Hydrolysis, Neurotransmitter, Acetylcholine, Neuron.*

Acetyl-CoA

Acetyl-coenzyme A. See *Ac-CoA.*

Acetyl-CoA Carboxylase

An enzyme that catalyzes the chemical reaction (i.e., conversion of Ac-CoA to malonyl-CoA via carboxylation), which is the first step in the series of chemical reactions via which some plants "manufacture" oils (e.g., soybean oil, canola oil). See also *Enzyme, Fats, Soybean oil, Canola.*

aCGH

Acronym for "array comparative genomic hybridization." See *Array comparative genomic hybridization.*

Acid

A substance that contains hydrogen atom(s) in its molecular structure, with a pH in the range from 0 to 6, which will react with a base to form a salt. Acids normally taste sour and feel slippery. For example, food product manufacturers often add citric acid, malic acid, fumaric acid, and itaconic acid in order to impart a "sharp" taste to food products. See also *Base (General), Citric acid, Fumaric acid* ($C_4H_4O_4$), *Gallic acid.*

Acidosis

A metabolic condition in which the capacity of the body to buffer changes in pH is diminished. Hence, acidosis is accompanied by decreased blood pH (i.e., the blood becomes more acidic than is normal). See also *Acid, Metabolism.*

Ac-P

Abbreviation for acetylphosphate.

ACP (Acyl Carrier Protein)

A protein that binds acyl intermediates during the formation of long-chain fatty acids. ACP is important in that it is involved in every step of fatty acid synthesis. See also *Fatty acid, Acyl-CoA, Fats.*

Acquired Immune Deficiency Syndrome (AIDS)

A disease in which a specific virus attacks and kills macrophages and helper T cells (thus causing collapse of the entire immune system). Once the immune system has been inactivated, other diseases, which under normal circumstances can be fought off, become fatal. See also *Human immunodeficiency virus type 1 (HIV-1), Human immunodeficiency virus type 2 (HIV-2), Helper T cells (T4 cells), Macrophage, Tumor necrosis factor (TNF).*

Acquired Mutation

A genetic change (i.e., mutation in DNA) that occurred within a somatic cell (i.e., cell *not* involved in the organism's reproduction), so the mutation is not passed down to subsequent generations. See also *Somatic cells, Mutation, Somatic variants, Cell, Organism, Deoxyribonucleic acid (DNA).*

Acrylamide Gel

See *Polyacrylamide gel.*

ACTH (Adrenocorticotropic Hormone [Corticotropin])

Adrenocorticotropic hormone (corticotropin) is a polypeptide secreted by the anterior lobe of the pituitary gland. This is an example of a protein hormone. See also *Polypeptide (protein), Endocrine glands, Endocrine hormones.*

Actin

First identified by Albert Szent-Gyorgyi, it is a contractile (i.e., periodically contracting) protein that is present within—or as part of the exterior of—eucaryotic cells. In mammalian cells, actin exists in two forms: G-actin (monomeric form) and F-actin (polymerized chain molecule form). The cell's actin is frequently polymerized (i.e., forms the molecular chains of F-actin) and depolymerized (i.e., the molecular chains break apart again into the individual "chain links" known as G-actin), as needed by the cell.

Via its contractions, F-actin is involved in eucaryote cells'

- Movement (e.g., when it "pulls" the cell to a new position) within an organism's body, much like a towrope. The "towrope" utilized is a long narrow structure extending from the exterior of cell, called a "filopodium," which is composed of actin surrounded by a layer of the cell's plasma membrane.
- *Note*: A protein known as N-WASP in breast cancer cells works in concert with actin to form branches bearing sharp external points on the cell surface by rearranging the cell's internal actin "skeleton." Those branches with sharp points are known as "pseudopodia," and the pseudopodia can grab onto and poke holes into the extracellular matrix (i.e., the supportive tissue located in between individual cells). Breast cancer cells attach enzymes to the pseudopodia and "dig into" the extracellular matrix, creating larger spaces for breast cancer to move into (i.e., beginning of metastasis process).
- Changes in the shape/morphology of the cell (e.g., in response to chemical signals such as a growth factor molecule encountering the exterior membrane of the cell).
- Separation of nuclear DNA during meiosis (dividing into two/reproductive step, in life of a cell).

In addition to being the raw material used by the cell for rapid formation of F-actin (filaments), G-actin also interacts with the cell's nucleus (thereby activating expression of certain genes involved in cell differentiation, cell growth, cell proliferation, and the F-actin filaments, which control cell shape/movement).

During an animal's development, some actin-rich protrusions form and extend from one individual muscle cell to an adjacent individual muscle cell. As part of the animal's maturation process, that leads those two muscle cells (and many other individual muscle cells) to fuse into long *myotubes* (*muscle fibers*) that contain multiple cell nuclei. Actin molecular analogues present in bacterial cells include *ParM, which separates DNA plasmids during meiosis*, and *MreB, which are located just beneath the outer membrane and determine cell shape in rod-shaped bacteria.* See also *Cell, Protein, Eucaryote, Deoxyribonucleic acid (DNA), Gene, Nucleus, Nuclear DNA, Meiosis, Extracellular matrix (ECM), Cytoskeleton, Analogue, Bacteria, Plasma membrane, Motor proteins, Cell motility, Listeria monocytogenes, Chemotaxis, MreB, ParM, Polymer, Morphology, Growth factor, Express, Differentiation, Metastasis.*

Activated Carbon

An adsorbent form of very finely divided carbon, which is produced by a high-temperature treatment of raw material (e.g., bone, coconut hulls) along with steam, air, or carbon monoxide.

One gram of the resultant activated carbon has an internal surface area (onto which many chemicals can tightly adsorb) of approximately 1200 m^2, so it is often utilized for purifying water, air, and other gases. It is sometimes utilized to remove hydrocarbon compounds from potable (e.g., drinkable) water or to remove polycyclic aromatic hydrocarbons from vegetable oils. See also *Meter.*

Activation Energy

The amount of energy (calories) required to bring all the molecules in one mole of a reacting substance to the transition state. More simply, it may also be viewed as the energy required to bring reacting molecules to a certain energy state from which point the reaction proceeds spontaneously. See also *Transition state (in a chemical reaction), Mole, Free energy.*

Activator (of Enzyme)

A small molecule that stimulates (increases) an enzyme's catalytic activity when it binds to an allosteric site. See also *Enzyme, Effector, Allosteric site, Catalyst.*

Activator (of Gene)

A protein molecule that increases the expression of a given gene, by binding to transcription control sites (e.g., within that gene or in adjacent intron). See also *Protein, Gene, Expressivity, Transcription activators, Signal transducers and activators of transcription (STATs), Transcription factors, Intron, Bursting.*

Active Site

The region of an enzyme surface that binds the substrate molecule and transforms the substrate molecule into the new (chemical) product (entity). This site is usually located not on a protruding portion of the enzyme but rather in a cleft or depression. This establishes a controlled environment in which the chemical reaction may occur. See also *Catalytic site, Agonists, Pharmacophore, Substrate (chemical), Enzyme, Antagonists, Meganuclease.*

Active Transport

Cell-mediated, energy-requiring translocation of a molecule across a membrane in the direction of increasing concentration (i.e., opposite of natural tendency). This is done via special membrane-bound proteins (i.e., protein molecules embedded in the cell's plasma membrane). See also *Osmotic pressure, Cell, Protein, Plasma membrane, Ion channels, G-proteins, Membrane transport.*

Activity Coefficient

The factor by which the concentration of a solute must be multiplied to give its true thermodynamic activity.

Activity-Based Screening

See *High-throughput screening (HTS).*

Acuron™ Gene

A gene, trademarked by Syngenta AG, that can be inserted into plants via genetic engineering techniques. When that gene is inserted into the genome (DNA) of a plant, it confers tolerance to herbicide(s) whose active ingredient is protoporphyrinogen oxidase (PPO) inhibitor (thus, such herbicides are known as PPO inhibitors). See also *Herbicide-tolerant crop, Gene, Genetic engineering, Genome, Deoxyribonucleic acid (DNA).*

Acute Myeloid Leukemia (AML)

One form of leukemia (cancer of the blood) that can be caused by fusion proteins. See also *Cancer, Fusion protein.*

Acute Transfection

Short-term infection of cells with DNA. See also *Transfection.*

Acyl Carrier Protein

See *ACP (acyl carrier protein).*

Acylcarnitine Transferase

An enzyme that converts the mitochondrial metabolite *acetyl carnitine* into carnitine. See also *Enzyme, Acetyl carnitine, Carnitine.*

Acyl-CoA

Acyl derivatives of coenzyme A (acyl-S-CoA). See also *Carnitine, Coenzyme A, Trypsin inhibitors.*

A

Acyltransferases

Refer to a class of enzymes that attach an acyl molecule (e.g., fatty acid) to the carbon-atom "skeleton" of a large organic molecule (e.g., a triacylglycerol). See also *Enzyme, Fatty acid, Triacylglycerols.*

AD

An acronym utilized to refer to the group of diseases known collectively as autoimmune disorders. These include diseases such as multiple sclerosis, lupus, and rheumatoid arthritis. See also *Autoimmune disease, Multiple sclerosis, Lupus.*

Adalimumab

A monoclonal antibody against tumor necrosis factor-alpha that was approved by the U.S. Food and Drug Administration (FDA) in 2003 for use as the pharmaceutical Humira™ to inhibit the structural damage (to body joints) of the autoimmune disease rheumatoid arthritis.

FDA subsequently approved adalimumab as a pharmaceutical treatment for

- Reducing signs and symptoms of moderately to severely active polyarticular juvenile idiopathic arthritis in patients 2 years of age and older
- Reducing signs and symptoms of active arthritis, inhibiting the progression of structural damage, and improving physical function
- Reducing signs and symptoms in patients with active ankylosing spondylitis disease
- Adult patients with moderate to severe chronic plaque psoriasis who are candidates for systemic therapy or phototherapy

In 2014, FDA approved adalimumab as a treatment for pediatric Crohn's disease in patients aged 6 years and older when other treatments have not worked well.

See also *Rheumatoid arthritis, Autoimmune disease, Food and Drug Administration (FDA), Monoclonal antibodies (MAb), Tumor necrosis factor (TNF), Phage display, Crohn's disease.*

Adaptation

Refers to the "adjustment" of a *population* of organisms to a changed environment.

For example, when the Industrial Revolution caused large amounts of black soot to be deposited onto the white bark of certain trees in England in the nineteenth century, it resulted in adaptation (e.g., via selective breeding) of the *population of a particular indigenous moth (Biston betularia), which had previously consisted of a mixture of all-white and all-black members.* Because the soot blackened the formerly white bark of the trees it rested on, predatory birds were able to easily catch and eat the all-white members of that moth population. Thus, there were fewer of the all-white moths present in the breeding population and a greater preponderance of all-black members.

In the twentieth century, antipollution efforts in England resulted in a cessation of the airborne soot, so that forest's tree bark returned to its original white color. Because the predatory birds were now able to more easily catch and eat the all-black members of that moth population, there were fewer of the all-black moths present in the breeding population and a greater preponderance of all-white members. See also *Organism.*

Adaptive Enzymes

See *Inducible enzymes.*

Adaptive Immune Response

See *CD8⁺ T cells, Immune response.*

ADBF

See *Azurophil-derived bactericidal factor (ADBF).*

ADC

Acronym for "antibody-drug conjugates." See *Immunoconjugate.*

ADCC

See *Antibody-dependent cellular toxicity.*

Additive Genes

Genes that interact but do not show dominance (in the case of alleles) or epistasis (if they are not alleles).

A single additive gene does not "show up" in the phenotype, but a *collective group* of additive genes can result in a trait that is evident in the phenotype. See also *Gene, Allele, Dominant allele, Epistasis, Phenotype, Trait, Additive variance.*

Additive Variance

Refers to the amount/percentage of an organism's genetic variance that results from a single given *additive gene.* See also *Gene, Genetics, Additive genes.*

Adenylate Cyclase

The enzyme (within cells) that catalyzes the synthesis (i.e., "manufacture") of cyclic AMP. See *Cyclic AMP.*

Adenine

A purine base, 6-aminopurine, occurring in ribonucleic acid (RNA) as well as in deoxyribonucleic acid (DNA) and a component of adenosine diphosphate (ADP) and adenosine triphosphate (ATP). Adenine pairs with thymine in DNA and uracil in RNA. See also *Base (nucleotide), Base pair (bp), Ribonucleic acid (RNA), Deoxyribonucleic acid (DNA).*

Adenosine

Refers to the *nucleoside* (i.e., hybrid with ribose or deoxyribose) *form* of adenine. See also *Adenine, Nucleoside.*

Adenosine Diphosphate (ADP)

A ribonucleoside 5′-diphosphate serving as a phosphate group acceptor in the cell energy cycle. See also *Catabolism, Adenosine triphosphate (ATP), Adenosine monophosphate (AMP).*

Adenosine Monophosphate (AMP)

A ribonucleoside 5′-monophosphate that is formed by hydrolysis of ATP or ADP. See also *Hydrolysis, Adenosine diphosphate (ADP), Adenosine triphosphate (ATP).*

Adenosine Monophosphate–Activated Protein Kinase

See *AMPK.*

Adenosine Triphosphate (ATP)

The major carrier of chemical energy in the cells of all living things on this planet. A ribonucleoside 5′-triphosphate functioning as a phosphate group donor in the energy cycle of the cell. ATP contains three phosphate/oxygen molecules linked together. When a phosphate–phosphate bond in ATP is broken (hydrolyzed), energy that the cell can use to carry out its functions is produced. Thus, ATP serves as the universal medium of biological energy storage and exchange, in living cells. See also *ATPase, ATP synthetase, Hydrolysis, Cyclic phosphorylation, Bioluminescence, ATP synthase, Adenosine monophosphate (AMP), Ubiquinone.*

Adenovirus

A category of virus that can infect humans, monkeys, rodents, cattle, and fowl. Like all viruses, adenoviruses can reproduce only inside living cells (of other host organisms). Adenovirus causes a protein (metabolite) to be made that disables the p53 gene. Because the p53 gene then cannot perform its usual function (i.e., prevention of uncontrolled cell growth caused by virus/DNA damage), the adenovirus thus "takes over" and causes the cell to make numerous copies of the virus until the cell dies (thus releasing the virus copies into the body of the host organism to cause further infection). See also *Virus, Retroviruses, Gene delivery, Gene therapy, Cell, Protein, p53 gene, Deoxyribonucleic acid (DNA).*

Adequate Intake (AI)

See *Choline.*

Adhesion Molecule

From the Latin *adhaerere* = "to stick to." The term "adhesion molecule" refers to a glycoprotein molecular "chain" that protrudes from the surface membrane of certain cells and causes cells (possessing "matching" adhesion molecules) to adhere to each other. For example, in 1952 Aaron Moscona observed that *(harvesting enzyme–separated)* chicken embryo cells did not remain separated but instead coalesced again into an (embryo) aggregate. In 1955, Philip Townes and Johannes Holtfreter showed that "like" amphibian (e.g., frog) neuron cells will rejoin together after being physically separated (e.g., with a knife blade), but "unlike" cells remain segregated (apart).

Adhesion molecules *per se* were formally discovered by Gerald M. Edelman in the 1970s.

Adhesion molecules also play a crucial role in guiding monocytes to sources of infection (e.g., pathogens) because adhesion molecules in the walls of blood vessels (after activation caused by pathogen invasion of adjacent tissue) adhere to like adhesion molecules in the membranes of monocytes in the blood. The monocytes pass through the blood vessel walls, become macrophages, and fight the pathogen infection (e.g., triggering tissue inflammation). See also *Monocytes, Macrophage, Polypeptide (protein), Cell, Pathogen, CD4 protein, CD44 protein, GP120 protein, Vaginosis, Harvesting enzymes, Harvesting, Signal transduction, Selectins, Lectins, Glycoprotein, Sugar molecules, Leukocytes, Lymphocyte, Neutrophils, Endothelium, Endothelial cells, P-selectin, ELAM-1, Integrins, Cytokines.*

Adhesion Protein

See *Adhesion molecule, Endothelial cells.*

Adipocytes

Specialized cells within an organism's lymphatic system, which store the triacylglycerols (i.e., also sometimes called "triglycerides"), after digestion of fats, and then later release fatty acids and glycerol into the bloodstream (e.g., when needed by the organism). See also *Cell, Triglycerides, Fatty acid, Digestion (within organisms), Fats.*

Adipocytokines

See *Adipokines.*

Adipokines

Refer to more than 50 protein hormones that are secreted by adipose cells and that act to help the body regulate its metabolism, immunity, homeostasis, etc.

Many adipokines promote inflammation and make cells resistant to the effects of insulin.

Production of adipokines can be decreased by consumption of conjugated linoleic acid. See also *Protein, Hormone, Cell, Adipose, Metabolism, Homeostasis, Visfatin, Leptin, Tumor necrosis factor-α, Resistin, Adiponectin, Interleukin-6, Conjugated linoleic acid (CLA), Insulin.*

Adiponectin

An adipokine that activates AMP-activated protein kinase and modulates signaling pathways controlled by NFκB. See also *Adipokines, AMP, Protein, Kinases, Signaling, Pathway, NFκB.*

Adipose

Utilized to refer to "energy storage" *tissues* within some animals, consisting of fat molecules. Adipose tissue tends to increase in animals' bodies if they consume more energy-dense food than needed for their level of energy expenditure (e.g., via exercise).

In humans older than 40, an increase in the body's amount of adipose tissue is correlated with an increased risk of premature death (e.g., from coronary heart disease).

Adipose tissue cells secrete a large number of compounds that impact the human body in a number of ways. For example,

- Leptin—a protein hormone signal to the brain that the body has "enough" energy stores, which also stimulates the body to consume calories faster.
- Visfatin—a protein that has some of the same effects as insulin (e.g., stimulates glucose uptake by the body, which lowers blood sugar levels).
- Tumor necrosis factor-α (TNF-α)—a cytokine protein that initiates changes (inflammation) in vascular tissues that result in monocytes adhering to internal walls of blood vessels, thereby becoming a macrophage and resulting in formation of a plaque deposit. Additionally and separately, TNF-α can also cause some tissues to become insulin resistant.
- Angiotensin—a precursor molecule that can become *angiotensin II* in the body. The hormone angiotensin II causes arteries to constrict (which can result in high blood pressure), promotes macrophage accumulation into plaque deposits on blood vessel walls, and enhances the metabolism of nitric oxide into free radical molecules.
- Adiponectin—a molecule that acts to inhibit the development of insulin resistance in tissues and acts to inhibit inflammation.

See also *Fats, Coronary heart disease* (*CHD*), *Leptin, Lecithin, Choline, Visfatin, Insulin, Cytokines, Monocytes, Adhesion molecule, Macrophage, Plaque, Metabolism, Nitric oxide, Free radical.*

Adipose Triglyceride Lipase

See *Lipase.*

Adjuvant (to a Herbicide)

From the Latin word for "aid," it refers to any compound that enhances the effectiveness (i.e., weed-killing ability) of a given herbicide. For example, adjuvants such as surfactants can be mixed (prior to application to weeds) with herbicide (in water), in order to hasten transport of the herbicide's active ingredient into the weed plant. That is because the herbicide must move from an aqueous (water) environment into one (i.e., the weed plant's cuticle or "skin") comprised of lipids/lipophilic molecules, before it can accomplish its task. See also *Surfactant, Lipids, Lipophilic.*

Adjuvant (to a Pharmaceutical)

From the Latin word for "aid," it refers to any compound that enhances the desired response by the body to that pharmaceutical. For example, adjuvants such as certain polysaccharides or surface-modified nanoparticles (e.g., pan-DR-binding epitope derivatized dendrimer nanoparticle acts as an adjuvant to amphotericin B, for treatment of leishmaniasis disease) can be injected along with (vaccine) antigen in order to increase the immune response (e.g., production of antibodies) to a given antigen.

Another example is that consumption of grapefruit juice by humans will increase the impact of certain pharmaceuticals. Those pharmaceuticals include some sedatives, antihypertensives, the antihistamine terfenadine, and the immunosuppressant cyclosporine. The adjuvant effect of grapefruit juice is thought to be caused via inhibition of the enzyme cytochrome P4503A4, which catalyzes reactions involved in the metabolism (breakdown) of those pharmaceuticals.

Another example is that consumption of the pharmaceutical known as clopidogrel (U.S. commercial name Plavix™) by people immediately following a mild heart attack (severe chest pain)—along with aspirin—greatly reduces the risk of death, strokes, and (new, additional) heart attacks versus taking aspirin alone after a mild heart attack.

See also *Cellular immune response, Humoral immunity, Polysaccharides, Nanotechnology, Antigen, Antibody, Enzyme, Metabolism, Histamine, Cyclosporine, Cytochrome P4503A4.*

ADME

Acronym for absorption, distribution (within the body), metabolism, and elimination of pharmaceuticals. See also *ADME tests, In silico screening, Scavenger receptor A.*

ADME Tests

Refer to *a*bsorption, *d*istribution (within the body), *m*etabolism, and *e*limination tests historically required by the U.S. Food and Drug Administration for approval of new pharmaceuticals or some new food ingredients.

Today, the relevant companies (e.g., pharmaceutical companies) are performing such tests at earlier stages in their screening and assessments of new compounds, so they can halt work on any compounds that are shown to be problematic.

To assess the *absorption* of a new pharmaceutical (candidate compound), scientists can test its permeability through an artificial membrane (e.g., hexadecane), its permeability through artificial lipid membranes, and/or its transport through a single layer of Caco-2 cells (in a cell culture vessel). Because such cultured Caco-2 cells act very much like human intestinal mucosa cells, such *Caco-2 tests* provide good prediction of pharmaceutical's active transport, passive transport, and also receptor-mediated efflux.

Distribution, which is related to bioavailability, is closely associated with compound's solubility (in body fluids) and the compound's ability to be "bound" by plasma proteins. Thus, distribution (e.g., of pharmaceutical compound) is assessed via a test to determine plasma protein binding.

Metabolism can sometimes lead to toxicity. For example, in people whose body cannot degrade pyrimidines, the (pyrimidine) metabolites of the anticancer drug 5-fluorouracil can build up in their body to lethal levels.

Elimination can lead to the pharmaceutical being removed from the bloodstream before it has the desired effect (e.g., entering and impacting the diseased tissue within the body). Note that applicable drugs or drug carriers (e.g., certain nanoparticles) bearing a slightly negative charge have a long circulation time. See also *Food and Drug Administration (FDA), Absorption, Caco-2, Plasma, Cell culture, Receptors, Active transport, Membrane transport, Protein, Metabolism, Intermediary metabolism, Metabolite, Pharmacokinetics, Pharmacogenomics, Codex Alimentarius Commission, ADME, ADMET, Haplotype, ADME/Tox, In silico screening, Microsomes, Nanoparticles.*

ADME/Tox

Refers to tests of the absorption, distribution (within the body), metabolism, elimination, and toxicity of a given compound (e.g., a pharmaceutical candidate).

To assess the *absorption* of a new pharmaceutical (candidate compound), scientists can test its permeability through an artificial membrane (e.g., hexadecane), its permeability through artificial lipid membranes, and/or its transport through a single layer of Caco-2 cells (in a cell culture vessel). Because such cultured Caco-2 cells act very much like human intestinal mucosa cells, such *Caco-2 tests* provide good prediction of a pharmaceutical's active transport, passive transport, and also receptor-mediated efflux.

Distribution, which is related to bioavailability, is closely associated with compound's solubility (in body fluids) and the compound's ability to be "bound" by plasma proteins. Thus, distribution (e.g., of pharmaceutical compound) is assessed via a test to determine plasma protein binding.

Metabolism can sometimes lead to toxicity. For example, in people whose body cannot degrade pyrimidines, the (pyrimidine) metabolites of the anticancer drug 5-fluorouracil can build up in their body to lethal levels.

Toxicity can be assessed via cell-based microarrays, for example, which reveal a compound's quantitative impact on specific cells' integrity and cellular functions (with regard to toxicity). See also *ADME tests, Absorption, Caco-2, Plasma, Cell, Cell culture, Receptors, Active transport, Membrane transport, Protein, Microarray (testing), Metabolism, Metabolite, Haplotype, Live cell array, Toxicogenomics, In silico screening.*

ADMET

Acronym for absorption, distribution (within the body), metabolism, elimination, and toxicity of pharmaceuticals. See also *ADME tests, In silico screening, ADME/Tox, Microsomes, Scavenger receptor A.*

A-DNA

A particular right-handed helical form of DNA (possessing 11 base pairs per turn), which is the form that DNA molecules exist in, when they are partially dehydrated. A-form DNA is found in fibers at 75% relative humidity and requires the presence of sodium, potassium, or cesium as the counterion. Instead of lying flat, the bases are tilted with regard to the helical axis and there are more base pairs per turn. The A-form is biologically interesting because it is probably very close to the conformation adopted by DNA–RNA hybrids or by RNA–RNA double-stranded regions. The reason is that the presence of the $2'^{2}$ hydroxyl group prevents RNA from lying in the B-form. See also *B-DNA, DNA–RNA hybrid, Deoxyribonucleic acid (DNA), Base pair (bp).*

Adoptive Cellular Therapy

The increase in immune response that is achieved by selectively removing certain immune system cells from a (patient's) body, multiplying them *in vitro* outside the body to greatly increase number, and then reinserting those (more numerous) immune system cells into the same body. See also *Cellular immune response, Cell culture, In vitro, Gene delivery, Gene therapy, Ex vivo (therapy).*

Adoptive Immunization

The transfer of an immune state from one animal to another by means of lymphocyte transfusions. See also *Lymphocyte.*

Ado-Trastuzumab Emtansine

An antibody–drug conjugate (a DM1 toxin molecule conjugated to trastuzumab) that was approved in 2013 by the U.S. Food and Drug Administration (FDA) as the pharmaceutical Kadcyla for the treatment of breast cancer caused by mutated HER-2 that has spread despite receiving other treatments. The monoclonal antibody in trastuzumab carries the conjugate to HER-2 (antiepidermal growth factor receptor-2) molecules located on the surface of applicable breast cancer tumors. See also *Antibody, Food and Drug Administration (FDA), Cancer, HER-2 receptor, Trastuzumab.*

ADP

See *Adenosine diphosphate (ADP).*

Adsorption

The tight adherence of an atom or molecule onto the surface of another substance (e.g., activated carbon). See also *Activated carbon.*

Adult Stem Cell

This term referred historically to a stem cell that is (extracted or) derived from the bone marrow tissue of adults, but scientists are now discovering additional types of "adult" stem cells within other tissues of older-than-infant humans. In addition to bone marrow, these adult stem cells have been found in liver, skin, adipose (fat), heart, testicle, intestine, menstrual blood, and brain tissue. Similar to human embryonic stem cells, adult stem cells can—under certain conditions—differentiate/proliferate into cell types specific to many of the human body's 210 different types of tissue.

For example, in 2013, Giuseppe Maria de Peppo and colleagues reprogrammed human skin cells (to act like stem cells) and inserted them into a bioreactor, where they grew into functional bone tissue. In 2014, Gordana Vunjak-Novakovic reprogrammed human fat cells to become mesenchymal stem cells that underwent a condensation stage and then grew into functional cartilage tissue.

For example, in 1980, Steven Teitelbaum and colleagues injected such cells from a donor into a 3.5-month-old girl who had the disease known as "osteopetrosis," which in that era almost always killed its victims in the first year of their life. Those adult stem cells differentiated/proliferated and that girl lived a long lifespan.

For example, in 2004, Nagy Habib injected such cells from a patient who was suffering from cirrhosis of the liver, into the patient's own hepatic artery in the liver. Those adult stem cells repopulated the liver and improved its function (i.e., reversed at least some of the liver cirrhosis).

Also, numerous researchers have reported that injecting adult stem cells taken from the bone marrow into the coronary arteries of patients who have "chronic ischemia" (a particular form of heart disease) results in neovasculogenesis (formation of new blood vessels, which lessens the chronic ischemia).

Adult stem cells are being utilized in treatments administered to humans for more than 100 diseases and conditions. In 2012, Sadia Mohsin discovered that when PIM-1 (a protein molecule that

A

promotes cell survival and growth) is added by man to adult stem cells that have been removed via biopsy from an (aged) heart, those adult stem cells are rejuvenated because telomerase enzyme activity is enhanced (resulting in elongation of the cells' telomeres) and stem cell proliferation is increased (when the PIM-1-modified stem cells are reinserted into the heart muscle).

See also *Cell, Stem cells, Multipotent adult stem cell, Stem cell growth factor (SCF), Protein, Differentiation, Bioreactor, Scaffolding (utilized in tissue engineering), Hematopoietic stem cells, Adipose, Mesenchymal stem cell (MSC), Coronary heart disease (CHD), Angiogenesis, Telomerase, Telomeres.*

Adventitious

From the Latin *adventitius* meaning "not properly belonging to." The term can be utilized to refer to

- Plant shoots emanating from sites other than typical ones (e.g., from a plant's leaves)
- A small amount of transgenic grain accidentally mixed into other grain, etc.

See also *Transgenic.*

Aerobe

A microorganism that requires oxygen to live (respire). See also *Microorganism.*

Aerobic

Exposed to air or oxygen. An oxygenated environment.

Affinity

Refers to the "attraction force" or "binding strength" between two entities (e.g., molecules). See also *Affinity chromatography, Antibody affinity chromatography, Affinity maturation.*

Affinity Chromatography

A method of separating a mixture of proteins or nucleic acids (molecules) by specific interactions of those molecules with a component known as a "ligand," which is immobilized on a support. If a solution of, say, a mixture of proteins is passed over (through) the column, *one* of the proteins binds to the ligand on the basis of specificity and high affinity (they fit together like a lock and key). If there is no naturally occurring "lock" inherent on the desired protein molecule (to go with the ligand's "key"), then the scientist can add an *affinity tag* during the synthesis of the protein (e.g., in a cell-free gene expression system) to act as that protein molecule's "key."

The other proteins in the solution wash through the column because they were not able to bind to the ligand. Once the column is devoid of the other proteins, an appropriate wash solution is passed through the column, which causes the protein/ligand complex to dissociate. The protein is subsequently collected in a highly purified form. See also *Chromatography, Protein, Nucleic acids, Antibody affinity chromatography, Ligand (in chromatography), Affinity, Affinity tag, Cell-free gene expression system.*

Affinity Maturation

See *B lymphocytes, Affinity.*

Affinity Tag

Refers to a particular sequence (of amino acids) added to a given protein molecule (e.g., produced via recombinant methods). That *affinity tag* can then later be utilized to make recovery of *that protein* (out of mixture of proteins, etc.) easier to accomplish (e.g., via chromatography, or as part of the "tandem affinity purification tagging" process). See also *Affinity chromatography, Affinity, Protein, Amino acid, Sequence (of a protein molecule), Ligand (in chromatography), Ribozymes, Tandem affinity purification tagging.*

AFGP

Abbreviation for "antifreeze glycoprotein." See *Thermal hysteresis proteins.*

Aflatoxin

The term that is used to refer to a group of related mycotoxins (i.e., metabolites produced by fungi that are toxic to animals and humans) produced by some strains of the fungi *Aspergillus flavus* and *Aspergillus parasiticus* and less often by *Penicillium puberulum. A. flavus* and *A. parasiticus* are common fungi that typically live on decaying vegetation.

Corn earworm (*Helicoverpa zea*) and European corn borer (*Ostrinia nubilalis*) can be vectors (carriers) of *A. flavus.*

Aflatoxin B_1 is the most commonly occurring aflatoxin and is one of the most potent carcinogens known to man. When ingested, it binds to DNA and interferes with replication and transcription.

When dairy cattle eat aflatoxin-contaminated feed, their metabolism process converts the aflatoxin (e.g., Aflatoxin B_1) into the mycotoxins known as Aflatoxin M_1 and Aflatoxin M_2, which soon appear in the milk produced by those dairy cows.

Consumption of aflatoxins by humans can also result in acute liver damage and/or interfere with the absorption/utilization of folic acid (a B vitamin) by a developing baby (i.e., in a pregnant woman), leading to the birth defect known as *spina bifida.* See also *Carcinogen, Toxin, Fungus, Mycotoxins, Stress proteins, P53 gene, Lipoxygenase (LOX), Peroxidase, Helicoverpa zea (H. zea), Beta-carotene, OH43, Bright greenish-yellow fluorescence (BGYF), Corn, European corn borer (ECB), Folic acid, Deoxyribonucleic acid (DNA), Replication (of DNA), Transcription.*

AFLP

Acronym for "amplified fragment length polymorphism." See *Amplified fragment length polymorphism.*

AFM

See *Atomic force microscopy.*

AFP

Acronym for "antifreeze protein." See *Thermal hysteresis proteins.*

AG

Abbreviation for the word "antigen." See *Antigen.*

Ag Biologicals

See *Crop biologicals.*

Agar

A complex mixture of polysaccharides obtained from marine red algae. It is also called agar-agar. Agar is used as an emulsion stabilizer in foods, as a sizing agent in fabrics, and as a solid substrate for the laboratory culture of microorganisms. Agar melts at 100°C (212°F) and when cooled below 44°C (123°F) forms a stiff and transparent gel. Microorganisms are seeded onto and grown (in the laboratory) on the surface of the gel. See also *Polysaccharides, Culture medium.*

Agarose

A highly purified form of agar. Used as a stationary phase (substrate) in some chromatography and electrophoretic methods. See also *Chromatography, Electrophoresis, Agar.*

Age-Related Macular Degeneration (AMD)

A disease in which the retina of the eye is damaged by either

- Abnormal deposits known as *drusen* in the macula (central region of the retina), which contain some complement proteins (complement factor H, factor B) and other materials (e.g., interleukin-18 [IL-18]). This form of age-related macular degeneration (AMD) is known as the "dry" form, and it often later progresses to the "wet" form (in the following text). Because IL-18 is anti-angiogenic (i.e., inhibits the formation/growth of new blood vessels), IL-18 helps to prevent or at least slow the progression to the "wet" form of AMD disease.
- Abnormal growth of new blood vessels in, and in front of, the retina/macula. This form of AMD is known as the "wet" form.

AMD typically afflicts people who are more than 60 years old.

In 2005, it was discovered that people whose DNA contains one particular SNP of *competent factor H* gene are much more likely to get AMD.

In 2006, it was discovered that people whose DNA contains one particular SNP of *factor B* gene are somewhat more likely to get AMD. See also *Protein, Complement, Complement cascade, Complement factor H gene, Vascular endothelial growth factor (VEGF), Ranibizumab, Deoxyribonucleic acid (DNA), Gene, Single-nucleotide polymorphisms (SNPS), Short interfering RNA (siRNA), Interleukin-18 (IL-18).*

Aging

The process, affecting organisms and most cells, whereby each cell division (mitosis) brings that cell (or organism composed of such cells) closer to its *final* cell division (i.e., death). Notable exceptions to this aging process include cancerous cells (e.g., myelomas) and the single-celled organism; both of which are "immortal." See also *Telomeres, Mitosis, Hybridoma, Myeloma, Cancer.*

Aglycon

A nonsugar component of a glycoside. See also *Glycoside.*

Aglycone

The biologically active (molecular) form of molecules of isoflavones. See also *Isoflavones, Biological activity.*

Agonists

Small protein or organic molecules that bind to certain cell proteins (i.e., receptors) at a site that is adjacent to the cell's "docking" site of protein hormones, neurotransmitters, etc. (i.e., receptor) to induce a conformational change in that cell protein hormone/neurotransmitter receptor molecule, thereby enhancing its activity (i.e., effect upon the cell).

For example, the resveratrol molecule acts as such an agonist binding partner with the estrogen receptor (without stimulating estrogenic cell proliferation) to beneficially control the body's inflammation response.

For example, obeticholic acid acts as such an agonist with the farnesoid X receptor, thereby making it one potential treatment for the disease nonalcoholic steatohepatitis with liver fibrosis. See also *Receptors, Farnesoid X receptor (FXR), Active site, Resveratrol, Chronic inflammation, Conformation, Cell, Hormone, Antagonists, Neurotransmitter, PPAR.*

Agraceutical

See *Nutraceuticals, Phytochemicals.*

Agriceuticals

See *Nutraceuticals, Phytochemicals.*

Agricultural Biologicals

See *Crop biologicals.*

Agrobacterium tumefaciens

A naturally occurring bacterium that is capable of inserting its DNA (genetic information) into plants, resulting in a type of injury to the plant known as crown gall. In 1980, Marc Van Montagu showed that *Agrobacterium tumefaciens* could alter the DNA of its host plant(s) by inserting its own ("foreign") DNA into the genome of the host plants (thereby opening the way for scientists to insert virtually any foreign genes into plants via use of *A. tumefaciens*).

In 1983, Luis Herrera-Estrella created the first man-made transgenic plant by inserting an antibiotic-resistance gene into a tobacco plant.

In 2000, Weija Zhou and Richard Vierling proved that *A. tumefaciens* is at least 10 times more effective (i.e., at "infecting" plants to insert DNA) *in space* (i.e., weightlessness/microgravity) than it is when on the surface of the Earth. Among others, Monsanto Company has developed a way to stop *A. tumefaciens* from causing crown gall, while maintaining its ability to insert DNA into plant cells, and now uses *A. tumefaciens* as a vehicle to insert desired genes into crop plants (e.g., the gene causing high production of CP4 EPSP synthase, thus conferring resistance to glyphosate-containing herbicide). See also *Bacteria, Deoxyribonucleic acid (DNA), Informational molecules, Genome, Transgenic (organism), Protoplast, EPSP synthase, CP4 EPSPS, "Shotgun Cloning Method," Biolistic® gene gun, Whiskers™, Genetic engineering, Gene, Bioseeds, Glyphosate, Glyphosate-trimesium, Glyphosate isopropylamine salt, NOS terminator.*

AHG

Antihemophilic globulin. Also known as factor VIII or antihemophilic factor VIII. See *Factor VIII, Gamma globulin.*

Al

Acronym for "adequate intake." See also *Choline.*

AIDS

See *Acquired immune deficiency syndrome* (*AIDS*).

Airlift Fermenter

A vessel utilized to grow cells in a liquid medium, in which those cells are mixed/stirred via air that is introduced at the bottom of the vessel, and the air rises through the culture medium. See also *Cell, Medium, Culture medium, Cell culture.*

aiRNA

See *Artificial interfering RNA* (*aiRNA*).

AKT1 Gene

See *Artificial interfering RNA* (*aiRNA*).

Alanine (Ala)

A nonessential amino acid of the pyruvic acid family. In its dry, bulk form, it appears as a white crystalline solid. See also *Essential amino acid.*

Albumin

A protein that the body's liver synthesizes (i.e., "manufactures"). Among its other functions, albumin stabilizes/protects antibodies and enzymes in the bloodstream, enhances the antibody–antigen reaction, and reversibly binds certain ligands (thereby enabling them to be carried by the blood to their destination within the body).

Most minerals and hormones utilized by the human body are first "attached" to a molecule of albumin before they are then transported within the bloodstream to where they are needed in the body. In 2005, the U.S. Food and Drug Administration approved albumin-stabilized paclitaxel nanoparticles as the pharmaceutical Abraxane™ for treatment of breast cancer, non-small-cell lung cancer, and pancreatic cancer. See also *Protein, Antibody, Enzyme, Enzyme denaturation, Antigen, Humoral immunity, Cellular immune response, Ligand* (*in biochemistry*), *Hormone, Supercritical carbon dioxide, Food and Drug Administration* (*FDA*), *Paclitaxel, Anoparticles, Cancer, Chemotherapy.*

ALCAR

Acronym for acetyl-L-carnitine. See *Acetyl carnitine.*

Aldose

A simple sugar in which the carbonyl carbon atom is at one end of the carbon chain. A class of monosaccharide sugars; the molecule contains an aldehyde group. See also *Monosaccharides.*

Aleurone

The layer ("skin") that covers the endosperm portion of a plant seed. See also *Endosperm.*

AlfAFP

Acronym for alfalfa antifungal peptide. See *Defensins.*

Algae

A heterogeneous (i.e., widely varying) group of photosynthetic plants, ranging from microscopic single-cell forms to multicellular, very large forms such as seaweed. All of them contain chlorophyll and hence most are green, but some of them may be different colors due to the presence of other, overshadowing pigments. See also *Photosynthesis.*

Algorithm (Bioinformatics)

Refers to a computational procedure that utilizes a combination of simple (e.g., mathematical) operations to process, analyze, and/or visualize (i.e., show pictorially) data about sequences (of DNA, RNA, proteins, etc.). See also *Bioinformatics, Sequence* (*of a DNA molecule*), *Sequence* (*of a protein molecule*), *Sequence MAP.*

Alicin

A compound that is produced naturally by the garlic plant (*Allium sativum*) when the cells within garlic bulbs are broken open (e.g., during food preparation or consumption). Enzymes present within cells convert alicin (precursor compound) to sulfenic acid, a powerful antioxidant that rapidly reacts with free radicals (to "quench" those radicals) in the body.

Research indicates that human consumption of alicin confers some specific health benefits (e.g., antithrombotic, reduce blood cholesterol levels, reduce/avoid coronary heart disease, enhance the immune system).

Alicin has also been shown to slow down the action of phase I detoxification enzymes (e.g., potentially reducing levels of some carcinogens within the human digestive system). See also *Cell, Phytochemicals, Enzyme, Inducible enzymes, Thrombosis, Coronary heart disease* (*CHD*), *Cholesterol, Carcinogen, Free radical, Phase I detoxification enzymes.*

Alkaline Hydrolysis

A chemical method of liberating DNA from a DNA–RNA hybrid. See also *Hydrolysis, Ribonucleic acid* (*RNA*), *DNA–RNA hybrid, Deoxyribonucleic acid* (*DNA*).

Alkaloids

A class of toxic compounds that are naturally produced by some organisms (e.g., certain ants, certain plants such as lupines and potatoes, and certain fungi such as ergot).

For example, the prickly yellow poppy (*Argemone mexicana*) naturally produces an alkaloid within the oil in its seeds.

For example, certain species of ants naturally produce alkaloids, as a self-defense mechanism. Poison dart frogs (*Dendrobates azureus*) and two species of New Guinea songbirds (*Pitohui dichrous* and *Ifrita kowaldi*) can tolerate those ant-produced alkaloids, so they also acquire that self-defense (toxin) by eating those particular ants.

Another example is the moth *Utetheisa ornatrix*, whose larvae (caterpillars) feed on certain plants (*Crotalaria* spp.) that contain pyrrolizidine alkaloids. Because those alkaloids are extremely bitter tasting and toxic, spiders that normally prey on them refuse to eat those *U. ornatrix*, even after they later become adult moths. If those moths (who consumed those pyrrolizidine alkaloids as larvae) get caught in the spider's web, the spider will cut it out of the web and release that particular (toxic) moth.

Vinca alkaloids isolated from the specific plants that produce them have been utilized as cancer-treating (i.e., antitumor) drugs.

A chlorine-containing alkaloid named epibatidine, isolated from the skin of the South American frog *Epipedobates tricolor*, has shown potential for use as a painkilling pharmaceutical (200 times more powerful than morphine). See also *Toxin, Fungus, Tremorgenic indole alkaloids, Ergotamine, Colchicine.*

Alkylating Agents

Refer to specific chemicals or other agents that result in chemical attachment of *alkyl molecular groups* (*e.g., methyl group, ethyl group*) to molecules such as DNA.

Some alkylating agents (e.g., ethyl methanesulfonate) are mutagens, for example, alkylating agent damage to DNA's guanine base to form O6-alkylguanine, a type of DNA damage that is particularly prevalent in colon or bowel cancer. See also *Methylated, Deoxyribonucleic acid (DNA), DNA methylation, Mutagen, Mutation breeding.*

Allele

From the Greek *allelon* meaning "mutually each other," the term refers to one of several alternate forms of a gene occupying a given locus on the chromosome, which controls expression (of product) in different ways. See also *Express, Gene, Chromosomes, Locus, Paramutation, Commensal, Wild type, DNA methylation.*

Allele Frequency

Refers to the relative amount of copies of a particular allele present within a population of organisms. For a specific allele, it is expressed as a fraction of *the grand total* number of copies of all alleles found at a given locus of the population's DNA. See also *Allele, Organism, Deoxyribonucleic acid (DNA), Locus.*

Allelic Exclusion

The expression in any particular manner of only one of the alleles in (the two inherited copies of each) chromosomes, due to chromosomal inactivation. For example, only one allele of an *antibody gene* within a B lymphocyte (blast cell) coding for the expressed antibody (in response to antigenic stimulus) is involved in expression, due to chromosomal inactivation during blast transformation. See also *Allele, Coding sequence, Gene, Chromosomes, B lymphocytes, Antibody, Immunoglobulin, Blast cell, Blast transformation, DNA methylation.*

Allelopathy

Refers to the secretion of certain chemicals (e.g., terpenoid compounds) by a plant, in order to hinder the growth or reproduction of other plants growing near it. For example,

- The crop plant known as sorghum secretes a chemical compound known as sorgoleone, which falls from that bush and inhibits growth of other plants near that bush.
- The *Callistemon citrinus* bush secretes a chemical compound known as leptospermone, which falls from that bush and inhibits growth of other plants near that bush.

In a similar manner, the fungus *Laccaria bicolor* secretes a chemical compound that paralyzes springtails (soil-dwelling insects) so the fungus can engulf and digest the springtails. The *L. bicolor* fungus lives symbiotically among the roots of the eastern white pine tree, so this results in the *nitrogen within the springtail's body* (i.e., fertilizer) being delivered to that eastern white pine tree. In return, the eastern white pine tree roots supply the *L. bicolor* fungus with certain carbohydrates that fungus requires. See also *Terpenoids, Fungus, Gallic acid.*

Allergies (Airborne)

See *Mast cells.*

Allergies (Foodborne)

Coined in 1906 by Clemens Freiherr von Piguet, it refers to an IgE-mediated (aggressive) immune system response to antigen(s) present on protein molecules—or (rarely) on a sugar molecule—in the particular food that a given person is allergic to. The antibodies (IgE) bind to those antigens and trigger a humoral immune response, which can cause vomiting, diarrhea, skin reactions (e.g., hives), blood pressure decline, wheezing, and respiratory distress. In severe cases, the immune response can cause death.

In some rare instances, the allergic reaction is mediated by sensitized T cells.

In some rare instances, the onset of a food allergy incident is induced by exercise (done before or after eating that particular food).

The U.S. Food and Drug Administration (FDA) requires testing in advance to determine if a genetically engineered foodstuff has the potential to cause allergic reactions in humans, before that genetically engineered foodstuff (e.g., a modified crop plant) is approved by the FDA.

In general, known food allergens (e.g., in peanuts, Brazil nuts, wheat) are protein molecules that are resistant to rapid digestion (e.g., because those protein molecules are too tightly "folded together" for digestive enzymes to access their chemical bonds to break down). One potential way to genetically engineer currently allergenic crops (e.g., wheat) to make them less allergenic is to insert gene(s) for extra production of thioredoxin.

Thioredoxin is a protein found in all living organisms, which "targets" and breaks down the chemical bonds holding together a tightly folded together protein molecule (thereby making those protein molecules easier to digest), so future crops engineered to contain more thioredoxin than traditional average level may be less allergenic. See also *Protein, Protein folding, Antibody, Antigen, Food and Drug Administration (FDA), Genetic engineering, Immunoglobulin, Humoral immunity, Mast cells, Leukotrienes, Digestion (within organisms), Organism, Reduced-allergen soybeans, Sugar molecules.*

Allicin

See *Alicin.*

Allogeneic

With a different set of genes (but same species). For example, an organ transplant from one nonrelated human to another is allogeneic. An organ transplant from a baboon to a human would be xenogeneic. See also *Gene, Species, Xenogeneic organs.*

A

Allosteric Enzymes

Regulatory enzymes whose catalytic activity is modulated by the noncovalent binding of a specific metabolite (effector) at a site (regulatory site) *other* than the catalytic site (on the enzyme). Effector binding causes a three-dimensional conformation change in the enzyme and is the root of the modulation.

The term (allosteric) is used to differentiate this form of regulation from the type that may result from the competition between substrate and inhibitors at the catalytic site. See also *Enzyme, Steric hindrance, Effector, Conformation, Active site.*

Allosteric Regulation

See *Allosteric site.*

Allosteric Site

The "site" on an (allosteric) enzyme molecule where, via noncovalent binding to the site, a given effector can increase or decrease that enzyme's catalytic activity. Such an effector is called an allosteric effector because it binds at a site on the enzyme molecule that is other (allo) than the enzyme's catalytic site. See also *Allosteric enzymes, Activator (of enzyme), Catalytic site, Effector, Conformation, Enzyme, Metabolite, Catalyst.*

Allosterism

Refers to the deformation of a protein molecule's conformation (and thus its activity) that is caused when a ligand (e.g., effector) binds to that protein molecule (e.g., enzyme) at a spot (e.g., "regulatory site" for enzyme) other than the protein's *active site* (i.e., which is "catalytic site" in an enzyme). See also *Protein, Conformation, Ligand, Effector, Enzyme, Active site, Catalytic site, Allosteric site, Allosteric enzymes.*

Allotypic Monoclonal Antibodies

Monoclonal antibodies that are isoantigenic. See also *Monoclonal antibodies (MAb), Antigen.*

Allozyme

Synonym for "allosteric enzyme." See *Sllosteric enzymes.*

Aloe vera L.

A "family" of short green plants (e.g., *Aloe barbadensis*) that grow in hot tropical climates, whose sap (juice) contains certain carbohydrates that naturally assist healing of human skin (wounds). Those carbohydrates "activate" macrophages, which causes those macrophages to produce cytokines (that regulate human immune system and inflammatory responses that promote healing). See also *Phytochemicals, Carbohydrates (saccharides), Macrophage, Cytokines.*

Alpha-Amylase Inhibitor-1

A protein that is naturally produced in the seeds of the plant known as the common bean *Phaseolus vulgaris*, which inhibits the amylase enzyme in the gut of the pest insect known as the pea weevil. Because the amylase enzyme (in its gut) is inhibited (i.e., prevented from helping digestion) by the alpha-amylase inhibitor-1, the seeds of the *P. vulgaris* plant are protected from depredation by the pea weevil. See also *Amylase inhibitors, Protein, Enzyme, Amylase, Weevils.*

Alpha Amylase T

A heat-tolerant (to greater than 300°F/149°C) amylase that is produced within the kernels of certain genetically engineered varieties of corn (maize). See also *Amylase, Enzyme, Corn, Genetic engineering.*

Alpha-Galactosides

Term utilized to refer to a "family" of polysaccharides (produced in plant seeds) composed (at the molecular level) of one sucrose unit linked by α 1,6 molecular bonds to several galactose units.

Alpha-galactosides include raffinose, stachyose, and verbascose. See also *Polysaccharides, Galactose (gal), Stachyose.*

Alpha Helix (α-Helix)

A highly regular (i.e., repeating) structural feature that occurs in certain large molecules. First discovered in protein molecules by Linus Pauling in the late 1940s. See also *A-DNA, Protein, Protein folding, Protein structure.*

Alpha Interferon

Also written as α-interferon. One of the interferons, it has been shown to prolong life and reduce tumor size in patients suffering from Kaposi's sarcoma (a cancer that affects approximately 10% of people with acquired immune deficiency syndrome). It is also effective against hairy cell leukemia and may work against other cancers. It has recently been approved by the U.S. FDA for use against certain types of sarcoma. Recent research indicates that injections of alpha interferon can limit the liver damage typically caused by hepatitis C, a viral disease. See also *Interferons, Cancer, Food and Drug Administration (FDA).*

Alpha-Linolenic (α-Linolenic) Acid

See *Linolenic acid.*

Alpha-Chaconine

See *Chaconine.*

Alpha-Rumenic Acid

See *Conjugated linoleic acid (CLA).*

Alpha-Solanine

See *Solanine.*

Alpha-Synuclein

A protein that is present within most cells of the brain. When something causes such protein molecules to aggregate (clump together), those aggregates can kill neurons in the brain, leading to diseases such as Parkinson's disease and Alzheimer's disease. Although there is no cure for those diseases, recent research indicates that molecular tweezers (e.g., CLR01) might be helpful, if developed into successful pharmaceuticals. Molecular tweezers are particular molecules shaped like the letter "C" that are capable of binding to specific proteins like alpha-synuclein, via wrapping around chains of lysine (an amino acid that is a constituent of alpha-synuclein proteins). Some molecular tweezers have been shown to be able to prevent α-synuclein from forming aggregates, prevent toxicity to neurons, and even break up existing aggregates.

Members of some human families, who inherit a (mutation) gene (SNP) that codes for a mutated form of alpha-synuclein, have a higher than average tendency to get Parkinson's disease. See also *Protein, Neuron, Cell, Parkinson's disease, Gene, Mutation, Single nucleotide polymorphisms (SNPs).*

ALS

A plant enzyme (also present in some microorganisms) known as "acetolactate synthase" or acetohydroxy acid synthase. ALS catalyzes (i.e., enables to occur) one of the early chemical reaction steps in the synthesis ("manufacturing") of branched-chain amino acids (isoleucine, leucine, valine), which are required by plants to sustain life (i.e., to make needed proteins).

Herbicides that deactivate/destroy ALS are effective at killing plants (e.g., weeds). See also *Enzyme, Gene, ALS gene, Microorganism, Catalyst, Amino acid, Isoleucine (ile), Leucine (leu), Valine (val).*

ALS

Acronym for the disease amyotrophic lateral sclerosis. See *Amyotrophic lateral sclerosis.*

ALS Gene

Gene that codes for (i.e., causes to be produced in microorganisms or plants' chloroplasts) the critical-to-plants enzyme "acetolactate synthase (ALS)." Also known as "acetohydroxy acid synthase," ALS catalyzes (i.e., enables to occur) one of the early chemical reaction steps in the synthesis ("manufacturing") of branched-chain amino acids (isoleucine, leucine, valine) by plants. Because those branched-chain amino acids are required by plants to sustain life, herbicides that deactivate/destroy ALS are effective at killing plants (e.g., weeds). See also *Gene, HTC, Microorganism, Chloroplasts, Enzyme, Catalyst, Amino acid, Isoleucine (ile), Leucine (leu), Valine, STS sulfonylurea (herbicide)-tolerant soybeans.*

Alternative mRNA Splicing

See *Alternative splicing.*

Alternative Splicing

The process (during transcription) via which alternative exons (i.e., portion of gene that codes for specific domain of a protein) within a given RNA molecule are combined (by RNA polymerase molecules known as snRNPs) to yield *different* mRNAs (messenger RNA molecules) from the same gene. Each such mRNA is known as a "gene transcript" and of course results in a different protein subsequently being produced by the cell.

Other causes/sources of alternative splicing include

- Varying translation start or stop site (on the mRNA during its translation), resulting in a given intron remaining in the mRNA transcript. For example, the COX-3 enzyme and the COX-1 enzyme are both produced from the COX-1 gene. The COX-3 enzyme results when *intron 1* is retained in the mRNA transcript.
- Frameshifting (i.e., different set of triplet codons in the mRNA/transcript is translated by the ribosome).

Different body tissues, some diseases, and some environmental stresses can cause alternative splicing (i.e., resulting in

- *Different* proteins being produced in *different* tissues of the organism
- *Different* proteins being produced in *diseased* tissues (versus healthy tissues)
- *Different* proteins being produced in a plant when that plant is under stress (i.e., disease infection) than when that plant is not under stress *from a given* gene

For example, alternative splicing results in a plant's ICS gene producing

- A protein that leads to the synthesis of salicylic acid (which helps the plant resist disease) during attack/infection by pathogens
- A protein that leads to the synthesis of vitamin K utilized by the plant in the photosynthesis process, when the plant is not under attack/infection by pathogens

See also *Transcription, Exon, Intron, Domain (of a protein), Protein, Gene, Enzyme, RNA polymerase, Transcript, Translation, Messenger RNA (mRNA), Coding sequence, Codon, Genetic code, Frameshift, Ribosomes, Transcriptome, Central dogma (new), COX-1, COX-3, Pathogen, Cyclooxygenase, Organism, Salicylic acid (SA), Vitamin, Read-through.*

Alu Family

A set of dispersed and related genetic sequences, each about 300 bp long, in the human genome. At both ends of these 300 bp segments, there is an AG-CT sequence. Alu 1 is a restriction enzyme that recognizes this sequence and cleaves (cuts) it between the G (guanine) and the C (cytosine). See also *Genome, Restriction endonucleases, Sequence (of a DNA molecule).*

Alu Sequences

See *Alu family.*

Aluminum Resistance

See *Citrate synthase (CSb) gene, Gene, Citric acid, Phenomics.*

Aluminum Tolerance

See *Citrate synthase (CSb) gene, Gene, Citric acid, Phenomics.*

A

Aluminum Toxicity

See *Citrate synthase (CSb) gene, Gene, Citric acid.*

Alzheimer's Disease

Named after Alois Alzheimer who first described the amyloid β-protein (AβP) plaques in the human brain that are caused by this disease, in 1906. Alzheimer's disease causes certain proteins to misfold and aggregate in the brain, resulting in progressive memory loss and dementia in its victims as it kills brain cells (neurons). AβPs are also harmful to the blood vessels that supply the brain—thereby further accelerating the cognitive decline of Alzheimer's disease by restricting the flow of blood-borne oxygen and nutrients.

Some human haplotypes are more susceptible to Alzheimer's disease than others. Women tend to get Alzheimer's disease significantly more often than men of the same age. Black people in North America tend to get Alzheimer's disease significantly more often than Caucasians in North America.

Alzheimer's disease begins within the brain's lateral entorhinal cortex (LEC). The LEC is the "gateway" to the brain's hippocampus, which plays a crucial role in the consolidation of long-term memory, among its other functions. So if the LEC is affected by Alzheimer's, other properties of the hippocampus will also inevitably be adversely impacted. The LEC is particularly vulnerable to Alzheimer's disease because it normally accumulates tau, which sensitizes the LEC to any accumulation there of amyloid precursor protein. Together, these two proteins damage neurons in the LEC.

With time, the Alzheimer's disease spreads from the LEC directly to other parts of the brain's cerebral cortex—in particular, to the parietal cortex, a region of the brain that is involved in the body's spatial orientation and navigation.

Some drugs (e.g., tacrine, donepezil) appear to slow the progression of Alzheimer's disease (by increasing the availability of acetylcholine in the brain), but there is currently no way to stop the disease. See also *Protein, Protein folding, Amyloid precursor protein, Amyloid β-protein (AβP), Amyloid β-protein precursor (AβPP), Neuron, Neurotransmitter, Acetylcholine, Oxidative stress, Curcumin, Haplotype, Rapid protein folding assay, Nitric oxide.*

Amber Stop Codon

See *Termination codon.*

AMD

Acronym for "age-related macular degeneration." See *Lutein, Angiogenesis, Lycopene, Beta-carotene, Complement factor H gene.*

American Type Culture Collection (ATCC)

An independent, nonprofit organization that was established in 1925 for the preservation and distribution of reference cultures. See also *Cell culture, Culture, Culture medium, Type specimen, Consultative Group on International Agricultural Research (CGIAR).*

Ames Test

A simple bacterial-based test for carcinogens that was developed by Bruce Ames in 1961. Although this test evaluates mutagenesis (i.e., causation of mutations) in the DNA of bacteria, its results have been utilized to approve or not approve certain compounds for consumption by humans. See also *Bioassay, Bacteria, Assay, Mutual recognition agreements (MRAs), Genotoxic carcinogens, Carcinogen, PARP.*

Amino Acid

A category of molecules that contain both "amino ($-NH_2$)" and "carboxyl ($-COOH$)" submolecular groups. There are 20 common natural amino acids and at least 2 uncommon natural amino acids, each specified by a different arrangement of three adjacent DNA nucleotides. These are the building blocks of proteins. Joined together in a strictly ordered chain, the sequence of amino acids determines the character of each protein (chain) molecule. The 20 common amino acids are alanine, arginine, aspartic acid, glutamic acid, glutamine, glycine, histidine, isoleucine, leucine, phenylalanine, proline, serine, threonine, tryptophan, tyrosine, valine, cysteine, methionine, lysine, and asparagine. Note that virtually all of these amino acids (except glycine) possess an asymmetric carbon atom and thus are potentially chiral in nature.

One of the least common amino acids is selenocysteine, which is made by humans, some animals, and some *archaea.* See also *Protein, Polypeptide (protein), Stereoisomers, Chiral compound, Messenger RNA (mRNA), Essential amino acids, Deoxyribonucleic acid (DNA), Absolute configuration, Archaea.*

Amino Acid Profile

Also known as "protein quality," this refers to a quantitative delineation of how much of each amino acid is contained in a given source of (livestock feed or food) protein. For example, the amino acid profile of soybean meal is matched closest to the profile of amino acids needed for human nutrition, of all the plant protein meals. See also *Ideal protein concept, Protein, Amino acid, Lipid rafts.*

Aminoacyl tRNA Synthetases (AARS)

See *Ribosomes, Aminoacyl-transfer RNA synthetases.*

Aminoacyl-Transfer RNA Synthetases

Also known as tRNA synthetases or aminoacyl tRNA synthetases, these are a family of 20 catalytic enzymes that play a key role in translation (the cellular process of producing proteins). The tRNA synthetases select the appropriate amino acid (in the sequence that is coded for by genes) and assign those amino acids to transfer RNAs (tRNAs) to manufacture a protein molecule comprised of the gene-specified amino acids within the cell's ribosome.

See also *Enzyme, Translation, Cell, Amino acid, Protein, Gene, Ribosomes.*

Aminocyclopropane Carboxylic Acid Synthase/Deaminase

See *ACC synthase, ACC.*

Amorph

See *Null mutation.*

AMP

See *Adenosine monophosphate (AMP)*.

Amphibolic Pathway

A metabolic pathway used in both catabolism and anabolism. See also *Anabolism, Catabolism.*

Amphipathic Molecules

Molecules bearing both polar and nonpolar domains (within the same molecule). Some examples of amphipathic molecules are wetting agents (SDS) and membrane lipids such as lecithin. See also *Micelle, Reverse micelle (RM), Polarity (chemical).*

Amphiphilic Molecules

Also known collectively as amphiphiles, they are molecules possessing distinct regions of hydrophobic ("water hating") and hydrophilic ("water loving") character within the same molecule. When dissolved in water above a certain concentration (known as the CMC), these molecules are capable of forming high-molecular-weight aggregates or micelles. See also *Critical micelle concentration, Hydrophobic, Hydrophilic, Micelle, Reverse micelle (RM).*

Amphoteric Compound

A compound capable of both donating and accepting protons and thus able to act chemically as either an acid or a base.

Ampicillin

An antibiotic of the penicillin class that interferes with the building of relevant bacteria's cell wall, thereby killing those bacteria. See also *Bacteria, Penicillin G (benzylpenicillin).*

AMPK

Acronym for "adenosine monophosphate–activated protein kinase" or "AMP-activated protein kinase." AMPK is a "master regulator" (enzyme) of cellular metabolism that is activated during times of reduced cell energy availability (e.g., during times of starvation). During such activation, AMPK helps to improve energy cells' lipid profile, energy homeostasis, the body's blood pressure, etc. AMPK is also involved in the activation of the p53 tumor suppressor gene. See *Enzyme, Metabolism, AMP, Protein kinases, Lipids, Homeostasis, p53 gene.*

Amplicon

A specific sequence of DNA that is produced by a DNA amplification technology such as the polymerase chain reaction technique. See also *Deoxyribonucleic acid (DNA), Sequence (of a DNA molecule), Polymerase chain reaction (PCR) technique, Nested PCR.*

Amplification

The production of additional copies of a chromosomal sequence, found in the form of either intrachromosomal or extrachromosomal DNA.

Some amplifications (of miRNA genes) can lead to cancer. See also *Whole genome amplification, In vitro selection, miRNA genes, Cancer, MicroRNAs.*

Amplified Fragment Length Polymorphism

Also known by its acronym "AFLP," it is a "DNA marker" utilized in a "genetic mapping" technique that utilizes the specific sequence of bases (nucleotides) in a piece of DNA (from an organism). Since the specific sequence of bases in their DNA molecules is different for *each* species, strain, variety, and individual (due to DNA polymorphism), AFLP can be utilized to "map" those DNA molecules (e.g., to assist and speed up plant breeding programs). See also *Genetic map, Sequence (of a DNA molecule), Deoxyribonucleic acid (DNA), Genome, Physical map (of genome), Marker (DNA sequence), Marker (genetic marker), Polymorphism (chemical), Nucleic acids, Nucleotide, Genetic code, Capillary electrophoresis, Anonymous DNA marker.*

Amplimer

See *Amplicon.*

Amylase

A term that is used to refer to a category of enzymes that catalyzes the chemical reaction in which amylose (starch) molecules are hydrolytically cleaved ("broken") to molecular pieces (e.g., the polysaccharides maltose, maltotriose, α-dextrin). For example,

- Amylases are produced within the digestive system of certain insects that feed upon starch-containing grains.
- α-amylase is used to break apart corn starch molecules, in the first step of manufacturing fructose (sweetener for soft drinks).

Since 1857, amylase has been utilized to remove (amylose) starch from woven fabrics in the textile industries, where the amylose was applied to cotton yarn as a "lubricant" in the weaving process. Modern uses of some amylases include

- Utilizing the enzyme to enable the substitution of barley grain for malt in the beer brewing process
- Utilizing the enzyme (via its addition to bread dough or flour) to slow the process of the subsequently produced bread from going stale
- Utilizing the enzyme within detergent products to remove starchy (e.g., food) stains from clothes or dishware

See also *Enzyme, Starch, Amylose, Barley, Hydrolytic cleavage, Polysaccharides, Amylase inhibitors, Alpha-amylase inhibitor-1, Digestion (within organisms).*

Amylase Inhibitors

Term that refers to a compound that chemically binds to an amylase (i.e., an enzyme that is produced within the digestive system of certain insects and other organisms) and thereby prevents the amylase from breaking down (i.e., digesting) any amylose (i.e., a carbohydrate present within many plants' seeds) consumed by the insect.

In 2004, South American researchers reported that coffee beans produced by coffee plants (*Coffea arabica*) that had been genetically

A

engineered to produce some amylase inhibitor in their beans were thereby protected from predation by the pest insect known as the coffee berry borer (*Hypothenemus hampei*). That particular amylase inhibitor's gene had been isolated from the South American bean plant known as *Phaseolus coccineus*, in which it is naturally produced. See also *Amylase, Enzyme, Protein, Digestion (within organisms), Amylose, Coffee berry borer, Gene, Genetic engineering, Alpha-amylase inhibitor-1.*

Amyloid AβP Protein (AβP)

A small protein that forms plaques in the brains and in the brain blood vessels of victims of Alzheimer's disease. AβP forms cation-selective ion channels in lipid bilayers (e.g., membranes surrounding cells). This ion channel formation disrupts calcium homeostasis, allowing (destructive) high concentrations of calcium ions in brain cells.

Amyloid β Protein Precursor (AβPP)

A (collective) set of protein molecules, from which are derived Amyloid b Protein (AβP).

Amyloid Plaques

See *Alzheimer's disease.*

Amyloid Precursor Protein

A transmembrane (i.e., extends through cells' plasma membrane) protein (prion) within brain cells of mammals. Under certain conditions, after being cut into pieces by gamma-secretase (γ-secretase) enzyme, it can form the molecular derivative known as amyloid β-protein, a cause of Alzheimer's disease. Amyloid β-proteins are also harmful to the blood vessels that supply the brain—thereby further accelerating the cognitive decline of Alzheimer's disease by restricting the flow of blood-borne oxygen and nutrients. See also *Protein, Cell, Plasma membrane, Alzheimer's disease, Docosahexaenoic acid (DHA), Gamma-secretase.*

Amylopectin

The water-insoluble form of starch (molecule) that consists of multibranched polymers, containing approximately 100,000 glucose units per molecule (polysaccharide). It is naturally present in "waxy" wheat and corn/maize varieties at 99%–100% of total kernel starch content, so as a result, "waxy corn starch" forms long cohesive pastes when cooked in water and subsequently cooled.

Among other uses, amylopectin is utilized by paper manufacturers to bind together pulp fibers and to "size" the surface of papers. See also *Starch, Polymer, Glucose (GLc), Polysaccharides, Waxy corn.*

Amylose

The form of starch (long-molecular-chain sugars) that consists of unbranched polymers, containing approximately 4000 glucose units per molecule (polysaccharide). It is naturally present in dent corn/maize at 24%–28% of total kernel starch content, so as a result, "dent corn starch" forms firm gels when cooked in water and subsequently cooled.

It is naturally present in wheat at 24%–26% of total kernel starch content and in potatoes at 23%–29% content (variation is thought to be caused by different growing conditions). Because amylose is slow to be digested (i.e., broken down to glucose molecules) by humans, its consumption does not cause a "spike" (i.e., sudden increase) in bloodstream levels of glucose, which can be hazardous for some people (e.g., diabetics). See also *Polymer, Starch, Glucose (GLc), Diabetes, Amylase, Polysaccharides, High-amylose corn, High-amylose wheat.*

Amyotrophic Lateral Sclerosis

Abbreviated as ALS, it is a disease that is characterized by progressive lethal damage to motor neurons, which results in a variety of symptoms that eventually includes total loss of voluntary muscle control of movement in ALS patients. Approximately 5% of ALS sufferers were discovered in 2011 to have within their DNA an altered version of a particular gene known as "C9orf72," which in those particular ALS patients contain hundreds of DNA repeat sequences *that are not present within the DNA of normal individuals.* See also *Deoxyribonucleic acid (DNA), Sequence (of a DNA molecule), Short tandem repeats, Gene, Neuron.*

Anabolic Pathway

Refers to a biosynthetic pathway via which a metabolite is made, within an organism. See also *Pathway, Metabolite, Organism.*

Anabolism

The phase of intermediary metabolism concerned with the energy-requiring biosynthesis of cell components from smaller precursor molecules. See also *Catabolism, Assimilation, Metabolism, Cell, Plasma membrane, Anabolic pathway.*

Anaerobe

An organism that lives in the absence of oxygen and generally cannot grow in the presence of oxygen. The catabolic metabolism of anaerobic microorganisms reduces a variety of organic and inorganic compounds in order to survive (e.g., carbon dioxide, sulfate, nitrate, fumarate, iron, manganese), and anaerobes produce a large number of end products of metabolism (e.g., acetic acid, propionic acid, lactic acid, ethanol, methane). See also *Catabolism, Metabolism, Metabolite, Reduction (in a chemical reaction), Anaerobic.*

Anaerobic

An environment without air or oxygen. See also *Anaerobe.*

Anaerobic Digestion

Refers to digestion (breakdown of materials into their subcomponents) in which molecular oxygen is unavailable. See also *Digestion (within chemical production plants), Anaerobe.*

Analog Gene

See *Ortholog.*

Analogue

(Analog) A compound (or molecule) that is a (chemical) structural derivative of a "parent" compound. The word is also used to describe a molecule that may be structurally similar (but not identical) to another and that exhibits many or some of the same biological functions of the other.

For example, the large class of antibiotics known as the sulfa drugs are all analogues of the original synthetic chemical drug (known as prontosil, cures streptococcal infections) discovered by the German biologist Gerhard Domagk. Domagk's and other discoveries made possible a program of further chemical syntheses based upon the original (sulfanilamide) molecular structure, which resulted in the large number of sulfonamide (also called "sulfa") drugs that are available today. All of the analogue (also analog) sulfa drugs that were patterned after the original sulfanilamide molecular structure may be called sulfanilamide analogues.

Today, analogues are known by man for various vitamins, amino acids, purines, sugars, growth factors, and many other chemical compounds. Research chemists produce analogues of various molecules in order to ascertain the biological role of, or importance of, certain structures (within the molecule) to the molecule's function within a living organism. See also *Biomimetic materials, Rational drug design, Heterology, Gibberellins, Quantitative structure–activity relationship (QSAR).*

Anaphase

The third of the four phases of eucaryotic mitosis (i.e., cell replication via division) during which the now split in half chromosome pairs (which had recently doubled via copying during S-phase of mitosis) are pulled toward the "north pole" and the "south pole" of the cell by actin-utilizing microtubules. See also *Mitosis, Eucaryote, Cell, Chromosomes, S-phase, Actin, Microtubules.*

Ancestral Trait Restoration

Refers to the creation of crop plants (today) whose genome (DNA) contains a particular trait that had been present within the genome of the wild-type ancestor of that crop plant but was subsequently lost during that crop plant's domestication process (e.g., 1000 years ago). For example, in 2014, Lijuan Qiu and Rongxia Guan discovered a salt tolerance gene present in the DNA of wild-type soybean plants that are the ancestors of today's domesticated soybean (*Glycine max* (L.) Merrill) varieties. Because today's domesticated soybean varieties do not possess that salt tolerance gene, a soybean breeder wanting to create a modern soybean variety that would grow well in salty soil could utilize a wide cross between a modern soybean variety (germplasm) and one of those salt-tolerant wild-type soybean species (i.e., retrieved from one of the seed banks utilized to store ancestral crop-plant relatives).

In addition to crop breeder use of a *wide cross* methodology, such crop "trait restoration" can be accomplished via certain other technologies. See also *Deoxyribonucleic acid (DNA), Gene, Genome, Trait, Soybean plant, Germplasm, Traditional breeding methods, Wide cross, Deletions, Trait restoration.*

ANDA (to FDA)

Abbreviated New Drug Application (to the U.S. Food and Drug Administration). See also *NDA, "Treatment" IND regulations, Food and Drug Administration (FDA).*

Anergy

Refers to a state of *inactivation (e.g., of a cell)*. For example, if a given human T lymphocyte cell's applicable receptor (T cell receptor) binds to a relevant protein molecule "presented" on the surface of a major histocompatibility complex but "that binding is *not* also accompanied by a required parallel series of chemical signals (to indicate the presented protein is a pathogen)," that T lymphocyte remains in a state of anergy.

This requirement for parallel/simultaneous chemical signals is one of the ways that the human immune system (usually) avoids attacking the body's own tissue/cells. See also *Cell, T lymphocytes, T cells, Receptors, T cell receptors, Major histocompatibility complex (MHC), Signaling.*

Aneuploidy

The process in which a new daughter cell gains chromosome(s) or loses chromosome(s). See also *Chromosomes, Cell mitosis.*

Angiogenesis

Formation/development of new blood vessels in the body. Discovered to be triggered and stimulated by angiogenic growth factors, in the early 1980s. For example, when heart arteries are clogged by arteriosclerosis, increased production of "granulocyte-macrophage colony-stimulating factor" can stimulate development of new blood vessels (to sometimes restore blood flow).

Angiogenesis is required for malignant tumors to metastasize (spread throughout the body), because it provides the (newly created) blood supply that tumors require. For example, the gene that codes for the production of vascular endothelial growth factor (VEGF) is greatly upregulated by chemical signals that are produced by hepatocarcinoma tumors. Some tumors produce transforming growth factor-beta to activate blood vessel formation.

Angiogenesis is crucial to the development of glaucoma and age-related macular degeneration (AMD), a major cause of blindness in older people. In the case of the "wet" form of AMD, the body's production of VEGF can cause creation/growth of blood vessels in front of the retina/macula, which eventually leads to blindness. Research indicates that siRNA-based drugs might be able to prevent "wet" AMD via prevention of (over)production of VEGF by the body.

Another pathway (independent of VEGF) that can cause creation/growth of blood vessels in front of the retina/macula can occur via activation of the molecule neuropilin 1 (NRP1) produced by certain connective tissue components within the eyeball. The NRP1 subsequently conveys signals inside the blood vessel cells via another molecule named ABL1 and thereby stimulates blood vessel growth.

Angiogenesis can also contribute to rheumatoid arthritis via development of new (inappropriate) blood vessels in the joints.

The drug thalidomide is also a potent inhibitor of angiogenesis, as are the monoclonal antibody pharmaceutical known as bevacizumab and the proteins "angiostatin" and "endostatin." See also *Angiogenic growth factors, Granulocyte-macrophage colony-stimulating factor (GM-CSF), Nitric oxide, Rheumatoid arthritis, Arteriosclerosis, Tumor, Cancer, Metastasis, Gene, Up-regulation, Vascular endothelial growth factor (VEGF), Transforming growth factor-beta (TGF-beta), Age-related macular degeneration (AMD), Short interfering RNA (siRNA), Anti-angiogenesis, Pathway, Signaling, Signaling molecule, Chiral compound, Angiostatin, Endostatin, Bevacizumab, Nanofibers.*

Angiogenesis Factors

See *Angiogenic growth factors.*

Angiogenesis Inhibitors

Refer to compounds (e.g., pharmaceuticals, interleukin-18) that work to inhibit/stop angiogenesis (i.e., formation/development of new blood vessels). Because angiogenesis is required for malignant tumors to grow and/or metastasize (spread), such compounds hold the potential to treat cancerous tumors.

For example, the biotechnology-derived angiogenesis inhibitor pharmaceutical known as Avastin® (bevacizumab) has been proven to be effective against metastatic colorectal cancer, some lung cancers, and some breast cancers. It acts by "starving" cancerous tumors of the blood supply (i.e., new blood vessels/feeders) those tumors need to survive and grow. See also *Angiogenesis, Cancer, Tumor, Metastasis, Anti-angiogenesis, Bevacizumab, Chiral compound, Interleukin-18 (IL-18).*

Angiogenic Growth Factors

Proteins that stimulate formation of blood vessels (e.g., in tissue being formed by the body to repair wounds). See also *Protein, Vascular endothelial growth factor (VEGF), Filler epithelial cells, Fibroblast growth factor (FGF), Mitogen, Angiogenin, Endothelial cells, Transforming growth factor-alpha (TGF-alpha), Transforming growth factor-beta (TGF-beta), Platelet-derived growth factor (PDGF), Angiogenesis.*

Angiogenin

One of the human angiogenic growth factors, it possesses potent angiogenic (formation of blood vessels) activity. In addition to stimulating (normal) blood vessel formation, angiogenin levels are correlated with placenta formation and tumor growth (tumors require new blood vessels). See also *Angiogenic growth factors, Angiogenesis, Tumor, Growth factor, Paneth cells.*

Angiostatin

An anti-angiogenesis (anti–blood vessel formation) human protein discovered by Judah Folkman. In combination with endostatin, it has been shown to cause certain cancer tumors in mice to shrink, via cutting off the creation of new blood vessels required to "feed" a growing tumor. Angiostatin acts to halt the creation of new blood vessels by binding to ATP synthase (an enzyme needed to initiate new blood vessels). See also *Protein, Anti-angiogenesis, Endostatin, Cancer, ATP synthase, Tumor.*

Angiotensin I

See *ACE, ACE inhibitors, Adipose, Insulin.*

Angiotensin II

See *ACE, ACE inhibitors, Adipose, Insulin.*

Angstrom (Å)

10^{-8} cm (3.937×10^{-9} in.).

Anion

See *Ion.*

Anneal

The process by which the complementary base pairs in the strands of DNA combine. See also *Base pair (bp), Deoxyribonucleic acid (DNA).*

Annotation (Bioinformatics)

Refers to analysis and commentary that is appended to DNA sequences, protein sequences, etc., data stored in databases. Such annotation can include

- Known information about a given sequence's coding (or noncoding)
- Known information about the protein(s) coded for *by an analogous gene* that has already been sequenced and delineated in a *model organism's* DNA
- Known (or predicted) protein structure coded for by a gene
- Known (or predicted) domain(s) of the protein
- Quaternary structure of the protein
- Known (or predicted) protein function (of protein coded for by a gene)
- Common posttranslational modifications of the protein (e.g., addition of carbohydrate moieties, phosphorylation, acetylation)
- Known clinically observed effect on organism (of the protein coded for by a gene, of that protein after *posttranslational modification of protein*, etc.)

See also *Sequence (of a DNA molecule), Sequence (of a protein), Bioinformatics, Coding sequence, Protein, Gene, Model organism, Homologous (chromosomes or genes), Functional genomics, Domain (of a protein), Phosphorylation, Quaternary structure, Posttranslational modification of protein.*

Anonymous DNA Marker

Refers to a DNA marker with a clearly identifiable sequence variation (i.e., it is detectable by the *specific variation* in its DNA sequence, whether or not it occurs in or near a coding sequence).

Examples of anonymous DNA markers include amplified fragment length polymorphisms and microsatellite DNA. See also *Deoxyribonucleic acid (DNA), Sequence (of a DNA molecule), Marker (DNA sequence), Microsatellite DNA, Amplified fragment length polymorphism.*

Antagonists

Molecules that bind to certain proteins (e.g., receptors, enzymes) at a specific (active) site on that protein. The binding suppresses or inhibits the activity (function) of that protein. See also *Receptors, Active site, Conformation, Agonists, Enzyme, Allosteric enzymes.*

Anterior Pituitary Gland

See *Pituitary gland.*

Anthocyanidins

Natural pigments (flavonoids) produced in blueberries (genus *Vaccinium*), blackberries (*Rubus fruticosus*), cranberries (*Vaccinium macrocarpon*), cherries (genus *Prunus*), black or purple carrots (*Daucus carota*), pomegranates (*Punica granatum* L.), and some types of grapes.

Consumption of anthocyanidins by humans has been shown to be beneficial to eyesight, via aiding the health of the retina. Within the human body, anthocyanidins act as antioxidants (i.e., "quenchers" of free radicals), so consumption of anthocyanidins apparently thereby reduces the risk of some cancers, coronary heart disease, eyesight loss, and cataracts. See also *Phytochemicals, Nutraceuticals, Carotenoids, Antioxidants, Oxidative stress, Cancer, Coronary heart disease (CHD), Insulin, Proanthocyanidins, FOSHU.*

Anthocyanins

See *Anthocyanidins.*

Anthocyanosides

Natural pigments (flavonoids) produced in bilberries (*Vaccinium myrtillus*) and certain other fruits. Consumption of anthocyanosides by humans has been shown to be beneficial to eyesight by aiding the health of retinal rhodopsin (a chemical within the retina).

Within the human body, anthocyanosides act as antioxidants (i.e., "quenchers" of free radicals), so consumption apparently reduces the risk of certain diseases (e.g., some cancers, eyesight loss, coronary heart disease). See also *Phytochemicals, Nutraceuticals, Carotenoids, Antioxidants, Oxidative stress, Cancer, Coronary heart disease (CHD), Insulin, Proanthocyanidins, FOSHU.*

Anti-Angiogenesis

Refers to impact of any compound that works to prevent angiogenesis (i.e., formation/development of new blood vessels). Because angiogenesis is required for malignant tumors to grow and/or metastasize (spread), anti-angiogenesis was proposed as a means to combat cancer, by Judah Folkman in 1970.

For example, the biotechnology-derived pharmaceutical known as Avastin® (bevacizumab) has been proven to be effective against metastatic colorectal cancer, some lung cancers, and some breast cancers. It acts by "starving" cancerous tumors of the blood supply (i.e., new blood vessels/feeders) those tumors need in order to survive and grow.

Because angiogenesis is required for the "dry" form of age-related macular degeneration (AMD) disease to progress to the "wet" form of AMD, some anti-angiogenesis pharmaceuticals have been used to treat AMD.

Because angiogenesis is required for embryonic development, anti-angiogenic drugs inhibit proper development/growth of infants in the womb. Drugs that have been found to possess anti-angiogenic properties include Avastin®, fumagillin, ovalicin, and thalidomide, as well as the human proteins angiostatin, endostatin, and interleukin-18. See also *Angiogenesis, Angiogenic growth factors, Tumor, Cancer, Metastasis, Angiostatin, Endostatin, Genistein, Receptor tyrosine kinase, Bevacizumab, Age-related macular degeneration (AMD), Interleukin-18 (IL-18).*

Antibiosis

Refers to the processes via which one organism produces a substance that is toxic or repellent to another organism (e.g., a parasite) that is attacking the first organism. For example, certain varieties of corn/maize (*Zea mays L.*) naturally produce chemical substances in their roots that are toxic to the corn rootworm. See also *Antibiotic, Bacillus thuringiensis (B.t.), Corn, Corn rootworm.*

Antibiotic

Coined by Selman Waksman during the 1940s, this term refers to organic compounds that are naturally formed and secreted by various species of microorganisms and/or plants. It has a defensive function and is often toxic to other species (e.g., penicillin, originally produced by bread mold, is toxic to numerous human pathogens). Antibiotics generally act by inhibiting protein synthesis, DNA replication, synthesis of cell wall (cytoskeleton) constituents, inhibition of required cell (e.g., bacteria) metabolic processes, and nucleic acid (DNA and RNA) biosynthesis, hence killing the (targeted bacteria) cells involved. Inorganic (e.g., certain metals) molecules may also have antibiotic properties. See also *Pathogen, Microorganism, Protein, Nucleic acids, Penicillin G (benzylpenicillin), Symbiotic, Gram stain, Gram-negative, Allelopathy, Bacteria, Gram-positive, Cell, Antibiosis, Aureofacin, Photorhabdus luminescens, Beta-lactam antibiotics, Metabolism, Deoxyribonucleic acid (DNA), Cytoskeleton, Plasma membrane, Ribonucleic acid (RNA), Nisin, Lantibiotics, Pharmacogenetics, Pharmacogenomics.*

Antibiotic Resistance

A property of a cell (e.g., pathogenic bacteria) that enables it to avoid the effect of an antibiotic that had formerly killed or inhibited that cell. Ways this can occur include the following:

- Changing the structure of the cell wall (plasma membrane).
- Oxidative stress.
- Synthesis (manufacture) of enzymes to inactivate the antibiotic (e.g., penicillinases, which inactivate penicillin).
- Synthesis of enzymes to prevent the antibiotic from entering cell.
- Active removal of the antibiotic from the cell. For example, the *membrane transporter protein* molecules known as "ABC transporters" are sometimes able to help pathogenic bacteria resist certain antibiotics by transporting out the antibiotic before it can kill the bacteria. The ABC transporter is a V-shaped molecule embedded in the (bacteria) cell's plasma membrane, with the "open end" of the "V" pointed toward the interior of the cell. When molecules of certain antibiotics (inside the cell) contact the ABC transporter molecule, the two "arms" of the ABC transporter *close around the antibiotic molecule; the ABC transporter flips over and thereby sends the antibiotic molecule out through the exterior of the cell's plasma membrane.*
- Replacing some critical cell metabolic processes, with (new) metabolic processes that bypass the antibiotic's (former) effect.
- Autophagy.
- Emission of hydrogen sulfide (H_2S), which inhibits oxidative stress.

See also *Cell, Pathogen, Pathogenic, Bacteria, Antibiotic, Plasma membrane, Enzyme, Penicillinases, Metabolism, ABC transporters, Mycobacterium tuberculosis, Glycoprotein remodeling, Oxidative stress, Hydrogen sulfide (H_2S), Autophagy.*

Antibody

Also called immunoglobulin (Ig). A large defense protein that consists of two classes of polypeptide chains, light (L) chains and heavy (H) chains. A single antibody molecule consists of two identical copies of the L chain and two of the H chain. They are synthesized (i.e., made) by the immune system (B lymphocytes) of the organism. The antibody is composed of four proteins linked together to form a Y-shaped bundle of proteins (looks somewhat like

A

a slingshot or two hockey sticks taped together at the handles). The amino acid sequence that makes up the stem (heavy chains or *constant region*) of the Y (i.e., the handles of the taped together hockey sticks) is similar for all antibodies. The stem is known as the Fc region of the antibody and it does not bind to antigen, but does have other regulatory functions.

The two arms of the Y are each made up of two side-by-side proteins called light chains and heavy chains (i.e., proteins are chains of amino acids), with identical antigen-binding (ab) sites (known as "variable region") on the tips of each "arm." The antibody is thus bivalent in that it has two binding sites for antigen. Taken together, the two arms of the Y are known as the Fab portions of the antibody molecule. The Fab portions can be cleaved from the antibody molecule with papain (an enzyme that is also used as a meat tenderizer), or the Fab portions can be produced via genetically engineered *Escherichia coli* (*E. coli*) bacteria.

When a foreign molecule (e.g., a bacterium, virus) enters the body, B lymphocytes are stimulated into becoming rapidly dividing blast cells, which mature into antibody-producing plasma cells. The plasma cells are triggered by the foreign molecule's epitope(s) (i.e., group or groups of specific atoms [also known as a hapten] that are recognized to be foreign by the body's immune system) into producing antibody molecules possessing antigen-binding (ab) sites (also called combining sites or determinants).

These fit into the foreign molecule's epitope. Thus, via the tips of its arms, the antibody molecule binds specifically to the foreign entity (antigen) that has entered the body. By this process, it inactivates that foreign molecule or marks it for eventual destruction by other immune system cells.

System marking of the foreign molecule (e.g., pathogen or toxin) for destruction is accomplished by the fact that the stem of the Y (i.e., the Fc) fragment hangs free from the combined antibody–antigen clump, thereby providing a receptor for phagocytes, which roam throughout the body ingesting and subsequently destroying such "marked" foreign molecules. This system is called "antibody-dependent cellular cytotoxicity."

Research published in 2001 indicates that antibodies may also *kill some pathogens themselves* by catalyzing the formation of hydrogen peroxide from *oxygen free radicals* (*singlet oxygen*) and water. Hydrogen peroxide is highly reactive and could potentially kill pathogens when generated by an (attached) antibody.

There are five classes of immunoglobulin: IgG, IgM, IgD, IgA, and IgE. See also *Humoral immunity, Immunoglobulin, Protein, Polypeptide* (*protein*), *Amino acid, B lymphocytes, Blast cell, Antigen, Hapten, Epitope, Combining site, Domain* (*of a protein*), *Sequence* (*of a protein molecule*), *Escherichia coliform* (*E. coli*), *Pathogen, Toxin, Phagocyte, Macrophage, Microphage, Monocytes, T cells, Polymorphonuclear leukocytes* (*PMN*), *Cellular immune response, Polymorphonuclear granulocytes, Genetic engineering, "Magic bullet", Engineered antibodies, Receptors, Oxygen free radical, Cytotoxic T cells.*

Antibody Affinity Chromatography

A type of chromatography in which antibodies are immobilized onto the column material. The antibodies bind to their target molecules, while the other components in the solution are not retained. In this way, a separation (purification) is achieved. See also *Antibody, Chromatography, Affinity chromatography, Affinity.*

Antibody Arrays

See *Protein microarrays.*

Antibody-Dependent Cellular Toxicity

Refers to the immune system mechanism whereby pathogens that are "marked" by antibodies subsequently get destroyed by other components of the immune system. See also *Pathogen, Antibody, Immune response, Cytotoxic T cells.*

Antibody–Drug Conjugate

See *Conjugate.*

Antibody-Laced Nanotube Membrane

See *Nanotube.*

Antibody-Mediated Immune Response

See *Humoral immune response.*

Anticoding Strand

Refers to the single strand of DNA (double helix) that is transcribed. Sometimes called the "antisense strand" or the "template strand." See also *Deoxyribonucleic acid* (*DNA*), *Transcription, Antisense* (*DNA sequence*).

Anticodon

A specific sequence of three nucleotides in a transfer RNA, complementary to a codon (also three nucleotides) for an amino acid in a messenger RNA. See also *Codon, Transfer RNA* (*tRNA*), *Amino acid, Messenger RNA* (*mRNA*), *Nucleotide.*

Antiepidermal Growth Factor Receptor Monoclonal Antibodies

Refer to monoclonal antibodies (e.g., the pharmaceutical products cetuximab, panituximab) that bind in a key into lock manner with the EGF receptor (i.e., receptor located on certain cells' surface, which receives and internalizes the epidermal growth factor molecule). Because the surfaces of some tumors (e.g., in colorectal cancer) are studded with EGF receptors, these monoclonal antibody pharmaceuticals can be utilized (alone or in combination with chemotherapy) to treat those tumors. See also *Receptors, Epidermal growth factor receptor, Monoclonal antibodies* (*MAb*), *Cell, Cancer, Tumor, Chemotherapy.*

Antifibrinolytic Drugs

Refer to pharmaceuticals that interfere with the action of fibrinolytic agents (i.e., those that dissolve blood clots) in the body. For example, the protein "aprotinin" is utilized during cardiac surgery to thereby help control excessive bleeding. See also *Fibrinolytic agents.*

Antifreeze Proteins

See *Thermal hysteresis proteins.*

Antigen

Also called an immunogen. Any large molecule or small organism whose entry into the body provokes synthesis of an antibody or immunoglobulin (i.e., an immune system response). See also *Hapten, Antibody, Epitope, Cellular immune response, Humoral immunity.*

Antigenic Determinant

See *Hapten, Epitope, Superantigens.*

Antihemophilic Factor VIII

Also known as factor VIII or antihemophilic globulin. See also *Factor VIII.*

Antihemophilic Globulin

Also known as factor VIII or antihemophilic factor VIII. See also *Factor VIII.*

Anti-Idiotype Antibodies

See *Anti-idiotypes.*

Anti-Idiotypes

Antibodies to antibodies. In other words, if a human antibody is injected into rabbits, the rabbit immune systems will recognize the human antibodies as foreign (regardless of the fact that they are antibodies) and produce antibodies against them. To the rabbit, the foreign antibodies represent just another invader or non-self to be targeted and destroyed. Anti-idiotypes mimic antigens in that they are shaped to fit into the antibody's binding site (in lock-and-key fashion). As such, anti-idiotypes can be used to create vaccines that stimulate production of antibodies to the antigen (that the anti-idiotype mimics). This confers disease resistance (to the pathogen associated with that antigen) without the risk that a vaccine using attenuated pathogens entails (i.e., that the pathogen "revives" to cause the disease). See *Antibody, Monoclonal antibodies (MAb), Antigen, Idiotype, Pathogen, Attenuated (pathogens).*

Anti-Interferon

An antibody to interferon. Used for the purification of interferons. See also *Antibody, Interferons, Affinity chromatography.*

Antioncogenes

See *Oncogenes, Antisense (DNA sequence).*

Antioxidants

Compounds (e.g., phytochemicals) or nanoparticles that act to prevent lipids from oxidizing (e.g., to plaque) and breaking down (e.g., to carcinogenic compounds), or that act to capture and halt singlet oxygen (O^-) free radicals, which can damage DNA in cells (i.e., causing mutations). Since oxidation of lipids in the blood is the initial step in atherosclerosis, consumption of large amounts of certain antioxidants (e.g., flavonoids, melanoidins) may help reduce the risk of atherosclerosis.

Because excessive oxidation reactions within the body often lead to formation of tissue-damaging free radicals (i.e., molecules containing an "extra" electron), consumption of applicable antioxidants can help to prevent such tissue damage.

Evidence indicates that tissue damage from free radicals may play a role in causing some arthritis, coronary heart disease, diabetes, and cancers.

Synthetic analogues have also been manufactured (e.g., synthetic vitamins), which perform a similar antioxidant function to naturally occurring antioxidant phytochemicals. See also *Oxidative stress, Phytochemicals, Lipids, Carcinogen, Cancer, Analogue, Oxidation, Coronary heart disease, Insulin, Lycopene, Mutagen, Mutation, Flavonoids, Isoflavones, Atherosclerosis, Astaxanthin, Human superoxide dismutase (hSOD), PEG-SOD (polyethylene glycol superoxide dismutase), Plaque, Phytate, Polyphenols, Ellagic acid, Beta-carotene, Vitamin E, Proanthocyanidins, Polyunsaturated fatty acids (PUFA), Conjugated linoleic acid (CLA), Catechins, Melanoidins, Nanoparticles, Nanoceria.*

Antiparallel

Describes molecules that are parallel but point in opposite directions. The strands of the DNA double helix are antiparallel. See also *Double helix.*

Antiporter

Refers to a membrane transport system in which the transport of one substance in one direction (across a cell's membrane) is coupled to the transport of a second substance in the opposite direction. For example, the

- *SOS1* gene in the *Arabidopsis thaliana* plant
- *AtNHX1* gene in the *A. thaliana* plant

code for a plasma membrane antiporter (ion channel) that transports NA^+ (sodium) ions *out* (of plant's xylem tissue) while transporting H^+ ions *in*. See also *Membranes (of a cell), Cell, Membrane transport, Membrane transporter protein, Ion channels, Gated transport, ABC transporters, Arabidopsis thaliana, Vacuoles, Salt tolerance.*

Antisense (DNA Sequence)

A strand of DNA that produces a messenger RNA (mRNA) molecule that (when reversed end for end) has the same sequence as (i.e., is complementary to) the unwanted ("bad") mRNA. The sense (i.e., forward) and antisense (i.e., backward) mRNA strands hybridize (i.e., tightly bond to each other), which prevents the bonded pair from leaving the cell's nucleus, so that bonded pair is rapidly degraded (destroyed) by nucleases within the cell nucleus.

In genetic targeting using antisense molecules (to block "bad" genes), antisense molecules are used to bind to a "bad" gene's (e.g., an oncogene) mRNA, thus cancelling the (cancer-causing) message of the gene and preventing cells from following its (tumor growth) instructions. Another example would be the use of antisense DNA to block the gene that codes for production of polygalacturonase (an enzyme that causes ripe fruit to soften).

Physically, *antisense* is accomplished by removing a given gene from an organism's genome, reversing it (end for end), and reinserting it back into the organism's genome. See also *Deoxyribonucleic acid (DNA), Coding sequence, Gene, Genome, Complementary DNA (cDNA), Messenger RNA (mRNA), Genetic targeting, Cancer, Polygalacturonase (PG), Oncogenes, Sense, Cosuppression, Gene silencing, Hybridization (molecular genetics), Nuclease, Anticoding strand.*

A

Antisense RNA

See *Antisense (DNA sequence).*

Antithrombogenous Polymers

Synthetic polymers (i.e., plastics) used to make medical devices that will be in contact with a patient's blood (e.g., catheters) and thus must not initiate the coagulation process as synthetic polymers usually do. The natural anticoagulant heparin is incorporated into the polymer and is gradually released into the bloodstream by the polymer, thus preventing blood coagulation on the surface of the polymer. See also *Polymer, Thrombosis.*

Antitoxin

See *Polyclonal antibodies, Diphtheria antitoxin.*

Antixenosis

Refers to the effect of a chemical compound (e.g., produced within a plant) that causes relevant predators (e.g., plant-chewing insects) to prefer to attack *other* plants. See also *Corn earworm, Corn rootworm, European corn borer (ECB).*

Aneuploid

Refers to cell/organism that possesses an abnormal number of chromosomes (for that particular species). From the Greek *ploos*, which means "fold" (i.e., referring to the visual appearance of the karyotype viewed in a microscope). See also *Cell, Chromosomes, Organism, Karyotyper, Diploid, Triploid.*

AP

Acronym for "atrial peptide." See *Atrial peptides.*

APC

Acronym for "antigen-presenting cells." See *Macrophage, Dendritic cells.*

APHIS

The Animal and Plant Health Inspection Service, which is the agency of the U.S. Department of Agriculture that is responsible for regulating the field (outdoor) testing of genetically engineered plants and certain microorganisms. See also *Coordinated Framework for Regulation of Biotechnology, Microorganism, Genetic engineering.*

Aplastic Anemia

An autoimmune disease of the bone marrow. See also *Autoimmune disease.*

Apo A-1 Milano

An apolipoprotein that was found to be naturally produced within the bodies of approximately forty related people living in an Italian town near Milan. It prevented any buildup of plaque in their arteries.

When synthetic (i.e., made by scientists) *apo A-1 milano* was injected into the bloodstreams of people who are not related to that Italian family, it caused a *reduction* in the buildup of plaque in their arteries. See also *Apolipoprotein, plaque.*

APO B-100

See *Low-density lipoproteins (LDLP), Apolipoproteins, Very low-density lipoproteins (VLDL).*

APO-1/Fas

See *CD95 protein.*

Apoenzyme

The protein portion of a holoenzyme. Many (but not all) enzymes are composed of functional "pieces," for example, a protein piece (chain) and another piece that is an organic and/or inorganic molecule. This other piece is known as a cofactor and it may be removed from the enzyme under certain conditions. When this is done, the resulting inactive enzyme is known as an apoenzyme. The inactive apoenzyme becomes functionally active again if it is allowed to recombine with its cofactor. See also *Cofactor, Enzyme, Holoenzyme.*

Apolipoprotein B

An apolipoprotein that is involved in human cholesterol metabolism. See also *Cholesterol, Low-density lipoproteins (LDLP), Apolipoproteins, Very low-density lipoproteins (VLDL).*

Apolipoproteins

The protein portion of lipoproteins (i.e., after the lipid portion is removed from those molecules). See also *Low-density lipoproteins (LDLP), Protein, Lipids, Very low-density lipoproteins (VLDL).*

Apomixis

A method of reproduction used by scientists to propagate (hybrid) plants without having to utilize sexual fertilization. By combining apomixis with tissue culture technology, Cai Detian, Ma Piugfu, and Yao Jialin were able to thus propagate rice varieties in 1994. In 1998, Dimitri Petrov, Phillip Sims, and Chester Deald were able to cause apomixis in corn (maize).

By "fixing" hybrid dominance, the need for (sexual) breeding is eliminated and the hybrid vigor is passed down via the seed from generation to generation. See also *Asexual, Germ cell, Hybrid vigor, Tissue culture, Hybridization (plant genetics), Corn, F1 hybrids.*

Apoptosis

From Greek syllables implying "dropping off." Also called "programmed cell death," it is a series of programmed steps that cause a cell to die via "self-digestion" without rupturing and releasing intracellular contents (e.g., nucleus, chromosomes, refractile bodies) into the local (i.e., surrounding tissue) environment. Manifestations of cell apoptosis include shrinking of the cell's cytoplasm and chromatin condensation, and the presence of phosphatidyl serine on the exterior surface of cell's plasma membrane.

If the normal cell apoptosis is prevented (e.g., by an enzyme that is present due to disease) in the body, cells can grow uncontrollably (i.e., causing cancer). For example, people with chronic myelogenous leukemia (*CML*, also known as chronic myeloid leukemia) typically have 10–25 times as many white blood cells as normal.

Certain compounds can help restore/induce some cells' apoptosis. For example, when ingested by humans, the chemical known as parthenolide (extracted from the herb plant known as *feverfew*) induces apoptosis in *human leukemia stem cells* (i.e., the progenitor cells from which all resultant leukemia cells in that body would be "descended"). In 2012, Teri W. Odom showed that certain aptamer-laden can penetrate the plasma membrane of cancer cells and restore the apoptosis pathway to those cells.

Research published by Duxin Sun and Max Wicha in 2010 indicated that when injected into mouse cell culture, the chemical known as sulforaphane (extracted from cruciferous plants such as broccoli, cabbage, kale) induces apoptosis in mouse *breast cancer stem cells* (i.e., the progenitor cells from which all resultant mouse breast cancer cells in that cell culture would be "descended").

Some fruits are broken down in the human digestive system to yield the short chain fatty acid known as butyrate within the human colon. Butyrate can act via an epigenetic mechanism (i.e., histone modification) to cause apoptosis of cancerous cells in the colon, thereby reducing the incidence of human colon cancer. See also *Cell, CD95 protein, Signal transduction, Signaling, Refractile bodies (RB), Nucleus, Chromosomes, Chromatin, Cytoplasm, Fusarium, Gene, p53 gene, Tubulin, Cancer, Selective apoptotic antineoplastic drug (SAAND), Replicon, Hypersensitive response, Signal transduction, Signal transducers and activators of transcription (STATS), MicroRNAs, Gene expression cascade, Enzyme, White blood cells, Philadelphia chromosome, GLEEVEC™, DNA fragmentation, RNAse 1, Gamma interferon, Caspases, Posttranslational modification of protein, Mitogen-activated protein kinase cascade, Nanostars, Aptamers, Plasma membrane, Phosphatidyl serine, NFκB.*

APP

Acronym for amyloid precursor protein. See *Amyloid precursor protein.*

Approvable Letter

From the Food and Drug Administration (FDA), one of the final steps in the U.S. FDA's review process for new pharmaceuticals. The letter precedes final FDA clearance for marketing of the new compound. See also *Food and Drug Administration (FDA), IND, IND exemption.*

Aptamers

Single-stranded nucleotides (RNA molecules) that form extended three-dimensional structures that bind (i.e., "stick to") other specific molecules (e.g., proteins, amino acids, sugars, or other ligands) and sometimes inactivate the molecules they "stick" to. The word "aptamer" is from the Latin *aptus* ("to fit").

In 2004, the U.S. Food and Drug Administration approved for use as a pharmaceutical the aptamer Macugen™ (pegaptanib sodium), which is "pegylated" (i.e., joined with polyethylene glycol—PEG—to camouflage the aptamer molecule from the body's immune system so it is not inactivated by immune response before it can do its work). Macugen™ binds to vascular endothelial growth factor (VEGF) when injected into eyeballs of people who are suffering from "wet" form of age-related macular degeneration (AMD). By doing so, Macugen™ prevents that VEGF from causing (more) growth of new blood vessels (i.e., in front of the retina) whose subsequent leakage of blood causes vision loss in AMD disease.

In 1992, Louis Bock and John Toole isolated aptamers that bind and inhibit the blood-coagulation enzyme "thrombin." Since thrombin is crucial to the formation of blood clots (coagulation), such aptamers may someday be useful for anticoagulant therapy (e.g., to prevent blood clots following surgery or heart attacks).

One current use of aptamers is as *capture agents* (i.e., ligands or other molecules that "bind" to proteins, which are attached to the microarray at specific/known locations). See also *Enzyme, Nucleotide, Ligand, Oligonucleotide, Protein, Inhibition, Thrombin, Thrombus, Thrombosis, Vascular endothelial growth factor (VEGF), Protein microarrays, Food and Drug Administration (FDA), Capture agent, Immune response, Pegylation.*

Aquaporins

A class of plasma membrane–spanning proteins that function as transport channels (for water movement into/out of cell) to allow cells to regulate cellular volume. Discovered by Peter C. Agre.

Recently, it was discovered that one plant cell aquaporin known as "nodulin 26-like intrinsic protein" will transport arsenic (dissolved in water) into or out of plant cells. See also *Plasma membrane, Protein, Cell.*

Arabidopsis thaliana

A small weed plant (*Cruciferae*) possessing 70,000 Kbp in its genome, with very little repetitive DNA. This makes it an ideal model for studying plant genetics. At least two genetic maps have been created for *Arabidopsis thaliana* (one using yeast artificial chromosomes). Because of this, a large base of knowledge about it has been accumulated by the scientific community.

A. thaliana was first genetically engineered in 1986. In 1994, researchers succeeded in transferring genes for polyhydroxylbutylate (PHB) ("biodegradable plastic") production into *A. thaliana*. Because production of PHB requires simultaneous expression of three genes (i.e., the PHB production process is "polygenic")—yet researchers have only been able to insert a maximum of two genes—they have to insert two genes into one plant and one gene into a second plant and then finally get the (total) three genes into (offspring) plants via traditional breeding.

In 2001, Eduardo Blumwald and Hong-Xia Zhang inserted a salt tolerance gene from *A. thaliana* into a tomato (*Lycopersicon esculentum*) and thereby made that tomato plant resistant to salt in concentrations up to 200 mM (i.e., far higher than it could previously survive). See also *Brassica, Gene, Express, Base pair (bp), Kilobase pairs (Kbp), Genome, Genetic code, Genetic map, Genetics, Trait, Polygenic, Deoxyribonucleic acid (DNA), Polyhydroxylbutylate (PHB), Yeast artificial chromosomes (YAC), Model organism, Tomato, Salt tolerance, Paramutation.*

Arachidonic Acid (AA)

Also known as eicosatetraenoic acid. Arachidonic acid (AA) is one of the "omega-6" (*n*-6) highly unsaturated fatty acids; AA is synthesized (i.e., "manufactured") by the human body from linoleic acid (e.g., obtained via consuming soybean oil). AA is present in human breast milk, and research indicates that it plays an important role in the brain and eye tissue development of infants. AA is

A

a crucial precursor for prostaglandins and other eicosanoids. The COX-1 enzyme converts AA to *constitutive prostaglandins*, and the COX-2 enzyme converts AA to *inducible prostaglandins*. See also *Cyclooxygenase, Polyunsaturated fatty acids (PUFA), N-6 fatty acids, Fatty acid, Unsaturated fatty acid, Linoleic acid, Soybean oil, Constitutive enzymes, Inducible enzymes, Leukotrienes, Essential fatty acids, Eicosanoids.*

Arbuscular Mycorrhizae

Refer to certain fungi (phylum Glomeromycota) that form filaments (hyphae) inside and around most crop plants' (80% of all land plant species) roots and symbiotically help those roots to better extract phosphates, nitrogen, sulfur, and other plant nutrients from the soil. Arbuscular mycorrhizae hyphae attach to the roots and radiate out into the surrounding soil, thereby enhancing those roots' access to nutrients and water, which increases that plant's tolerance of drought. In return, arbuscular mycorrhizae receive plant-synthesized carbon.

Roots of the corn (maize) plant (*Zea Mays* L.) that need phosphate allow *arbuscular mycorrhizae* to form microscopic structures known as arbuscules within the cells of their root tissues. Those arbuscules improve nutrient (e.g., phosphate) entry into the roots. See also *Fungus, Symbiotic, Mycorrhizae, Glomalin.*

Archaea

Single-celled life-forms that can live at extreme ocean depths (i.e., high pressure) and elsewhere in the absence of oxygen. *Archaea* were delineated/named by Carl Woese. Enzymes robust (i.e., sturdy) enough for industrial process utilization have been isolated by scientists from some strains of *Archaea*.

Other *Archaea* strains are sometimes present in the rumen (i.e., "first stomach") of cattle and sheep. Those *Archaea* produce methane gas by breaking down some of the feed consumed by the cattle and sheep. See also *Enzyme, Extremozymes, Cell, Anaerobe, Anaerobic, Strain, Sulfolobus solfataricus.*

Arginine (Arg)

An amino acid, commonly abbreviated arg. In dry, bulk form arginine is colorless, crystalline, and water soluble. It is an essential amino acid of the α-ketoglutaric acid family.

When consumed, arginine helps the walls of blood vessels to relax. See also *Amino acid, Essential amino acids, Nitric oxide synthase.*

Argonaute Proteins

See *piRNAs.*

ARM

Acronym for "antibiotic resistance marker." See *Marker* (*genetic marker*).

Armchair Form of Carbon Nanotubes

Refers to one specific chiral form of carbon nanotubes that behaves like a pure metal (in terms of high electrical conductivity) and is the ideal quantum wire. See also *Quantum wire.*

ARMD

Acronym for age-related macular degeneration. See *Lutein.*

Armed Antibody

See *Immunoconjugate.*

ARMG

Acronym for antibiotic resistance marker gene. See also *Antibiotic, Antibiotic resistance, Gene, Marker* (*genetic marker*), *Recombinase.*

Armyworm

Caterpillars (pupae) of the lepidopteran insect *Pseudaletia unipuncta* species, most of which are harmful to crops (e.g., wheat, corn/maize) grown by humans.

Armyworms are susceptible to some of the "cry" proteins (e.g., they are killed if they eat plants genetically engineered to contain Cry1A(b), Cry9C, or Cry1F proteins).

Armyworms are preyed upon by some species of ground beetles, sphecid wasps, toads, birds, etc. See also *Protein, Volicitin, CRY proteins, CRY1A(b) protein, CRY1F protein, CRY9C protein, Corn, Wheat, Fall armyworm.*

AroA

Refers to the transgene (cassette) that was initially isolated/extracted from the genome of the *Agrobacterium* bacteria species (strain CP4) and inserted via "genetic engineering" techniques into a crop plant (e.g., soybean, *Glycine max* L.), in order to make that (soybean) plant tolerant to glyphosate-based herbicides (and also sulfosate-based herbicides). See also *Gene, Transgene, Cassette, Genome, Agrobacterium tumefaciens, EPSP synthase, mEPSPS, CP4 EPSPS, Soybean, Herbicide-tolerant crop, Genetic engineering, Soybean plant, Glyphosate, Sulfosate.*

Array Comparative Genomic Hybridization

Refers to a researcher's use of microarrays to sift through the DNA of a chromosome, searching for aberrations from normal such as copy number variation (e.g., which can sometimes be a source of disease). The capture molecules attached to the surface of such microarrays are numerous different known segments of the targeted chromosome.

The chromosome's DNA segments (and *numbers* of those segments) that hybridize to (i.e., adhere) the applicable capture molecules on the microarray reveal the presence (and copy number) of each relevant gene within the chromosome being evaluated. See also *Microarray* (*testing*), *Genome, Hybridization* (*molecular genetics*), *Hybridization surfaces, Capture molecules, Deoxyribonucleic acid* (*DNA*), *Chromosomes, Gene, Copy number variation, Sequence* (*of a DNA molecule*).

ARS

See *ARS element.*

ARS Element

A sequence of DNA that will support autonomous replication (sequence, ARS). See also *Deoxyribonucleic acid* (*DNA*), *Sequence* (*of a DNA molecule*).

Arteriosclerosis

A group of diseases (including atherosclerosis) that is characterized by deposits of plaques on the inside of blood vessel walls (beginning in late teenage years), a decrease in elasticity (i.e., "stretchiness"), and a thickening of the walls of the body's arteries. See also *Atherosclerosis, Coronary heart disease (CHD), Plaque, Granulocyte-macrophage colonystimulating factor (GM-CSF), C-reactive protein (CRP), Homocysteine.*

Arthritis

Refers to a group of approximately 100 different diseases that adversely affect tissues in and near joints. Those diseases include osteoarthritis, juvenile arthritis, rheumatoid arthritis, lupus erythematosus, and gout. See also *Osteoarthritis, Rheumatoid arthritis, Autoimmune disease, Lupus, Prostaglandins, Ascorbic acid.*

Artificial Interfering RNA (aiRNA)

A term coined in 2013 by Xiaowei Wang, to refer to a molecule created via combination of an siRNA and a microRNA, both possessing a similar seed region RNA sequence. Such an aiRNA against the AKT1 gene (which encourages uncontrolled cell division) has been shown to more effectively inhibit (mutagenesis) migration and proliferation of cancer cells, than either siRNA or microRNA alone. See also *Short interfering RNA (siRNA), MicroRNAs, Ribonucleic acid (RNA), Gene, Cancer, Mutagenesis breeding.*

Ascites

Liquid accumulations in the peritoneal cavity. Used as an input in one of the methods for producing monoclonal antibodies. See also *Monoclonal antibodies (MAb), Peritoneal cavity/membrane, Antibody.*

Ascorbic Acid

A water-soluble vitamin and antioxidant that is produced by plants in response to adverse factors in their environment, such as extremely bright sunlight or drought. Also known as vitamin C, ascorbic acid is a powerful antioxidant, to protect the plants from damage (e.g., the damaging oxidative process that can occur within cells of a dehydrated plant).

Humans who consume significant amounts of ascorbic acid are less likely than others to get the diseases known as iron deficiency anemia, scurvy, or gout (a form of arthritis which can be caused by elevated levels of uric acid in the blood).

Ascorbic acid is also utilized by man to synthesize some nanorods. See also *Antioxidants, Nanorods, Arthritis.*

Asexual

Denotes fertilization and/or reproduction by *in vitro* means. Without sex. See also *In vitro, Apomixis, Germ cell.*

Asian Corn Borer

Also known by its Latin name *Ostrinia furnacalis*, it is an insect (originally from Asia) whose larvae (caterpillars) eat and bore into the corn/maize (*Zea Mays* L.) plant. In doing so, they can act as vectors (i.e., carriers) of the fungi known as *Aspergillus flavus* (a source of aflatoxin), *Fusarium moniliforme* (a source of fumonisin), or *Aspergillus parasiticus* (a source of aflatoxin). See also *European corn borer (ECB), Corn, Fungus, Aflatoxin, Fusarium, Fusarium moniliforme.*

Asparagine (asp)

An amino acid, commonly abbreviated asp. In dry, bulk form, asparagine appears as a white, crystalline solid. It is found in high amounts in many plants. Asparagine's name arose from the fact that it was first extracted from the asparagus plant and identified, in 1806, by Louis Nicolas Vauquelin and Pierre-Jean Robiquet.

A crop (e.g., potato) bred/engineered to contain lower levels of asparagine, would have a lower risk for production of acrylamide, a potential human carcinogen, when such potatoes are cooked at high temperatures. See also *Amino acid.*

Aspartic Acid

A dicarboxylic amino acid found in plants and animals, especially in molasses from young sugarcane and sugar beets. See also *Amino acid.*

Aspergillus flavus

See *Aflatoxin, Peroxidase, Beta-carotene.*

Aspergillus fumigatus

See *High-phytase corn and soybeans.*

Assay

A test (specific technique) that measures a response to a test substance or the efficacy (effectiveness) of the test substance. See also *Immunoassay, Bioassay, Luminescent assays, Hanging drop assays, Multiplexed (assay), Multiplex assay, Hybridization surfaces.*

Assimilation

The formation of "self" cellular material from small molecules derived from food. See also *Insulin-like growth factor-1 (IGF-1), Ribosomes, Messenger RNA (mRNA).*

Association Mapping

See *Haplotype, Haplotype map.*

Association of Biotechnology Companies (ABC)

An American trade association of companies involved in biotechnology and services to biotechnology companies (e.g., accounting, law). Formed in 1984, the ABC tended to consist of the smaller firms involved in biotechnology (and service firms that worked for all biotechnology companies). In 1993, the Association of Biotechnology Companies was merged with the Industrial Biotechnology Association to form the Biotechnology Industry Organization. See also *Industrial Biotechnology Association (IBA), Biotechnology Industry Organization (BIO), Biotechnology.*

A

Astaxanthin

A carotenoid pigment that is responsible for the characteristic pink coloring of salmon, trout, and shrimp and the red color of lobsters. It is produced by the microorganisms in the natural (wild) diets of those aquatic animals.

Research has shown that astaxanthin (an antioxidant) helps to boost the immune systems of humans that consume it. Research has shown that astaxanthin helps to reduce oral cancer in rats and inhibits breast cancer in mice, when those rodents consume it. See also *Carotenoids, Antioxidants, Oxidative stress.*

Astrocytes

Refer to one category of glial cells that are present in the brain and spinal cord tissues. Most astrocyte cells have a central star-shaped mass, with numerous long "tubes" (processes) extending from that central mass. During early brain development and also during brain repair after traumatic injuries, neurons (which are only produced in a few distant portions of the brain) migrate to where they need to be located throughout the brain, via those long astrocyte tubes. Following a stroke, astrocytes protect neurons from oxidative stress in addition to rebuilding the neural circuits that improve learning and memory.

The neurons themselves help control the formation and location of astrocyte tubes via secretion of the protein "Slit1." See also *Cell, Neuron, Protein.*

ATCC

See *American type culture collection (ATCC), Type specimen, Accession (germplasm).*

Atherosclerosis

One form of arteriosclerosis; it is characterized by deposition and buildup of fatty deposits (plaque) on the internal walls of the body's arteries, in addition to the decrease in elasticity of arteries' walls that characterizes all forms of arteriosclerosis. When a piece of plaque breaks off, a blood clot generally forms, and that clot often blocks blood flow through the artery, which causes "heart attack" or stroke in the person. See also *Arteriosclerosis, Coronary heart disease (CHD), Cholesterol, High-density lipoproteins (HDLPs), Thrombosis, Thrombus, Flavonoids, Oxidative stress, Antioxidants, Plaque, Nanoparticles.*

Athymic "Nude" Mouse

See *"Nude" mouse.*

AT-III

A human blood factor that promotes clotting. A deficiency of AT-III can be inherited, resulting from certain surgical procedures, certain illnesses, and sometimes use of certain oral contraceptives. See also *Factor VIII.*

AtNHX1 Gene

See *Antiporter, Salt tolerance.*

Atomic Force Microscopy

Refers to one type of scanning probe microscopy that is particularly utilized for the study of biological systems. Developed in 1986, atomic force microscopy (AFM) can produce high-resolution three-dimensional images of a (biological) surface in aqueous environments without the need to stain the biological specimen.

In AFM, a very sharp probe (stylus) is carefully suspended in near proximity to the specimen surface via a high precision device such as a cantilever, piezoelectric, or magnetic/electrostatic process. By carefully moving the probe over the entire specimen surface, the force between the probe and the atoms of the specimen keeps the probe just above the specimen surface and thereby delineates the surface topography of the specimen, at atomic-scale resolution.

AFM can also be utilized for "dip-pen nanolithography." See also *Dip-pen nanolithography, Piezoelectric effect.*

Atomic Weight

The total mass of an atom, it is equal to the sum of the isotope's number of protons and neutrons (in the atom's nucleus). The atomic weights of the earth's elements are based on the assignment of exactly 12.000 as the atomic weight of the carbon-12 isotope (variation of atom). The atomic (weight) theory was established as a framework in 1869 by Meyer and Mendeléev, but standard precise values were not adopted internationally until an "international commission on atomic weights" was formed in 1899 in response to an initiative by the German Chemical Society.

An element's atomic weight averages in the presence of all isotopes present on Earth, so it does not come out to a whole number (with the exception of carbon), because of the existence of small amounts of isotopes, which differ slightly with respect to the number of neutrons each contains. See also *Molecular weight, Isotope.*

ATP

See *Adenosine triphosphatase (ADP).*

ATP Synthase

An enzyme complex that forms ATP from ADP and phosphate during oxidative phosphorylation in the inner mitochondrial membrane (in animals), in chloroplasts (in plants), and in cell membranes (in bacteria). This is an energy-producing reaction in that ATP is a high-energy compound used by cells to maintain their living condition. One Janssen Biotech, Inc. pharmaceutical, SIRTURO™ (bedaquiline), controls pulmonary multidrug-resistant tuberculosis via specifically inhibiting mycobacterial ATP (adenosine 5′-triphosphate) synthase, an enzyme that is essential for the generation of energy in *Mycobacterium tuberculosis.*

ATP synthase is also present on the surface of endothelial cells (lining of blood vessels) where it helps to build new blood vessels (e.g., to replace tissue damaged by injury or disease). Under certain circumstances, this also creates new blood vessels that provide blood supply to tumors.

When separated from cell's membrane, ATP synthase hydrolyzes (i.e., breaks down) ATP via a chemical process in which one subunit (designated γ) of ATP synthase rotates within the other (hollow) part of ATP synthase. See also *Enzyme, Chloroplasts, Adenosine triphosphate (ATP), Hydrolysis, Adenosine diphosphate (ADP), Mitochondria, Tumor, Endothelial cells, Angiostatin.*

ATP Synthetase

See *ATP synthase.*

ATPase

Adenosine triphosphatase, an enzyme that hydrolyzes (clips the bond between two phosphates in) ATP to yield ADP, phosphate, and energy. The reaction is usually coupled to an energy-requiring process. ATP is hydrolyzed in the act of shivering and the energy produced is converted into heat to increase body temperature. This type of heat production involves what is known as a futile cycle because the energy is converted to (and wasted as) heat rather than used in motion, etc. See also *ATP synthase, Enzyme, Adenosine triphosphate (ATP), Adenosine diphosphate (ADP), Futile cycle, Hydrolysis, Hydrolyze.*

Atrial Natriuretic Factor

An atrial peptide hormone that may regulate blood pressure and electrolyte balance within the body. An example is a peptide hormone. See also *Hormone, Atrial peptides, Peptide.*

Atrial Peptides

Endocrine components (proteins) that act to regulate blood pressure, as well as water and electrolyte homeostasis within the body. Atrial peptides are made by the heart in response to elevated blood pressure levels, and they stimulate the kidneys to excrete water and sodium into the urine, thus lowering blood pressure. They also slow the heartbeat. An example is a peptide hormone. See also *Endocrine hormones, Homeostasis, Electrolyte.*

Attenuated (Pathogens)

Inactivated, rendered harmless (e.g., killed viruses used to make a vaccine). Some of the ways in which viruses and other pathogens may be attenuated are by heat, chemical, or radiation treatment. See also *Pathogen.*

Attenuation (of RNA)

Premature termination of an elongating RNA chain. See also *Ribonucleic acid (RNA).*

Aureofacin

An antifungal antibiotic produced by a strain of *Streptomyces aureofaciens.* At least one company has incorporated the gene for this antibiotic (which acts against wheat take-all disease) into a *Pseudomonas fluorescens*, to be used to confer resistance to wheat take-all disease. This is done by allowing the bacteria to colonize the wheat's roots. In this way, the plant obtains the benefits of the antibiotic because the bacteria become a part of the plant. See also *Pseudomonas fluorescens, Endophyte, Antibiotic, Bacillus thuringiensis (B.t.).*

Autoantibody

See *HSE.*

Autogenous Control

The action of a gene product (a molecule) that either inhibits (negative autogenous control) or activates (positive autogenous control) expression of the *gene that codes for it* (Greek *auto* = "self"). The presence of the product either causes or stops its own production. See also *Gene, Express.*

Autoimmune Disease

First proven to be a cause of disease by Ivan Roitt in 1956, it refers to a disease in which the body produces an immunogenic (i.e., immune system) response to some constituent of its own tissue. In other words, the immune system loses its ability to recognize some tissue or system within the body as "self" and targets and attacks it as if it were foreign. Autoimmune diseases can be classified into those in which predominantly one organ is affected (e.g., hemolytic anemia and chronic thyroiditis) and those in which the autoimmune disease process is diffused through many tissues (e.g., multiple sclerosis, systemic lupus erythematosus, and rheumatoid arthritis).

For example, multiple sclerosis is thought to be caused by T cells attacking acetylcholine receptors in the sheaths (myelin) that surround the nerve fibers of the brain and spinal cord. This eventually results in loss of coordination, weakness, and blurred vision. Arthritis is caused by immune system cells attacking joint tissues.

In 2012, some researchers utilized receptor binding mapping to link vitamin D deficiency to an increased risk for cancer and the autoimmune diseases rheumatoid arthritis, multiple sclerosis, and systemic lupus erythematosus.

Certain bacterial infections (e.g., Lyme disease, *Salmonella*) are followed by arthritis in approximately 10% of cases. The antigen (on surface of those bacteria) that is targeted by the human immune system is similar (in its molecular shape) to a protein that is located on the surface of cells in human joint tissue(s). See also *Thymus, Superantigens, T cells, Tumor necrosis factor (TNF), Multiple sclerosis, Myoelectric signals, Acetylcholine, Lupus, Insulin-dependent diabetes mellitis (IDDM), Diabetes, Antigen, Bacteria, Salmonella typhimurium, Protein, Cell, Rheumatoid arthritis, Glutamic acid decarboxylase (GAD), Commensal, Anergy, HSE, Crohn's disease, Immunomodulating agent, Receptor binding mapping, Graves' disease.*

Autoinducer

Refers to a signaling molecule that is utilized by applicable microorganisms for quorum sensing. Quorum sensing is the collective determination that "enough" of that microorganism are present, to initiate a *collective* action. Such collective actions can include

- "Turning on" one or more pathways for production of specific product(s) from certain substrate(s). For example, certain pathogenic bacteria (e.g., *Vibrio cholerae*) will often live benignly within the digestive system of an animal until "enough" of them are present, as determined via quorum sensing (e.g., utilizing an acyl homoserine lactone signaling molecule). At that point in time, those bacteria collectively turn on a pathway for production of their particular enterotoxin.
- "Differentiating" into specialized subtypes of cells, which perform different needed functions (for the biofilm/colony to survive).

A

- "Infecting" another (host) organism, if pathogenic bacteria.
- "Sporulating" (creation of spores for survival and/or reproduction).
- "Bioluminescing" (creation of light).

See also *Quorum sensing, Signaling, Signaling molecule, Microorganism, Pathway, Substrate (chemical), Bacteria, Pathogen, Enterotoxin, Differentiation, Cell, Biofilm, Bioluminescence.*

Autologous

Refers to two things that are derived from the same organism. See also *Organism.*

Autolysis

Refers to the reaction in which some enzymes catalyze their own "digestion" (lysis) at room temperature. See also *Enzyme, Lysis.*

This can be minimized by man (e.g., for enzymes utilized in industrial applications) via encapsulation of enzymes inside nanospheres. See also *Enzyme.*

Autonomous Replicating Segment

See *ARS element.*

Autophagosome

See *Autophagy.*

Autophagy

From the Greek words meaning "self eating," this refers to the cellular processes involved in cleaning up damaged cell parts (e.g., parts of old protein molecules, organelles), viruses, bacteria, etc. That detritus is gathered up in the cell's cytoplasm by phagophores (open-ended globules composed of sheets of proteins and lipids) that enclose the detritus by forming themselves into a sphere around it. Once the sphere is formed, it is known as an autophagosome. The autophagosomes carry the detritus to digestive organelles (e.g., lysosomes) within the cell, where the detritus is broken down and recycled.

Autophagosomes also gather up and carry pathogens or parts of pathogens (e.g., antigens, epitopes) to the cell's toll-like receptors, which thereby activate the organism's innate immune system while also preventing inappropriate inflammation.

Autophagosomes also sometimes gather up and carry certain pharmaceuticals (e.g., chemotherapy agents introduced into cancer cells) to lysosomes within the cell, where those pharmaceuticals are broken down and/or excreted (e.g., by efflux pumps). See also *Ubiquitin, Cell, Protein, Lipids, Lysosome, Cytoplasm, Toll-like receptors, Antigen, Pathogen, Epitope, Innate immune system, Cancer, Antibiotic resistance, ABC transporters, Chemotherapy, Efflux pump, Chronic inflammation.*

Autoradiography

A technique to detect radioactively labeled molecules by creating an image on photographic film. The slab of gel or other material in which the molecules are held (suspended) is placed on top of a piece of photographic film. The two are then securely fastened together such that movement is eliminated and the film is exposed for a period of time. The exposed (to the radiation) film is subsequently developed and the radioactive area is seen as a dark (black) area. Among other uses, autoradiography has been used to track the spread of (radioactively labeled) viruses in a living plant. After treatment (i.e., the radioactive labeling process), the whole plant (in a slab) is placed on top of a piece of photographic film. When the film is subsequently developed, the "picture" seen is of a plant, with darker areas indicating regions of greater virus concentration. See also *Label (radioactive), Virus.*

Autosomes

All chromosomes except the sex chromosomes. A diploid cell has two copies of each autosome. See also *Cell, Chromosomes, Diploid.*

Autotroph

An organism that can live on very simple carbon and nitrogen sources, such as carbon dioxide and ammonia. See also *Heterotroph.*

Aux/IAA

See *Auxins.*

Auxins

From the Greek *auxein*, "to increase," this term refers to a "family" of chemical compounds (chemically related to indole-3-acetic acid) that regulate plant growth (e.g., stimulate cell enlargement, cell division, initiate roots/growth, flowering, fruit growth and ripening).

Auxins also control plant root growth, plant leaf reorientation response to light, the location where a given plant grows a new stem or a new leaf, fruit ripening, the flowering of some plants, how a plant defends itself against attacking pathogens (e.g., by shutting down some pathogenic bacteria–promoted plant part growth, which is harmful to the plant), etc.

An auxin molecule functions by binding to a protein molecule (known as TIR1) present in plants; that itself is a component of a molecular complex that attaches ubiquitin molecules to *protein molecules that are destined for recycling by the cell.* Those "tagged" protein molecules are said to have been "ubiquitinated" or "ubiquitylated."

Upon the binding to it by an auxin molecule, that molecular complex (with auxin bound to its TIR1 component) attaches ubiquitin molecules to a class of protein molecules known as Aux/IAA. Because Aux/IAA normally represses plant genes that are triggered by auxins, this ubiquitination of Aux/IAA results in their removal (for recycling elsewhere in the cell) and thereby *activation of the auxin-triggered genes.* See also *Hormone, Plant hormone, Indole-3-acetic acid, Phytohormone, Cell, Rice blast, Bacteria, Pathogen, Protein, PAMPs, Ubiquitin, Ubiquitinated, Gene, Repression (of gene transcription/translation).*

Auxotroph

Auxotrophic mutant. A mutant defective in the synthesis of a given biomolecule. The biomolecule must be supplied to the organism if normal growth is to be achieved. See also *Mutation, Gene, Gene delivery (gene therapy), Essential fatty acids.*

Avena Gene

A gene that (when present within the genome of a plant) confers resistance to mesotrione herbicide. See also *Gene, Genome, Herbicide-tolerant crop.*

Avidin

A protein that is naturally present in egg white, oilseed protein (e.g., soybean meal), and grain (e.g., corn/maize). The protein is 70 kDa in mass (weight) and has a high affinity for biotin (i.e., it "sticks" tightly to the biotin molecule). Since grain-eating insects require biotin (a B-complex vitamin) to live, adding extra avidin to grain (e.g., via inserting a gene to cause overproduction of avidin in the grain kernels) may be a way to protect grain from insects (e.g., weevils in stored corn/maize). See also *Protein, Soy protein, Corn, Kilodalton (KD), Biotin, Weevils, Vitamin, Streptavidin.*

Avidity (of an Antibody)

The "tightness of fit" between a given antibody's combining site and the antigenic determinant that it combines with. The firmness of the combination of antigen with antibody. See also *Antigenic determinant, Antibody, Antigen, Combining site, Polyclonal response, Catalytic antibody.*

Avirulence Gene

See *R genes.*

A_w

See *Water activity* (A_w).

Axon

From the Greek word *axis*; axons are also known as nerve fibers. An axon is a slender long projection from surface of a neuron (nerve cell) that usually conducts electrical impulses away from the neuron's cell body. The typical function of axons is to transmit information from neuron to different neurons, muscles, and glands. See also *Neuron, Cell, Oligodendrocytes.*

Azadirachtin

The pharmacophore (i.e., active ingredient) in secretions of the tropical neem tree, which resists insect depradation. See also *Pharmacophore, Neem tree, Crop biologicals.*

Azelaic Acid

A dicarboxylic acid that is produced by plants when the plant is infected by pathogenic bacteria. The azelaic acid acts as a signaling molecule (to distant plant tissues), resulting in those distant tissues' production of salicylic acid to combat the bacterial infection. See also *Acid, Signaling molecule, Bacteria, Pathogenic, Salicylic acid (SA).*

Azurophil-Derived Bactericidal Factor (ADBF)

Potent antimicrobial protein produced by neutrophils (a type of white blood cell). See also *Leukocytes.*

α-Helix

See *Alpha helix.*

α-Linolenic Acid

See *Linolenic acid.*

αβ–Tubulin

See *Tubulin.*

B

B Cells

B lymphocytes. See *Lymphocyte, B lymphocytes, T cells, Blast cell.*

B Lymphocytes

A class of white blood cells originating in the bone marrow and found in the blood, spleen, and lymph nodes. They are the precursors of (blood) plasma cells (B cells) that secrete antibodies (IgG) directed against invading antigens (e.g., of pathogenic bacteria).

Via a complex "gene splicing" process, the B cells of the human body are able to produce more than one billion different IgG antibodies (i.e., able to bind onto and neutralize a billion different antigens). Via a natural process known as affinity maturation, the immune system selects those B cell–producing antibodies with greater affinity for the antigen of the invading pathogen, to combat the invader.

Sometimes, B cells can go awry and contribute to causing the disease rheumatoid arthritis (e.g., by producing antibodies against the body's own tissue). The pharmaceutical Rituxan™ (rituximab) is a monoclonal antibody that can be utilized to inhibit the structural damage (to body joints) of rheumatoid arthritis. See also *Antigen, Antibody, Blast cell, Lymphocyte, Pathogen, Bacteria, Gene splicing, Immunoglobulin, Allelic exclusion, Rheumatoid arthritis, Monoclonal antibodies (MAb).*

B.t.

See *Bacillus thuringiensis (B.t.).*

B.t. israelensis

One of the approximately 30 subspecies groupings within the approximately 20,000 different strains of the soil bacteria known (collectively) as *Bacillus thuringiensis (B.t.).* When eaten (e.g., due to their presence on food), the protoxin proteins produced by *B.t. israelensis* are toxic to mosquitoes and black fly (Diptera) larvae. See also *Bacillus thuringiensis (B.t.), Protoxin, Ion channels.*

B.t. k.

See *B.t. kurstaki.*

B.t. kumamotoensis

One of the approximately 280 subspecies groupings within the approximately 50,000 different strains of the soil bacteria known (collectively) as *Bacillus thuringiensis (B.t.).* When eaten (e.g., due to their presence on/in their food plants), the protoxin proteins produced by *B.t. kumamotoensis* are toxic to larvae of the insect known as the corn rootworm (*Diabrotica virgifera virgifera*). See also *Bacillus thuringiensis (B.t.), Protoxin, Ion channels, Corn, Corn rootworm, Strain, Bacteria.*

B.t. kurstaki

One of the approximately 30 subspecies groupings within the approximately 20,000 different strains of the soil bacteria known (collectively) as *Bacillus thuringiensis (B.t.).* When eaten (e.g., as part of a genetically engineered plant), the protoxin proteins produced by *B.t. kurstaki* are toxic to certain caterpillars (*Lepidoptera* larvae), such as the European corn borer (pyralis). See also *Bacillus thuringiensis (B.t.), Protoxin, CRY1A (b) protein, Ion channels, European corn borer (ECB).*

B.t. tenebrionis

One of the approximately 30 subspecies groupings within the approximately 20,000 different strains of the soil bacteria known (collectively) as *Bacillus thuringiensis (B.t.).* When eaten (e.g., as part of a genetically engineered plant), the protoxin proteins produced by *B.t. tenebrionis* are toxic to certain insects. See also *Bacillus thuringiensis (B.t.), Protoxin, Genetic engineering, Ion channels.*

B.t. tolworthi

One of the approximately 30 subspecies groupings within the approximately 20,000 different strains of the soil bacteria known (collectively) as *Bacillus thuringiensis (B.t.).* When eaten (e.g., as part of a genetically engineered crop plant), the protoxin proteins produced by *B.t. tolworthi* are toxic to certain caterpillars (*Lepidoptera* larvae), such as the European Corn Borer (pyralis). See also *Bacillus thuringiensis (B.t.), Protoxin, CRY9C protein, Genetic engineering, Ion channels.*

BAC

Acronym for "bacterial artificial chromosomes." See also *Bacterial artificial chromosomes (BAC).*

Bacillus

Rod-shaped bacteria. See also *Bacteria, Bacillus subtilis (B. subtilis), Bacillus thuringiensis (B.t.), Bacillus licheniformis.*

Bacillus licheniformis

A (rod-shaped) bacterium that dwells in soil.

Various biotechnology companies have extracted a number of enzymes (e.g., amylases) from the wild type *Bacillus licheniformis*, modified the relevant enzymes' genes via mutagenesis (to improve the activity or other properties of that enzyme), and today sell that improved enzyme produced via genetically engineered *B. licheniformis.* See also *Bacteria, Enzyme, Amylase, Gene, Wild type, Mutation breeding, Genetic engineering.*

B

Bacillus subtilis

A (rod-shaped) aerobic bacterium commonly used as a host in recombinant DNA experiments. When they are grown in a liquid (e.g., nutrient broth in a fermentation vat) and the *Bacillus subtilis* population exceeds a particular density threshold, their group *collective swimming* behavior makes a "liquid flow" in one direction.

During the 1990s, research showed that corn (maize) plant tissues infected with the endophyte *B. subtilis* were less likely to become infected with *Fusarium moniliforme* fungus.

Another research has indicated the potential for prior infection of corn (maize) plant tissues to hinder any subsequent aflatoxin production in that plant by *Aspergillus flavus* fungus. See also *Bacteria, Host vector (HV) system, Deoxyribonucleic acid (DNA), Corn, Endophyte, Fungus, Fusarium moniliforme, Aflatoxin, cspB gene.*

Bacillus thuringiensis

Discovered by bacteriologist Ishiwata Shigetane on a diseased silkworm in 1901. Later discovered on a dead Mediterranean flour moth, and first named *Bacillus thuringiensis*, by Ernst Berliner in 1915.

Today, *B. thuringiensis* refers to a group of rod-shaped soil bacteria found all over the earth that produce "cry" proteins, which are indigestible by—yet still "bind" to—specific insects' gut (i.e., stomach) lining (epithelium cell) receptors, so those "cry" proteins are thereby toxic to certain classes of insects (corn borers, corn rootworms, mosquitoes, black flies, some types of beetles, etc.), but which are harmless to all mammals. At least 20,000 strains of *B. thuringiensis* are known.

Genes that code for the production of these "cry" proteins that are toxic to insects have been inserted by scientists since 1989 into vectors (i.e., viruses, other bacteria, and other microorganisms) in order to confer insect resistance to certain agricultural plants (e.g., via expression of those *B.t.* proteins by one or more tissues of the transgenic plant). For example, the *B.t.* strain known as *B.t. kurstaki*, which is fatal when ingested by the European corn borer was first (genetically) inserted into a corn plant (via vector) in 1991. *B.t. kurstaki* kills borers via perforation of that insect's gut by cry ("crystal-like") proteins that are coded for by the *B.t. kurstaki* gene. The vectors as listed earlier are entities that can take up and carry the DNA into plant or other cells. Vectors are DNA-carrying vehicles. See also *Endophyte, Corn, Gene, Pseudomonas fluorescens, Agrobacterium tumefaciens, Aureofacin, European corn borer* (ECB), *Cowpea trypsin inhibitor* (CpTI), *Protein, "Shotgun" method, Coding sequence, Fusarium, Vector, Express, Genetic engineering, "Explosion" method, Biolistic® gene gun, CRY proteins, CRY1A (b) protein, CRY1A (c) protein, CRY9C protein, B.t. kurstaki, B.t. tenebrionis, B.t. israelensis, B.t. tolworthi, Ion channels.*

Back Mutation

Reverse the effect of a mutation that had inactivated a gene; thus, it restores wild phenotype. See also *Phenotype, Mutation.*

Bacteria

From the Greek *bakterion* = "stick," since the first bacteria viewed by man (via crude microscopes) appeared to be stick shaped. Bacteria were first observed in 1676 by Anton van Leeuwenhoek, the inventor of the microscope.

Any of a large group of microscopic organisms having round, rodlike, spiral, or filamentous unicellular or noncellular bodies that are often aggregated into colonies are enclosed by a cell wall or membrane (procaryotes), and lack fully differentiated nuclei. Bacteria may exist as free-living organisms in soil, water, and organic matter, or as parasites in the live bodies of plants and animals. See also *Bacteriology, Cell, Membranes (of a cell).*

Bacterial Artificial Chromosomes

Pieces of DNA (e.g., plant DNA) that have been cloned (made) inside living bacteria (e.g., by plant researchers who need to "manufacture" some pieces of plant DNA). They can be utilized as vectors (for genetic engineering) to carry (inserted) genes into certain organisms.

Some potential uses of bacterial artificial chromosomes include

- The "manufacture" of probes (i.e., sequences of DNA utilized to "find" complementary sequences within large pieces of DNA) via hybridization
- The "manufacture" of "DNA sequence markers" for use in marker-assisted selection (e.g., to guide choices made by commercial crop breeders, so they can more quickly select plants bearing gene(s) for a particular trait) to develop future improved crop varieties faster than was previously possible

See also *Bacteria, Clone (a molecule), Synthesizing (of DNA molecules), Chromosomes, Yeast artificial chromosomes (YAC), Human artificial chromosomes (HAC), Probe, Marker assisted selection, Complementary DNA (c-DNA), Hybridization (molecular genetics), Deoxyribonucleic acid (DNA), Sequence (of a DNA molecule), Marker (DNA sequence), Gene, Trait, Genetic engineering, Vector.*

Bacterial Expressed Sequence Tags

Expressed sequence tags (ESTs) that are based on sequenced/mapped bacterial genes instead of the genes of ("traditional" EST) *Caenorhabditis elegans* nematode. They are utilized to "label" a given gene (i.e., in terms of that gene's function/protein). See also *Best, Expressed sequence tags (EST), Bacteria, Sequencing (of DNA molecules), Sequence (of a DNA molecule), Mapping, Caenorhabditis elegans (C. elegans).*

Bacterial Two-Hybrid System

See *Two-hybrid systems.*

Bactericide

See *Microbicide, Biocide, Antibiotic.*

Bacteriocide

See *Bactericide.*

Bacteriocins

Proteins produced by many types of bacteria that are toxic (primarily) to other closely related strains of the particular bacteria that produce those proteins. Bacteriocins hold promise (e.g., after genetic engineering of the DNA responsible for their production) for future possible use as food preservatives (i.e., acting against bacteria species that cause food spoilage).

Examples include

- The bacteriocin known as curvaticin 13, which is produced by *Lactobacillus curvatus* bacteria, inhibits the food-poisoning bacteria *Listeria monocytogenes.*
- The bacteriocin known as sakacin K, which is produced by *Lactobacillus sakei* bacteria, inhibits the food-poisoning bacteria *Listeria monocytogenes.*

However, the effectiveness of both curvaticin 13 and sakacin K are lessened by the presence of salt (e.g., in processed meat products), so salt resistance would be a desired property that may someday be engineered into those bacteriocins.

Nisin is also referred to as a bacteriocin, but it is active against many more species of bacteria than most of the bacteriocins. See also *Protein, Bacteria, Bacteriology, Bifidus, Strain, Toxin, Genetic engineering, Deoxyribonucleic acid (DNA), Coding sequence, Colicins, Listeria monocytogenes, Extremophilic bacteria, Nisin.*

Bacteriology

The science and study of bacteria, a specialized branch of microbiology. The bacteria constitute a useful and essential group in the biological community. Although some bacteria prey on higher forms of life, relatively few are pathogens (disease-causing organisms). Life on earth depends on the activity of bacteria to mineralize organic compounds and to capture the free nitrogen molecules in the air for use by plants. Also, bacteria are important industrially for the conversion of raw materials into products such as organic chemicals, antibiotics, and cheeses. Genetically engineered bacteria are often used to produce high-value-added pharmaceuticals and specialty chemicals. See also *Escherichia coliform (E. coli).*

Bacteriophage

Discovered in 1917 by Felix d'Herelle (fr. "bacteria eaters"), a bacteriophage is a virus that attaches to, injects its DNA into, and multiplies inside bacteria, which eventually causes the bacteria to die. Upon the bacteria's death, 10–200 more new bacteriophages are released to infect other bacteria. Often abbreviated as simply *phage.* Phage is also another name sometimes utilized for virus.

Bacteriophages were first used to treat human skin infections in 1921. Some countries used bacteriophages to treat bacterial infections for a half century.

During 2007, the U.S. Department of Agriculture (USDA) approved the use of a *bacteriophage that targets E. coli 0157:H7 bacteria* in spray or wash-product form to be applied to livestock prior to their slaughter.

In recent decades, some bacteriophages have been utilized as tools by biological researchers. As an example, bacteriophage lambda is commonly used as a vector in rDNA experiments in *Escherichia coliform* strains (*not 0157:H7*) and attaches to a specific receptor, which in that bacteria also normally functions in sugar transport across the cell wall. See also *Escherichia coliform (E. coli), Receptors, Virus, Transduction (gene), Transduction (signal), Transfection, Lambda phage, Holins, Escherichia coliform 0157:H7, Pro-phage.*

Bacterium

See *Bacteria.*

Baculovirus

A class of virus that infects lepidopteran insects (e.g., cotton bollworm or gypsy moth larva). Baculoviruses can be modified via genetic engineering to insert new genes into the larva, causing those larvae to then produce proteins desired by man (e.g., pharmaceuticals).

Baculoviruses are potentially very useful for pharmaceutical production, because

- The protein molecules produced are *glycosylated (i.e., have relevant oligosaccharides attached to them).*
- Baculoviruses cannot infect vertebrate animals; thus, such pharmaceuticals are not even a theoretical pathogenic risk to humans.

See also *Virus, Genetic engineering, Gene, Protein, Glycosylation, Baculovirus expression vectors (BEVs), Pathogen.*

Baculovirus Expression Vector

Refers to vectors (used by researchers to carry new genes into insect cells) in which the agent is a baculovirus (i.e., a virus that infects certain types of insect cells only). A genetically engineered baculovirus expression vector (BEV) is commonly utilized to carry a new gene into the insect cells within a baculovirus expression vector system (BEVS) to induce cell culture production of a protein desired by man.

A BEV could conceivably be used to make a genetically engineered "insecticide" that is specific to a targeted insect (i.e., wouldn't harm anything but that insect). For example, a BEV might be used to cause a cotton bollworm *adult* protein to be expressed when the bollworm is a *juvenile,* thus killing the bollworm before it has a chance to damage a cotton crop. See also *Baculovirus, Virus, Vector, Gene, Protein, Cell, Genetic engineering, Insect cell culture, Baculovirus expression vector system (BEVS).*

Baculovirus Expression Vector System

Refers to an insect cell culture system, invented in 1982 by Gale Smith and Max Summers, in which a genetically engineered baculovirus expression vector (BEV) is utilized to carry into the insect cells a gene that codes for a protein desired by man. See also *Insect cell culture, Cell, Gene, Genetic engineering, Coding sequence.*

Bakanae

See *Fusarium moniliforme.*

BAR Gene

A dominant gene from the *Streptomyces hygroscopicus* bacterium, which codes for (i.e., causes production of) the enzyme "phosphinothricin acetyl transferase (PAT)." When the BAR gene is inserted into a plant's genome (i.e., its DNA), it imparts resistance to glufosinate-ammonium-based herbicides.

Because the glufosinate-ammonium herbicides act via inhibition of glutamine synthetase (an enzyme that catalyzes the synthesis

B

of glutamine), this inhibition (of enzyme) kills plants (e.g., weeds). That is because glutamine is crucial for plants to synthesize critically needed amino acids.

The BAR gene is often utilized by genetic engineers as a marker gene. See also *Gene, Genome, Genetic engineering, Marker (genetic marker), Dominant allele, Essential amino acids, Herbicide-tolerant crop, GTS, Soybean plant, Canola, Corn, Glutamine, Glutamine synthetase, Phosphinothricin, Phosphinothricin acetyltranserase (PAT), PAT gene.*

Barley

The grain of the domesticated plant *Hordeum vulgare* is utilized by man for various purposes:

- Feed barley varieties, utilized for feeding of livestock.
- Malting barley varieties (containing beta-amylase in their seeds) were created via mutation breeding (i.e., bombardment of the seeds by ionizing radiation to cause random genetic mutations, followed by selection of the particular mutation in which maltose is produced by that barley plant in its seeds).

See also *Traditional breeding methods, Mutation, Mutation breeding, Amylase.*

Barnase

Abbreviation for the ribonuclease "*Bacillus amyloliquefaciens* RNase"; it is an enzyme that catalyzes destruction of nucleic acids (which thus kills the cell that the barnase is in).

When the gene that codes for barnase is inserted via genetic engineering into a given plant and activated only in that plant's anther tissue (i.e., the barnase is produced only in its anther pollen-producing cells), that plant's male parts become sterile. For crop plants possessing both male and female parts (i.e., monoecious plants), such "male sterility" facilitates the development of hybrids, because self-pollination does not occur. See also *Enzyme, Nucleic acids, Cell, Gene, Genetic code, Coding sequence, Genetic engineering, Genetics, Hybridization (plant genetics), F1 hybrids, Monoecious, RNase, Barstar.*

Barnase–Barstar Gene System

See *Barnase* and *Barstar.*

Barstar

Refers to a gene (extracted by man from the bacterium *Bacillus amyloliquefaciens*) that codes for the production (e.g., in the anther parts of plants) of an enzyme that inhibits barnase. See also *Gene, Genetic code, Coding sequence, Bacteria, Enzyme, Inhibition, Barnase.*

Base (General)

A substance with a pH in the range 7–14 that will react with an acid to form a salt. Mild bases normally taste bitter and feel slippery to the touch. See also *Acid.*

Base (Nucleotide)

A segment of the DNA (and RNA) molecules—one of the four (repeating) chemical units that comprise DNA/RNA that, according to their order and pairing (i.e., on the parallel strands of DNA/RNA molecules), represent the different amino acids (i.e., within the protein molecule that each gene in the DNA codes for). The four bases that comprise DNA are adenine (A), cytosine (C), guanine (G), and thymine (T).

The four bases that comprise RNA are adenine (A), cytosine (C), guanine (G), and uracil (U). See also *Deoxyribonucleic acid (DNA), Ribonucleic acid (RNA), Polymer, Coding sequence, Control sequences, Expression, Amino acid, Protein, Gene, Adenine, Cytosine, Guanine, Thymine, Uracil, Base pair (bp).*

Base Calls

Refers to the raw data resultant from a DNA sequencing process. See also *Sequencing (of DNA molecules)*, and *Base (nucleotide).*

Base Excision Repair

See *DNA repair, Nucleotide, Base (nucleotide).*

Base Excision Sequence Scanning

A method that can be utilized to detect a "point mutation" in DNA (via rapid DNA sequence scanning). See also *Base pair (bp), Nucleotide, Deoxyribonucleic acid (DNA), Mutation, Point mutation, Excision, Sequencing (of DNA molecules), Sequence (of a DNA molecule).*

Base Pair

Two nucleotides that are in different strands of a given nucleic acid molecule (DNA or RNA) and whose bases pair (interact) by hydrogen bonding. In DNA, the nucleotide bases are adenine (which pairs with thymine) and guanine (which pairs with cytosine).

In RNA, the nucleotide bases are adenine (which pairs with uracil) and guanine (which pairs with cytosine). See also *Deoxyribonucleic acid (DNA), Nucleotide, Genetic code, Informational molecules, Ribonucleic acid (RNA), Hydrogen bonding, Lesion.*

Base Pairing

Refers to the hydrogen-bonding-driven matching of complementary base pairs (e.g., components of DNA molecules) during replication of DNA in cells, or during hybridization (e.g., of a DNA sample fragment to a DNA probe). See also *Deoxyribonucleic acid (DNA), DNA probe, Replication (of DNA), Mismatch repair, Hybridization (molecular genetics).*

Base Substitution

Replacement of one base (within a DNA molecule) by another base. See also *Base (nucleotide), Transition, Transversion, Oligonucleotide-mediated mutagenesis, Lesion.*

Basic Fibroblast Growth Factor

See *Fibroblast growth factor (FGF).*

Basophilic

Refers to something that stains strongly with basic dye. For example, basophil leukocytes are polymorphonuclear leukocytes that stain strongly with (take up a lot of) basic dyes. See also *Polymorphonuclear leukocytes (PMN)*.

Basophils

Also called "basophilic leukocytes."

A type of white blood cell (leukocyte) produced by stem cells within the bone marrow that synthesizes and stores histamine and also contains heparin. When two IgE molecules of the same antibody "dock" at adjacent receptor sites on a basophil cell, the two IgE molecules capture an allergen between them. A chemical signal is sent to the basophil causing the basophil cell to release histamine, serotonin, bradykinin, and "slow-reacting-substance." Release of these chemicals into the body causes the blood vessels to become more permeable, which consequently causes the nose to run. These chemicals also cause smooth muscle contraction, resulting in sneezing, coughing, wheezing, etc. See also *Mast cells, Antigen, Antibody, Histamine, White blood cells, Basophilic, Leukocytes, Polymorphonuclear leukocytes (PMN), Stem cells.*

BB T.I.

See *Trypsin inhibitors.*

BBB

See *Blood–brain barrier (BBB).*

BCA

Acronym for "*bio-barcode amplification.*" See *Bio-bar codes.*

Bce4

The name of a promoter (region of DNA) that controls/enhances an oilseed plant's gene(s) that code for components (e.g., fatty acids, amino acids) of that plant's seeds. The Bce4 promoter causes such genes to be expressed during one of the earliest stages of canola plants' seed production, for instance. See also *Promoter, Deoxyribonucleic acid (DNA), Gene, Polygenic, Plastid, Express, Canola, Soybean plant, Transcription.*

Bcr-abl Protein

Refers to a particular tyrosine kinase that is coded for by the abnormal chromosome that bears the Bcr-abl gene (present in the DNA of approximately 95% of people who get the disease known as chronic myelogenous leukemia). See also *Protein, Enzyme, Kinases, Tyrosine kinase, Deoxyribonucleic acid (DNA), Gene, Genetic code, Coding sequence, Chromosomes, Bcr-Abl genetic marker, Bcr-Abl gene.*

Bcr-Abl Gene

The gene (SNP) that causes the blood cancer *chronic myelocytic leukemia (CML)* in humans that possess it. See also *Gene, Single-nucleotide polymorphisms (SNPs), Cancer, Gleevec™.*

Bcr-Abl Genetic Marker

See *Genetic marker, Fluorescence in situ hybridization (FISH).*

B-DNA

A helical form of DNA. B-DNA can be formed by adding back water to (dehydrated) A-DNA. B-DNA is the form of DNA of which James Watson and Francis Crick first constructed their model in 1953. It is found in fibers of very high (92%) relative humidity and in solutions of low ionic strength. This corresponds to the form of DNA that is prevalent in the living cell.

Irradiation by gamma rays of B-DNA (in copper-containing solution) converts B-DNA to Z-DNA. See also *Deoxyribonucleic acid (DNA), A-DNA, Ion, Cell, Z-DNA.*

BDNF

Acronym for brain-derived neurotrophic protein. See *Brain-derived neurotrophic protein.*

Beige Fat

See *Brown adipocytes.*

BER

Acronym for "*base excision repair.*" See *Base excision repair, DNA repair.*

BESS Method

See *Base excision sequence scanning (BESS).*

BESS T-Scan Method

See *Base excision sequence scanning (BESS).*

Best Linear Unbiased Prediction

A statistical (data) technique that is utilized by livestock breeders to determine the breeding (genetic trait) value of animals in a breeding program. See also *Genetics, Trait, Phenotype, Genotype, Expected progeny differences (EPD).*

Beta Carotene

A phytochemical (vitamin precursor) that is naturally produced in carrots, other orange vegetables, apricots, cantaloupe, kiwi, papaya, and the endosperm portion of the corn (maize) kernel. If the corn kernel seed coat is torn (e.g., via insect chewing), the beta carotene inhibits growth of *Aspergillus flavus* fungi in the endosperm region of the kernel.

In 1970, an orange (-fruited) cauliflower was discovered growing in Bradford Marsh in Canada. It was the result of a natural mutation that caused beta carotene to be produced in that cauliflower plant at a level that was approximately 100 times higher than normal for cauliflower.

Beta carotene has been found to aid eyesight and to strengthen the immune system in people who consume it and may help prevent lung cancer and heart disease.

Because beta carotene is processed into vitamin A by the human body, consumption of this phytochemical can help avoid human diseases (e.g., in developing countries where vitamin A is scarce) that result from deficiency of vitamin A, for example,

- Coronary heart disease
- Certain cancers (e.g., cancer of prostrate, lungs)
- Childhood blindness
- Age-related macular degeneration, a leading cause of blindness in older people
- Various childhood diseases that often result in death, due to weakened immune system

See also *Vitamin, Golden rice, Aflatoxin, Fungus, OH43, Phytochemicals, Nutraceuticals, Carotenoids, Cancer, Coronary heart disease* (*CHD*), *Antioxidants, AMD, Desaturase.*

Beta Cells

Insulin-producing cells in the pancreas. If these cells are destroyed, childhood (also known as early-onset or Type I) diabetes result. See also *Islets of Langerhans, Insulin, Type I diabetes, Serotonin.*

Beta Conformation

An extended, zigzag arrangement of a polypeptide (molecule) chain. See also *Polypeptide* (*protein*).

Beta Interferon

One of the interferons, it is a protein that binds to a receptor located on the surface of T cells. That binding slows down the body's immune response (e.g., normally after the immune response has defeated an infection).

Man-made beta interferon was approved by America's Food and Drug Administration in 1993 to be used to treat multiple sclerosis, a disease that is caused by an immune response that fails to ever shut down. See also *Interferons, Food and Drug Administration* (*FDA*), *Protein, Receptors, T cell receptors, T cells, Immune response, Multiple sclerosis.*

Beta Oxidation

See *Carnitine.*

Beta Sitostanol

See *Sitostanol.*

Beta Sitosterol

See *Sitosterol.*

Beta-Conglycinin

Abbreviated β-conglycinin. One of the (structural) categories of proteins that is produced in seeds of legumes. For example, it constitutes approximately 5% of soybeans. In general, β-conglycinin contains one-quarter to one-third as much cysteine and methionine per unit of protein as does glycinin.

β-Conglycinin has greater emulsifying capacity (in water) and emulsion stability than does glycinin, so its presence can assist the manufacture of better protein-based (emulsion) drinks. Human consumption of a moderate amount of beta-conglycinin helps to reduce inflammation. Human consumption of significant amounts of beta-conglycinin causes a reduction in bloodstream levels of cholesterol and triglycerides. See also *Protein, Cysteine* (*cys*), *Methionine* (*met*), *Glycinin, Emulsion, Chronic inflammation, Cholesterol.*

Beta-D-Glucuronidase

See *GUS gene.*

Beta-Glucan

Refers to a type of water-soluble fiber produced in oat bran (and some other grain kernels) that is a polysaccharide composed entirely of glucose (molecular) units. Wheat and barley also contain some beta-glucan.

Beta-glucan passes intact through the stomach and small intestine and is fermented to produce short-chain fatty acids in the large intestine.

In 1997, the U.S. Food and Drug Administration approved a (label) health claim that associates consumption of beta-glucan oat fiber with reduced blood cholesterol content and with reduced coronary heart disease. Research indicates that consumption of beta-glucan also helps prevent colorectal cancer and diabetes. See also *Water soluble fiber, Polysaccharides, Glucose* (*GLc*), *Short-chain fatty acids, Food and Drug Administration* (*FDA*), *Type II diabetes.*

Beta-Glucuronidase

See *Beta*-D-glucouronidase.

Beta-Lactam Antibiotics

A category of antibiotics (e.g., penicillin G, ampicillin) that kill targeted bacteria by altering their essential cellular function of enzymatic controls that keep cell wall (peptidoglycan) synthesis (i.e., creation/repair) in balance with cell wall degradation, thereby causing cell wall breakdown and death of those bacteria (pathogens). See also *Antibiotic, Penicillin G, Bacteria, Cell, Enzyme, Pathogen, bla gene.*

Beta-Secretase

An enzyme that (in the human brain) is linked to the presence of Alzheimer's disease. See also *Enzyme, Alzheimer's disease, Amyloid* β *protein precursor* (*A*β*PP*).

Bevacizumab

An antiangiogenic monoclonal antibody that was approved by the U.S. Food and Drug Administration for use as a pharmaceutical in conjunction with specific types of applicable chemotherapy for people with metastatic colorectal cancer and for women with persistent, recurrent, or metastatic carcinoma of the cervix.

Bevacizumab used along with applicable types of chemotherapy is effective for the treatment of nonsquamous, non–small cell lung cancer and for treatment of recurrent ovarian cancer that is resistant to platinum-containing chemotherapy.

Research indicates that bevacizumab would be effective in helping to treat certain other cancers. See also *Cancer, Angiogenesis,*

Antiangiogenesis, Angiogenesis inhibitors, Monoclonal antibodies (MAb), Metastasis, Food and Drug Administration (FDA), Cell.

BEVS

See *Baculovirus, Baculovirus expression vector (BEV), Baculovirus expression vector system (BEVS).*

BFGF

Basic fibroblast growth factor. See *Fibroblast growth factor (FGF).*

BGYF

See *Bright greenish-yellow fluorescence (BGYF).*

BHK Cells

Abbreviation for *baby hamster kidney cells.* This refers to cell lines propagated/grown in cell culture (e.g., in petri dishes) that were originally removed from the kidney of baby hamster(s). See also *Cell, Cell culture, Mammalian cell culture.*

Bifidobacteria

See *Bifidus.*

Bifidus

A "family" of bacteria species that live within the digestive systems of certain animals (e.g., humans, swine). Examples include *Bifidobacterium bifidum, Bifidobacterium longum, Bifidobacterium infantis, Bifidobacterium adolescentis,* and *Bifidobacterium acidophilus.* In general, *Bifidus* bacteria help to promote good health of the host animals by several means:

- They produce organic acids (e.g., propionic, acetic, lactic), which make the host animal's digestive system more acidic. Because most pathogens (i.e., disease-causing microorganisms) grow best at a neutral pH (i.e., neither acidic nor base/caustic), the growth rates of pathogens are thereby inhibited.
- They "crowd out" enteric pathogens, since Bifidus bacteria grow fast in the acidic environment created by those organic acids.
- Some of the organic acids (e.g., propionic) produced by Bifidus bacteria are able to pass-through the outer cell membrane of pathogenic bacteria and fungi. Once inside those pathogens' cells, these acids dissociate and acidify the cell interior (which disrupts protein synthesis, growth, and replication of that pathogen).
- They produce bacteriocins, which are proteins that suppress growth of the pathogenic bacteria.
- They produce certain short-chain fatty acids, which are absorbed by the host animal (e.g., in the colon) and thereby result in a reduction of triglyceride (fat) levels in the host animal's bloodstream. That (triglyceride reduction) lowers the risk of coronary heart disease and thrombosis.

See also *Bacteria, Species, Probiotics, Acid, Base (general), Pathogen, Cell, Plasma membrane, Microorganism, Fungus, Protein, Ribosomes, Growth (microbial), Fructose oligosccharides, Fatty acid, Triglycerides, Coronary heart disease (CHD), Thrombosis, Prebiotics, Bacteriocins, Inulin, Transgalacto-oligosaccharides.*

Bile

A liquid (mixture) made by the liver to help digest fats (in the intestine) and facilitate intestinal absorption of certain fat-soluble vitamins and minerals. Bile consists primarily of water, cholesterol, lipids (fat), "natural detergents" (i.e., salts of bile acids such as cholic acid, chenodeoxycholic acid) that help break up fat globules in the intestines, and bilirubin. See also *Bile acids, Bilirubin, Fats, Digestion (within organisms), Farnesoid X receptor (FXR), Enterocytes.*

Bile Acids

A "family" of acids (chenodeoxycholic acid, cholic acid, etc.) that are derived by the human liver from dietary cholesterol (i.e., from foods), and excreted into the bile by the liver. They help to emulsify (food source) fats in the small intestine, as part of crucial first step in the digestion of fats. See also *Cholesterol, Digestion (within organisms), Lecithin, Fats, Lipids, Farnesoid X receptor (FXR), Enterocytes.*

Bilirubin

A component (pigment) of red blood cells (i.e., erythrocytes), which is recovered (from old red blood cells) via oxidative degradation of heme and recycled via making bile (a liquid that aids the digestive process) by the liver.

During 2008, Cary Pirone and David W. Lee discovered that the bird of paradise tree (*Strelitzia nicolai*) also manufactures bilirubin and stores it in its seed arils. See also *Erythrocytes, Heme, Oxidation (chemical reaction), Bile, Digestion (within organisms), Endothelium, BOXes.*

Binning

A term utilized within metagenomics (a methodology for simultaneously assessing the DNA of multiple microorganisms within a sample or a series of samples taken from a given environment). After sequencing of the sample's DNA *en masse* yields its (mixed) information, binning refers to how computational tools (bioinformatics) are utilized to do data assembly and the subsequent assignment of DNA fragments to each of the respective microorganisms present. See *Deoxyribonucleic acid (DNA), Metagenomics, Microorganism, Sequence (of a DNA molecule), Sequencing (of DNA molecules), Gene, Genetic code, Shotgun sequencing, Sequence map.*

BIO

See *Biotechnology industry organization (BIO).*

Bio Brick

See *Synthetic biology.*

Bioassay

Determination of the relative strength or bioactivity of a substance (e.g., a drug). A biological system (such as living cells, organs,

B

tissues, or whole animals) is exposed to the substance in question and the effect on the living test system is measured. See also *Biological activity, Assay, Hanging drop assays, Biochip, Multiplex assay.*

Bioavailability (of Plant Nutrients)

See *Precision agriculture, Crop biologicals.*

Bio-Bar Codes

Refers to oligonucleotides (DNA segments) located on the surface of the following:

- *Nanoparticle probes*—These are gold particles of 30-nm-size dimensions to which have been attached one *antibody specific to target protein molecule*, plus thousands of (hybridized) single strands of a *specific DNA sequence.* Because each antibody binds to only one protein, these specific DNA sequences thereby serve as a *bar code–like label specific to that protein.* During 2003, Chad Mirkin, Jwa-Min Nam, and C. Shad Thaxton created such "bar-coded" nanoparticle probes whose attached antibody was specific to the protein known as prostate-specific antigen (PSA). When utilized in conjunction with magnetic particles whose antibodies are also themselves specific to the same protein (i.e., PSA), these nanoparticle probes *jointly* attach to that protein molecule along with the magnetic particles. A magnetic field was utilized to remove the magnetic particle/nanoparticle probe agglomeration from (solution mixture); then a dehybridization solution was used to remove the specific-to-PSA-molecule DNA segments for subsequent identification (e.g., via DNA microarray). Because the identification segments (of DNA) are thousands of times more numerous than the analyte (i.e., protein molecules), Mirkin/Nam/Thaxton named this process *bio-bar code amplification.*
- This nanoparticle probe/magnetic particle system can be utilized to simultaneously detect and identify *numerous different proteins* within a given sample or *numerous different DNA segments* within a given sample.
- *"Phage-displayed library" peptides*—See the entry within this glossary for PHAGE DISPLAY.

See also *Oligonucleotides, Nanometers* (nm), *Nanotechnology, Deoxyribonucleic acid (DNA), Antibody, Protein, Sequence (of a DNA molecule), Peptide, Prostate-specific antigen (PSA), Phage display, Magnetic particles, Hybridization (molecular genetics), DNA microarray, Target (of a therapeutic agent).*

Biochemistry

The study of chemical processes that comprise living things (systems). The chemistry of life and living matter. Despite the dramatic differences in the appearances of living things, the basic chemistry of all organisms is strikingly similar. Even tiny one-celled creatures carry out essentially the same chemical reactions that each cell of a complex organism (such as man) carries out. See also *Molecular biology, Molecular diversity, Cell.*

Biochips

A term first used with regard to an electronic device that utilizes biological molecules as the "framework" for other molecules that act as semiconductors and functions as an integrated circuit

1. During the 1990s, this term also became commonly used to refer to various "laboratories on a chip" (e.g., to analyze very small samples of DNA, to assess the impact of pharmaceuticals—or pharmaceutical drug candidate molecules—on specific cells [i.e., attached to the biochip's surface] or on specific cellular receptors [ligand–receptor response of a cell], to size and sort DNA fragments [genes] via the [proportional] fluorescence of dyes intercalated in the DNA molecules, to detect the presence of a specific DNA fragment [gene] via hybridization to a probe [which was fabricated onto the "chip"], to size and sort protein molecules [via various cells fabricated onto "chip"], to assess pharmaceuticals via adhesion molecules attached to a "chip," to detect specific pathogens or cancerous cells in a blood sample [e.g., by applying controlled electrical fields to cause those cells to collect at electrodes on the "chip"], to screen for compounds that act against a disease [e.g., by applying antibodies linked to fluorescent molecules and then measuring electronically the fluorescence that is triggered by antibody binding], to conduct gene expression analysis by measuring fluorescence of messenger RNA [specific to which particular gene is "turned on"] when that mRNA hybridizes with DNA [from genome] on the hybridization surface of a chip).
2. Shortly after the 1990s, several companies began manufacturing "biochips" capable of sequencing (i.e., determining the sequence of) DNA samples. Such biochips have—attached to their surfaces—all possible "DNA probes" (i.e., short sequences of DNA). The sample (i.e., the unknown DNA molecule) is passed over the probe-covered surface of the biochip, where each relevant segment (within the large unknown DNA molecule) hybridizes with (i.e., "pairs" with) the short "DNA probe" attached to a known location on the surface of the biochip. Because the sequence of each DNA probe—at each specified location on the biochip—is known, that information (i.e., the probes' sequences that the unknown DNA molecule hybridized to) is then utilized to "assemble the complete sequence" of the unknown DNA molecule.
3. Sometimes refers to an electronic device that uses biological molecules as the framework for other molecules that act as semiconductors and functions as an integrated circuit. The future working parts of the science of bioelectronics, biochips may consist of 2D or 3D arrays of organic molecules used as switching or memory elements.

One application will be to shrink currently existing biosensors in size. This would enable the biosensors to be implanted in the body or in organs and tissues for the sake of monitoring and controlling certain bodily functions. A future possibility is to try to provide sight for the blind using light-sensitive (e.g., protein-covered electrode) biochips implanted in the eyes to replace a damaged retina. For example, during 2001, Alan Chow implanted such biochips into several men whose retinas had been damaged by the disease "retinitis pigmentosa."

See also *Bioelectronics, Bionics, Biosensors (electronic), Deoxyribonucleic acid (DNA), Ribonucleic acid (RNA), Gene,*

Receptors, High-throughput screening (HTS), Bioinorganic, Target-ligand interaction screening, Antibody, Characterization assay, Bioassay, Assay, Luminescent assay, Protein, Ligand (in biochemistry), Microfluidics, Probe, Proteomics, Proteome chip, Bioreceptors, Hybridization (molecular biology), Fluorescence, Adhesion molecule, Gene expression analysis, Pathogen, Bioinformatics, Microarray (testing), Hybridization surfaces, Messenger RNA (mRNA), Genomics, Quantum dot, Quantum wire, Nanocomposites, Sequencing (of DNA molecules), Carbohydrate microarrays, ChIP, Peptide nucleic acid, Nanosheets.

Biocide

Any chemical or chemical compound that is toxic to living things (systems). Literally "biokiller" or killer of biological systems. Includes insecticides, bactericides, fungicides, etc.

Most bactericides accomplish their task (i.e., killing bacteria) via massive lysis (disintegration) of bacteria cell walls (membranes). However, one (i.e., triclosan) kills bacteria by inhibiting enoyl-acyl protein reductase, a crucial enzyme utilized by bacteria in their synthesis of fatty acids. See also *Bactericide, Microbicide, Lysis, Bacteria, Cell, Fatty acid, Enzyme, Protein, Essential fatty acids, Essential nutrients.*

Bioconversions

See *Feedstock, Fermentation, Substrate* (chemical), *Ionic liquids.*

Biodegradable

Describes any material that can be broken down by biological action (e.g., dissimilation, digestion, denitrification). The breakdown of material (e.g., animal carcasses, dead plants, even man-made chemicals) by microorganisms (bacteria, fungus, etc.).

The biodegradation process is often assisted (i.e., first step) by the actions of animals and insects (e.g., feeding on dead carcasses, which physically breaks down those carcasses to make their materials more available for microorganisms to "feed" upon). For example, the vulture and the yellow swallowtail butterfly often are the first to feed on the carcasses of dead alligators in the U.S. state of Florida, which helps to make the alligator's material (body tissue) more readily available to microorganisms (e.g., in the dung excreted by those "first step" carcass feeders).

In the medical field, the term "biodegradable" refers to materials (e.g., polylactic acid-polymer-based sutures left in body tissues following a surgery operation) that naturally break down within body tissues without leaving any harmful residues. See also *Digestion (within organisms), Microorganisms, Bacteria, Fungus, Glycolysis, Metabolism, Nitrification, Polyhydroxyalkanoic acid (PHA), Polyhydroxylbutylate (PHB).*

Biodesulfurization

The removal of organic and inorganic sulfur (a pollution source) from coal by bacterial and soil microorganisms. See also *Bioleaching, Biorecovery,* and *Biosorbents.*

Biodiversity

Defined to be "the variability among living organisms from all sources including terrestrial, marine/aquatic, and the complexes of which they are a part" by the Convention on Biological Diversity. See also *Convention on biological diversity.*

Bioelectronics

Also called biomolecular electronics. It is the field where biotechnology is crossed with electronics. The branch of biotechnology that deals with the electroactive properties of biological materials, systems, and processes together with their exploitation in electronic devices. For example, during 2003, Susan L. Lindquist utilized yeast prions (which self-assemble into 60–300 nm long fibers) to create *nanowires* by subsequently coating those fibers with gold or silver.

Bioelectronics will attempt to replace traditional semiconductor materials (e.g., silicon or gallium arsenide) with organic materials such as proteins (e.g., in biochips) and/or "hybrid" materials such as nanowires. See also *Biochips, Biosensors (electronic), Bioinorganic, Bionics, Quantum wire, Self-assembly (of a large molecular structure), Nanowire, Prion, nanoFET.*

Biofertilizer

Refers to a microorganism that either *mobilizes* a soil-borne chemically bound plant nutrient/mineral (i.e., makes nutrient/mineral bioavailable to crop plant roots) or itself *produces* (e.g., nitrate from the nitrogen in atmosphere) a plant nutrient. See also *Crop biologicals, Nitrates, Nitrogen fixation, Rhizobium (bacteria), Nitrogenase system.*

Biofilm

Refers to an integral layer of living microorganisms (e.g., on the surface of a vessel, on the surface of teeth, on the surface of an artificial joint/implant, and on the surface of the heart valve) that are held together within the biofilm by a polymer (e.g., certain proteins, cellulose and other sugars) they synthesize. Those microorganisms within a biofilm often differentiate in order for certain of the microorganisms to perform different tasks necessary for the survival of the overall biofilm. For example, the microorganisms within the "bottom" layer might specifically differentiate/change in a manner that enables them to better adhere the entire biofilm onto underlying substrate.

For example, *Streptococcus mutans* bacteria can form a biofilm on the surface of teeth. Biofilms are often responsible for a large number of human diseases and can sometimes help pathogenic bacteria to be resistant to antibiotics, because it is difficult for antibiotics to penetrate the biofilm. Approximately 80% of human pathogenic bacteria form biofilms during at least one part of their life cycle.

Sometimes, the microorganisms that constitute a biofilm will *act collectively* to do something (e.g., "turning on" one or more pathways for production of specific chemical product(s) from certain substrate(s)). When "enough" of that species of microorganism are present, as determined via quorum sensing, those microorganisms collectively turn on a pathway for the production of a product (e.g., a toxin, in the case of cholera disease).

During 2009, Erik Taylor and Thomas Webster discovered that specific nanoparticles (i.e., made of iron oxide with an average diameter of 8 nm) would penetrate and kill the *Staphylococcus epidermidis* bacteria within a biofilm constructed by those bacteria on the surface of a man-made medical implant in the body. See also *Microorganism, Bacteria, Cholera toxin, Pathogen, Substrate (structural), Streptococcus mutans, Quorum sensing, Pathway, Substrate (chemical), Polymer, Cellulose, Nanoparticles.*

B

Biogenesis

The theory that living organisms are produced only by other living organisms. That is, the theory of generation from preexisting life. It is the opposite of abiogenesis, or spontaneous generation.

Biogeochemistry

A branch of geochemistry that is concerned with biological materials and their relation to earth's chemicals in an area.

Bioinformatics

This term refers to the generation/creation, collection, storage (in databases), and efficient utilization of data/information from genomics (functional genomics, structural genomics, etc.), combinatorial chemistry, high-throughput screening, proteomics, DNA-sequencing research efforts, capillary electrophoresis (e.g., to determine the molecular structure of glycan molecule), etc. in order to accomplish a (research) objective (e.g., to discover a new pharmaceutical or a new herbicide).

Examples of the data/information that are manipulated and stored include gene sequences, biological activity/function, pharmacological activity, biological structure, molecular structure, protein–protein interactions, and gene expression products/amounts/timing. See also *Genomics, Functional genomics, Pharmacogenomics, Structural genomics, Combinatorial chemistry, High-throughput screening, Proteomics, Biochip, Gene, Genetic map, Genetic code, Sequencing (of DNA molecules), In silico biology, In silico screening, Gene expression analysis, Metamodel methods (of bioinformatics), Capillary electrophoresis, Glycoinformatics.*

Bioinorganic

This term refers to the combination of "organic" (life) materials with inorganic materials to create (useful materials). For example, Abalone shellfish make their shells via a combination of protein and calcium carbonate.

The scientific discipline known as *bioinorganic chemistry* arose during the 1970s after a series of Gordon conferences brought together scientists from the field of biology and from the field of inorganic chemistry. For example, researchers are working on making bioinorganic devices such as semiconductor devices (chips) containing peptides, etc. attached to silicon or gallium arsenide. See also *Protein, Biochip, Peptide, Biosensors (electronic), Nanocomposites.*

Bioleaching

The biomediated recovery of precious metals from their ores. In the recovery of gold, for example, the microorganism *T. ferrooxidans* may be used to cause the gold to leach out of the ore so it may then be concentrated and smelted. Aluminum may be similarly bioleached from clay ores, using heterotropic bacteria and fungi. See also *Biorecovery, Biogeochemistry, Bacteria, Biosorbents.*

Biolistic Gene Gun

The word "biolistic" was coined from the words "biological" and "ballistic" (pertaining to a projectile fired from a gun). Used to shoot tiny gold or tungsten pellets that are coated with genes (e.g., for desired crop traits) into plant seeds or plant tissues in order to get those plants to then express the new genes. The gun uses an actual explosive (.22 caliber blank) to propel the material. Compressed air or steam may also be used as the propellant. The Biolistic Gene Gun was invented in 1983–1984 at Cornell University by John Sanford, Edward Wolf, and Nelson Allen. It and its registered trademark are now owned by E. I. du Pont de Nemours and Company. See also *Whiskers™, "Shotgun" method, Genetic engineering, Gene, Bioseeds, Microparticles.*

Biologic Response Modifier Therapy

Refers to patient treatments (e.g., certain pharmaceuticals) that impact biological responses within an organism. For example, Avastin (bevacizumab) is a monoclonal antibody used in the treatment of certain cancers, which acts by inhibiting angiogenesis (formation of blood vessels within the body that "feed" a growing tumor, in response to chemical signals sent out by that tumor). See also *Organism, Cancer, Angiogenesis, Tumor, Antiangiogenesis, Monoclonal antibodies (MAb), Signaling.*

Biological Activity

The effect (e.g., change in metabolic activity inside living cells) caused by specific compounds, molecules, or other agents. For example, the drug aspirin causes the blood to thin, that is, to clot less easily. See also *Glycoform, Metabolism, Nitrosylation, Bioassay, Pharmacophore, Retinoids.*

Biological Crop Protection Products

See *Crop biologicals.*

Biological Nitrogen Fixation

See *Nitrogen fixation.*

Biological Oxygen Demand

The oxygen used in meeting the metabolic needs of aerobic organisms in water containing organic compounds. Numerically, it is expressed in terms of the oxygen consumed in water at a temperature of 68°F (20°C) during a 5-day period. The biological oxygen demand is used as an indication of the degree of water pollution. See also *Metabolism.*

Biological Pesticides

See *Crop biologicals.*

Biological Seed Treatments

See *Crop biologicals.*

Biological Vectors

See *Vectors.*

Biologicals

See *Crop biologicals.*

Biology

From the two Greek words *bios* (life) and *logos* (word), it is the field of science encompassing the "study of life." See also *Genetics, Cladistics, Organism, Species.*

Bioluminescence

The enzyme-catalyzed production of light by living organisms, typically during mating or hunting. This word literally means "living light." Bioluminescence was first identified/analyzed in 1947, by William McElroy.

For example, bioluminescence results when the enzyme luciferase comes into contact with adenosine triphosphate (ATP)/luciferin inside the photophores (organs which emit the light) of the organism. Such production of light by living organisms is exemplified by fireflies, South America's railroad worm, and by many deep ocean marine organisms.

Bioluminescence has been utilized by man as a genetic marker (e.g., to cause a genetically engineered plant to glow as evidence that a gene was successfully transferred into that plant).

Another use of bioluminescence by man is for the rapid detection of foodborne pathogenic bacteria (e.g., in a food processing factory). One rapid test for bacteria uses two chemical reagents that first break down bacteria cell membranes and then cause the ATP from those broken cells to luminesce. Another rapid test uses electrophoresis to first separate the sequences of bacteria's DNA (following its extraction from cell and enzymatic fragmentation), causing those separated sequences to luminesce, and then a camera is used to record the sequence-pattern light emission and compare that pattern to patterns of pathogenic bacteria previously stored in a database. See also *Enzyme, Marker (genetic marker), Bacteria, Toxin, Pathogenic, Escherichia coliform 0157:H7 (E. coli 0157:H7), Cell, Luminescent assay, Adenosine triphosphate (ATP), Genetic engineering, Electrophoresis, Polyacrylamide gel electrophoresis (PAGE), Sequence (of a DNA molecule), Photorhabdus luminescens, Restriction endonucleases, Nitric oxide, LUX gene, Quorum sensing.*

Bioluminescence Resonance Energy Transfer

Abbreviated BRET; it is a method to monitor dynamic protein–protein molecular interactions within living cells via molecular tags (e.g., molecules chemically attached to protein) that serve to indicate when a particular biological interaction occurs (e.g., a protein–protein interaction). BRET utilizes a bright red fluorescent dye molecule attached to one protein (e.g., a receptor or a signaling protein) and a bioluminescent enzyme from deep-sea shrimp (*Oplophorus*) attached to the other protein (e.g., a biopharmaceutical candidate). When those two proteins interact (i.e., the biopharmaceutical exerts the desired effect on a receptor or on a signaling protein), the bioluminescent enzyme excites the red dye so the dye fluoresces in a manner that scientists can detect in BRET test. See also *Bioluminescence, Protein, Cell, Reporter molecules, Signaling proteins.*

Biomarkers

Refers to various *proteins, metabolites and other compounds, genes,* or *biological events* that are indicative of a relevant biological condition (e.g., disease, predisposition to a disease, disease progression, disease regression, inflammation).

For example, the presence of a specific antigen (e.g., the prostate-specific antigen or PSA) in the bloodstream for some time (e.g., several years in the case of PSA) prior to its specific disease (e.g., prostate cancer, in the case of PSA) makes that antigen useful as a *biomarker for the presence of that disease.*

The presence in a diabetes patient's bloodstream of a molecule known as "hemoglobin AlC" is a biomarker indicative of how well a given pharmaceutical is controlling that patient's blood glucose levels. Similarly, the Brownian motion of *intra*cellular water within certain cancers' solid tumors (as measured by diffusion MRI) can indicate whether a given chemotherapy or radiation treatment regime is decreasing the tumor.

The presence of a protein known as EMP-1 within tumors of non–small cell lung cancer is a biomarker for tumors' resistance to the pharmaceutical Iressa (gefitinib).

Certain molecules (e.g., *C-reactive protein* or *epidermal growth factor receptor*) can function as biomarkers in pharmacogenomics (i.e., indicating whether a given pharmaceutical will be efficacious in a specific person's body) due to that person's haplotype.

The molecule *thiopurine S-methyl transferase* can be utilized as a biomarker in toxicogenomics (e.g., indicating likelihood for one haplotype of *pediatric leukemia patients* to suffer severe/life-threatening reactions to certain leukemia treatment drugs). See also *Gene, Protein, Metabolite, Diabetes, Hemoglobin, Pharmacogenomics, Pharmacogenetics, Haplotype, Toxicogenomics, ADME tests, ADME/TOX, C-reactive protein (CRP), Prostate-specific antigen (PSA), Cancer, Tumor, Magnetic particles, BOXes, Western blot.*

Biomass

All organic matter grown by the photosynthetic conversion of solar energy (e.g., plants) and organic matter from animals.

See *Photosynthesis, Low-tillage crop production, No-tillage crop production.*

BioMEMS

Refers to MEMS that are designed to work within biological systems/organisms.

Examples include microfluidic cell sorters, or a "biochip" possessing diverging nanometer-scale etched channels and a fluorescence detector. Via an electrical field that would drive electrophoretic separation of DNA (fragments), samples of DNA could be separated/sorted/identified via fluorescence.

See *MEMS (nanotechnology), Organism, Electrophoresis, Microfluidics, Cell sorting, Nanometers (nm), Fluorescence, Biochip, Nanotechnology.*

Biomimetic Materials

From the Greek *bios* (life) and *mimesis* (to imitate). Refers to synthetic (i.e., man-made) molecules or systems that are analogs of natural (i.e., made by living organisms) materials. For instance, molecules have been synthesized by man that act chemically like natural proteins but are not as easily degraded by the digestive system (as are those natural protein molecules). Other systems such as reverse micelles and/or liposomes exhibit certain properties that mimic certain aspects of living systems. See also *Protein, Digestion (within organisms), Reverse micelle (RM), Liposomes, Analogue, Bionics, Biopolymer.*

B

Biomimicry

From the Greek *bios* (life) and *mimesis* (to imitate). Refers to the creation of man-made or man-designed molecules, systems, etc. whose design was inspired by an example observed in nature. See also *Biomimetic materials*.

Biomolecular Electronics

See *Bioelectronics*.

Biomotors

Refers to biologically based technologies/techniques utilized to "power" nanometer-size "machines" (e.g., "nanobots") in one way or another. For example, in 2000, Bernard Yurke and colleagues created a molecular-machine "tweezers" (grasper) consisting of three separate strands of DNA (i.e., two of them were hybridized separately to small complementary sequences near the two ends of the first DNA strand). The "tweezers" can then be closed (and/or opened) by sequentially adding other DNA strands (to the three) that

- Hybridize to small complementary sequences on second and third strands
- Hybridize to #4 strand, causing it to unhybridize from #2 and #3 strands

See also *Nanotechnology*, *Biology*, *Nanometers (nm)*, *Molecular machines*, *Deoxyribonucleic acid (DNA)*, *Hybridization (molecular genetics)*, *Sequence (of a DNA molecule)*, *Complementary (molecular genetics)*, *Self-assembly (of a large molecular structure)*.

Bionanotechnology

Refers to the application of biotechnology within the fields of nanotechnology. For example

- Using genetic engineering to create a "molecular template" on which is subsequently formed a nanotechnology device (e.g., a *nanowire*)
- Using genetic engineering to create specific molecules that will subsequently self-assemble into a nanotechnology tool or device (e.g., a *nanofiber*)
- Using genetic engineering to create *nanobodies*, which could be utilized to "coat" an acid-sensitive pharmaceutical molecule, to enable that pharmaceutical to be orally administered

See also *Nanotechnology*, *Genetic engineering*, *Biotechnology*, *Template*, *Nanowire*, *Self-assembly (of a large molecular structure)*, *Nanofibers*, *Directed self-assembly*, *Nanobodies*, *Nanocapsules*, *Orally administered*, *Nanotube*, *Nanoscience*, *Nanospheres*, *BioNEMs*.

BioNEMS

Acronym for biomedical nanoelectromechanical systems.

Bionics

An interscience discipline for constructing artificial systems that resemble or have the characteristics of living systems. Bionics can encompass (in whole, or in part) bioelectronics, biosensors, biomimetic materials, biophysics, biomotors, and self-assembly (of a large molecular structure). See also *Biology*, *Bioelectronics*, *Biomimetic materials*, *Biosensors (electronic)*, *Biophysics*, *Biomotors*.

Bioorthogonal Chemistry

Refers to chemistry conducted within an organism (e.g., use of click chemistry to "assemble" a pharmaceutical compound inside a living cell from two smaller precursor chemical molecules) that doesn't interfere with any of the normal biological processes in the organism. See also *Organism*, *Cell*, *Click chemistry*.

Biopanning

Refers to certain screening/searching methodologies (e.g., phage display) in which applicable interactions (e.g., selective binding/hybridization) among numerous proteins, peptides, pathogens, etc. are utilized to find useful candidates (e.g., a pharmaceutical compound that is active against a given disease). See also *Phage display*, *Hybridization (molecular genetics)*, *Protein*, *Peptide*, *Pathogen*.

Biopesticides

See *Crop biologicals*.

Biophysics

An area of scientific study in which physical principles, physical methods, and physical instrumentation are used to study living systems or systems related to life. It overlaps with biophysical chemistry, which is more specialized in scope since it is concerned with the physical study of chemically isolated substances found in living organisms. See also *ET*, *Metalloproteins*, *Tryptophan (trp)*.

Biopolymer

A high-molecular-weight organic compound found in nature, whose structure can be represented by a repeated small unit (i.e., monomer [links]). Common biopolymers include cellulose (long-chain sugars found in most plants and the main constituent of dried woods, jute, flax, hemp, cotton, etc.) and proteins in general and specifically collagen and gelatin. See also *Molecular weight*, *Protein*, *Polymer*.

Biorationals

See *Crop biologicals*.

Bioreactor

Refers to a vessel in which living cells are grown or maintained, while needed nutrients/oxygen, and scaffolding (if needed) are provided by the bioreactor to those cells. See also *Cell*, *Scaffolding (utilized in tissue engineering)*, *Adult stem cell*.

Bioreceptors

Refers to fragments of DNA, antibodies, protein molecules, and cellular probes (e.g., adhesion molecule) when those are attached to a man-made surface (e.g., biochip) for purposes of analyzing biological substances. See also *Hybridization surfaces*, *Biochips*, *Antibody*, *Deoxyribonucleic acid (DNA)*, *Protein*, *Adhesion molecule*, *Orphan receptors*, *Microarray (testing)*.

Biorecovery

The use of organisms (including bacteria, plants, fungi, and algae) in the recovery of (collecting of) various metals and/or organic compounds from ores or garbage (other matrices). See also *Bioleaching*, *Consortia*, *Biosorbents*, *Phytoremediation*, *Metabolic engineering*, *Bacteria*, *Fungus*.

Bioremediation

The use of organisms (e.g., plants, ferns, bacteria, fungi) to consume or otherwise help remove (e.g., biorecovery) unwanted materials (e.g., toxic chemical wastes, metals) from the soil, water, etc. of a contaminated site (e.g., remove toluene from the land and ponds on site of an old refinery, remove arsenic from the surface soils surrounding an old mine). See also *Biorecovery*, *Phytoremediation*, *Metabolic engineering*, *Bioleaching*, *Biodesufurization*, *Organism*, *Bacteria*, *Fungus*, *Endophyte*.

Biosafety

See *Convention on Biological Diversity (CBD)*.

Biosafety Protocol

See *Convention on Biological Diversity (CBD)*, *International Plant Protection Convention (IPPC)*.

Bioseeds

Plant seeds produced via genetic engineering of existing plants. See also *Genetic engineering*, *Biolistic® gene gun*, *Herbicide-tolerant crop*, *PAT gene*, *EPSP synthase*, *ALS gene*, *CP4 EPSPS*, *Glyphosate oxidase*, *Cholesterol oxidase*, *High-lysine corn*, *Acuron™ gene*, *High-methionine corn*, *High-phytase corn and soybeans*, *High-stearate soybeans*, *Low-stachyose soybeans*, *LOX null*, *Plant's novel trait (PNT)*, *"Shotgun" method* [to introduce foreign (new) genes into plant cells], *Bacillus thuringiensis (B.t.)*, *B.t. kurstaki*, *B.t. tenebrionis*, *B.t. israelensis*, *CRY proteins*, *CRY1A (b) protein*, *CRY1A (c) protein*, *CRY9C protein*.

Biosensors (Chemical)

Chemically based devices that are able to detect and/or measure the presence of certain molecules (e.g., DNA, antigens, glucose, active ingredients of pesticides). These devices are currently created in the following forms:

- A two-part diagnostic test that can detect the presence of trace amounts of specific chemicals (e.g., pesticides). The (chemical) biosensor consists of an immobilized enzyme (to bind the trace chemical) combined with a color reagent (to indicate visually the presence of the trace chemical).
- Carbon nanotubes onto which have been deposited a layer of glucose oxidase, with a layer of potassium ferricyanide adsorbed onto the glucose oxidase. Such coated nanotubes are placed into a tiny (permeable to glucose) dialysis capillary tube whose ends are then sealed. When that capillary tube is inserted beneath the skin—for example, of a person with diabetes, and subsequently illuminated with near-infrared light (which can pass through human tissue)—those carbon nanotubes fluoresce in a specific manner that is directly dependent on the glucose concentration (i.e., telling the diabetic when he needs to inject his insulin). That is because his body's glucose enters the semipermeable capillary; the glucose oxidase (enzyme) acts on the glucose to produce hydrogen peroxide, which then complexes with the ferricyanide in a way that changes the fluorescence properties of the nanotubes in a manner that is directly dependent on the concentration of glucose.
- Nanosheets of peptoids onto which have been attached biologically active ligands (e.g., which result in one or more types of signal being generated when those ligands bind to the specific molecule sought).
- A one-part test that can detect specific DNA segments in complex ("dirty," multiple component) samples. The biosensor consists of 13 nm gold particles onto which are attached numerous nucleotide "molecular chains." Each "nucleotide chain" contains 28 nucleotides. The 13 nucleotides that are closest to each gold particle serve as a "spacer," and solutions containing such (spaced) randomly distributed gold particles appear red in color when illuminated by appropriate light.

The 15 nucleotides that are farthest from each gold particle are chosen to be complementary to, and thus bind to (complementary), nucleotide sequences in the target (e.g., DNA) molecule. In the presence of the specific target molecule, a closely linked network of gold particles and double-stranded nucleotide molecular chains forms (overcoming the 13-nucleotide "spacer" that previously held apart the gold particles). When double-stranded chains form (i.e., target molecule is present), the distance between gold particles becomes less than the size of those particles, which makes the solution containing (bound) particles appear blue in color when illuminated by appropriate light. See also *Enzyme*, *Immunoassay*, *Nanocrystal molecules*, *Nanotechnology*, *Deoxyribonucleic acid (DNA)*, *Nanometers (nm)*, *Antigen*, *Glucose (GLc)*, *Dialysis*, *Sequence (of a DNA molecule)*, *Nucleotide*, *Polymer*, *Complementary DNA (c-DNA)*, *Double helix*, *Duplex*, *Self-assembly*, *Carbon nanotubes*, *Glucose oxidase*, *Fluorescence*, *Biochips*, *Nanosheets*, *Ligand (in biochemistry)*.

Biosensors (Electronic)

Electronic sensors that are able to detect and measure the presence of biomolecules such as sugars or DNA segments. Some of these devices are currently created by the following:

- Fusing organic matter (e.g., enzymes, antibodies, receptors, or nucleic acids) to tiny electrodes, yielding devices that convert natural chemical reactions into electric current to measure blood levels of certain chemicals (e.g., glucose or insulin), control functions in an artificial organ, monitor some industrial processes, act as a robot's "nose," etc.
- Fusing organic matter (e.g., segment of DNA, antibody, enzyme) onto the surfaces of etched silicon wafers, yielding devices that convert supramolecular interactions

B

(e.g., nucleotide hybridization, enzyme–substrate binding, lectin–carbohydrate [sugar] interactions, antibody–antigen binding, host–guest complexation) into electric current via a charge-coupled device detector that measures the shift in interference pattern caused by change in refractive index that results when a (sensed) molecule tightly binds to the fused (electronic) organic matter. For such an etched-silicon-wafer biosensor, the nucleotide hybridization (binding) enables the detection of femtomolar (10^{-15} mole or 0.000000000000001) concentrations of DNA. If the (sensed) DNA segment is not complementary to the fused DNA segment, there is no significant change in the interference pattern.

A major future goal is to build future generations of biosensors directly into computer chips. (Researchers have discovered that proteins can replace certain metals in semiconductors.) This would enable low-cost mass production via processes similar to those now used for existing semiconductor chips with circuits built right into the sensor to process data picked up by the biological matter on the chip. See also *Biochips, Quartz crystal microbalances, Bioelectronics, Enzyme, Genosensors, Receptors, Antibody, Bioinorganic, Insulin, Combinatorial chemistry, Substrate (chemical), Lectins, Sugar molecules, Carbohydrates (saccharides), Glucose (GLc), Deoxyribonucleic acid (DNA), Nucleotide, Hybridization (molecular genetics), Hybridization surfaces, Antigen, Complementary DNA (c-DNA), Gene, Nanotechnology, Template, Charge coupled device, Nanosheets, Field effect transistor, nanoFET.*

Biosensors (Light-Based)

Sensors that are able to detect and measure the presence of biomolecules such as sugars or DNA segments. Some of these devices are currently created by the following:

- Making arrays consisting of apertures (i.e., small holes) approximately 200–350 nm in size through thin metallic films. When a particular pathogen (e.g., a virus) within a sample solution (e.g., in blood) binds to this array's surface near an aperture, surface plasmon resonance causes directly proportional changes in the refractive index of reflected light striking the metallic film. By shining a highly focused beam of light (e.g., laser, polarized light) on the metallic surface and measuring the readily detectable change (i.e., shift) in the resonance frequency of light transmitted through the nanoholes, the mass changes (resulting from the pathogen binding to the metallic film) and thus the presence/identity and concentration of the pathogen within the solution can be determined.
- Forming multiple cloaks into a device that can positively identify biological materials based on the amount of light they absorb and then subsequently emit (i.e., fluorescence spectroscopy), because the cloaks slow down that light, and slowed-down light has a more pronounced interaction with molecules than does light travelling at normal speed, thereby enabling a more thorough analysis of the biological materials.

See *Plasmonic nanohole arrays, Surface plasmon resonance (SPR), Metamaterials.*

Biosilk

A biomimetic, man-made fiber produced by

1. Sequencing the "dragline silk" protein that is produced by the orb-weaving spider (*Nephila clavipes*)
2. Synthesizing gene to code for that "dragline silk" protein (components), which are mostly glycine and alanine
3. Expressing the gene in a suitable host organism (e.g., yeast, bacteria, plants) to cause production of the protein
4. Dissolving the protein in a suitable solvent and then "spinning" the protein into fiber form by passing the liquid (dissolved protein) through a small orifice, followed by drying to remove the solvent

This results in biosilk fibers that are extremely strong. See also *Biomimetic materials, Biopolymer, Protein, Sequencing (of protein molecules), Gene, Gene machine, Synthesizing (of DNA molecules), Deoxyribonucleic acid (DNA), Express, Glycine (gly), Alanine (ala), Supercritical carbon dioxide.*

Biosorbents

Microorganisms that, either by themselves or in conjunction with a support/substrate system (e.g., inert granules), effect the extraction (e.g., from ore) and/or concentration of desired (precious) metals or organic compounds by means of selective retention of those entities. Retention of organic compounds (e.g., gasoline) may be for the purpose of cleaning polluted soil. See also *Biorecovery, Bioleaching, Consortia.*

Biosphere

All living matter on or in the earth, the oceans and seas, and the atmosphere. The area of the planet in which life is found to occur.

Biostimulants

Refers to a crop biological (agent) applied to crop plants by a farmer that induces the crop plants to grow faster/yield more. May be applied alone or in combination with other crop stimulants (e.g., certain herbicides). For example, the application of certain diphenyl-ether-based herbicides to soybean (*Glycine max* (L.) Merrill) plants at an appropriate point during growing season (after emergence of the main bud) has been shown to trigger the soybean plant to do more branching and setting more nodes and more pods and shorten the ultimate plant height achieved, each of which potentially increases the soybean yield. See also *Crop biologicals, Soybean plant, Harpin.*

Biosynthesis

Production of a chemical compound or entity by a living organism.

Biotechnology

The means or way of manipulating life forms (organisms) to provide desirable products for man's use. For example, beekeeping and cattle breeding could be considered to be biotechnology-related endeavors. The word biotechnology was coined in 1919 by Karl Ereky, to apply to the interaction of biology with human technology.

However, usage of the word biotechnology in the United States has come to mean all parts of an industry that knowingly create, develop, and market a variety of products through the willful manipulation, on a molecular level, of life forms or utilization of knowledge pertaining to living systems.

A common misconception is that biotechnology refers only to recombinant DNA (rDNA) work. However, recombinant DNA is only one of the many techniques used to derive products from organisms, plants, and parts of both for the biotechnology industry. A list of areas covered by the term biotechnology would more properly include recombinant DNA, plant tissue culture, rDNA or gene splicing, enzyme systems, plant breeding, meristem culture, mammalian cell culture, immunology, molecular biology, fermentation, and others. See also *Genetic engineering, Biorecovery, Recombinant DNA (rDNA), Recombination, Deoxyribonucleic acid (DNA), Bioleaching, Gene splicing, Mammalian cell culture, Fermentation.*

Biotechnology Industry Organization

An American trade association composed of companies and individuals involved in biotechnology and in services to biotechnology companies (e.g., accounting, law). Formed in 1993, the Biotechnology Industry Organization (BIO) was created by the merger of its two predecessor trade associations: the Association of Biotechnology Companies and the Industrial Biotechnology Association. The BIO works with the government and the public to promote safe and rational advancement of genetic engineering and biotechnology. See also *Biotechnology, Association of Biotechnology Companies (ABC), Industrial Biotechnology Association (IBA), Japan Bioindustry Association, Senior Advisory Group on Biotechnology (SAGB).*

Biotic Stresses

The stress (e.g., to crop plants) caused by insects, bacteria, viruses, fungi, nematodes, and/or other living things that attack plants. See also *Nematodes, Fungus, Virus, Bacteria.*

Biotin

A B-complex vitamin, also known as *vitamin H*, that is essential (i.e., required) for life of many grain-eating insects and is also essential for many of the metabolic pathways (i.e., series of chemical reactions) involved in milk production by cattle.

All of the predominant cellulolytic bacteria (i.e., those that break down cellulose molecules) within the rumen (first stomach) of cattle require biotin for them to be able to grow. Biotin (within certain molecules) acts as a coenzyme in carboxylation reactions, thereby playing a critical role in gluconeogenesis, fatty acid synthesis ("manufacture"), and protein synthesis reactions occurring within all animals.

Biotin binds very tightly to streptavidin (avidin). This property is utilized by some scientists to attach various molecules such as antibodies (e.g., to quantum dots, probes), utilizing the biotin streptavidin to make a *molecular bridge*.

Biotin enzymes are inhibited (i.e., blocked) by the protein avidin. Since insects must have biotin to live, avidin might be a useful ingredient to add to grain in order to protect it during storage from insects such as weevils. See also *Vitamin, Metabolism, Intermediary metabolism, Pathway, Bacteria, Cellulose, Lysis, Enzyme, Coenzyme, Weevils, Gluconeogenesis, Fatty acid, Protein, Streptavidin, Molecular bridge, Quantum dot, Antibody, Probe.*

Biotinylation

See *Streptavidin.*

Biotransformation (of a Biosynthesized Product)

See *Post-translational modification of protein.*

Biotransformation (of an Introduced Compound)

See the *biological means* portion of definition of PERSISTENCE.

BIR

Acronym for break-induced replication; it is one of the living cell's DNA repair mechanisms. See also *DNA repair, Deoxyribonucleic acid (DNA), Cell.*

bla Gene

A gene that confers resistance to β-lactam (beta-lactam) antibiotics (e.g., ampicillin). See also *Gene, Beta-lactam antibiotics, Marker (genetic marker).*

Black Layer

A layer of tissue within corn (maize) kernels near the tip at which each kernel is embedded in corn cob. That tissue layer conveys sugar molecules and other materials from the corn plant into the kernel during the growing season, then the layer of cells collapses, stops functioning, and turns black in color when the kernel is mature.

It thus serves as an indicator of the corn's maturity. It refers to a distinctive dark line that forms within each corn kernel at maturity (i.e., the point in time near end of the growing season at which the kernels have achieved their full weight). See also *Corn.*

Black-Lined (Corn)

See *Black-layered (corn).*

Blast Cell

A large, rapidly dividing cell that develops from a B cell (B lymphocyte) in response to an antigenic stimulus. The blast cell then becomes an antibody-producing plasma cell. See also *Antigen, Antibody, B lymphocytes, Lymphocyte.*

Blast Transformation

The process via which a B cell (B lymphocyte) becomes a blast cell. See also *Antibody, Lymphocyte, Blast cell.*

Blinatumomab

A recombinant, single-chain monoclonal antibody that contains antigen-recognition sites for CD3 and CD19 surface proteins on B lymphocytes. CD3 is a complex of T cell surface glycoproteins, while CD19 is a tumor-associated antigen present on the surface of some B cells. Because blinatumomab possesses both recognition sites, it (when administered) promotes cytotoxic T cell and helper T cell activity against B lymphocyte malignancies, particularly cancerous CD19-expressing B lymphocytes. See also *B lymphocytes, Antigen, Monoclonal antibodies (MAb), T cells, Helper T cells, Cytotoxic T cells, T cell receptors, Tumor, Tumor-associated antigens, Cancer.*

B

Blood Clotting

See *Fibrin*, *Serotonin*.

Blood Derivatives Manufacturing Association

A trade organization of firms involved in producing pharmaceuticals from collected blood. See also *Serum*, *Buffy coat* (*cells*), *Serology*.

Blood Plasma

See *Plasma*.

Blood Platelets

See *Platelets*.

Blood Serum

See *Serum*.

Blood–Brain Barrier

The specialized layer of endothelial cells that line all blood vessels in the brain. The blood–brain barrier (BBB) prevents most organisms (e.g., bacteria) and toxins from entering the brain via the bloodstream. However, the BBB does allow oxygen and needed nutrients (e.g., iron, glucose, tryptophan) to enter the brain from the bloodstream. For example, transferrin receptors that line BBB cell surfaces (on the bloodstream side of the BBB) "latch onto" transferrin molecules (which contain iron molecules) as those transferrin molecules pass by in the bloodstream. These transferrin receptors first bind to the (passing) transferrin molecules, transport those transferrin molecules through the BBB via a process called receptor-mediated transcytosis, and then release those transferrin molecules (in order to supply needed iron to the brain cells). Factors such as aging, trauma, stroke, multiple sclerosis, and some infections will cause an increase in the permeability of the BBB. See also *Endothelial cells*, *Toxin*, *Transferrin*, *Transferrin receptor*, *Chelating agent*, *Glucose*, *Receptors*, *Vaginosis*, *Heme*, *Bacteria*, *Tryptophan* (*trp*), *Serotonin*.

Blue Biotechnology

Term utilized in some countries to refer to *environmental improvement* applications of genetic engineering. One example would be bioremediation. See also *Genetic engineering*, *Bioremediation*.

Blunt-End DNA

A segment of DNA that has both strands terminating at the same base pair location, that is, fully base-paired DNA. No sticky ends. See also *Sticky ends*.

Blunt-End Ligation

A method of joining blunt-ended DNA fragments using the enzyme T4 ligase that can join fully base-paired, double-stranded DNA. See also *Ligase*, *Deoxyribonucleic acid* (*DNA*), *Base pair* (*bp*), *Blunt-end DNA*.

BLUP

See *Best linear unbiased prediction* (*BLUP*).

BNF

Acronym for biological nitrogen fixation. See *Nitrogen fixation*.

BOD

See *Biological oxygen demand* (*BOD*).

Boletic Acid

See *Fumaric acid* ($C_4H_4O_4$).

Bollworms

See *Heliothis virescens* (*H. virescens*), *Helicoverpa zea* (*H. zea*), *Pectinophora gossypiella*, *B.t. kurstaki*.

Bone Morphogenetic Proteins

A family of proteinaceous growth factors (nine identified as of 1994) for bone tissue formation (e.g., at the site where a bone has been broken, cut, etc.). In humans, bone morphogenetic proteins (BMPs) stimulate a "recruitment" of bone-forming cells (e.g., to the site of bone injury) that first form cartilage and then that cartilage is mineralized to form bone.

During 2010, Ken Muneoka discovered that in mice (which have the ability to regenerate a "fingertip" after one gets cut off), the regeneration-capable injury sites will release enough BMPs to cause full regeneration of the cutoff digit. The expression of MSX genes, which "turn on" production of BMPs, increases during the mouse digit-regeneration process. See also *Growth factor*, *Periodontium*, *Protein*, *Express*, *Expressivity*.

Bone Morphogenetic Protein-Signaling Pathway

Refers to the biochemical (signaling) pathway that is utilized by the body to cause bone formation (e.g., at the site where a bone has been broken). See *Bone morphogenetic proteins* (*BMP*), *Pathway*, *Signaling*, *Protein signaling*, *Pathway feedback mechanisms*.

Bortezomib

A boronic acid–containing dipeptide proteasome inhibitor that has been approved by the U.S. Food and Drug Administration as the pharmaceutical Velcade™ for the treatment of multiple myeloma and mantle cell lymphoma (MCL) and for treating some patients with relapsed or refractory MCL. See also *Peptide*, *Proteasomes*, *Proteasome inhibitors*, *Multiple myeloma*, *MCL*, *Food and Drug Administration* (*FDA*).

Bovine Somatotropin

Also called bovine growth hormone. A protein hormone, produced in a cow's pituitary gland, that increases the efficiency of the cow in converting its feed into milk. Increases milk production in cows, and promotes cell growth in healing tissues of all ages of cattle. Promotes body growth of young cattle. See also *Protein*, *Growth hormone (GH)*, *Hormone*, *Somatomedins*, *Species specific*.

Bowman–Birk Trypsin Inhibitor

See *Trypsin inhibitors*.

BOXes

Bilirubin oxidation molecules, or *bilirubin oxidation products*. These are a group of molecules that can serve as biomarkers of some blood vessel diseases (e.g., vasospasm or other vasoconstrictions), which result from oxidative damage to bilirubin inside the body.

Research indicates that BOXes can also exacerbate tissue damage caused by stroke and *compartment syndrome* (i.e., a compression of certain nerves and blood vessels). See also *Bilirubin*, *Oxidation (chemical reaction)*, *Oxidative stress*, *Biomarkers*.

bp

Common abbreviation for base pair. See *Base pair (bp)*.

Bradyrhizobium japonicum

A nitrogen-fixing strain of bacteria that lives symbiotically among the roots of the soybean plant and provides almost all of the nitrogen needed by the soybean plant. See also *Bacteria*, *Symbiotic*, *Nodulation*, *Nitrogen fixation*, *Soybean plant*, *Isoflavones*, *Rhizobium (bacteria)*.

Brain-Derived Neurotropic Protein

Abbreviated BDNF, it is a protein that is produced as a result of a mammal engaging in endurance exercise. As a result of that exercise, the muscles produce FNDC5 protein. The increase of FNDC5 protein in the body in turn boosts the expression within the brain of BDNF in the *dentate gyrus* of the hippocampus (i.e., portion of the brain involved in learning and memory).

The hippocampus is one of only two portions of the adult human brain that can generate new neurons/nerve cells. BDNF promotes the development of new nerves and synapses (i.e., the connections between brain neurons that allow learning and memory to be stored) and also helps to preserve existing brain cells. See also *Protein*, *Expression*, *Neuron*, *Synapse*.

Brassica

A fast-growing category of the mustard plant family, which also produces sulfur-based gases (a natural defense against certain fungi, nematodes, and insect pests). For example, Australian CSIRO scientists discovered in 1994 that sulfur-based isothiocyanates emitted by *Brassica* actively combat Wheat Take-All Disease (a fungal disease that attacks the roots of the wheat plant). Those isothiocyanates also combat parasitic soybean cyst nematodes (*Heterodera glycines*), thereby benefitting soybean crop planted in rotation after *Brassica* species (e.g., canola, oilseed rape). See also *Arabidopsis thaliana*, *Wheat*, *Wheat take-all disease*, *Canola*, *Glucosinolates*, *Allelopaty*, *Fungus*, *Nematodes*, *Soybean cyst nematodes (SCN)*, *Crop rotation*.

Brassica campestre

See *Brassica*.

Brassica campestris

See *Canola*, *Brassica*.

Brassica napus

See *Canola*, *Brassica*.

Brassinosteroids

A category of steroid hormones that are active in certain plants. Brassinosteroids act to control specific plant developmental processes such as stem elongation, seed size, differentiation of vasculatory (i.e., liquid carrying) tissues, and flowering time.

Brassinosteroids act to control specific physiological processes such as fertility, resistance to abiotic stresses (e.g., drought, cold), softening of certain fruits (e.g., wine grapes), and resistance to biotic stresses (e.g., attacks by insects to the plant). See also *Hormone*, *Steroid*, *Differentiation*, *Cell differentiation*, *Biotic stresses*, *Abiotic stresses*.

Brazzein

A protein that imparts a sweet taste to foods that contain it. See *Protein*.

BRCA 1 Gene

A particular tumor suppressor gene present within some humans' DNA (e.g., is in the DNA of approximately 2% of women who are of Northern European ancestry, most Caucasian women in the United States, and Ashkenazi Jews whose ancestors are from Central and Eastern Europe). The acronym BRCA stands for breast cancer.

Normally, the BRCA 1 gene helps prevent cancer by

- Helping (breast tissue) cells repair DNA damage that can occasionally occur in those cells (e.g., as a result of ultraviolet radiation)
- Helping cells in their natural degradation of the cell surface receptor molecules (i.e., *progesterone receptor*) via which the female hormone progesterone accomplishes its growth-promoting effect on breast tissue cells

When mutated, the loss of the two aforementioned BRCA 1 gene functions can lead to the development of breast cancer. Poly (ADP-ribose) polymerase (PARP) inhibitors have been shown to act against cancers when used in women with breast and ovarian cancers linked to BRCA mutations. See also *Gene*, *Tumor suppressor genes*, *Oncogenes*, *Cancer*, *BRCA genes*, *Protein*, *Tumor suppressor proteins*, *Cell*, *Deoxyribonucleic acid (DNA)*, *Genetic code*, *Mutation*, *Hormone*, *Progesterone*, *Receptors*, *PARP*, *PARP inhibitors*.

BRCA 2 Gene

See *BRCA genes*.

B

BRCA Genes

Tumor suppressor genes (sometimes act as oncogenes) that, when mutated, can allow development of breast cancer or ovarian cancer. All humans possess BRCA genes of one sort or another (the acronym "BRCA" stands for breast cancer). However, the two specific BRCA genes most likely to lead to breast cancer (i.e., *BRCA 1*, discovered by Mary-Claire King, and *BRCA 2*) are present in only 2% of women who are of Northern European ancestry, most Caucasian women in the United States, and Ashkenazi Jews whose ancestors are from Central and Eastern Europe.

Those women possessing the *BRCA 1* gene in their genome (DNA) have a 20%–60% chance of developing ovarian cancer (and a 36%–85% chance of developing breast cancer) in their lifetime. Those women possessing the *BRCA 2* gene in their genome (DNA) have a 15%–20% chance of developing ovarian cancer (and a 36%–85% chance of developing breast cancer) in their lifetimes. Women in the general population have a 1.7% chance of developing ovarian cancer, and a 13% cancer of developing breast cancer in their lifetime.

Poly (ADP-ribose) polymerase (PARP) inhibitors have been shown to act against these cancers when used in women with breast and ovarian cancers linked to BRCA mutations.

See also *Gene*, *Mutation*, *Cancer*, *Tumor suppressor genes*, *Oncogenes*, *BRCA 1 gene*, *HER2 gene*, *PARP*, *PARP inhibitors*, *Olaparib.*

Break-Induced Replication

Refers to one of the living cell's DNA repair mechanisms. See also *DNA repair*, *Deoxyribonucleic acid* (*DNA*), *Cell.*

Breeder's Rights

See *Plant breeder's rights.*

BRET

See *Bioluminescence resonance energy transfer.*

Bright Greenish-Yellow Fluorescence

An indication of the presence of fungus (e.g., in a sample of grain), when light of an appropriate wavelength is shone on sample. For example, when the fungus *Aspergillus flavus* infects cottonseed during boll development on the cotton plant, the resultant seed (when harvested) shows bright greenish-yellow fluorescence on its lint and linters. That fungus gains entry into the bolls typically via holes made by the pink bollworm (*Pectinophora gossypiella*). See also *Mycotoxins*, *Aflatoxin*, *Fungus*, *Pectinophora gossypiella*, *Fluorescence.*

Brinjal

One of the common names of the crop known as eggplant (*Solanum melongena* L.).

Broad Spectrum

See *Gram stain.*

Bromoxynil

An active ingredient in some herbicides, it kills certain types of plants (weeds). See also *Nitrilase.*

Broth

A fluid culture medium (for growing microorganisms). See also *Medium*, *Culture medium.*

Brown Adipocytes

Adipocyte (body fat) cells that specialize in converting fat/lipid molecules into heat (e.g., to maintain body temperature). Brown adipocytes contain many tiny droplets of lipids and the most mitochondria (i.e., "energy factories" containing pigmented cytochromes that bind iron; so appear brown visually) of any cell type.

Under certain conditions (organism in warm environment), brown adipocytes can interconvert to become white adipocytes (i.e., highly flexible energy "storehouses," which are "filled up" in times of calorie abundance for an organism). Under certain conditions (organism in cold environment), white adipocytes can interconvert to become brown adipocytes via the mTORC1 molecular pathway, which is regulated by the Grb10 protein. See also *White adipocytes*, *Lipids*, *Mitochondria.*

Brown Adipose Tissue

See *Brown adipocytes.*

Brown Fat Cells

See *Brown adipocytes.*

Brown Stem Rot

A plant disease that can be caused by the soilborne fungus *Phialaphora gregata* in the soybean plant (*Glycine max* L. Merrill). Some soybean varieties are genetically resistant to brown stem rot. See also *Fungus*, *Soybean plant*, *Genotype*, *Gene*, *Pathogenic.*

BSA

Acronym for *bovine serum albumin.* See *Albumin*, *Serum.*

BSE

Bovine spongiform encephalopathy. A neurodegenerative disease of cattle. See *Prion.*

BSP

Biosafety protocol. See *Convention on Biological Diversity* (*CBD*).

BSR

See *Brown stem rot* (*BSR*).

BST

See *Bovine somatotropin* (*BST*).

BtR-4 Gene

See *Toxicogenomics.*

Buffy Coat (Cells)

The layer of white blood cells (leukocytes) that separates out when blood is subjected to centrifugation. See also *Ultracentrifuge*, *Leukocytes*, *Plasma*, *Blood Derivatives Manufacturing Association*.

Bundesgesundheitsamt

German Federal Health Organization. The German government agency that must approve new pharmaceutical products for sale within Germany; it is the equivalent of the U.S. Food and Drug Administration. See also *Food and Drug Administration* (*FDA*), *Koseisho*, *Committee for Proprietary Medicinal Products* (*CPMP*), *Committee on Safety in Medicines*, *Medicines Control Agency* (*MCA*), *European Medicines Evaluation Agency* (*EMEA*).

Bursting

Refers to the "turning on" and "turning off" of individual genes in a cell, resulting in the synthesizing of proteins in bursts (e.g., like periodic "bursts" of machine gun fire by soldiers during a war). See also *Gene*, *Cell*, *Protein*, *Activator* (*of gene*), *Transcription activators*, *Transcription Factors*, *Transactivation*.

Butyrate

See *Short-chain fatty acids*.

BXN Gene

See *Nitrilase*.

β-Conglycinin

See *Beta-conglycinin*.

β-Sitostanol

See *Beta sitostanol* (β-sitostanol).

C

C Terminus

See *Carboxyl terminus (of a protein molecule).*

C Value

The total amount of DNA in a haploid genome. See also *Deoxyribonucleic acid (DNA), Haploid, Genome.*

C. elegans

See *Caenorhabditis elegans.*

C3 Pathway

See *C3 Photosynthesis.*

C3 Photosynthesis

Refers to the particular photosynthesis chemical system utilized by most green plants, in which an enzyme known as RuBisCO helps (along with sunlight of course) chemically combine carbon dioxide with a two-carbon molecule to initially yield a three-carbon molecule. Subsequent to that initial step, numerous other carbohydrate (sugar) molecules and other chemicals needed by the plant are synthesized from the initial three-carbon molecule.

Other plants utilize C4 photosynthesis, so called because it chemically combines carbon dioxide with a three-carbon molecule to initially yield the four-carbon molecule known as oxaloacetic acid. Subsequent to that initial step, numerous other carbohydrate (sugar) molecules and other chemicals are synthesized from the oxaloacetic acid.

C4 photosynthesis is much more efficient than C3 photosynthesis. For example, the C3 photosynthesis crop plant known as rice (i.e., domesticated form of *Oryza sativa* and/or *Oryza glaberrima*) produces approximately half the carbohydrate amount of the C4 photosynthesis crop plant maize (*Zea mays* L.). See also *Photosynthesis, Enzyme, Carbohydrates, Sugar molecules, Rice.*

C4 Pathway

See *C4 Photosynthesis.*

C4 Photosynthesis

Refers to the particular photosynthesis chemical system (pathway) utilized by some green plants, in which several enzymes plus the enzyme known as RuBisCO help (along with sunlight of course) to chemically combine carbon dioxide with a three-carbon molecule to initially yield the four-carbon molecule known as oxaloacetic acid. Subsequent to that initial step, numerous other carbohydrate (sugar) molecules and other chemicals needed by the plant are synthesized from the oxaloacetic acid.

Other plants utilize C3 photosynthesis, so called because it chemically combines carbon dioxide with a two-carbon molecule to initially yield a three-carbon molecule. C4 photosynthesis is much more efficient than C3 photosynthesis. For example, the C3 photosynthesis crop plant known as rice (i.e., domesticated form of *Oryza sativa* and/or *Oryza glaberrima*) produces approximately half the carbohydrate amount of the C4 photosynthesis crop plant maize (*Zea mays* L.).

Corn/maize (*Zea mays* L.) and sugarcane are 2 of the approximately 40 plant species that utilize C4 photosynthesis. See also *Photosynthesis, Enzyme, Carbohydrates, Sugar molecules, Corn, Rice, Hydrilla verticillata.*

Caco-2

Developed during the late 1980s and early 1990s by Ronald Borchardt and Ismael Hidalgo, it refers to an immortal cell line (i.e., cells propagated over time in cell culture) of human colon adenocarcinoma cells that is utilized by research scientists. When Caco-2 cells are grown on suitable surfaces (in cell culture vessel), those cells differentiate and assume properties akin to intestinal mucosa cells.

Such cultured Caco-2 cells are used to assess *absorption* of pharmaceutical candidate (chemical) compounds (e.g., the likelihood and rate for a given candidate compound to be absorbed into the body through cell membranes from the gastrointestinal tract).

Enough is now known of Caco-2's absorption of each major category/type of chemical) for such absorption to often be predicted *in silico* (i.e., via computer modeling). See also *Cell, Differentiation, ADME Tests, Absorption, Plasma membrane, Pharmacokinetics, Pharmacogenomics, Cell culture, ADME, ADMET, In silico screening, Structure-activity models, ADME/Tox, Efflux pump.*

Cadherins

A class of (cell surface) adhesion molecules that causes cells (e.g., in the lining of the intestine known as the epithelium) to *stick together* to form a continuous lining; plus cadherins sometimes function as cellular adhesion receptors.

For example, the (food poisoning) pathogenic bacteria *Listeria monocytogenes* is able to infect humans via its use of the E-cadherin receptor located on the surface of intestinal epithelium cells. That bacteria's *key* (a bacterial membrane surface protein known as internaulin) is *inserted* into the E-cadherin (*lock*), which opens up the otherwise closed-to-bacteria intestinal epithelium. The *L. monocytogenes* bacteria then leaves the intestine and infects the human body tissues. See also *Adhesion molecule, Cell, Receptors, Listeria monocytogenes, Epithelium.*

Caenorhabditis elegans (C. elegans)

The name of a nematode (microscopic roundworm) that is commonly utilized by scientists in genetics experiments. Because of this, a large base of knowledge about *C. elegans* genetics has been accumulated by the world's scientific community.

For example, of the nearly 300 *disease-causing genes* in the human genome, more than half of them have an analogous gene within the *C. elegans* genome. *C. elegans* was one of the first animals to have its entire genome sequenced by man.

Thus, one of the methodologies utilized by researchers to rapidly screen large numbers of chemical compounds for their potential use as pharmaceuticals is to

- Expose large numbers of *C. elegans* to the various chemical compounds that the researcher wants to investigate for potential pharmaceutical activity
- Pass those large numbers of previously exposed *C. elegans*, suspended in liquid such as water, through a small transparent chamber where a focused laser beam is shined upon the roundworm's side (for its full length, as the roundworm passes by)
- Utilize expression of fluorescent protein, autofluorescence, lectin (in the fluid) binding detected via laser reflectance, antibody (in the fluid) binding detected via laser reflectance, etc. as the basis for individual *C. elegans* to be *sorted* via tiny jets of air that blow into a container those *C. elegans* that show thus-visible sign(s) of having been changed by the particular chemical compound they were exposed to
- Evaluate in detail (e.g., via conventional gene expression analysis) the specific impact of that particular chemical compound on those *C. elegans* that had indicated an apparent change, so were sorted into the *likely target* receptacle

See also *Nematodes, Genetics, Gene, Genome, Gene expression, Gene expression markers, Expressed sequence tags (EST), Sequencing (of DNA molecules), High-throughput screening (HTS), High-throughput identification, Gene expression analysis, Target-ligand interaction screening, Target (of a therapeutic agent), Fluorescence, Lectins, Model organism.*

Caffeine

A chemical [$C_8H_{10}N_4O_2$] that is naturally produced in some plants (e.g., coffee tree) to repel predatory insects. It also acts as a stimulant (when consumed by humans), so it is classified as a *phytochemical.* Caffeine was first isolated chemically and named in 1819 by Friedlieb Ferdinand Runge.

Research done by Seymour Diamond last 2000 showed that caffeine consumption causes interactions within the human body, with the synthetic chemical painkiller known as ibuprofen. Consuming both together was shown to be more effective in relieving pain than was consuming ibuprofen alone and brought pain relief faster than consumption of ibuprofen alone. See also *Phytochemicals, Coffee tree.*

Calcium Channel Blockers

Refers to

- Drugs (e.g., verapamil, amlodipine, diltiazem, nifedipine) that are used to slow down calcium movement through cell membranes. This leads to dilation of the blood vessels and reduces the heart's workload. Blood vessels need calcium to contract (causing flow constriction and hence an increase in blood pressure), so the drug-induced shortage of available calcium causes the body's blood vessels to remain dilated (which results in lower blood pressure).
- Drugs such as Prialt™/ziconotide (an *N-type calcium channel blocker*), or Neurontin™ and Lyrica™ (GABAergic calcium channel blockers) that act as powerful painkillers by slowing down calcium movement through certain cell membranes

See also *Cell, Ion channels, Membrane transport.*

Calcium Oxalate

A crystalline salt that is normally deposited in the cells of some species of plants. In spinach, the presence of such oxalate inhibits absorption of the calcium (present in the spinach) by humans eating that spinach. In many animals, calcium oxalate is excreted in the urine, or retained by the animal's body in the form of urinary calculi. See also *Absorption, Oxalate, Cell.*

Callipyge

(Means *beautiful buttocks* in the Greek language) An inherited trait in livestock (e.g., sheep) that results in thicker, meatier hindquarters. First identified as a genetic trait in 1983, this desirable trait results in a higher meat yield per animal. See also *Trait, Genotype, Phenotype, Wild type.*

Callus

An undifferentiated cluster of plant cells that is a first step in

- Repair of a physical wound in some plants
- Regeneration of plants from excised sample (explant) placed into tissue culture medium
- Plant cell fermentation (in which calluses are propagated/kept alive in a water-based system containing needed amino acids, sugars, vitamins, trace elements, and other nutrients while they produce a desired substance such as paclitaxel)

See also *Cell, Somaclonal variation, Tissue culture, Culture medium, Plant cell fermentation, Amino acids, Vitamin, Paclitaxel.*

Calorie

The amount of heat (energy) required to raise the temperature of 1 g of water from 14.5°C (58°F) to 15.5°C (60°F) at a constant pressure of one standard atmosphere. This unit measure of energy (i.e., one calorie) is also frequently utilized to express the amount of energy contained within certain foods or animal feeds. See also *Carbohydrates (saccharides), Fats, TME (N).*

Calpain-10

A gene that increases the likelihood for development of diabetes disease, in humans whose DNA carries that gene (i.e., approximately 80% of humans carry that gene). See also *Diabetes, Insulin, Insulin-dependent diabetes mellitus (IDDM), Gene, Deoxyribonucleic acid (DNA).*

CAM

Acronym for crassulacean acid metabolism. See also *Crassulacean acid metabolism (CAM)*.

CAM

Acronym for cell adhesion molecule. See *Adhesion molecule.*

Campesterol

A phytosterol that is produced within the seeds of the soybean plant (*Glycine max* L.), among others. Evidence shows that human consumption of campesterol helps to reduce total serum (blood) cholesterol and low-density lipoprotein levels and thereby lowers the risk of coronary heart disease.

Evidence indicates that certain phytosterols (including campesterol) interfere with the absorption of cholesterol by the intestines and decrease the body's recovery and reuse of cholesterol-containing bile salts, which causes more (net) cholesterol to be excreted from the body. See also *Phytosterols, Phytochemicals, Sterols, Soybean plant, Cholesterol, Stigmasterol, Beta-sitosterol (B-Sitosterol), Coronary heart disease (CHD).*

Campestrol

See *Campesterol.*

Campsterol

See *Campesterol.*

Camptothecins

See *Rubitecan.*

CaMV

See *Cauliflower mosaic virus 35S promoter (CaMV 35S).*

CaMV 35S

See *Cauliflower mosaic virus 35S promoter (CaMV 35S).*

Canavanine

An uncommon amino acid. It is used in biology as an arginine (another amino acid) analogue. It is a potent growth inhibitor of many organisms. See *Amino acid, Biomimetic materials.*

Cancer

The name given to a group of diseases that are characterized by uncontrolled cellular growth (e.g., formation of tumor) without any differentiation of those cells (i.e., into specialized and different tissues).

Causes include consumption of carcinogens (e.g., certain mycotoxins), mutagens (e.g., certain radiation), some viruses (e.g., approximately 70% of human cervical cancers and 30% of oropharyngeal cancers are caused by the human papilloma virus), etc. During 2010, Stuart Gordon discovered that people infected with hepatitis C were twice as likely to develop kidney cancer, as non-infected people.

During 1930, Otto Warburg discovered that most cancer cells utilize glycolysis to generate energy via oxidation of sugar molecules, instead of utilizing the cell mitochondria as normal cells do, for energy generation. That utilization of glycolysis enables cancer cells to better survive hypoxia (shortage of oxygen due to lack of good blood supply to a growing tumor) and to better avoid apoptosis (i.e., *programmed cell death*, initiated by mitochondria in cells whose DNA is damaged). See also *Carcinogen, Oncogenes, Tumor-suppressor genes, ras gene, Tumor, Virus, Telomeres, Retinoids, Mutagen, Cell, Telomerase, Neoplastic growth, Chemotherapy, Differentiation, Oropharyngeal cancer, Oral cancer, Mycotoxins, RNase 1 gene, Regulatory T cells, Phosphorylation, Oncolytics, Chronic inflammation, Glycolysis, Hypoxia, Sugar molecules, Mitochondria, Apoptosis, miRNA gene, Receptor binding mapping.*

Cancer Epigenetics

See *Epigenetic, Micro-RNAs.*

Cancer Immunotherapy

Refers to cancer treatments that target the body's immune system rather than tumors directly. When effective, these treatments induce the body's T cells and other immune system cells to combat the cancer/tumors. See also *Tumor, Cancer, Cellular immune response, Immunogen, Checkpoint blockade.*

Cancer Stem Cells

See *Apoptosis.*

CANDA

Computer-assisted new drug application. An application to the U.S. Food and Drug Administration (FDA) seeking approval of a drug that has undergone Phase 2 and Phase 3 clinical trials. A CANDA is submitted in the form of computer-readable (e.g., clinical) data that provides the FDA with a sophisticated database that allows the FDA reviewers to evaluate (e.g., statistically) the data themselves, directly. See also *NDA (to FDA), NDA (to Koseisho), Food and Drug Administration (FDA), MAA Marketing authorization application, Phase I clinical testing.*

Canola

Historically, this term has referred to *Brassica napus* or *Brassica campestris/rapa* strains of the rapeseed plant (oilseed rape), which were developed by plant breeders after the 1960s. This was because oil produced from rapeseed grown prior to 1971 contained 30%–60% erucic acid.

By 1974, canola varieties producing oil containing less than 5% erucic acid constituted virtually all of that year's Canadian rapeseed crop, and Canadian breeders continued to develop new canola varieties with ever-lower erucic acid content (e.g., oil from *double-zero* canola varieties contains less than 0.1% erucic acid).

In 1982, Canada filed with the U.S. Food and Drug Administration (FDA) to have low-erucic-acid rapeseed (LEAR) oil affirmed to be Generally Recognized As Safe, which the FDA did. LEAR was one of the first foodstuffs to be determined to be *substantially equivalent* under the OECD-defined criteria for *substantial equivalence* because LEAR was shown (in OECD petition) to be very similar to, and composed of the same basic components as, traditional rapeseed oil (and other commonly consumed vegetable oils) except for a lower level of erucic acid (the component of earlier concern earlier).

In 2002, a *Brassica juncea* canola variety was introduced for the first time ever in Canada. Genomic differences among the three species sometimes result in certain diseases causing more damage in one canola species than the other. For example, the fungal disease *Alternaria* black spot tends to cause more damage in *juncea* or *rapa* canola varieties than in *napus* canola varieties. For example, fungal *Albugo candida* staghead disease tends to cause more damage in *rapa* canola varieties than in *napus* or *juncea* canola varieties.

Because it is a *Brassica* plant, canola also produces sulfur-based gases (a natural defense against certain fungi, nematodes, and insect pests). For example, Australian CSIRO scientists discovered in 1994 that sulfur-based isothiocyanates emitted by *Brassica* actively combat Wheat Take-All Disease (a fungal disease that attacks the roots of the wheat plant). Those isothiocyanates also combat parasitic soybean cyst nematodes (*Heterodera glycines*), thereby benefitting soybean crop planted in rotation after *Brassica* species such as canola or oilseed rape. See also *Strain, Fats, Laurate, Fatty acid, Oleic acid, Gras list, Organization for Economic Cooperation and Development (OECD), Glucosinolates, Brassica, High-stearate canola, Nematodes, Soybean cyst nematodes (SCN), Fungus, Wheat, Wheat take-all disease, Crop rotation.*

CAP

Catabolite gene-activator protein, also known as catabolite regulator protein (CRP) or cyclic AMP receptor protein. The protein mediates the action of cyclic AMP (cAMP) on transcription in that cAMP and CAP must first combine. The cAMP–CAP complex then binds to the promoter regions of *Escherichia coli* and stimulates transcription of its operon. Since a cell component increases rather than inhibits transcription, this type of regulation of gene expression is called "positive transcriptional control." See also *Escherichia coliform (E. coli), Catabolite repression, Transcription, Operon, Transcription activators.*

Capillary Electrophoresis

A research technology/methodology that is utilized to electrophoretically separate ions (e.g., DNA/RNA/nucleic acids, protein molecules). That separation occurs inside a tiny capillary tube, when a powerful electrical field is applied across the (length of) capillary tube, because the ions (in solution inside capillary tube) move at different speeds through the tube depending on their charges and their molecular size/weight. Optical detection systems are typically utilized to determine each of the ions (each molecule) as they emerge from the capillary tube.

Capillary electrophoresis is utilized to perform DNA sequencing, biowarfare (pathogen) detection, heterozygote detection, mutation analysis (e.g., in site-directed mutagenesis efforts), single-nucleotide polymorphism analysis, gene expression analysis, amplified fragment length polymorphism analysis/"fingerprinting," quantitation of PCR (products), quantitation of RT-PCR (products), the identity of glycan molecular structures (which can be determined with the assistance of glycoinformatics based on their electrophoretic migration-time-based glucose unit [GU] values), etc. See also *Electrophoresis, Ion, Protein, Isoelectric focusing (IEF), Molecular weight, Nucleic acids, Deoxyribonucleic acid (DNA), Protein, Ribonucleic acid (RNA), Sequence (of a DNA molecule), Gene, Sequencing (of DNA molecules), Pathogen, Heterozygote, Mutation, Site-directed mutagenesis (SDM), Single-nucleotide polymorphisms (SNPs), Amplified fragment length polymorphism, Gene expression analysis, PCR, RT-PCR, Isotachophoresis, Glycoinformatics, Structural biology, Glucose unit (GU) values.*

Capillary Isotachophoresis

See *Isotachophoresis.*

Capillary Isotechophoresis

See *Isotachophoresis.*

Capillary Zone Electrophoresis

See *Capillary electrophoresis.*

Capsid

The external protein coat of a virus particle that surrounds the nucleic acid. The individual proteins that make up the capsid are called "capsomers" or protein subunits. It has been discovered that resistance to certain viral diseases may be imparted to some plants by inserting the gene for production of the capsid protein coat into the plants (thereby preventing the virus from uncoating, which it must first do in order to infect the plants). See also *Tobacco mosaic virus (TMV), Virus, Protein.*

Capsule

An envelope surrounding many types of microorganisms. The capsule is usually composed of polysaccharides, polypeptides, or polysaccharide–protein complexes. These materials are arranged in a compact manner around the cell surface. Capsules are not absolutely essential cellular components. See also *Microorganism, Polysaccharides, Polypeptide (protein), Protein, Cell, Gram-negative (G–), Mannanoligosaccharides (MOS), Gram-positive (G+).*

Capture Agent

Also known as a *capture molecule.* See *Capture molecule.*

Capture Molecule

Also known as a *capture agent.* Refers to molecules such as ligands, receptors, aptamers, DNA segments, enzymes, antigens, antibodies, etc. that bind to specific molecules sought by a scientist (e.g., within a sample being analyzed via microarray testing). See also *Microarray (testing), Protein microarrays, Deoxyribonucleic acid (DNA), Hybridization (molecular genetics), DNA chip, Biochip, Magnetic particles, Ligand (in biochemistry), Receptors, Aptamers, Enzyme, Antigen, Antibody, Nanosheets.*

CARB

See *Center for Advanced Research in Biotechnology (CARB).*

Carbetimer

An antineoplastic (i.e., anticancer) low-molecular-weight polymer that acts against several types of cancer tumors, perhaps via stimulation of the patient's immune system. It has minimal toxicity. See also *Polymer, Cancer.*

Carbohydrate Engineering

The selective, deliberate alteration/creation of carbohydrates (and the oligosaccharide side chains of glycoprotein molecules) by man. See also *Gluconeogenesis, Glycobiology, Glycoform, Glycolipid, Glycolysis, Glycoprotein, Glycosidases, Restriction endoglycosidases, Glycoside, Glycosylation, Carbohydrate microarrays.*

Carbohydrate Microarrays

Refers to a piece of glass, plastic, or silicon onto which has been placed a large number of specifically known sugar molecules (also known as oligosaccharides, polysaccharides, carbohydrates, or glycans) in specific locations. These microarrays can then be utilized to test a single biological sample for a variety of carbohydrate-specific attributes or effects.

The sugar molecules can be bound to the chip (glass, plastic, or silicon) via *use of thiol molecular groups, use of biotinylation* (to subsequently adhere the biotinylated sugar molecules onto streptavidin-coated chip surface), *conversion of the sugar molecules to glycolipids* (to subsequently adhere those glycolipids via hydrophobic adsorption onto the chip surface), other glycoconjugates (e.g., glycosaminoglycans), etc.

For example, during 2002, Denong Wang attached numerous pathogen-applicable sugar molecules onto chips (thereby creating a carbohydrate microarray) and then utilized those microarrays to evaluate the *specificity* of various antibodies and other immune system proteins in binding to those *polysaccharides typically attached to surfaces of pathogens.* See also *Microarray (testing), Oligosaccharides, Biochips, High-throughput screening (HTS), Target (of a therapeutic agent), Assay, Bioassay, Thiol group, Biotinylation, Streptavidin, Glycolipid, Glycobiology, Pathogen, Antibody, Combining site, Glycoprotein, Glycoform, Glycoconjugates.*

Carbohydrates (Saccharides)

A large class of carbon–hydrogen–oxygen compounds. Monosaccharides are called "simple sugars," of which the most abundant is D-glucose. It is both the major fuel for most organisms and constitutes the basic building block of the most abundant polysaccharides, such as starch and cellulose.

While starch is a fuel source, cellulose is the primary structural material of plants. Carbohydrates are produced by photosynthesis in plants. Most, but not all, carbohydrates are represented chemically by the formula $Cx(H_2O)n$, where n is 3 or higher. On the basis of their chemical structures, carbohydrates are classified as polyhydroxy aldehydes, polyhydroxy ketones, and their derivatives.

The term *carbohydrates* was originally utilized to apply to any compounds (i.e., saccharides) whose molecular formula *could* be written in a form implying an equal number of moles of carbon and water. See also *Glucose (GLc), Glycogen, Monosaccharides, Oligosaccharides, Polysaccharides, Sialic acid, C4 photosynthesis, Mole.*

Carbon Nanohorns

Refers to tiny tubes composed of carbon *that are closed off on one end via a cone-shaped cap,* whose diameter is measured in nanometers.

Groups of carbon nanohorns self-assemble into spherical structures with the capped (i.e., *horn*) ends pointing outward in all directions. These spherical nanohorn structures have a diameter of less than 100 nm, so are able to penetrate the plasma membrane (i.e., *outer skin*) of cells, but not the nucleus of cells.

When certain natural polysaccharides or gum arabic is applied to the surfaces of carbon nanohorns, these nanohorn structures can subsequently be utilized to carry certain pharmaceutical compounds into cells. This can be a means to carry specific pharmaceuticals into cells, which would otherwise not penetrate the plasma membrane of the (diseased) cells. See also *Nanometers (nm), Nanoscience, Nanotechnology, Self-assembly (of large molecular structure), Cell, Nucleus, Plasma membrane, Polysaccharides.*

Carbon Nanotubes

Refers to any tiny tube composed of carbon, whose diameter is measured in nanometers (nm). There are several potential applications for the utilization of carbon nanotubes (CNTs) within fields of nanobiotechnology.

For example, during 2003, Bruce J. Hinds and coworkers were able to incorporate numerous CNTs into a polymer membrane (film) in a manner such that the CNTs served as *pores* through which molecules possessing 1–10 nm *diameters* could pass from one side of the membrane to the other. Such *nanotube membranes* hold significant potential utility as *molecular sieves* (e.g., to screen certain biochemicals out of solution/mixture), as the contact surface for certain biosensors (e.g., allowing-in only the molecules *sought to be sensed).*

During 2009, Stuart Lindsay and colleagues utilized single-walled carbon nanotubes (SWNTs) to construct a *nanopore sequencer.*

During 2012, Marshall, Brown, and Jimmy Xu synthesized carbon nanotubes that have a diameter of approximately 40 nm (i.e., large enough to carry anticancer drug molecules suspended in a temperature-sensitive hydrogel). After injecting these carbon nanotubes and waiting for them to be taken up by tumor cells, an alternating magnetic field was applied (external to the body), which induced an alternating electric current within the nanotubes. The nanotubes' electrical resistance to that current generated heat that liquefied the hydrogel, thereby releasing the drugs into the tumor cells.

During 2004, Hongjie Dai and Paul A. Wender utilized single-walled carbon nanotubes to *ferry*-specific proteins (e.g., streptavidin) across the plasma membrane of certain cells, where the protein was able to then act upon the cell's interior. Dai and Wender showed that those specific proteins (bound to biotin-*coated* carbon nanotubes) entered the cells via endocytosis.

Certain carbon nanotubes can form into bundles that rotate (spin) rapidly in the presence of a rotating magnetic field. Via precise manipulation of that magnetic field, these *nanodrills* could be utilized to drill into specific cells (e.g., cancer cells) inside the body to deliver drugs into those cancer cells or to kill them outright.

Some carbon nanotubes can also act as an *antenna* to receive electromagnetic radiation possessing wavelengths of several hundred nanometers length (i.e., visible light). That visible light's energy is converted by the carbon nanotubes into either electricity or thermal energy (i.e., to drive a chemical reaction in adjacent substrate). Thus, these carbon nanotubes may be utilized in the future to construct light-sensing biosensors.

Some single-walled carbon nanotubes are fluorophores. See also *Nanoscience, Nanobiotechnology, Nanometers (nm), Nanotechnology, Self-assembly (of a large molecular structure), Single-stranded DNA, Polymer, Biosensors (chemical), Biosensors (electronic), Nanofibers, Protein, Streptavidin, Cell, Plasma membrane, Biotin, Endocytosis, Substrate (chemical), Activation energy, Fluorophores, Single-walled carbon nanotubes, Nanopore sequencing, Nanobionics, Nanodrills, Tumor, Cancer, Modulatory nanotechnologies.*

Carboxyl Terminus (of a Protein Molecule)

Refers to the carboxyl group [–COOH] that is attached to one end of some protein molecules, or one end of some amino acid molecules. See also *Protein, Amino acid.*

Carcinogen

A cancer-causing agent. See also *Mutagen, Proto-oncogenes, Aflatoxins, Antioxidants.*

Carnitine

A *vitamin-like* nutrient that occurs naturally in the cells within animals and that is needed for the body to convert fatty acids to energy (which can then be used by the body's cells). Carnitine is essential to facilitate the transport of acyl-CoA enzyme (attached to a fatty acid molecule) into the cell's mitochondria, where the beta-oxidation of fatty acids occurs (thereby providing energy to the cell).

Before fatty acids can enter the mitochondria, they must be *activated* by a chemical reaction (which occurs on the outer mitochondrial membrane), in which acyl-CoA is attached to the fatty acid molecule via a chemical reaction that is driven by adenosine triphosphate and is catalyzed by acyl-CoA synthetase. Adenosine monophosphate is a by-product of that chemical reaction. See also *Fatty acids, Metabolism, Acyl-CoA, Enzyme, Acetyl carnitine, Acetylcarnitine transferase, Mitochondria, Plasma membrane, Activation energy, Adenosine triphosphate (ATP), Synthase, Adenosine monophosphate (AMP).*

Carotenoids

A general term for a group of plant-produced and microorganism-produced pigments ranging in color from yellow to red and brown that act as protective antioxidants in photosynthetic plants and in animals that consume carotenoids.

Approximately 600 carotenoids have been discovered and studied by man. The carotenes and the xanthophylls, orange to yellow in color, are the most common. Carotenoids are responsible for the coloration of certain plants (e.g., the carrot) and of some animals (e.g., the lobster). The carotenoid pigments are transferred to animals as an element in their foods. Carotenoids are composed of isoprene units (usually eight) that may be modified by the addition of other chemical groups on the molecule. The carotenes are of importance to higher animals because they are utilized in the formation of vitamin A.

Carotenoids act as antioxidants (*quenchers* of free radicals), so consumption of carotenoids apparently thereby reduces the risk of some cancers, coronary heart disease, eyesight loss, and cataracts. See also *Vitamin, Beta carotene, Cancer, Coronary heart disease (CHD), Astaxanthin, Lycopene, Antioxidants, Free radical, Oxidative stress, Insulin, Lutein, Zeaxanthin, Golden rice, Photosynthesis, Microorganism.*

Cartilage-Inducing Factors A and B

Compounds produced by the body that also have immunosuppressive activity. See also *Immunosuppressive.*

Cas Proteins

Abbreviation for *CRISPR-associated protein nuclease.* It is a family of RNA-guided bacterial nucleases (i.e., DNA-cutting enzymes) that targets a particular DNA sequence (e.g., of a specific virus that had earlier invaded a bacterium) for destruction, thereby protecting the bacterium from later reinvasion by that virus. See *Bacteria, CRISPR, CRISPR/CAS9 gene-editing systems, Deoxyribonucleic acid (DNA), Ribonucleic acid (RNA), Enzyme, Nuclease, Virus, Site-directed nucleases.*

Cas1

See *CRISPR.*

Cas1–Cas2 Complex

See *CRISPR.*

Cas2

See *CRISPR.*

Cas9

Abbreviation for *CRISPR-associated protein nuclease.* It is a family of RNA-guided bacterial nucleases (i.e., DNA-cutting enzymes) that targets a particular DNA sequence (e.g., of a specific virus that had earlier invaded a bacterium) for destruction, thereby protecting the bacterium from later reinvasion by that virus. See *Bacteria, CRISPR/CAS9 gene-editing systems, Deoxyribonucleic acid (DNA), Ribonucleic acid (RNA), Enzyme, Nuclease, Site-directed nucleases, Virus.*

Cascade

A sequential series of events (e.g., gene expressions, chemical reactions, immune responses), which is initiated (i.e., *set off*) by a specific first event (e.g., a signaling molecule *docking* at a receptor molecule, an antibody–antigen complex forming in the body, thrombin cleaving fibrinogen). See also *Signaling molecule, Signal transduction, Receptors, Protein signaling, Systemic acquired resistance (SAR), Harpin, Complement (component of immune system), Complement cascade, Thrombin, Fibrin, Gene expression cascade, R genes, Transactivating protein, Viral transactivating protein, Kinases, Mitogen-activated protein kinase cascade.*

Casimir Force

A small naturally occurring attractive force that acts between two (uncharged) small objects (e.g., nanoparticles) that are near each other.

In certain instances (e.g., gold nanoparticles located very close to a small silicon plate), the shining of appropriate wavelength light onto the silicon plate increases the number of electrons present at the surface of the silicon, which thereby increases the Casimir force acting on the gold nanoparticle. This ability to remotely control the Casimir force impacting on nanometer-scale objects (e.g., tiny machines) points to the future ability to remotely power/control *nanobots* (e.g., micromachines designed to do specific tasks within the body's bloodstream, cells, etc.). See also *Nanometers (nm), Nanoparticles, Nanobots, Nanoscience, Self-assembling molecular machines, Nanotechnology, Nanoelectromechanical systems (NEMS).*

Caspases

Refers to a *family* of cysteine proteases, which are coded for by certain genes (e.g. the **ced-3** gene) during cell apoptosis.

As a result of apoptosis causing synthesis of several caspases, the (dying) cells cleave/destroy a variety of cellular proteins (i.e., the cell undergoes *self-digestion*). See also *Enzyme, Protease, Cell, Gene, Apoptosis, Protein, PARP.*

Cassette

A *package* of genetic material (containing more than one gene) that is inserted into the genome of a cell via gene splicing techniques. May include promoter(s), leader sequence, termination codon, etc. See also *Gene splicing, Leader sequence, Promoter, Genetic code, Termination codon (sequence), Genetic engineering, Transgene, Genome.*

Catabolism

Energy-yielding pathway. The phase of metabolism involved in the energy-yielding breakdown of nutrient (food) molecules. See also *Dissimilation, Metabolism, Pathway, Sterols.*

Catabolite Activator Protein

See CAP.

Catabolite Repression

Common in bacteria. The decreased expression of catabolic enzymes as brought about by a catabolite such as glucose. For example, glucose is the preferred fuel source for certain bacteria, and when it is present in the culture medium, it represses the formation of enzymes that are required for the utilization of other fuel sugars such as β-galactosidase. Since glucose or other catabolites (other molecules derived from glucose) cause the repression, it is known as catabolite repression. See also *CAP, Operon, Glucose (GLc), Adenosine monophosphate (AMP), Pathway feedback mechanisms.*

Catalase

An enzyme that catalyzes the very rapid decomposition of hydrogen peroxide to water and oxygen. Catalase is in the group of enzymes known as metalloenzymes because it requires the presence of a metal in order to be catalytically active. The metal (known as a cofactor) is, in the case of catalase, iron, which is found in both plants and animals.

For example, Karin U. Schallreuter showed that during the early decades of a human's life, the production of the applicable catalase within human hair prevents hydrogen peroxide (also produced within human hair) from bleaching that hair. Later in life, the catalase production within hair declines, and the hair of older humans gets bleached white/gray by the hydrogen peroxide. See also *Hydrolysis, Human superoxide dismutase (hSOD), PEG-SOD (polyethylene glycol superoxide dismutase).*

Catalysis

Coined by Jons J. Berzelius in 1838, this term refers to the act of increasing the rate of a given chemical reaction via use of a catalyst. Almost all chemical reactions in biological systems (e.g., within an organism) are catalyzed by molecules known as enzymes. Enzymes typically increase the rate of a given biological/chemical reaction by at least a millionfold. See also *Catalyst, Catalytic site, Enzyme, Metalloenzyme.*

Catalyst

From the Greek word "katalyein," which means "to dissolve." Any substance (entity), either of protein or of nonproteinaceous nature, that increases the rate of a chemical reaction, without being consumed itself in the reaction. In biosciences, the term "enzyme" is used for a proteinaceous catalyst. Enzymes catalyze biological reactions. See also *Enzyme, Catalytic site, Active site, Catalytic antibody, Semisynthetic catalytic antibody, Metalloenzyme.*

Catalytic Antibody

Discovered by Richard A. Lerner and Peter G. Shultz during the 1980s, these are antibodies produced by an organism's body in order to help catalyze certain chemical reactions (e.g., needed for certain body functions).

Scientists have subsequently been able to cause organisms to produce an antibody in response to a carefully selected antigen (e.g., target molecule in bloodstream, or molecule involved in chemical reaction of interest), which itself catalyzes the *splitting* of a molecule in the bloodstream (e.g., heroin into two harmless small molecules) or mimics

- Restriction endonucleases that cleave (cut) proteins or DNA molecules precisely at specific locations on those molecules
- Restriction endoglycosidases that are capable of cleaving oligosaccharides or polysaccharide molecules precisely at specific locations on those molecules
- Transition state chemical complex in the chemical reaction that is to be catalyzed—resultant antibody acts both as an antibody (to the selected transition-state-complex antigen) and as a catalyst (for the chemical reaction possessing that selected transition state chemical complex)

This catalyst (enzyme) thus possesses the remarkable specificity of an antibody (i.e., specific only to the desired transition-state reactant) that holds the potential to yield chemical reaction products of greater purity than those achieved via current (less specific) catalysts.

Because the immune system will (in theory) produce an antibody to virtually every molecule of sufficient size to be detected by the immune system (i.e., 6–34 Å), it should be possible to raise catalytic antibodies for a large number of industrial chemical reactions that are currently catalyzed via conventional (less specific) catalysts. Commercial quantities of such antibodies would be produced via monoclonal antibody techniques (e.g., in bioreactors/fermentation vats). See also *Oligosaccharides, Catalyst, Antibody, Organism, Restriction endonucleases, Restriction endoglycosidases, Cell, Monoclonal antibodies (MAb), Antigen, Transition state, Protein, Activation energy, Semisynthetic catalytic antibody, Angstrom (Å), Abzymes.*

Catalytic Domain

See *Domain (of a protein).*

C

Catalytic RNA

Discovered by Thomas R. Cech and Sidney Altman in 1983, this refers to an RNA (ribonucleic acid) molecule that acts to cleave (*cut*) any other RNA. See also *Ribozymes*, *Ribonucleic acid* (*RNA*).

Catalytic Site

The site (geometric area) on an enzyme molecule (or other catalyst) that is actually involved in the catalytic process. The catalytic site usually consists of a small portion of the total area of the enzyme. See also *Catalyst*, *Enzyme*, *Active site*, *Catalytic antibody*.

Catechins

Refers to a *family* of polyphenol chemical compounds (phytochemicals) that are naturally produced in most teas, red wines, apples, grapes, chocolates, etc.

When consumed by humans, catechins have been shown to have beneficial antioxidant, anti-inflammatory, and antithrombotic effects in the human body. See also *Polyphenols*, *Phytochemicals*, *Antioxidants*, *Thrombosis*.

Catecholamines

Hormones (such as adrenalin, dopamine) that are amino derivatives of a base structure known as catechol. Some catecholamines (e.g., endorphins) are released into the bloodstream by exercise and act as natural *tranquilizers*. See also *Endorphins*, *Dopamine*, *Hormone*.

Cation

See *Ion*, *Chelation*, *Chelating agent*.

Cationic Lipids

Refers to lipid molecules that possess a positive charge on one end. See also *Lipids*, *Cation*.

Cauliflower Mosaic Virus 35S Promoter (CaMV 35S)

A promoter (sequence of DNA) that is often utilized in genetic engineering to control expression of (inserted) gene, that is, synthesis of desired protein in a plant. See also *Virus*, *Promoter*, *Deoxyribonucleic acid* (*DNA*), *Gene*, *Genetic engineering*, *Protein*.

Caveolae

Discovered during the 1950s by Eichi Yamada and George Palade. Named by Yamada, who felt they looked like small caves (Latin *caveola*, meaning "small caves"). See *Plasma membrane*.

CBA

Acronym for *cell-based assay*. See *Cell-based assays*.

CBD

See *Convention on Biological Diversity* (*CBD*).

CBF Proteins

A *family* of cold (temperature)-regulated transcriptional activators (transcription factors). See *CBF1*.

CBF/DREB1 Pathway

See *CBF proteins*, *DREB proteins*, *Pathway*.

CBF1

A transcription factor (i.e., special protein that acts upon genes) that is synthesized (i.e., manufactured) within certain plants (e.g., *Arabidopsis thaliana*) when that plant is exposed to cold temperatures. CBF1 then interacts with certain portions of the plant's DNA (i.e., COR genes, regulatory sequences) to thus "switch on" the process of cold hardening (via proteins coded for by that plant's COR genes). See also *Transcription factors*, *Protein*, *Gene*, *Synthesizing* (*of proteins*), *Arabidopsis thaliana*, *Genetic code*, *Coding sequence*, *Regulatory sequence*, *Deoxyribonucleic acid* (*DNA*), *Cold hardening*, *Acclimatization*, *COR genes*.

C-C Chemokine Receptor Type 5

See *CCR5 protein*.

CCC DNA

A covalently linked circular DNA molecule, such as a plasmid. See also *Deoxyribonucleic acid* (*DNA*), *Plasmid*.

CCD

See *Charge-coupled device*.

CCR5 Protein

Also referred to as C-C chemokine receptor type 5, it is a G-protein-coupled receptor (embedded in surface membrane of cell) to facilitate entry of certain chemokines into cell. The CCR5 protein is expressed in the plasma membrane of macrophages, T cells, and dendritic cells. See also *Protein*, *Receptors*, *G-protein-coupled receptors*, *Plasma membrane*, *Chemokines*.

CCR5-Delta 32

A deletion mutation within the human CCR5 gene that confers resistance to certain pathogenic viruses such as bubonic plague (*Yersinia pestis*) and HIV/AIDS. The delta-32 mutation causes the cell surface receptor coded for by the CCR5 gene to be "turned off," thereby preventing entry into cell of the bubonic plague or HIV/AIDS virus. See also *CCR5 protein*, *Cell*, *Deletions*, *Mutation*, *Gene*, *Pathogenic*, *Virus*, *Receptors*, *Plasma membrane*, *Human immunodeficiency virus type 1*, *Human immunodeficiency virus type 2*, *Acquired immune deficiency syndrome* (*AIDS*), *Coding sequence*.

CD19

See *Blinatumomab*.

CD20 Protein

A cell-membrane-spanning glycosylated phosphoprotein expressed on the surface of all human B cells. It is coded for by the human MS4A1 gene and plays a role in the development and differentiation of B cells into plasma cells. See also *Protein*, *Cell*, *Plasma membrane*, *Transmembrane proteins*, *B cells*, *Plasma cell*, *Glycosylation (to glycosylate), Posttranslational modification of protein, Gene, Coding sequence, Differentiation, Humanized antibody.*

CD3

See *Blinatumomab.*

CD4 EPSP Synthase

See *EPSP synthase, CP4 EPSPS.*

CD4 EPSPS

See *EPSP synthase, CP4 EPSPS.*

CD4 Protein

An adhesion molecule (protein) imbedded in the outer wall (envelope) of human immune system and brain cells that functions as the receptor (door to entry into the cell) for the HIV (AIDS) virus. The gp120 envelope glycoprotein of the HIV (i.e., AIDS virus) directly interacts with the CD4 protein on the surface of helper T cells to enable the virus to invade the helper T cells. See also *T cell receptors, Adhesion molecule, GP120 protein, Soluble CD4.*

CD44 Protein

One of the adhesion molecules (embedded in the surface of the linings of blood vessels) that assists the neutrophils on their journey from the bloodstream through the walls of blood vessels (e.g., to combat pathogens into adjacent tissues). Tumor cells also exploit CD44 molecules in order to metastasize (spread throughout the body's tissue from a single beginning tumor) via a similar (tumor cell) through-blood vessel-wall adhesion molecule mechanism. See also *Adhesion molecule, CD4 protein, Protein, Neutrophils, Pathogen, Tumor, Cancer, Soluble CD4.*

CD4-PE40

A pharmaceutical discovered in 1988 by Ira Pastan and Bernard Moss that has indicated potential to combat acquired immune deficiency syndrome (AIDS). CD4-PE40 is a conjugated protein (fusion protein) consisting of a CD4 protein (molecule) attached to *Pseudomonas* exotoxin (a substance produced by *Pseudomonas* bacteria that is toxic to certain living cells). The gp 120 glycoprotein on the surface of the HIV (i.e., AIDS) virus attaches preferentially to the CD4 portion of this immunoconjugate, and the virus is inactivated by the *Pseudomonas* exotoxin portion of this immunoconjugate. See also *Protein, CD4 protein, Fusion protein, GP120 protein, Soluble CD4, Immunotoxin, Conjugated protein, Acquired immune deficiency syndrome (AIDS), Human immunodeficiency virus type 1 (HIV-1), Human immunodeficiency virus type 2 (HIV-2), Ricin, Abrin.*

CD8$^+$ T Cells

One class of T cells (i.e., part of the immune system) that is triggered by an infection (or a vaccination) to differentiate into two different populations of cells. One population (i.e., effector T cells) attacks the pathogen to clear it from the body. Effector T cells have a short lifespan.

The second population (i.e., memory T cells) constitute a long-lived reservoir of cells that *remember* that pathogen's specific antigenic determinant and mediate the immune system's response to any future infection by the same pathogen (faster/more effective response). That future memory, T-cell-mediated response is known as the *adaptive immune response.*

Memory T cells also contain elevated levels of a protein known as the Signal Transducer and Activator of Transcription #4 (STAT4), which makes them much more responsive to cytokines. Thus, these STAT4-containing CD8$^+$ T cells are also part of the future cytokine-mediated *innate immune response.* See also *Cell, T cells, Effector T cells, Memory T cells, Differentiation, Pathogen, Antigenic determinant, Cellular immune response, Innate immune system, Innate immune response, Signal transducers and activators of transcription (STATs).*

CD95 Protein

Also called "APO-1/Fas," it is a transmembrane protein (embedded within the surface membrane of the cell) that transmits apoptosis ("programmed" cell death) "signal" into cells. Transduction of that apoptosis signal occurs when certain ligands or antigens (i.e., the APO-1/Fas antigen) bind to the extracellular (i.e., portion outside of cell membrane) part (i.e., receptor) of the CD95 protein. See also *Apoptosis, Protein, Cell, Signal transduction, Signaling, Nuclear receptors, Antigen, Receptors, Fusarium.*

cDNA

See *Complementary DNA (cDNA).*

C-DNA

Also known as copy DNA. A helical form of DNA. It occurs when DNA fibers are maintained in 66% relative humidity in the presence of lithium ions. It has fewer base pairs per turn than B-DNA. See also *B-DNA, Deoxyribonucleic acid (DNA), Base pair (bp), Complementary DNA (cDNA).*

cDNA Array

See *Microarray (testing).*

cDNA Clone

A DNA molecule synthesized (*made*) from an mRNA sequence via sequential use of reverse transcriptase (acting on mRNA) and DNA polymerase.

A collection of such cloned molecules that represents all the genetic information expressed by a given cell or by a given tissue type is referred to as a "cDNA library." See also *Deoxyribonucleic acid (DNA), Messenger RNA (mRNA), Complementary DNA (cDNA), Sequence (of a DNA molecule), Reverse transcriptase, DNA polymerase, Clone (a molecule), Cell, Genetic code.*

cDNA Library

See *cDNA clone.*

cDNA Microarray

See *Microarray (testing), Complementary DNA (cDNA).*

C

CDR

Acronym for *complementarity-determining regions*, or acronym for *commonly deleted region*. See the links. See also *Deoxyribonucleic acid (DNA)*, *Complementarity (molecular genetics)*, *Deletions*.

CDx

Abbreviation for companion diagnostic. See *Companion diagnostic*.

CE

Acronym for capillary electrophoresis. See *Capillary electrophoresis*.

Cecrophins

(Lytic proteins) Proteins produced by certain white blood cells (called cytotoxic T lymphocytes [CTL] or killer T cells). The proteins allow lysis (i.e., bursting) of infected cells. Cecrophins are amphipathic (i.e., contain both a hydrophobic region and a hydrophilic region) and work by "worming" the hydrophobic portion into the cell membrane (so the hydrophobic portion of the cecrophin molecule is out of the water). This creates a transmembrane pore (i.e., a hole in the membrane) that is lined with the cecrophin's hydrophilic portion. Membranes function simply to separate various components. This separation is required for life to exist. When holes are introduced into cell membranes, water rushes into the targeted cell due to differences in osmotic pressure and the cell ruptures (explodes). T cecrophins are only able to lyse (i.e., burst) infected cells because only "sick" cells have a weakened cytoskeleton (located just inside the cell membrane), which cannot prevent the contents of the cell from spilling out through the pores (created by cecrophins). See also *Helper T cells (T4 cells)*, *Pathogen*, *Complement*, *Hydrophobic*, *Hydrophilic*, *Complement cascade*, *Lyse*, *Lysis*.

Cecropin A

See *Cecropin A peptide*.

Cecropin A Peptide

See *Cecrophins*, *Peptide*.

C-Extein

See *Intein*.

Celiac Disease

From the Greek word *koelia* (abdomen), it is a genetic-susceptibility-resultant disease of the small intestine. People possessing the relevant genes (especially common in Italians and people descended from Italians) are unable to tolerate gluten (a component of cereal grains) or gluten-similar proteins (secalin and hordein) in rye and barley. Consumption of gluten (e.g., in wheat-containing products) causes villi (small fingerlike projections on the surface of the inner wall of the small intestine) to become chronically inflamed. An enzyme named tissue transglutaminase is released by the damaged cells (enterocytes). In genetically susceptible individuals, the B cells of the immune system make and release antibodies against the tissue transglutaminase. When those antibodies encounter molecules of tissue transglutaminase on the surfaces of villi, they can cause degeneration of villi in the small intestine and thereby result in a reduction in the body's subsequent ability to absorb nutrients from foods. In severe cases, this may result in malnutrition occurring while an otherwise adequate diet is later consumed.

Tissue transglutaminase can also modify undigested gluten (peptides) present in the small intestine, thereby enabling the gluten to tightly bind to certain histocompatibilty leukocyte antigens (HLA-DQ2 and HLA-DQ8). When that occurs, applicable T cells are activated to release into the bloodstream cytokines and chemokines. See also *Gluten*, *Heredity*, *Genetics*, *Enterocytes*, *Peptide*, *Antigen*, *Antibody*.

Cell

From the Latin word "cella," which means "small room." Discovered by Robert Hooke in 1665, the cell is the fundamental self-containing unit of life. The living tissue of every multicelled organism is composed of these fundamental living units. Certain organisms may consist of only one cell, such as yeast or protein bacteria, protozoa, some algae, and gametes (the reproductive stages) of higher organisms. Larger organisms are subdivided into organs that are relatively autonomous but cooperate in the functioning of that plant or animal. Unicellular (i.e., single-cell) organisms perform all life functions within the one cell.

In a higher organism (i.e., a multicellular organism), entire populations of cells (i.e., an organ) may be designated a particular specialized task (e.g., the heart to facilitate circulation). The cells of muscle tissue are specialized for movement and those of bone and connective tissue, for structural support.

The surface of some cells bear extensions (e.g., extended "arms"). For example, such extensions from the bone cells known as osteocytes are called "dendrites."

While most cells are too small to be seen with the unaided eye, the egg yolk of birds is a single cell, so the egg yolk of an ostrich is the world's largest cell. See also *Plasma membrane*, *Gamete*, *Germ cell*, *Microbiology*, *Oocytes*, *Dendrites (in bone)*.

Cell Adhesion Molecule

See *Adhesion molecule*.

Cell Adhesion Proteins

From the Latin *adhaerere* meaning "to stick to." The term "cell adhesion protein" refers to a glycoprotein molecular "chain" that protrudes from the surface membrane of certain cells and causes cells (possessing "matching" adhesion proteins) to adhere to each other. For example, in 1952 Aaron Moscona observed that (harvesting enzyme-separated) chicken embryo cells did not remain separated but instead coalesced again into an (embryo) aggregate. In 1955, Philip Townes and Johannes Holtfreter showed that "like" amphibian (e.g., frog) neuron cells will rejoin together after being physically separated (e.g., with a knife blade); but "unlike" cells remain segregated (apart).

Cell adhesion proteins *per se* were formally discovered by Gerald M. Edelman during the 1970s.

Adhesion proteins also play a crucial role in guiding monocytes to sources of infection (e.g., pathogens) because adhesion molecules in the walls of blood vessels (after activation caused by pathogen invasion of adjacent tissue) adhere to like adhesion molecules in the membranes of monocytes in the blood. The monocytes pass through the blood vessel walls, become macrophages, and fight the pathogen infection (e.g., triggering tissue inflammation). When certain cell adhesion proteins are absent from tissues (where they are normally present), it can lead to cancers of the endometrium, bladder, prostate, skin, breast, pancreas, colon, etc.

See also *Monocytes, Macrophage, Polypeptide (protein), Cell, Pathogen, CD4 protein, CD44 protein, GP120 protein, Vaginosis, Harvesting enzymes, Harvesting, Signal transduction, Selectins, Lectins, Glycoproteins, Sugar molecules, Leukocytes, Lymphocytes, Neutrophils, Endothelium, Endothelial cells, P-selectin, ELAM-1, Integrins, Cytokines, Cancer, Chronic inflammation.*

Cell Culture

The *in vitro* (i.e., outside of body, in a test tube or vat) propagation of cells isolated from living organisms. See also *Mammalian cell culture, Insect cell culture, Dissociating enzymes, Harvesting enzymes, Vero, MDCK, Plant cell culture.*

Cell Cytometry

See *Cell, Cell sorting, Fluorescence activated cell sorter (FACs), Magnetic particles.*

Cell Differentiation

The process whereby descendants of a common parental cell achieve and maintain specialization of structure and function. In humans, for instance, all the different types of cells (e.g., muscle cells, bone cells) differentiate from the zygote (itself formed by union of the simple sperm and egg). In humans, the various blood cell types (e.g., red blood cells, white blood cells) differentiate from stem cells in the bone marrow. Cell differentiation is caused/triggered/assisted by micro-RNAs, colony-stimulating factors, growth factors, and certain other proteins (e.g., hedgehog proteins, mediator, cohesin). See also *Stem cells, Stem cell one, Differentiation, Protein, Hedgehog signaling pathway, Hedgehog proteins, Erythrocytes, Leukocytes, Colony-stimulating factors, Growth factors, Mitogen-activated protein kinase cascade, Micro-RNAs, Tetraspanin proteins, Mediator, Cohesin.*

Cell Fusion

The combining of cell contents of two or more cells to become a single cell. Fertilization is such a process (fusing of gametes' cells). See also *Gamete, Cell.*

Cell Motility

Refers to *cell movement* (e.g., during an organism's early development, during repair of some tissues, during cancer metastasis).

For example, during some stages of a human baby's development in the womb, entire "sheets" of cells will suddenly move significant distances to a new location on the baby's body.

For example, when blood vessels get injured, the harmed cells release a signal. That signal causes some of the endothelial smooth muscle cells to "transform" from *contractile* phenotype (i.e., normal state, in which they help to control blood pressure) to *synthetic* phenotype (i.e., which can move). The cells in synthetic state move to the site of the injury, where they repair it via growing/dividing, and then they return to the contractile state.

During metastasis, transforming growth factor-beta (TGF-beta) exuded by a cancerous tumor causes epithelial cells to "transform" to a mesenchymal phenotype (thereby enabling subsequent cell motility).

The molecule *n-cofilin* plays a critical role in the body's regulation of cell motility (e.g., helps to break down actin fibers, resulting in cells moving during development). See also *Cell, Pathway, Signaling, Hedgehog signaling pathway, Embryology, Endothelium, Cancer, Metastasis, Transforming growth factor-beta (TGF-beta), Phenotype, Cholesterol, Actin.*

Cell Recognition

See *Adhesion molecule, Signal transduction, Receptors.*

Cell Signaling

See *Signaling.*

Cell Sorting

A process utilized (e.g., by researchers) to sort/separate different cells (e.g., pathogens, cancerous vs. normal cells, sperm that are bearing chromosomes for male vs. female).

Some automated means of cell sorting include *biochips* (utilizing controlled electrical fields to collect specific cell types onto electrodes in the biochip), fluorescence-activated cell sorter machines using a laser to light up cell surface proteins, magnetic particles (e.g., attached to antibodies that themselves attach to cell surface proteins), level(s) of RNA resultant from gene expression in the cell, etc. See also *Cell, Pathogen, Cancer, Chromosome, Biochip, BioMEMS, Fluorescence activated cell sorter (FACs), Magnetic particles, Gene expression.*

Cell State

See *Mediator, Cohesin.*

Cell-Based Assays

Refers to assays in which whole cells (generally living) are probed. See also *Assay, Bioassay cell, Multiplexed (assay), High-content screening, Flow cytometry, Whole-cell patch-clamp recording, Molecular beacon.*

Cell-Differentiation Proteins

The various growth factors and other proteins that cause/assist in cell differentiation. See also *Cell differentiation, Mediator, Cohesin, Hedgehog proteins.*

Cell-Free Gene Expression System

Refers to a (science researcher's) system of carefully prepared compounds/vessels for the *expression of a given gene* in crude cell extract, without the use of any cells.

For example, a given gene can be transcribed in a research vessel (test tube) via addition of the proper RNA polymerase. The resultant RNA is then translated via the proper lysate (e.g., extracted from rabbit reticulocytes or from wheat germ). See also *Cell, Gene, Express, Transcription, Translation.*

Cell-Free Translation System

See *Cell-free gene expression system.*

C

Cell-Free Fermentation

Discovered by Eduard Buchner in 1896, this refers to a (science researcher's) system of carefully prepared compounds/vessels for the *fermentation of a particular substrate* (e.g., glucose) without the use of any cells. See also *Fermentation, Cell, Substrate (chemical), Glucose (GLc).*

Cell-Mediated Immunity

See *Cellular immune response.*

Cell-Penetrating Peptide

See *Peptide-oligonucleotide conjugates.*

Cellular Adhesion Molecule

See *Adhesion molecule.*

Cellular Adhesion Receptors

See *Cell, Adhesion molecule, Receptors, Integrins, Selectins, Cadherins.*

Cellular Affinity

Tendency of cells to adhere specifically to cells of the same type. This property is lost in some cancer cells. See also *Cell, Adhesion molecule, Cell differentiation.*

Cellular Immune Response

Also called "cell-mediated immunity." The immune response that is carried out by specialized cells, in contrast to the response carried out by soluble antibodies. The specialized cells that make up this group include cytotoxic T lymphocytes, helper T lymphocytes, macrophages, and monocytes. This system works in concert with the humoral immune response. See also *Humoral immunity, T cells, $CD8^+$ T cells, T cell receptors, Phagocyte, Helper T cells (T4 cells), Cytokines, Macrophage.*

Cellular Oncogenes

See *Proto-oncogenes.*

Cellular Pathway Mapping

Refers to the process of determining each of the pathways and pathway feedback mechanisms within a given cell's vital processes.

Cellular pathway mapping can be utilized to identify targets of therapeutic agents, identify cross talk between some of the cell's pathways, and identify "branched" pathways that—when perturbed by a potential therapeutical agent (e.g., pharmaceutical)—could result in toxic side effects. See also *Pathway, Cell, Pathway feedback mechanism, Target (of a therapeutic agent), Validation (of target), Toxicogenomics, Metabolite profiling, High-content screening, RNA interference (RNAi), Metabonomics.*

Cellulase

An enzyme that digests cellulose to simple sugars such as glucose. Commercial cellulases (e.g., for biofuel production) have been extracted from some fungi and from archaea. See also *Enzyme, Cellulose, Digestion (within chemical production plants), Fungus, Archaea.*

Cellulose

A polymer of glucose units found in all plant matter; it comprises 40%–55% of the cell wall in plant cells. Because of its presence in all plant cells, cellulose is the most abundant biological compound on earth.

Cellulose is also synthesized (made) by some single-celled organisms such as cyanobacteria. See also *Carbohydrates, Glucose (GLC), Cell, Van der Waals forces, Cortical microtubules.*

CenH3 Gene

See *CENH3 protein, Doubled-haploid breeding program.*

CenH3 Protein

One of the histone proteins (i.e., around which are "wound" the cell's DNA to make chromosomes). CENH3 is found only within the centromere (i.e., the part of the chromosome that controls how it is passed to the next generation). See also *Chromosomes, Histones, Protein, Meiosis, Mitosis, Deoxyribonucleic acid (DNA), Centromere, Doubled-haploid breeding program.*

Center for Advanced Research in Biotechnology (CARB)

A protein engineering research consortium that was established in Rockville, Maryland, during 1989 by the U.S. Government, the University of Maryland, and the local government. See also *Protein engineering.*

Central Dogma (New)

Coined by Shankar Subramaniam during 1999, it is a restatement of the (old) former "central dogma" to include the fact that an organism's environment/activity also impacts *when* and *how* and *how much* some of its genes are expressed (e.g., to cause certain proteins to be "manufactured"). Environmental factors impacting gene expression include temperature, sunlight, humidity, consumption of some vitamins, industrial chemicals, the presence of certain bacteria, the presence of signal transducers and activators of transcription (STATs), etc.

In addition, epigenetics cause organisms to preferentially express certain alleles (e.g., those inherited from the mother or from the father). For example, in mice, more maternal-origin alleles are expressed within the developing brain, and more paternal-origin alleles are expressed within the adult mouse brain than would occur from a simple random 50/50 contribution of parental alleles to the offspring's DNA.

For example, the eggs of the saltwater crocodile (*Crocodylus porosus*) yield a larger fraction of male offspring when those eggs are incubated in the nest (made of rotting vegetation) at temperatures above 90°F (32°C) than when those eggs are incubated at temperatures below 90°F (32°C).

Recent research indicates that *physical exercise* changes the expression levels of some genes (within human skeletal muscles) involved in the body's metabolism of carbohydrates.

That (central dogma) restatement also expressly includes the fact that *more than one protein* can result from each gene in an organism's genome (e.g., due to interactions *between* genes, interactions between genes and their protein products [e.g., STATs], interactions between genes and histones, and interactions between genes and some environmental factors). Mechanistically, this results in (different) proteins via the following:

- Alternative splicing of the mRNA transcript. For example, the COX-3 enzyme is produced in the human body when *intron 1* is retained in the mRNA transcript during transcription of the COX-1 gene. For example, a single intronic base substitution that is present within the IKAP gene (the allele responsible for the human disease known as *familial dysautonomia*) affects the splicing of the IKAP transcript (i.e., the mRNA segment that determines which protein is subsequently *manufactured* by relevant cells).
- Varying translation start or stop site (on the gene).
- Frameshifting (i.e., different set of triplet codons in the mRNA is translated).
- Contiguous genes.
- Recombination of some gene segments.

See also *Central dogma (OLD), Organism, Molecular genetics, Complementary DNA (c-DNA), Gene, Allele, Protein, Enzyme, Replication (of virus), Gene, Transcription, Translation, Deoxyribonucleic acid (DNA), Genome, Ribonucleic acid (RNA), Messenger RNA (mRNA), Transcription factors, Ribosomes, Signal transduction, Signal transducers and activators of transcription (STATs), Photoperiod, Gene expression, Alternative splicing, Gene splicing, Splicing, Splice variants, Frameshift, Codon, Intron, Pharmacoenvirogenetics, Cyclooxygenase, Metabolism, Carbohydrates, Transcriptome, Activator (of gene), Contiguous genes, Epigenetic, Vitamin, Histones, Epigenetic.*

Central Dogma (Old)

The historical organizing principle of molecular genetics; it states that genetic information flows from DNA to RNA to protein, or stated in another way, DNA makes RNA, which makes protein. This principle was first stated by Watson and Crick. It is, however, not rigorously accurate as illustrated by the following facts:

- DNA (i.e., genes) "information flow" is influenced (e.g., timing, amounts) by some environmental factors (e.g., temperature, humidity).
- The enzyme reverse transcriptase produces ("makes") DNA using an RNA template.
- Prions do not contain any DNA.

See also *Molecular genetics, Complementary DNA (c-DNA), Protein, Enzyme, Replication (of virus), Transcription, Translation, Deoxyribonucleic acid (DNA), Ribonucleic acid (RNA), Messenger RNA (mRNA), Prion, Template, Central dogma (new), Reverse transcriptases, RT-PCR, Activator (of gene).*

Centrifuge

A machine that is used to separate heavier from lighter molecules and cellular components and structures. See also *Ultracentrifuge.*

Centriole

An organelle within cells that helps organize the cell's microtubules (e.g., to move certain protein molecules around the cell's interior, from where they were manufactured to where they are needed by the cell). See also *Microtubules, Cell, Protein vesicular transport (of a protein).*

Centromere

A constricted region of a chromosome that includes the site of attachment to the mitotic or meiotic spindle. Due to that role, the centromere is a crucial segment of DNA for ensuring that the right numbers of chromosomes are delivered to the correct location within each "daughter cell" during cell division. See also *Deoxyribonucleic acid (DNA), Chromosomes, Meiosis, Chromatin, Mitosis, Karyotype, Karyotyper.*

Centrosome

Refers to the *pair of centromeres* in a cell that (during interphase) organize microtubules within the cell (e.g., to pull apart the paired chromosomes, prior to separation of the "parent" cell into two daughter cells). Centrosomes also form the base of cilium—a tail-like protrusion on the surface of certain cells (e.g., epithelial cells that line the inner surface of the trachea/bronchial tubes) that "sweep" debris (e.g., bacteria) out to where it can be expelled via coughing, etc. See also *Cell, Centriole, Meiosis, Microtubules, Cilia.*

Cerebrose

See *Galactose (gal).*

Cessation Cassette

A three-gene cassette (genetic sequence construct) that, *when inserted into a plant* and when activated via tetracycline antibiotic, prevents the seeds produced by that plant from germinating. That is because the *cessation cassette* stops those resultant seeds from synthesizing a specific protein needed for seed germination. See also *Cassette, Gene, Genetic engineering, Protein, Synthesizing (of proteins), Sequence (of a DNA molecule), antibiotic.*

Cetuximab

A monoclonal antibody approved by the U.S. Food and Drug Administration as the pharmaceutical Erbitux™ to treat colon cancer and head and neck cancers. By attaching itself to epidermal growth factor receptors (EGF receptors) on the surface of those cancer cells, it thereby interferes with inappropriate (malignant) signaling and thus slows or even stops growth of those cancerous cells. See also *Monoclonal antibodies (MAb), Food and Drug Administration (FDA), Cancer, Cell, Receptor, Epidermal growth factor (EGF), EGF receptor, Signaling.*

CFH Protein

Abbreviation for complement factor H protein. See *Complement factor H gene.*

CFP

Acronym for cyan fluorescent protein. See *Visible fluorescent proteins.*

C

CFTR

See *Cystic fibrosis transmembrane regulator protein (CFTR).*

CGE

Acronym for control of gene expression. See *Genetic use restriction technologies.*

CGIAR

See *Consultative Group on International Agricultural Research (CGIAR).*

cGMP

Current Good Manufacturing Practices. The set of current, up-to-date methodologies, practices, and procedures mandated by the Food and Drug Administration (FDA) that are to be followed in the testing and manufacture of pharmaceuticals. The set of rules and regulations promulgated and enforced by the FDA to ensure the manufacture of safe clinical supplies. The cGMP guidelines are more fine-tuned and up to date (technologically speaking) than the more general GMP. See also *Phase I clinical testing, IND, Good manufacturing practices (GMP).*

Chaconine

A neurotoxin that is naturally present at low levels within potatoes. As a result of that, chaconine is present at detectable levels in the bloodstream of humans that consume potatoes.

When consumed by humans, chaconine acts as a plasma cholinesterase inhibitor. See also *Toxin, Solanine, Plasma, Cholinesterase, Inhibition.*

Chakrabarty Decision

Diamond vs. Chakrabarty, U.S. Department of Commerce, 1980, a landmark case in which the U.S. Supreme Court held that the inventor of a "new" microorganism (*Burkholderia cepacia* bacteria) whose invention otherwise met the legal requirements for obtaining a patent could not be denied a patent solely because the invention was alive. It essentially allowed the patenting of life forms.

The scientist Ananda Chakrabarty had modified the genes of *Burkholderia cepacia*, making it better able to break down petroleum and digest it. See also *U.S. Patent and Trademark Office (USPTO), Microorganism, Bacteria.*

Chalcone Isomerase

An enzyme present within some plants (e.g., tomato) that can catalyze/increase production of certain flavonols (e.g., naringenin chalcone, quercetin glycosides), which act as antioxidants in the human body when they are consumed by humans.

Because oxidation of certain lipids (e.g., low-density lipoproteins) in the bloodstream is the initial step in atherosclerosis disease, consumption of large amounts of such flavonols may help to prevent atherosclerosis (and some other diseases caused by oxidative stress). See also *Flavonols, Oxidation, Lipids, Oxidative stress, Antioxidants, Atherosclerosis, Quercetin.*

Channel Blockers

See *Calcium channel blockers.*

Chaotrope

See *Chaotropic agent.*

Chaotropic Agent

From the Greek meaning "disorder maker," it is a substance that yields ions (in water solution) that can increase unfolding/denaturation of protein molecules in solution, dissolve biological membranes/proteins, and/or denature nucleic acids. One example of a chaotropic agent is guanidium isothiocyanate. See also *Ion, Plasma membrane, Protein, Protein folding, Denaturation, Denatured DNA, Nucleic acids.*

Chaperone Molecules

See *Chaperones.*

Chaperone Proteins

See *Chaperones.*

Chaperones

Protein molecules inside living cells of organisms that assist with

- Correct protein folding as the protein molecule emerges from the cell's ribosomes
- Correct RNA molecule folding

Also, they help to convey those protein(s) and RNA(s) to their ultimate destination(s) in the organism.

Later, when cellular protein molecules begin to "unfold" due to age, heat, viruses, or exposure to certain chemicals or ultraviolet light, chaperones often cause those unfolded protein molecules to return to their correct (initial) conformation.

Examples of such chaperone molecules include heat-shock proteins (e.g., heat-shock protein 70, heat-shock protein 40), certain cold-shock proteins, GroEL protein, and GroES protein. See also *Leader sequence (protein molecule), Protein folding, Heat shock proteins, Protein, Ribosomes, Cell, Conformation, Virus, Congo red, Cold-shock protein, Ribonucleic acid (RNA).*

Chaperonins

Protein molecules inside living cells that facilitate proper folding of the (new) protein molecules that are synthesized (i.e., *manufactured*) in the cell's ribosomes. Chaperonins also facilitate proper folding of some (old) proteins that have been stress-denatured (e.g., they lost their proper folded structure as a result of stress or aging of the cell). Chaperonins accomplish this by encapsulating the applicable protein molecules inside a protective chamber that is formed from two rings of molecular complexes stacked back to back. See also *Protein, Co-chaperonin, Chaperones, Molecular chaperones, Protein folding, Ribosomes, Cell, Conformation.*

Characterization Assay

See *Assay*, *High-throughput screening (HTS)*, *Bioassay*, *Biochips*.

Charge-Coupled Device

A matrix of photosensitive electronic circuits, which functions akin to a camera in the recording of an image (*photograph*) electronically. See also *Dynamic light scattering*.

CHARM

Acronym for comprehensive high-throughput arrays for relative methylation. This particular microarray technology allows scientists to analyze the whole genome of an organism at once. See also *Microarray (testing)*, *Cell array*, *Multiplexed (assay)*, *Gene expression analysis*, *Genome*, *Genomics*, *Organism*, *DNA methylation*, *Methylated*, *Whole-genome association*, *Whole-genome shotgun sequencing*.

Chassis

See *Synthetic biology*.

CHD

Acronym for coronary heart disease. See *Coronary heart disease (CHD)*, *Atherosclerosis*, *Low-density lipoproteins (LDLPs)*, *Carotenoids*.

Checkpoint Blockade

Refers to how certain cancer tumors avoid being destroyed by the human immune system by synthesizing and secreting certain protein molecules that bind to receptors located on the surface of applicable immune system cells, thereby "blockading" those immune system cells. See also *Tumor*, *Cancer*, *Cancer immunotherapy*, *Protein*, *Receptors*, *Synthesizing (of proteins)*.

Chelating Agent

A molecule capable of "binding" metal atoms. The word "chelated" is Greek for *claw/binding together.* The chelating agent/metal complex is held together by coordination bonds that have a strong polar character. One example of a common chelating agent is ethylenediamine tetraacetate (EDTA), which tightly and reversibly binds Mg^{2+} and other divalent cations (positively charged ions). If a chelate is allowed to bind to metal ions required for enzyme activity, the enzyme will be inactivated (inhibited). Cobalamin (vitamin B_{12}), EDTA, and the iron–porphyrin complex of heme (which provides the red color of blood) are other examples of chelates. See also *EDTA*, *Phytate*, *Low-phytate soybeans*, *Low-phytyate corn*, *Chelation*, *Heme*, *Transferrin*.

Chelation

From the Greek for *claw/binding together.* The binding of metal cations (metal atoms or molecules possessing a positive electrical charge) by atoms possessing unshared electrons (thus the electrons can be "donated" to a bond with a cation). The binding of the metal (cation) to the (electron excess) chelator atom (ligand) results in formation of a chelator/metal cation complex. The intra-atom bonds thus formed are given the name of coordination bonds.

The properties of the chelator/metal cation complex frequently differ markedly from the "parent" cation. Both carboxylate and amino (molecular) groups readily bind metal cations. One of the most widely used chelators is ethylenediamine tetraacetate (EDTA). It has a strong affinity for metal cations possessing two (bi) or more positive (electrical) charges. Each EDTA molecule binds one metal cation. The EDTA molecule can be visualized as a "hand" (having only four fingers) that grasps the metal cation. Some enzymes (which require metal cations for their activity) are inactivated by EDTA (and other chelators) in that the chelators preferentially remove the metal from the enzyme. See also *Ion*, *EDTA*, *Ligand (in biochemistry)*, *Carbohydrates*, *Enzyme*, *Heme*, *Chelating agent*, *Transferrin*, *Phytate*, *Low-phytate corn*, *Low-phytate soybeans*.

Chemical Genetics

Coined by Rebecca Ward and Tim Mitchison, this term refers to the creation and use of synthetic chemicals that act to either change the sequence (of amino acids), change the conformation, block, or enhance the activity of a protein (or gene that codes for protein). This enables scientists to then determine the specific function(s) of specific protein molecules.

For example, during 2002, Henning D. Mootz and Tom W. Muir devised a methodology to use a dimerizer ligand to initiate protein splicing. When carefully devised (e.g., the dimerizer ligands are each bound to one-half of an *intein*), which are themselves bound to *exteins*, the intein is thereby "popped out" of the center (of the protein molecule), and the two exteins (i.e., the two "end sequences" of protein molecule after intein is removed) are spliced together in a manner that is controlled (to yield desired *new net protein* molecule). See also *Genomics*, *Functional genomics*, *Protein*, *Gene*, *Genetic code*, *Zinc finger proteins*, *Combinatorial chemistry*, *Conformation*, *Genomic sciences*, *Gene function analysis*, *Sequence (of a protein molecule)*, *Ligand (in biochemistry)*, *Intein*, *Extein*.

Chemiluminescence

See *Luminescent assays*, *Chemiluminescent immunoassay (CLIA)*.

Chemiluminescent Immunoassay (CLIA)

An immunoassay (i.e., an antibody-based bioassay) that utilizes a signal that is generated by light-releasing chemical reactions (e.g., triggered by the binding of antibody to analyte). See also *Immunoassay*, *Antibody*, *Luminescent assays*.

Chemoautotroph

A microorganism that obtains its energy from reactions between it and inorganic (chemical) compounds. For example, certain *Archaea* microorganisms derive energy from chemicals present on the ocean floor where they live (e.g., in/near volcanoes, lava vents). See also *Autotroph*, *Archaea*.

Chemokines

Refers to a category of small cytokines (approx. 8–10 kDa in mass) that are able to cause nearby cells to undergo chemotaxis (i.e., those cells move toward or away from the source of the chemokines).

That directed cell migration is an important part of an organism's growth/development. Because of that, the word "chemokines" is created from the phrase *chemo*tactic cyto*kines*. See also *Cytokines, CCR5 protein, Cell, Chemotaxis, Kilodalton (kDa)*.

Chemometrics

An empirical methodology utilized to (inexpensively) infer a chemical quantity/value from (indirect) measurement(s) of other physical/chemical values (which can be obtained inexpensively).

The term "chemometrics" was coined in 1975 by Bruce Kowalski. One example of the use of chemometrics is to infer the "true metabolizable energy" (TME [N]) of high-oil corn from that corn's protein and oil (fat) content. See also *High-oil corn, TME (N), Protein, Fats*.

Chemopharmacology

Therapy (to cure disease) by chemically synthesized drugs. See also *Pharmacology, Cisplatin*.

Chemotaxis

Sensing of, and movement toward or away from, a specific chemical agent by living, freely moving cells (e.g., bacteria, macrophages, neutrophils).

For example, the *Clostridium botulinum* bacteria can sense and move away from nitric oxide (which can kill *Clostridium botulinum*). See also *Cell, Bacteria, Macrophage, Neutrophils, Actin, Nodulation, Nitric oxide, Chemokines*.

Chemotherapy

When this term was first coined by Paul Ehrlich in 1905, it was defined as any therapy (to cure diseases) via chemically synthesized drugs.

Over time, the term "chemotherapy" has increasingly been utilized to refer to only application of such therapy to treat cancers.

Note that autophagosomes sometimes gather up and carry certain pharmaceuticals (e.g., chemotherapy agents introduced into cancer cells) to lysosomes within the cell, where those pharmaceuticals are broken down and/or excreted (e.g., by efflux pumps).

See also *Chemopharmacology, Cancer, Cisplatin, Taxol, Paclitaxel, Toxicogenomics, Autophagy, Lysosome, Efflux pump*.

Chimera

An organism consisting of tissues or parts of a diverse genetic constitution. An example of a chimera would be a centaur, the half-man, half-goat figure in Greek mythology.

The word "chimera" is from the mythological creature by that name that possessed the head of a lion, the body of a goat, and the tail of a serpent. The word "chimera" is very general and may be applied to any number of entities. For example, chimeric antibodies may be produced by cell cultures in which the variable, antigen-binding regions are of murine (mouse) origin while the rest of the molecule is of human origin. It is hoped that this combination will lead to an antibody that, when injected into patients, would not elicit *rejection* and not give rise to a lesser immune response by the host against disease(s) the antibody is *aimed* at. See also *Deoxyribonucleic acid (DNA), Genetic engineering, Chimeric DNA, Chimeric proteins, Chimeric antibody, Chimeraplasty, Organism, Antibody, Engineered antibodies*.

Chimeraplasty

A method utilized by man to introduce a gene (from the same or another species) into the DNA of a living organism or cell, via *gene repair* mechanism. Scientists add the desired DNA (gene) to a cell, along with RNA, in a paired-group known as a chimeraplast. The chimeraplast attaches itself to the cell's DNA at the site of the specific gene (to be changed) and *repairs* it utilizing its (new) chimeraplast DNA as a *template*. See also *Gene repair (done by man), Gene, Species, Deoxyribonucleic acid (DNA), DNA repair, Organism, Cell, Chimera, Template, Ribonucleic acid (RNA), Oligonucleotide-mediated mutagenesis*.

Chimeric Antibody

A (genetically engineered) antibody that combines characteristics of antibodies from two different sources. For example, the complementarity-determining (i.e., antigen-binding) portion of an animal antibody (e.g., raised against a specific antigen) with human monoclonal antibody.

The pharmaceutical rituximab (Rituxan™) is a chimeric antibody utilized to treat non-Hodgkin's lymphoma. Its complementarity-determining portion binds to CD20, a receptor found on the surface of B cells in humans who have non-Hodgkin's lymphoma (a cancer of the bone marrow/spleen/lymph nodes). That binding to CD20 induces death of those B cells via apoptosis or humoral immune response. Because bone marrow stem cells (progenitors to B cells) do not have CD20 receptors, rituximab does not bind to them, so after the treatment has ended, those stem cells will again make (noncancerous) B cells.

The pharmaceutical cetuximab (Erbitux™) is a chimeric antibody used to treat certain metastatic colorectal cancers and head and neck cancer. Its complementarity-determining portion binds to an EGF receptor, a receptor found in abundance on the surface of those tumors' cells. That binding to EGF receptors induces tumor cell death via apoptosis or humoral immune response. See also *Antibody, Antigen, Avidity, Monoclonal antibodies (MAb), Chimera, Chimeric proteins, Genetic engineering, Humanized antibody, Cancer, Humoral immune response, EGF receptor, Anti-epidermal growth factor receptor monoclonal antibodies, Tumor, Rituximab*.

Chimeric DNA

(Recombinant) DNA containing spliced genes from two different species.

Transcription/translation of chimeric DNA results in synthesis (by ribosome) of a *chimeric protein* (also known as a fusion protein). See also *Deoxyribonucleic acid (DNA), Gene, Transcription, Translation, Ribosome, Protein, Chimeric proteins, Gene splicing, Species, Recombinant DNA (rDNA), Genetic engineering, Gene fusion*.

Chimeric Molecule

A molecule consisting of diverse constituents (e.g., a peptide and an oligonucleotide). See also *Chimera, Chimeric antibody, Chimeric DNA, Chimeric proteins, Peptide-oligonucleotide conjugates*.

Chimeric Oligonucleotide-Dependent Mismatch Repair

See *Oligonucleotide-mediated mutagenesis, Genome editing*.

Chimeric Proteins

Fused proteins from different species that are produced from the chimeric DNA template. See also *Chimera, Chimeric DNA, Deoxyribonucleic acid (DNA), Antibody, Engineered antibodies, Chimeric antibody, Gene fusion, Peptide-oligonucleotide conjugates.*

Chinese Hamster Ovary Cells

See *Cho cells.*

CHIP

Acronym for "chemical inkjet printer." Such CHIPs are sometimes utilized to manufacture certain microarrays by depositing precise amounts of chemicals (e.g., DNA segments) onto microarray surface (e.g., slide) at specific location(s). See also *Microarray (testing), Deoxyribonucleic acid (DNA), DNA chip, Probe, Hybridization (molecular biology), Hybridization surfaces, Biochips, High-throughput screening (HTS).*

ChIP

Acronym for "Chromatin ImmunoPrecipitation method" (test). It is a test methodology utilized to determine which protein molecules (e.g., transcription factors) bind to specific DNA segments (e.g., regulatory sequence).

The test device (biochip) is created by attaching DNA segments of known sequence to a substrate and then determining which protein molecules (e.g., from a solution passed over the substrate) attach themselves to which DNA segment. See also *Protein, Deoxyribonucleic acid (DNA), Transcription factors, Sequence (of a DNA molecule), Chromatin, Chromatin immunoprecipitation, Substrate (structural), Biochips, cis-acting protein, trans-acting protein, Surface plasmon resonance (SPR), Proteomics, Genomics.*

Chiral Compound

A chemical compound that contains an asymmetrical center and is capable of occurring in two nonsuperimposable mirror images. This phenomenon was first described by Louis Pasteur. "Chiral" is a word derived from the Greek *cheir* (meaning "hand").

For example, human hands may be used to illustrate chirality in that when the left and right hand are held one on top of the other, one thumb sticks out on one side while the other thumb sticks out on the other side. The point is that the same number and type of fingers and thumbs exist in both hands, but their arrangement in space may be different. So it is with the arrangement of a given molecule's (e.g., a drug's) atoms in 3D space. The two are designated as "R" for right-handed and "S" for left-handed (S is from the Latin "sinistro").

Approximately 40% of drugs on the market today consist of chiral compounds. In many chiral drugs, only one type of the molecule is beneficially biologically active (i.e., acts beneficially to control disease, reduce pain, etc.), while the other type of the drug molecule is either inactive or else causes undesired impacts (called "side effects" of the drug mixture). For example, one enantiomer of the drug thalidomide is a potent angiogenesis inhibitor (e.g., halts multiple myeloma and leprosy), but the other enantiomer causes birth defects in babies of pregnant women taking it. See also *Stereoisomers, Angiogenesis, Optical activity, Enantiomers, cis/trans isomerism.*

Chitin

A water-insoluble polysaccharide polymer composed of *N*-acetyl-D-glucosamine molecular units, which is a major constituent of the cell walls of fungi and also forms the exoskeletons and some other parts of arthropods (insects) and crustacea. Shellac is produced from chitin.

Because the lining of the midgut (*stomach*) of certain insect pests is composed at least partially of chitin, genetically engineering a crop plant to produce within its applicable tissues some chitinase (an enzyme that degrades chitin) or the lectin known as HFR-3 (which tightly latches-onto chitin molecules) can help such crop plants to resist being attacked by that particular insect pest. See also *Polysaccharides, Polymer, Chitinase, Fungus, Cell, Lectins, PAMPs, Genetic engineering.*

Chitinase

An enzyme that degrades (breaks down) chitin. It is one of the pathogenesis-related proteins produced by certain plants as a disease-fighting response to entry into plant of pathogenic (i.e., disease-causing) fungi.

Because the lining of the midgut ("stomach") of certain insect pests is composed at least partially of chitin, genetically engineering a crop plant to produce within its applicable tissues some chitinase can help such crop plants to resist being attacked by that particular insect pest.

It (chitinase) is also sometimes produced by certain fungi and actinomycetes that destroy the eggs (i.e., chitin-containing shells) of harmful roundworms. See also *Chitin, Enzyme, Stress proteins, Pathogenesis related proteins, Fungus, Aflatoxin, Genetic engineering.*

Chloroplast Transit Peptide (CTP)

A transit peptide that, when fused to a protein, acts to transport that protein into chloroplast(s) in a plant. Once (both are) inside the chloroplast, the transit peptide is cleaved off the protein and that protein is then free (to do the task it was designed for). For example, the CP4 EPSPS enzyme in genetically engineered glyphosate-resistant soybean [*Glycine max* (L.) *Merrill*] plant is transported into the soybean plant's chloroplasts by the CTP known as "N-terminal petunia chloroplast transit peptide." After (both) reaching the chloroplast, the CTP is cleaved and degraded, so the CP4 EPSPS is then free to do its task (i.e., confer resistance to glyphosate). See also *Peptide, Chloroplasts, Gated transport, Vesicular transport, Transit peptide, Fusion protein, Protein, Soybean plant, CP4 EPSPS, EPSP synthase, Herbicide-tolerant crop.*

Chloroplasts

Specialized chlorophyll-containing photosynthetic organelles (plastids) in eucaryotic cells (i.e., the sites where photosynthesis takes place in plants).

Because there are approximately 100 chloroplasts within each plant cell, and each chloroplast contains approximately 100 copies of the plant's DNA, it is theoretically possible to have 10,000 copies (e.g., of a gene inserted via genetic engineering) coding for a given protein. See also *Eucaryote, Organelles, Cell, Photosynthesis, Chloroplast transit peptide (CTP), Transit peptide, Deoxyribonucleic acid (DNA), Gene, Genetic engineering, Coding sequence, Protein.*

C

CHO Cells

Abbreviation for "Chinese hamster ovary cells." This refers to cell line(s) propagated/grown in cell culture (e.g., in petri dishes) that were originally removed from a Chinese hamster. Such cell culturing of CHO cells has been done by scientists since the 1960s to study genetics, gene expression, nutrition, etc.

Some pharmaceutical proteins (e.g., etanercept) and some enzymes (e.g., PARP) are produced by CHO cells via large-scale cell culture (e.g., fermentation vats, which have internal substrates for the CHO cells). See also *Cell, Cell culture, Mammalian cell culture, Gene, Gene expression, Fusion protein, Etanercept, Substrate (structural), PARP.*

Cholera Toxin

The toxin that is produced by the *Vibrio cholerae* (Latin America) bacteria, a source of food/water-borne gastrointestinal disease.

The cholera toxin has a strong affinity for certain receptors that are present on the surface of gastrointestinal cells. See also *Toxin, Enterotoxin, Conjugate, Immunoconjugate, Receptors, G-proteins.*

Cholesterol

From the Greek word *chole* (bile), it is a sterol (sterol–lipid) that is an essential material for the creation of cell membranes, cell differentiation, and cell proliferation and is a *building block* for certain hormones (progesterone, estrogens, etc.), sterols, and acids used by the body. For example, the bile acids are made in the liver from cholesterol.

Cholesterol is also vital for normal embryonic development (e.g., of humans in the uterus) because it comprises a crucial portion of the *hedgehog proteins* that direct tissue differentiation (of the mammal embryo into various organs, limbs, etc.).

In addition to getting some via dietary intake, cholesterol is synthesized by the human body using the enzyme HMG-CoA reductase. However, deposition of (excess) oxidized cholesterol on the interior walls of blood vessels (in the form of plaque) can result in atherosclerosis and/or coronary heart disease, two often fatal diseases. See also *High-density lipoproteins (HDLPS), Low-density lipoproteins (LDLPs), Cell, Sterols, Phytosterols, Hormone, Sitostanol, Fructose oligosaccharides, Enzyme, Cholesterol oxidase, Coronary heart disease (CHD), High-oleic oil soybeans, Steroid, Lipids, Hedgehog, Differentiation proteins, Campesterol, Stigmasterol, Sitosterol, Sitostanol, Resveratrol, Bile acids, Atherosclerosis, Plaque, CYP46, APOE4, Alzheimer's disease.*

Cholesterol Oxidase

An enzyme that catalyzes the breakdown of cholesterol molecules (causing oxygen consumption in the breakdown process). Because cholesterol molecules are essential for creation and maintenance of cell membranes and some hormones, an excess of cholesterol oxidase can be harmful (e.g., to certain insects).

When the gene (which codes) for cholesterol oxidase is inserted into the genome of the corn (maize) plant, it can enable that plant to resist many of the worm pests (e.g., corn earworm, European corn borer, corn rootworm, black cutworm, armyworm) that attack corn (maize) in the field.

When the gene (which codes) for cholesterol oxidase is inserted into the cotton plant, it can enable that plant to resist weevils and other sucking insects that attack cotton plants in the field. See also *Enzyme, Gene, Genetic engineering, Genome, Corn, Cholesterol, Helicoverpa zea (H. zea), Corn rootworm.*

Choline

Formerly known as vitamin B_4, choline is an essential nutrient that takes part in many of the metabolism processes in the human body. Naturally present in egg yolks, organ meats, dairy products, soybean lecithin, spinach, and nuts, choline

- Is a major component of cell membranes
- Is required by the body to make phospholipids
- Promotes fat metabolism in the liver
- Is used by the liver to make certain choline-based compounds (necessary for the transport of fat from the liver to the rest of the body)
- Is used for the synthesis of high-density lipoproteins (i.e., HDLP, also known as "good" cholesterol) by the liver

It is also utilized by the body in order to synthesize (i.e., *manufacture*) acetylcholine, an important neurotransmitter (substance that transmits nerve impulses).

Because significant choline deficiency can cause liver carcinogenesis, cirrhosis, coronary heart disease, and hypertension and can impair cell signaling, the U.S. government has defined choline to be an essential nutrient and has formally established an *adequate intake* level per day for choline (550 mg/day for men and 425 mg/day for women, per NAS, 1998). In 1998, the U.S. Food and Drug Administration authorized a formal *nutrient content claim (on labels)* for food products and dietary supplements containing appropriate amounts of choline (e.g., those containing soybean lecithin).

One active metabolite of choline is the platelet-activating factor, which is involved in the body's hormonal and reproductive functions. Choline is so important in proper infant development/growth that it is included in manufactured infant formula at the rate of at least 7 mg/100 kcal. See also *Lecithin, Metabolism, Metabolite, High-density lipoproteins (HDLPS), Essential nutrients, Cell, Plasma membrane, Phospholipids, Hormone, Soybean oil, Vitamin, Acetylcholine, Cholinesterase, Neurotransmitter, Fats, Cancer, Coronary heart disease (CHD), Signaling, Homocysteine.*

Cholinesterase

An enzyme that catalyzes the chemical reaction in which the neurotransmitter (i.e., substance that transmits nerve impulses) molecule *acetylcholine* is synthesized (i.e., *manufactured*) from Ac-CoA and choline. See also *Enzyme, Neurotransmitter, Ac-CoA, Choline, Lecithin, Alzheimer's disease, Solanine, Chaconine.*

Chromatids

Copies of a chromosome produced by replication within a living eucaryotic cell during the prophase (i.e., the first stage of mitosis). They are compact cylinders consisting of DNA coiled around flexible rods of histone protein. See also *Chromatin, Eucaryote, Mitosis, Chromosomes, Replication (of virus), Histones, Protein.*

Chromatin

From the Greek word *chroma* meaning color. Named by Walter Flemming in 1882, due to the fact that chromatin's band-like structures stained darkly, chromatin is the complex of DNA and (histone) protein of which the chromosomes are composed. Consisting of fibrous swirls of unraveled DNA molecules in the nucleus of the interphase (i.e., the prolonged period of cell growth between cell division phases) eucaryote cell, chromatin DNA gradually coils itself around flexible rods of histone protein during the prophase (i.e., the first stage of mitosis), forming two parallel compact cylinders (called "chromatids") connected by a knot-like structure (called a "centromere") at their middles. In appearance, they are sort of like two rolls of carpeting standing side by side that are tied together with rope at their middles.

These (recently replicated) cylinders (that are joined at their middles) are homologous chromosomes (i.e., the genes of the two chromosomes are linked in the same linear order within the DNA strands of both chromosomes). While they are still joined at their middles, these paired chromosomes appear X-shaped when photographed by a karyotyper to produce a karyotype.

Chromatin is usually not visible during the interphase of a cell but can be made more visible during all phases by reaction with basic stains (dyes) specific for DNA.

Chromatin modification is a term that refers to any (epigenetic) change in a cell's chromatin that impacts how (or if) a given gene is expressed. See also *Cell, Basophilic, Deoxyribonucleic acid (DNA), Protein, Histones, Chromatids, Chromosomes, Mitosis, Replication (of virus), Centromere, Karyotype, Eucaryote, Karyotyper, Epigenetic, Express, Chromatin remodeling, Chromatin remodeling elements, Short interfering RNA (siRNA), Differentiation pathways.*

Chromatin Immunoprecipitation

Refers to the use in genomics/proteomics of antibodies created to adhere to a given DNA-binding protein (e.g., transcription factor, DNA repair proteins) to find sites where a particular DNA-binding protein will bind to the DNA.

The researcher treats applicable cells with a chemical such as formaldehyde, which causes cross-linking of a cell's DNA and relevant protein-binding molecules. The cells are then broken open, their chromatin is separated out, and the DNA molecule within that chromatin is cut into small fragments. A selected antibody (against one selected DNA-binding protein) is added to the mixture in order to precipitate that protein *along with the DNA (fragment) it is bound to.* Following a chemical reaction that breaks the protein–DNA cross-links, the now-liberated DNA fragments are analyzed to determine precisely the locations on an organism's DNA where each DNA-binding protein attaches. This thereby reveals all DNA points (e.g., genes, promoters) impacted by that DNA-binding protein. See also *Deoxyribonucleic acid (DNA), Gene, Antibody, Organism, Protein, Transcription factors, Promoter, Sequence (of a DNA molecule), Genomics, Proteomics, Chromatin, Histone, Cell, Organism, DNA repair, Gene repair (natural), Sliding clamps.*

Chromatin Immunoprecipitation Method

See *Chromatin immunoprecipitation.*

Chromatin Modification

See *Chromatin, DNA methylation.*

Chromatin Remodeling

Refers to the reshaping (at molecular scale) of chromatin (i.e., organism's complex of DNA and histone protein) that *alters which specific genes in that organism's DNA subsequently get expressed.* Can be caused by short interfering RNA, certain transcription activators, acetylation of histone, methylation of histone, sumoylation of histone, etc. See also *Chromatin, Deoxyribonucleic acid (DNA), Gene, Gene expression, Epigenetic, Epigenetic marks, Gene silencing, Gene splicing, Short interfering RNA (siRNA), Silencing, Transcription activators, Histones, Methylated, Repression (of gene transcription or translation), Small ubiquitin-related modifier, Differentiation pathways, Apoptosis.*

Chromatin Remodeling Elements

See *Chromatin remodeling, Transcription activators, Short interfering RNA (siRNA).*

Chromatography

Coined by Mikhail S. Tswett in 1906, this word refers to a process by which complex mixtures of different molecules may be separated from each other. This is accomplished by subjecting the mixture to many repeated partitionings between a flowing phase and a stationary phase. Chromatography constitutes one of, if not, *the* most fundamental separation techniques used in the biochemistry/biotechnology arena to date. See also *Polyacrylamide gel electrophoresis (PAGE), Substrate (in chromatography), Affinity chromatography, Monolithic chromatography substrates, Biotechnology, Agarose, Gel filtration.*

Chromosomal Packing Unit

See *Nucleosome.*

Chromosomal Translocation

See *Jumping genes.*

Chromosome Map

See *Linkage map.*

Chromosome Painting

See *Fluorescence in situ hybridization (FISH).*

Chromosome Walking

A methodology for determining the location of, and sequencing of, *a given gene* (within an organism's DNA) by sequencing (specific DNA sequences that overlap and span collectively) that gene's location within the organism's DNA. See also *Gene, Deoxyribonucleic acid (DNA), Sequence (of a DNA molecule), Organism, Chromosomes.*

Chromosomes

Discrete units of the genome carrying many genes, consisting of (histone) proteins and a very long molecule of DNA. Found in the nucleus of every plant and animal cell. See also *Genome, Gene, Genetic code, Chromatin, Chromatids, Karyotype, Karyotyper, "Designer" chromosome, Philadelphia chromosome.*

Chronic Heart Disease

See *Coronary heart disease (CHD).*

C

Chronic Inflammation

Refers to inflammation (i.e., the body's natural response to infection or injury) that does not stop after the initial cause (infection/injury) has disappeared. Chronic inflammation is present in some diseases such as coronary heart disease, gastric ulcers, diabetes, Crohn's disease, periodontitis (periodontal disease), and chronic kidney disease and results when certain immune cells (e.g., neutrophils, macrophages) release reactive chemicals such as hypochlorous acid (HOCl), reactive oxygen species, CO_3, NO_2, HOBr, and N_2O_3. Although those reactive chemicals are intended to kill invading pathogens during an infection, their *chronic* release by neutrophils/macrophages (e.g., after an infection is over, or as a result of plaque deposition on walls of blood vessels) can harm body tissues by causing inflamed blood vessel linings, swollen joints, and damaged DNA (potentially leading to certain cancers such as breast cancer).

In people who have an applicable cancer, their body often overproduces interleukin-6 (IL-6) (i.e., a cytokine that normally stimulates several different types of immune system cells in response to infections or injuries) because those cancerous cells cause decreased presence of the protein known as SOCS3. SOCS3 is present in normal cells, where it functions as an *off-switch* in a feedback loop involving the IL-6, thereby normally halting that inflammation promoter after IL-6's work (e.g., combatting the infection) is completed. However, in the case of late-stage/metastatic triple-negative breast cancer, IL-6 levels are 40 times higher than normal (thereby adding to the chronic inflammation aspect of the cancer).

Consumption of adequate amounts of certain anti-inflammatory nutrients such as linolenic acid helps dampen inflammatory reactions within the human body via blocking the formation of certain compounds that promote inflammation such as omega-6 (*n*-6)-derived eicosanoids, cytokines, platelet-activating factor, and C-reactive protein.

Consumption of resveratrol helps dampen inflammatory reactions within the human body via the resveratrol molecule acting as such an agonist-binding partner with the estrogen receptor (without stimulating estrogenic cell proliferation) to beneficially control the body's inflammation response. See also *Diabetes, Crohn's disease, Neutrophils, Macrophage, Reactive oxygen species, Protein, Cell adhesion proteins, Cancer, Interleukin-6 (IL-6), Beta-conglycinin, Tight junction proteins, N-3 fatty acids, Resveratrol, Curcumin, Atheroserosis, Plaque, Coronary heart disease (CHD), Nanoparticles, C-reactive protein (CRP), Agonists.*

Chymosin

Also known as rennin. It is an enzyme used to make cheeses (from milk). Chymosin occurs naturally in the stomachs of calves and is one of the oldest commercially used enzymes. Chymosin (rennin) is chemically similar to renin, an enzyme that plays an important role in regulating blood pressure in humans. See also *Renin.*

Cilia

Protein-based structures that occur in certain cells of both the plant and animal world. Cilia are very tiny hair-like structures and occur in large numbers on the outside of certain cells. In higher organisms such as man, they usually function to move extracellular material along the cell surface. An example is the *sweeping out of foreign matter* action of cilia in the bronchial tubes in which very small particles are moved into the throat to be expelled or swallowed.

In man, the cilia on cartilage cells swiftly increase in length by 50% when exposed to the inflammatory protein (cytokine) known as interleukin-1 (IL-1). Thus, these cartilage cells' primary cilia are linked to the inflammatory response.

Some lower organisms use their cilia for locomotion (swimming). Cilia are used in the swimming motion of bacteria toward sources of nutrients in a process called "chemotaxis." Cilia are shorter and occur in larger numbers per cell than flagella. Singular: cilium. See also *Chemotaxis, Microtubules, Centrosomes, Flagella, Dynein, Interleukin-1 (IL-1), Inflammatory response.*

Ciliary Neurotrophic Factor (CNTF)

A human protein that has been shown to help the survival of those cells in the nervous system that act to convey sensation and control the function of muscles and organs. CNTF was approved by the U.S. FDA to treat amyotrophic lateral sclerosis (also known as Lou Gehrig's disease) in 1992. Amyotrophic lateral sclerosis causes the victim's muscles to degenerate severely, and it affects approximately 30,000 people per year in the United States. CNTF *might* prove useful for treating Alzheimer's Disease and/or other human neurological diseases.

Research published in 2006 reported that CNTF also activates an enzyme within muscles that increases the metabolism off fats and sugars. See also *Protein, Cell, Nerve growth factor (NGF), Food and Drug Administration (FDA), Enzyme, Fats, Sugar molecules, Metabolism.*

cis/trans Isomerism

A type of geometrical isomerism found in alkenic systems in which it is possible for each of the doubly bonded carbons to carry two different atoms or groups. Two similar atoms or groups may be on the same side (i.e., *cis*) or on opposite sides (i.e., *trans*) of a plane bisecting the alkenic carbons and perpendicular to the plane of the alkenic systems. See also *Isomer, Chiral compound, Trans fatty acids.*

cis/trans Test

Assays (determines) the effect of relative configuration on the expression of two (gene) mutations. In a double heterozygote, two mutations in the same gene show mutant phenotype in the *trans* configuration and wild (phenotype) in the *cis* configuration. The phenotypic distinction is referred to as the position effect.

See *Gene, Phenotype, cis-acting protein, Position effect, Heterozygote, Mutation.*

cis-Acting Protein

A *cis*-acting protein has the exceptional property of acting only on the molecule of DNA from which it was expressed. See also *trans-acting protein, Deoxyribonucleic acid (DNA).*

Cisgenesis

Refers to the genetic modification of an organism via insertion of a gene(s) from a sexually compatible (i.e., crossable) organism that is the same species or a closely related species. The inserted gene would include its (native) introns, promoter, and terminator in their normal SENSE orientation. See also *Gene, Intron, Promoter, Terminator, Sense, Intragenesis.*

Cisplatin

First synthesized by Michel Peyrone during 1845, it is a platinum-containing drug that is used in chemotherapy regimens against certain types of cancer tumors (e.g., testicular cancer, ovarian cancer, bladder cancer, lung cancer).

Cisplatin works against (tumor) cells by binding to the cell's DNA and generating intrastrand cross-links (between the two strands of the DNA molecule). These intrastrand cross-links prevent replication and cause cell death. See also *Chemopharmacology, Chemotherapy, Cancer, Deoxyribonucleic acid (DNA), Replication fork, Replication (of DNA).*

Cistron

Synonymous with *gene*, it refers to a specific DNA sequence that codes for the synthesis (by ribosome) of a single protein (polypeptide molecular chain). See also *Gene, Deoxyribonucleic acid (DNA), Protein, Ribosome.*

Citrate Synthase

The enzyme that is utilized (e.g., by plants) to synthesize (i.e., create) citric acid. See also *Enzyme, Citric acid.*

Citrate Synthase (CSb) Gene

A bacterial gene that is utilized by certain bacteria (e.g., *Pseudomonas*) to code for (i.e., cause to be produced by bacterium possessing that gene) the enzyme known as citrate synthase. That enzyme is utilized to synthesize (i.e., create) citric acid.

In 1996, Luis Herrera-Estrella discovered that inserting the CSb gene from *Pseudomonas aeruginosa* into certain plants caused those plants to produce up to 10 times more citrate in their roots and to release up to 4 times more citric acid from those roots into the surrounding soil (thus decreasing aluminum toxicity via chemically "binding" aluminum ions that are present in some soils). Such soil aluminum, which slows plant growth and decreases crop yields, is present to a certain degree in approximately one-third of the Earth's arable land (e.g., in the country of Colombia, it affects 70% of the arable land). See also *Gene, Enzyme, Express, Citrate synthase, Ion, Citric acid.*

Citrate Synthase Gene

A gene that codes for (i.e., causes to be produced by an organism possessing that gene) the enzyme known as "citrate synthase." See also *Gene, Enzyme, Express, Citrate synthase, Citric acid.*

Citric Acid

A tricarboxylic acid occurring naturally in plants, especially citrus fruits. It is used as a flavoring agent, as an antioxidant in foods, as an animal feed ingredient, and as a sequestering agent. The commercially produced form of citric acid melts at 153°C (307°F). Citric acid is found in all cells, its central role is in the metabolic process.

Some plants naturally release citric acid from their roots into the surrounding soil in order for that citric acid to chemically *bind* aluminum ions that are present in some soils. Such aluminum, which slows plant growth and decreases crop yields, is present to a certain degree (which causes at least some crop yield reduction) in approximately one-third of the world's arable land. For example, 70% of the agricultural land in the country of Colombia possesses harmful amounts/conditions of aluminum to damage crops.

Corn (maize) yields are reduced up to 80% by such aluminum in soils. Soybeans, cotton, and field bean yields are also reduced. See also *Metabolism, Acid, Cell, Citrate synthase, Citrate synthase gene, Citrate synthase (Csb) gene, Citric acid cycle, Metabolite, Cell, Ion, Soybean plant, Corn, Probiotics.*

Citric Acid Cycle

Also known as the tricarboxylic acid cycle (TCA cycle because the citric acid molecule contains three [tri] carboxyl [acid] groups). Also known as the Krebs cycle after Hans A. Krebs, who first postulated the existence of the cycle in 1937 under its original name of "citric acid cycle." A cyclic sequence of chemical reactions that occurs in almost all aerobic (air requiring) organisms. A system of enzymatic reactions in which acetyl residues are oxidized to carbon dioxide and hydrogen atoms and in which formation of citrate is the first step. See also *Citric acid, Citrate synthase, Citrate synthase gene, Citrate synthase (Csb) gene, Acid, Aerobic, Metabolism, Enzyme, Oxidation.*

Citrinin

A mycotoxin initially isolated in 1931 from a culture of *Penicillium citrinum*. It has since been found to be produced by a number of other fungal species that are found or used in the production of human foods, such as grain, cheese, sake, and certain red pigments. Those fungal species include *Aspergillus niveus, Aspergillus ochraceus, Aspergillus oryzae, Aspergillus terreus, Monascus ruber, Monascus purpureus*, and *Penicillium camemberti*. See also *Mycotoxins.*

c-kit Genetic Marker

See *Genetic marker, Fluorescence in situ hybridization (FISH).*

CKR-5 Proteins

See *Human immunodeficiency virus type 1 (HIV-1), Human immunodeficiency virus type 2 (HIV-2), Receptors, Protein.*

CLA

Abbreviation for conjugated linoleic acid. See *Conjugated linoleic acid (CLA).*

Clades

The taxonomic subgroups within cladistics. See also *Cladistics.*

Cladistics

Initially popularized by Willi Hennig's 1950 book entitled *Phylogenetic Systematics*, cladistics is a system of taxonomic classification of organisms (and/or their specimens) that is based upon (determined) similar lines of selected shared traits. See also *Clades, Type specimen, Genetics, Biology, Species, Systematics, American type culture collection (ATCC), Trait.*

C

Clathrin

A protein that forms itself into a lattice-like structure on the surface of the cell membrane forming a vesicle within a cell during the process of endocytosis. See *Endocytosis*.

Clathrin-Dependent Endocytosis

See *Endocytosis*.

Cleistogamous

Refers to *self-pollinating* plants. See also *Monoecious*.

CLIA

Acronym for **c**hemi**l**uminescent **i**mmuno**a**ssay. See *Chemiluminescent assay (CLIA)*.

Click Chemistry

Invented by K. Barry Sharpless, it is a category of chemistry (e.g., a *family* of related chemical reactions) that utilizes heteroatom links to hook together specific molecular units (modules) into longer molecular structures, in a modular manner.

Click chemistry can be utilized to attach fluorescent labels (i.e., molecular units that fluoresce when illuminated by light of applicable wavelength) to the surface of living cells. That facilitates subsequent imaging of those cells by scientists who are investigating functions of cells.

Click chemistry can be utilized to make certain small molecule *modules* capable of passing through the blood–brain barrier (BBB) that Matthew Disney and colleagues developed in 2014 to bind to adjacent portions of the RNA defect known as a "tetranucleotide repeat" in which a series of four nucleotides is repeated more times than normal within an individual's genetic code. When that tetranucleotide repeat causes applicable RNA splicing abnormalities, it results in the progressive muscle-weakening disease known as myotonic dystrophy type 2. However, via this use of click chemistry, the Disney-created small molecule modules are able to pass through the BBB, bind to the defective RNA, and thereby reverse the effect of the disease.

Click chemistry can be utilized to attach a targeting molecule (e.g., folic acid) onto a nanoparticle (e.g., nanocapsules, nanoshells) that has been filled with an applicable pharmaceutical (e.g., tumor necrosis factor). Because many cancerous tumors consume very large amounts of folic acid during their rapid growth, folic acid can be utilized as a *targeting molecule* (e.g., attached to the surface of such therapeutic nanoparticles) to deliver the tumor necrosis factor (TNF) directly to the tumor. The TNF can then act to disrupt formation of the new vasculature (blood vessels) needed by the tumor for blood supply. See also *Fluorescence, Fluorescence mapping, Label (fluorescent), Target (of a therapeutic agent), Blood-brain barrier (BBB), Ribonucleic acid (RNA), RNA splicing, Folic acid, Cancer, Nanoparticles, Nanocapsules, Tumor necrosis factor (TNF)*.

Clinical Trial

One of the final stages in the collection of data (for drug approval prior to commercialization) in which the new drug is tested in human subjects. Used to collect data on effectiveness, safety, and required dosage. See also *Phase I clinical testing, Food and Drug Administration (FDA), Koseisho, Bundesgesundheitsamt (BGA), Committee on Safety in Medicines, Committee for Proprietary Medicinal Products (CPMP)*.

CLL

Acronym for chronic lymphocytic leukemia. See *Rituximab, Ibrutinib*.

Clone (a Molecule)

To create copies of a given molecule via various methods. See also *Polymerase chain reaction (PCR), Monoclonal antibodies (MAb), Cocloning, Antibody, cDNA clone*.

Clone (an Organism)

A group of individual organisms (or cells) produced from one individual cell through asexual processes that do not involve the interchange or combination of genetic material. As a result, members of a clone have identical genetic compositions.

For example, many plants reproduce asexually (i.e., without sex) via a process known as apomixis.

For example, man has reproduced numerous trees via grafting of a branch from a valuable tree into a less-valuable tree (which subsequently provides nutrients, etc. to that ingrafted branch so it can continue to grow, flower, and reproduce). Via such grafting, every navel (seedless) orange tree on Earth is an exact genetic copy of one bud mutation that occurred in 1820 on a sour orange tree in Bahia, Brazil.

Protozoa, bacteria, and some animals (e.g., the anemone *Anthopleura elegantissima*) can reproduce asexually (i.e., without sex) by a process called "binary fission." In binary fission a single-celled organism undergoes cell division. The result is two cells with identical genetic composition. When these two identical cells undergo division, the result is four cells with identical genetic composition. These identical offspring are all members of a clone. The word "clone" may be used either as a noun or a verb.

Scientists have cloned some adult mammals via nuclear transfer. In that process, the nucleus of an oocyte is removed and replaced with a nucleus taken out of another conventional somatic (adult's body) cell. That oocyte can then grow up to become a clone of the (adult) animal. See also *Organism, Apomixis, Bacteria, Cell, Oocytes, Somatic cells, Mutation, Nuclear transfer, Reprogramming*.

Clostridium

A genus of bacteria. Most are obligate anaerobes, and form endospores. See also *Anaerobe, Endospore*.

CMC

See *Critical micelle concentration*.

CML

Abbreviation for *chronic myelogenous leukemia* (also known as *chronic myeloid leukemia*, or *chronic myelocytic leukemia*). See *Gleevec™*.

CMV

See *Cytomegalovirus (CMV)*.

CNHs

Acronym for *carbon nanohorns*. See *Carbon nanohorns*.

CNP

Acronym for *copy number polymorphisms*. See *Copy number polymorphisms*.

CNTF

See *Ciliary neurotrophic factor (CNTF)*.

CNTs

Acronym for *carbon nanotubes*. See *Carbon nanotubes*.

CNV

Acronym for *copy number variant* or *copy number variation*. See *Copy number variant*, *Copy number variation*.

CoA

See *Coenzyme A*.

Co-Chaperonin

A protein molecule inside living cells that *works together* with applicable chaperonin(s) to help ensure proper folding of the (new) protein molecules that are synthesized (*manufactured*) in the cell's ribosomes. See *Chaperonins*, *Protein*, *Protein folding*, *Cell*, *Ribosomes*, *Conformation*.

Coccus

A spherical-shaped bacterium. See also *Bacillus*.

Cocloning (of Molecules)

The additional (accidental) cloning (i.e., copying) of extramolecular fragments, other than the desired one, that sometimes occurs when a scientist is attempting to clone a molecule. See also *Clone (a molecule)*, *Polymerase chain reaction (PCR)*, *Q-beta replicase technique*.

Codex Alimentarius

See *Codex alimentarius commission*.

Codex Alimentarius Commission

An international regulatory body that is part of the United Nations' Food and Agriculture Organization (FAO), it is one of the three international SPS (sanitary and phytosanitary) standard-setting organizations that is recognized by the World Trade Organization. It was created in 1962 by the UN's FAO and the World Health Organization (WHO). It has 165 member nations.

In the Latin language, *Codex Alimentarius* means "food law" or "food code." The Codex Alimentarius Commission is responsible for execution of the Joint FAO/WHO Food Standards Program. The Codex Alimentarius standards are a set of international food mandates that have been adopted by the commission. The commission is composed of delegates from member country governmental agencies. The Codex Secretariat is headquartered in Rome, Italy.

The commission periodically determines and then publishes a list of food ingredients and maximum allowable levels that it deems safe for human consumption (known as the *Codex Alimentarius*). See also *Maximum residue level (MRL)*, *SPS*, *International Plant Protection Convention (IPPC)*, *International Office of Epizootics (OIE)*, *World Trade Organization (WTO)*.

Coding Region

See *Coding sequence*.

Coding Sequence

The region within a DNA molecule (i.e., between the start and stop codons) that encodes the amino acid sequence of a protein, or for a specific microRNA. See also *Genetic code*, *Informational molecules*, *Gene*, *Messenger RNA (mRNA)*, *Base (nucleotide)*, *Control sequences*, *Codon*, *MicroRNAs*.

Codon

A triplet of nucleotides (three nucleic acid units [residues] in a row) within either DNA or messenger RNA that codes for an amino acid (triplet code) or a termination signal. See also *Genetic code*, *Deoxyribonucleic acid (DNA)*, *Termination codon (sequence)*, *Amino acid*, *Nucleotide*, *Informational molecules*, *Messenger RNA (mRNA)*, *Leader sequence (mRNA)*.

Coenzyme

A nonproteinaceous organic molecule required for the action of certain enzymes. The coenzyme contains as part of its structure one of the vitamins. This is why vitamins are so critically important to living organisms. Sometimes the same coenzyme is required by different enzymes that are involved in the catalysis of different reactions. By analogy, a coenzyme is like a part of a car, such as a tire, that can be identified in and of itself and that can, furthermore, be removed from the car. The car (enzyme), however, must of necessity have the tire in order to carry out its prescribed function. Coenzymes have been classified into two large groups: fat soluble and water soluble. Examples of a few water-soluble vitamins are: thiamin, biotin, folic acid, vitamin C, and vitamin B_{12}. Examples of fat-soluble vitamins are: vitamins A, D, E, and K. See also *Enzyme*, *Catalyst*, *Holoenzyme*, *Vitamin*, *Polypeptide (protein)*, *Biotin*.

Coenzyme A

A water-soluble vitamin known as pantothenic acid. A coenzyme in all living cells. It is required by certain condensing enzymes and functions in acyl-group transfer and in fatty acid metabolism. Abbreviated CoA. See also *Enzyme*, *Fats*, *Fatty acid*.

Coenzyme Q10

A name sometimes utilized for ubiquinone, as dietary ingredient. See also *Ubiquinone*.

Cofactor

A nonprotein component required by some enzymes for activity. The cofactor may be a metal ion or an organic molecule called a "coenzyme." The term "cofactor" is a general term. Cofactors are generally heat stable. See also *Coenzyme, Holoenzyme, Molecular weight*.

Cofactor Recycle

The regeneration of a spent cofactor by an auxiliary reaction such that it may be reused many times over by a cofactor-requiring enzyme during a reaction. See also *Cofactor, Holoenzyme, Enzyme*.

Coffee Berry Borer

Refers to the pest insect *Hypothenemus hampei*, which attacks berries of the coffee tree (Coffee Arabica, Coffea canephora). See also *Amylase inhibitors, Coffee tree*.

Coffee Tree

Refers to the specific plants:

- *Coffea canephora*, whose berries are utilized to make approximately 30% of the world's coffee production
- *Coffee arabica*, whose berries are utilized to make the majority of the world's coffee, and which make a resultant coffee possessing a less acidic taste and lower caffeine than *Coffea canephora*

See also *Coffee berry borer, Caffeine*.

Cohesin

Refers to a protein molecule that, together with another protein known as *mediator*, forms a protein complex (structure) that helps a cell's DNA form into the specific loop that is necessary for the applicable gene(s) in the DNA to be activated that control that particular cell's state (e.g., the tissue it has differentiated into, if the cell is no longer in its embryonic state).

Additionally, Cohesin also helps to hold chromosomes together. See also *Protein, Cell, Deoxyribonucleic acid (DNA), Cohesin, Loop, Gene, Activator (of gene), Expressivity, Cell differentiation, Embryonic stem cells, Pluripotent stem cells, Differentiation, Chromosomes*.

Cohesive Ends

See *Sticky ends*.

Cohesive Termini

See *Sticky ends*.

Colchicine

Discovered in 1937, it is a chemical (alkaloid) that can be extracted from certain members of the lily family of plants (e.g., *Colchicum autumnale*, autumn crocus, or meadow saffron). It has sometimes been used as an anti-inflammatory, to try to treat gout in humans.

Colchicine has also been used by some plant breeders to induce mutations in crop plants (e.g., by soaking seeds in it) in order to create crop plant varieties with new traits. That happens because colchicine prevents chromosomes from separating during the anaphase of mitosis, thereby causing the cell to become tetraploid (i.e., four copies of each chromosome). Such induced polyploidy (i.e., extra copies of chromosomes in the cells of breeding "parents") can be utilized in crop-breeding programs to speed up the rate at which new crop varieties are produced/commercialized. See also *Alkaloids, Mutation breeding, Traditional breeding methods, Trait, Chromosomes, Induced polyploidy, Mitosis, Tetraploid*.

Cold Acclimation

See *Cold hardening*.

Cold Acclimatization

See *Cold hardening*.

Cold Hardening

A process of acclimatization in which certain organisms produce specific proteins that protect them from freezing to death during the winter. Among other organisms, the common housefly, the *Arabidopsis thaliana* plant, the fruit fly *Drosophila*, and *no-see-ems* (i.e., *Culicoides variipennis*) can produce these proteins (e.g., during the gradually decreasing temperatures of a typical autumn season in North America). The amount of such proteins produced within their bodies is proportional to the severity and duration of the cold experienced.

For example, prior to cold hardening, *Culicoides variipennis* insects usually die after exposure for 2 h to a temperature of 14°F (−10°C). If those insects are first exposed for 1 h to a temperature of 41°F (5°C), approximately 98% of these insects can then survive exposure for 3 days to a temperature of 14°F (−10°C).

In certain plants, such exposure to cold causes oxidative stress. That oxidative stress then can initiate the activation of the mitogen-activated protein kinase cascade, resulting in production of several *stress responsive proteins* (e.g., heat-shock proteins). Those *stress proteins* help protect such plants from cold temperatures. See also *Acclimatization, Protein, Low-tillage crop production, No-tillage crop production, Drosophila, Arabidopsis thaliana, CBF1, Transcription factors, Linoleic acid, Mitogen-activated protein kinase cascade, Oxidative stress, Stress proteins*.

Cold-Shock Protein B

Refers to a cold-shock protein (naturally produced in the bacterium *Bacillus subtilis*) that, when inserted into the DNA of a corn (*Zea mays* L.) plant, confers resistance to drought and other environmental stresses. See also *Cold-shock protein, Protein, Deoxyribonucleic acid (DNA), Corn*.

Cold Tolerance

See *Cold hardening*.

Cold-Shock Protein

Refers to particular chaperone protein molecules that are expressed by cells in an organism exposed to low-environmental temperatures, to protect living cells (from freeze damage).

For example, at low temperatures, *Escherichia coli* bacteria sometimes express CspA, a cold-shock protein that protects those *E. coli* bacteria from (some) freeze damage.

For example, at low temperatures, *Bacillus subtilis* bacteria sometimes express CspB, a cold-shock protein that protects those *B. subtilis* bacteria from (some) freeze damage. See also *Cold hardening, Chaperones, Protein, Organism, Bacteria, cspB gene, Bacillus subtilis (B. subtilis), Cold shock protein B.*

Colicins

Proteins produced by *Escherichia coli* (*E. coli*) that are toxic (primarily) to other closely related strains of bacteria. The particular *E. coli* that produce a given colicin are generally unaffected by the colicin that they produce. See also *Bacteriocins, Bacteriology, Strain, Bacteria, Protein, Toxin, Escherichia coliform (E. coli).*

Colinearity

See *Co-linearity.*

Collagen

The major structural protein in connective and bone tissue. It is instrumental in wound healing (stimulated by fibroblast growth factor, platelet-derived growth factor, and insulin-like growth factor-1). See also *Protein, Fibroblast growth factor (FGF), Platelet- derived growth factor (PDGF), Insulin-like growth factor-1 (IGF-1), Inhibition.*

Collagenase

An enzyme that catalyzes the cleavage of collagen. One example of this is when bacteria in the mouth cause production of collagenase that then cleaves (i.e., breaks down) the collagen that holds teeth in place. Some cancers use collagenase to break down connective tissues in the body they inhabit, to enable the cancers to form the (new) blood vessels that nourish those cancers and help those cancers to spread through the body. Collagenase may also be responsible indirectly for certain autoimmune diseases such as arthritis, via breaking down the protective proteoglycan coat that covers cartilage in the body. See also *Stromelysin (MMP-3), Proteolytic enzymes, Enzyme, Collagen, Cancer, Autoimmune disease.*

Collective Swimming

See *Bacillus subtilis (B. subtilis).*

Co-Linearity

Refers to when the DNA segment(s) that are common to two different organisms (e.g., rice and maize/corn) are present in the same linear order within their respective DNA molecules (i.e., when one overlooks other *inserted/deleted segments*, sometimes called "indels" or "in/dels"). See also *Deoxyribonucleic acid (DNA), Sequence (of a DNA molecule), Organism, Corn, Gene.*

Colony

A growth of a group of microorganisms derived from one original organism. After a sufficient growth period, the growth is visible to the eye without magnification. See also *Microorganism.*

Colony Hybridization

A technique using *in situ* hybridization to identify bacterial colonies carrying inserted DNA that is homologous with some particular sequence (probe). See also *DNA probe, Homology, In situ, Regulatory sequence.*

Colony-Stimulating Factors (CSFs)

Specific glycoprotein growth factors required for the proliferation and differentiation of hematopoietic progenitor cells. Different CSFs stimulate the growth of different cells. See also *Macrophage colony-stimulating factor (M-CSF), Granulocyte colony-stimulating factor (G-CSF), Granulocyte-macrophage colony-stimulating factor (GM-CSF), Epidermal growth factor (EGF), Fibroblast growth factor (FGF), Hematologic growth factors (HGF), Insulin-like growth factor-1 (IGF-1), Megakaryocyte stimulating factor (MSF), Nerve growth factor (NGF), Platelet-derived growth factor (PDGF), Transforming growth factor-alpha (TGF-alpha), Transforming growth factor-beta (TGF-beta).*

Combinatorial Biology

A term used to describe the set of DNA technologies that are utilized to generate a large number of samples of new chemicals (metabolites) via creation of nonnatural metabolic pathways. This collection of samples thus generated is called a "library," and the samples are then tested for potential use (e.g., for therapeutic effect, in the case of pharmaceutical). These technologies enable greater efficiency in a pharmaceutical researcher's screening process for drug discovery. See also *Combinatorial chemistry, Target, Molecular diversity, Metabolism, Intermediary metabolism, Metabolite, Receptors.*

Combinatorial Chemistry

A term used to describe the set of technologies that are utilized to generate a large number of samples of (new) chemicals, which are then tested (screened) for potential use (e.g., for therapeutic effect, in the case of a pharmaceutical). These large numbers of chemical samples, thus generated, are called a "library" and are screened (e.g., for therapeutic effect) via a variety of laboratory, biosensor, computational, receptor, or animal tests.

Combinatorial chemistry was made feasible by the fact that, during the 1980s, H. Mario Geysen developed a methodology to synthesize arrays of peptides on pin-shaped solid supports. In addition, Richard A. Houghten developed a technique for the creation of peptide libraries in small mesh "bags" by solid-phase parallel synthesis, thereby enabling automation of the process (in the early 1990s).

For a library that is used for new drug (candidate) screening, high diversity in molecular structure among the chemicals in the library is desired, to increase the efficiency of the screening process. One method used to measure diversity of the molecular structure among samples in a library is called "molecular fingerprinting." If two samples are identical in molecular structure, the "fingerprint" coefficient is 1.0. If two samples are totally dissimilar in molecular structure, the coefficient is 0. The diversity of a library is measured by

C

comparing each sample's molecular structure to that of all the others in the library. See also *Combinatorial biology, Target, Molecular diversity, Receptors, Biosensors (electronic), Peptide, Synthesizing (of proteins), Biochips, High-throughput screening, Target–ligand interaction screening.*

Combinatorics

See *Combinatorial chemistry.*

Combining Site

The site on an antibody molecule that locks (binds) onto an epitope (hapten). See also *Antibody, Epitope, Engineered antibodies, Nanobodies, Hapten, Catalytic antibody.*

Commensal

A term that literally means *eating at the same table*; it is often used to refer to organisms such as

- The house mouse (*Mus musculus*), etc. that tends to thrive alongside/among humans
- The *Bacteroides thetaiotaomicron* bacteria that live within the human gastrointestinal tract and induce appropriate glycosylation (fucosylation) of intestinal epithelial cells
- Certain strains of the bacteria *Pseudomonas fluorescens* that thrive living on the surfaces of plants and can even help protect those plants against pathogens via the production of antibiotics, etc.

The commensal aspect tends to be species specific. For example, numerous strains of *Salmonella* bacteria can live within the intestine of an adult cow without harming that cow, but would be pathogenic (i.e., disease-causing) in a human's intestine.

Another example is that the *E. coli 0157:H7* strain of *Escherichia coliform* bacteria can live within the digestive system of an adult cow without harming that cow, but would be pathogenic (i.e., disease-causing) in a human's digestive system. However, hundreds of *other* strains of *Escherichia coliform* bacteria live within the digestive system of humans, without causing harm to the human body (i.e., those hundreds of strains of *Escherichia coliform* bacteria are commensal).

In some people, commensal intestinal bacteria can lead to an autoimmune disease. For example, certain individual humans possessing specific alleles of the *NOD2/CARD15* gene have been shown to be likely to mount an inappropriate immune system response against their own bowel tissues, which are in intimate contact with commensal intestinal bacteria. That can result in such autoimmune diseases as Crohn's disease or ulcerative colitis. See also *Organism, Microorganism, Bacteria, Glycosylation (to glycosylate), Pseudomonas fluorescens, Salmonella typhimurium, Salmonella enteritidis, Pathogen, Pathogenic, Strain, Escherichia coliform (E. coli), Escherichia coliform 0157:H7, Autoimmune disease, Gene, Allele, Immune response, Antibiotic, Helicobacter pylori.*

Commission E Monographs

Documents published by the government of Germany, which detail the proven safety and efficacy of certain phytochemical-containing herbs (approved by the German government).

For example, consumption of *Saint John's wort* (a plant native to Europe) is approved in Germany for treatment of depressive mood disorders, anxiety, and nervous unrest. See also *Phytochemicals, Saint John's wort.*

Commission of Biomolecular Engineering

An agency of the French government, established to oversee and regulate all genetic engineering activities in the country of France. See also *Genetic engineering, IOGTR, Recombinant DNA Advisory Committee (RAC), ZKBS (Central Committee on Biological Safety), Indian Department of Biotechnology, Gene technology regulator (GTR), Gene Technology Office.*

Committee for Proprietary Medicinal Products (CPMP)

The European Union's (EU's) scientific advisory organization dealing with new human pharmaceuticals approval. Its recommendations (e.g., to either approve or not approve a new product) are usually adapted by the European Medicines Evaluation Agency (EMEA), to which the CPMP reports.

Within 60 days of a CPMP "approval for recommendation" being adopted by the EMEA, each of the EU's member countries must advise the EMEA of its progress toward a regulatory decision on that pharmaceutical's submission for approvals. See also *Food and Drug Administration (FDA), Koseisho, European Medicines Evaluation Agency (EMEA), Committee on Safety in Medicines, Bundesgesundheitsamt (BGA).*

Committee for Veterinary Medicinal Products (CVMP)

The European Union's scientific advisory organization dealing with approvals of new medicinal products intended for use in animals. Its recommendations (e.g., to either approve, or not approve a new product) are usually adopted by the European Medicines Evaluation Agency. See also *Committee for Proprietary Medicinal Products (CPMP), Food and Drug Administration (FDA), Koseisho, Committee on Safety in Medicines, Medicines Control Agency (MCA), EMEA, Bundesgesundheitsamt (BGA).*

Committee on Safety in Medicines

The British Government agency that must approve new pharmaceutical products for sale within the United Kingdom. In concert with the Medicines Control Agency, it regulates all pharmaceutical products in the United Kingdom. It is the equivalent of the U.S. Food and Drug Administration. See also *Food and Drug Administration (FDA), Medicines Control Agency (MCA), Committee for Proprietary Medicinal Products (CPMP), Koseisho, NDA (to koseisho), IND, Bundesgesundheitsamt (BGA), EMEA.*

Community Plant Variety Office

An agency of the European Union that was established by Council Regulation 2100/94 and is located in Angers, France. It applies the Union for Protection of New Varieties of Plants rules across all countries of the European Union when a plant breeder registers a new plant variety at the Community Plant Variety Office. Thus, it confers/protects the plant breeder's rights across the entire European Union in a manner analogous to the way the European Patent Office confers patent rights (for patented inventions) across the entire European Union. See also *Union for Protection of New*

Varieties of Plants (UPOV), Plant breeder's rights (PBR), European Patent Office (EPO), Plant Variety Protection Act (PVP).

Companion Diagnostic

Abbreviated **CDx**, this term refers to diagnostic tests that determine in advance the likelihood that a particular pharmaceutical will benefit a given patient, based on that patient's gene(s) or biomarker(s) applicable to a specific disease or condition.

For example, the U.S. Food and Drug Administration (FDA) has approved as a companion diagnostic the BioMerieux THxID-BRAF test, a polymerase chain reaction test that can pinpoint the particular melanoma (skin cancer) patients whose tumors are driven by specific mutations (V600E and V600K) in the BRAF gene. Those melanoma tumors thus identified by that BioMerieux companion diagnostic are susceptible to treatment by Tafinlar (dabrafenib) and Mekinist (trametinib), two GlaxoSmithKline melanoma drugs.

For example, the U.S. FDA has approved as a companion diagnostic the Myriad Genetics, Inc. BRACAnalysis test, to be used in conjunction with the AstraZeneca's drug Lynparza (olaparib). Lynparza is the first FDA-approved poly ADP-ribose polymerase (PARP) inhibitor for patients with germline mutations in BRCA1/2 advanced ovarian cancer who have had three or more lines of chemotherapy. See also *Gene, HER-2 gene, Biomarkers, Food and Drug Administration (FDA), Polymerase chain reaction (PCR), Tumor, Mutation, Cancer, PARP inhibitors, Chemotherapy.*

Comparative Analysis

See *Homologous (chromosomes or genes).*

Competence Factor

See *Platelet-derived growth factor (PDGF).*

Complement (Component of Innate Immune System)

A group of more than 15 soluble proteins found in blood serum that interacts in a sequential fashion, in which a precursor molecule is converted into an active enzyme. Each enzyme uses the next molecule in the system as a substrate and converts it into its active (enzyme) form. This cascade of events and reactions leads ultimately to the formation of an attack complex that forms a transmembrane channel in the cell membrane (e.g., of a pathogen). It is the presence of the channel that leads to lysis (rupturing) of the cell. See also *Innate immune system, Innate immune response, Plasma membrane, Cell, Pathogen, Cascade, Complement cascade, Complement factor H gene, Cecrophins, Humoral immunity, Lyse, Lysis.*

Complement Cascade

The precisely regulated, sequential interaction of proteins (in the blood) that is triggered by a complex of antibody and antigen to cause lysis of infected cells. The triggering of lysis by multivalent antibody–antigen complexes is mediated by the classical pathway, beginning with the activation of C1, the first component (protein) of the pathway. This activation step, in which C1 undergoes conversion from a zymogen to an active protease, results in sequential cleavage of the C4, C2, C3, and C5 components (proteins). C5b, a fragment of C5, and then joins C6, C7, and C8 to penetrate the (cell) membrane bearing the antigen. Finally, the binding of some 16 molecules of C9 to this *bridgehead* produces large pores in the (cell) membrane, which cause the lysis and destruction of the target cell. See also *Antibody, Antigen, Lysis, Cell, Plasma membrane, Complement, Complement factor H gene, Zymogens, Cecrophins, Cascade, Pathway.*

Complement Factor H Gene

A gene that codes for the production of complement factor H, a protein also known as "CFH protein" that helps regulate the complement cascade of the human immune system. For example, CFH protein can bind (inflammation) initiation factors such as C-reactive protein and can inactivate certain components of the complement cascade.

Certain variants (alleles) of this gene in humans increase the probability of that person developing age-related macular degeneration disease. See also *Gene, Protein, Coding sequence, Allele, Complement, Complement cascade, Initiation factors, C-reactive protein (CRP), Age-related macular degeneration (AMD).*

Complementary (Molecular Genetics)

Refers to strands of DNA that will hybridize (*bind*) to each other, due to one-for-one matchup of each strand's sequence of nucleotides. Any sequence (within the two strands) that does *not* match up one for one will not hybridize to the respective sequence (in adjacent strand). See also *Molecular genetics, Hybridization (molecular genetics), Deoxyribonucleic acid (DNA), Double helix, Nucleotide, Microarray (testing), Biomotors, Southern blot analysis.*

Complementary DNA (cDNA)

A single-stranded DNA that is complementary to a strand of mRNA. The DNA is synthesized *in vitro* by an enzyme known as reverse transcriptase. Then, a second DNA strand is synthesized via the enzyme known as *DNA polymerase*.

Complementary DNA is often utilized in hybridization studies and in microarrays (e.g., to detect/identify genes) because cDNAs usually do not contain *regulatory sequences* of DNA, since the cDNA was copied from mRNA. Because cDNA is a DNA copy of mRNA (messenger RNA), it is an exception to the (old) central dogma. See also *Deoxyribonucleic acid (DNA), Messenger RNA (mRNA), Central dogma (old), Enzyme, DNA polymerase, Hybridization (molecular genetics), Microarray (testing), Gene expression analysis, Regulatory sequence, Reverse transcriptases.*

Compound Q

See *Trichosanthin.*

Computational Biology

See *Bioinformatics, In silico biology, Rational drug design, Docking (in computational biology).*

Computer-Assisted New Drug Application

(also called "computer-assisted NDA") See *CANDA.*

Computer-Assisted Drug Design (CADD)

See *Rational drug design, Pharmacophore searching.*

C

Configuration

The 3D arrangement in space of substituent groups in stereoisomers.

Confocal Microscopy

Invented by Marvin Minsky in 1957, this refers to the use of a special microscope that is utilized to scan (e.g., in tissue) *a 2D plane* at varying depths. Today, this is typically done using the following:

- Laser beams that rapidly raster scan the sample via galvomirrors. The resultant images can then be put together via a process known as volume rendering, in order to yield a 3D overall image.
- Light that has been passed through a pattern of tiny slits or pinholes in a specially designed (and often rotating) disk, resulting in that light being confined to (and illuminating) the desired 2D sample plane.

Today, using visible fluorescent proteins to "label" some protein molecules of interest, it is possible to watch the movement/interactions of labeled proteins inside living cells via confocal microscopy.

Some confocal microscopes utilize fluorescence resonance energy transfer to achieve better resolution and/or 4D images. See also *Volume rendering, Multiplex assay, Fluorescence, Protein, Label (fluorescent), Visible fluorescent proteins, Green fluorescent protein, Fluorescence resonance energy transfer (FRET).*

Conformation

The 3D arrangement of substituent groups in a protein or other molecular structure (e.g., aptamer molecule) that is free to assume different positions. The geometric form or shape of a protein in 3D space. See also PROTEIN.

Some protein molecules (e.g., receptors) change their conformation when applicable ligands bind to those protein molecules. See also *Native conformation, Tertiary structure, Aptamers, Effector, Protein, Protein folding, Proteomics, Unfoldases, Transcriptome, Disulfide bond, Structure-activity models, Raman optical activity spectroscopy, Receptors, Ligand (in biochemistry), Target–ligand interaction screening, Nuclear proteins.*

Congo Red

A chemical dye that adheres to β-amyloid protein (which can lead to Alzheimer's disease when clumped together inside neurons). At high concentrations, Congo red can inhibit such clumping.

Research indicates that when molecules of Congo red are chemically linked to relevant ligands for FKBP (a large cellular *chaperone protein*), that linked-together chemical entity *recruits* an FKBP protein molecule to insert itself between β-amyloid proteins, which could prevent clumping. See also *Alzheimer's disease, Protein, Cell, Neuron, Ligand (in biochemistry), Chaperones.*

Conjugate

A molecule created by fusing together (e.g., via recombination or chemically) two unlike (different) molecules. The purpose of this is to create a molecule in which one of the original molecules has one function, for example, a toxic, cell-killing function, while the other original molecule has another function, such as targeting the toxin to a specific site in the body, which might be cancerous cells.

For example, molecules of interleukin-2 (IL-2) have been fused with molecules of diphtheria toxin to create a conjugate that does the following:

- It enters leukemia and lymphoma cells. Because these two types of cancer cells possess IL-2 receptors on their surfaces, the IL-2 (targeting function) binds to that receptor and is internalized by the cell.
- The diphtheria toxin (killing function) then shuts down protein synthesis within the cancer cells.
- It then kills the cancerous cells.

This type of approach is widespread and there are many different types of this category of conjugate.

Another type of conjugate consists of enzymes used in the treatment of certain molecular diseases attached covalently to polyethylene glycol (PEG). In this case, the PEG greatly diminishes both the immunogenicity (the tendency to induce the body's immune reaction) and the antigenicity (the ability to react with preformed antibodies).

Another type of conjugate consists of various molecules (e.g., fluorophores, toxins) or nanoparticles (e.g., quantum dots) attached to antibodies. Such conjugated antibodies may be utilized as vectors to carry either small molecules of destructive toxins or imaging proteins (e.g., green fluorescent protein) or imaging particles (e.g., quantum dots) to specific sites (cells) within the body. Antibodies may be coupled to enzymes, toxins, and/or ribosome-inhibiting proteins, as well as radioisotopes. These conjugates are known collectively as immunoconjugates. See also *Immunoconjugate, Conjugated protein, "Magic bullet," Fusion protein, Molecular bridge, Recombination, Toxin, Interleukin-2 (IL-2), Ricin, Abrin, Receptors, Ribosomes, Messenger RNA (mRNA), Diphtheria toxin, Antibody, Fluorphore, Green fluorescent protein, Quantum dot, Enzyme, Nanoparticles, Chimeric molecules, Peptide-oligonucleotide conjugates.*

Conjugated Linoleic Acid (CLA)

Also known as *alpha-rumenic acid* or *9-cis, 11-trans C 18:1*, it is a naturally occurring *n*-6 polyunsaturated fatty acid (PUFA) discovered in 1979 by Michael W. Pariza whose consumption by humans has been linked to

- Reduction in risk for atherosclerosis
- Reduction in blood triglyceride levels
- Reduction in blood pressure
- Reduction in body fat (adipose tissue) in obese humans
- Increase in lean body mass
- Reduction in risk for breast cancer, skin cancer, and some other types of cancer

CLA inhibits angiogenesis (i.e., formation of new blood vessels, such as the ones needed for tumors to be able to grow), and CLA exhibits powerful antioxidant properties (i.e., it "quenches" free radicals). Chemically, CLA consists of two linoleic acid molecules linked together by a chemical bond, so it is a dimer.

Foods that are naturally highest in CLA content include beef, lamb, full-fat milk, butter, cheese, some creams, and full-fat yogurt. However, that natural level (3–7 mg/g of fat) is too small to exert much beneficial impact. Feeding of soybean oil (in feed rations) to livestock has been proven to increase CLA content in the resultant meat. In 1998, T.R. Dhiman showed that feeding of soybean oil (i.e., whole) to dairy cattle did also increase the content of CLA in their milk.

Research conducted during the 1990s indicated that consumption of CLA (e.g., by humans, swine, rats) causes the bodies of those animals to change the way they utilize and store energy. Thus, the body requires less food to perform at the same level. The body also tends to produce less body fat (adipose tissue) and more lean protein (e.g., muscle) tissue. See also *Polyunsaturated fatty acids (PUFA), Fats, Linoleic acid, Atherosclerosis, Oxidative stress, Antioxidants, Soybean oil, Adipose, Adipokines, Cancer, Volicitin, Oligomer, Tumor, Angiogenesis.*

Conjugated Protein

A protein containing a metal or an organic prosthetic group (e.g., heme group, carbohydrate, lipid group), or both. For example, a glycoprotein is a conjugated protein bearing at least one oligosaccharide group. See also *Prosthetic group, Glycoprotein, Protein, Oligosaccharides, Conjugate, CD4-PE40.*

Conjugation

A process akin to sexual reproduction occurring in bacteria, mating in bacteria. A process that involves cell-to-cell contact and the one-way transfer of DNA from the donor to the recipient. In contrast to some other DNA-transfer processes of bacteria, conjugation may involve the transfer of large portions of the genome. The discovery caused considerable controversy at the time. See also *Transformation, Bacteria, Transduction (gene), Transduction (signal), Deoxyribonucleic acid (DNA), Genome, Sexual conjugation.*

Consensus Sequence

The nucleotide sequence (within a DNA molecule) that gives the *most common* nucleotide at each position (along that sequence of that DNA molecule), for those instances (in certain organisms) where a (usually small) number of variations in nucleotide sequences can occur (e.g., for a given nucleotide sequence such as a promoter sequence). See also *Nucleotide, Deoxyribonucleic acid (DNA), Sequence (of a DNA molecule), Genetic code, Gene, Promoter, Pharmacogenomics.*

Conservation Tillage

Refers to crop production (farming) techniques/practices such as *low-tillage crop production, no-tillage crop production*, etc. that avoid or minimize the disturbance of topsoil. The field's topsoil is protected from soil erosion by the decomposing leftover crop residue on the field surface resulting from low or no tillage. Via shading of the field's topsoil and via reducing field surface wind speed to near zero, the leftover crop residue also minimizes evaporation of moisture from the field's soil. See also *Low-tillage crop production, No-tillage crop production, Drought tolerance, Glomalin.*

Conserved

A term used to describe the following:

- The number of genes that are present within the DNA of more than one species. For example, approximately 25% of the genes found within the human genome (DNA) are also found within the DNA of plants.
- A particular domain (region) of a molecule on the surface of a rapidly mutating microorganism (e.g., the influenza virus, the AIDS virus) that remains the same in all, or most, variations of that microorganism. If that *conserved* region is suitable to act as an antigen (hapten, epitope), it may be possible to create a successful vaccine against that microorganism that would otherwise be unsuccessful due to the fact that the rapid mutation would cause it (e.g., the AIDS virus) to appear to be *different* than the one (antigen) the vaccine was designed against.

See also *Domain (of a protein), GP120 protein, Superantigens, Mutation, Acquired immune deficiency syndrome (AIDS), Antigen, Hapten, Epitope, Virus, Gene, Deoxyribonucleic acid (DNA), HIV-1 and HIV-2.*

Consortia

Microorganisms that interact with each other (or at least *coexist peacefully*) when growing together. An example of such interaction/coexistence would be bioleaching. See also *Bioleaching, Biorecovery, Biodesulfurization, Biosorbents.*

Constant Region

See *Antibody.*

Constitutive Enzymes

Enzymes that are part of the basic, permanent enzymatic machinery of the cell. They are formed at a constant rate and in constant amounts regardless of the metabolic state of the organism. For example, enzymes that function in the production of cell-usable energy (such as ATP) might be good candidates. And this, in fact, is the case with the enzymes of the glycolytic sequence, which is the most ancient energy-yielding catabolic pathway. See also *Enzyme, Metabolism, Cell, Pathway.*

Constitutive Genes

Expressed as a function of the interaction of RNA polymerase with the promoter, without additional regulation. They are sometimes also called "household genes" in the context of describing functions expressed in all cells at a low level. See also *Gene, RNA Polymerase, Promoter.*

Constitutive Heterochromatin

The inert state of permanently nonexpressed sequences, usually satellite DNA. See also *Express, Coding sequence, Deoxyribonucleic acid (DNA), Chromatin.*

Constitutive Mutations

Mutations (changes in DNA) that cause genes that are nonconstitutive (have controlled protein expression) to become constitutive (in which state the protein is expressed all of the time). See also *Constitutive genes, Mutation, Regulatory sequence, Protein.*

C

Constitutive Promoter

Refers to a promoter that is (present/acts at) high level in all cells of an organism. See also *Promoter, Cell, Organism.*

Construct

See *Cassette, Transgene.*

Consultative Group on International Agricultural Research (CGIAR)

An organization that is cosponsored by the Rome-based United Nations Food and Agriculture Organization (FAO), the United Nations Development Programme, and the World Bank. The CGIAR is an association of 58 public and private donors that jointly support 16 international agricultural research centers that are located primarily in developing countries. Twelve of the research centers have collectively assembled 500,000 different preserved samples (i.e., germplasm) of major food, forage, and forest plant species into a gene bank. This, the world's largest internationally held collection of genetic resources, was legally placed under the auspices of the FAO in 1994 in order "to hold the collection in trust for the international community." Since 1970, CGIAR's collection has supported research efforts to develop better varieties of staple foods consumed primarily in developing countries of the world. See also *American type culture collection (ATCC), Type specimen, Germplasm.*

Contaminant

By definition, it is any unwanted or undesired organism, compound, or molecule present in a controlled environment. Unwanted presence of an entity in an otherwise clean or pure environment. See also *Organism.*

Contiguous Genes

A *group* of genes that are situated together on an organism's chromosome and that often function *together as a unit* to express a trait in that organism. See also *Linkage, Organism, Gene, Chromosome, Trait.*

Con-Till

An abbreviation that refers to *conservation tillage* farming practices. See also *Conservation tillage, Low-tillage crop production, No-tillage crop production, Glomalin.*

Continuous Perfusion

A type of cell culture in which the cells (either mammalian or otherwise) are immobilized in a part of the system, and nutrients/oxygen are allowed to flow through the stationary cells, thus effecting nutrient/waste exchange. Ideally the system incorporates features that retard the activity of proteolytic enzymes and reduce the need for anti-infective agents (e.g., antibiotics) and fetal bovine serum, which are required by most other cell culture systems. Continuous perfusion is used because, among other things, it eliminates the need to separate the cells from the culture medium when fresh medium is exchanged for old. See also *Mammalian cell culture, Enzyme, Proteolytic enzymes.*

Control Sequences

Those sequences of DNA that are adjacent to a gene (in genome) and "turn on" and/or "turn off" that gene. See also *Sequence (of a DNA molecule), Gene, Genome, Promoter, Termination codon (terminator sequence), Base (nucleotide), Coding sequence.*

Convention on Biological Diversity (CBD)

The international treaty governing the conservation and use of biological resources around the world, which was signed by more than 150 countries at the 1992 United Nations Conference on Environment and Development.

Article 19.4 of the CBD called for the establishment of a *protocol on biosafety* to govern the transnational-boundary movement of nonindigenous living organisms. See also *MEA, Consultative Group on International Agricultural Research (CGIAR), International Plant Protection Convention* (IPPC), *Biodiversity, Introduction.*

Convergent Improvement

See *Transgressive segregation.*

Coordinated Framework for Regulation of Biotechnology

The regulatory *framework* via which the United States evaluates/approves new products derived via biotechnology. The Coordinated Framework assigns specific regulatory tasks to each of the U.S. government's applicable agencies (see the following).

For example, the U.S. Environmental Protection Agency is assigned to evaluate/regulate all genetically modified pest-protected new plants, in terms of their impact on pests. The U.S. Food and Drug Administration is assigned to evaluate/regulate all new food crops derived via biotechnology, in terms of their potential food safety impact (e.g., allergenicity, toxicity). The U.S. Department of Agriculture is assigned to evaluate/regulate all new plants derived via biotechnology, in terms of field (i.e., outdoor) testing, in terms of potential environmental impacts such as weediness. See also *Biotechnology, Food and Drug Administration (FDA), Genetically modified pest-protected (GMPP) plants, Allergies (foodborne), APHIS.*

Coordination Chemistry

See *Chelation.*

Copy DNA (C-DNA)

See *C-DNA.*

Copy Number (Plasmid or Plastid)

The number of molecules (copies) of an individual plasmid or plastid that is typically present in a single (e.g., bacterial for *plasmid*, plant for *plastid*) cell. Each plasmid has a characteristic copy number value ranging from 1 to 50 or more. Higher copy numbers result in a higher yield of the protein encoded for by the plasmid gene in each cell. See also *Plasmid, Plastid, Protein, Gene, Extranuclear genes, Genetic code, Multi-copy plasmids.*

C

Copy Number (Protein Molecules)

The number of protein molecules coded for/produced by a specified gene within the DNA of an organism, as a result of *copy number variation.* Higher copy numbers (of that gene, within the DNA) result in more protein molecules being synthesized. See also *Protein, Deoxyribonucleic acid (DNA), Gene, Organism, Copy number variation, Multi-allelic copy number variation loci.*

Copy Number Polymorphisms

Abbreviated CNP, it refers to the genotypic variations (e.g., among a population of organisms of the same species) resulting from the loss, gain, or duplication of specific segments (sequences) of their DNA. See also *Copy number variant, Polymorphism (genetic), Genotype, Organism, Species, Deoxyribonucleic acid (DNA), Sequence (of a DNA molecule), Multi-allelic copy number variation loci.*

Copy Number Variant

Abbreviated CNV, it refers to different members of the same species possessing differing copy number variations—that is, fragments of the organism's DNA that are either missing or existing in extra copies (e.g., fewer or more copies of a given gene)—within their genome. CNVs can perturb (affect the function of) many genes within a genome simultaneously. See also *Copy number variation, Copy number polymorphisms, Species, Deoxyribonucleic acid (DNA), Genome, Multi-allelic copy number variation loci.*

Copy Number Variation

Refers to differing numbers of a specific protein molecule produced from the gene that codes for that protein. This copy number variation (CNV) is typically caused by *extra* copies of that gene inserted in an organism's DNA or *deleted* copies of the gene (although it is sometimes due to another source such as *deleterious mutations* within a gene or within the regulatory sequence for that gene, which prevent proper function of that gene) and may be a cause of

- Response of the body to certain pharmaceuticals (e.g., pharmacogenetics/pharmacogenomics)
- Rate of cancer progression/metastasis
- Susceptibility to certain genetic diseases

For example, the heritable disorder known as "congenital generalized hypertrichosis terminalis" results when large fractions of DNA are absent (due to mutations) from four specific human genes.

For example, some research is indicative that copy number variation in the gene(s) that code for one or another subunit of the *myelin protein structure (which is composed of multiple of such subunits)* may cause or contribute to the human disease known as "multiple sclerosis."

For example, a 2012 research by David E. Cook et al. showed that overexpression (10X higher copy variant) of a set of three specific soybean genes together at locus *Rhg1* conferred enhanced resistance to soybean cyst nematode (*Heterodera glycines* or SCN, a parasitic roundworm that attacks soybean plants).

Copy number variation is even found in DNA of monozygotic (*identical*) twins. It typically results when both strands of the applicable DNA molecule break, and the DNA repair process inserts extra copies of a gene or leaves out some genes.

People in some human cultures that consume a lot of starch-containing foods tend to have a higher copy number for the enzyme amylase (which helps digestion of starch) in their saliva.

Copy number variations occur in less than 10% of human genes. See also *Protein, Deoxyribonucleic acid (DNA), Gene, Organism, Multiple sclerosis, aCGH, Mutation, Multi-allelic copy number variation loci, DNA repair, Pharmacogenetics, Pharmacogenomics, Cancer, Metastasis, Amylase, Aneuploidism, Regulatory sequence, Soybean cyst nematodes (SCN).*

CoQ10

See *Coenzyme Q10.*

COR Genes

Refer to a category of plant genes that, when activated, express proteins that protect plant cells from membrane damage and other cold (temperature)-induced damage. See *CBF1.*

Core Histones

See *Histones, Epigenetic marks.*

Corepressor

A small molecule that combines with the repressor to trigger repression (the shutting down) of transcription. See also *Transcription.*

Corn

The domesticated plant *Zea mays* L. also known as maize. A green, leafy (grain) plant that is one of the world's largest providers of edible starch and fructose (sugar) for mankind's use. This summer annual plant varies in height from 2 ft (0.5 m) to more than 20 ft (6 m) tall. The seeds (kernels) are borne in cobs, ranging in size from 2 ft long to smaller than a man's thumb.

Due to genetic variation (i.e., of different hybrids/varieties), the fraction of kernel that consists of recoverable starch varies between 42% and 73% for different corn varieties.

Due to genetic variation (i.e., of different hybrids/varieties), the fraction of kernel that consists of protein varies between 8% and 10%, but that protein content can be increased by 10% via insertion into corn plant of the *glutamate hydrogenase gene.*

Due to genetic variation (i.e., of different hybrids/varieties), the fraction of kernel that consists of oil varies between 3.5% and 8.5% for different corn varieties.

Grown widely in the world's temperate zones, corn is grown as far north as latitude 58° in Canada and Russia and as far south as latitude 40° in the Southern Hemisphere.

During the 1980s, scientists were able to insert genes from the *Bacillus thuringiensis* bacteria into the corn plant, to make that plant resistant to certain insects. During the 1990s, scientists were able to insert genes into the corn plant, to make it tolerant to certain herbicides and to cause the corn plant to produce monoclonal antibodies.

Some of the major economic pests of corn include the European corn borer (*Ostrinia nubilalis*), corn earworm/soybean podworm (*Helicoverpa zea*), corn rootworm (*Diabrotica virgifera virgifera*), and beet armyworm (*Pseudaletia unipuncta*). See also *Hybridization (plant genetics), Bacillus thuringiensis (B.t.), Protein, Stress proteins, Maysin, Cry proteins, CRY1A (b) protein, CRY1A*

C

(c) protein, CRY9C protein, Gene, "Stacked" genes, Opague-2, High-methionine corn, High-lysine corn, B.t. kurstaki, Value-enhanced grains, Helicoverpa zea (H. zea), Chloroplast transit peptide (CTP), Herbicide-tolerant crop, High-oil corn, European corn borer (ECB), Aflatoxin, Fusarium, Corn rootworm, Volicitin, GA21, Transposable element, Transposon, Glutamate dehydrogenase, Black-layered (corn), Monoclonal antibodies (MAb), Photorhabdus luminescens, Cholesterol oxidase, MIR1-CP, cspB gene.

Corn Borer

See *European corn borer (ECB), Asian corn borer, Southwestern corn borer.*

Corn Earworm

Also known as soybean podworm (when found on soybean plants) and as the tomato fruitworm (when it is on tomato plants). See *Helicoverpa zea (H. zea), Corn.*

Corn Rootworm

A complex of several strains of beetles, it refers to the larva stage of the corn rootworm beetle (*Diabrotica virgifera virgifera*), which historically has laid its eggs on corn/maize (*Zea mays* L.) plants. When they hatch, the larva must feed on the roots of the corn/maize plant in order to live. Adult corn rootworm beetles also feed on leaves and silks of corn/maize plants.

Some strains of *Bacillus thuringiensis* have proven to be effective against the corn rootworm, when sprayed onto them or genetically engineered into the corn/maize plant.

In 1992, a new genetic variant of corn rootworm known as the "Western phenotype" or Western corn rootworm (*Diabrotica virgifera virgifera* LeConte) was discovered in the United States. It prefers to lay its eggs on soybean plants instead of corn plants. During 2007, James Baum et al. showed that RNA interference (RNAi) could potentially be utilized to control this insect pest via double-stranded RNA (dsRNA) via oral delivery to the larvae (e.g., via corn plant tissues genetically engineered to contain relevant dsRNA). The relevant dsRNA is taken up by the larvae's midgut cells and processed by its cells' native RNAi machinery, which leads to specific knockdown of the applicable mRNA (e.g., targeted mRNA that encodes a protein required for an essential function in the insect's cells).

Other genetic variants of the corn rootworm include the "Northern phenotype" or Northern corn rootworm (*Diabrotica barberi*) and the Mexican corn rootworm (*Diabrotica virgifera zeae*). See also *Corn, Phenotype, Soybean plant, Strain, Bacillus thuringiensis (B.t.), Genetic engineering, CRY3B(b) protein, B.t. kumamotoensis, Antibiosis, Ribonucleic acid (RNA), RNA interference (RNAi), Double-stranded RNA (dsRNA).*

Coronary Heart Disease (CHD)

A disease of the heart and arteries, in which (among other effects) cholesterol is deposited on the interior walls (lumen endothelium), where it can sometimes later break off and cause death (e.g., via *heart attack*).

Risk factors (i.e., increased risk) for CHD include high blood levels of triglycerides, high levels of apolipoprotein B, high levels of LDLPs/VLDLs (the two lipoproteins that are most likely to deposit cholesterol on artery walls), and/or low levels of HDLPs (the lipoproteins that help to clear-away cholesterol deposits from artery walls).

A human diet containing a large amount of certain phytosterols (e.g., CAMPESTEROL, BETA-SITOSTEROL, and/or STIGMASTEROL) has been shown to lower total serum (blood) cholesterol and low-density lipoprotein (LDLP) levels by approximately 10% and thereby lower the risk of CHD.

A human diet containing a large amount of oleic acid causes lower blood cholesterol levels, and can thus lower risk of CHD and atherosclerosis. See also *Cholesterol, Low-density lipoproteins (LDLPs), Sitosterol, Very-low-density lipoproteins (VLDLs), High-oleic oil soybeans, Phytosterols, Sterols, Campesterol, High-density lipoproteins (HDLP), Beta-sitosterol (B-sitosterol), Stigmasterol, Serum lifetime, Lycopene, Atherosclerosis, Resveratrol, Lumen, Endothelium, Triglycerides, Endothelin, Adipose, Homocysteine.*

Cortical Microtubules

The microtubules that are located on the inner face of the plasma membrane of a cell. In plant cells, within the hypocotyl (i.e., stem of a germinating seedling), these microtubules are arranged with a predominant orientation that is perpendicular to the axis of expansion of the growing stem. In response to sunlight striking the stem, the cortical microtubules swiftly reorient themselves 90°, to become parallel to the axis of the growing stem. That new orientation of the cell's cortical microtubules causes cellulose deposition (i.e., building of the structural-strength members of plant cell wall) to occur in a manner that causes the plant to *grow in the direction of the sunlight (a phenomenon known as phototropism)*. See also *Microtubules, Cell, Plasma membrane, Cellulose.*

Corticotropin

See *ACTH.*

Cortisol

A steroid hormone that is utilized by the human body to regulate blood pressure (e.g., via increased water retention, by decreasing the kidney's water-excretion rate).

A deficiency of cortisol causes *Addison's disease*. See also *Transcription activators, Glycyrrhizic acid, Steroid, Hormone, Homeostasis.*

Cosuppression

A significant decrease ("silencing") in the expression of a gene (within an organism's genome/DNA) that (often) results when man *inserts and causes to be expressed* a homologous gene.

For example, *high-oleic oil soybeans* result when the *GmFad2-1* gene (which codes for native Δ12 desaturase enzyme) is inserted and expressed in traditional varieties of soybeans. That is because the inserted gene "silences" itself and the endogenous *GmFad2-1 gene* (i.e., the one naturally/originally present in the soybean plant), which thus prevents formation of the Δ12 desaturase enzyme (which normally causes most oleic acid within soybeans to be converted into polyunsaturated linolenic acid/linoleic acid). See also *Gene silencing, Oleic acid, Linoleic acid, Linolenic acid, Express, Gene, Post-transcriptional gene silencing (PTGS), Knockout, Genome, Homologous (chromosomes or genes), Soybean plant, High-oleic oil soybeans, Δ12 desaturase, Antisense (DNA sequence), FAD3 gene, RNA interference (RNAi).*

Cowpea Mosaic Virus (CpMV)

A virus that infects cowpea (*Vigna unguiculata*) plants (which are known as black-eyed peas in the United States), but does not infect animals. Researchers have discovered how to cause CpMV to express certain animal virus proteins (i.e., antigens) on its surface, via genetic engineering. These virus antigens hold potential to replace the antigens currently used in vaccines, which are fraught with problems due to their production in animal cells, bacterial cells, or yeast cells. In addition, CpMV acts as an intrinsic natural adjuvant to the (animal virus) antigens, since it provokes an immune response itself. See also *Virus, Cowpea trypsin inhibitor (CpTI), Express, Protein, Adjuvant (to a pharmaceutical), Immune response, Antigen.*

Cowpea Trypsin Inhibitor (CpTI)

A chemical that is naturally coded for by a certain cowpea (*Vigna unguiculata*) plant gene. It kills certain insect larvae by inhibiting digestion of ingested trypsin by the larvae, thereby starving the larvae to death. See also *Trypsin, Trypsin inhibitors, Gene, Coding sequence.*

COX

See *Cyclooxygenase.*

COX Gene

It refers generally to any gene that codes for cyclooxygenase (COX), itself a term that refers to a *family* of enzymes (isozymes) that convert arachidonic acid to prostaglandins in the human body. Some of the different forms of cyclooxygenase cause the body to produce compounds that promote inflammation.

For example, the bodies of men whose DNA has a COX-2 gene variant called "rs4647310" have cyclooxygenase that produces inflammatory compounds (and a higher risk of developing advanced prostate cancer).

During 2009, John S. Witte showed that consumption of long-chain omega-3 fatty acids reduces the risk of prostate cancer, even in men possessing the COX-2 gene variant called "rs4647310." See also *Gene, Coding sequence, Cyclooxygenase, Enzyme, Isozymes, Arachidonic acid (AA), Deoxyribonucleic acid (DNA), Prostate, Cancer, Omega-3 fatty acids, N-3 fatty acids.*

COX-1

See *Cyclooxygenase.*

COX-2

See *Cyclooxygenase.*

COX-2 Gene

See *Cox gene, Cyclooxygenase.*

COX-3

See *Cyclooxygenase.*

CP4 EPSP Synthase

See *CP4 EPSPS.*

CP4 EPSPS

The enzyme 5-enolpyruvyl-shikimate-3-phosphate synthase, which is naturally produced by an *Agrobacterium* species (strain CP4) of soil bacteria. CP4 EPSPS is essential for the functioning of that bacterium's metabolism biochemical pathway. CP4 EPSPS happens to be unaffected by glyphosate-containing or sulfosate-containing herbicides, so introduction of the CP4 EPSPS gene into crop plants (e.g., soybeans) makes those plants essentially impervious to glyphosate-containing or sulfosate-containing herbicides. See also *Enzyme, Metabolism, Gene, Genetic engineering, EPSP synthase, Glyphosate, Sulfosate, Soybean plant, Glyphosate oxidase, Bacteria, Chloroplast transit peptide (CTP), Herbicide-tolerant crop, Pathway.*

CpDNA

See *Cytoplasmic DNA.*

CPMP

See *Committee for Proprietary Medicinal Products (CPMP).*

CpMV

See *Cowpea mosaic virus (CPMV).*

CPP

Acronym for cell-penetrating peptide. See *Peptide-oligonucleotide conjugates.*

CpTI

See *Cowpea trypsin inhibitor (CPTI).*

CR

Acronym for *complete remission.*

Crassulacean Acid Metabolism (CAM)

See *Metabolism, C4 photosynthesis.*

C-Reactive Protein (CRP)

Discovered in 1929 by Oswald Avery, C-reactive protein (CRP) is a general inflammation *biomarker* (protein molecule) produced in humans in the liver in response to certain bacterial infections or certain physical trauma (which cause inflammation). Elevated blood levels of CRP are related to the degree of risk of arteriosclerosis, coronary heart disease, and heart attack.

Typical healthy humans (e.g., not suffering an infection) tend to have blood CRP levels of less than 3 mg/L. Blood levels of CRP increase 1000-fold or more when the individual becomes infected/inflamed.

C

Aspirin and statin-type pharmaceuticals (e.g., pravastatin, simvastatin, atorvastatin) help to lower inflammation and bloodstream CRP levels. Consumption of α-linolenic acid causes a decline in bloodstream CRP levels. Blood levels of CRP also decline when (overweight) people lose weight.

Blood levels of CRP are increased by a person's

- Aging
- Obesity
- Type II diabetes
- Smoking and excess alcohol consumption
- Gum disease (periodontal disease)

See also *Biomarkers, Protein, Bacteria, Arteriosclerosis, Coronary heart disease (CHD), Chronic inflammation, Humoral immune response, Interleukin-6 (IL-6), Type II diabetes, Linolenic acid, Statins.*

Cre-Lox System

Refers to the use of a particular phage/enzyme system to accomplish a site-specific (on organism's or virus's DNA) insertion or deletion of a specific DNA fragment. *Cre* is the name of an enzyme that specifically joins *LoxP* sites (on DNA molecule) that were earlier engineered into the DNA of both a shuttle vector (i.e., plasmid in this case) and the DNA of the "target" organism or virus.

Certain "knockouts" in transgenic organisms can be created via use of the Cre-Lox System. The Cre-Lox system is particularly useful for knocking-out (removing) a particular gene from only a specific subset of the organism's tissues. See also *Phage, Organism, Virus, Deoxyribonucleic acid (DNA), Deletions, Enzyme, Shuttle vector, Knockout.*

CRISPR

Acronym for *clustered regularly interspaced short palindromic repeats*, which is one method for scientists to do genetic editing today. It was first described in 1987 for the bacterium *Escherichia coli*, in which it is a naturally occurring immune system process that confers some resistance to harmful exogenous genetic elements (*infectious agents*) such as plasmids and phages/viruses. It was later shown to also be present within archaea.

Bacteria and archaea utilize CRISPR in combination with *Cas proteins* (i.e., CRISPR-**as**sociated proteins) known as Cas1 and Cas2 to silence crucial segments of an invading infectious agent's genetic element content via a mechanism analogous to RNA interference and also to thereby retain immunity against invasion by that same infectious agent in the future. The latter is accomplished via a molecular complex (Cas1–Cas2) that can latch onto a specific segment of an invader's genetic content, copy it as DNA segment, and then insert that DNA segment into the bacteria's (or archaea's) DNA in the form of a *spacer* (which functions as an immune system memory to facilitate a swifter and stronger response to that particular invader next time). When, for example, a later-invading virus' DNA binds to such CRISPR segments in the bacteria's DNA, the Cas9 enzyme cuts up that viral DNA (thereby halting the viral invasion).

After initially being called other things, it was renamed CRISPR or CRISPR-Cas in 2002. See also *Gene, Gene editing, Palindrome, Bacteria, Archaea, Escherichia coliform (E. coli), Plasmid, Virus, Phage, Gene silencing, CRISPR/CAS9 gene-editing systems.*

CRISPR/Cas9 Gene-Editing Systems

Acronym for transcription activator-like effector nucleases and *clustered regulatory interspaced short palindromic repeats*, a technology system for scientists to do precise genetic editing today. The CRISPR/Cas9 system utilizes tailored segments of man-made short *single guide RNA* (known as sgRNA or gRNA) to guide Cas9 nuclease (a DNA-cutting enzyme) to virtually any desired site on a DNA molecule.

Via the scientist supplying a carefully selected gene (DNA sequence) for the CRISPR system to use in its repair of the damage of that "cut," the CRISPR/Cas9 gene-editing system can be utilized to insert genes (e.g., to create a genetically engineered crop plant) to cure certain animal disorders/diseases caused by a single genetic mutation, to impart a new trait, etc. The Cas9-sgRNA recognizes targeted DNA based on complementarity between a sgRNA spacer (i.e., the leading sequence of sgRNA) and its (targeted) DNA.

In addition to inserting a gene (specific DNA sequence replacement), this sgRNA-guided endonuclease technology can be used to induce in/dels (indel mutations), insertions and large deletions, or genomic rearrangements at any specific selected location in the organism's genome, knock-in of a particular gene, knockout of a particular gene, and knockdown of a particular gene, or be utilized to mediate up- or downregulation of specific gene(s) within the organism's genome. CRISPR/Cas9 gene-editing systems can also be utilized to alter histone modifications or DNA methylation within the organism's genome.

See also *Deoxyribonucleic acid (DNA), Gene, Sequence (of a DNA molecule), Genome, Gene editing, Gene silencing, Organism, Palindrome, Ribonucleic acid (RNA), sgRNA, CAS9, CRISPR, Enzyme, Nuclease, Genomic surgery, Genetic engineering, Mutation, Indel mutations, Knockin, Knockout, Knockdown, Histones, Histone modifications, DNA methylation, Mutagenic chain reaction, Gene drive.*

CRISPR/Cas9 Genome-Editing Systems

See *CRISPR/Cas9 gene-editing systems.*

CRISPR-Cas

See *CRISPR.*

CRISPR-Cas Immune System

See *CRISPR.*

Critical Micelle Concentration

Also known as the CMC of a surfactant. It is the lowest surfactant concentration at which micelles are formed. That is, the CMC represents that concentration of surfactant at which the individual surfactant molecules aggregate into distinct, high-molecular-weight spherical entities called "micelles." Or from another viewpoint, it represents the concentration of a surfactant, above which micelles or reverse micelles will spontaneously form through the process of self-aggregation (self-assembly).

For example, liposomes in a water solution will self-assemble into micelles/vesicles if their concentration is higher than that liposome's CMC. See also *Micelle, Reverse micelle (RM), Liposomes.*

Crohn's Disease

An intestinal disease of humans that can cause inflammation of the colon, abdominal pain, diarrhea, and weight loss and decrease the body's ability to absorb dietary-source vitamin D. See also *Commensal, Dendritic cells, Vitamin, Immunomodulating agent.*

Crop Biologicals

Refers to microorganisms that are utilized by man (e.g., applied to crop seeds in the form of a seed treatment coating, sprayed onto plants, inserted into field topsoil) to enhance the growth of those crop plants (e.g., via increasing expression of applicable plant genes), to help those crop plants to better absorb plant nutrients (e.g., phosphorous, nitrogen) from soil, to resist pests (e.g., certain insects, phytoparasitic nematodes), to resist pathogens (e.g., harmful bacteria, harmful fungi, certain viruses), to fix nitrogen from the atmosphere for a crop plant's roots to absorb, etc. It can be applied to crops alone or in combination with other agents (e.g., biostimulants).

For example, the fungal pathogen *Nomuraea rileyi* is an effective biological control agent for the soybean podworm (*Helicoverpa zea*) insect pest in soybean plants.

For example, the parasitic-to-soybean cyst nematode (SCN) *Pasteuria* sp. bacteria can be applied to soybean seeds (e.g., as a crop biological coating) prior to planting, in order to help control soybean cyst nematodes. The *Pasteuria* bacteria must attach their spores (for reproduction) to the juvenile nematodes, so that the *Pasteuria* offspring can consume the SCN when those spores later germinate.

For example, the fungal pathogen *Coniothyrium minitans* (*strain CON/ M/91-08*) is an effective biological control agent for the soilborne pathogenic fungi *Sclerotinia* spp.

Other crop biologicals include *rhizobia*, mycorrhizae, penicillium, trichoderma, and bacillus species/compounds. See also *Microorganism, Microbiology, Gene, Express, Expressivity, Bacteria, Pasteuria, Nitrogen fixation, Rhizobium* (*bacteria*), *Bradyrhizobium japonicum, Penicillium, Fungus, Pathogen, Nematodes, Soybean cyst nematodes* (*SCN*), *Biofertilizers, Biostimulants, Pharmacoenvirogenetics, Azadirachtin, Neem tree, Seed treatments, Systemic acquired resistance* (*SAR*), *Helicoverpa zea* (*H. zea*), *Sclerotinia* spp.

Crop Rotation

Refers to the alternate growing of *different species* of crops in a given farm field during subsequent growing seasons (e.g., canola during the first growing season, soybean during the second growing season, maize/corn during the third growing season). In addition to decreasing the field populations of crop pests (e.g., certain crop-chewing insects such as the European corn borer, certain parasitic roundworms such as the soybean cyst nematode) via denying them their preferred food/host plant in the field, such crop rotation also increases the yields of the crops that are grown in such a (rotated) field. That is because crop rotation increases the biodiversity of microorganism within the field's topsoil, which benefits the heath of subsequently grown in field crops' roots. See also *Canola, Soybean plant, Corn, Microorganism, Brassica, European corn borer* (*ECB*), *Soybean cyst nematodes* (*SCN*).

Cross-Reaction

When an antibody molecule (against one antigen) can combine with (bind to) a different (second) antigen. This sometimes occurs because the second antigen's molecular structure (shape) is very similar to that of the first antigen. See also *Antibody, Antigen.*

Cross-Reactivity

See *Cross-reaction.*

Crossing Over

The reciprocal exchange of material between chromosomes that occurs during meiosis. The event is responsible for genetic recombination. The process involves the natural breaking of chromosomes, the exchange of chromosome pieces, and the reuniting of DNA molecules. See also *Linkage, Deoxyribonucleic acid* (*DNA*), *Chromosomes, Recombination.*

Crown Gall

See *Agrobacterium tumefaciens.*

CRP

Acronym for *catabolite regulator protein*. See *CAP.*

CRP

Acronym for *C-reactive protein*. See *C-reactive protein* (*CRP*).

CRTL Gene

See *Golden rice, Gene.*

Cruciferae

A taxonomic group (*family*) of plants that includes canola, mustard, oilseed rape, etc. See also *Brassica.*

Cry Proteins

A class of proteins produced by *Bacillus thuringiensis* (*B.t.*) bacteria (or plants into which a *B.t.* gene has been inserted). Cry (i.e., *crystal-like*) proteins are toxic to certain categories of insects such as corn borers (e.g., *Ostrinia nubilalis*), corn rootworms (*Diabrotica virgifera virgifera*), armyworms (e.g., *Spodoptera frugiperda*), black cutworms (*Agrotis ipsilon*), velvetbean caterpillar (*Anticarsia gemmatalis*), mosquitoes, black flies, tobacco hornworm, and some types of beetles, but harmless to mammals and most beneficial insects. See also *Bacillus thuringiensis* (*B.t.*), *Protein, Bacteria, Gene, Protoxin, Corn, European corn borer* (*ECB*), *Corn rootworm, Armyworm, Tobacco hornworm, CRY1A*(*b*) *protein, CRY1A*(*c*) *protein, CRY3B*(*b*) *protein, CRY9C protein, mCRY3AA protein, Ion channels, Cotton, Toxicogenomics.*

Cry1A (b) Protein

One of the *cry* (i.e., *crystal-like*) proteins, it is a protoxin that—when eaten by certain insects (e.g., *Lepidoptera* larvae such as the armyworm or tobacco hornworm or European corn borer)—is toxic to those crop-pest insects. However, if eaten by a mammal, the Cry1A(b) protein is digested within 1 min, harmlessly. See also

C

Cry proteins, Protein, B.t. kurstaki, Protoxin, European corn borer (ECB), Armyworm, Tobacco hornworm, Ion channels.

Cry1A (c) Protein

One of the *cry* (i.e., *crystal-like*) proteins. See also *Cry proteins, Ion channels.*

Cry1F Protein

One of the *cry* (i.e., *crystal-like* proteins), it is a protoxin that—when eaten by the European corn borer (*Ostrinia nubilalis*), southwestern corn borer (*Diatraea grandiosella*), black cutworm (*Agrotis ipsilon*), fall armyworm (*Spodoptera frugiperda*), and Western bean cutworm—is toxic to those insects. See also *Cry proteins, Bacillus thuringiensis (B.t.), Protoxin, Protein, European corn borer (ECB), Armyworm, Ion channels.*

Cry3A (a) Protein

One of the *cry* (i.e., *crystal-like*) proteins, it is a protoxin that—when eaten by certain insects (e.g., larvae of corn rootworm [*Diabrotica virgifera virgifera*])—is toxic to those insects. See also *Protein, Cry proteins, Protoxin, Corn rootworm, Ion channels, B.t. kumamotoensis.*

Cry3B (b) Protein

One of the cry (i.e., *crystal-like*) proteins, it is a protoxin that—when eaten by certain insects (e.g., larvae of corn rootworm [*Diabrotica virgifera virgifera*])—is toxic to those insects. See also *Protein, Cry proteins, Protoxin, Corn rootworm, Ion channels, B.t. kumamotoensis.*

Cry9C Protein

One of the *cry* (i.e., *crystal-like*) proteins, it is a protoxin that—when eaten by European corn borer (*Ostrinia nubilalis*), southwestern corn borer (*Diatraea grandiosella*), black cutworm (*Agrotis ipsilon*), and some species of armyworm (e.g., *Spodoptera frugiperda*)—is toxic to those insects. See also *Cry proteins, Bacillus thuringiensis (B.t.), B.t. tolworthi, Protoxin, Protein, European corn borer (ECB), Armyworm, Ion channels.*

CSF

See *Colony-stimulating factors (CSFs).*

CspB Gene

A naturally occurring gene within *Bacillus subtilis* bacteria that causes the production of a particular cold-shock (stress response) protein that acts as an RNA chaperone (i.e., helps to convey RNA molecule(s) to their ultimate destination(s) in the cell).

When inserted (via genetic engineering) into the DNA of a corn/maize (*Zea mays* L.) plant, the production of that cold-shock protein helps that plant cope better with the stress of drought conditions. That is because it decreases the rate at which the plant absorbs water from soil in dry conditions. See also *Corn, Gene, Genetic engineering, Drought tolerance, Drought tolerance trait, Cold-shock protein, Chaperones, Ribonucleic acid (RNA).*

CT

Refers to conservation tillage practices of crop production. See also *Low-tillage crop production, No-tillage crop production, Glomalin.*

CTAB

See also *Hexadecyltrimethylammonium bromide (CTAB).*

CTC

Acronym for *circulating tumor cell* (i.e., those that are shed by tumors as part of the metastasis process). See also *Cell, Cancer, Tumor, Metastasis.*

CTNBio

Acronym for Brazil's National Technical Commission on Bio-safety, which is the Brazilian government's regulatory body for granting formal approval to a new genetically engineered plant (e.g., a genetically engineered crop to be planted).

CTNBio is analogous to Germany's ZKBS (Central Commission on Biological Safety), Australia's GMAC (Genetic Manipulation Advisory Committee), Kenya's Biosafety Council, and India's Department of Biotechnology. See also *GMAC, Recombinant DNA Advisory Committee (RAC), ZKBS (Central Commission on Biological Safety), Genetic engineering, Kenya Biosafety Council, Indian Department of Biotechnology.*

CTP

See *Chloroplast transit peptide (CTP).*

Culture

Any population of cells (e.g., bacteria, algae, protozoa, virus, yeasts, plant cells, mammalian cells) growing on, or in, a medium that supports their growth. Typically used to refer to a population of the cells of a single species or a single strain. A medium that contains only one specific organism (e.g., *E. coli* bacteria) is known as a pure culture. A culture may be preserved (i.e., stored alive) via freezing, drying (in which the cells go dormant), subculturing on an agar medium, or other preservation methods. See also *Culture medium, Type specimen, Lyophilization, American type culture collection (ATCC), Species, Strain, Cell culture, Mammalian cell culture.*

Culture Medium

Any nutrient system for the artificial cultivation of bacteria or other cells. It usually consists of a complex mixture of organic and inorganic materials. For example, the classic culture (growth) medium used for bacteria consists of nutrients (required by that bacteria) plus agar to solidify or semisolidify the nutrient-containing mass. See also *Medium, Agar, Cell culture, Mammalian cell culture, Airlift fermenter.*

Curcumin

A polyphenol compound (naturally found in some plants) that acts as an antioxidant in the body's tissues when consumed by humans. Research has shown that it also acts to prevent inflammation of neurological tissues.

For example, curcumin is naturally produced in the roots of turmeric plant (*Curcuma longa*).

Research indicates that lifelong consumption of curcumin might help to prevent or delay symptoms of Alzheimer's disease, because of its amyloid β-protein-binding properties (inside applicable human tissues). See also *Oxidative stress, Antioxidants, Alzheimer's disease, Chronic inflammation, Protein, Amyloid precursor protein.*

Curing Agent

A substance that increases the rate of loss of plasmids during bacterial growth. See also *Growth (microbial), Plasmid.*

Current Good Manufacturing Practices

See *cGMP.*

CUS

See *HSE.*

Cut

An enzyme-induced, highly specific break in both strands of a DNA molecule (opposite one another). The enzymes involved are called "restriction enzymes." See also *Restriction endonucleases, Enzyme, Deoxyribonucleic acid (DNA).*

CVD

Acronym for cardiovascular disease. See *Atherosclerosis, Arteriosclerosis.*

Cyclic AMP

A molecule of AMP (adenosine monophosphate) in which the phosphate group is joined to both the 3′ and the 5′ positions of the ribose, forming a cyclic (ring) structure. When cAMP binds to CAP, the complex is a positive regulator of procaryotic transcription. See also *Adenosine monophosphate (AMP), CAP, Procaryotes, Transcription, Adenilate cyclase.*

Cyclic Phosphorylation

Synthesis (i.e., *manufacturing*) of adenosine triphosphate (chemical reaction) that occurs during photosynthesis in plants. Also called "photosynthetic phosphorylation" (photophosphorylation).

See also *ATP synthase, Adenosine triphosphate (ATP), Photosynthesis, Photosynthetic phosphorylation.*

Cyclodextrin

A macrocyclic (doughnut-shaped) carbohydrate ring produced enzymatically from starch. The external surface is hydrophobic while the interior is hydrophilic in nature. The hole of the doughnut is large enough to accommodate guest molecules. Uses include solubilization, separation, and stabilization of molecules in the interior cavity of or in association with the cyclodextrin molecules.

For example, during 2005, Timothy Triche utilized cyclodextrins to carry some short interfering RNA (siRNA) into mouse tumors (after he attached a molecular tag specific to tumors to the exterior of those cyclodextrins). After entry to the tumors, the siRNA inhibited growth of those tumors. See also *Carbohydrates, Short interfering RNA (siRNA), Tumor.*

Cycloheximide

Also called Actidione. A chemical that inhibits protein synthesis by the 80S eucaryotic ribosomes; it does not, however, inhibit the 70S ribosomes of procaryotes. The chemical blocks peptide bond formation by binding to the large ribosomal subunits. See also *Protein, Ribosomes.*

Cyclooxygenase

Abbreviated COX, it refers to a *family* of enzymes (isozymes) that convert arachidonic acid to prostaglandins in the human body. There are at least three forms of cyclooxygenase:

- COX-1 (also known as PGHS-1) and COX-3, which convert arachidonic acid to *constitutive* prostaglandins, which help to maintain the tissues of the stomach, kidneys, and intestines. COX-1 is present in nearly all tissues of the body.
- COX-2 (also known as PGHS-2), which converts arachidonic acid to *inducible* prostaglandins, which can cause pain and inflammation in the body's joints when they accumulate in those joints. COX-2 is generally not present in body tissues until those tissues are *inflamed* by monocytes (macrophages)/mast cells or *injured* (via mechanical shear/abrasion of endothelial cells). Research indicates that overexpression of the *COX-2 gene* is one of the causative factors in onset of breast cancer. COX-2 also mediates the transformation of omega-3 fatty acids into electrophilic fatty acid oxidation products that help reduce oxidation and inflammation.
- COX-3, which results when *intron 1* is retained in the mRNA transcript during transcription of the COX-1 gene (i.e., alternative slicing).

Aspirin and some other pain-relieving drugs (e.g., ibuprofen, indomethacin) chemically block the earlier-described activity of COX-1 and COX-2. Long-term use of aspirin causes a reduction in incidents of colorectal cancer.

Nexrutine (an extract from the *Phellodendron amurense* tree) and certain pain-relieving drugs (celecoxib, rofecoxib, etc.) chemically block the earlier-described activity of COX-2, while *not* blocking the (beneficial) COX-1.

The pain-relieving drug acetaminophen chemically blocks the activity of COX-3, without blocking COX-1 or COX-2. See also *Enzyme, Isozymes, Arachidonic acid, Platelets, Inducible enzymes, Selective apoptotic anti-neoplastic drug (SAAND), Eicosanoids, Monocytes, Mast cells, Prostaglandins, Endothelial cells, PGHS, Intron, Transcription, Messenger RNA (mRNA), Gene, Cyclooxygenase, Gene, Expressivity, Alternative splicing, Cancer, COX gene, Omega-3 fatty acids, Oxidative stress, Macrophage.*

Cyclosporin A

An immune-system-suppressing drug that was isolated from a mold in the mid-1970s by the Swiss firm F. Hoffmann-LaRoche & Co. AG. The drug is used to prevent (organ recipient's) immune system

C

from *rejecting* a transplanted organ and typically must be taken by the organ recipient for the duration of his/her lifetime.

Cyclosporin's mechanism of action is to prevent the divalent calcium cation (Ca^{2+}) from entering T lymphocytes to activate certain genes within those T lymphocytes (which trigger the *rejection* process).

In 1996, Thomas Eisner reported that the mold *Tolypocladium inflatum*, from which cyclosporin is harvested, prefers a natural (wild) substrate of a deceased dung beetle.

During 2000, it was discovered that cyclosporin inhibits growth of the parasitic microorganism *Toxoplasma gondii* (which can cause loss of sight and neurological disease in humans). See also *T-lymphocytes*, *Fungus*, *Xenogeneic organs*, *Cation*, *Gene*, *Graft-versus-host disease (GVHD)*, *Human leukocyte antigens (HLA)*, *Major histocompatibility complex (MHC)*, *Microorganism*, *Growth (microbial)*.

Cyclosporine

See *Cyclosporin A*.

CYP

See *Cytochrome P450 (CYP)*.

CYP3A4

See *Cytochrome P4503A4*.

CYP46 Gene

A human gene that codes for a protein (within the brain) involved in the brain's usage/processing of cholesterol.

People whose DNA has a mutated version of the CYP46 gene are at a higher than average risk of getting Alzheimer's disease. See also *Gene*, *Protein*, *Genetic code*, *Cholesterol*, *Mutation*, *APOE4*, *Deoxyribonucleic acid (DNA)*, *Haplotype*, *Alzheimer's disease*.

Cysteine (cys)

An amino acid of molecular weight (mol wt) 121 Da. It is incorporated in many proteins. It possesses a sulfhydryl group that makes cysteine a mild reducing agent. Cysteine can cross-link with another cysteine located on the same or on a different polypeptide chain to form disulfide bridges. The *free* cysteine group is called a "thiol group."

High levels of cysteine content in certain genetically engineered corn (maize) kernels have been shown to inhibit in-field production of mycotoxins in corn (e.g., by several species of fungi that can be carried into corn plants by insects). See also *Amino acid*, *Cystine*, *Disulfide bond*, *Homocysteine*, *Polypeptide (protein)*, *Protein*, *Mycotoxins*, *Reduction (in a chemical reaction)*.

Cystic Fibrosis

See *Cystic fibrosis transmembrane regulator protein (CFTR)*.

Cystic Fibrosis Transmembrane Regulator Protein (CFTR)

A protein, also known as *CF transmembrane conductance regulator*, that regulates proper chloride ion transport across the cell membranes of human lung airway epithelial cells.

When the gene that codes for the CFTR protein is damaged/mutated, the (mutant) CFTR protein fails to function properly (i.e., conducts chloride ions at a much slower rate, or not at all), which causes mucous (and bacteria) to accumulate in the lungs. This lung disease is known as Cystic Fibrosis, and more than ten different mutations of the gene that codes for CFTR protein can cause it. During 2014, the U.S. Food and Drug Administration approved KALYDECO® (ivacaftor) as a pharmaceutical to treat people ages 6 and older who have the R117H, G551D, G178R, S549N, S549R, G551S, G1244E, S1251N, S1255P, or G1349D mutations of that gene.

The SNP for cystic fibrosis was identified in 1989. See also *Protein*, *Gene*, *Mutation*, *Ion*, *Ion channels*, *Deoxyribocycleic acid (DNA)*, *Informational molecules*, *Genome*, *Genetic code*, *Ribosomes*, *Transcription*, *Single-nucleotide polymorphisms (SNPs)*, *Genomic surgery*.

Cystine

Two cysteine amino acids that are covalently linked via a disulfide bond. These units are important in biochemistry in that disulfide bridges represent one important way in which the conformation of a protein is maintained in the active form. Cystine bridges lock the structure of the proteins in which they occur in place by disallowing certain types of (molecule) chain movement. When the disulfide bond is with a *free* cysteine (i.e., one that is not a part of the same protein molecule's amino acid backbone), the *free* cysteine is known as a *thiol group*. Cystine can be metabolized from methionine by certain animals (e.g., swine), but not vice versa. See also *Cysteine (cys)*, *Amino acid*, *Conformation*, *Protein*, *Methionine (met)*, *Metabolism*, *Disulfide bond*.

CystX

Refers to a naturally occurring group of genes present in the genome (DNA) in some varieties of soybean plant, which confers on those particular soybean varieties (some) resistant to the soybean cyst nematode. Discovered via marker-assisted breeding and developed during the 1990s by Jamal Faghihi, John Ferris, Virginia Ferris, and Rick Vierling. See also *Soybean plant*, *Soybean cyst nematodes (SCN)*, *Gene*, *Marker assisted breeding*.

Cytochrome

Any of the complex protein respiratory pigments (enzymes) occurring within plant and animal cells. They usually occur in mitochondria and function as electron carriers in biological oxidation. Cytochromes are involved in the *handing off* of electrons to each other in a stepwise fashion. In the process of *handing off*, other events take place, which result in the production of energy that the cell needs and is able to use. See also *Protein*, *Enzyme*, *Mitochondria*, *Cell*.

Cytochrome P450

A *family* of enzymes within the liver that contain an iron-heme cofactor. They catalyze many different biological hydroxylation reactions (e.g., metabolism of certain compounds), epoxidation reactions, oxidative ring-coupling reactions, and heteroatom oxygenation/release reactions. Essentially, the enzyme renders fat-soluble (hydrophobic) molecules water soluble or *more water soluble* (by introduction of the hydrophilic hydroxyl group) so that the molecules may be removed (i.e., filtered/washed) from the body's bloodstream via the kidneys and excreted.

These enzymes are being investigated for their potential as catalysts in the hydroxylation of specific (valuable) industrial chemicals.

In some plants such as sorghum (*Sorghum bicolor*), their cytochrome P450 molecules can help them to respond to certain kinds of stress (e.g. drought) if those cytochrome P450 molecules are present in high enough abundance. See also *Cytochrome*, *Enzyme*, *Cofactor*, *Heme*, *Hydroxylation reaction*, *Metabolism*, *Cytochrome P4503A4*, *Cytochrome P450(CYP)*, *Microsomes*, *Pro-drug therapy*.

Cytochrome P450 (CYP)

Refers to a class of liver enzymes (approximately 4000 known so far) that are responsible for the metabolism (breakdown) of more than 50% of human pharmaceuticals, when those pharmaceuticals enter the bloodstream.

For example, *cytochrome P4503A4* catalyzes the breakdown of some pharmaceutical sedatives, the antihistamine terfenadine, antihypertensives, and the immunosuppressant cyclosporin. *CYP2D6* catalyzes such rapid breakdown of the pain reliever codeine that patients within the haplotype whose liver contains large amounts of CYP2D6 derive virtually no benefit from taking the standard dose of codeine.

Another example is that consumption of the pharmaceuticals tolbutamide, warfarin, or phenytoin can be riskier for people who possess a mutation (i.e., an SNP that codes for less or no expression of *CYP2C9*) within their liver tissue. That is because CYP2C9 enzyme causes rapid metabolism of tolbutamide, warfarin, and phenytoin (and some other pharmaceuticals); so the *typical dose* could result in higher-than-expected bloodstream levels of those pharmaceuticals in people possessing that particular SNP. See also *Cytochrome P450*, *Cytochrome P4503A4*, *Cytochrome*, *Enzyme*, *Metabolism*, *Haplotype*, *Mutation*, *SNP*, *Coding sequence*, *Express*, *Expressivity*, *Pharmacogenetics*, *Pro-drug therapy*.

Cytochrome P4503A4

An enzyme within the liver that, in humans, catalyzes reactions involved in the metabolism (breakdown) of estrogen plus approximately half of all modern pharmaceuticals. Those pharmaceuticals include some sedatives, antihypertensives, birth control pills, antihistamine terfenadine, and immunosuppressant cyclosporin.

Prior consumption of grapefruit juice decreases the activity of cytochrome P450 3A4 (lessening its ability to break down estrogen and some pharmaceuticals). Prior consumption of Saint John's wort (a plant native to Europe) increases its activity. See also *Enzyme*, *Cytochrome P450*, *Metabolism*, *Histamine*, *Cyclosporin*, *Metabolic pathway*, *Cytochrome*, *Saint John's wort*.

Cytokines

A large class of glycoproteins similar to lymphokines but produced by nonlymphocytic cells such as normal macrophages, fibroblasts, keratinocytes and a variety of transformed cell lines. They participate in regulating immunological and inflammatory processes and can contribute to repair processes and to the regulation of normal cell growth and differentiation.

Although cytokines are not produced by glands, they are hormone-like in their intercellular regulatory functions. They are active at very low concentrations and for the most part appear to function nonspecifically. For example, the cytokines stimulate the endothelial cells to express (synthesize and present) P-selectins and E-selectins on the internal surfaces (of blood vessels). These selectins protrude into the bloodstream, which causes passing white blood cells (leukocytes) to adhere to the selectins, and then leave the bloodstream by *squeezing* between adjacent endothelial cells. Cytokines are exemplified by the interferons. See also *Interleukin-1 (IL-1)*, *Lymphokines*, *Interferons*, *Glycoprotein*, *Protein*, *T cells*, *Interleukin-6 (IL-6)*, *Macrophage*, *Lectins*, *Fibroblasts*, *Hormone*, *Endothelial cells*, *Endothelium*, *Selectins*, *P-selectin*, *ELAM-1*, *Leukocytes*, *Adhesion molecule*, *Erythropoietin (EPO)*.

Cytokinins

A widely occurring (i.e., in many species) category of plant hormones, most of which promote growth/cell division in plants.

During 2009, John Burke discovered that application of applicable man-made cytokinins to young cotton plants caused those plants' roots to quickly grow and spread (deeper in soil) faster than they otherwise would. This had the effect of making those cotton plants more resistant to drought. See also *Okant hormone*, *Stress hormones*, *Pink pigmented facultative methylotroph (PPFM)*.

Cytolysis

The dissolution of cells, particularly by destruction of their surface membranes. See also *Lysis*, *Cecrophins*, *Lysozyme*, *Magainins*, *Complement*, *Complement cascade*.

Cytomegalovirus (CMV)

A virus that infects different groups of people in varying amount, depending on their behavior. For example, 40%–90% of American heterosexuals and about 95% of homosexuals are infected with CMV. CMV normally produces a latent (nonclinical, nonobvious) infection because a healthy immune system's neutrophils produce a protein known as TRAIL that causes death of CMV-infected cells. However, when AIDS or other events (e.g., organ transplant) result in immune system suppression, CMV produces a febrile (fever-causing) illness that is usually mild in nature but can become retinitis (eye infection).

CMV can result in babies born to CMV-infected individuals being born blind, deaf, or brain damaged.

CMV can be treated (to halt life- and sight-threatening infection) in immunocompromised patients (i.e., transplant patients and AIDS victims) with Ganciclovir™, an antiviral compound developed by Syntex, or Foscarnet™, a compound developed by Astra Pharmaceuticals.

In 1996, Stephen E. Epstein found that latent CMV may cause changes in artery wall cells that aid clogging of arteries in adults (especially following balloon angioplasty). See also *Virus*, *Acquired immune deficiency syndrome (AIDS)*, *Neutrophils*.

Cytopathic

Damaging to cells. See also *Cell*.

Cytoplasm

From the Greek words *kytos*, which means "vessel to hold liquid," and *plasma*, which means "form." Cytoplasm refers to the protoplasmic contents of the cell (e.g., plastids, mitochondria) not including the nucleus. See also *Cell, Nucleus, Protoplasm, Cytoplasmic DNA, Plasma membrane, Plastid, Mitochondria, Chloroplasts.*

Cytoplasmic DNA

The DNA within an organism (e.g., plant) that is not inside cell's nucleus. Cytoplasmic DNA (i.e., located in the cells' mitochondria and the chloroplasts) is not transferred from plant to plant via pollen as *nuclear DNA* is. See also *Deoxyribonucleic acid (DNA), Organism, Cell, Cytoplasm, Nucleus, Mitochondria, Mitochondrial DNA, Chloroplasts.*

Cytoplasmic Genes

See *Cytoplasmic DNA.*

Cytoplasmic Membrane

See *Plasma membrane.*

Cytosine

A pyrimidine occurring as a fundamental unit (one of the bases) of nucleic acids. See also *Nucleic acids, Base (nucleotide).*

Cytoskeleton

This term refers to the *structural framework* of a cell/cytoplasm. Some antibiotics work (e.g., kill a bacterial cell) via inhibition of *cytoskeleton building/repair* by relevant bacterial cells.

During the late 1970s, scientists utilized fluorescent-labeled monoclonal antibodies to *show* (visually, under microscope) the existence of the cytoskeleton within cells.

Components of a cell's cytoskeleton include microtubules, actin, etc. See also *Cell, Cytoplasm, Plasma membrane, Bacteria, Antibiotic, Label (fluorescent), Monoclonal antibodies (MAb), Label (fluorescent), Microtubules, Actin.*

Cytotoxic

Poisonous to cells. See also *Cell.*

Cytotoxic Killer Lymphocyte

See *Cytotoxic T cells.*

Cytotoxic T Cells

Also called "killer T cells." T cells that have been created by stimulated helper T cells. The T refers to cells of the cellular system rather than to cells of the humoral system (B cells). Cytotoxic T cells detect and destroy infected body cells by use of a special type of protein. The protein attaches to the infected cell's membrane and forms holes in it. This allows the uncontrolled leakage of ions out of and water into the cell, causing cell death. In general, the loss of the integrity of the cell membrane leads to death. The cytotoxic T cells also transmit a signal to the (leaking) infected cells that causes the cell to *chew up* its DNA. This includes its own DNA as well as that of the virus. See also *Cecrophins, Magainins, Interleukin-4 (IL-4), Helper T cells (T4 cells), Virus, T cells, Suppressor T cells, Protein, Interleukin-2 (IL-2), Deoxyribonucleic acid (DNA), Plasma membrane, Insulin-dependent diabetes mellitus.*

CZE

Acronym for capillary zone electrophoresis. See *Capillary zone electrophoresis.*

D

"Designer" Chromosome

Refers to a chromosome that has been entirely synthesized (manufactured) by man. Potential uses include inserting it into the DNA of a microorganism to cause that microorganism to produce certain new medicines, to produce biofuels, to produce industrial raw materials, to produce food, etc. See also *Chromosomes*, *Deoxyribonucleic acid (DNA)*, *Microorganism*, *Synthetic biology*.

D Loop

A region within mitochondrial DNA in which a short stretch of RNA is paired with one strand of DNA, displacing the original partner DNA strand in this region. The same term is used also to describe the displacement of a region of one strand of duplex DNA by a single-stranded invader in the reaction catalyzed by RecA protein. See also *Deoxyribonucleic acid (DNA)*, *Mitochondria*, *Ribonucleic acid (RNA)*, *Duplex*, *Displacement loop*.

Daffodil Rice

See *Golden rice*.

Daffodils

Refers to the approximately 80 species of flowering plants within the genus *Narcissus*. Native to Southern Europe and Northern Africa, they are the source of *golden rice* and the Alzheimer's disease treatment compound *galantamine hydrobromide*. See also *Golden rice*, *Alzheimer's disease*.

Daidzein

See *Isoflavones*.

Daidzin

The β-glycoside form (isomer in which glucose is attached to the molecule at the seven position of the A ring) of the isoflavone known as daidzein (aglycone form). See also *Isoflavones*, *Isomer*, *Daidzein*.

Dalton

A unit of mass very nearly equal to that of a hydrogen atom (precisely equal to 1.0000 on the atomic mass scale). Named after John Dalton (1766–1844) who developed the atomic theory of matter. It is 1.660×10^{-24} g. See also *Kilodalton (kDa)*.

Dark Genome

Refers to noncoding DNA within a genome. See also *Deoxyribonucleic acid (DNA)*, *Genome*, *Intron*.

Data Mining

Refers to a computational methodology utilized to search for relationships between and *overall patterns among* the myriad data within a (bioinformatics) database.

Some *data mining* techniques include neural network analysis, genetic algorithms (which improve themselves over time), volume rendering, etc. See also *Bioinformatics*, *In silico biology*, *In silico screening*, *Volume rendering*.

DBT

An acronym that is used by some to designate the Indian Department of Biotechnology. See *Indian Department of Biotechnology*.

DC

Acronym for dendritic cells. See *Dendritic cells*.

ddPCR

Acronym for *droplet digital polymerase chain reaction*. See also *Polymerase chain reaction (PCR)*.

ddRNAi

Acronym for *DNA-directed RNA interference*. See also *DNA-directed RNA interference*.

De novo Sequencing

The sequencing of protein or DNA molecules via techniques that do not depend on having in your possession some preexisting knowledge of what the sequence of that particular molecule is. See also *Sequencing (of protein molecules)*, *Sequencing (of DNA molecules)*, *Sequence (of a protein molecule)*, *Sequence (of a DNA molecule)*.

Deamidation

See *Posttranslational modification of protein*.

Deamination

The removal of amino groups from molecules (e.g., in an animal's food) via the energy-consuming metabolism of *excess* amino acids eaten by that animal. For example, when livestock are fed more lysine (amino acid) than their body needs in a given day (i.e., animals' bodies can only utilize the essential amino acids in precise amounts/ratios of their daily diet), that excess lysine is metabolized to urea and then excreted in the animal's urine. See also *Metabolism*, *Amino acid*, *Essential amino acids*, *Lysine (lys)*, *Ideal protein*, *Ideal protein concept*, *PDCAAS*, *ACC synthase*.

D

Defective Virus

A virus that, by itself, is unable to reproduce when infecting its host (cell), but that can grow in the presence of another virus. This other virus provides the necessary molecular machinery that the first virus lacks. See also *Virus, Cell, Symbiotic.*

Defensins

A class of proteins that inhibits certain fungi- and bacteria-caused diseases. These defensin proteins are produced as a natural defense by some plants and some animals.

For example, the alfalfa plant produces a defensin known as "alfAFP" (alfalfa antifungal peptide). In addition to protecting the alfalfa plant from certain diseases, the alfAFP also inhibits a fungal disease known as "potato early dying complex" (also called "Verticillium wilt"), which is caused by the fungus *Verticillium dahliae.*

For example, the flowers of the ornamental tobacco plant produce a defensin known as NaD1, which inhibits a wide variety of fungal diseases.

For example, the immune system of honeybees (*Apis mellifera*) produces a defensin known as "defensin-1 protein," which is added by bees to honey. See also *Protein, Peptide, Fungus, Paneth cells.*

Deficiency

Refers to the insufficiency (or total absence) of relevant form(s) of nutrients, enzymes, or environmental inputs (e.g., temperature) required for physiological functions. Such a deficiency can prevent or hinder an organism's development, growth, metabolism, or other physiological functions. See also *Enzyme, Metabolism, Flux, Digestion (within organisms), Essential nutrients, Essential amino acids, Essential fatty acids, Isozymes, Iron deficiency anemia (IDA), Ergotamine, Insulin.*

Degenerate Codons

Two or more codons that code for the same amino acid. For example, isoleucine is specified by the AUU, AUC, and AUA triplets. Since in this case more than one triplet codes for isoleucine, the codons are called degenerate. See also *Genetic code, Codon.*

Dehydrogenases

Enzymes that catalyze the removal of pairs of hydrogen atoms from their substrates. See also *Substrate (chemical), Glutamate dehydrogenase, Enzyme, Dehydrogenation.*

Dehydrogenation

The removal of hydrogen atoms from molecules. When those molecules are the components of vegetable oils/fats, this results in a lower content percentage of *saturated* fats. See also *Fats, Monounsaturated fats, Saturated fatty acids (SAFA), Fatty acid.*

Deinococcus radiodurans

A species of bacteria that is capable of surviving 1.5 million rads of gamma radiation (i.e., 3000 times the lethal radiation dose for humans), surviving long periods of dehydration, and surviving high doses of ultraviolet radiation.

Deinococcus radiodurans was discovered in 1956, in some canned meat. See also *Bacteria, Extremophilic bacteria.*

Delaney Clause

Formerly part of American federal law (1959 Delaney amendment to Food, Drug, and Cosmetic Act); it was eliminated during 1996. The Delaney Clause had set a zero-risk tolerance level for carcinogenic pesticide residues in processed foods. See also *Carcinogen.*

Deleterious Mutations

See *Copy number variation.*

Deletion Mutation

See *Deletions.*

Deletions

Loss of a section of the genetic material from a chromosome. The size of a deleted material can vary from a single nucleotide to sections containing a number of genes.

Some deletions (in miRNA genes) can lead to cancer. See also *Gene, Chromosomes, Nucleotide, CRE-LOX system, Mutation, Mutation breeding, miRNA gene, Cancer, TALENs.*

Delta 12 Desaturase

An enzyme that is present within the soybean plant and in other oilseed crops (e.g., sunflower, canola, maize/corn).

Delta 12 desaturase ($\Delta 12$) is involved in the synthesis "pathway" utilized by oilseed crops to synthesize (i.e., *manufacture*) polyunsaturated fatty acids (e.g., linoleic acid, linolenic acid) from monounsaturated fatty acids (e.g., oleic acid) in seeds (while those seeds are developing). See also *Enzyme, Desaturase, Fatty acid, Unsaturated fatty acid, Monounsaturated fatty acids (MUFA), Polyunsaturated fatty acids (PUFA), Pathway, Oleic acid, Linoleic acid, Linolenic acid, Soybean plant, High-oleic sunflowers, Corn, Canola, Cosuppression.*

Delta 32

See *CCR5-delta 32.*

Delta Endotoxins

See *CRY proteins, Protein.*

Demethylase

See *Methylated.*

Demethylating Agent

Refers to pertinent enzymes, chemicals, etc. that cause demethylation of DNA. See *Demethylation.*

Demethylation

The enzymatic removal of methyl submolecule groups from DNA inside living cells (e.g., due to epigenetic gene programming) via

DNA-repair dioxygenase enzyme, LSD1 enzyme, certain pharmaceuticals, etc. See also *Methylated, DNA methylation, Enzyme, Deoxyribonucleic acid (DNA), Cell, Epigenetic.*

Denaturation

The loss of the native conformation of a macromolecule resulting, for instance, from heat, extreme pH (i.e., by acidity or basicity) changes, chemical treatment, etc. It is accompanied by loss of biological activity. See also *Conformation, Configuration, Macromolecules, Laser inactivation, Biological activity, Structural biology, Unfoldases.*

Denatured DNA

DNA that has been converted from double-stranded to single-stranded form by a denaturation process such as heating the DNA solution. In the case of heat denaturation, the solution becomes very gelatinous and viscous. See also *Denaturation, Deoxyribonucleic acid (DNA), Duplex.*

Denaturing Gradient Gel Electrophoresis

One particular method of gel electrophoresis, which can be utilized to separate different segments of double-stranded DNA or RNA from each other.

In denaturing gradient gel electrophoresis, a steadily increasing level of a *denaturing agent* such as urea and formamide is present in the "path" of the DNA or RNA molecules as they are moved under the influence of the applied electrical field within the gel matrix (e.g., polyacrylamide gel). The denaturing agent(s) causes the double-stranded DNA or RNA molecules to denature (e.g., "unwind" into single strands), which changes those DNA/RNA molecules' individual rates of movement through the gel, thereby enhancing their separation and ease of identification via the gel electrophoresis process. See also *Deoxyribonucleic acid (DNA), Double helix, Polyacrylamide gel electrophoresis (PAGE), Denaturation, Denatured DNA, Denaturing polyacrylamide gel electrophoresis, Short interfering RNA (siRNA).*

Denaturing Polyacrylamide Gel Electrophoresis

The use of polyacrylamide gel electrophoresis in order to separate and analyze DNA fragments (sequences) after that DNA is first denatured. This methodology can be utilized to scan DNA in order to detect point mutations. See also *Polyacrylamide gel electrophoresis (PAGE), Point mutation, Denaturing gradient gel electrophoresis, Deoxyribonucleic acid (DNA), Denatured DNA, Base excision sequence scanning (BESS).*

Dendrimers

Polymers (i.e., molecules composed of repeating atomic units within the molecule) that repeatedly branch (while "growing" due to addition of more atoms in a repeating pattern) until that branching is stopped by the physical constraint of contacting itself (i.e., having formed a complete, hollow sphere). Named based on the Greek word for *tree.*

Research indicates that some dendrimers can encapsulate certain pharmaceuticals and subsequently deliver the pharmaceuticals into a person's bloodstream when those drug-containing dendrimers are spread upon the surface of the skin.

Discovered during 1979 by Donald A. Tomalia, dendrimers possess sites on their exterior surface to which genetic material (e.g., genes or other portions of DNA) can be "attached." Dendrimers bearing such genetic material have been shown to be able to successfully transfer that genetic material into more than 30 types of living animal cells. See also *Polymer, Dendritic polymers, Nanocapsules, Gene, Genetic engineering, Gene delivery, Informational molecules, Coding sequence, Tumor-suppressor genes, Deoxyribonucleic acid (DNA), Genetic targeting, Genetics, Dendrimersomes.*

Dendrimersomes

Refers to self-assembling nanostructures that are formed when *Janus* dendrimers (i.e., polymers composed of many repeating, highly branched, bifunctional atomic units within the molecule) are added to water.

These nanostructures take the form of tiny bubbles (spherical), tubes, disks, etc. that can encapsulate certain pharmaceuticals or genes and subsequently deliver those payloads into a person's bloodstream or targeted cells. See also *Self-assembly (of a large molecular structure), Dendrimers, Polymer, Nanocapsules, Gene, Gene delivery.*

Dendrites (in Bone)

Refers to small extensions (protruding "arms") of the bone cells known as osteocytes. Among other functions, these dendrites sense mechanical loading (e.g., impact of exercise on the bones), which thereby promotes maintenance of good bone health. See also *Osteoporosis.*

Dendrites (in Brain)

Highly branched structures that extend from the (nucleus of) neurons to (synapse junctions with) other neurons (e.g., in human brain tissue). The primary purpose of dendrites is to "process" signals that are generated/received at the synapses (e.g., from the dendrites of adjoining neurons).

Neuron ribosomes are located in the dendritic spines, the dendrite projections that form synapses (i.e., the junctions between dendrites where "signal transfer" between neurons takes place). Thus, those ribosomes make the proteins that are crucial to learning and memory (e.g., accomplished via growth/changes of dendrites).

Messenger RNAs are synthesized (i.e., "manufactured") in the nucleus of the neuron and then transported on microtubules (filaments within neuron cell) to the ribosomes in the dendrites, where they cause manufacture of proteins (e.g., enzymes) in response to synapse activity (i.e., signals). See also *Neuron, Cell, Neurotransmitter, Ribosomes, Protein, Enzyme, Messenger RNA (mRNA), Microtubules, Synapse.*

Dendritic Cells

Discovered in 1973 by Ralph Steinman, these are rare white blood cells that act to stimulate the human immune system lymphocytes (i.e., *naive* T cells, or *naive* B cells) to become *effector T cells* or *effector B cells* and combat certain pathogens by "presenting" the *antigens of those pathogens* to those naive T cells or naive B cells.

Dendritic cells are present within the lymph system (and some nonlymphoid tissues such as the external wall of intestines), and their action of recognizing and "presenting" *antigens (of some tumors)* can cause the immune system to also halt certain types of cancer.

Within the external wall of the intestines, dendritic cells act to keep T cells unresponsive (i.e., naive) so that the immune system does not mount a harmful response against antigens in food or against the commensal intestinal bacteria. Such an immune response can result in autoimmune diseases (e.g., Crohn's disease or ulcerative colitis). See also *Cell, White blood cells, Lymphocyte, T cells, Pathogen, Antigen, Immune response, Cancer, Leukocytes, Major histocompatibility antigen—Class II, Major histocompatibility complex (MHC), Gut-associated lymphoid tissues (GALT), Commensal, Autoimmune disease, Crohn's disease, Nanovaccine.*

Dendritic Langerhans Cells

A type of cell, located in the mucous membranes of the mouth and genital areas, that during inflammation (i.e., start of infection process) extends narrow cellular protrusions known as dendrites through cellular seams in those body-surface membranes. The dendrites have receptor molecules on their tips, to bind to antigens/pathogens and then bring them in to the Langerhans cells to be processed and presented to the body's immune system to facilitate that immune system combating the pathogen.

Unfortunately, that action permits the human immunodeficiency virus (i.e., the virus that causes AIDS) to enter and infect the body—even when there are no cuts or abrasions through those mucous membranes. See also *Innate immune response, Cellular immune response, Human immunodeficiency virus type 1 (HIV-1), Human immunodeficiency virus type 2 (HIV-2), Acquired immune deficiency syndrome (AIDS), Adhesion molecule, Dendritic polymers.*

Dendritic Polymers

Polymers (i.e., molecules composed of repeating atomic units within the molecule) that repeatedly branch (while "growing" due to the addition of more atoms in a repeating pattern) until that branching is stopped (e.g., by physical constraints, for those polymers within living tissues). In the absence of physical constraints, dendritic polymers can continue branching (and growing) until they form a complete tiny (hollow) sphere.

Such spheres are potentially useful for

- Protecting and "delivering" a fragile pharmaceutical molecule to specific tissue(s) within the body
- Delivery of high-potency (toxic) pharmaceutical(s) to specific targeted sites within the body (where that pharmaceutical is then released) without harming nontargeted tissues

See also *Polymer, Dendrimers, DNA buckyballs.*

Denitrification

Reduction of nitrate to nitrites or into gaseous oxides of nitrogen or even into free nitrogen by organisms. See also *Reduction (in a chemical reaction).*

Denitrification

The process (i.e., internal respiration) via which denitrifying bacteria (e.g., in soil) convert nitrates to gaseous nitrogen/nitrous oxide, which then enters the atmosphere. See also *Nitrates, Bacteria, Respiration.*

Denitrifying Bacteria

See *Denitrification.*

Dent Corn

See *Amylose.*

Deoxynivalenol

A mycotoxin (i.e., toxin that is naturally produced by a fungus under certain conditions) that, under specific temperature/moisture conditions, is sometimes produced by certain *Fusarium* fungi (e.g., *Fusarium graminearum* and *Fusarium culmorum*) growing in some grains (e.g., wheat *Triticum aestivum* or corn/maize *Zea mays* L.).

Deoxynivalenol is also known as DON, and/or "vomitoxin," because certain animals (especially swine) will often vomit after they have consumed grain that contains deoxynivalenol, due to its toxicity. See also *Toxin, DON, Mycotoxins, Fungus, Fusarium.*

Deoxyribonucleic Acid (DNA)

Discovered by Frederick Miescher in 1869, it is the chemical basis for genes. The chemical building blocks (molecules) of which genes (i.e., paired nucleotide units that code for a protein to be produced by a cell's machinery, such as its ribosomes) are constructed.

Every inherited characteristic has its origin somewhere in the code of the organism's complement of DNA. The code is made up of subunits, called nucleic acids. The sequence of the four nucleic acids is interpreted by certain molecular machines (systems) to produce the proteins required by an organism.

The structure of the DNA molecule was elucidated in 1953 by James Watson, Francis Crick, and Maurice Wilkins. The DNA molecule is a linear polymer made up of deoxyribonucleotide repeating units (composed of the sugar 2-deoxyribose, phosphate, and a purine or pyrimidine base). The bases are linked by a phosphate group, joining the 3′ position of one sugar to the 5′ position of the next sugar. Most molecules are double stranded and antiparallel, resulting in a right-handed helix structure that is held together by hydrogen bonds between a purine on one chain and pyrimidine on the other chain. DNA is the carrier of genetic information, which is encoded in the sequence of bases; it is present in chromosomes and chromosomal material of cell organelles such as mitochondria and chloroplasts and also present in some viruses. See also *A-DNA, B-DNA, cDNA, Z-DNA, Transcription, Antiparallel, Double helix, Messenger RNA (mRNA), Nucleotide, Protein, Ribosomes, Genetic code, Gene, Chromosomes, Chromatids, Chromatin, Mitochondrial DNA, Cytoplasmic DNA, Nuclear DNA, Hydrogen bonding.*

Deprotection (of a Peptide)

See *HF cleavage.*

Derepression

The opposite of *repression (of gene transcription/translation).* It typically occurs via removal of the *repressor protein* from the applicable site on relevant DNA (or RNA) molecule. See also *Repression (of gene transcription/translation), Repressor (protein), Gene, Transcription, Translation, Deoxyribonucleic acid (DNA).*

Desaturase

An enzyme (group) "family" that is present within the soybean plant and other oilseed crops (e.g., sunflower, canola, corn/maize). One or more desaturases are involved in the synthesis "pathway" via which oilseed crops produce unsaturated fatty acids (e.g., linoleic acid).

A desaturase is also involved in the production of beta carotene (in some plants). See also *Enzyme, Fats, Stearoyl-ACP desaturase, Delta 12 desaturase, Soybean plant, High-oleic sunflowers, Pathway, Linoleic acid, Fatty acid, Unsaturated fatty acid, Golden rice, Beta carotene.*

Desert Hedgehog Protein (Dhh)

See *Hedgehog proteins.*

Desferrioxamine Manganese

An iron-chelating agent (i.e., it chemically binds to iron atoms in the blood, thus trapping the iron atoms). The molecule also acts as a human superoxide dismutase mimic by capturing harmful oxygen free radicals in the blood before they damage the walls of blood vessels. Recent research indicates that desferrioxamine manganese may be useful in blocking the onset of cataracts. See also *Human superoxide dismutase (hSOD), Xanthine oxidase, Lazaroids.*

Desulfovibrio

A genus of bacteria that reduces sulfate to H_2S (hydrogen sulfide). Energy is obtained by oxidation of H_2 or organic molecules. Not a strict autotroph because CO_2 cannot be used as a sole carbon source. See also *Reduction (in a chemical reaction), Autotroph.*

Dextran

A polysaccharide produced by yeasts and bacteria as an energy storage reservoir (analogous to fat in humans). Consists of glucose residues, joined almost exclusively by alpha-1,6 linkages. Occasional branches (in the molecule) are formed by alpha 1,2, alpha 1,3, or alpha 1,4 linkages. Which linkage is used depends on the species of yeast or bacteria producing the dextran. See also *Polysaccharides.*

Dextrorotary (D) Isomer

A stereoisomer that rotates the plane of plane-polarized light to the right. Dextro means right. See also *Stereoisomers, Levorotary* (L) *isomer, Polarimeter.*

DGGE

Acronym for denaturing gradient gel electrophoresis. See *Denaturing gradient gel electrophoresis.*

DHA

See *Docosahexaenoic acid (DHA).*

dHPLC

Acronym for *denaturing high-pressure liquid chromatography.* See *HPLC, Denaturation.*

Diabetes

A grouping of diseases in which the body either does not synthesize (i.e., *manufacture*) insulin or else its tissues are insensitive to the insulin that it does synthesize.

Approximately 5%–10% of all people with diabetes are unable to synthesize insulin (e.g., because their insulin-making tissue was destroyed by autoimmune disease). Approximately 90%–95% of all people with diabetes are insensitive to the insulin their body synthesizes.

One of the many impacts of diabetes is an increase in the sorbitol content within cells, which causes swelling of certain cells due to osmotic pressure. When cells thus swell in the lens of the eye, this can lead to creation of diabetic cataracts. See also *Pancreas, Insulin, Insulin-dependent diabetes mellitis (IDDM), Flux, Cell, Autoimmune disease, Beta cells, N-3 fatty acids, Calpain-10, Type I diabetes, Type II diabetes, Haptoglobin, Osmotic pressure.*

Diacylglycerols

Molecules that consist of two fatty acids attached to a glycerol "backbone." Research during the 1990s indicated that consumption of vegetable oils (e.g., used in frying foods) containing primarily diacylglycerols (versus typical triacylglycerols) is less likely to result in it being deposited as body fat (adipose tissue). See also *Fatty acid, Saturated fatty acids (SAFA), Unsaturated fatty acid, Adipose, Triacylglycerols.*

Dialysis

The separation of low molecular weight compounds from high molecular weight components in solution by diffusion through a semipermeable membrane. Frequently utilized to remove salts and to remove biological effectors (such as nicotinamide adenine dinucleotides, nucleotide phosphates, etc.) from polymeric molecules such as protein, DNA, or RNA. Commonly used membranes have a molecular weight cutoff (threshold) of around 10,000 Da, but other membrane pore sizes are available. See also *Hollow fiber separation, Active transport.*

Diamond vs. Chakrabarty

See *Chakrabarty decision.*

Diastereoisomers

Four variations of a given molecule consisting of a pair of stereoisomers about a second asymmetric carbon atom for each of the two isomers of the first asymmetric carbon atom. See also *Stereoisomers, Chiral compound.*

Diced siRNA

See *d-siRNA.*

DICER

See *RNA interference (RNAi).*

Dicer Enzymes

A "family" of RNAse III dsRNA-specific nucleases (i.e., nucleic acid–digesting enzymes). See *RNA interference (RNAi).*

D

Differential Display

A technique of *gene expression analysis* in which two different tissues (or *same* tissue under two different conditions) are compared in terms of proteins expressed. See also *Gene, Gene expression, Gene expression analysis, Genetic code, Express, Protein, Gene expression profiling, Microarray* (*testing*), *Epigenetic.*

Differential Splicing

A cellular process in which numerous mRNA molecules can be created by the joining of different exons (i.e., RNA sequence fragments) within a single RNA molecule. See also *Cell, Transcription, Messenger RNA* (*mRNA*), *Splicing, Ribonucleic acid* (*RNA*), *Splice variants, Splicing junctions.*

Differentiation

Refers to processes via which a single type of cells (e.g., stem cells, embryonic stem cells) become multiple, different types of (specialized) cells. Among many other cues to guide these cells regarding the specific type of cell they become, the physical stiffness of the extracellular matrix touching them is a major factor.

Another factor is biomechanical stress, or force, which living cells within tissue exert on one another.

For example, when preexisting mechanical force disappears—such as at a wound site where cells have been destroyed, leaving an empty cell-free space—a protein molecule known as DII4 is secreted by remaining cells, and it coordinates nearby cells to migrate to the wound site and collectively cover it with new tissue. See also *Cell, Cell differentiation, Epigenetic, Long noncoding RNAs, Stem cells, Stem cell one, Stem cell growth factor* (*SCF*), *Totipotent stem cells, Colony stimulating factors* (*CSFs*), *Embryonic stem cells, Human embryonic stem cells, Hedgehog signaling pathway, Hedgehog proteins, Granulocyte-macrophage colony stimulating factor* (*GM-CSF*), *Multipotent, Adult stem cell, Multipotent adult stem cell, Retinoid X receptor* (*RXR*), *Micro-RNAs, Brassinosteroids, Morphogenetic, Differentiation pathways, Extracellular matrix, PIM-1 protein, CD8*$^{+}$ *T cells, Extracellular matrix.*

Differentiation Pathways

Refers to chemical/gene expression pathways responsible for causing a single type of cells (e.g., stem cells, embryonic stem cells) to become multiple, different types of (specialized) cells.

Expression of each of the specific genes responsible for (i.e., the genes that code for the particular proteins that cause) differentiation is itself controlled by exquisite methylation and acetylation of the histone proteins adjacent to those genes. See also *Differentiation, Epigenetic, Pathway, Deoxyribonucleic acid* (*DNA*), *Gene, Cell, Protein, Express, Gene expression cascade, Histones, Stem cells, Coding sequence, Genetic code, Methylated, DNA methylation.*

DIGE

Acronym for two-dimensional *difference gel electrophoresis.* See also *Two-dimensional* (*2D*) *gel electrophoresis.*

Digestion (within Chemical Production Plants)

Breakdown of feedstocks by various processes (chemical, mechanical, and biological) to yield their desired building block components for inclusion as raw materials in subsequent chemical or biological processes. See also *Anaerobic digestion.*

Digestion (within Organisms)

The enzyme-enhanced hydrolysis (breakdown) of major nutrients (food) in the gastrointestinal system to yield their building block components (to the organism), such as amino acids, fatty acids, or other essential nutrients. See also *Hydrolysis, Fats, Protein, Amino acid, Essential amino acids, Essential nutrients, Fatty acid, Essential fatty acids, Lipase, Ideal protein concept, Enzyme, Proteases, Proteolytic enzymes, Absorption, Trypsin, Lecithin, Protein digestibility-corrected amino acid scoring* (*PDCAAS*).

Diglycerides

See *Triglycerides.*

Dll4 Protein

See *Differentiation.*

Dimeric RNase III Ribonucleases

Also sometimes known as dicer enzymes. See *RNA interference* (*RNAi*).

Dioecious

A category of plants that is characterized by the male and female reproductive structures *not* being located on the same plant (i.e., male on one plant and female on a separate plant).

Diphtheria Antitoxin

Discovered by Emil von Behring in 1900. See *Antitoxin, Enterotoxin.*

Diphtheria Toxin

Refers to the toxin that is produced by the pathogen (*Corynebacterium diphtheriae* bacteria) that causes the disease known as diphtheria. That toxin is an exotoxin that inhibits normal cell protein synthesis, plus it kills cells.

Scientists have been able to make and attach that toxin to *guidance* molecules (e.g., antibodies), which guide/steer the toxin to an intended target inside the body. For example, the pharmaceutical known as Ontak™ is an immunoconjugate consisting of the diphtheria toxin attached to an antibody that binds to the CD25 cell surface protein (i.e., on the surface of *regulatory T cells* that are overactive). That immunoconjugate (i.e., immunotoxin) thereby kills [only] those targeted regulatory T cells. See also *Toxin, Exotoxin, Diphtheria antitoxin, Protein, Antibody, Magic bullet, Immunoconjugate, Immunotoxin, Cancer, Tumor, Cell, T cells, Regulatory T cells, Synthesizing* (*of proteins*).

Diploid

The state of a cell in which each of the chromosomes, except for the sex chromosomes, is always represented twice (46 chromosomes in humans). In contrast to the haploid state in which each chromosome is represented only once. See also *Diplophase, Chromosomes, Homozygous, Triploid.*

Diplophase

A phase in the life cycle of an organism in which the cells of the organism have two copies of each gene. When this state exists the organism is said to be diploid. See also *Diploid, Gene, Homozygous, Cell.*

Dip-Pen Lithography

See *Dip-pen nanolithography.*

Dip-Pen Nanolithography

Refers to the use of atomic force microscopy to apply very small amounts of specific molecules to very precise locations (e.g., probes on the surface of a *microarray*, or DNA/thiol molecules on the individual "pieces" of a *self-assembling molecular structure*). See also *Atomic force microscopy, Microarray (testing), Self-assembly (of a large molecular structure), Directed self-assembly, Probe, Deoxyribonucleic acid (DNA), Thiol group.*

Direct Transfer

Refers to methods of inserting a gene directly into a cell's DNA without the use of a vector. One example of direct transfer is electroporation. See also *Gene, Genetic engineering, Vectors, Cell, Deoxyribonucleic acid (DNA), Electroporation.*

Directed Evolution

See *DNA shuffling.*

Directed Evolution

See *Synthetic biology, DNA shuffling.*

Directed Self-Assembly

Refers to man's use of

- Carefully preplanned man-made molecular components that can be caused to self-assemble via affinity/hybridization to each other of DNA or thiol molecular segments (attached to the relevant man-made molecular components)
- An atomic force microscope stylus tip to apply specific molecular (e.g., thiol, DNA) segments to surfaces such as metals, oxides, etc. in order for those preapplied DNA segments to thereby direct (via hybridization to each other) the assembly of nanometer-scale structures such as gene chips, catalysts, nanoscale circuits, etc.

See also *Deoxyribonucleic acid (DNA), Hybridization (molecular genetics), Dip-pen nanolithography self-assembling molecular machines, Nanometer (nm), Template, Catalyst, Gene chips, Self-assembly (of a large molecular structure), Thiol group.*

Disaccharides

Carbohydrates consisting of two covalently linked monosaccharide units—hence "di" for "two." See also *Oligosaccharides, Monosaccharides, Polysaccharides.*

Disease-Sensitivity Gene

See *Marker-assisted selection.*

Displacement Loop

The unique DNA molecular structure that is created when the (normally) double-stranded DNA molecule takes up/incorporates an inserted third strand of DNA or RNA. See also *Deoxyribonucleic acid (DNA), Duplex, D loop, Ribonucleic acid (RNA).*

Dissimilation

The breakdown of food material to yield energy and building blocks for cellular synthesis. See also *Digestion (within organisms).*

Dissociating Enzymes

See *Harvesting enzymes.*

Distribution

See *ADME tests, Pharmacokinetics.*

Disulfide Bond

An important type of covalent bond formed between two sulfur atoms of different cysteines in a protein molecule (or one each, in two different protein molecules).

Disulfide bonds (linkages, bridges) contribute to holding proteins together and also help provide the internal structure (conformation) of the protein molecule. See also *Protein, Cysteine (cys), Cystine, Conformation, Tertiary structure.*

Disulphide Bond

See *Disulfide bond.*

Diversity (within a Species)

Refers to the genetic variation that exists within a population (of organisms) in a species. For example, black cattle and white cattle; or both toxic and nontoxic strains/serotypes of *Escherichia coliform* (*E. coli*) bacteria. This diversity is due to one or more single-nucleotide polymorphisms in each individual's genome (DNA) within the population of organisms. See also *Species, Single-nucleotide polymorphisms (SNPs), Polymorphism (genetic), Nucleotide, Organism, Strain, Serotypes, Escherichia coliform (E. coli), Escherichia coliform 0157:H7 (E. coli 0157:H7).*

Diversity Biotechnology Consortium

A nonprofit U.S. organization that was formed in August of 1994 by a group of research institutions and companies. The consortium's first president was Stuart A. Kauffman of the Santa Fe Institute. The consortium's purpose is to further the use of molecular diversity as a tool in drug design and in the study of mutating viruses. See also *Molecular diversity, Rational drug design, Diversity estimation (of molecules), Molecular biology, Virus, Mutation, Mutant, Site-directed mutagenesis (SDM), Combinatorial chemistry, Combinatorial biology.*

D

Diversity Estimation (of Molecules)

See *Combinatorial chemistry.*

DLS

Acronym for *dynamic light scattering.* See *Dynamic light scattering.*

DMPK

See *Pharmacokinetics.*

DNA

See *Deoxyribonucleic acid (DNA).*

DNA Acetylation

See *Epigenetic.*

DNA Adduct

Refers to certain chemicals covalently bound to DNA molecules (e.g., within the cells of an organism). For example, following consumption by humans of aristolochic acids (i.e., carcinogenic compounds found in *Aristolochia* plants), adducts of aristolochic acids covalently bound to DNA can be found in cells found in the urine of those people. See also *Deoxyribonucleic acid (DNA), Cell, Organism.*

DNA Analysis

See *DNA profiling.*

DNA Bridges

Large segments of DNA whose sequence (i.e., composition) is known/mapped in total. Those sequences are then utilized by scientists to piece together (i.e., "bridging" the DNA segments) and assemble a (more) complete map (e.g., of an organism's chromosome or genome). See also *Deoxyribonucleic acid (DNA), Genetic map, Sequence (of a DNA molecule), Chromosome, Genome, Sequence map, Shotgun sequencing.*

DNA Buckyballs

Refers to self-assembling hollow balls of approximately 400 nm diameter, composed of selected pieces of branched DNA fused to polystyrene plastic molecules. Because approximately 70% of the volume of these "balls" is hollow, water (and water-soluble pharmaceuticals) can enter the balls' interiors.

Potential uses include

- Protecting and "delivering" a fragile pharmaceutical molecule to specific tissue(s) within the body
- Delivery of high-potency (toxic) pharmaceutical(s) to specific targeted sites within the body (where that pharmaceutical is then released) without harming nontargeted tissues

See also *Deoxyribonucleic acid (DNA), Self-assembly (of a large molecular structure), Nanometers (nm), Nanocapsules, Nanotechnology, Nanoshells, Dendrimers.*

DNA Chimera

One DNA molecule composed of DNA from two different species. See also *Chimera.*

DNA Chip

See *Biochips, Multiplexed assay, Gene expression analysis, ChIP, Proteomics, Nanoparticles.*

DNA Codon

See *Codon.*

DNA Demethylation

See *Demethylation.*

DNA Fingerprinting

See *DNA profiling.*

DNA Fragmentation

The cleavage (i.e., *chewing up*) of DNA (within a cell) at internucleosomal sites on that DNA molecule.

DNA fragmentation during cellular apoptosis prevents (aberrant) a cell's DNA from causing any further problems in the organism's body. See also *Deoxyribonucleic acid (DNA), Cell, Nucleosome, Apoptosis.*

DNA Glycosylase

Refers to a category of enzymes (within cells) that initiate repair of a cell's (damaged) DNA under certain circumstances. See also *Enzyme, Cell, Deoxyribonucleic acid (DNA).*

DNA Gyrase

An enzyme, also known as "helix unwinding protein," that works to "relax" the tension within a supercoiled DNA molecule. See also *Enzyme, deoxyribonucleic acid (DNA), Helix, Double helix, Supercoiling, Positive supercoiling, DNA topoisomerase, Protein.*

DNA Helicase

See *Helicase.*

DNA Hybridization

See *Hybridization (molecular genetics), Hybridization surfaces.*

DNA Ladder

See *Molecular-weight size marker.*

DNA Ligase

Discovered during the 1960s by Baldomero Olivera, it is an enzyme that creates a phosphodiester bond between the 3′ end of one DNA segment and the 5′ end of another, while they are base paired to a template strand. The enzyme seals (joins) the ends of single-stranded DNA in a duplex DNA chain. DNA ligase constitutes a part of the DNA-repair mechanism available to the cell. See also *Nick*, *Ligase*, *Deoxyribonucleic acid (DNA)*, *Sliding clamps*, *Gene repair (natural)*, *Duplex*.

DNA Looping

Refers to a process in which certain DNA sequence(s) within some introns in an organism's DNA can interact with (e.g., "turn on," "turn off") a specific gene that is located a long distance away (within the same DNA molecule) from that initial DNA segment. This physical interaction (i.e., creation of a large loop in the organism's DNA molecule, to cause the applicable intron to "touch" the relevant gene) results in an apparent genetic effect. For example, via such DNA looping, aberrant DNA sequences in fourteen different introns result in increased risk of bowel cancer for those people whose DNA contains one or more of those aberrant DNA sequences. See also *Deoxyribonucleic acid (DNA)*, *Gene*, *Intron*, *Cancer*, *Sequence (of a DNA molecule)*.

DNA Marker

See *Marker (DNA marker)*.

DNA Melting Temperature

See *Melting temperature (of DNA)* (T_m).

DNA Methylase

Refers to a category of enzymes (within a cell) that catalyze the addition of methyl groups ($-CH_3$) to DNA molecules. The methyl groups ($-CH_3$) thereby inactivate relevant genes in the cell's DNA. See also *Enzyme*, *Cell*, *Catalyst*, *Gene*, *DNA methylation*, *DNA methyltransferases*.

DNA Methylation

Refers to a process resulting in a DNA molecule that either has one methyl submolecular group attached to that DNA molecule, has several methyl submolecular groups attached to that DNA molecule, or is saturated with methyl groups (i.e., methyl submolecular groups, CH_3, have attached themselves to the DNA molecule's "backbone" at all possible locations on that DNA molecule). DNA methylation is used by healthy cells to perform the following:

- "Turn off" certain genes when those particular genes are not needed (e.g., they turn off genes involved in juvenile development after the organism reaches adulthood).
- Preferentially express certain alleles (e.g., those inherited from the mother or from the father). For example, in mice, more *maternal-origin* alleles are expressed within the *developing* brain and more *paternal-origin* alleles are expressed within the *adult* mouse brain than would occur from a simple random 50/50 contribution of parental alleles to the offspring's DNA.
- Stabilize genome/prevent the spread/activation of potentially harmful nucleic acids (e.g., certain transposable elements) within the organism's genome. For example, the mouse pigment gene known as *agouti* can be rendered defective when a certain transposable element embeds itself in that gene's nearby regulatory sequence, resulting in yellow or mottled mouse fur, which is less effective in camouflaging the mouse from predators.
- Enable some genomes to render their organism more adaptive to its environment. For example, the black truffle fungi (*Tuber melanosporum*), whose genome consists of approximately 58% "jumping genes" (also known as transposable genetic elements, or transposons), utilize reversible DNA methylation to "turn on" and "turn off" the transposable genetic elements as needed due to changes in its environment.
- Inactivate X chromosome (required in humans and some other species for creation of males).

DNA methylation causes *inbreeding depression* (i.e., offspring resulting from the mating of two closely related individuals are less fit and less fertile than offspring from mating of individuals who are not closely related). DNA methylation can be impacted by environmental factors including diet, exercise, and stress. DNA methylation (e.g., of tumor suppressor genes that would normally prevent inappropriate cell division/proliferation) also occurs in some cancers.

The specific DNA methylation resulting from the addition of a methyl group to the particular DNA base cysteine is vital to the regulation of many cellular processes such as embryonic development, gene transcription, and the cell's chromatin structure.

Some DNA methylations are reversible and can impact an organism's behavior. For example, during the lifetime of honeybees (*Apis mellifera* L.), specific patterns of DNA methylation in 155 genes found within their brain cells determine which individuals become *nurse bees* (who feed and clean bee larvae) versus *forager bees* (who travel out from hive to find and bring back nectar from flowers). If there is a sudden shortage of *nurse bees* in the hive, some of the *forager bees* will revert to become *nurse bees* via changes in the methylation patterns of 57 of those 155 genes.

Some DNA methylations (e.g., accomplished via feeding of a small amount of a *methyl group donor* compound such as folic acid) can correct certain genetic defects. For example, feeding of folic acid to pregnant mice from a line bearing the earlier-described defective *agouti* gene typically results in at least half of her offspring having normal brown color fur (making them more likely to remain hidden from predators). See also *Deoxyribonucleic acid (DNA)*, *Gene*, *Allele*, *Genome*, *DNA methyltransferases*, *Methylated*, *Epigenetic marks*, *Cell*, *Imprinting*, *Cancer*, *Transcription*, *Genetic code*, *Messenger RNA (mRNA)*, *P53 gene*, *Tumor-suppressor genes*, *Epigenetic*, *Short interfering RNA (siRNA)*, *Transposable element*, *Regulatory sequence*, *Differentiation pathways*, *Chromatin*, *Chromatin modification*, *Chromosome*, *X chromosome*, *Genomic imprinting*, *Alkylating agents*.

DNA Methyltransferases

Enzymes that catalyze methylation of the C-5 position of cytosine (within an organism's DNA). See also *Deoxyribonucleic acid (DNA)*, *Enzyme*, *Cytosine*, *Organism*, *DNA methylation*, *DNA methylase*, *Epigenetics*, *Methylated*.

DNA Microarray

Initially developed by Patrick Brown during the 1980s, these microarrays enable analysis of the levels of expression of genes in an organism or comparison of gene expression levels (e.g., between diseased and nondiseased tissues) via hybridization of messenger RNA (mRNA) to its counterpart DNA sequence when biological samples containing DNA (e.g., in liquid) are passed over the array surface.

To manufacture the DNA microarray, cellular mRNA is used to make segments of complementary DNA (cDNA) in lengths of approximately 500–5000 base pairs long, using the reverse transcriptase polymerase chain reaction. These cDNA segments are then attached to a nylon or glass surface at *known* spots, so when hybridization of sample DNA occurs, the location of the spot tells what DNA was in sample.

Another way to manufacture another type of DNA microarray is to similarly attach oligonucleotides or peptide nucleic acids of known sequence (composition) at known spots on the nylon or glass surface and pass the biological sample containing DNA (e.g., in liquid) over that surface to identify the DNA in the sample via which spot it hybridizes to. See also *Gene, Organism, Biochips, ChIP, Microfluidics, Deoxyribonucleic acid (DNA), Messenger RNA (mRNA), Hybridization (molecular genetics), Express, Gene expression analysis, Proteomics, Microarray (testing), Multiplexed assay, Oligonucleotide, Nucleic acids, Sequence (of a DNA molecule), Bioinformatics.*

DNA Origami

Invented by Paul Rothemund in 2005, this refers to one method of creating man-made molecular-scale structures or devices that comprise nucleic acids (i.e., DNA strands and sometimes including RNA strands). Via careful design, the creator encodes sequence complementarity into DNA strands in such a way that those strands (sometimes along with analogously preplanned RNA strands) self-assemble into the desired devices or structures. For example, coupling short individual strands of DNA to hold a long single strand of the nucleic acid in a predetermined shape (e.g., a structure that resembles a gridiron).

Scientists have thereby created tubes, gridirons/lattices, ribbons, nanopores, circuits, switches, nanobots, etc. See also *Deoxyribonucleic acid (DNA), Nucleic acids, Ribonucleic acid (RNA), Sequence (of a DNA molecule), Self-assembly (of a large molecular structure), Self-assembling molecular machines, Nanobots, Nanoscience, Nanotechnology, Nanopore, Complementary (molecular genetics), RNA origami.*

DNA Origami Robots

See *Nanobots, DNA origami.*

DNA Phosphorylation

See *Epigenetic.*

DNA Polymerase

Discovered in 1956 by Arthur Kornberg, it refers to a "family" of enzymes that catalyzes the synthesis of DNA. They do this by catalyzing the addition of deoxyribonucleotide residues to the free 3′-hydroxyl end of a DNA molecular chain, starting from a mixture of the appropriate triphosphorylated bases, which are dATP, dGTP, dCTP, and dTTP. This chemical reaction is reversible and, hence, DNA polymerase also functions as an exonuclease.

DNA polymerases include

- DNA polymerase alpha—which helps manufacture short RNA–DNA primers
- DNA polymerase delta—which replicates the lagging DNA molecular strand during the DNA-repair or manufacturing process
- DNA polymerase epsilon—which replicates the leading DNA molecular strand during the DNA-repair or manufacturing process
- *Taq* DNA polymerase—a DNA polymerase that was originally isolated from the thermophilic archaean *Thermus aquaticus* and is today utilized by man to catalyze DNA strand manufacturing in PCR reactions due to its heat resistance (needed for thermal cycles utilized in the PCR technique)

See also *Enzyme, Exonuclease, Deoxyribonucleic acid (DNA), Primer (DNA), TAQ DNA polymerase, Proof-reading, Gene repair (natural), Gene repair (synthetic), Synthesizing (of DNA molecules), Therophilic, PCR, Polymerase chain reaction (PCR) technique, Archaea.*

DNA Polymorphism

See *Polymorphism (genetic).*

DNA Probe

Also called gene probe or genetic probe. Short, specific (complementary to desired gene), artificially produced segments of DNA used to combine with and detect the presence of specific genes (or shorter DNA segments) within a chromosome.

If a DNA probe of known composition and length is mingled with pieces of DNA (genes) from a chromosome, the probe will cling to its exact counterpart in the "chromosomal DNA pieces" (genes), forming a stable double-stranded hybrid. The presence of this (now) "labeled" probe is detected visually or with the aid of another detection instrument.

Because the composition of the DNA probes is known, scientists can riffle through a chromosome, spotting segments of DNA (i.e., genes) that seem to be linked to genetic diseases. See also *Muscular dystrophy (MD), Probe, Polymerase chain reaction (PCR), Gene, Polymerase chain reaction (PCR) technique, Chromosomes, Double helix, Duplex, Hybridization (molecular genetics), Hybridization surfaces, Deoxyribonucleic acid (DNA), Base pairing, Homeobox, Rapid microbial detection (RMD), Southern blot analysis.*

DNA Profiling

Invented in 1985 by Alec Jeffreys, it is a technique used by forensic (i.e., crime-solving) chemists to match biological evidence (e.g., a blood stain) from a crime scene to the person (e.g., the assailant) involved in that particular crime. DNA profiling involves the use of restriction fragment length polymorphism (RFLP) analysis or allele-specific oligonucleotide/polymerase chain reaction (ASO/PCR) analysis to analyze the specific sequence of bases (i.e., nucleotides) in a piece of DNA taken from the biological evidence. Since the specific sequence of bases in DNA molecules is different for each individual (due to DNA polymorphism), a criminal's DNA can be matched to that of the evidence to prove guilt or innocence. Biological evidence may include among other things blood, hair, nail fragments, skin, and sperm. See also *Deoxyribonucleic acid (DNA), Restriction fragment length polymorphism (RFLP) technique, Polymorphism (chemical), Polymerase chain reaction (PCR) technique, Allele, Nucleotide, Nucleic acids, Oligomer, Genetic code, Informational molecules, Oligonucleotide, Codon, Nanoparticles.*

DNA Repair

Refers to the several ways in which damaged DNA gets repaired within living cells. Some naturally occurring examples include

- The *mismatch repair system*
- The *"SOS" repair system*
- Base excision repair
- Nucleotide excision repair
- Double-strand break repair
- The *CHK1 and CHK2 signaling pathways*
- The *break-induced replication* pathway
- Nonhomologous end joining
- MAPKAP kinase-2 (sometimes called "CHK3 signaling pathway")
- DNA photolyase enzymes that harness light energy to repair DNA that has been damaged by ultraviolet light
- Genome editing

Some examples of what scientists are doing for DNA repair include

- Targeted gene repair
- Chimeraplasty
- Targeted nucleotide exchange
- Therapeutic nucleic acid repair
- Oligonucleotide-mediated gene editing
- Oligonucleotide-mediated gene repair
- Oligodeoxynucleotide-directed gene modification

See also *Cell, Deoxyribonucleic acid (DNA), Double-strand breaks (in DNA), DNA ligase, Sliding clamps, Mismatch repair, SOS repair system, SOS response (in Escherichia coli bacteria), Gene repair (natural), Gene repair (done by man), Break-induced replication, Proof-reading, Ubiquitin, Signaling, Pathway, Photolyases, Editing, p53 protein, PARP, Copy number variation, Zinc finger proteins, CRISPR, CRISPR/Cas9 editing systems, p53 gene, Chimeraplasty, Oligonucleotide-mediated mutagenesis, Nonhomologous end joining.*

DNA Shuffling

Refers to a process in which man breaks apart a DNA segment (e.g., a gene), shuffles (i.e., changes the order of) the order of the relevant nucleotides within that sequence, and then recombines those nucleotides into an intact DNA segment.

When repeated DNA shuffling is coupled to a gene expression and assessment/improvement process (e.g., each new "shuffled" DNA segment is expressed and the resultant protein is evaluated against a desired goal) incorporating feedback to the DNA-shuffling process, the process is sometimes called "directed evolution."

For example, during 2003, Linda A. Castle and coworkers utilized this process to increase the activity of a GAT enzyme 10,000-fold. See also *Deoxyribonucleic acid (DNA), Gene, Genetic code, Nucleotide, Sequence (of a DNA molecule), Express, Protein, Gene expression analysis, Enzyme, Active site, Turnover number, GAT.*

DNA Synthesis

See *Synthesizing (of DNA molecules).*

DNA Typing

See *DNA profiling.*

DNA Vaccines

Products in which "naked" genes (i.e., pieces of bare DNA) are used to stimulate an immune response (e.g., either a cellular immune response, humoral immune response, or otherwise raising antibodies against the pathogen from which the naked genes have arisen/been derived). See also *Deoxyribonucleic acid (DNA), Immune response, Cellular immune response, Humoral immunity, Antibody, Naked gene, Pathogen, DNA vector.*

DNA Vector

A vehicle (such as a virus) for transferring genetic information (DNA) from one cell to another. See also *Bacteriophage, Retroviruses, Vector.*

DNA-Binding Proteins

A term utilized to refer to all protein molecules (e.g., transcription factors, DNA-repair proteins, histone-binding proteins, etc.) that attach to specific sites on the DNA of a given organism in order to control a cell's DNA repair, transcription, replication, chromosome segregation, etc. See also *Protein, Deoxyribonucleic acid (DNA), Transcription, Cell, Transcription factors, DNA repair, Sliding clamp, Chromosome, Replication.*

DNA-Dependent RNA Polymerase

See *RNA polymerase.*

DNA-Directed RNA Interference

Abbreviated *ddRNAi*, this refers to when a scientist makes RNA interference occur via causing cellular gene(s) to code for production of the relevant *shRNA* (i.e., *short hairpin RNA*, a dsRNA that the cell's dicer enzymes turn into the short interfering RNA strands that cause RNA interference). See also *RNA interference (RNAi), Cell, Gene, Coding sequence, Ribonucleic acid (RNA), Short interfering RNA (siRNA), Dicer enzymes, dsRNA, Short hairpin RNA.*

DNA-Protein Interaction Testing

See *Chromatin immunoprecipitation, ChIP.*

DNA-RNA Hybrid

A double helix that consists of one chain of DNA hydrogen bonded to a chain of RNA by means of complementary base pairs.

See also *Hybridization (molecular genetics), Hybridization (plant genetics), Double helix.*

DNase

Deoxyribonuclease, an endonuclease enzyme "family" that degrades (cuts up) DNA molecules.

DNase I is produced and secreted by the salivary glands, intestines, liver, and pancreas of animals. It has optimal activity (i.e., greatest ability to cut up DNA molecules) at neutral pH (i.e., neither acidic nor basic).

DNase II has optimal activity between pH 4.6 and 5.5 (i.e., in slightly acidic solutions). See also *Enzyme, Deoxyribonucleic acid (DNA), Endonucleases, Pancreas, Acid, Base (general).*

D

Docking Proteins

Refers to certain specific protein molecules that help (enable) other specific protein molecules to interact on a molecular scale (e.g., via docking with each other). For example, the plant gene known as Early Flowering 3 (*ELF3*) serves as an important docking protein that enables the "ELF4" and "LUX" protein molecules to dock with each other (which does not occur in the absence of ELF3). Only thus joined, do those protein molecules (collectively known as the "evening complex") cause plants to not grow (e.g., via elongation of stems, leaves, etc.) during the *early* hours of the evening. Then (shortly before each day's dawn during the growing season), two genes important in promoting plant growth (*PIF4* and *PIF5*) are released to promote plant growth, because levels of the evening complex have decreased by that time of the day. See also *Protein, Gene.*

Docosahexaenoic Acid (DHA)

One of the "omega-3" (*n*-3) highly unsaturated fatty acids (HUFA), DHA is important in the development of the human infant's brain, spinal cord, and retina tissues. DHA aids optimal brain and nervous system development in human infants and is required for optimal brain function throughout life. DHA comprises 40% of the polyunsaturated fatty acids in human brain tissue, and 60% of all fatty acids in human eye tissue. DHA is naturally present in human breast milk and fish oil from fatty fish species such as sardines, salmon, mackerel, and herring.

The human body converts linolenic acid (e.g., from consumption of soybean oil) to the two highly unsaturated fatty acids (HUFA) *DHA* and *eicosapentaenoic acid.*

Research indicates that consumption of docosahexaenoic acid also helps

- To reduce the risk of heart disease (by lowering blood pressure)
- To reduce depression (via its effect in the brain)
- Decrease the size of cancerous solid tumors, plus increase the potency of the chemotherapy drug cisplatin
- To cause brain cells to manufacture more LR11, a brain protein (also known as SorLA) that helps to clear the brain of amyloid precursor protein

See also *Polyunsaturated fatty acids (PUFA), Highly unsaturated fatty acids (HUFA), N-3 fatty acids, Fatty acids, Unsaturated fatty acids, Essential fatty acids, Linolenic acid, Soybean oil, Eicosanoids, Eicosapentaenoic acid (EPA), Amyloid precursor protein, Cancer, Tumor, Chemotherapy, Cisplatin.*

Domain (of a Chromosome)

May refer either to a discrete structural entity defined as a region within which supercoiling is independent of other domains, or to an extensive region, including an expressed gene that has heightened sensitivity to degradation by the enzyme DNase I. See also *Gene, Express, Enzyme.*

Domain (of a Protein)

A discrete continuous part of the amino acid sequence that can be equated with a particular function. See also *Exon, Combining site, Epitope, Idiotype, Protein, p53 protein, Minimized proteins.*

Dominant (Gene)

See *Dominant allele.*

Dominant Allele

Discovered by Gregor Mendel in the 1860s, it is a gene that produces the same phenotype when it is heterozygous as it does when it is homozygous (i.e., trait, or protein, is expressed even if only one copy of the gene is present in the genome). See also *Genetics, Recessive allele, Heterozygote, Homozygous, Phenotype, Genotype, Genome.*

Domoic Acid

A neurotoxin that is sometimes naturally produced by certain ocean coastal algae. Also known as "amnesic shellfish poison" due to its memory impairment of people who consume it by eating seafood that contains it, this heat-resistant neurotoxin toxin can accumulate in mussels, clams, scallops, and fish.

Because the kidneys try to filter domoic acid from the body, the kidneys can be damaged by it (additional to the brain damage it causes). See also *Toxin.*

DON

Abbreviation for mycotoxin *deoxynivalenol,* which is produced by certain *Fusarium* fungi (e.g., *Fusarium graminearum*). DON is also known as "vomitoxin," because it can cause some animals to vomit if they consume it. See also *Mycotoxins, Deoxynivalenol, Fusarium, Fungus, Vomitoxin.*

Donor Junction Site

The junction between the left 5′ end of an exon and the right 3′ end of an intron. See also *Exon, Intron, Acceptor junction site.*

Dopamine

A catecholamine neurotransmitter that is produced within several different regions of the brain. It has many functions, and when it is absent or in low supply (e.g., due to injury or Parkinson's disease), control of voluntary body movements is disrupted.

During 2011, Andras Simon showed that the absence/low supply of dopamine in salamander brains resulted in brain stem cells becoming new neurons.

See also *Catecholamines, Neurotransmitter, Parkinson's disease, Neuron, Stem cells.*

Double Helix

The natural coiled conformation of two complementary, antiparallel DNA chains. This structure was first put forward by Watson and Crick in 1953. See also *Deoxyribonucleic acid (DNA).*

Doubled-Haploid Breeding Program

Refers to a commercial crops breeding program (e.g., to produce hybrid corn/maize) in which the "parent" corns (i.e., the two that will be bred together to produce the hybrid seed that is subsequently sold to farmers) are created through the following methods:

1. First pollinating elite crop germplasm (e.g., optimized to the applicable growing climate/latitude and after insertion of desired new gene(s) via genetic engineering) with a haploid-inducer parent.
2. When seeds resultant from that pollinating are harvested, their genome contains only half the number of chromosomes as normal plants of that species (e.g., 10 single chromosomes instead of 10 pairs of chromosomes, for corn/maize). These haploid organisms are then treated with a special compound that results in each single chromosome being copied, so the result are plants bearing 10 identical pairs of chromosomes in the case of corn/maize. It is referred to as a *doubled-haploid* plant.
3. Those doubled-haploid plants (i.e., the "parent" plants) are then bred together to produce the hybrid seed that is subsequently sold to farmers.

One way to create a haploid-inducer parent in soybean plants is via disrupting and modifying the activity of a gene named CenH3. The net result of a doubled-haploid breeding program is creation of commercial seed in less time than required by a conventional crop-breeding program. See also *Elite germplasm, Gene, CenH3 gene, CenH3 protein, Genome, Hybrid, Corn, Haploid, Genetic engineering, Chromosomes.*

Double-Strand Break Repair

See *DNA repair.*

Double-Strand Breaks (in DNA)

Refers to the breaking of both strands of the double helix DNA molecule at one location in a cell's chromosome. This can be caused by the DNA's contact with certain metabolites (e.g., reactive oxygen species) of that cell, ultraviolet light, some chemical mutagens (e.g., colchicine, sodium azide), or ionizing radiation. Double-strand breaks can potentially lead to cell death if not repaired or lead to genomic rearrangements (i.e., gene deletions, translocations, and/or fusions) resulting in the cell becoming cancerous.

See also *Deoxyribonucleic acid (DNA), Double helix, Cell, Chromosome, Metabolite, Reactive oxygen species, Mutation, Mutagen, Colchicine, Mutation breeding, Cancer, Gene fusion.*

Double-Stranded DNA

See *dsDNA.*

Double-Stranded RNA

See *dsRNA.*

Double-Zero Canola

See *Canola.*

Dough Conditioner

Refers to any ingredient that is added to (e.g., wheat-based) bread dough to strengthen it, so it traps more of the carbon dioxide produced by yeast during the dough-rising process to make the resultant bread's texture finer or to otherwise improve the bread in some way. Examples of commercially utilized dough conditioners include lecithin, potatoes, gluten, ascorbic acid, ammonium chloride, enzymes, milk, and calcium salts such as calcium iodate.

For example, lecithin complex added to wheat-based flour dough at a rate of 0.25%–0.6% acts as a dough conditioner. It disperses fat evenly throughout the dough, thereby enabling that conditioned dough to trap more of the carbon dioxide produced by yeast during fermentation (rising) process. Such lecithin-conditioned doughs also tend to produce a bread of fine-grain, larger baked volume, and improved slicing properties.

See also *Wheat, Fermentation, Ascorbic acid, Lecithin (crude, mixture), Enzyme.*

Down Promoter Mutations

Those mutations that decrease the frequency of initiation of transcription. Down promoter mutations lead to the production of less mRNA than is the case in the nonmutated state. See also *mRNA, mutation, Transcription, Down regulating.*

Downregulating

Phrase utilized to refer to regulatory sequences, chemical compounds (e.g., transcription factors), mutations (e.g., down promoter mutations), etc. that cause a given gene to express *less* of the protein that it normally codes for. See also *Gene, Gene expression, Regulatory sequence, Transcription factors, Down promoter mutations, Protein, Coding sequence, Transcriptional repressor, Negative control, Riboswitches, Micro-RNAs, Methylated, NFkB.*

DPN

Acronym for dip-pen nanolithography. See also *Dip-pen nanolithography.*

DREB Proteins

A "family" of cold (temperature)-regulated transcriptional activators (transcription factors). Discovered by Kazuo Shinozaki, they are also known as *CBF proteins.* See *CBF1.*

DREs

Acronym for DNA regulatory elements. See also *Deoxyribonucleic acid (DNA), Regulatory sequence, Regulatory genes, Down regulating.*

Drosha

See *Micro-RNAs.*

Drosophila

The name of a type of fly (*Drosophila melanogaster*)that reproduces rapidly and that is commonly utilized in genetics experiments due to its short life cycle (14 days) and simple genome (four chromosome pairs). Because of this, a large base of knowledge about *Drosophila* genetics has been accumulated by the world's scientific community. For example, of the nearly 300 *disease-causing* genes in the human genome, more than half of them have an analogous gene in the *Drosophila* genome.

D

Drosophila was one of the first organisms to have its entire genome sequenced by man. See also *Genetics, Genome, Genetic code, Genetic map, Chromosomes, Cold hardening, Homeobox, Sequencing (of DNA molecules), Gene.*

Drought Tolerance

Refers to a given crop/plant's ability to survive a prolonged period of little or no rainfall. This may result from the following:

- A plant possessing a *drought tolerance trait* either inherently or because it was genetically engineered (e.g., to activate CBF/DREB1 pathway, ABA pathway, etc.). For example, insertion of gene(s) to produce trehalose or to increase the level of vitamin C in stomatal pore-containing leaf tissues.
- A crop possessing greater drought tolerance as a result of it being genetically engineered to possess a new trait that allows farmers to utilize husbandry practices that conserve topsoil moisture. For example, genetically engineered herbicide-tolerant soybeans enable U.S. farmers to utilize conservation tillage practices (e.g., no-tillage or low-tillage crop production) on most U.S. soybean hectares/acres, which conserves topsoil moisture and results in the overall U.S. soybean crop being more drought-tolerant than before.

In some plants such as sorghum (*Sorghum bicolor*), their cytochrome P450 molecules can help them to respond to certain kinds of stress (e.g., drought) if those cytochrome P450 molecules are present in high enough abundance. See also *Drought tolerance trait, Gene, Trait, Genetic engineering, Trehalose, Quercetin, Herbicide-tolerant crop, Soybean plant, Conservation tillage, No-tillage crop production, Low-tillage crop production, CBF/DREB1 pathway, Phenomics, cspB gene, Abscisic acid, Cytokinins, Stomatal pores, Ion channels, PARP, Abiotic stresses.*

Drought Tolerance Trait

Refers to the genetic trait whereby a given plant is able to survive a prolonged period of little or no rainfall.

For example, during the 1990s, Monty Jones crossed the Asian rice variety *Oryza sativa* with the African variety *Oryza glaberrima*. The result was "New Rice for Africa" (NERICA) variety, a drought-resistant rice.

Some of the strategies being pursued to create new drought-tolerant crop varieties have been the following:

- Insertion into soybean plant of the HAHB4 gene (also known as "HB4 gene") from the sunflower (*Helianthus annuus* L.) plant, resulting in the soybean plant tolerating drought stress longer by repressing its ethylene signaling pathway (thereby avoiding senescence during drought).
- Insertion into soybean plant of the flavodoxin gene, resulting in the soybean plant tolerating drought stress longer.
- Insertion into crop plant of a promoter to increase expression of the gene for aldehyde dehydrogenase (ALDH), an enzyme that detoxifies (metabolites incurred via drought).
- Breeding for enhanced leaf survival so that the crop can survive intermittent severe drought stress and still produce a harvestable yield. This has involved selecting for those with extreme low epidermal (leaf) water conductance, moderate osmotic adjustment and low critical relative water content.
- Breeding for the SLOW WILTING trait (in which higher ascorbic content of leaf stomata cells results in them closing faster in times of water stress to minimize leaf water loss).
- Breeding for increased pinitol (a cyclic-molecule carbohydrate) content in leaves, which enables photosynthesis activity to continue longer during times of drought stress.

In some plants such as sorghum (*Sorghum bicolor*), their cytochrome P450 molecules can help them to respond to certain kinds of stress (e.g., drought) if those cytochrome P450 molecules are present in high enough abundance.

Another possibility is the use of genetic engineering to activate a plant's CBF/DREB1 pathway, ABA pathway, etc. See also *Trait, Gene, Promoter, Protein, Enzyme, Polygenic, Genetic engineering, Corn, Trehalose, Drought tolerance, CBF/DREB1 pathway, Cytochrome P450, Abscisic acid, cspB gene, Stomatal pores, Ion channels, PARP, PARP inhibitors, ERU maps, Abiotic stresses, Ethylene, Senescence.*

DSB

Acronym for double-strand breaks. See also *Double-strand breaks (in DNA).*

DSBR

Acronym for double-strand break repair. See also *Double-strand breaks (in DNA), DNA repair.*

dsDNA

Acronym for the double-stranded structure of DNA molecule. See also *Deoxyribonucleic acid (DNA), Double helix.*

d-siRNA

Abbreviation for *diced RNA*. Refers to the products (i.e., short interfering RNA segments) resulting from the dicer enzymes cleaving applicable long segments of double-stranded RNA (dsRNA). See also *Dicer enzymes, RNA interference (RNAi), dsRNA.*

dsRNA

Acronym for the double-stranded structure of RNA molecule. Among its other functions, dsRNA can induce degradation of its counterpart (i.e., *matching*) mRNA, thereby causing RNA interference.

It has been shown that dsRNA can induce methylation of DNA in some species. Long dsRNA is specific to viral infections (i.e., the presence of this dsRNA within a human or animal cell indicates that the host cell has been invaded by a virus, which subsequently is in the process of making more viruses). See also *Ribonucleic acid (RNA), Double helix, RNA interference (RNAi), Virus, DNA methylation, Corn rootworm.*

Duchenne Muscular Dystrophy (DMD) Gene

See *Muscular dystrophy (MD).*

Duplex

The double-helical structure of DNA (deoxyribonucleic acid). See also *Double helix*, *Deoxyribonucleic acid (DNA)*.

Dx

Acronym for diagnostic (test). See also *Companion diagnostic*.

Dynamic Light Scattering

A technology utilized for rapid detection and characterization of very small, *soft* particles (e.g., protein molecule aggregates) in a solution. Output data from dynamic laser light scattering instruments (which shine the laser into the particle-containing solution) such as how the intensity of reflected light changes with time enable the characteristics (e.g., size) of the particles to be determined.

Collection/measurement of that reflected light (e.g., via a charge-coupled device) enables calculation of the hydrodynamic radius of the aggregates (which increases with protein solution stress such as high heat) and intensity of scattered light (which increases with greater protein molecule aggregation). See also *Protein*, *Charge-coupled device*.

Dynamics

Term used to refer to the study of changes in (a given population of organisms') genetics over time. See also *Genetics*, *Organism*.

Dynamin

A protein molecule within cells that assists in cellular endocytosis (engulfing of external molecules such as nutrients, growth factors, etc. by the cell). The external entity under consideration binds to receptor(s) located in the plasma (cell) membrane, which then invaginates (enfolds) hence taking up the entity via *endosomes* (formed by cell's dynamin molecules pinching-off of the enfold to form a "bag") into vesicles located within the cell. See also *Protein*, *Cell*, *Endocytosis*, *Receptors*.

Dynein

A molecular motor protein that "powers" within cells the movement of vesicles, organelles, certain steps of cellular mitosis, and also the *beating action* of cilia and flagella. See also *Motor proteins*, *Cell*, *Vesicle*, *Organelles*, *Mitosis*, *Cilia*, *Flagella*.

Dynein–Dynactin Complex

See *Dynein*.

Δ12 Desaturase

One of the desaturases (enzymes). See also *Delta 12 desaturase*, *Cosuppression*, *Enzyme*, *Desaturase*.

Δ15 Desaturase

One of the desaturases (enzymes). See also *Enzyme*, *Desaturase*, *Delta 12 desaturase*.

δ Endotoxins

See *Delta endotoxins*.

E

"Explosion" Method

To introduce foreign (new) genes into plant cells. A technique for gene-into-cell introduction in which the gene (genetic material) is driven into plant cells by the force of an explosion (vaporization) of a drop of water (to which the gene and gold particles have been added). The explosion is caused by application of high-voltage electricity to the drop of gene-laden water; the water then vaporized explosively, driving the "shot" (gold particles) and genetic material through the cell membrane. The plant cell then heals itself (reseals the hole where the gene entered), incorporates the new gene into its genetic complement, and produces whatever product (e.g., a protein) that the newly introduced gene codes for. See also *Agrobacterium tumefaciens, Coding sequence, Genetic engineering, Vector, "Shotgun cleaning method," Gene, Genome, Ribosomes.*

E. coli

See *Escherichia coliform* (*E. coli*).

E. coli 0157:H7

See *Escherichia coliform 0157:H7* (*E. coli 0157:H7*).

EA

Acronym for endocrine activity. See *Endocrine disruptors.*

EAA

See *Excitatory amino acids* (*EAAs*).

EAA

See *Essential amino acids.*

Early Development

This refers to the period of a phage infection before the start of DNA replication. See also *Phage, Bacteriophage, Deoxyribonucleic acid* (*DNA*).

Early Flowering3 Protein

See *Docking proteins.*

Early versus Late Genes

Those genes transcribed early in a bacteriophage-mediated infection process as compared to those genes transcribed some time later. May require different "p factors" (sigma) for recognition of promoters. See also *Gene, Promoter.*

Early versus Late Proteins

During viral infection, viral-specific proteins are synthesized at characteristic times after infection. They are called "early" and "late." Often under the positive control of bacterial and viral sigma factors. See also *Early versus late genes, Protein.*

Earthworms

Refer to *Lumbricus terrestris, Aporrectodea caliginosa, Eisenia fetida*, etc. These worms live in the soil, often come to the soil's surface to feed at night and consume up to 10 tons of organic matter (e.g., old crop plant stalks, husks, etc. lying on the soil's surface) per acre (i.e., approximately 0.4 hectare) per year. In so doing, earthworms make the soil more fertile, since the process breaks down that organic matter into soil (i.e., excreted underground by those earthworms) and humic acids. Earthworm tunnels also help aerate and drain soil, which encourages/fosters healthy plant root systems, enables soil to absorb more rainwater, etc. See also *Humic acids, Low-tillage crop production, Glomalin, No-tillage crop production.*

ECB

See *European corn borer* (*ECB*).

ECM

Acronym for *extracellular matrix.* See also *Extracellular matrix.*

Ecology

The study of the interrelationships between organisms and their environment. See also *Habitat.*

Ectodermal Adult Stem Cells

Certain stem cells present within (adult) bodies of organisms that can be differentiated (via chemical signals) to give rise to cells of skin, hair, tooth enamel, mucous membranes, and some glandular tissues. See also *Stem cells, Multipotent adult stem cells, Cell, Organism, Signaling.*

ED

Acronym for endocrine disruptors. See *Endocrine disruptors.*

Edible Vaccines

Edible substances, bearing antigens, that cause activation of an animal's immune system via that animal's gut-associated lymphoid tissues (GALT). These "edible vaccines" are derived from transgenic plants (e.g., grains, tubers, fruits) or eggs (i.e., via the activation

E

of the hen's immune system to cause that hen to secrete desired molecule(s) into the eggs it lays). See also *Gut-associated lymphoid tissues (GALT), Peyer's patches, Antigen, Cellular immune response, Molecular Pharming™, Humoral immunity, Plantigens.*

Editing

A term with several different meanings:

- In *transcription*—process that removes the intron sequences during synthesis of mRNA from DNA and joins together the exon sequences
- During *DNA recombination*—process of ligating two segments of DNA together
- During *genome editing*—for example, the process of utilizing zinc finger proteins (coupled with relevant nucleases), TALENs, CRISPR/Cas9 gene editing, or oligonucleotide-mediated mutagenesis to insert a desired mutation (e.g., impart drought tolerance in a crop plant) or to silence a nondesired DNA sequence within cells of a living organism
- During *some types of gene therapy*—for example, the process of utilizing zinc finger proteins (coupled with relevant nucleases) or CRISPR/Cas9 gene editing to "correct" a "wrong" DNA sequence (e.g., a disease-causing SNP) within cells of a living organism

See also *Deoxyribonucleic acid (DNA), Gene, Transcription, Messenger RNA (mRNA), Sequence (of a DNA molecule), Intron, Exon, Spliceosomes, Genetic code, DNA repair, Gene repair, Recombination, Recombinant DNA, Ligation, Gene silencing, Genome editing, Zinc finger proteins, Gene therapy, Single-nucleotide polymorphisms (SNPs), Liposomes, CRISPR, CRISPR/CAS9 gene-editing systems, TALENs, Oligonucleotide-mediated mutagenesis, Genomically recoded organisms (GROs).*

EDTA

Ethylenediamine tetraacetate. An organic molecule that, due to the chemical groups it contains and their juxtaposition within that molecule, is able to chelate (bind) certain other molecules such as divalent metal cations. EDTA thus inhibits some enzymes requiring such ions for activity. See also *Chelation, Cofactor, Chelating agent, Ion, Enzyme.*

EETI

Acronym for *Ecballium elaterium trypsin inhibitors*, a category of trypsin inhibitors naturally present in some plants (e.g., squash). See also *Trypsin inhibitors, Knottins.*

EFA

See *Essential fatty acids.*

Effector

A class of (usually small) molecules that regulates the activity of a specific protein (e.g., enzyme, messenger RNA) molecule by binding to a specific site on the protein. Control of (existing) enzyme molecules may be achieved by combination of the effector with the enzyme. The effector molecule may either physically block the active site on the enzyme molecule or alter the three-dimensional conformation of the enzyme molecule. That conformation change results in a change in the enzyme's catalytic activity.

A special class of effector, known as an allosteric effector, binds to the enzyme molecule at a site other than the enzyme's active site (thereby activating or inhibiting).

Effector is a general term. Effector molecules may be activators (cause an *increase* in the enzyme's catalytic activity) or inhibitors (cause a *decrease* in the enzyme's catalytic activity).

In the case of RNA interference, the applicable effector molecules are short interfering RNAs (siRNA). See also *Protein, Enzyme, Conformation, Allosteric enzymes, Allosteric site, Active site, Feedback inhibition, Catalytic site, Messenger RNA (mRNA), Short interfering RNA (siRNA).*

Effector B Cells

See *Gut-associated lymphoid tissues (GALT).*

Effector T Cells

See *Dendritic cells, $CD8^+$ T cells.*

Effectors (Bacterial)

Refer to a class of protein molecules protruding from the surface of certain (pathogenic) bacteria such as *Erwinia amylovora*, a rod-shaped bacterium that attacks apple and pear trees, causing the disease known as fire blight. The effectors bid to the surface of applicable plant cells and alter those cells to facilitate the infection of the tree. See also *Protein, Cell, Pathogen, Bacteria.*

Effectors (Fungal)

Refer to a class of protein molecules protruding from the surface of (pathogenic) fungi, which enable that pathogen to penetrate and thereby infect healthy cells of the host (plant) organism. Specific regions of the effector protein's "molecular chain" bind to the lipid known as phosphatidylinositol 3-phosphate (located in lipid rafts on the surface of host cell). Because a lipid raft is a portion of the cell's outer membrane that can be internalized by the cell, that lipid acts as an entry way for entry of the fungal pathogen into the cell.

Once inside, the effectors disable the plant's immune system via RNA interference. See also *Protein, Fungus, Pathogen, Cell, Organism, Lipids, Lipid rafts, Phosphatidylinositol, RNA interference.*

Efflux Pump

Refers to a particular class of molecule within the cell membrane of some cells, which "pumps" out certain pharmaceutical compounds. For example, Caco-2 cell membranes contain the efflux pump known as P-glycoprotein (P-gp), which pumps out toxins that enter the cell.

Autophagosomes also sometimes gather up and carry certain pharmaceuticals (e.g., chemotherapy agents introduced into cancer cells) to efflux pumps in the cell, where those pharmaceuticals are pumped out. See also *Cell, P-glycoprotein, Plasma membrane, Caco-2, Cancer, Autophagy.*

EFOX

See *Cyclooxygenase, Macrophage.*

EGF

See *Epidermal growth factor (EGF)*.

EGF Receptor

A protein embedded in the surface of the membranes of epidermal (skin) cells and epithelial cells (e.g., lining the interior of the intestines/stomach). The receptor consisting of (1) an outside (of the cell membrane) enzyme that recognizes epidermal growth factor (EGF) and binds to it and (2) an enzyme on the inside of the cell membrane, which is of the tyrosine kinase class. When free EGF comes in contact with an EGF receptor, they bind (in a lock-and-key fashion) and then enter the cell (through the cell membrane) together (where EGF then stimulates growth/division of cell via *ras* protein and *ras* gene). The EGF receptor (and receptors in general) is like a butler who allows the EGF (a guest) to enter the cell (home).

The EGF receptor is also (over)expressed in the plasma membranes of the cells of some tumors in colorectal cancer, head and neck cancers, lung cancer, pancreatic cancer, and some other cancers.

Scientists discovered via gene expression analysis that humans of Asian ancestry are more likely to express the particular mutation of the *gene for EGF receptor* that can result in lung cancer. See also *Oncogenes, Protein, Plasma membrane, Transmembrane proteins, ras gene, ras protein, Receptors, Signal transduction, Mutation, Cancer, Gene expression analysis*.

EGFR

See *EGF receptor*.

EHEC

See *Enterohemorrhagic E. coli*.

EIA

See *Enzyme immunoassay (EIA)*.

Eicosanoids

A group of chemical compounds, containing 20 carbon atoms in their "molecular backbone," which the human body synthesizes (i.e., "manufactures") from eicosapentaenoic acid, arachidonic acid, docosahexaenoic acid, or other n-3 and n-6 fatty acid starting materials. The term "eicosanoids" is from the Greek *eicosa* meaning 20.

One subgroup of eicosanoids is that of the prostaglandins (cyclic fatty acids that act as hormones in the body). For example, the COX-1 enzyme converts arachidonic acid to *constitutive prostaglandins*, and the COX-2 enzyme converts arachidonic acid to *inducible prostaglandins*.

Another subgroup of eicosanoids is that of the leukotrienes (lipid mediator molecules involved in the body's inflammation processes).

Another subgroup of eicosanoids is that of the prostacyclins.

Another subgroup of eicosanoids is that of the thromboxanes. See also *Eicosapentaenoic acid (EPA), Arachidonic acid (AA), Docosahexaenoic acid (DHA), Cyclooxygenase, Constitutive enzymes, Inducible enzymes, Prostaglandins, Hormone, COX-1, COX-2, Leukotrienes, Fatty acid, N-3 fatty acids, N-6 fatty acids*.

Eicosapentaenoic Acid (EPA)

One of the "omega-3" (n-3) polyunsaturated fatty acids, eicosapentaenoic acid (EPA) is important for the development of the human brain and retina tissue and prevention of high blood pressure, coronary heart disease (CHD), and some cancers.

The human body converts linolenic acid (e.g., from consumption of soybean oil) to the two highly unsaturated fatty acids—EPA and docosahexaenoic acid (DHA). EPA is a precursor of (i.e., human body converts EPA to) several eicosanoids. EPA is protective against heart attack via its antithrombosis properties and is protective against osteoporosis via its anti–bone resorption properties. See also *N-3 fatty acids, Polyunsaturated fatty acids (PUFA), Unsaturated fatty acid, Essential fatty acids, Coronary heart disease (CHD), Cancer, Highly unsaturated fatty acids (HUFA), Linolenic acid, Soybean oil, Stearidonic acid, Eicosanoids, Thrombosis, Thrombus, NFκB*.

Eicosatetraenoic Acid

See *Arachidonic acid (AA)*.

ELAM-1

Also known as E-selectin, it is a selectin molecule that is synthesized by endothelial cells after (adjacent) tissue is infected. ELAM-1 molecules then help leukocytes to leave the bloodstream to fight the infection. See also *Selectins, Lectins, Adhesion molecules, Leukocytes*.

Elastase

An enzyme secreted by neutrophils (white blood cells that engulf pathogens) that catalyzes the cleavage (breakdown) of specific proteins that function to provide elasticity to certain tissues. May be indirectly responsible for some autoimmune diseases, such as arthritis (which results from breakdown of cartilage tissue). Elastase may also be indirectly responsible for the emphysema (caused by loss of lung elasticity) that results from prolonged smoke inhalation. When α-1 antitrypsin (antielastase) efficacy is reduced (via smoke), the now-unrestrained excess elastase destroys alveolar walls in the lungs by digesting elastic fibers and other connective tissue proteins. See also *Leukocytes, Neutrophils, Proteolytic enzymes*.

Electrolyte

Any compound (e.g., salt, acid, base) that in aqueous solution dissociates into ions (charged atom-sized particles). Electrolytes may be either strong (completely or nearly completely dissociated) or weak (only partially dissociated). See also *Ion*.

Electron Carrier

A protein, such as flavoprotein or a cytochrome, that can gain and lose electrons reversibly and functions in the transfer of electrons from one carrier to another until the electron is taken up by a final molecule or atom such as oxygen. See also *Protein, Cytochrome*.

Electron Microscopy (EM)

A technique for greatly magnifying and visualizing very small entities such as viruses and even large molecules. The technique

E

uses beams of electrons instead of light rays. Because of the physics involved, beams of electrons permit much greater magnification than is possible with a light microscope. Electron microscopes have been used to examine the structures of viruses, bacteria, pollen grains, molecules, etc. See also *Virus, Bacteria, Label (radioactive).*

Electropermeabilization

See *Electroporation.*

Electrophoresis

A technique for separating molecules based on the differential movement of charged particles through a matrix when subjected to an electric field. The term is usually applied to large ions of colloidal particles dispersed in water. The most important use of electrophoresis (currently) is in the analysis of proteins, and then a technique known as gel electrophoresis is used. Since the proportion of proteins varies widely in different diseases, electrophoresis can be used for diagnostic purposes.

Electrophoresis, through agarose or other gel matrices, is a common way to separate, identify, and purify plasmid DNA, DNA fragments resulting from digestion (of DNA) with restriction endonucleases, and RNA. Electrophoresis is also used to study bacteria and viruses, nucleic acids, and some types of molecules, including amino acids. See also *Protein, Amino acid, Bioluminescence, Polyacrylamide gel electrophoresis (PAGE), Two-dimensional (2D) gel electrophoresis, Capillary electrophoresis, Chromatography, Gel, Agarose, Plasmid, Deoxyribonucleic acid (DNA), Restriction endonucleases, Ribonucleic acid (RNA), Bacteria, Virus, BioMEMS.*

Electroporation

A process that can be utilized to introduce a foreign gene into the genome of an organism. Examples are the following:

- In 1995, the U.S. company Dekalb Genetics Corp. received a patent for producing genetically engineered corn by introduction of a foreign gene into corn plant cells via electroporation.
- During 2005, Richard Heller and Adil Daud were able to deliver into human melanoma (skin cancer) tumor cells a gene that codes for production of interleukin-12 (IL-12) via electroporation. Because IL-12 helps stimulate the human immune system to try to combat melanoma, it is hoped that this will someday become a way to treat melanoma.

Electroporation, also called electroporesis or electropermeabilization, uses a brief direct current electrical pulse to cause formation of "micropores" (tiny holes) in the surface of cells (or protoplasts, in the case of plants, e.g., suspended in a solution containing DNA sequences [genes]). After the gene(s) enters a cell via the temporarily created micropores, the electrical pulse ceases, and the micropores close so that the gene(s) cannot depart the cell. The cell then incorporates (some of) the new genetic material (genes) into its genetic complement (genome) and produces whatever product (i.e., a protein) the newly introduced gene codes for. See also *Coding sequence, Genetic engineering, Vector, Biolistic® gene gun, "Explosion" method, Agrobacterium tumefaciens, Gene, Genome, Cell, Corn, Protoplast, Deoxyribonucleic acid (DNA), Protein, Interleukin-12 (IL-12), Cancer, Tumor, Melanoma, Gene delivery.*

ELF3 Protein

See *Docking proteins.*

ELF4 Protein

See *Docking proteins.*

Eliminylation

See *Posttranslational modification of protein.*

ELISA (Test for Proteins)

An enzyme-linked immunosorbent assay (hence the acronym) that can readily measure less than a nanogram (10^{-9} g) of a protein. This assay is more sensitive than simple immunoassay (tests) because one of the two antibodies used to bind and quantitate (measure) the protein's antigen, based on two concurrent epitopes within the protein, is attached to an enzyme. The enzyme can rapidly convert an added colorless substrate into a colored product, or a nonfluorescent substrate into an intensely fluorescent product (thus enabling finer quantitation). See also *Absorbance (A), Immunoassay, Protein, Antigen, Enzyme, Nanogram (ng), Fluorescence.*

Elite Germplasm

Refers to germplasm that is adapted (selectively bred) and optimized to new surroundings (i.e., environment). For example, corn/maize (*Zea mays* L.), which is native to the country of Mexico, has been adapted and optimized to grow in field conditions in many of the world's countries. See also *Germplasm, Introgression, Marker-assisted selection, Corn.*

Ellagic Acid

A naturally occurring plant phenol (phytochemical) that has been shown to possess general antioxidant properties and to help inhibit some cancers when consumed by humans.

Research also indicates that human consumption of ellagic acid inhibits growth of certain pathogenic bacteria such as *Salmonella* and *Campylobacter.*

Ellagic acid is naturally present in red raspberries, strawberries, pomegranate (*Punica granatum* L.), etc. See also *Phytochemicals, Polyphenols, Antioxidants, Oxidative stress, Cancer, Bacteria, Pathogen, Salmonella.*

Ellagic Tannin

See *Ellagic acid.*

EM

See *Electron microscopy (EM).*

EMAS

Acronym for Eco-Management and Audit Scheme.

Embryo Rescue

Refers to the tissue culture techniques/technologies that are utilized to enable the fertilized embryo resulting from a "wide cross" (between two non–sexually compatible plant species) to grow and mature into a seed-producing plant. See also *Traditional breeding methods*, *Wide cross*, *Tissue culture*.

Embryology

From the Latin *embryon* meaning embryo. The study of the early stages in the development of an organism. In these stages, a single highly specialized cell, the egg, is transformed into a complex many-celled organism resembling its parents. See also *Cell*, *Organism*, *Anti-angiogenesis*, *Gamete*, *Hedgehog signaling pathway*, *Imprinting*, *Micro-RNAs*, *Insulin-like growth factor-2* (*IGF-2*), *DNA methylation*.

Embryonic Induction

Refers to the processes involved in the initiation of cell differentiation within the embryo of complex, multicellular organisms. In embryonic induction, the action of one type of cells on an adjacent type of cells is what leads to the establishment of the developmental pathway in the responding tissue. See also *Cell*, *Organism*, *Differentiation*.

Embryonic Stem Cells

See *Human embryonic stem cells*.

EMEA

See *European Medicines Evaluation Agency* (*EMEA*).

EMP-1 Protein

See *Biomarkers*.

Emulsion

From the Latin *emulgere* (= "to milk out"), it is a stable dispersion of one liquid in a second, immiscible (i.e., nonmixable) liquid. For example, milk is an emulsion of oil (fat) in water and latex paint is an emulsion of paint resin in water.

Certain ingredients (e.g., β-conglycinin protein) help enable a greater content of the first liquid to be dispersed in the second liquid.

Certain ingredients (e.g., β-conglycinin protein) make a given liquid/liquid emulsion more stable (i.e., prevents the two liquids from separating over an extended period of time). See also *Protein*, *β-conglycinin*.

Enantiomers

From the Greek word *enantios*, which means "opposite." Enantiomers are a pair of nonidentical, mirror-image molecules. This means that both molecules are made up of the same atoms, that is, they have the same molecular formula, but the constituent groups that are attached to a carbon atom can be arranged in two different ways (forms) around the carbon atom. This gives rise to an asymmetric molecule that can exist in either of two mirror-image forms whose mirror images are not superimposable. A pair of these molecules is known as enantiomers. The four attached groups are all different from each other. See also *Racemate*, *Optical activity*, *Chiral compound*, *Enantiopure*.

Enantiopure

Refers to a compound (e.g., a pharmaceutical) that consists of only *one* of that compound's two possible enantiomers. Sometimes expressed in relative terms. For example, 98% enantiopure would refer to a compound that consists of 98% (of) desired enantiomer. See also *Enantiomers*, *Chiral compound*, *Racemate*, *Optical activity*.

Endergonic Reaction

A chemical reaction with a positive standard free energy change (i.e., an "uphill" reaction). A (heat) energy-requiring reaction. A nonspontaneous reaction at ambient temperature. See also *Exergonic reaction*, *Free energy*.

Endocrine Activity

See *Endocrine disruptors*.

Endocrine Disruptor Chemicals

See *Endocrine disruptors*.

Endocrine Disruptors

Refer to substances that are defined by three simultaneous criteria—the presence of (1) an adverse impact on an organism or a (sub) population of organisms, (2) an endocrine activity (e.g., entry of the substance into an organism causes a change in the organism's production/utilization of endocrine hormones), and (3) a causal relationship between #1 and #2. For example, research indicates that lavender oil (extracted from *Lavandula hybrida* and marketed as an ingredient in some personal care products) may be an endocrine disruptor. See also *Endocrine hormones*, *Hormone*, *Organism*.

Endocrine Glands

Glands that secrete their products (hormones) into the blood, which then carries them to their specific target organs. For example, adrenalin, produced in the adrenal glands, is carried to the heart (and other muscles) when needed during periods of stress. The endocrine glands are the pituitary, thyroids, adrenals, pancreas, ovaries (in females), and testes (in males). Endocrine glands are found in some invertebrates as well as in vertebrates. See also *Hormone*, *Endocrine hormones*.

Endocrine Hormones

These are the products secreted by the endocrine glands. These help control long-term bodily processes, such as growth, lactation, sex cycles, and metabolic adjustment. The endocrine system and the nervous system are interdependent and often referred to collectively as the neuroendocrine system. For example, the juvenile hormone, found in insects and annelids, affects sexual maturation. There is currently great interest among scientists in the potential use of such hormones in the control of destructive insects (e.g., by preventing larvae from maturing into adults). See also *Endocrine glands*, *Hormone*, *Pheromones*.

E

Endocrinology

The branch of science that studies the endocrine glands, hormones, and hormonelike substances. See also *Endocrine glands, Hormone, Endocrine hormones, Endocrine disruptors.*

Endocytosis

Also called "receptor-mediated endocytosis." The import of substances (e.g., hormones, growth factors, nutrients, viruses, and toxins) into a cell via specific receptor/ligand binding. The chemical entity under consideration binds to a receptor(s) located in the plasma (cell) membrane, which then uses clathrin protein to form a pit that subsequently invaginates (infolds), hence taking up the entity via "endosomes" (formed by cell's dynamin molecules pinching off of the infold to form a "bag") into vesicles located within the cell.

It is one route to deliver essential metabolites into cells (e.g., low-density lipoprotein), and it is a means to modulate the cell's responses to many protein hormones and growth factors (e.g., insulin, epidermal growth factor, and nerve growth factor). It is a route by which certain proteins targeted for destruction can be taken up and delivered to the cell's lysosomes. For example, phagocytic cells have receptors enabling them to take up *antigen–antibody complexes* for subsequent destruction by the phagocytic cell. This route is also a means exploited by certain viruses and toxins to gain entry into cells through the otherwise impervious cell membranes (e.g., used by the AIDS virus and the Semliki Forest virus). Disorders of endocytosis can lead to disease states (e.g., high cholesterol levels in the blood of people whose low-density lipoprotein receptors are impaired).

Drugs (e.g., certain painkillers) can be targeted to specific receptors via receptor mapping and receptor fitting for greater efficacy.

Certain drugs (e.g., streptavidin) can be delivered into the interior of specific cells via biotin-*coated* carbon nanotubes (to which the streptavidin binds, until the cell is entered, whereupon it is released). See also *Cell, Invasin, Adhesion molecule, CD4 protein, Exocytosis, T cell receptors, Signal transduction, Vaginosis, Receptors, Receptor fitting (RF), High-density lipoproteins (HDLPS), Low-density lipoproteins (LDLP), Receptor mapping (RM), Signaling, Nuclear receptors, Streptavidin, Carbon nanotubes, Dynamin, Clathrin.*

Endodermal Adult Stem Cells

Certain stem cells present within (adult) bodies of organisms, which can be differentiated (via chemical signals) to give rise to cells of tongue, tonsils, the bladder/urethra, digestive tract, liver, pancreas, lung tissues, etc. See also *Stem cells, Multipotent adult stem cells, Cell, Organism, Signaling.*

Endoglycosidase

An enzyme capable of hydrolyzing (i.e., breaking) interior bonds in the oligosaccharide molecular branches of a glycoprotein molecule. That is, the enzyme is capable of cutting a sugar-to-sugar bond anywhere within the sugar polymer molecule (depending, of course, on the specificity of the enzyme). This is in contrast to an exoglycosidase, which must cut away at the polymer from the outside, that is, from the free end, one unit (or section as the case may be) at a time. See also *Exoglycosidase, Glycoprotein, Enzyme, Oligosaccharides, Restriction endoglycosidases, Hydroxylation reaction.*

Endometrium

The lining of the uterus.

Endonucleases

A class of enzymes capable of hydrolyzing (breaking) the interior phosphodiester bonds of DNA or RNA chains. As opposed to cleavage (by exonucleases) at the terminal bonds (ends) of the molecular chain. See also *Enzyme, DNase 1, DNase 2, Exonuclease, Endoglycosidase, Meganuclease, Deoxyribonucleic acid (DNA), Ribonucleic acid (RNA).*

Endophyte

A microorganism (fungus or bacteria) that lives inside vascular tissues of plants (e.g., in spaces between plant cells). The *Metarhizium* fungus species live within both plant tissues and soil (where they parasitize certain insects in that soil). As those insects thus die and are consumed by the *Metarhizium* fungi, some of the nitrogen is transferred to the plants.

At least one biotech seed company has incorporated the gene for a protein toxic to insects (taken from *Bacillus thuringiensis*) into an endophyte to confer insect resistance to a crop plant.

During 2002, Regina Redman and Russell Rodriguez discovered that *Curvularia protuberata* fungi (which live inside *Dichanthelium lanuginosum* grass that grows in hot soils adjacent to magma-heated geysers) impart heat tolerance to the grass they live in, when those fungi are themselves infected with *Curvularia* thermal tolerance virus (CThTV). Redman, Rodriguez, and Joan Henson were later able to show that when these CThTV-infected fungi were inserted into tomato and watermelon seedlings, those plants/roots were also able to withstand far higher temperatures than before.

During 2004, Daniel van der Lelie and other researchers incorporated a gene enabling certain bacteria to break down (biodegrade) toluene into the endophytic bacteria that naturally colonize yellow lupine. Such colonized plants could be useful for bioremediation of toluene-polluted land sites.

When endophyte-infested fescue grass is fed to cattle, sheep, horses, or rabbits, it is generally toxic to those animals, due to mycotoxin(s) or ergot alkaloid toxins produced by that endophyte. See also *Microorganism, Bacteria, Bacillus thuringiensis (B.t.), Fungus, Protein, Thermoduric, Mycotoxins, Tremorgenic indole alkaloids, Bioremediation, Virus.*

Endoplasmic Reticulum (ER)

Discovered in 1963 by George Palade, the endoplasmic reticulum (ER) is a highly specialized, complex network of branching, intercommunicating tubules (surrounded by membranes) found in the cytoplasm of most animal and plant cells. The two types of ER recognized are the rough ER and smooth ER. The ER that is covered with many ribosomes is called rough and the ER without or with fewer ribosomes attached is called smooth. This nomenclature comes about because of the appearance of the ER under a high magnification microscope. The rough ER is very well developed to facilitate cells carrying on abundant protein synthesis, because proteins are synthesized (manufactured) in ribosomes.

MHC class II molecules are synthesized in the ER of applicable dendritic cells and lymphocytes. See also *Cell, Cytoplasm, Ribosomes, Fats, Lipids, Plasma membrane, Protein, Dendritic cells, Lymphocyte, Phospholipids.*

E

Endorphins

Discovered during the 1970s by U.S. and Scottish scientists, these are hormones produced in the brain, which act as natural painkillers. For example, runners and long-distance walkers achieve something of a "high" due to endorphins released by the brain during long runs or walks. See also *Enkephalins*, *Catecholamines*, *Hormone*.

Endosome

See *Endocytosis*.

Endosperm

The interior portion of a plant seed, beneath the outer hull (e.g., it is the portion that people tend to eat, in food crops). In grains (e.g., rice or corn/maize), the endosperm consists primarily of starch (carbohydrate). In legumes (e.g., beans), the endosperm contains mainly protein, a small amount of carbohydrates, and sometimes vegetable oil. See also *Starch*, *Corn*, *Soybean plant*, *Carbohydrates (saccharides)*, *Soybean oil*, *Aleurone*.

Endospore

A highly resistant, dormant inclusion body formed within certain bacteria. To kill spores, temperatures above boiling point are usually needed. For this, pressure cookers and autoclaves are required. Endospores have survival value since the spore can remain for long periods of time in a nongrowing state and then, under appropriate conditions, can be induced to germinate and regenerate the original cell. Endospore formation may be viewed as being akin to hibernation, that is, a kind of "bacterial hibernation." See also *Bacteria*, *Cell*.

Endostatin

An *anti-angiogenesis* human protein discovered by Judah Folkman. In concert with angiostatin, it causes certain cancer tumors in mice to shrink. See also *Protein*, *Anti-angiogenesis*, *Angiostatin*, *Cancer*.

Endothelial Cells

These are the flat, sort of plate-shaped cells that line the surface of all blood vessels, heart, and lymphatics within the body. A blood vessel's endothelial cells must fit tightly together to form a solid tubular structure so that blood can flow through. The protein molecule known as *Syx* normally ensures the junctions between endothelial cells are tight.

Endothelial cells possess transmembrane (i.e., through the cell membrane) molecules known as adhesion molecules, which selectively allow the passage (from bloodstream to tissues) of some molecules (e.g., leukocytes, monocytes, hormones). Endothelial cells are packed much tighter together in the capillaries that provide blood to the brain. This tighter packing limits the size and kind of molecules that can pass into the brain. This blood–brain barrier serves to protect the sensitive brain tissue from pathogens or harmful molecules (e.g., toxins). See also *Endothelium*, *Vascular endothelial growth factor (VEGF)*, *Adhesion molecules*, *Monocytes*, *Mitogen*, *Selectins*, *Blood–brain barrier (BBB)*, *Lectins*, *ELAM-1*, *ATP synthase*, *Oxidative stress*, *Cyclooxygenase*, *von Willebrand factor*, *Estrogen receptors*.

Endothelial Nitric Oxide Synthase (eNOS)

An enzyme within certain endothelial cells, which synthesizes ("manufactures") nitric oxide in response to a number of different stimuli such as exercise, certain hormones, certain neurotransmitters (e.g., bradykinin, acetylcholine), or the stress imparted on blood vessels by high blood pressure. See also *Nitric oxide synthase*, *Endothelial cells*, *Enzyme*, *Nitric oxide*, *Acetylcholine*.

Endothelin

A "family" of peptides that cause arteries to contract (which consequently causes blood pressure to increase).

Research indicates that overproduction of *endothelin-1* can combine with plaque deposits (on interior walls of arteries) to "clog" those arteries. See also *Peptide*, *Atrial peptides*, *Plaque*, *Coronary heart disease (CHD)*, *Polyphenols*.

Endothelium

The layer of epithelial cells that line blood vessels throughout the body. The layer selectively allows the passage (from bloodstream to tissues) of nutrients, hormones, and other molecules that are essential for tissue growth and function. The endothelium is involved in the recovery and recycling of old red blood cells. It also produces

- Nitric oxide, which causes neighboring smooth-muscle (blood vessel) cells to relax so that those (neighboring) blood vessels dilate and the body's blood pressure is lowered
- Two compounds that prevent blood clotting—prostacyclin and von Willebrand factor

See also *Endothelial cells*, *Vascular endothelial growth factor (VEGF)*, *Selectins*, *Lectins*, *Adhesion molecules*, *Nitric oxide*, *Nitric oxide synthase*, *Bilirubin*.

Endotoxin

A lipopolysaccharide (fat/sugar complex; poison, also known as LPS) that forms an integral part of the cell wall of gram-negative bacteria. It is only released when the cell is ruptured. It can cause, among other things, septic shock and tissue damage. Pharmaceutical preparations are routinely tested for the presence of endotoxins. This is one reason why pharmaceuticals must be prepared in a sterile environment. See also *Sepsis*, *Bacteria*, *Lipids*, *Polysaccharides*, *Toxin*, *Cholera toxin*, *Gram-negative (G–)*, *Good manufacturing practices (GMP)*.

Engineered Antibodies

Chimeric monoclonal antibodies, produced via genetic engineering of human antibody-producing cells (clones). For example, the genes coding for antilymphoma binding sites from a rat have been inserted into human antibody-producing cells to yield rat (antigen) binding sites mounted on human antibody "stems." See also *Chimeric proteins*, *Monoclonal antibodies (MAb)*, *Antibody*, *Genetic engineering*, *Combining site*, *Lymphocyte*, *Semisynthetic catalytic antibody*.

Engineered Nanoparticle

See *Nanoparticles*, *Nanoshells*, *Nanocrystals*, *Nanocrystal molecules*, *Nanocapsules*, *Nanocomposites*, *Nanowire*.

E

Enhanced Nutrition Crops

See *Nutrient enhanced*™.

Enhancer

Refers to particular regulatory DNA segments within introns (i.e., nongene portions) of the DNA of a eukaryotic organism that act to either increase/activate/"turn on" a given gene or that act to decrease/silence/"turn off" a given gene. The genes that enhancers thereby regulate may be located up to several thousand base pairs distant from that enhancer. See also *Deoxyribonucleic acid (DNA), Sequence (of a DNA molecule), Intron, Gene silencing, Gene, Promoter, Eucaryote, Organism, Transcription, Expressivity, Base pair (bp).*

Enkephalins

A class of hormones produced in the brain that act as natural painkillers. Discovered by John Hughes and Hans Kosterlitz in 1975, they are some of the endorphins. See also *Endorphins.*

Enolpiruvil Shikimate

See *EPSP synthase.*

Enolpyruvil Shikimate

See *EPSP synthase.*

eNOS

Acronym for endothelial nitric oxide synthase. See *Endothelial nitric oxide synthase (eNOS).*

Enoyl-Acyl Protein Reductase

An enzyme that is utilized by bacteria in their synthesis ("manufacture") of fatty acids. See also *Enzyme, Protein, Bacteria, Fatty acid, Essential fatty acids.*

ENP

Acronym for engineered nanoparticle. See *Engineered nanoparticle.*

Ensiling

The fermentation of (usually chopped-up) agricultural vegetation in order to preserve it. It is carried out for 1–2 weeks, using either indigenous microorganisms (e.g., *Lactobacillus* spp.) or introduced microorganisms (to speed up the process, yield product containing more nutrients for livestock, etc.), in the absence of oxygen (to prevent the growth of aerobic mold fungi). When indigenous microorganisms are used, *Lactobacillus* spp. become the dominant microorganisms present, and heat is generated by the microorganisms within the vegetative mass (optimum temperature is 25°C–30°C, which is 77°F–86°F). Lactic acid is produced by the microorganisms, which inhibits the growth of bacteria that would normally putrefy the vegetation. See also *Fermentation, Microorganism, Aerobic, Fungus, Optimum temperature.*

Enterobactin

See *Siderophore.*

Enterocytes

Specialized cells within the ileum (lining the intestines) that recover and reclaim bile acids from the intestinal (food) mass, which occurs via the following:

- Bile acids trigger the expression within enterocytes of ileal bile acid–binding protein (IBABP), a cytosolic binding protein.
- IBABP causes translocation/transport of the bile acid molecules into the portal circulation (carried to the liver).
- When these bile acids reach the liver, they activate farnesoid X receptors, which represses transcription of the genes involved in bile acid creation. That prevents overproduction of bile acids in the body.

See also *Cell, Digestion (within organisms), Bile, Bile acids, Receptors, Farnesoid X receptor (FXR), Celiac disease, Gene, Transcription, Repression (of gene transcription/translation).*

Enterohemorrhagic *E. coli*

The several dozen (approximately 60 known) serotypes (strains) of *E. coli* bacteria that cause internal hemorrhaging in humans that ingest those bacteria. The toxin produced by these particular *E. coli* bacteria attacks the human kidney, which often leads to kidney failure and/or death of infected humans. See also *Escherichia coli 0157:H7 (E. coli 0157:H7), Toxin, Serotypes, Enterotoxin.*

Enterotoxin

The category (i.e., intestinally active) of toxins, produced by certain bacterial strains and/or serotypes, which attack the body's internal organs. For example, enterotoxins secreted by the serotype of *Escherichia coliform* bacteria known as *E. coli 0157:H7* attack the kidneys and other internal organs of humans, also causing internal bleeding and sometimes death.

A different enterotoxin known as STa, which is secreted by certain other serotypes of *Escherichia coliform* bacteria, stimulates the buildup of fluids in the human intestine, thereby causing diarrhea. See also *Toxin, Bacteria, Escherichia coliform 0157:H7, Enterohemorrhagic E. coli, Serotypes, Cholera toxin.*

Environmental Response Unit Maps

See *ERU maps.*

Environmental RNAi

See *eRNAi.*

Enzyme

A protein-based catalyst that is not itself used up in the chemical reaction (that it catalyzes). It is naturally produced by living cells to catalyze biochemical reactions. Enzymes were first determined to be proteins in 1926 by James B. Sumner.

Each enzyme is highly specific with regard to the type of chemical reaction that it catalyzes and to the substances (called substrates) upon which it acts. This specific catalytic activity and its control by other biochemical constituents are of primary importance in the physiological functions of all organisms. Although all enzymes are proteins, they may, and usually do, contain additional nonprotein components called coenzymes that are essential for catalytic activity. See also *Apoenzyme*, *Catalyst*, *Coenzyme*, *Holoenzyme*, *Substrate (chemical)*, *Protein*, *Hormone*, *Extremozymes*, *Turnover number*, *Cell.*

Enzyme Denaturation

The loss of enzyme (catalytic) activity due to loss of the correct functional structure of the protein. Denaturation may be caused by factors such as exposure to heat and organic solvents, degradation of the enzyme molecule by proteases, oxygen, and acid or alkaline pH. See also *Enzyme*, *Conformation*, *Denaturation*, *Extremozymes*, *Structural biology.*

Enzyme Derepression

Commonly known as induction (of an enzyme). Initially a repressor protein is bound to a specific region of DNA. This binding inhibits transcription to mRNA, thus blocking the synthesis of the protein (enzyme) specified by the mRNA. When present, the inducer molecule binds to the repressor protein and inactivates it. Thus, the inhibition caused by the repressor protein is overcome and mRNA can be synthesized, which consequently leads to synthesis of the mRNA-specified protein (enzyme). The word derepression is sometimes used because the repressor protein is, by itself, active in repressing protein (enzyme) synthesis. Its repressive action is mitigated (derepressed) by the inducer molecule. Hence, derepression (or unrepression) of repression equals induction. See also *Continuous perfusion*, *Enzyme repression*, *Enzyme*, *Repression (of an enzyme)*, *Inducible enzymes.*

Enzyme Evolution

See *Directed evolution.*

Enzyme Immunoassay (EIA)

See *ELISA.*

Enzyme Repression

Inhibition of enzyme synthesis caused by the availability of the product of that enzyme. On a molecular level, a repressor molecule (which could be, e.g., the amino acid arginine) combines with a specific repressor protein that is present in the cell. This repressor molecule/repressor protein complex is then able to bind to a specific region of DNA at the initial end of the gene, which is called the operator region. It is in this region where the synthesis of mRNA is initiated. The repressor "roadblock" thus stops the synthesis of mRNA, and therefore the synthesis of the protein is also blocked. See also *Enzyme*, *Repression (of an enzyme)*, *Enzyme derepression.*

Enzyme-Linked Immunosorbent Assay

See *ELISA.*

Eosinophils

Polymorphonuclear leukocytes made in the bone marrow. They circulate in the blood for a number of hours (3–8) and then migrate into the tissue where they reside. They kill parasites too large to be phagocytized by secreting substances that kill the parasites (hookworms, trichinosis, etc.). They also inhibit histamine release from mast cells and secrete chemicals that neutralize histamine. Allergy causes an increase in eosinophils. GM-CSF stimulates eosinophil production. See also *Polymorphonuclear leukocytes (PMN)*, *Basophils*, *Antigen*, *Cellular immune response.*

EpCAM

Acronym for "epithelial cell adhesion molecule," a particular adhesion molecule that is produced on the surface of most carcinoma (cancerous) cells. See also *Adhesion molecule*, *Epithelium*, *Cancer.*

EPD

See *Expected progeny differences.*

Epidermal Growth Factor (EGF)

A protein of 53 amino acids that greatly increases growth/reproduction of epidermal (skin) cells and epithelial cells (e.g., lining the interior of the intestines/stomach). This protein also increases

- Growth of wool in sheep
- Growth in more than 50% of human tumors

High concentrations of epidermal growth factor (EGF) are found in human tears. EGF was discovered by Stanley Cohen. See also *Protein*, *EGF receptor*, *Growth factor*, *Nerve growth factor (NGF)*, *Amino acid*, *Filler epithelial cells*, *Tumor.*

Epidermal Growth Factor Receptor

See *EGF receptor*, *HER-2 receptor*, *HER-2 gene.*

Epigenetic

Refers to mechanisms "in control" of changing gene expression (and interaction) primarily during development (e.g., of embryonic organism), which *do not require changes in actual gene/DNA sequences to occur.* While the actual gene/DNA sequences (genetic information) constitute the "blueprint" for the synthesis of all the proteins needed by a living organism, the epigenetic mechanism (epigenetic information) provides the "instructions" *to cell via when, where, and how* (much) *that genetic information is utilized* (expressed). The epigenetic mechanism (epigenome) can also be impacted by environmental factors including diet, exercise, and stress.

Coined from the Greek *epi* meaning "upon" by Conrad H. Waddington in the 1940s, this term essentially means "above the genome" or "on top of the genome," but more broadly refers to *all* of the "*non*classical genetic/heredity" sources of a given organism's phenotype. Examples are as follows:

- As a female mammal embryo (including humans) develops, epigenetics causes random inactivation of one X chromosome.

- Preferential expression of certain alleles (e.g., those inherited from the mother or from the father). In mice, more *maternal-origin* alleles are expressed within the *developing* brain and more *paternal-origin* alleles are expressed within the *adult* mouse brain, than would occur from a simple random 50/50 contribution of parental alleles to the offspring's DNA.
- As a plant embryo develops into a seedling (juvenile in the case of an animal), different genes are expressed (or silenced) in that organism, analogous to mammal embryo development. For example, during 2002, researchers discovered that feeding small amounts of methionine, folic acid, zinc, betaine, choline, and vitamin B_{12} to certain pregnant inbred (yellow-colored) laboratory mice will cause their offspring to have *brown*-colored fur via an *epigenetic change caused by DNA methylation* because the folic acid donates methyl molecular groups to the mouse's chromatin, thereby *silencing* the applicable gene (known as "agouti" gene). When epigenetic change results from consumption of certain foods/compounds, it is referred to as "nutritional epigenetics."
- As a result of certain disease(s) afflicting the organism, different genes within the organism's DNA are silenced (or expressed). For example, hypoxia (e.g., resulting during tuberculosis or certain cancers) has been shown to result in epigenetic events within the bodies of relevant patients. When it results from cancer, it is referred to as "cancer epigenetics."
- As an organism (especially during its embryo stage) interacts with its environment, some of those environmental impacts (e.g., temperature) cause certain genes to be expressed (or silenced). For example, in the mouse-ear cress plant (*Arabidopsis thaliana*), a gene known as "FLC" normally represses flowering. However, exposure of that plant to prolonged cold will epigenetically silence that gene (thereby enabling flowering to begin when spring's warm weather arrives). For example, the eggs of the saltwater crocodile (*Crocodylus porosus*) yield a larger fraction of male offspring when those eggs are incubated (in the nest) at temperatures above 90°F (32°C) than when those eggs are incubated at temperatures below 90°F (32°C).
- As an organism passes from juvenile stage to later life stage(s). For example, the female honeybee (*Apis mellifera*) "switches on"/off approximately 40% of her total genes as she matures from being a "nurse bee" (taking care of pupae while she is a juvenile) to being a "forager bee." During times in which the honeybee colony is in need of more forager bees (i.e., adults), the juveniles *mature faster* (to "forager bees") than during normal times.

Epigenetic events include impacts of microRNAs, short interfering RNA (siRNA), gene silencing, DNA methylation, histone methylation, DNA/histone acetylation, DNA/histone phosphorylation, DNA/histone ubiquitination, histone biotinylation, chromatin remodeling, chromatin modification, histone modification, paramutation, nucleolar dominance, and gene imprinting.

At least one type of epigenetic regulation is caused by changes in the (molecular-scale) shape of an organism's chromatin. Such "chromatin remodeling" can be caused by introduction (into cell) of certain siRNA, certain transcription activators, certain short-chain fatty acids, etc.

Some epigenetic changes are heritable. For example, research indicates that a girl infant who experiences "famine" in the womb (i.e., her mother endures near starvation while pregnant) is much more likely to bear grandchildren who live shorter-than-average life spans and both children and grandchildren who become diabetic when they are adults.

For example, if a male arachnid known as the pseudoscorpion (*Cordylochernes scorpioides*) is exposed to the antibiotic tetracycline, both it and its male offspring will produce sperm bearing 25% lower viability.

See also *Gene, Expression, Phenotype, Deoxyribonucleic acid (DNA), Heredity, Gene silencing, Differentiation, DNA methylation, Alkylating agents, Chromatin, Chromatin remodeling, Histones, Short interfering RNA (siRNA), MicroRNAs, Gene imprinting, Imprinting, Organism, Central dogma (new), Cancer, Cell, Hedgehog signaling pathway, Paramutation, Nucleolar dominance, Epigenetic variation, Epigenomics, Histone deacetylase inhibitors, Epigenetic therapy, Short-chain fatty acids, Genomic imprinting, X chromosome, Ubiquitination, Biotinylation, Posttranslational modification of protein, long noncoding RNAs, Vernalization.*

Epigenetic Inheritance

Refers to the process via which epigenetic changes are expressed in the offspring of the organism to which the initial epigenetic changes (accomplished via epigenetic marks on its DNA, such as DNA methylation) occurred. The epigenetic inheritance process is a two-step process, with changes first occurring to maintain a chromatin (molecular) state that facilitates later actual silencing of any particular given genetic locus.

Later, heritable epigenetic chemical marks (e.g., DNA methylation, acetylation) are added to the genes/chromatin at the applicable genetic locus (i.e., specific location on a chromosome where the gene is located). These chemical tags serve as a form of "molecular memory," allowing cells to recognize the applicable genes and remember to silence them again in each new generation. See also *Epigenetic, DNA methylation, Epigenetic marks, Gene, Chromatin.*

Epigenetic Marks

Refer to the reshaping (at molecular scale) of chromatin (i.e., an organism's complex of DNA and histone protein) that alters which specific genes in that organism's DNA subsequently get expressed. It is caused by

- RNA-directed DNA methylation (RdDM) via short interfering RNAs (siRNAs) that guide the addition of methyl (one carbon) molecular groups to matching DNA strands/histone
- Acetylation of the histone (i.e., addition to histone of a two-carbon molecular group)
- Certain transcription activators
- Sumoylation of histone, etc.

The particular molecular group (e.g., methyl, acetyl) that gets attached to a specific gene's DNA as a result of the epigenetic regulation process thereby impacts the subsequent expression of that gene in the offspring. See also *Epigenetic, Epigenetic inheritance, Deoxyribonucleic acid (DNA), Gene, Expression, DNA methylation, Ribonucleic acid (RNA), Short interfering RNA (siRNA).*

Epigenetic Therapy

Refers to use of a therapeutic agent (e.g., certain pharmaceuticals) to reverse specific epigenetic changes in an organism (e.g., earlier methylation of tumor suppressor genes). See also *Epigenetic*, *Gene*, *Tumor Suppressor genes*, *DNA methylation*, *Deoxyribonucleic acid (DNA)*, *Long noncoding RNAs*.

Epigenetic Variation

Refers to the (nonhereditary) genetic variation arising via epigenetic changes. Among other sources, epigenetic changes can result from environmental factors including diet, exercise, and stress.

For example, studies of identical human twins have shown that over a lifetime, their different post-birth epigenetic changes result in increasing epigenetic variation (between the two twins) with age.

For example, the somaclonal variation that can result from *in vitro* culturing of land plants is sometimes caused by epigenetic variation. See also *Epigenetic*, *DNA methylation*, *Nucleolar dominance*, *Paramutation*, *Somaclonal variation*.

Epigenome

See *Epigenetic*, *Epigenetic variation*.

Epigenomics

Study of epigenetic variation on a genome-wide scale. See also *Epigenetic*, *Epigenetic variation*.

Epimerase

An enzyme capable of the reversible interconversion of two epimers. See also *Enzyme*, *Epimers*.

Epimers

Two stereoisomers differing in configuration. See also *Configuration*, *Stereoisomers*.

Episome (of a Bacterium)

An independent genetic element (DNA) that occurs inside bacterium in addition to the normal bacterial cell genome. The episome can replicate either as an autonomous unit or as one integrated into the host genome. The F (fertility) factor is an episome. See also *Genome*, *Plasmid*, *Bacteria*, *Deoxyribonucleic acid (DNA)*.

Epistasis

Interaction between nonallelic genes in which the presence of a certain allele at one locus prevents expression of an allele at a different locus. See also *Allele*, *Gene*, *Express*, *Locus*.

Epistatic Genetic Interactions

Refer to one in which multiple genes interact, resulting in a nonadditive phenotype. See also *Gene*, *Phenotype*.

Epithelial Projections

Refer to projections that anchor the epidermis (surface skin) to the dermis (subsurface tissue). Growth of these projections is increased by epidermal growth factor during the wound healing process. See also *Epidermal growth factor (EGF)*.

Epithelium

The prefix "epi-" means on, above, or upon. The membranous cellular tissue that covers a free surface or lines a tube or cavity of an animal body. It serves to enclose and protect the other tissues, to produce secretions and excretions, and to function in assimilation. See also *Assimilation*, *Cadherins*, *Ion channels*, *Commensal*.

Epitope

Also called antigenic determinant. The specific group of atoms (on an antigen molecule) that is recognized by (that antigen's) antibodies (thereby causing an immune response). See also *Antibody*, *Antigen*, *Idiotype*, *Humoral immune response*.

EPO

See *Erythropoietin*, *European Patent Office*.

EPPO

See *European Plant Protection Organization*.

EPSP Synthase

Enolpyruvyl-shikimate phosphate synthase. An enzyme produced by virtually all plants and internally transported into their cells' chloroplasts; it is essential in a plant's metabolism biochemical pathway and for the biosynthesis (i.e., creation) of the aromatic (ring-shaped molecule) amino acids tyrosine, phenylalanine, and tryptophan, which are needed for plants to live.

Some (glyphosate-containing and sulfosate-containing) herbicides kill unwanted plants (e.g., weeds) by inhibiting EPSP synthase. By incorporating a gene that causes (over)production of CP4 EPSP synthase into several crops (e.g., soybeans, cotton), scientists have been able to help those crops to survive postemergence application(s) of glyphosate-containing herbicide. Additional resistance to glyphosate-containing and sulfosate-containing herbicides can be conferred to plants via incorporation into plants of a gene (GO) that causes those plants to produce glyphosate oxidase. See also *Enzyme*, *Metabolism*, *Gene*, *PAT gene*, *BAR gene*, *Genetic engineering*, *Soybean plant*, *Corn*, *Glyphosate*, *Glyphosate oxidase*, *CP4 EPSPS*, *Herbicide-tolerant crop*, *Sulfosate*, *mEPSPS*, *Chloroplasts*, *Chloroplast transit peptide (CTP)*, *Target (of a herbicide or insecticide)*.

EPSPS

See *EPSP synthase*, *CP4 EPSPS*, *mEPSPS*.

eQTLs

Acronym for "expression quantitative trait loci." See also *Quantitative trait loci (QTL)*, *Express*, *Expressivity*, *Expression analysis*, *Expression profiling*.

ER

See *Endoplasmic reticulum*.

erb B-2 Gene

A designation that is sometimes utilized for the "HER-2/neu" gene. See also *HER-2 gene*.

E

ERBB2 Gene

A designation that is sometimes utilized for the "HER-2/neu" gene. See also *HER-2 gene*.

Ergotamine

A mycotoxin (i.e., metabolite produced by a fungus, which is toxic to animals and humans) produced by the fungus (*Claviceps purpurea*) known as ergot. Ergotamine is an alkaloid vasoconstrictor, so consumption of it can lead to severe constriction of blood vessels in the brain and extremities, causing hallucinations and dry gangrene. Humans whose bodies are deficient in vitamin A are especially vulnerable to ergotism ("ergot poisoning"). See also *Mycotoxins*, *Toxin*, *Fungus*, *Vitamin*.

eRNAi

Abbreviation for "environmental RNA interference," that is, the sequence-specific knockdown of gene expression (RNA interference) within an organism that results via ingestion by the organism of double-stranded RNA.

eRNAi has been shown to occur in nematodes and certain insect species (e.g., ingestion of double-stranded RNA by certain pest insects of *crops that have been genetically engineered to contain specific double-stranded RNA that causes the pest insect's digestive system to be unable to digest that crop plant*). For example, to control a pest known as the soybean cyst nematode (microscopic roundworm), scientists can design and insert a gene into a soybean plant that, when expressed in the plant:

- That gene's resultant RNA is chopped into pieces by dicer molecule, as usual.
- One or more of the small RNA pieces enter into the soybean cyst nematode when it chews on the soybean plant, where they silence gene(s) responsible for SCN feeding. The resultant nematode starvation results in control of the nematode. See also *RNA interference (RNAi)*, *Gene*, *Express*, *Organism*, *Short interfering RNA (siRNA)*, *Ribonucleic acid (RNA)*, *Soybean cyst nematodes (SCN)*.

ERU Maps

Acronym for "environmental response unit maps." Refer to precision agriculture software products that utilize a combination of public soil databases (including soil depth, texture, soil organic matter, and water-holding capacity), company proprietary (analytics, high-resolution field elevation data, field topography, and watersheds/hydrogeology), and electrical conductivity soil testing (basis: topsoil depth, pH, salt concentrations, and available water-holding capacity) to divide farm fields into contiguous ERU areas (subportions of field, also sometimes known as management zones or yield environments) where crops planted there will respond positively to farmer management decisions regarding crop inputs such as amounts and timing of fertilizer applied to field, amounts and timing of irrigation water applied to field, and number of crop seeds planted per hectare.

For example, such precision agriculture software might thereby recommend far less irrigation water be applied by the farmer to an ERU located in a low-elevation, heavy-soil-type area of farm field that is naturally wet throughout the growing season due to the field's hydrogeology, plus a higher number of crop seeds planted per hectare (of a wet-environment-tolerant crop variety). The precision agriculture software would recommend different inputs/rates for an ERU located in a sandy soil, higher-elevation area of the farm field (e.g., more irrigation water applied, lower number of crop seeds planted per hectare) of a drought-tolerant crop variety. Such properly integrated precision management of ERUs would maximize the field's crop yield while minimizing its consumption of inputs such as fertilizer, irrigation water, and crop seeds. See also *Drought-tolerance trait*.

Erwinia caratovora

A species of bacteria that can cause significant postharvest losses to potato farmers, when it infects potatoes and causes "soft rot" (spoilage). See also *Bacteria*, *Species*.

Erwinia uredovora

See *Golden rice*.

Erythroblasts

See *Macrophage*.

Erythrocytes (Red Blood Cells)

Hemoglobin-containing cells (manufactured in the bone marrow) that transport the oxygen from the lungs to the body tissues where it is needed. See also *Hemoglobin*.

Erythropoiesis

The formation of red blood cells (erythrocytes) from pluripotent stem cells. Stimulated by the protein erythropoietin, which is secreted by the kidneys.

People with damaged kidneys or with myeloma typically do not produce enough erythropoietin, so they often become anemic. See also *Stem cells*, *Erythropoietin (EPO)*, *Macrophage*, *Pluripotent stem cells*.

Erythropoietin (EPO)

A glycoprotein cytokine produced in the kidneys that stimulates pluripotent stem cells in the bone marrow to differentiate and then increase the number of red blood cells. Erythropoietin can be used to help correct a variety of anemias. See also *Glycoprotein*, *Cytokines*, *Erythrocytes*, *Stem cells*, *Differentiation*, *Pluripotent stem cells*.

Escherichia coli

See *Escherichia coliform (E. coli)*.

Escherichia coli 0157:H7 (E. coli 0157:H7)

See *Escherichia coliform 0157:H7*.

Escherichia coliform (E. coli)

Named after Theodor Escherich, who isolated it in 1885, it is a bacterium that commonly inhabits the human intestine as well as the intestine of other vertebrates (i.e., animals possessing a skeleton). The most thoroughly studied of all bacteria, *Escherichia coli* is used in many microbiological experiments. It has historically been considered the workhorse of genetic engineering research, and genetically engineered versions have been used to produce human proteins (e.g., insulin).

One of the more exotic uses of genetically engineered *E. coli* was to make indigo dye (originally discovered in 1983, using indole or tryptophan as starting materials). In 1993, Burt D. Ensley and coworkers at Amgen discovered a way to genetically engineer *E. coli* to produce indigo from glucose starting material. *E. coli* has 4288 genes. See also *Bacteria, Genetic engineering, Gene, Recombinant DNA (rDNA), Escherichia coliform 0157:H7, Tryptophan (trp).*

Escherichia coliform 0157:H7

The particular strain (serotype) of *Escherichia coliform* (*E. coli*) bacteria that causes often-fatal diarrhea, internal bleeding, and kidney damage in humans, via the Shiga toxins they produce. Children are more susceptible to *E. coli 0157:H7* than adults, because children possess more of the receptors (on cells inside the digestive tract) that are utilized by *E. coli 0157:H7* to enter the body from the digestive tract.

Although cattle were susceptible to *E. coli 0157:H7*'s toxins prior to the 1980s, they eventually developed resistance. That meant that the cattle could carry these bacteria without getting sick and transmit *E. coli 0157:H7* to humans whenever conditions allow (e.g., when *E. coli 0157:H7*–infected cattle are slaughtered and people consume the meat without first heating it to a high enough temperature to kill the *E. coli 0157:H7*). Some varieties of *E. coli 0157:H7* are resistant to the antibiotics tetracycline and streptomycin.

In 1996, researchers at Cornell University in New York, USA, discovered that nonambulatory cows (which could not walk) were approximately four times as likely as other cows to test positive for *E. coli 0157:H7.* Other research in Canada indicates that fasting of cattle (common occurrence for nonambulatory cows) tends to alter the pH inside the cow's rumen (stomach) in a way that encourages the proliferation of *E. coli 0157:H7* instead of the bacteria that normally populate the rumen.

E. coli 0157:H7 is the most common Shiga toxin–producing strain in North America. In Europe, *E. coli* 026 is the most common Shiga toxin–producing strain. In South America, *E. coli* 0111 is the most common Shiga toxin–producing strain. See also *Escherichia coliform (E. coli), Bacteria, Serotypes, Toxin, Shigellosis, Receptors, Bioluminescence, Strain, Enterotoxin, Commensal, Bacteriophage.*

E-Selectin

See *ELAM-1.*

ESI

See *Mass spectrometer.*

Essential Amino Acids

Those amino acids that cannot be synthesized by humans and most other vertebrates, and therefore must be obtained from the diet. They are phenylalanine, valine, threonine, tryptophan, isoleucine, methionine, histidine, arginine, leucine, and lysine (glycine and proline for poultry). See also *Amino acid, Lysine (lys), Methionine (met), Soy protein, Opague-2, Protein digestibility-corrected amino acid scoring (PDCAAS).*

Essential Fatty Acids

The group of polyunsaturated fatty acids of plants that are required in the human diet, because the human body cannot synthesize (i.e., "manufacture") them yet must have them for proper functioning (e.g., of the body's metabolism, immune system function). These include linoleic acid, linolenic acid, arachidonic acid, and docosahexaenoic acid.

If humans and other higher animals do not consume enough essential fatty acids per day, they suffer decreased growth rates, increased susceptibility to infection, impaired reproduction, kidney damage, and other adverse physiological effects. See also *Fatty acid, Soybean oil, Lecithin, Fats, Essential nutrients, Polyunsaturated fatty acids (PUFA), Linoleic acid, Linolenic acid, Docosahexaenoic acid (DHA), Arachidonic acid (AA).*

Essential Nutrients

Chemical compounds in foods that are required for (consuming organism's) life, growth, or tissue repair, and cannot be synthesized by that organism. See also *Essential amino acids, Essential fatty acids, Essential polyunsaturated fatty acids, Vitamin.*

Essential Polyunsaturated Fatty Acids

See *Essential fatty acids.*

EST

See *Expressed sequence tags (EST).*

Establishment Potential

Refers to a formal estimate of the likelihood for a given pest (e.g., weed, insect, disease) to successfully establish a permanent (reproducing) population within a specific "pest-free area" (e.g., country or region where *that pest* has not previously been established).

Determination of *establishment potential* requires reliable biological information (e.g., life cycle, host range if any, climate tolerance, epidemiology). For example, a pest that requires tropical temperatures/humidity to reproduce is not likely to have a significant *establishment potential* in a cold and dry country. See also *International Plant Protection Convention (IPPC), Pest risk analysis (PRA), Quarantine pest, Introduction.*

Estradiol

See *Estrogen.*

Estrogen

Refers to a "family" of female sex hormones, secreted by the ovaries and the brain, that promotes estrus and helps to regulate the pituitary gland's production of luteinizing hormone (LH) and follicle-stimulating hormone (FSH). Estrogens cause proliferation of breast tissue (cells) and a number of other cell types. Estrogens are responsible for the development of female secondary sex characteristics (e.g., smaller body size, lack of facial hair, higher pitch voice in humans). Most of the estrogens (e.g., estradiol, a primary hormone that controls the body's menstrual cycle) are manufactured in the ovaries. When estradiol subsequently circulates throughout the body (i.e., including in the brain and pituitary gland), it influences reproduction, body weight, learning, and memory.

More generally, estrogens (which are also synthesized in small amounts in males) regulate reproductive behavior, promote tissue survival, and protect cells from apoptosis, cardiovascular disease, osteoporosis, diabetes, and dementia.

Research indicates that lack of estrogen (e.g., in postmenopausal women) makes human more prone to colon cancer and heart disease but less prone to the "hormone-dependent" cancers (e.g., ovarian cancer, uterine cancer). See also *Hormone, Pituitary gland, Follicle stimulating hormone (FSH), Selective estrogen effect, Testosterone, Luteinizing hormone (LH), Hypothalamus, Cancer, Cell, Estrogen receptors.*

Estrogen Receptors

Receptor molecules that are specific to the estrogen hormone. Estrogen receptors are generally located in the nucleus of applicable cells, except some are also located on the surfaces of certain cells, especially the endothelial cells that line arteries and veins. See also *Receptors, Selective estrogen effect, Resveratrol, Hormone, Estrogen, Cell, Nucleus, Endothelial cells.*

ET

Acronym for "electron transfer." See also *Metalloproteins, Biophysics, Tryptophan (trp).*

Etanercept

A soluble fusion protein consisting of the Fc region of an antibody plus p75-TNF receptor protein. It is produced in CHO cells and acts to inhibit tumor necrosis factor. It was approved as the pharmaceutical Enbrel™ by the U.S. Food and Drug Administration (FDA) for the treatment of several types of arthritis and psoriasis. See also *Fusion protein, TNF blockers, CHO cells, Antibody, Fc region, Receptors, Arthritis, Tumor necrosis factor (TNF).*

Ethylene

A plant hormone that is synthesized ("manufactured") by most plants to induce ripening (of their fruit/seeds) and by some plants as a stress hormone (as described in the following paragraph).

When certain plants are under stress conditions such as drought, flooding, cold, wounding (e.g., by chewing insects), or pathogen attack (e.g., bacteria, viruses), ethylene is synthesized to instruct applicable plant cells to make some needed adjustments for the plant to defend against those stresses. When applicable plant cells sense such ethylene, they trigger a cascade of events. Ethylene results in a signal going to the cell's nucleus, which initiates gene transcription so the plant adjusts according to the particular stress condition that it faces. Those adjustments include leaves wilting or rolling up (to reduce loss of water via leaf transpiration/evaporation) and premature leaf senescence (aging). For crop plants, that unfortunately results in reduced photosynthetic efficiency, loss of chlorophyll, poor pollination, and/or flower, fruit, and seed loss.

Ethylene is also synthesized when needed to instruct petioles (i.e., the part of the plant that connects a leaf to the plant's stem) to grow/elongate (e.g., if another nearby plant causes that petiole's leaf to be in shade).

In some plants (e.g., petunias, carnations, orchids), ethylene causes their flowers to shed their petals (i.e., abscission) in the autumn, via activating genes within the flower that cause production of hydrolytic enzymes. See also *Plant hormone, Abiotic stresses, Stress hormones, Enzyme, ACC synthase, ACC, Gene, SAM-K gene, Hydrolytic cleavage, Abscission, Signaling, Bacteria, Pathogen, Virus, Cell, Nucleus, Transcription, Petiole.*

Etiological Agent (of a Disease)

The microorganism (or other agent) that causes the disease. See also *Pathogen, Etiology.*

Etiology

The science (study) of the cause (source) of a disease. See also *Pathogen, Etiological agent.*

Eucaryote

Also spelled eukaryote. A cell characterized by compartmentalization (by membranes) of its extensive internal structures or an organism made up of such cells. For example, eucaryotes possess a distinct membrane-surrounded nucleus containing the DNA. Eucaryotic cells (e.g., human cells) are much larger and more complex than procaryotic cells (e.g., bacteria). The cells of all higher organisms, both plant and animal, are eucaryotic, so those higher (complex) organisms are often referred to as eucaryotes.

Most eucaryotic organisms cannot survive temperatures greater than 131°F (55°C). However, one called the Pompeii worm (*Alvinella pompejana*) can withstand long-term exposure in water up to a temperature of 176°F (80°C). See also *Procaryotes, Cell, Thermophile, Deoxyribonucleic acid (DNA), Plasma, Membranes, Microtubules.*

Eugenics

First formulated by Francis Galton, who was a cousin of Charles Darwin, eugenics is the concept that a species can be "improved" by encouraging reproduction of only those organisms in that species that possess "desired" traits.

This belief became popular in a number of countries during the early twentieth century. Margaret Sanger, founder of America's Planned Parenthood organization, referred to African Americans as "human weeds" and called for "more children from the fit, less from the unfit." Based upon Charles Darwin's written assertion that "the civilized races of man will almost certainly exterminate and replace the savage races," a number of large genocides were committed by some national governments. See also *Genetics, Gene, Trait, Genotype, Heredity, Heritability, Genome.*

Eukaryote

See *Eucaryote.*

Euploid

A cell carrying an exact multiple of the haploid chromosome number. For example, a diploid possesses twice the haploid number of chromosomes. See also *Haploid, Diploid, Chromosomes.*

European Corn Borer (ECB)

Also known as pyralis (*Pyralidae*). Latin (Linnaean) name *Ostrinia nubilalis*, it is an insect whose larvae (caterpillars) eat and bore into the corn/maize plant (*Zea mays* L.). In doing so, they can act as vectors (i.e., carriers) of the fungi known as *Aspergillus flavus* (a source of aflatoxin), *Fusarium moniliforme* (a source of fumonisin), or *Aspergillus parasiticus* (a source of aflatoxin).

Full-grown ECB larvae overwinter by sheltering inside a variety of vegetative materials (e.g., plant stalks lying on top of soil in some fields).

ECB control can be effected via some of the following methods:

- Spraying of conventional synthetic chemical pesticides
- Spraying of pesticides produced via promulgation of *Bacillus thuringiensis* (*B.t.*) bacteria
- Incorporating a (protoxin) gene from *Bacillus thuringiensis* (*B.t.*) into the DNA of the corn plant, so that the plant itself produces B.t. protoxin

As part of integrated pest management, farmers can utilize the following:

- Corn possessing *Bacillus thuringiensis* (*B.t.*) gene(s) to control populations of ECB without applying insecticides.
- The parasitic *Euplectrus comstockki* wasp to help control the ECB. When that wasp's venom is injected into ECB larva, it stops the larva from molting (and thus maturing).
- Crop rotation.
- Other additional methods, alone or in concert with the aforementioned.

See also *Corn, Fungus, Aflatoxin, Integrated pest management* (*IPM*), *Crop rotation, Bacillus thuringiensis* (*B.t.*), *B.t. kurstaki, Fusarium, Fusarium moniliforme, Asian corn borer, Protoxin, Volicitin.*

European Medicines Evaluation Agency (EMEA)

A London-based agency of the European Union (EU) that began operation in 1995. It coordinates drug licensing and safety matters throughout the nations of the EU. Its licensing/approval process is compulsory throughout the EU. See also *Committee for Proprietary Medicinal Products* (*CPMP*), *Medicines Control Agency* (*MCA*), *Food and Drug Administration* (*FDA*), *Koseisho, Bundesgesundheitsamt* (*BGA*), *Committee on Safety in Medicines, Committee for Veterinary Medicinal Products* (*CVMP*).

European Patent Convention

An international patent treaty signed in 1973, by which the countries of Europe agreed to recognize and honor the patents granted by each country, plus those patents granted by the European Patent Office. Plant varieties or animal breeds were initially excluded from patentability by the European Patent Convention. In 1998, the European Parliament removed that exclusion. See also *European Patent Office* (*EPO*), *U.S. Patent and Trademark Office* (*USPTO*), *Plant's novel trait* (*PNT*), *Plant breeder's rights* (*PBR*), *Union for Protection of New Varieties of Plants* (*UPOV*).

European Patent Office (EPO)

The agency of the European Union (EU) based in Munich, Germany—established in 1977—that is responsible for common patent protection matters for all of the (EU) member countries, plus the non-EU countries of Switzerland and Liechtenstein. The European Patent Office originally did not allow a "plant or animal breed" to be patented, whereas its U.S. counterpart—the U.S. Patent and Trademark Office (USPTO)—does allow patenting of microbes, plants, and animals (e.g., those which have been genetically engineered by man). In 1998, the European Parliament removed that exclusion, and in 1999 the European Patent Court issued a ruling that caused the European Patent Convention to allow patents on novel plants, thus making the two patent systems compatible. See also *European Patent Convention, Microbe, Genetic engineering, Biotechnology, American type culture collection* (*ATCC*), *U.S. Patent and Trademark Office* (*USPTO*), *Plant's novel trait* (*PNT*), *Plant breeder's rights* (*PBR*), *Union for Protection of New Varieties of Plants* (*UPOV*), *Community Plant Variety Office.*

European Plant Protection Organization (EPPO)

One of the international SPS standard–setting organizations that develops plant health standards, guidelines, and recommendations (e.g., to prevent transfer of a plant disease or plant pest from one country to another). Its secretariat is in Paris, France.

EPPO is one of the organizations within the International Plant Protection Convention, and it covers the countries of Europe. See also *International Plant Protection Convention* (*IPPC*), *North American Plant Protection Organization* (*NAPPO*), *SPS, Plant's novel trait* (*PNT*), *Plant breeder's rights* (*PBR*).

Evanescent Wave

See *TIRF microscopy.*

Evening Complex

See *Docking proteins.*

Event

Refers to each instance of a genetically engineered organism. For example, the same gene inserted by man into a given plant genome at two different locations (i.e., loci) along that plant's DNA would be considered two different "events." Alternatively, two different genes inserted into the same locus of two same-species plants would also be considered two different "events."

Generally speaking, the world's regulatory agencies confer new biotech-derived product approvals in terms of events. See also *Genetic engineering, Genetically engineered organism* (*GEO*), *Gene, Deoxyribonucleic acid* (*DNA*), *Locus, Loci, Genome, Mutual recognition agreements* (*MRAs*).

Ex Vivo (Testing)

The testing of a substance by exposing it to (excised) living cells (but not to the whole, multicelled organism) in order to ascertain the effect of the substance (e.g., pharmaceutical) on the biochemistry of the cell. See also *In vitro, In vivo.*

Ex Vivo (Therapy)

Removal of cells (e.g., certain blood cells) from a patient's body and alteration of those cells in one or more therapeutic ways, followed by reinsertion of the altered cells into the patient's body. See also *In vitro, In vivo.*

E

Excision

The cutting out of a piece of damaged or defective DNA by enzymes. DNA damage might be constituted by the presence of a thymine dimer that inactivates that part of the DNA. The region of the dimer is cut out and it is then repaired. See also *Recombination, Genome, Informational molecules.*

Excitatory Amino Acids (EAAs)

Amino acids present in the brain (when released by certain immune system cells), which can kill brain cells when in excess (e.g., results from strokes, which cause the release of too many EAAs in the brain). Another source of harmful EAAs (e.g., glutamate) is the disease known as multiple sclerosis.

Some spiders paralyze their prey with venom that contains a substance that blocks the action of EAAs (thus pharmaceuticals based on an active ingredient in that venom may someday be used to prevent brain damage in stroke and in multiple sclerosis victims). See also *Amino acid, Multiple sclerosis, Cell, Immune response.*

Exclusion Chromatography

See *Gel filtration.*

Exergonic Reaction

A chemical reaction with a negative standard free energy change (i.e., a "downhill" reaction). A reaction that releases energy (exothermic; in the form of heat). See also *Endergonic reaction, Free energy.*

Exobiology

Extraterrestrial biology.

Exocytosis

The releasing of an entity that was bound inside an "endosome" (e.g., inside a cell). See also *Endocytosis.*

Exoglycosidase

An enzyme that hydrolyzes (cuts) only a terminal (i.e., end) bond in the oligosaccharide (molecular) branch(es) of a glycoprotein. See also *Endoglycosidase, Glycoprotein, Restriction endoglycosidases.*

Exome

Refers to all the exons (i.e., the segment of each gene that is transcribed into a messenger RNA molecule; it codes for a specific domain of a protein) in a eucaryotic organism. See also *Deoxyribonucleic acid (DNA), Gene, Exon, Messenger RNA, Protein, Sequence (of a dna molecule), Organism.*

Exome Sequencing

Refers to sequencing of all the exons (i.e., the segment of each gene that is transcribed into a messenger RNA molecule; it codes for a specific domain of a protein) in a eucaryotic organism. See also *Deoxyribonucleic acid (DNA), Gene, Exon, Messenger RNA, Protein, Sequence (of a DNA molecule), Sequencing (of DNA molecules), Organism.*

Exon

The segment of a eucaryotic gene that is transcribed into a messenger RNA molecule; it codes for a specific domain of a protein. See also *Deoxyribonucleic acid (DNA), Protein, Eucaryote, Messenger RNA (mRNA), Gene, Homeobox, Transcription, Editing, Exome sequencing.*

Exonuclease

An enzyme that hydrolyzes (cuts) only a terminal phosphodiester bond of a nucleic acid. See also *Hydrolyze.*

Exotic Germplasm

Germplasm that has not been adapted (selectively bred) to the environment intended (for its offspring, via selective breeding by man). See also *Germplasm, Introgression, Hybridization (plant genetics).*

Exotoxin

Proteins (toxins) produced by certain bacteria that are released by the bacteria into their surroundings (growth medium). Produced by primarily gram-positive bacteria. Diphtheria toxin was the first one discovered. Other exotoxins cause botulism, tetanus, gas gangrene, and scarlet fever. Exotoxins are generally more potent and specific in their actions than endotoxins. See also *Toxin, Diphtheria toxin, Endotoxin, Bacteria, Gram-positive (G+), Protein.*

Expanded Repeat (within DNA)

See *Trinucleotide repeat.*

Expected Progeny Differences (EPDs)

Numerical rankings of (livestock) parental genetics, in terms of an animal's genetic impact on progeny's four following commercial traits:

1. Number of progeny born alive
2. Weight of progeny at weaning age
3. Number of days required to reach slaughter weight, when fed adequately
4. Carcass lean meat versus fat percentages

EPDs allow a farmer to estimate differences in performance of future offspring (of a given parent) versus offspring produced by parents of average genetic value. For example, a boar (male pig) possessing an EPD of -4 for "number of days required to reach slaughter weight" produces offspring that reach slaughter weight in four fewer days (of feeding time) versus offspring that are sired by a boar possessing an EPD of 0. See also *Genetics, Trait, Phenotype, Genotype, Best linear unbiased prediction (BLUP).*

E

Express

To translate the cell's genetic information stored in the DNA (gene) into a specific protein (synthesized by the cell's ribosome system) or a specific miRNA gene.

When present in cells, certain proteins such as transcription factors and/or certain RNAs (e.g., long noncoding RNAs) regulate the expression (e.g., increase/decrease/timing) of some genes. See also *Gene expression cascade, Ribosomes, Gene, Deoxyribonucleic acid (DNA), Cell, Transcription, Transcription factors, Long noncoding RNAs, Translation, Messenger RNA (mRNA), Transcription unit, Protein, Cosuppression, Gene expression analysis, Functional genomics, Splice variants, Coding sequence, miRNA genes, Genomic imprinting, Polymery, Polygenic, Expressivity.*

Expressed Sequence Tags (ESTs)

Molecular tags consisting of a DNA sequence approximately 200–500 bp long, which are utilized to "label" a given gene (i.e., in terms of that gene's function/protein). Physically, the EST was historically composed of mRNA (i.e., the gene's "message" after the "junk DNA" [introns] have been *edited out*)—produced by the analogous gene in (simple) *model organisms* such as (traditionally) *Caenorhabditis elegans* nematode—which has been sequenced/mapped. *Functions* of the "labeled" genes were (at least initially) inferred from (known function) *C. elegans* genes.

Since ESTs are derived from mRNA, ESTs can be utilized to assess (all of) the genes *being expressed* in a specific tissue at a particular stage of an organism's development. For example, such gene expression can be assessed by attaching numerous (all possible) ESTs to a microarray substrate and then passing a tissue sample over the EST-covered microarray. See also *Gene, Intron, Protein, Complementary DNA (cDNA), Deoxyribonucleic acid (DNA), Messenger RNA (mRNA), Junk DNA, Caenorhabditis elegans (C. elegans), Sequencing (of DNA molecules), Sequence (of a DNA molecule), Mapping, Model organism, Express, Bacterial expressed sequence tags (BEST), Microarray (testing), Gene expression analysis, Serial analysis of gene expression (SAGE), Substrate (structural), Organism.*

Expression

See *Express.*

Expression Analysis

See *Gene expression analysis, Microarray (testing).*

Expression Array

See *Microarray (testing).*

Expression Elements

See *Promoters, Control sequences, Leader sequence.*

Expression Profiling

See *Gene expression analysis.*

Expressivity

The intensity with which the effect of a gene is realized in the phenotype. The degree to which a particular effect is expressed by individuals. For example, expression of the gene for NFκB protein in humans is increased 500% after a short period of weightlessness (space flight). See also *Phenotype, Gene, Express, Ribosomes, Long noncoding RNAs, NFκB.*

Extein

See *Intein.*

Extension (in Nucleic Acids)

The nucleic acid strand elongation (lengthening) that occurs in a polymerization reaction. See also *Nucleic acids, Polymer.*

Extracellular Matrix

Abbreviated "ECM," this refers to the structural connective tissue located between cells within an organism. The constituents of the ECM are manufactured inside the cells that subsequently reside within the ECM.

The ECM regulates signaling between those cells, provides a structural substrate for those cells to adhere to, and can store/release, when needed, various compounds that cause those cells to migrate or otherwise change cellular function. Among many other cues to guide stem cells regarding the specific type of cell they become, the physical stiffness of the extracellular matrix touching them is a major factor. See also *Substrate (structural), Cell, Stem cells, Signaling, Cell differentiation, Tissue engineering.*

Extranuclear Genes

Genes that reside within the cell but *outside* the nucleus. Generally, extranuclear genes reside in the organelles such as mitochondria and chloroplasts. See also *Gene, Cell, Nucleus, Copy number, Organelles, Chloroplasts, Mitochondria.*

Extremophilic Bacteria

Bacteria that live and reproduce outside (either colder or hotter) the typical temperature range of 40°F (4°C) to 140°F (60°C) that bacteria tend to be found in, on Earth. Other extremes are high pressure (e.g., at the ocean bottom), salt saturation (e.g., the Dead Sea), pH lower than 2 (e.g., coal deposits), pH higher than 11 (e.g., sewage sludge), high levels of radiation, etc. See also *Bacteria, Thermophilic bacteria, Thermophile, Thermoduric, Deinococcus radiodurans.*

Extremozymes

Enzymes within the microorganisms (e.g., extremophilic bacteria) that populate extreme environments. Because extremozymes can catalyze reactions under high pressure, high temperatures, etc., they are increasingly being used as catalysts for industrial processes. See also *Extremophilic bacteria, Enzyme, Archaea, Phyto-manufacturing.*

EZH2

Acronym for "enhancer of zeste homolog 2." See *MicroRNAs.*

F

F Protein

See *Nanovaccine.*

F1 Hybrids

The first-generation offspring of crossbreeding; also known as *first filial* hybrids. They tend to be more healthy, productive, and uniform than their parents. See also *Genetics*, *Hybridization* (*plant genetics*).

FABP

Acronym for *fatty acid–binding protein.* See *Fatty acid–binding protein.*

Facilitated Folding

Refers to the improvement in protein molecular folding (e.g., into its native conformation) of certain protein molecules, when those molecules

- Fuse with other relevant partner "folding facilitator" molecules such as small ubiquitin-related modifier (SUMO)
- Partner with chaperone molecules, chaperonins, etc. as the newly formed protein molecules emerge from ribosomes within the cell
- Partner with certain heat-shock proteins (e.g., HSP 90, HSP 70)—also called stress proteins—after stresses such as heat, age, exposure to ultraviolet light, certain viruses, or exposure to certain chemicals cause some protein molecules in cell to begin to "unfold"

See also *Cell, Protein, Fusion protein, Protein folding, Conformation, Native conformation, Small ubiquitin-related modifier, Chaperones, Chaperonins, Ribosomes, Heat-shock proteins, Stress proteins.*

FACS

See *Fluorescence-activated cell sorter* (*FACS*).

F-Actin

See *Actin.*

Factor IX

A protein factor in the blood serum that is instrumental in the *cascade* of chemical reactions (involving 17 blood components) that leads to clot formation, following a cut or other wound to body tissue.

A deficiency of factor IX is the cause of the disease known as "hemophilia B" (approximately 15% of all hemophilia patients). See also *Fibrin, Fibronectin, Protein, Cascade, Factor VIII.*

Factor VIII

Also known as antihemophilic globulin (AHG) or antihemophilic factor VIII. A protein factor in the blood serum that is instrumental in the "cascade" of chemical reactions (involving 17 blood components in the *intrinsic pathway*) that leads to clot formation following a cut or other wound to body tissue. Also, a deficiency of AHG is the cause of the classical type of hemophilia sometimes known as "hemophilia AM" (approximately 85% of all hemophilia patients). See also *Fibrin, Protein, Fibronectin, Cascade, Pathway, Factor IX.*

Facultative Anaerobe

An organism that will grow under either aerobic or anaerobic conditions. See also *Aerobe, Anaerobe, Organism.*

Facultative Cells

Cells that can live either in the presence or absence of oxygen. See also *Aerobe, Anaerobe.*

FAD

See *Flavin adenine dinucleotide* (*FAD*).

FAD Enzymes

Acronym for *fatty acid desaturase* enzymes (located in plastids in applicable plants) that catalyze the production of desaturated vegetable oils. See also *Enzyme, Fatty acid, Plastid.*

Fad Genes

Refers to genes that code for the synthesis ("manufacturing") in plants of a specific *fatty acid desaturase* enzyme. See also *Gene, Fad3 gene, Enzyme, Fatty acid, Desaturase, Δ12 desaturase, Delta 12 desaturase.*

Fad2 Gene

See *Plastid.*

Fad3 Gene

A gene naturally found with the soybean plant (*Glycine max* L.) DNA, which codes for (i.e., causes production of) the ω-desaturase; an enzyme within the pathway in which the soybean plant synthesizes (*manufactures*) oleic acid within the vegetable oil produced in the soybean.

Specifically, the ω-desaturase enzyme converts linoleic acid to linolenic acid in that pathway, so by increasing expression of the *Fad* 3 gene in a (genetically engineered) soybean plant, a scientist could cause that soybean plant to produce soybean oil comprised of as much as 50% linolenic acid. See also *Gene, Express, Fad genes, Soybean plant, Deoxyribonucleic acid (DNA), Enzyme, Pathway, Soybean oil, Fatty* acid, *Oleate, Oleic acid, High-oleic oil soybeans, Cosuppression, Desaturase, Δ12 desaturase, Delta* 12 *desaturase, Linolenic acid, Linoleic acid, Genetic engineering, High-linolenic oil soybeans.*

Fall Armyworm

Caterpillars (pupae) of the Lepidopteran insect *Spodoptera frugiperda*, which are harmful to certain crops grown by humans.

Fall armyworms are susceptible to *some* of the "cry" proteins. See also *Protein, Cry proteins, Armyworm, Insect cell culture.*

FAME

Acronym for *fatty acid methyl esters*, that is, the compounds that result when fatty acid molecules are reacted with compounds containing methyl submolecule groups ($-CH_3$).

The FAME acronym is also utilized to refer to a method of identifying species of microorganisms (e.g., during the investigation of a disease outbreak) based on the fatty acid composition of the microorganism's cell wall (membrane). Such cell walls can contain up to 115 different fatty acids in varying amounts/ratios of each one.

The cell wall of the microorganism of interest is broken down, and its fatty acids are converted to FAMEs as detailed in the first paragraph. Then, the FAMEs are analyzed using a gas chromatograph to yield a "fingerprint" unique to each species/strain of microorganism. See also *Fatty acid, Microorganism, Species, Strain.*

FAO

Food and Agriculture Organization of the United Nations. See also *Consultative Group on International Agricultural Research (CGIAR), Codex Alimentarius Commission.*

Farnesene

A 15-carbon alkene that can be produced via fermentation of sugars by certain strains of baker's yeast that have been modified via insertion of specially designed synthetic biology farnesene-producing modules. See *Fermentation, Synthetic biology.*

Farnesoid X Receptor (FXR)

Refers to nuclear receptors primarily in the enterohepatic system (e.g., cells within the liver) for bile acids. When certain bile acids (e.g., chenodeoxycholic acid, cholic acid) "dock" at the FXR, they "turn on" the bile salt efflux pump (BSEP), a bile acid transporter that increases the flow of bile acids into the body's bile (where they are utilized in the digestive system to help render fats and fat-soluble vitamins more easily absorbed). See also *Bile acids, Nuclear receptors, Vitamin, Agonists.*

Farnesyl Transferase

An enzyme utilized by the *ras* gene (to help "signal" certain cells to divide/grow). See also *ras gene, Gene, Enzyme, Cell, Signaling molecule, Farnesyl transferase inhibitors.*

Farnesyl Transferase Inhibitors

Refers to compounds that inhibit the enzyme farnesyl transferase. For example, these compounds inhibit the tendency of farnesyl transferase to

- Modify the precursor molecule (i.e., prelamin A) of lamin A protein
- Direct that prelamin A to the nuclear envelope of cells
- Modify certain other proteins via prenylation (e.g., farnesyl transferase inhibitor prevents hepatitis delta virus [HDV] infection replication inside hepatic cells and blocks the ability of the HDV virus to multiply)

In humans possessing certain mutation(s) of the gene coding for production of lamin A protein, the farnesyl transferase–directed excess of (modified due to mutation) prelamin A protein at the cells' nuclear envelopes causes distortion in shapes of those nuclei and the disabling/fatal effects of *Hutchinson–Gilford progeria syndrome.*

During 2014, the U.S. Food and Drug Administration (FDA) granted to lonafarnib (an orally active farnesyl transferase inhibitor) an Orphan Drug designation (i.e., approval for pharmaceutical use) for the treatment of HDV infection in humans. See also *Enzyme, Farnesyl transferase, Prenylation inhibition, Cell, Nucleus, Nuclear envelope, Mutation, Gene, Coding sequence, Protein, Hutchinson–Gilford progeria syndrome, Food and Drug Administration (FDA), Orphan drug.*

Fats

Energy storage substances produced by animals and some plants (e.g., soybeans), which consist of a combination of fatty acids and glycerol that form predominantly triglyceride molecules (although some diglyceride molecules are also often present in fats). The structure of triglyceride molecules consists of three fatty acids attached to a glycerol *molecular backbone*, so "triglyceride" molecules are more accurately called "triacylglycerides," but the term "triglyceride" is most often used.

Two separate components of plant cells are involved in the synthesis (i.e., *manufacturing*) of plant fats (lipids): the *plastid* and the *endoplasmic reticulum.* Synthesis of fatty acids begins in the plastid, where Ac-CoA is first carboxylated (thereby becoming malonyl CoA) via the enzyme *acetyl-CoA carboxylase.* Next, a group of seven related enzymes (known as "fatty acid synthetases") catalyzes synthesis of palmitoyl-CoA (which is a long molecule possessing 18 carbon atoms in its "molecular backbone"), although shorter-length molecules result when a specific *acyl carrier protein (ACP)* thioesterase enzyme is present in plastid (e.g., $C_{16:0}$ACP), which results in fatty acids of various "carbon chain" *length.* After the palmitoyl-CoA is elongated (i.e., made a longer molecule via addition of carbons to its *molecular backbone*) to become the (stearate-like) molecule "oleoyl-ACP" in a chemical reaction catalyzed by a *palmitoyl elongase enzyme*, the oleoyl-ACP is transported to the plant's endoplasmic reticulum. In the endoplasmic reticulum, the oleoyl-ACP is either further elongated (via the addition of more carbon atoms to the fatty acid's *molecular carbon chain* "backbone") or further desaturated (i.e., via desaturase-catalyzed removal of hydrogen atoms from that fatty acid molecule). Stearic acid (also known as stearate) is desaturated to become oleic acid, which can be desaturated to become linoleic acid, which can be desaturated to become linolenic acid.

Three of the resultant fatty acid molecules are then chemically attached to a *glycerol-3-phosphate* molecule (with the cleaved-off phosphate atom "recycled" in the endoplasmic reticulum for further utilization in the energy cycle of the cell).

The content levels of individual fatty acids vary somewhat with the diet of the animal (i.e., for animal fat) and vary somewhat with the plant's growing conditions (i.e., for *plant fat* also known as vegetable oil). No natural fat is either totally saturated or unsaturated.

When eaten, fats are generally not absorbed directly through the intestinal wall. They are first emulsified and then hydrolyzed by the lipase enzyme. The components (i.e., fatty acids, cholesterol, monoacylglycerol, phospholipids, etc.) form micelles that pass through the intestinal wall and are absorbed by the body. Such emulsification/micelle formation is aided by the nutrient lecithin (a component in soybeans).

When fats are oxidized in cells, they provide energy for the body. Some of the energy is released as heat and some is stored in the form of adenosine triphosphate, which "fuels" metabolic processes. See also *Fatty acid, Hydrolysis, Hydrolytic cleavage, Hydrolyze, Lipase, Monounsaturated fats, Saturated fatty acids, Triglycerides, Triacylglycerols, Diacylglycerols, Micelle, Cell, Metabolism, Digestion (within organisms), Cholesterol, Lipids, Lecithin, Soybean oil, Free fatty acids, Oxidative stress, Plastid, ACP, Oxidation (of fats/oils/lipids), Plasma membrane, Enzyme, Ac-CoA, Endoplasmic reticulum, Fatty acid synthetase, Thioesterase, Desaturase, Mitochondria, Lauroyl-ACP thioesterase, Stearoyl-ACP desaturase, Adipocytes, Adenosine triphosphate (ATP), Bile acids, Phosphate transporter genes, Photosynthesis, Oleosomes, Stearate (stearic acid), Oleic acid, Linoleic acid, Linolenic acid (α-linolenic acid), Conjugated linoleic acid (CLA).*

Fatty Acid

A long-chain aliphatic acid found in natural fats and oils. Fatty acids are abundant in cell membranes and (after extraction/purification) are widely used as industrial emulsifiers, for example, phosphatidylcholine (lecithin). In general, fats possessing the highest levels of saturated fatty acids tend to be solid at room temperature, and those fats possessing the highest levels of unsaturated fatty acids tend to be liquid at room temperature. That rule of thumb was the original "dividing line" between compounds called *fats* and *oils*, respectively. In general, saturated fatty acids tend to be more stable (resistant to oxidation and thermal breakdown) than unsaturated fatty acids.

In plant cells, fatty acids are synthesized (manufactured) in plastids.

Fatty acids in biological systems (e.g., produced by plants in oilseeds) tend to contain an even number of carbon atoms in their molecular "backbone," typically between 14 and 24 carbon atoms. The molecular backbone (alkyl chain) may be saturated (no double bonds) or it may contain one or more double bonds. The configuration of the double bonds in most unsaturated fatty acids is *CIS*. See also *Essential fatty acids, Laurate, Phytochemicals, Saturated fatty acids, Lecithin, Soybean oil, Unsaturated fatty acid, Monounsaturated fats, Polyunsaturated fatty acids (PUFA), LPAAT protein, Stearoyl-ACP desaturase, Soybean oil, Canola, Fats, Oleic acid, Trans fatty acids, Enoyl-acyl protein reductase, Oxidation (of fats/oils/lipids), Lipids, Mitochondria, Adipocytes, Oleosomes, Delta 12 desaturase, Linoleic acid, Linolenic acid, Fatty acid synthetase, Carnitine, Biotin, Ac-CoA, Acyltransferases, Plastids.*

Fatty Acid–Binding Protein

Refers to certain protein molecules present within animal cells that bind to specific fatty acids (after they come through the cell's plasma membrane) and help transport those fatty acids to their needed destinations within the cell.

For example, oleic acid molecules are transported to the nucleus of some cells (e.g., in human breast tissue), where the oleic acid molecules bind to the cell's DNA in a manner that reduces over-expression of the Her-2/neu gene (in those women whose breast cells overexpress the Her-2/neu gene), and thereby confer some protection against breast cancer. See also *Protein, Transport proteins, Fatty acid, Cell, Nucleus, Deoxyribonucleic acid (DNA), Plasma membrane, Intracellular transport, Membrane transport, Transcription factors, Express, Expressivity, Cancer, Her-2/neu gene, Downregulating.*

Fatty Acid Methyl Esters

Abbreviated FAME. See *FAME.*

Fatty Acid Synthase

Acronym FAS, it refers to an enzyme involved in synthesis within the body of long-chain aliphatic fatty acids, which are abundant in cell membranes of the body. Because cancerous (e.g., tumor) cells multiply very rapidly, inhibition of FAS is one means to slow the cancer spread. See also *Enzyme, Fatty acid, Cell, Plasma membrane, Cancer, Tumor.*

Fatty Acid Synthetase

A group of seven related enzymes that catalyze synthesis (i.e., "manufacturing") of fatty acids within the soybean plant (*Glycine max* (L.) Merrill). See also *Enzyme, Catalyze, Fatty acid, Soybean plant, Desaturase, Fats, Oleosomes, Pathway, Delta 12 desaturase.*

F-Box Proteins

Proteins produced ("manufactured") within some eucaryotic cells that play an essential role in the degradation (i.e., breakdown) of cellular regulatory proteins, after those regulatory proteins have "completed their job" in the cell. Sometimes some F-box proteins go awry. In some cases of pneumonia, the infecting bacteria activate an F-box protein known as *Fbxo*3 to form a molecular complex that degrades another protein called *Fbxl*2 that is needed to suppress the body's inflammatory response. If that occurs, the result is a harmfully overactive inflammatory response that can cause further damage of the lung tissue, multiple organ failure, and septic shock (sepsis).

See also *Protein, Cell, Eucaryote, Sepsis.*

FC

Acronym for *flow cytometry*. See *Flow cytometry.*

F

Fc Region

See *Antibody.*

FD

See *FT protein.*

Federal Coordinated Framework for Regulation of Biotechnology

The legal framework created by the United States' government in 1986, which divided regulation of biotechnology among the U.S. Department of Agriculture, the U.S. Environmental Protection Agency, and the U.S. Food and Drug Administration. See also *Food and Drug Administration.*

Federal Insecticide Fungicide and Rodenticide Act (FIFRA)

A law enacted by the United States Congress in 1972. During 1994, the U.S. Environmental Protection Agency (EPA) proposed that the substances produced by plants (e.g., genetically engineered crops) for their defense against pests and diseases would be regulated by EPA under FIFRA. See also *Toxic Substances Control Act (TSCA), Genetically engineered microbial pesticides (GEMP), Wheat take-all disease, Bacillus thuringiensis (B.t.).*

Feedback Inhibition

Inhibition of the first enzyme in a metabolic pathway by the end product of that pathway. This is a method of shutting down a metabolic pathway that is producing a product that is no longer needed. See also *Metabolism, Enzyme, Effector.*

Feedstock

Raw material(s) used for the production of chemicals or growth substrates of microbes (e.g., yeasts or bacteria that require a solid phase to attach themselves to). See also *Fermentation, Ionic liquids, Bacteria, Yeast.*

Fermentation

A term first used with regard to the foaming that occurs during the manufacture of wine and beer. The process dates back to at least 6000 B.C. when the Egyptians made wine and beer by fermentation. From the Latin word *fermentare*, "to cause to rise."

The term "fermentation" is now used to refer to so many different processes that fermentation is no longer accepted for use in most scientific publications. Three typical definitions are given:

1. A process in which chemical changes are brought about in an organic substrate through the actions of enzymes elaborated (produced) by microorganisms.
2. The enzyme-catalyzed, energy-yielded pathway in cells by which "fuel" molecules such as glucose are broken down anaerobically (in the absence of oxygen). One product of the pathway is always the energy-rich compound adenosine triphosphate. The other products are of many types: alcohol, glycerol, and carbon dioxide from yeast fermentation of various sugars; butyl alcohol, acetone, lactic acid, and acetic acid from various bacteria; and citric acid, gluconic acid, antibiotics, and vitamin B_{12} and B_2 from mold fermentation. The Japanese utilize a bacterial fermentation process to make the amino acid, L-glutamic acid, a derivative of which is widely used as a flavoring agent.
3. An enzymatic transformation of organic substrates (feedstocks), especially carbohydrates, generally accompanied by the evolution of gas. A physiological counterpart of oxidation, permitting certain organisms to live and grow in the absence of air; used in various industrial processes for the manufacture of products such as alcohols, acids, and cheese by the action of yeasts, molds, and bacteria. Alcoholic fermentation is the best known example. It is also known as zymosis. The leavening of bread depends on the alcoholic fermentation of sugars. The dough rises due to production of carbon dioxide gas that remains trapped within the viscous dough.

See also *Zymogens, Substrate (chemical), Adenosine triphosphate (ATP), Microorganism, Enzyme, Feedstock, Carbohydrates (Saccharides), Cell-free fermentation, Plant cell fermentation.*

Ferritin

An iron–protein complex (a metalloprotein) that occurs in living tissues. Functions in iron storage in the spleen. Dietary sources include the soybean plant. See also *Hemoglobin, Metalloprotein, Soy protein.*

Ferrobacteria

Also called iron bacteria. Any of a group of bacteria that oxidize iron as a source of energy. The oxidized iron in the form $Fe(OH)_3$ is then deposited in the environment by secretion from the bacterium. The energy obtained from these reactions is used to carry on processes in which the basic substances needed by the bacterium are manufactured. These bacteria are commonly found in seepage waters of coal and iron mining areas where iron compounds abound. Ferrobacteria are not disease producers (i.e., pathogenic), but they are important as scavengers. Sometimes they create a nuisance by multiplying so profusely in iron water pipes that they stop the flow of water. Ferrobacteria have been active through long periods of geologic time. For example, the great Mesabi iron (ore) seam of America's Lake Superior region is thought to be a product of ferrobacteria activity. See also *Pathogen.*

Ferrochelatase

A mitochondrial enzyme that catalyzes the incorporation of iron into the protoporphyria molecule. See also *Mitochondria, Enzyme, Catalyst, Porphyrins.*

Ferrodoxin

An iron- and sulfur-containing protein important in the electron transfer processes of photosynthesis in plants. It also plays a role in the metabolism of some bacteria and was first found in an anaerobic bacterium. See also *Photosynthesis, Metabolism.*

Fertility Factor (F)

A type of transmissible (i.e., can enter other cells) plasmid that is often found in *Escherichia coli (E. coli).* See also *Plasmid, Vector, Escherichia coliform (E. coli).*

Fertilization

The union of the (haploid) male and (haploid) female germ cells (sex cells or gametes) to produce a diploid zygote. Fertilization marks the start of development of a new individual (organism), the beginning of cell differentiation. See also *Cell*, *Germ cell*, *Stigma*, *Self-pollination*, *Organism*.

FFA

Acronym for free fatty acids. See *Free fatty acids*.

FGF

See *Fibroblast growth factor (FGF)*.

FGMP

See *Food good manufacturing practice (FGMP)*.

FHB

Acronym for *Fusarium head blight*. See *Fusarium*.

FIA

Refers to immunodiagnostic tests that are based on fluorescence tracers (labels). See also *Immunoassay*, *Fluorescence*, *Radioimmunoassay*.

Fibrin

The ordered fibrous array of fibrin monomers, called a fibrin-platelet clot (blood clot), which spontaneously assembles from fibrin monomers (which themselves are formed by the thrombin-catalyzed conversion of fibrinogen into fibrin). Fibrinogen itself is the product of a controlled series of zymogen activation steps (enzymatic cascade) triggered initially by substances that are released from body tissues as a consequence of trauma (harm) to them. See also *Fibronectin*, *Zymogens*, *Cascade*, *Lipoprotein-associated coagulation (clot) inhibitor (LACI)*.

Fibrinogen

See *Fibrin*, *Lipoprotein-associated coagulation (clot) inhibitor (LACI)*.

Fibrinolytic Agents

Blood-borne compounds that activate fibrin in order to dissolve blood clots. See also *Tissue plasminogen activator (tPA)*, *Thrombolytic agents*, *Fibrin*.

Fibroblast Growth Factor (FGF)

First described in the mid-1970s by Dr. Gospodarowicz and fellow researchers at the University of California, San Francisco. It is a protein that stimulates the formation/development of blood vessels and fibroblasts (precursors to collagen, the connective tissue "glue" that holds cells together).

Fibroblast growth factor (FGF) also is mitogenic (causes cells to divide and multiply) for both fibroblasts and endothelial cells and attracts those two cell types (i.e., is chemotactic).

Dr. Gospodarowicz named the FGF originally derived from bovine (cow) brain tissue to be acidic FGF. Dr. Gospodarowicz named the FGF originally derived from bovine pituitary tissue to be basic FGF (BFGF). This was due to their identical *biological* activity but differing isoelectric points (i.e., the former being acidic and the latter being basic). BFGF is, however, 10 times more "potent" than acidic FGF in most bioassays.

Researchers recently discovered that exposure to BFGF can also be utilized to coax certain mature (differentiated) cells to become "self-renewing" multipotent stem cells. See also *Angiogenic growth factors*, *Protein*, *Cell*, *Fibroblasts*, *Pituitary gland*, *Collagen*, *Mitogen*, *Endothelial cells*, *Chemotaxis*, *Biological activity*, *Bioassay*, *Acid*, *Base*, *Nanofibers*, *Differentiation*, *Cell differentiation*, *Stem cells*, *Multipotent adult stem cell*.

Fibroblasts

Cells that are precursors to the connective tissue cells found in the skin. They make structural proteins like collagen, which gives skin its strength. Because fibroblasts do not express antigens on their cell surfaces (free standing, separated), fibroblasts possess potential for use in making artificial organs (e.g., artificial pancreas for diabetics), since recipient immune system cannot recognize the fibroblast cells as foreign. See also *Cellular immune response*, *Humoral immunity*, *Graft-versus-host disease (GVHD)*, *Xenogeneic organs*, *Cell*, *Multipotent*, *Fibroblast growth factor (FGF)*, *Collagen*.

Fibronectin

An adhesive glycoprotein that forms a link between the epithelial cells and the connective tissue matrix (essential for blood clotting). Research has indicated that fibronectin may solve the problem of getting new cells to stick to existing tissue, once a growth factor has caused them to grow (e.g., when growth factor is administered after a serious wound to tissue). See also *Fibrin*, *Glycoprotein*, *Growth factor*, *Organogenesis*.

Field Effect Transistor

Abbreviated FET, it refers to the electronics portion of certain miniature biosensors, in which the transistor "gate" (which controls current flow through the transistor) is replaced by a biochemical (and/or nanowire) that serves as a *sensing material of the biosensor*. A change in the microenvironment (e.g., liquid within tissue) immediately surrounding the sensing material results in a *field effect* (impact on the device's electrical field) that drains electrical current off the transistor. Thus, the FET can be utilized to continually monitor/measure that microenvironment for any changes. See also *Biosensors (electronic)*, *Bioelectronics*, *Nanowire*.

Field Inversion Gel Electrophoresis (FIGE)

A chromatographic procedure for the separation of a mixture of molecules by means of a 2D electrical field, applied across a gel matrix containing those molecules. For example, field inversion gel electrophoresis (FIGE) is commonly used to separate mixtures of large DNA molecules by their size and (electrical) charge. FIGE can be used to separate (resolve) DNA molecules up to 2000 kbp in length.

See also *Two-dimensional* (2D) *gel electrophoresis*, *Chromatography*, *Electrophoresis*, *Kilobase pairs* (*kbp*), *Polyacrylamide gel electrophoresis* (*PAGE*), *Deoxyribonucleic acid* (*DNA*).

FIFRA

See *Federal Insecticide Fungicide and Rodenticide Act* (*FIFRA*).

Filgrastim

A commercial biopharmaceutical (Amgen Corporation's Neupogen®, Sandoz's Zarzio®) version of granulocyte colony stimulating factor, which is approved for reducing the rate of infection that can result from neutropenia (a low white blood cell count) in cancer patients receiving chemotherapy. See *Granulocyte colony stimulating factor* (*G-CSF*).

Filler Epithelial Cells

Skin cells that initially form under a scab in the wound healing process, in response to stimulation by epidermal growth factor. See also *Epidermal growth factor* (*EGF*).

Filopodia

See *Motor proteins*, *Actin*.

Finger Proteins

See *Zinc finger proteins*.

Fingerprinting

See *Peptide mapping* (*"fingerprinting"*), *Combinatorial chemistry*.

FIONA

Acronym for *fluorescence imaging with one nanometer accuracy*. Developed by Ahmet Yildiz. See *Fluorescence*, *Fluorescence mapping*, *Multiplexed assay*, *Nanometers* (*nm*).

Firefly Luciferase–Luciferin System

See *Fluorescence*, *Luciferase*, *Luciferin*.

First Filial Hybrids

See *F1 hybrids*.

FISH

Acronym for fluorescence *in situ* hybridization. See also *Fluorescence in situ hybridization* (*FISH*), *In situ*.

Flagella

A protein-based, flexible, whiplike organ of locomotion found on some microorganisms. With these, microorganisms are able to swim. Flagella are usually very long and there are usually only one or two per cell. The tails of sperm cells are examples of flagella. Flagella are used in the swimming motion of bacteria toward sources of nutrients in a process called chemotaxis. Singular: flagellum. See also *Microtubules*, *Cilia*, *Chemotaxis*, *Bacteria*, *Protein*, *Dynein*.

Flanking Sequence

A segment of DNA molecule that either precedes or follows the region of interest on the molecule. See also *Deoxyribonucleic acid* (*DNA*).

Flavin

Also known as lyochrome. One of a group of pale yellow, greenly fluorescing biological pigments widely distributed in small quantities in plant and animal tissues. Flavins are synthesized only by bacteria, yeast, and green plants; for this reason, animals are dependent on plant sources for riboflavin (vitamin B_2), the most prevalent member of the group.

Flavin Adenine Dinucleotide (FAD)

The coenzyme of some adenine dinucleotide (FAD) oxidation–reduction enzymes; it contains riboflavin. See also *Flavin*, *Enzyme*, *Coenzyme*, *Oxidation–reduction reaction*.

Flavin Mononucleotide (FMN)

Riboflavin phosphate, a coenzyme of certain oxidoreduction enzymes. See also *Coenzyme*.

Flavin Nucleotides

Nucleotide coenzymes (FMN and FAD) containing riboflavin. See also *Flavin mononucleotide* (*FMN*), *Flavin adenine dinucleotide* (*FAD*).

Flavin-Linked Dehydrogenases

Dehydrogenases are enzymes (involved in removing hydrogen atoms from their substrate) that require one of the riboflavin coenzymes, FMN or FAD, in order to function. See also *Dehydrogenases*, *Flavin mononucleotide* (*FMN*), *Flavin adenine dinucleotide* (*FAD*), *Substrate* (*chemical*).

Flavinoids

See *Flavonoids*.

Flavonoids

A category consisting of approximately 7000 phytochemicals that perform various functions within the plants that produce them. Some of those functions are aiding pollen production, resistance to some plant diseases, flower color, resistance to damage from ultraviolet radiation, signaling by soybean plant to rhizobia to form nitrogen-fixing nodules in its roots, and so on.

Flavonoids are typically beneficial to the health of humans that consume them (e.g., help lower blood cholesterol levels). Hundreds of flavonoids are naturally produced (by plants) in common human foods. For example, the three isoflavones (genistein, daidzein, and glycitein) produced in seeds of the soybean plant (*Glycine max* (L.) Merrill) are flavonoids, and they confer several health benefits to humans that consume them.

Coffee, tea, and chocolate products contain a number of antioxidant flavonoids (i.e., polyphenols). Because oxidation of lipids (e.g., low-density lipoproteins) in the bloodstream is the initial step in atherosclerosis disease, consumption of large amounts of coffee may help to prevent atherosclerosis. Research conducted by Joe Vinson in 1999 indicated that high coffee consumption by humans reduced oxidation of lipids in the bloodstream by 30%.

Cranberries (*Vaccinium macrocarpon*) contain a number of antioxidant flavonoids, and research indicates that consumption of large amounts on a regular basis may inhibit development of breast cancer. Blueberries (*Vaccinium ashei*, *Vaccinium corymbosum*, etc.) contain a number of flavonoids, and research indicates that consumption of large amounts on a regular basis helps to prevent urinary tract infections, strengthen eyesight, improve memory, inhibit certain cancers, and inhibit some physical aspects of the aging process.

Other subcategories of flavonoids are flavones, flavonols, flavanols, aurones, chalcones, and so on. See also *Phytochemicals, Isoflavones, Soybean plant, Rhizobium (bacteria), Quercetin, Atherosclerosis, Oxidation, Antioxidants, Oxidative stress, Cancer, Lipids, Anthocyanidins, Proanthocyanidins, Flavonols, Polyphenols, Cholesterol.*

Flavonols

A group of phytochemicals, consisting of a subcategory of the *flavonoid* "family" of phytochemicals. Flavonols are typically beneficial to the health of humans that consume them and are typically found in citrus fruits such as grapefruit, oranges, and so on. However, at least two flavonols (quercetin glycoside and naringenin chalcone) are found in tomato peels. See also *Phytochemicals, Flavonoids, Chalcone isomerase.*

Flavoprotein

An enzyme containing a flavin nucleotide as a prosthetic group. See also *Prosthetic group.*

Flesh-Eating Fungus

Refers to the fungus *Apophysomyces* spp., which ordinarily live within soil, wood, or outdoor water (e.g., ponds), but can rapidly attack living human flesh if it is introduced deep into the body through a blunt trauma puncture wound (e.g., if human is struck by high-velocity debris during a tornado). See also *Fungus.*

Flesh-Eating Infection

A colloquialism for *necrotizing fasciitis*. See also *Streptococcus, Flesh-eating fungus.*

FLIPR™

See *Fluorometric imaging plate readers.*

FLK-2 Receptors

See *Totipotent stem cells.*

Flora

The microorganisms found in a given situation, for example, reservoir flora (the microorganisms present in a given municipal water reservoir) or intestinal flora (the microorganisms found in the intestines). See also *Commensal.*

Floury-2

A gene in corn/maize (*Zea mays* L.) that (when present in the DNA of a given plant) causes plant to produce seed that contains higher than traditional levels of the amino acids methionine and tryptophan. See also *Gene, Corn, Methionine (met), High-methionine corn, Essential amino acids, Value-enhanced grains, Deoxyribonucleic acid (DNA).*

Flow Cytometry

See *Cell sorting, Fluorescence-activated cell sorter (FACs), Magnetic particles.*

Fluorescence

The reaction of certain molecules (known as fluorophores) upon absorption of a specific amount/wavelength of light, in which those molecules emit (reradiate) light energy possessing a longer wavelength than the original light absorbed. All cells will naturally fluoresce, at least a bit.

Human colon cancer cells, and precursor cells, fluoresce much more (and emit much more red light when they fluoresce) than noncancerous cells, which may lead to a new and better means of early detection. See also *Fluorophore, Cell, Fluorescence mapping, Cancer, FIA, Bright greenish-yellow fluorescence (BGYF), Immunosensor, Biochip, Near-infrared spectroscopy (NIR).*

Fluorescence-Activated Cell Sorter (FACs)

A machine or MEMS/lab-on-a-chip that is used to sort specific cells from a mixed group of cells (e.g., to remove only the cells of one type of tissue, or cells into which a new gene has been inserted via genetic engineering techniques). Initially invented by Leonard A. Herzenberg during the late 1960s.

The desired cells are first labeled with a specific fluorescent dye, or a gene for a fluorophore (e.g., green fluorescent protein) is inserted; then the cells (e.g., as part of a mixture) are passed through a flow chamber that is illuminated by a laser beam, which causes the labeled cells to fluoresce (i.e., glow).

The FACs can separate out those cells possessing the green fluorescent protein, or those cells bearing the molecules of the fluorescent dye that "stick" to only one type of cell in the mixture and that contain chromophores that can be elevated to an excited, unstable state via irradiation with specific wavelength(s) of light. Those chromophores remain in that excited state for a maximum of 10^{-9} s before releasing their energy by emitting light and returning to their unexcited "ground" state. This fluorescence ("glow") is a measurable property and the FACS machine utilizes it to separate the desired cells from the rest of the mixture. See also *Basophilic, Gene, Genetic engineering, Cell, Fluorescence, Fluorophore, Cell sorting, Label (fluorescent), Green fluorescent protein, Lab-on-a-chip, MEMS (nanotechnology).*

Fluorescence *In Situ* Hybridization (FISH)

A method for detecting the presence of particular genes (e.g., in a biological sample), which utilizes a number of fluorescein-"tagged" DNA probes. When those DNA probes hybridize to each of their

F

respective particular genes (i.e., that they were selected to be complementary to), each DNA probe's "tag" fluoresces at a different wavelength (different "color"), thereby indicating positively the presence in sample of that particular gene.

During August, 2002, the U.S. Food and Drug Administration (FDA) approved use of *information from a FISH test* (for detection of the overexpression of HER-2 gene in individual women) to guide the administration of the humanized monoclonal antibody (trastuzumab) that FDA had approved for use in conjunction with chemotherapy, and so on, against metastatic breast cancers. That FISH test (an application of pharmacogenomics) helps to detect the approximately 35% of patients for whom trastuzumab will be effective.

Another use of such *genetic markers* is for the selection of the human haplotypes (patient groups) utilized in Phase I and/or Phase II clinical tests of new pharmaceutical candidate compounds. For example, the pharmaceutical known as Gleevec™ showed near 100% efficacy in clinical trials for

- *Chronic myelogenous leukemia* disease when all patients in the patient group consisted of the haplotype possessing the genetic marker (gene) known as "bcr-abl."
- *Gastrointestinal stromal tumors* when all patients in the patient group consisted of the haplotype possessing the genetic marker (gene) known as "c-kit."

See also *Gene, Genetic marker, Haplotype, Fluorescence, Probe, DNA probe, Complementary (molecular genetics), Hybridization (molecular genetics), HER-2 gene, HER-2 receptor, Cancer, Pharmacogenomics, Pharmacogenetics, Monoclonal antibodies (MAb), Metastasis, Label (fluorescent), Food and Drug Administration (FDA), Trastuzumab, Humanized antibody, Phase I clinical testing, Phase* II *clinical tests, Gleevec™.*

Fluorescence Mapping

Refers to use of a special microscope/light of selected wavelength (i.e., to induce fluorescence of "targets") in order to scan (e.g., in tissue) *2D planes* at varying depths, in order to thoroughly "map" *in three dimensions* all of the molecules of interest that fluoresce (e.g., when a pharmaceutical compound binds to each "target" molecule, such as a cell receptor). See also *Fluorescence, Multiplexed assay, Confocal microscopy cell, Receptors, FIONA, Microfluidics, Phenomics, Click chemistry.*

Fluorescence Multiplexing

See *Fluorescence mapping, Multiplexed assay.*

Fluorescence Polarization (FP)

A technology that can be utilized to detect the presence or the behavior of single molecules—or a single molecular species—within

- Living cells (without killing the cell).
- Biological fluids (without disrupting/destroying other compounds in those biological fluids). For example, FP immunoassays have been extensively utilized since 1980 to measure the concentration of drugs in biological samples being evaluated in clinical laboratories.

In FP, plane-polarized light of *specific (to relevant molecule) wavelength* is utilized to cause that specific molecule to fluoresce. If that molecule remains stationary, that (fluorescing) molecule emits light in the *same* plane as the original light. See also *Cell, Immunoassay, Gene expression profiling, High-throughput screening (HTS), Single-nucleotide polymorphisms (SNPs).*

Fluorescence Resonance Energy Transfer (FRET)

Refers to (fluorescence-induced) resonance that occurs when two different molecular (fluorescent) labels are in very close proximity to each other. That resonance causes the fluorescence-excitation energy to transfer from one to the other (causing the second one to fluoresce) or else (the two in combination) to emit a third color (wavelength); but the two revert (to original two colors) when some event moves the two labels apart, such as

1. A ligand "docking" at a cell's receptor (e.g., resulting in signal transduction that releases a labeled chemical signal molecule within the cell, from the receptor)
2. A change in ion concentration (e.g., thereby causing two fluorophore-labeled molecules to move apart)

FRET can be utilized as a microscopy tool by scientists to obtain quantitative information about the binding or other molecular interactions between enzymes, other proteins, lipids, DNA, and RNA. Via fluorescent labeling (e.g., with green fluorescent protein), FRET microscopy has been used to trace the movement of protein molecules inside living cells and to delineate the functioning/organization within cells. See also *Fluorescence, Fluorophore, Label (fluorescent), Green fluorescent protein, Ligand (in biochemistry), Cell, Receptors, Signal transduction, Ion, Enzyme, Protein, Lipids, Deoxyribonucleic acid (DNA), Ribonucleic acid (RNA).*

Fluorescent Real-Time PCR

See *Real-time PCR (testing).*

Fluorogenic Probe

See *Molecular beacon.*

Fluorometric Imaging Plate Readers

Abbreviated FLIPR, this is a fluorescence-detection-based testing system, which is based on readouts of images produced by sample cells fluorescing in the bottom of shallow wells located in plates (i.e., a flat plastic research tool utilized by researchers).

The most common use of FLIPR is measurement of intracell calcium levels (e.g., in a researcher's high-throughput plate-based assay system for assessing potential new pharmaceutical compounds based upon their impact on calcium levels in cells), for example, when a pharmaceutical compound acts as an agonist or an antagonist to certain G-protein-coupled receptors (resulting in rapid changes to cells' calcium levels).

FLIPR can also be utilized to screen for potential new pharmaceutical compounds that act either to close or to open specific ion channels (e.g., thereby changing the amount of relevant fluorescent dye(s) on both sides of a cell membrane in the plate-based assay). See also *Assay, Fluorescence, Fluorophore, Label (fluorescent), Cell, Plasma membrane, Ion channels, Calcium channel blockers, High-throughput screening (HTS), Agonists, Antagonists, G-proteins.*

Fluorophore

Refers to any substance that is fluorescent (e.g., certain molecules, single-walled carbon nanotubes). See also *Fluorescence*, *Carbon nanotubes*.

Flux

Refers to the specific biochemical reactions/cascades (and amounts/rate of metabolite production) in a given metabolic pathway.

For example, the human disease diabetes results in elevated levels of glucose in the bloodstream. Such elevated glucose levels cause the body's metabolism to switch increasing flux through the *polyol pathway* (of metabolism). In nondiseased individuals, the polyol pathway is used very little, if at all. However, this massive increase in flux in the polyol pathway results in accumulation of the *reduced form* of the cofactor nicotinamide adenine dinucleotide, reduced (NADH) and thereby an increase in the ratio of NADH to NAD (its oxidized form). That higher ratio of NADH/NAD adversely impacts many of the biochemical pathways that NADH has a critical role in.

Some pathogens (e.g., bacteria) are able to resist certain antibiotics via a change in their flux.

Metabolic flux analysis refers to the research methodology utilized to comprehensively evaluate an organism's metabolic biochemical pathways and their responses to environmental and genetic inputs. See also *Metabolism*, *Metabolic pathway*, *Metabolite*, *Metabolic engineering*, *Cascade*, *Feedback inhibition*, *Cofactor*, *Deficiency*, *Oxidation (chemical reaction)*, *Reduction (in a chemical reaction)*, *Oxidation–reduction reaction*, *Glucose (GLc)*, *Antibiotic resistance*.

Folic Acid

One of the B complex vitamins, it is needed by the body for synthesis of nucleic acids. A shortage of folic acid can lead to anemia.

Because many cancerous tumors consume very large amounts of folic acid during their rapid growth, folic acid can be utilized as a "targeting molecule" (e.g., attached to the surface of therapeutic nanoparticles) for nanoparticle-delivered pharmaceuticals. See also *Nucleic acids*, *Aflatoxin*, *Vitamin*, *Cancer*, *Tumor*, *Nanoparticles*, *Click chemistry*.

Follicle Stimulating Hormone (FSH)

A protein hormone used in conventional medical therapy in an attempt to increase production of sperm in men (inside the follicles of the testes). See also *Thyroid stimulating hormone (TSH)*, *Grave's disease*, *Protein*, *Hormone*, *Pituitary gland*.

Follicular Helper T Cells

See *T cells*.

Food and Drug Administration (FDA)

The federal agency charged with approving all pharmaceutical and food ingredient products sold within the United States.

In 1992, prior to approval of any of the biotechnology-derived food crop plants, the FDA decided that food crops produced via "biotechnological (i.e., recombinant) technologies" must meet the same rigorous safety standards as those created via "traditional breeding methods," both categories of which are regulated by the FDA.

Historically, new food crops created via "traditional breeding technologies" (e.g., crossing with wild type in order to confer disease resistance, increased yield, and so on, on the resultant domesticated plant varieties/strains) have sometimes contained unexpectedly high levels of known (and naturally occurring) toxins (e.g., solanine, a naturally occurring toxin in potatoes and some other plants; psoralene, a naturally occurring toxin in celery). See also *Koseisho*, *Committee for Proprietary Medicinal Products (CPMP)*, *Committee for Veterinary Medicinal Products (CVMP)*, *Committee on Safety in Medicines*, *Wild type*, *Strain*, *"Treatment" IND regulations*, *Kefauver rule*, *IND*, *IND exemption*, *Recombinant DNA (rDNA)*, *Phase I clinical testing*, *European Medicines Evaluation Agency (EMEA)*, *Medicines Control Agency (MCA)*, *Bundesgesundheitsamt (BGA)*, *Traditional breeding methods*, *Solanine*, *Psoralene*.

Food Good Manufacturing Practice (FGMP)

The Food and Drug Administration's (FDA's) approval mechanism for a process to manufacture a given food or food additive. It is implemented instead of specific regulations (such as those used to dictate processes in simple food manufacture, as in beef packing), due to the newness of the technology, and may later be superseded (due to further advances in the technology). See also *Food and Drug Administration (FDA)*.

Footprinting

A technique used by researchers to determine precisely *where* (on DNA molecule) certain DNA-binding proteins make specific contact with that DNA molecule. For example, certain types of drugs act by binding tightly to certain DNA molecules in specific locations (e.g., in order to halt cancerous growth of cells). See also *Deoxyribonucleic acid (DNA)*, *Protein*, *Genotoxic*.

For Treatment IND

See *"Treatment" IND regulations*.

Force Spectroscopy

See *Atomic force microscopy*.

Formaldehyde Dehydrogenase

An enzyme that catalyzes the oxidation of formaldehyde to formic acid (formate at intracellular pH). It requires NAD (i.e., nicotinamide-adenine dinucleotide) as an electron acceptor. It is important in the metabolism of methanol. See also *Metabolism*, *Enzyme*, *NAD (NADH, NADP, NADPH)*, *Catalyst*.

Forster Resonance Energy Transfer

See *Fluorescence resonance energy transfer (FRET)*.

Forward Mutation

A mutation from the wild (natural) type to the mutant (type). See also *Mutation*, *Wild type*.

FOS

See *Fructose oligosaccharides*.

F

FOSHU

A Japanese government designation meaning "Foods of Specified Health Use." Introduced in the early 1980s, these are foods or food ingredients that meet the following specific criteria:

- Must improve human nutrition and health.
- A benefit to human health and nutrition must be proven for that food/ingredient.
- An appropriate daily dose (i.e., amount to be consumed) must be confirmed by doctors or dieticians.
- The food/ingredient must guarantee balanced nourishment.
- The active component (e.g., phytochemical) must be scientifically confirmed regarding (#1) its quantitative and qualitative definition and (#2) its chemical and/or physical features.
- The active component must not lower nutritional value (e.g., of the food it is added to).
- The food/ingredient must be consumed in a normal fashion (i.e., eaten or drank, not as pill or powder form).
- The active component must be of natural origin.

Some of the foods/ingredients designated "FOSHU" have been those containing polyphenols, anthocyanins, and diacylglycerols. See also *Nutraceuticals, Phytochemicals, Mannanoligosaccharides, Fructose oligosaccharides, Anthocyanins, Polyphenols, Diacylglycerols.*

Foundation on Economic Trends

A small organization that lobbies against agricultural biotechnology. See also *Biotechnology.*

FP

Acronym for *fluorescence polarization.* See *Fluorescence polarization (FP).*

Fragile X Syndrome

A genetic disease that can cause mental retardation, autism, and anxiety.

The gene for fragile X was discovered in 1991 by Stephen Warren. See also *Gene.*

Frameshift

A shift (displacement) of the reading frame in a DNA or RNA molecule. Frameshifts generally result from the addition or deletion of one or more nucleotides to/from the DNA or RNA molecule. See also *Reading frame, Codon, Genetic code, Mutation, Deoxyribonucleic acid (DNA), Nucleotide, Ribonucleic acid (RNA), Central dogma (new), Nonhomologous end-joining.*

Free Energy

The component of the total energy of a system that can do work at a constant temperature and pressure. Also known as Gibbs free energy.

Free energy is a key variable calculated and monitored for different (proposed) drug molecules or drug/target interactions during *rational drug design* activities (e.g., molecular modeling). See also *Rational drug design, Target (of a therapeutic agent), Activation energy.*

Free Fatty Acids (FFA)

Individual fatty acid molecules within a vegetable oil, which exist in an *uncombined with glycerine* molecular state. The presence of free fatty acids (FFA) can be caused by naturally occurring noncombination (e.g., in some varieties of oilseeds), sprouting of the oilseeds prior to processing into vegetable oil, or breakdown of the fat (oil) during processing or usage. See also *Fats, Fatty acid, Saturated fatty acids, Unsaturated fatty acid.*

Free Radical

Sometimes called *reactive oxygen species, singlet oxygen,* or *oxygen free radical.* Term utilized to refer to an oxygen (atom) bearing an "extra" electron. Because of that, it possesses a large amount of energy, and in a biological system (i.e., inside the body of an organism), excessive amounts of it can damage body tissues when it "discharges" that energy.

For example, during 2001, researchers showed that an excess of free radicals within tissues of diabetic organisms is a major factor in the development of the vascular and nerve damage typically found in late stage diabetes. See also *Oxidative stress, Antioxidants, Human superoxide dismutase (hSOD), Carotenoids, Conjugated linoleic acid (CLA), Alicin, Diabetes, Insulin, Haptoglobin, Neutrophils, Ubiquitin, Nanoceria.*

FRET

Acronym for fluorescence resonance energy transfer. See *Fluorescence resonance energy transfer (FRET).*

Fructan

A general term utilized to refer to any carbohydrate in which *fructosyl–fructose (molecule)* linkages constitute the majority of the molecule's glycosidic bonds (i.e., between atoms in the molecule). See also *Carbohydrates (saccharides), Oligosaccharides, Fructose oligosaccharides, Glycoside.*

Fructooligosaccharides

See *Fructose oligosaccharides.*

Fructose Oligosaccharides

A "family" of oligosaccharides, some of which help to foster the growth of bifidobacteria in the lower colon of monogastric animals (e.g., humans, swine). Those bifidobacteria generate certain short-chain fatty acids, which are absorbed by the colon and result in a reduction of triglyceride (fat) and cholesterol levels in the bloodstream, thereby lowering the risk of coronary heart disease and thrombosis. Research indicates that they also promote absorption of calcium from foods (in the large intestine).

Fructose oligosaccharides are classified as a "water-soluble fiber" (e.g., by the European Union's government food regulatory agencies) because humans cannot digest them. See also *Bifidobacteria, Bifidus, Inulin, FOSHU, Oligosaccharides, Nutraceuticals, Cholesterol, High-density lipoproteins (HDLPs),*

Low-density lipoproteins (LDLPs), Bacteria, Fatty acid, Prebiotics, Mannanoligosaccharides (MOS), Coronary heart disease (CHD), Triglycerides, Thrombosis.

FT Gene

A gene that codes for production of FT protein in plants. See *Gene, Protein, FT protein.*

FT Protein

A protein that is produced within plant leaves each year, just prior to the point in the growing season at which that plant produces flowers. When light-day length (at the latitude where that plant grows) becomes long enough, the FT protein migrates to growth tip location (of flower) where the FT protein triggers the transcription factor FD. FD is specific to growth tip cells and initiates the growth that creates that plant's flower. See also *Protein, FT gene, Transcription factors.*

FTO Gene

See *Ghrelin.*

FtsZ

A contractile (i.e., periodically contracting) stringlike protein that is present within bacteria as the principal cytoskeleton component of the *Z-ring* that first constricts and then divides the bacterial (parent) cell into two different cells during mitosis.

FtsZ is a homologous protein to the protein tubulin in eucaryotic cells. See also *Protein, Cell, Bacteria, Cytoskeleton, Mitosis, Homologous protein, Eucaryote.*

Fumarase (fum)

An enzyme that catalyzes the hydration (addition of hydrogen atoms) of fumaric acid to maleic acid, as well as the reverse dehydration reaction (removal of hydrogen atoms). See also *Enzyme, Catalyst.*

Fumaric Acid ($C_4H_4O_4$)

A dicarboxylic organic acid produced commercially by chemical synthesis and fermentation; the *trans* isomer of maleic acid; colorless crystals, melting point 87°C (191°F); used to make resins, paints, varnishes, and inks, in food, as a mordant (dye fixer/stabilizer), and as a chemical intermediate. Also known as boletic acid. See also *Acid, Isomer, Boletic acid.*

Fumonisins

A "family" of mycotoxins that are primarily produced by the fungi *Fusarium moniliforme* and *Fusarium verticillioides* and *Fusarium proliferatum* (e.g., in insect-damaged corn/maize and wheat).

Consumption of fumonisins by horses and swine can be fatal to those animals. Consumption of fumonisins by other animals (including humans) can result in tumors (e.g., cancer of the esophagus, in humans). See also *Mycotoxins, Fungus, Fusarium, Fusarium moniliforme, European corn borer (ECB), Cancer, P53 gene.*

Functional Foods

Refers to foods that provide health benefit beyond basic nutrition. See also *Nutraceuticals, Phytochemicals, FOSHU.*

Functional Gene

See *Joining segment.*

Functional Genomics

Study of, or discovery of, what traits/functions (generally via proteins expressed) are conferred to an organism by given (gene) sequences. The timing and location of the expression of those genes is also impacted by external/environmental factors sometimes, such as temperature, sunlight, humidity, the presence of signal transducers and activators of transcription, and so on. Also impacting the functions/traits are interactions among genes, signaling cascades, and response/reaction mechanisms within the body of that organism.

Typically, functional genomic study follows after discovery of gene sequences found via structural genomics study.

Some methods utilized to determine which traits/functions result from which gene(s) are as follows:

- Site-directed mutagenesis, to compare two same-species organisms possessing two different genes at the same site on the genome
- Antisense DNA sequence, to compare two same-species organisms (one of which has gene at the same site "turned off" via antisense DNA)
- Reporter gene, to compare two same-species organisms (with two different genes at the same site on genome) via a "reporter" gene adjacent to the gene/site, to detect the presence of a desired trait/function
- Chemical genetics, to compare two same-species organisms (one of which has a gene at the same site on a DNA molecule at least partially inactivated by a specific chemical)
- "Silencing" or "knocking out" a particular gene via other methods than antisense or chemical genetics, to compare

See also *Genomics, Trait, Gene, Genotype, Phenotype, Polygenic, Express, Structural gene, Structural genomics, Deoxyribonucleic acid (DNA), Sequence (of a DNA molecule), Pleiotropic, Genetic code, Expressed sequence tags, Informational molecules, Point mutation, Site-directed mutagenesis (SDM), Antisense (DNA sequence), Reporter gene, Methylation, zinc finger proteins, DNA methylation, Positional cloning, Chemical genetics, Gene silencing, Drosophila, Caenorhabditis elegans, Central dogma (new), Transcription factors, Signal transducers and activators of transcription (STATs), Gene expression analysis, Gene function analysis, Pathway, Pathway feedback mechanisms, Cascade, Intein.*

Functional Group

A molecule, or portion of a molecule, that will react with other molecule(s). For example, "hedgehog proteins" must first add a cholesterol molecule (to themselves) before they can carry out their task of directing/controlling tissue differentiation during mammal embryo development (into various organs, limbs, etc.).

An "acetyl (functional) group" must be added to a choline molecule in order for the body to have the critical neurotransmitter acetylcholine. See also *Protein, Peptide, Hedgehog proteins, Cholesterol, Acetyl choline, Neurotransmitter, Signal transduction.*

Functional Protein Microarrays

Refers to the category of protein microarrays in which the *capture agents* are themselves protein molecules, so that such microarrays can evaluate

- Protein–protein interactions
- Protein–ligand interactions
- How some protein molecules modify other proteins (e.g., how tyrosine kinases will phosphorylate some proteins)

See also *Protein microarrays, Protein, Capture agent, Protein interaction analysis, Protein–protein interactions, Target–ligand interaction screening, Protein tyrosine kinases, Phosphorylation.*

Fungicide

Any chemical compound that is toxic to fungi. See also *Biocide, Fungus.*

Fungus (Plural: Fungi)

Any of a major group of saprophytic and parasitic plants that lack chlorophyll and flowers, including molds, toadstools, rusts, mildews, smuts, ergot, mushrooms [*Agaricus bisporus*], and yeasts.

Under certain conditions (e.g., temperature, humidity), some fungi can produce mycotoxins via their metabolism. See also *Rusts, Aspergillus flavus, Mycotoxins, Fusarium, Fusarium graminearum, Aflatoxin, Fumonisins, Vomitoxin, DON, Ergotamine, Metabolism, Rice blast, Melanin.*

Furanocoumarins

See *Psoralene.*

Furanose

A sugar molecule containing the five-membered furan ring. See *Sugar molecules.*

Furocoumarins

A term that is sometimes utilized to refer to furanocoumarins. See *Furanocoumarins.*

Fusaric Acids

See *Fusarium moniliforme.*

Fusarium

A genus of fungus that infests certain grains (e.g., wheat [*Triticum aestivum*], corn or maize [*Zea mays* L.]) during growing seasons in which climate (e.g., high humidity, cool weather) and other conditions combine to enable rapid growth/proliferation of the fungus.

In wheat, the (*Fusarium graminearum* head blight) fungus infestation, also known as "scab," causes the wheat plant to weaken and to produce empty seed heads, which reduces yield.

In corn (maize), the (*Fusarium graminearum*) fungus infestation, also known as *Gibberella zeae* or "Gibberella ear rot," ruins grain kernels, which reduces yields. As a by-product of their metabolism, some of the *Fusarium* types (species) produce deoxynivalenol (also known as DON or "vomitoxin"—produced by *Fusarium graminearum*), zearalenone (ZEA), and fumonisins (a group of very potent mycotoxins that are produced by *Fusarium moniliforme* and *Fusarium proliferatum* and *Fusarium verticillioides* fungi). Fumonisin B_1 is the most prevalent *Fusarium*-produced mycotoxin in corn (maize). Its presence can cause livestock to refuse to eat infested feed, decrease reproductive efficiency in swine, and even kill horses (via equine leukoencephalomalacia).

When consumed by humans, fumonisin B_1 induces cell death via apoptosis, and the tissues that are adjacent to killed cells respond with cell replication/proliferation to replace the lost cells.

Fumonisin B_1 inhibits the enzyme ceramide synthetase (which is crucial to the biosynthetic pathway for the creation of sphingolipids in cells), resulting in accumulation of sphinganine in cells, and decreases ceramides and complex sphingolipids. These internal changes signal the cells to die via apoptosis ("programmed cell death"), especially liver and kidney cells.

Maximum fumonisin content allowed in flour (for U.S. bread) is one part per million. Maximum fumonisin content allowed in U.S. malting barley (*Hordeum vulgare*) is zero.

In 1997, Iowa State University research showed that *B.t.* corn varieties (which express the *B.t.* protoxin in the corn ears) have significantly less ear mold caused by *Fusarium* fungi. That is because the European corn borer is a vector (carrier) of *Fusarium.* See also *Fungus, Mycotoxins, Toxin, Metabolism, Fumonisins, Zearalenone, Apoptosis, Enzyme inhibition, Lipids, Vomitoxin, DON, Deoxynivalenol, Bacillus thuringiensis (B.t.), European corn borer (ECB), CD95 protein, Soybean cyst nematodes (SCN), Fusarium moniliforme, Fusarium graminearum.*

Fusarium graminearum

A fungus, also sometimes known as *Gibberella zeae,* that can infect wheat (*Triticum aestivum*) or corn/maize (*Zea mays* L.), under certain growing season conditions. In wheat, this fungus infestation—also known as "scab"—causes the wheat plant to weaken and to produce empty seed heads, which reduces yield.

In corn/maize, this fungus infestation—also known as "Gibberella ear rot"—ruins grain kernels, which reduces yield.

As a by-product of its metabolism, this fungus can sometimes produce the mycotoxins *deoxynivalenol* (also known as DON or "vomitoxin") and *zearalenone.* See also *Fungus, Fusarium, Corn, Wheat, Toxin, Metabolism, Mycotoxins, Deoxynivalenol, DON, Vomitoxin, Zearalenone.*

Fusarium moniliforme

One of the *Fusarium* fungi; therefore it can produce one or more *fumonisins* (a group of mycotoxins) under certain environmental conditions, when it grows in some grains (see the entry for *Fusarium*).

When *Fusarium moniliforme* grows within growing plants of domesticated rice (*Oryza sativa*), it can cause the plant disease known as "bakanae" (also known as "foolish seedling" disease). Symptoms of bakanae include rice plants that are much taller than normal rice plants and leaves that are much longer than normal. That abnormal growth (of rice plant/leaves) is caused by a gibberellin compound that is excreted by the *Fusarium moniliforme* fungus. The fungus also excretes fusaric acids, which can stunt or kill rice plants. See also *Fusarium, mycotoxins, Fumonisins, Fungus, Gibberellins.*

Fusion Gene

Refers to two protein-encoding genes, joined together end to end (which causes the cell's ribosome to subsequently produce the resultant fusion protein). Although it can occur spontaneously, a fusion gene can also be manmade. When done by man, this fusion is generally done to

- Put the expression of one of the (fused) genes under the control of the strong promoter for the first gene.
- Allow the gene of interest (which is difficult to assay) to be more easily studied via substituting some of the (gene) protein with a more easily measured (assayed) function, for example, fusing a difficult-to-study gene with the β-galactosidase gene, the (protein) product of which can easily be measured (assayed) using chromatography. Another example is to fuse the gene of a fluorescent protein to that of a gene coding for a given protein being assayed regarding its folding inside a living cell (i.e., fluorescence then indicates that protein to have folded properly).
- Create a pharmaceutical consisting of relevant domains of two different proteins. For example, the pharmaceutical Enbrel™ (etanercept) is a fusion protein consisting of the *extracellular (i.e., portion sticking out of cell's plasma membrane) sequence* of human tumor necrosis factor receptor (TNFR) and the *Fc sequence of the human antibody IgG1.* When injected, the TNFR segment of the etanercept fusion protein binds to a tumor necrosis factor molecule, and the Fc segment of the etanercept protein marks that molecule for removal by other immune system cells, thereby reducing the structural damage (to body joints) caused by excess tumor necrosis factor in the autoimmune disease rheumatoid arthritis.

Naturally occurring fusion genes can lead to some tumorigenesis (tumor creation) via overactivating certain proto-oncogenes, deactivating tumor suppressors, or altering the regulation and/or splicing of other genes that lead to defects in key signaling pathways. For example, fusion gene formation from the PAX3 and MAML3 genes leads to biphenotypic sinonasal sarcoma (tumor in nose/facial tissues). See also *Fusion protein, Gene, Tumor, Proto-oncogene, Tumor suppressor, Splicing, pathway.*

Fusion Inhibitors

See *CD4-PE40, Soluble CD4.*

Fusion Protein

A protein consisting of all or part of the function-applicable amino acid sequences (known as the "domain") of two or more proteins. Fusion proteins are formed by the following:

- Some natural cellular processes. One example of a naturally occurring fusion protein results when the ubiquitin protein fuses with certain (degraded/misfolded) protein molecules inside cells to "mark" those degraded protein molecules for destruction by the cell's proteasomes. In some specific instances, the fusion of ubiquitin to certain protein molecules in cells causes that "partner protein" to be expressed in larger amounts than previously. When fusion proteins are produced uncontrollably, it can lead to cancer (e.g., acute myeloid leukemia).
- Scientists fusing the two protein-encoding genes that causes the cell's ribosome to subsequently produce the desired fusion protein. This fusion is generally done to
  - Put the expression of one of the (fused) genes under the control of the strong promoter for the first gene.
  - Allow the gene of interest (which is difficult to assay) to be more easily studied via substituting some of the (gene) protein with a more easily measured (assayed) function. For example, fusing a difficult-to-study gene with the β-galactosidase gene, the (protein) product of which can easily be measured (assayed) using chromatography. Another example is to fuse the gene of a fluorescent protein to that of a gene coding for a given protein being assayed regarding its folding inside a living cell (i.e., fluorescence then indicates that protein to have folded properly).
  - Create a pharmaceutical consisting of relevant domains of two different proteins. For example, the pharmaceutical Enbrel™ (etanercept) is a fusion protein consisting of the *extracellular (i.e., portion sticking out of cell's plasma membrane) sequence* of human tumor necrosis factor receptor (TNFR) and the *Fc sequence of the human antibody IgG1.* When injected, the TNFR segment of the etanercept fusion protein binds to a tumor necrosis factor molecule, and the Fc segment of the etanercept protein marks that molecule for removal by other immune system cells, thereby reducing the structural damage (to body joints) caused by excess tumor necrosis factor in the autoimmune disease rheumatoid arthritis.

See also *Protein, Amino acid, Sequence (of a protein molecule), Gene, Express, Cell, Ribosomes, Promoter, Assay, Coding sequence, Domain (of a protein), Ubiquitin, Proteasomes, Protein folding, Gene fusion, Fluorescence, Visible fluorescent proteins, Green fluorescent protein, Tumor necrosis factor (TNF), Plasma membrane, Receptors, Antibody, Sequence (of a protein molecule), Rapid protein folding assay, CD4-PE40, Fusion inhibitors, Metalloproteins, Acute myeloid leukemia (AML).*

Fusion Toxin

A fusion protein that consists of a toxic protein (domain) plus a cell receptor-binding region (protein domain). The cell receptor portion (of the total fusion toxin molecule) delivers the toxin directly to the (diseased) cell, thus sparing other healthy tissues from the effect of the toxin. See also *Fusion protein, Toxin, Ricin, Protein, Protein engineering, Domain (of a protein), Receptors, Endocytosis.*

Fusogenic Agent

Any compound, virus, and so on that causes cells to fuse together. For example, one of the effects of the HIV (i.e., AIDS-causing) viruses is to cause the T cells of the human immune system to fuse (causing collapse of the immune system). See also *Acquired immune deficiency syndrome (AIDS), Human immunodeficiency virus type 1 (HIV-1), Human immunodeficiency virus type 2 (HIV-2), Helper T cells (T4 cells), Adhesion molecule.*

Futile Cycle

An enzyme-catalyzed set of cyclic reactions that results in release of thermal energy (heat) through the hydrolysis of adenosine triphosphate (ATP). The hydrolysis of ATP is normally coupled to other cycles and reactions in which the energy released is metabolically used. However, futile cycles would appear to waste the energy of ATP as heat—except when one is shivering to keep warm. The production of heat by shivering is an example of the futile cycle. See also *Adenosine triphosphate* (*ATP*), *Enzyme*, *Hydrolysis*.

FXR

Acronym for farnesoid X receptor. See *Farnesoid X receptor* (*FXR*).

G

G Proteins

See *G-proteins.*

G–

See *Gram-negative* (*G*–).

G+

See *Gram-positive* (*G*+).

GA21

A naturally occurring gene (i.e., expressed at low levels in some plants) that confers resistance to glyphosate-containing herbicides. When the "GA21 gene" is inserted by man into crop plants (e.g., maize/corn) in a way that causes high expression, those crop plants are subsequently unaffected when glyphosate-containing herbicides are applied to fields to control weeds in those crops. See also *Gene, Express, Expressivity, Protein, Genetic engineering, Corn, Herbicide-tolerant crop, Glyphosate.*

G-Actin

See *Actin.*

GAL4

See *Two-hybrid systems.*

Galactomannan

See *High-mannogalactan soybeans.*

Galactose (gal)

A monosaccharide occurring in both levo (L) and dextro (D) forms as a constituent of plant and animal oligosaccharides (lactose and raffinose) and polysaccharides (agar and pectin). Galactose is also known as cerebrose. See also *Stereoisomers, Dextrorotary* (*D*) *isomer, Levorotary* (*L*) *isomer.*

Gall

See *Ti plasmid.*

Gallic Acid

The chemical 3,4,5-trihydroxybenzoic acid. It is produced in gallnuts, oak bark, sumac, witch hazel, and some other plants (e.g., to deter predators from eating those plants).

The marsh plant known as *Phragmites australis* also exudes gallic acid from its roots, as an allelopathic compound (to deter any competitor plants growing nearby). The gallic acid dissolves tubulin, a structural protein that imparts strength to the roots of competitor plants. The resultant loss of plant root strength causes any nearby competitor plants to die. See also *Acid, Allelopathy, Tubulin.*

GalNAc

Abbreviation for *N*-acetyl-D-galactosamine.

GALT

See *Gut-associated lymphoid tissue* (*GALT*).

Gamete

A germ or reproductive cell. In animals (and humans) the functional, mature, male gamete is called a spermatozoon; in plants it is called a spermatozoid. In both animals and plants the female gamete is called the ovum, or egg. See also *Oocytes.*

Gamma Globulin

A type of blood protein that plays a major role in the process of immunity (immune system response). Sometimes the term "gamma globulin" refers to a whole group of blood proteins that are known as antibodies or immunoglobulins (Ig). Most often, however, it applies to a particular immunoglobulin, designated as IgG, believed to be the most abundant type of antibody in the body. See also *Antibody, Gut-associated lymphoid tissue* (*GALT*), *Protein, Immunoglobulin.*

Gamma Interferon

Produced by T lymphocytes. See also *Interferons, T lymphocytes.*

Gamma-Secretase

Abbreviated (γ-secretase). An enzyme (e.g., within the brains of mammals) that, under certain conditions, will cut into pieces the transmembrane protein known as amyloid precursor protein, thereby allowing those pieces to subsequently form the molecular derivative known as amyloid β-protein, a cause of Alzheimer's disease. See also *Enzyme, Cell, Protein, Plasma membrane, Alzheimer's disease.*

GAP

A double-stranded DNA is said to be "gapped" when one strand is missing over a short region of the molecule. See also *Deoxyribonucleic acid* (*DNA*).

G

GAT

Acronym for *glyphosate N-acetyltransferase*. See *Glyphosate N-acetyltransferase*.

Gated Transport (of a Protein)

One of three means for a protein molecule to pass between compartments within eucaryotic cells. The compartment "wall" (membrane) possesses a "sensor" (receptor) that detects the presence of a correct protein (e.g., after that protein has been synthesized in the cell's ribosomes) and then opens a "gate" (pore) in the membrane to allow that protein to pass from the first compartment to the second compartment. See also *Protein*, *Eucaryote*, *Cell*, *Ribosomes*, *Signaling*, *Vesicular transport*.

GDH Gene

See *Glutamate dehydrogenase*.

gDNA

Abbreviation for *genomic DNA*. See also *Deoxyribonucleic acid (DNA)*, *Genome*, *Genomic sciences*.

GDNF

See *Glial derived neurotrophic factor*.

GEAC

The country of India's Genetic Engineering Approval Committee. The GEAC must approve an rDNA product (e.g., a genetically engineered crop plant that earlier received its "biosafety clearance" from the Indian Department of Biotechnology) before that rDNA product is allowed to be commercially planted in the country of India. See also *Genetic engineering*, *rDNA*, *Indian Department of Biotechnology*.

Gel

A colloid where the dispersed phase is liquid and the dispersion medium is solid. See also *Denaturing gradient gel electrophoresis*, *Denaturing polyacrylamide gel electrophoresis*, *Field inversion gel electrophoresis (FIGE)*, *MALDI-TOF-MS*, *Mass spectrometer*, *Northern blotting*, *Polyacrylamide gel electrophoresis (PAGE)*, Sequencing (*of DNA molecules*).

Gel Electrophoresis

See *Two-dimensional (2D) gel electrophoresis*, *Polyacrylamide gel electrophoresis (PAGE)*, *Electrophoresis*, *Denaturing gradient gel electrophoresis*.

Gel Filtration

Also known as exclusion chromatography. An effective technique for separating molecules (such as peptide mixtures) on the basis of size. This is accomplished by passing a solution of the molecules to be separated over a column of Sephadex®, for example, which is a polymerized carbohydrate derivative that contains tiny holes. The holes are of such a size that some of the smaller molecules diffuse into them and are in this way retained (held back) while the larger molecules are not able to get into the holes and pass on by the solid phase (Sephadex, in this example). This, simplistically, is how separation is effected. See also *Electrophoresis*, *Chromatography*, *Field inversion gel electrophoresis*.

Gel Permeation Chromatography (GPC)

Also known as size exclusion chromatography, it separates mixtures of different-sized molecules based on their differential rates of passage through a structured gel. GPC is the most common analytical technique for the measurement of molecular weight distribution, structure, and viscosity of a wide variety of polymers. See also *Polymer*.

GEM

A project conducted under the auspices of the United States Department of Agriculture, in concert with 16 American universities and 20 corn (maize) seed companies. This acronym stands for Germplasm Enhancement for Maize. GEM's intent is to cross exotic (not in current use) germplasm with commercial maize lines in order to increase corn yield. See also *Corn*, *Germplasm*, *Hybridization (plant genetics)*, *Pleiotropic*.

Geminivirus

Refers to a category of DNA viruses that transcribe their genes in the nucleus of plant cells that they have infected.

See also *Virus*, *Deoxyribonucleic acid (DNA)*, *Gene*, *Cell*, *Transcription*.

GEMM

Acronym for genetically engineered mouse model. For example, a laboratory mouse that is engineered to get "human cancer" would allow researchers to study cancer in a way that more naturally simulates how human tumors exist within live tissue in a body. See also *Genetic engineering*, *Cancer*, *Tumor*.

GEMP (Genetically Engineered Microbial Pesticide)

See *Genetically engineered microbial pesticide*, *Integrated pest management (IPM)*.

Gene

A natural unit of the hereditary material, which is the physical basis for the transmission of the characteristics of living organisms from one generation to another. The basic genetic material is fundamentally the same in all living organisms: it consists of chain-like molecules of nucleic acids—deoxyribonucleic acid (DNA) in most organisms and ribonucleic acid (RNA) in certain viruses—and is usually associated in a linear arrangement that (in part) constitutes a chromosome.

The segment of DNA that is involved in producing a polypeptide chain, or in producing a microRNA. It includes regions preceding and following the coding region (leader and trailer) as well as intervening sequences (introns) between individual coding segments (exons).

More than one protein can be expressed (made) from a given gene, that is, the particular protein expressed is determined by factors such as follows:

- The cell's temperature or other environmental variable, which can trigger alternative splicing
- The presence of STATs (some of which *themselves* are proteins)
- Other factors

See also *Informational molecules, Deoxyribonucleic acid (DNA), Ribonucleic acid (RNA), Gene expression, Chromosomes, Express, Messenger RNA (mRNA), Codon, Intron, Exon, Coding Sequence, Gene expression cascade, Central dogma (new), Signal transducers and activators of transcription (STATs), Micro-RNAs, Alternative splicing, miRNA genes.*

Gene "Stacking"

See *"Stacked" genes.*

Gene Amplification

The copying of segments (e.g., genes) within the DNA or RNA molecule. This can be done by man (e.g., polymerase chain reaction [PCR]), can be caused by certain chemical carcinogens (e.g., phorbol ester), or occur naturally (e.g., in procaryotes and certain lower eucaryotes). The five primary techniques that are used by man to perform gene amplification are as follows: (1) PCR, (2) ligase chain reaction, (3) self-sustained sequence replication, (4) Q-beta replicase technique, and (5) strand displacement amplification. See also *Gene, Q-beta replicase technique, Polymerase chain reaction (PCR), Carcinogen, Procaryotes, Eucaryote.*

Gene Array Systems

See *Biochips, Proteomics, Gene expression analysis.*

Gene Chips

See *Biochips, Gene expression analysis, Proteomics.*

Gene Correction

See *Gene editing.*

Gene Deletions

See *Double-strand breaks (in DNA).*

Gene Delivery (Gene Therapy)

The insertion of genes (e.g., via retroviral vectors, liposomes) into selected cells in the body in order to

- Cause those cells to produce specific therapeutic agents (e.g., growth hormone in livestock, factor VIII in hemophiliacs, insulin in diabetics). A potential way of curing some genetic diseases, in that the inserted gene will produce the protein and/or enzyme that is missing in the body due to a defective gene (thus causing the genetic disease). Approximately 4000 genetic diseases are known to man. Examples of genetic diseases include cystic fibrosis, sickle cell anemia, Huntington's disease, phenylketonuria, Tay–Sachs disease, adenosine deaminase enzyme deficiency (ADA deficiency), and thalassemia.
- Cause those cells to produce a specific compound (e.g., interleukin-12) that will result in the body's immune system becoming more active against a specific disease (e.g., melanoma).
- Cause those cells to become (more) susceptible to a conventional therapeutic agent that previously was ineffective against that particular condition/disease (e.g., insertion of Hs-tk gene into brain tumor cells to make those tumor cells susceptible to the Syntex drug Ganciclovir).
- Cause those cells to become less susceptible to a conventional therapeutic agent (e.g., insert genes into healthy tissue in order to enable that healthy tissue to resist the harmful effects of such conventional chemotherapy agents as vincristine).
- Counter the effects of abnormal (damaged) tumor suppressor genes via insertion of normal tumor suppressor genes.
- Cause expression of ribozymes that cleave oncogenes (cancer-causing genes).
- Be used for other therapeutic uses of genes in cells.

See also *Gene, Tumor suppressor genes, Oncogenes, Cancer, p53 gene, Tumor, Melanoma, Proto-oncogenes, Retroviral vectors, Retroviruses, Huntington's disease, Genetic code, Informational molecules, Deoxyribonucleic acid (DNA), Chromosomes, Hormone, Enzyme, Protein, Genetic targeting, Polycation conjugate, Electroporation, Liposomes, Zinc finger Proteins, Spiral* polypeptides.

Gene Dosage Variation

See *Multiallelic copy number variation loci.*

Gene Drive

Refers to either of the following:

1. A naturally occurring phenomenon that results in the spread of a given genetic element through a population of organisms via causing that genetic element to be inherited more often than classical genetics would predict.
2. A technology/methodology utilized by scientists to spread a (new) gene to virtually all organisms of a targeted population/species. For example, the mutagenic chain reaction can be utilized (i.e., via CRISPR/Cas9 genome editing system) for creating heterozygous autocatalytic (self-catalyzed/powered) mutations to generate homozygous loss-of-function mutations in somatic and germline cells of organisms.

See also *Gene, Organism, Mutagenic chain reaction, CRISPR/Cas9 gene-editing systems.*

Gene Editing

See *Genome editing, Editing, CRISPR, CRISPR/Cas9 gene-editing systems, TALENs.*

G

Gene Expression

Conversion of the *genetic information* within a gene into an actual protein (or cell process).

Note that many genes are only expressed at specific times during the lifetime of a cell/organism. Some genes are expressed in a "cascade" of related expressions. See also *Gene, Genetic code, Informational molecules, Express, Gene expression analysis, Biochips, Gene expression cascade, Micro-RNAs, Central dogma (new), CHO cells, Enhancer, Long noncoding RNAs.*

Gene Expression Analysis

Generally done via use of real-time PCR, two-dimensional gel electrophoresis, "biochips" (i.e., which have numerous detection/analysis devices fabricated onto their silicon surface), or "microarrays" (e.g., with specific cDNA molecules attached to surface). In whole or in part, gene expression analysis involves evaluation of the expression (and expression *levels*) of numerous genes in a biological sample, to analyze/compare any differences between gene expression/products in

- Normal cells versus diseased cells
- Normal cells versus those responding to a stimulus
- Cells from same organism, at different stages of development (e.g., embryo vs. adult)
- Normal (historic wild-type) cells versus genetically engineered cells (e.g., those that have been engineered to cure a disease, resist a herbicide)
- Normal cells versus those same cells treated with a given pharmaceutical or nutraceutical (candidate)

Analysis generally involves measurement of *gene expression markers* (i.e., molecules synthesized or cellular consequences such as apoptosis) to determine which genes are expressed (and when/*how much*, etc.). For example, during 2002, Yiwei Li utilized a cDNA-based microarray to show that genistein (an isoflavone obtained from soybeans) downregulated the expression levels of 11 human genes (which have been shown to be involved in angiogenesis and/or cancer metastasis) and upregulated (i.e., increased the expression levels of) 2 human genes associated with connective tissue cell signaling.

Scientists discovered via gene expression analysis that humans of Asian ancestry are more likely to express the particular mutation of the *gene for epidermal growth factor receptor* that can result in lung cancer.

See also *Gene, Gene expression, Gene expression profiling, Capillary electrophoresis, Microarray (testing), Genomics, Functional genomics, Express, Expressed sequence tags (EST), Zinc finger proteins, Biochips, High-throughput screening (HTS), Microfluidics, Herbicide-tolerant crop, Gene delivery (Gene therapy), Hormone, Proteomics, Promoter, Gene expression markers, Gene expression cascade, Apoptosis, Real-time PCR, RT-PCR, Differential display, Isoelectric focusing (IEF), Two-dimensional (2D) gel electrophoresis, Genistein (GEN), Nutraceuticals, Angiogenesis, Metastasis, Anti-angiogenesis, Signaling, Serial analysis of gene expression (SAGE), Mutation, Cancer, EGF receptor.*

Gene Expression Cascade

A sequential series of *individual gene expressions* (i.e., each gene causing a separate/different protein to be "manufactured") that is initiated (i.e., "set off") by the first gene expression.

For example, a *gene expression cascade* is often initiated by the first gene causing expression of a transcription factor (i.e., protein that *itself* interacts with cell's DNA to either cause or speed up yet *another* gene expression). The protein resulting from that second gene expression could be yet *another* transcription factor that triggers another (i.e., third) gene expression, and so on. See also *Gene, Express, Gene expression, Cascade, Protein, Cell, Deoxyribonucleic acid (DNA), Promoter, Transcription factors, Apoptosis, Enhancer.*

Gene Expression Markers

Refers to molecules (e.g., synthesized due to a specific gene's expression) or consequences (e.g., cell apoptosis due to a specific gene's expression) that can be measured as *proof of gene's expression* in gene expression analysis. See also *Gene expression, Gene, Gene expression analysis, Express, Expressed sequence tags (EST), Biochips, Protein, Cell, Apoptosis, Green fluorescent protein, Kusabira Orange, Fluorescence, Fluorescence polarization (FP), Nutraceuticals.*

Gene Expression Profiling

Determination of specifically *which genes are "switched on" (e.g., in a cell)*, thereby enabling the precise definition of the phenotypic condition of that cell (i.e., the phenotype of that cell at that moment).

Typical uses (i.e., comparison of such tissue phenotypes) include

- Comparing *diseased* cell with *normal* cell
- Defining quantitatively the "normal" state
- Comparing a given drug's impact—that is, *treated* cell with *untreated* cell
- Comparing the impact of a given *nutraceutical's consumption* (i.e., treated cell vs. untreated cell)
- Comparing *old* cell with *young* cell

In subsequent gene expression analysis, the quantitative *amounts of each protein being expressed* can be determined via the use of such technologies as two-dimensional (2D) gel electrophoresis, Southern blot analysis, fluorescence tagging, radiolabeling, RT-PCR, QPCR, fluorescence polarization (FP), plane polarimetry, and so on. See also *Gene, Gene expression, Protein, Cell, Phenotype, Gene expression analysis, Two-dimensional (2D) gel electrophoresis, Southern blot analysis, Radiolabeled, RT-PCR, QPCR, Gene expression markers, Microarray (testing), Multiplexed assay, Fluorescence, Fluorescence polarization (FP), Nutraceuticals.*

Gene Function Analysis

The determination of which protein is expressed (i.e., caused to be "manufactured") by each gene in an organism's genome/DNA. Typically, gene function analysis follows after discovery of gene sequences found via *structural genomics* study. Some methods utilized to determine which proteins result from which gene(s) are

- Site-directed mutagenesis to compare two same-species organisms possessing two different genes at the same site (SNP) on the genome (i.e., on organism's DNA)
- Antisense DNA sequences to compare two same-species organisms—one of which has gene at the same site "turned off" (silenced) via antisense DNA
- Reporter gene, to compare two same-species organisms (possessing two different genes at same site on genome/DNA)

via a *reporter gene* adjacent to the gene/site, to detect the presence or absence of the desired trait/function
- Comparison of the same organism (e.g., crop plant) when one of the two is "challenged" by a specific plant disease
- Chemical genetics, to compare two same-species organisms (one of which has gene at the specific site at least partially *inactivated by a specific chemical*)
- "Silencing" or "knocking out" a particular gene via *other methods than antisense or chemical genetics*, to compare
- Use of already known "model organisms" (e.g., *Drosophila* for comparing insect genes, *Arabidopsis thaliana* for plant genes, *Caenorhabditis elegans* for animal genes), and so on

See also *Gene, Gene expression, Genetic code, Informational molecules, Express, Protein, Genome, Genomics, Structural genomics, Functional genomics, Zinc finger proteins, Trait, deoxyribonucleic acid (DNA), Sequence (of a DNA molecule), Point mutation, Site-directed mutagenesis (SDM), Antisense (DNA sequence), Gene silencing, Reporter gene, Methylation, Positional cloning, DNA methylation, Chemical genetics, Model organism, Drosophila, Arabidopsis thaliana, Caenorhabditis elegans (C. elegans), Central dogma (old), Central dogma (new), Transcription factors, Transwitch®, Single-nucleotide polymorphisms (SNPs).*

Gene Fusion

Refers to the technology/methods utilized to fuse together two or more genes. When such a "fused gene" is then inserted into a genome (e.g., the DNA of a plant), it causes production (in plant's ribosomes) of protein(s) consisting of all or part of the amino acid sequences (known as the "domain") of the two proteins typically coded for by those two genes. This fusion is often done in order to put expression of the "second" (fused) gene under the control of the (strong) promoter of the "first" gene.

During 2001, Rajbir Sangwan and colleagues inserted a fused gene into a potato plant (*Solanum tuberosum*), a major source of plant starch. That fused gene coded for the production of the two proteins *α-amylase* and *glucose isomerase*—both are enzymes. α-amylase catalyzes the conversion of potato starch into glucose (a sugar), and glucose isomerase catalyzes conversion of glucose to fructose (a more valuable sugar). See also *Gene, Genome, Deoxyribonucleic acid (DNA), Double-strand breaks (in DNA), Genetic engineering, Ribosomes, Coding sequence, Protein, Amino acid, Sequence (of a protein molecule), Fusion protein, Rapid protein folding assay, Express, Promoter, Enzyme, Amylase, Glucose, Isomerase.*

Gene Gun

See *Biolistic gene gun.*

Gene Imprinting

See *Imprinting.*

Gene Machine

An instrument that, when fed information on the amino acid sequence of a protein (usually via a protein sequencer), will automatically produce polynucleotide gene segments to code for that protein. See also *Sequencing (of DNA molecules), Synthesizing (of DNA molecules), Gene, Amino acid, Protein.*

Gene Manipulation

See *Genetic engineering.*

Gene Map

See *Linkage map, Genetic map, Physical map (of genome).*

Gene Mapping

See *Sequencing* (of *DNA molecules*), *Genetic map, Linkage map, Physical map (of genome).*

Gene Probe

See *DNA probe.*

Gene Repair (Done by Man)

A term with several different meanings:

- One of the natural modes of *DNA repair* when engendered by man.
- The "repair" of a *damaged gene (e.g., mutation)* or replacement of a given gene via a process invented by Eric Kmiec in 1993. The desired DNA (gene) is added to a cell, along with RNA, in a paired group known as a chimeraplast. The chimeraplast attaches itself to the cell's DNA at the site of the specific gene (i.e., the one that is to be changed) and "repairs" it using its (new) chimeraplast DNA as a "template."
- The gene therapy form of *editing.*

See also *DNA repair, Gene, Chimeraplasty, Mutation, Deoxyribonucleic acid (DNA), Double-strand breaks (in DNA), Ribonucleic acid (RNA), Cell, Template, Editing, Gene therapy, Zinc finger proteins.*

Gene Repair (Natural)

Refers to the natural processes via which all cells in an organism are continually repairing their DNA (which can be damaged by ultraviolet light, various mutagenic chemicals, etc.). In these natural cell (gene repair) processes:

- First, an enzyme complex detects the damaged DNA (e.g., on one of the two strands of the DNA molecule).
- Next, an enzyme cuts out the damaged portion of the DNA (on that one strand, leaving the other—good—strand intact).
- Then, a DNA polymerase enzyme enters the gap and synthesizes (i.e., "manufactures") the new DNA (to replace the portion that was cut out), using the intact—good—DNA strand as a *template.*
- Finally, the new DNA is joined to the "old" DNA via the help of DNA ligase enzyme.

See also *Cell, Enzyme, Deoxyribonucleic acid (DNA), Double-strand breaks (in DNA), Mutagen, DNA repair, DNA polymerase, DNA ligase, Sliding clamps, Template, Editing, Zinc finger proteins, p53 protein.*

G

Gene Replacement Therapy

See *Gene delivery*.

Gene Silencing

The suppression of gene expression in cells or cell-invading infectious elements via

- The cell's natural gene regulation (e.g., occurs with some genes in an organism as the organism matures (e.g., from an embryo to a seedling/juvenile) via formation of heterochromatin, and so on
- Cosuppression
- A scientist's use of CRISPR/Cas9 gene editing system
- Natural epigenetic regulation
- Infection of plant cells by a geminivirus (which commandeers the cell's nucleus to instead transcribe the genes of that particular DNA virus)
- Genetic engineering done by man (e.g., silencing of a fruit plant's gene for polygalacturonase that causes the fruit to ripen, of the *FLC* gene in *Arabidopsis thaliana* plant, which results in it subsequently flowering, of the gene for allergenic P34 protein in soybeans) via a variety of methods (e.g., via RNA interference, chemical genetics, effect of certain viruses, CRISPR, "zinc finger proteins," sense or antisense genes, epigenetic silencing)

See also *Gene, Cell, Express, Gene expression, Transcription, RNA interference (RNAi), Knockout, Short interfering RNA (siRNA), Genetic code, Informational molecules, Protein, Chemical genetics, Zinc finger proteins, Virus, Gene function analysis, Gene silencing, Cosuppression, Antisense (DNA sequence), Transwitch®, Sense, Polygalacturonase (PG), GPA1, Reduced-allergen soybeans, Heterochromatin, Enhancer, Histone deacetylase inhibitors, Posttranscriptional gene silencing (PTGS), Epigenetic, Geminivirus, Imprinting, Micro-RNAs, Vernalization, Long noncoding RNAs, CRISPR, CRISPR/Cas9 gene-editing system.*

Gene Splicing

The enzymatic attachment (joining) of one gene (or part of a gene) to another; also removal of introns and splicing of exons during mRNA synthesis.

Another category of gene splicing occurrence is when chromatin remodeling results in two *recombination signal sequences* that flank a relevant gene/sequence within the DNA that is looped around histones in the chromatin becoming close enough/exposed so that cellular recombinase enzymes recognize them and catalyze their splicing together (along with relevant genes/sequences). For example, B lymphocyte cells are thereby able to splice together large numbers of gene segments that enable those B lymphocytes to collectively possess antigen receptors (Ig) specific to a huge number of antigens on innumerable pathogens. See also *Splicing, Central dogma (new), Messenger RNA (mRNA), Cell, Gene, Chromatin, Histones, Chromatin remodeling, B lymphocytes, Enzyme, Recombinase, Recombination, Sequence (of a DNA molecule), Antigen, Receptors, Pathogen.*

Gene Switching

See *Gene, Genetic code, Coding sequence, Deoxyribonucleic acid (DNA), Sequence (of a DNA molecule), Regulatory sequence, Transcription factors, CBF1, Enhancer, Cold hardening, Cessation cassette, Systemic acquired resistance (SAR).*

Gene Targeting

See *Genetic targeting, Gene splicing, Gene delivery, Genetic engineering.*

Gene Taxi

A term used in some regions (e.g., Europe) to refer to a vector (e.g., *Agrobacterium tumefaciens*) that is utilized to carry a gene/cassette into an organism and insert that gene/cassette into the organism's DNA. See also *Vector, Agrobacterium tumefaciens, Organism, Gene, Cassette, Deoxyribonucleic acid (DNA).*

Gene Technology Office

An agency of the Australian government, established in 1997, to oversee and regulate all genetic engineering activities conducted in the country of Australia. Replaced/superceded by Australia's newly formed Interim Office of the Gene Technology Regulator (IOGTR) in 1999. See also *IOGTR, Gene Technology Regulator (GTR), Genetic engineering, Recombinant DNA Advisory Committee (RAC), ZKBS (Central Committee on Biological Safety), Indian Department of Biotechnology, Commission of Biomolecular Engineering.*

Gene Technology Regulator (GTR)

The regulatory body of Australia's government that is responsible for approvals of new rDNA products (e.g., new genetically engineered crops) before they can be introduced into Australia. GTR replaced Australia's IOGTR (Interim Office of the Gene Technology Regulator) in this role on June 21, 2001. See also *Interim Office of the Gene Technology Regulator (IOGTR), Gene Technology Office, Genetic Manipulation Advisory Committee (GMAC), rDNA, Deoxyribonucleic acid (DNA), Genetic engineering, Recombinant DNA Advisory Committee (RAC), Commission of Biomolecular Engineering, Indian Department of Biotechnology.*

Gene Therapy

Refers broadly to any therapy that is accomplished/delivered to patient via insertion of a gene or genes into one or more of that patient's tissues. See also *Gene, Gene delivery, Optogenetics, Editing.*

Gene Transcript

See *Transcript.*

Gene Translocations

See *Double-strand breaks (in DNA).*

Generation Time

The time required for a population of cells to double. The average time required for a round of cell division. See also *Cell, Mitosis.*

Genestein

See *Genistein (gen).*

Genetic Code

The set of triplet code words in DNA coding for all of the amino acids. There are more than 20 different amino acids and only four bases (adenine, thymine, cytosine, and guanine). The mRNA code is a triplet code, that is, each successive "frame" of three nucleotides (sometimes called a codon) of the mRNA corresponds to one amino acid of the protein. This rule of correspondence is the genetic code. The genetic code consists of 64 entries—the 64 triplets possible when there are four possible nucleotides, each of which can be at any of three places ($4 \times 4 \times 4 = 64$). A triplet code was required because a doublet code would have only been able to code for ($4 \times 4 = 16$) 16 amino acids. A triplet code allows for the coding of 64 theoretical amino acids. Since only a little over 20 exist, there is some redundancy in the system. Hence some certain amino acids are coded for by two or three different triplets. See also *Messenger RNA (mRNA), Deoxyribonucleic acid (DNA), Informational molecules, Coding sequence, Codon.*

Genetic Drift

See *Adaptation.*

Genetic Editing

See *Gene editing.*

Genetic Engineering

Coined in 1951 by John Stewart Williamson, it is the selective, deliberate alteration of genes (genetic material) by man. This term has come to have a very broad meaning including the manipulation and alteration of the genetic material (constitution) of an organism in such a way as to allow it to produce endogenous proteins with properties different from those of the traditional (historic/typical) or to produce entirely different (foreign) proteins altogether. Some other words often applicable to the same process are gene splicing, gene manipulation, or recombinant DNA technology (techniques). See also *Gene, Informational molecules, Chromosomes, Gene editing, Gene amplification, Vector, Plasmid, Agrobacterium tumefaciens, Gene splicing, Deoxyribonucleic acid (DNA), Transgenic (organism), Biolistic® gene gun, Whiskers™, "Shotgun" method, Nuclear transfer, GMO, Recombinant DNA (rDNA), Recombination, Recombinant adeno-associated virus-based genome editing, CRISPR/Cas9 gene-editing systems, Heterokaryon, Heredity, Messenger RNA (mRNA), Heteroduplex, Positive and negative selection (PNS), Polymerase chain reaction (PCR) technique, Biotechnology, Metabolic engineering, Phenomics, Systems biology.*

Genetic Engineering Approval Committee

See *GEAC.*

Genetic Event

See *Event.*

Genetic Fingerprinting

Another name for **DNA profiling**. See *DNA profiling.*

Genetic Linkage

See *Linkage, Linkage group.*

Genetic Manipulation

See *Genetic engineering.*

Genetic Manipulation Advisory Committee (GMAC)

A body that advises the Australian government on matters pertaining to genetic engineering (e.g., new rDNA product approvals).

The GMAC is analogous to Germany's ZKBS (Central Commission on Biological Safety), Brazil's CTNBio (National Technical Biosafety Commission), and the Kenya Biosafety Council. See also *GMAC, ZKBS (Central Commission on Biological Safety), Recombinant DNA Advisory Committee (RAC), Genetic engineering, rDNA, Deoxyribonucleic acid (DNA), CTNBio, Kenya Biosafety Council, Gene Technology Office, Gene Technology Regulator (GTR).*

Genetic Map

A diagram showing the relative sequence and position of specific genes along a chromosome (DNA) molecule. *Markers* utilized as "signposts"/guideposts in such maps include single-nucleotide polymorphisms (SNPs), restriction sites (i.e., the specific locations where each restriction endonuclease "cuts" a DNA strand), and microsatellites. Such *markers* located in or close to the *gene of interest* (e.g., a disease-causing gene within a chromosome) *to a researcher* are more likely to be inherited along with that gene. See also *Position effect, Gene, Genome, Chromosomes, Deoxyribonucleic acid (DNA), Physical map (of genome), Single-nucleotide polymorphisms (SNPs), Restriction site, Microsatellite DNA, Marker-assisted selection.*

Genetic Marker

Refers to a segment of DNA (e.g., gene) within an organism's overall DNA, which can be detected by man (e.g., via use of FISH) and is a reliable indicator that that particular organism possesses a specific trait of interest. See also *Marker (DNA sequence), Gene, Deoxyribonucleic acid (DNA), Fluorescence in situ hybridization (FISH), Trait, HER-2 gene, Gleevec™, Marker-assisted selection, Whole-genome shotgun sequencing.*

Genetic Probe

See *DNA probe.*

Genetic Targeting

The insertion of antisense DNA molecules *in vivo* into selected cells of the body in order to block the activity of undesirable genes. These genes might include oncogenes or genes crucial to the life cycle of parasites such as trypanosomes (which cause sleeping sickness). See also *Antisense (DNA sequence), Gene,* Gene *delivery, Oncogenes, Dendrimers, Short hairpin RNA.*

G

Genetic Use Restriction Technologies (GURTs)

A general term utilized to refer to several different technologies intended to control the expression (or nonexpression) of the gene(s) for specific (e.g., valuable) traits. See also *Cessation cassette, Gene, Trait, Express, Value-enhanced grains.*

Genetically Engineered Microbial Pesticides (GEMP)

One or more microbes that have been genetically engineered in such a way as to cause them to be effective in combating pest(s) that attack crops or livestock. For example, a microbe that naturally attacks a crop pest could be genetically engineered to make the microbe more potent or more durable in the field environment when applied to the field via selected methods of microbe application. See also *Microbe, Genetic engineering, Wheat take-all disease, Baculovirus, Bacillus thuringiensis* (*B.t.*), *Federal Insecticide Fungicide and Rodenticide Act* (*FIFRA*), *Toxic Substances Control Act* (*TSCA*).

Genetically Engineered Organism (GEO)

See *GEO.*

Genetically Manipulated Organism (GMO)

See *GMO.*

Genetically Modified Microorganism (GMM)

See *GMM.*

Genetically Modified Organism (GMO)

See *GMO.*

Genetically Modified Pest Protected (GMPP) Plants

Plants that have been genetically engineered so that they resist (or are more tolerant to) attacks by pests (e.g., insects). See also *Genetic engineering, Bacillus thuringiensis* (*B.t.*), *Cowpea trypsin inhibitor* (*CpTI*), *Cry proteins, CRY1A* (*b*) *protein, CRY1A* (*c*) *protein, CRY9C protein, B.t. kurstaki, B.t. tenebrionis, B.t. israelensis, Pathogenesis-related proteins, Photorhabdus luminescens.*

Genetics

Coined by William Bateson in 1905, this word refers to the study of the patterns of inheritance of specific traits, by organisms. See also *Genetic code.*

Genistein (Gen)

One of several phytochemicals produced by the soybean plant as a defense against certain plant diseases and to signal *Rhizobium japonicum* bacteria (to produce nitrogen for the soybean plant via colonization of its roots, followed by nitrogen fixation from the air). Genistein can also be produced as a by-product of mycobacterium fermentation (process used to produce commercial amounts of certain antibiotics). Genistein is an isoflavone, a steroid-like compound that can be lethal to certain animal cells via its kinase-inhibiting and other properties. Genistein fights cancer (tumor cells) by inhibiting protein tyrosine kinase and topoisomerase II. Genistein also exhibits the property of anti-angiogenesis (i.e., inhibition of tumor growth via prevention of the formation/development of new blood vessels in tumors).

Attached to a pharmaceutical "guided missile" such as a monoclonal antibody or the CD4 protein, genistein is potentially useful for treatment against some tumors and has been investigated as a possible treatment against B-cell precursor leukemia. A human diet containing a large amount of genistein has been shown to increase bone density and to decrease total serum (blood) cholesterol, thereby lowering risk of osteoporosis and coronary heart disease.

Research indicates that human consumption of genistein can help to prevent breast cancer, help prevent prostate cancer/metastasis, prevent adverse increases in blood platelet aggregation, and inhibit the proliferation of *smooth muscle cells* in plaque deposits (inside blood vessels).

Research also indicates that human consumption of genistein can enhance the effectiveness of both radiation and of chemotherapy drugs (e.g., cisplatin) used in the treatment of cancer in those humans. See also *Immunotoxin, Monoclonal antibodies* (*MAb*), *CD4 protein, Genetic engineering, Nitrogen fixation, Nodulation, Phytochemicals, Fusion protein, Fusion toxin, Soluble CD4, Isoflavones, Soybean plant, Ricin, Tyrosine* (*tyr*), *Steroid, Cancer, Inhibition, Stress proteins, "Magic bullet," tyrosine kinase, Coronary heart disease* (*CHD*), *Cholesterol, Osteoporosis, Selective estrogen effect, Anti-angiogenesis, Protein tyrosine kinase inhibitor, Gene expression analysis, Plaque, Metastasis, NFκB.*

Genistin

The β-glycoside form (isomer in which glucose is attached to the molecule at the seven position of the A ring) of the isoflavone known as genistein (aglycone form). See also *Genistein* (*Gen*), *Isoflavones, Isomer.*

Genome

Coined in 1920 by Hans Winkler, this term refers to the entire hereditary material (which was proven by Oswald Avery in 1944 to be DNA) in a cell. In addition to the DNA contained in cell nucleus (known as nuclear DNA), an organism's cells contain DNA in other locations within those cells:

- Bacteria also contain some DNA in PLASMIDS.
- Plants also contain some DNA in PLASTIDS.
- Animals also contain some DNA in MITOCHONDRIA.

An organism's nuclear DNA is composed of one or more chromosomes, depending on the complexity of the organism. See also *Deoxyribonucleic acid* (*DNA*), *Chromosomes, Plastid, Plasmid, Mitochondria, Mitochondrial DNA.*

Genome Editing

Refers to several techniques utilized by scientists to correct or to introduce specific mutations/corrections at a particular site (locus) within the DNA of an organism. In each of those techniques, the mutations are induced via the oligonucleotide (chosen to be specific to the selected DNA locus). The techniques used to accomplish these site-specific corrections or directed mutations (base substitution, addition, or deletion) include CRISPR/Cas9 gene editing, TALENs, targeted gene repair, chimeraplasty, targeted nucleotide exchange, therapeutic nucleic acid repair approach, oligonucleotide-mediated gene editing, oligonucleotide-mediated gene repair, oligonucleotide-directed gene modification, oligodeoxynucleotide-directed gene modification, chimeric oligonucleotide-dependent mismatch repair, triplex-forming oligonucleotide-induced recombination, and so on. See also *Editing, CRISPR, CRISPR/Cas9 gene-editing systems, TALENs, Oligonucleotide-mediated mutagenesis, Mutation, Nucleotide, Oligonucleotide, Deoxyribonucleic acid (DNA), Locus, Base substitution, Organism, DNA repair, Mismatch repair, Chimeraplasty, Gene, Gene repair (done by humans), Genomically recoded organisms (GROs), Meganucleases.*

Genome Walking

Refers to several procedures utilized by scientists for the direct identification of unknown nucleotide sequences (e.g., a gene) from a purified genome (e.g., the DNA of a eucaryote). See also *Genome, Deoxyribonucleic acid (DNA), Eucaryotes.*

Genome-Wide Association Studies

See *GWAS.*

Genomic Imprinting

A cellular epigenetic process that occurs via DNA methylation in which certain alleles (e.g., those inherited from the mother or from the father) within an organism's cells are preferentially expressed. For example in mice, more *maternal-origin* alleles are expressed within the *developing* brain, and more *paternal-origin* alleles are expressed within the *adult* mouse brain, than would occur from a simple random 50/50 contribution of parental alleles to the offspring's DNA. See also *Epigenetic, Gene, Imprinting, Allele, Cell, Express, DNA methylation.*

Genomic Rearrangements

See *Double-strand breaks (in DNA).*

Genomic Sciences

An encompassing term utilized to refer to all knowledge of and attempts to decipher/understand the structure and function of the genomes of organisms. See also *Genomics, Genome, Structural genomics, Functional genomics, Genotype, Gene, Genetics, Genetic map, Genetic targeting, Genetic code, Sequencing (of DNA molecules), Informational molecules, Deoxyribonucleic acid (DNA), Gene amplifications, Coding sequence, Chemical genetics.*

Genomic Surgery

Refers to potential future use of certain genome-repair/editing methodologies (e.g., CRISPR/Cas9 Gene-editing System) to correct a mutated gene that causes disease (e.g., single mutation that causes cystic fibrosis). See also *Genome, Gene, Mutation, CRISPR/Cas9 gene-editing systems, Cystic fibrosis.*

Genomically Recoded Organisms (GROs)

See *Oligonucleotide-mediated mutagenesis, Genome editing.*

Genomics

Coined in 1986 by Tom Roderick, by combining *gene* and "-omics" (from the Greek word for "all"), this term refers to the scientific study of all the genes and their roles in an organism's structure, growth, health, disease (and/or resistance to disease, etc.). For example, how the (approximately) 3000 genes in a given strain of bacteria, or the (approximately) 6000 genes in a given strain of yeast, contribute to the shape, function, and the development of those whole organisms.

Some tools/methods utilized in genomics include the following:

- *Structural genomics*: The study or discovery of what particular gene sequences are present and where they are located within an organism's DNA.
- *Gene function analysis*: The determination of which protein is expressed (i.e., caused to be "manufactured") by each gene in an organism's genome. Typically, gene function analysis follows after structural genomics study.
- *Functional genomics*: The study or discovery of what traits/functions are conferred to an organism by given gene sequence(s).
- *Chemical genetics*: Used to compare two same-species organisms (one of which has a given gene, or genes, inactivated by a specific chemical or site mutation).
- *Gene expression analysis*: Used to determine product(s) produced (such as an enzyme or other critical protein) when a given gene is "switched on," by measuring fluorescence of individual messenger RNA (mRNA) molecules (specific to which particular gene is "switched on" at the time), when that mRNA hybridizes (with DNA pieces corresponding to proteins produced/analyzed that were attached to hybridization surface on biochip).
- *Chromatin immunoprecipitation*: Used to determine all points on an organism's DNA that a given protein (e.g., transcription factor) binds to.
- *Receptor-binding mapping*: *Receptor-fitting* assessment of all molecular receptors in the body, regarding which of them bind (in a lock-and-key manner) a given entity (e.g., hormone, vitamin, antibody). For example, during 2012, some researchers utilized receptor-binding mapping to link vitamin D deficiency to an increased risk for cancer and autoimmune diseases (e.g., rheumatoid arthritis, multiple sclerosis, lupus).

One example of genomics was scientists' discovery via gene expression analysis that humans of Asian ancestry are more likely to express the particular mutation of the *gene for epidermal growth factor receptor (shown to be present in their DNA via structural genomics)* that can result in lung cancer. See also *Genotype, Gene, Genetic map, Genetic targeting, Genetics, Genetic code, Sequencing (of DNA molecules), Informational molecules, Deoxyribonucleic acid (DNA), Functional genomics, Gene amplification, Coding sequence, Structural genomics, Genomic sciences, Bacteria, Yeast, Strain, Chemical genetics, Fluorescence, Enzyme, Protein, Messenger RNA (mRNA), Biochips, Express, Expressed sequence tags (EST), Hybridization surfaces, Gene expression, Gene expression analysis, Gene function analysis, Organism, Protein,*

G

Transcription factors, Chromatin immunoprecipitation, Chromatin, Metagenomics, Receptor-binding mapping.

Genosensors

Biosensors (electronic) that can detect the individual nucleotides that comprise a genome (DNA) molecule. Automated genosensors enable rapid, nondestructive sequencing of DNA molecules. See also *Genome, Nucleotide, Deoxyribonucleic acid (DNA), Sequencing (of DNA molecules), Template, Biosensors (Electronic), Footprinting, Nanotechnology, Biochips.*

Genotoxic

Refers to compounds that interfere with normal functioning of genetic material (i.e., DNA). For example, the antitumor antibiotic family of duocarmycin drugs. See also *Deoxyribonucleic acid (DNA), Genotoxic carcinogens, Footprinting.*

Genotoxic Carcinogens

Compounds that act directly on the genetic material (i.e., DNA) of an organism, thus causing cancer in that organism. Of the numerous chemicals that have been documented to be human carcinogens, the majority of them are genotoxic. See also *Carcinogen, Cancer, Gene, Deoxyribonucleic acid (DNA).*

Genotype

The total genetic, or hereditary, constitution that an individual receives from its parents. An individual organism's genotype is distinguished from its phenotype, which is its appearance or observable character. See also *Trait, Phenotype, Wild type, Copy number polymorphisms.*

Gentechnik Gesetz (Gene Technology Law)

The 1990 law that governs recombinant DNA research and development in the country of Germany. It was amended January 1, 1994 to make it somewhat less restrictive. See also *ZKBS (Central Commission on Biological Safety), Recombinant DNA Advisory Committee (RAC), Genetic engineering, Recombinant DNA (rDNA), Recombination, Biotechnology, Bundesgesundheitsamt (BGA), Indian Department of Biotechnology.*

Genus

A group of closely related species. See also *Species, Clades.*

GEO

Genetically engineered organism. See also *Genetic engineering, GMO, Gene, Gene splicing, GMM.*

Geomicrobiology

Applications of microbiological knowledge to an understanding of geological phenomena. See also *Ferrobacteria.*

GEP

Acronym for *gel electrophoresis*. See *Gel electrophoresis.*

Germ Cell

The sex cell (sperm or egg). It differs from other cells in that it contains only half (haploid) the usual number of chromosomes. See also *Gamete, Haploid.*

Germplasm

The total genetic variability to an organism, represented by the total available pool of germ cells or seed. See also *Organism, Cell, Germ cell, GEM.*

German Gene Law

See *Gentechnik Gesetz (Gene Technology Law).*

GFP

Acronym for *green fluorescent protein*. See *Green fluorescent protein.*

GH

See *Growth hormone.*

Ghrelin

Discovered by Masayasu Kojima in 1999, ghrelin is an appetite-stimulating hormone that is sometimes produced by the stomach. Ghrelin also slows down fat metabolism. In response to elevated levels of ghrelin, people and animals feel hungry. Research indicates that reduced food intake and sleep deprivation (e.g., less than 8 h per night, for most people) can lead to increased ghrelin bloodstream levels.

During 2013, research showed that people possessing one particular variant of the FTO gene have both higher bloodstream levels of ghrelin plus their brain is more sensitive to ghrelin; so those people are more prone to obesity. See also *Hormone, Metabolism, Fats.*

Giant Vesicles

Large hollow spheres in which the exterior (membrane) is composed of a film of lipids (i.e., water-insoluble fats). These can function as carriers of certain pharmaceuticals to deliver those pharmaceuticals to specific targeted tissues in the body, as synthetic mimics of the plasma membrane that envelops living cells, and so on. See also *Lipids, Lipid bilayer, Cell, Plasma membrane.*

Gibberella Ear Rot

See *Fusarium graminearum.*

Gibberella zeae

See *Fusarium graminearum.*

Gibberellins

Plant hormones that, among other functions, regulate the growth of grass species, including rice (after the relevant gibberellin is activated by an enzyme). A gene known as "sd1" controls the amount of that enzyme produced (e.g., less of the enzyme = less gibberellin = shorter plant stalk).

Following a time of environmental stress (e.g., drought) to the plant, gibberellins break down the proteins (e.g., stress proteins) that the plant synthesized to repress growth during the environmental stress.

In 1996, Lew Mander and Richard Pharis discovered an analog (i.e., a chemical that is similar) to grass gibberellin that does *not* cause grass to grow. When this analog is sprayed onto grass, it mixes into the naturally occurring grass gibberellin and significantly slows grass growth (thus potentially reducing the amount of mowing required for lawns, golf courses, etc.). See also *Hormone, Plant hormone, Enzyme, Analogue, Fusarium moniliforme, Stress proteins, Synthesizing (of proteins).*

GIST

See *Gleevec™.*

Gleevac™

See *Gleevec™.*

Gleevec™

A pharmaceutical (imatinib mesylate, also known as STI571), developed and trademarked by Novartis AG, that is used to treat the blood cancer known as "chronic myelogenous leukemia" or "chronic myeloid leukemia" or "chronic myelocytic leukemia (CML)." CML results from a genetic defect (SNP) that causes excessive production of white blood cells in the body of the affected (human). That excessive production of white blood cells results when the defective gene (i.e., SNP) causes excessive production of the enzyme *Bcr-Abl tyrosine kinase.*

Because Gleevec™ is a protein tyrosine kinase *inhibitor*, it inhibits excessive production of white blood cells (and induces apoptosis—cell death—in the cells that have the Bcr-Abl gene/SNP).

Gleevec can also be utilized to treat GIST (gastrointestinal stromal tumors), where it targets the receptor tyrosine kinase known as KIT.

Research indicates Gleevec is also effective for treating the cancers known as

- Hypereosinophilic syndrome, one type of blood cancer
- Dermatofibrosarcoma protuberans, one type of skin cancer
- Systemic mastocytosis, one type of blood cancer

See also *Cancer, White blood cells, Gene, Mutation, Single-nucleotide polymorphisms (SNPs), Enzyme, Apoptosis, Protein, Bcr-abl protein, bcr-abl genetic marker, Protein tyrosine kinase inhibitor, Fluorescence in situ hybridization (FISH), Kinase assays, Receptors, Receptor tyrosine kinase, Eosinophils, Mast cells, Cytopathic.*

Glia Cells

From the Greek word for "glue," glia cells are abundant within the brain, where they

- Physically hold the brain's neurons together
- Regulate the synapses and help to sort information, for the brain to store

Glia cells are especially numerous in the brain's hippocampus and the cortex, the two parts of the brain that have the most control over the brain's ability to process information, learn, and memorize.

Glial-Derived Neurotrophic Factor (GDNF)

A neurotrophic factor that assists the survival and functional activity of the brain's dopaminergic neurons. Because dopaminergic neurons typically deteriorate and die in brains of the victims of Parkinson's disease, it is possible that GDNF may someday be used in treatment of Parkinson's disease. See also *Neurotransmitter, Parkinson's disease.*

Globular Protein

A soluble protein in which the polypeptide chain is tightly folded in three dimensions to yield a globular (roughly oval, circular) shape. See also *Protein folding, Polypeptide (protein), Conformation, Tertiary structure.*

Glomalin

A "sticky" protein molecule that is naturally produced by certain fungi (arbuscular mycorrhizal fungi) that grow on most plant roots (in the soil). It was discovered and named by Sara F. Wright in 1996. As plant roots grow, glomalin is sloughed off into the surrounding soil.

Glomalin acts like a sort of glue, thereby improving soil stability by "gluing" soil into clumps. Proper soil "clumping" (i.e., glomming together) allows air and water to pass through the soil more easily, increases the amount of carbon contained within the soil (thereby removing the "greenhouse gas" carbon dioxide from the atmosphere), increases the number of ("healthy") bacteria in the soil, and improves the soil's overall fertility (i.e., its ability to produce high-yield crops or a large amount of biomass per hectare/acre). The glomalin (and thus carbon) content of soil in a field is increased by farmer utilization of low-tillage or "no-tillage" methods of crop production. See also *Protein, Fungus, Mycorrhizae, Arbuscular mycorrhizae, Bacteria, Biomass, Conservation tillage, Low-tillage crop production, No-tillage crop production.*

GLP

See *Good laboratory practices (GLPs).*

GLQ223

See *Trichosanthin.*

GLS

Abbreviation for *glucosinolates*. See *Glucosinolates.*

Glucagon

A hormone produced by the pancreas that causes the breakdown of glycogen in the liver. Glycogen is a form of storage sugar and its breakdown releases glucose for energy production. See also *Glycogen, Hormone, Glucose, Pancreas.*

G

Glucan

See *Water soluble fiber, Polyphenols.*

Glucanases

See *Thermal hysteresis proteins.*

Glucocerebrosidase (Trade Name Ceredase)

An enzyme used in the treatment of inherited Gaucher's disease in which there is abnormal deposition of glucocerebrosides (hydrophobic lipid molecules that contain a hydrophilic sugar head group). Gaucher's disease is an enzyme deficiency disease that may be amenable to cure by incorporation of the gene coding for glucocerebrosidase into the patient's genome via gene delivery techniques. See also *Enzyme, Gene delivery.*

Glucogenic Amino Acid

Amino acids whose carbon chains can be metabolically converted by cells into glucose or glycogen. See also *Gluconeogenesis, Cell, Amino acid, Metabolism.*

Gluconeogenesis

The net biosynthesis (formation) of new glucose from noncarbohydrate precursors such as pyruvate, lactate, glycerol, acetyl-CoA (in plants), certain amino acids, and intermediates of the citric acid cycle. See also *Carbohydrates, Glucose (GLc), Citric acid cycle, Ac-CoA, Biotin.*

Glucose (GLc)

A prime fuel for the generation of energy by organisms. It is broken down (to obtain energy) via a metabolic process called glycolysis. Glucose is a hexose, a sugar possessing six carbon atoms in its molecule. The six carbon atoms are connected to each other to form a closed ring structure known as a hexose (6) ring.

Animal cells store glucose in the form of glycogen (sometimes called animal starch), a large branched polymer of glucose units (GUs). Plant cells store glucose in the form of starch, a large polymer of GUs.

Yeasts and bacteria store glucose in the form of dextran, a polymer of GUs. The difference between the forms of storage glucose is

- In the size (molecular weight) of the final polymer formed
- In the type of linkages that connect the single GUs together in the branched molecule
- In the degree of branching that occurs in the polymer

Note that a glucose polymer does not consist of just a single long straight chain. The backbone chain has other polymer chains branching off of it. The whole molecule may be visualized as looking somewhat like a tree without the trunk. The other very abundant polymer formed by GUs is structural in nature and is called cellulose. It is the most abundant cell wall and structural polysaccharide in the plant world. Hence, glucose is used not only as an energy source but also as a structural material. See also *Amylose, Amylopectin, Glycolysis, Gluconeogenesis, Glycogen, Starch, Dextran, Cellulose.*

Glucose Isomerase

An enzyme that catalyzes the conversion of glucose to fructose. A molecule of fructose contains the same atoms as a molecule of glucose (but in a different arrangement). See also *Enzyme, Glucose, Gene fusion.*

Glucose Oxidase

An enzyme that breaks down sugar molecules (causing oxygen consumption in an organism). Industrial uses include to remove dissolved oxygen from certain food products (e.g., sugar-containing drink products). See also *Enzyme, Glucose (GLc), Glycolysis, Sugar molecules, Organism.*

Glucose Unit (GU) Value

Refers to the molecular mass of a single glucose molecule. See *Glycans.*

Glucosin

See *Glucosinolates.*

Glucosinolates

Toxins (neurotoxic phytotoxins) that are naturally produced in the seeds and certain tissues of some plants (e.g., rapeseed, wild mustard (*Brassica juncea/Brassica rapa, Sinapis arvensis*), grass pea (*Lathyrus sativus*), and so on, in order to dissuade wild animals or insects from eating those plants' seeds. When hydrolyzed (e.g., during digestion), the resultant isothiocyanate compounds can be toxic to certain pests.

For example, when large amounts of grass pea (*Lathyrus sativus*) are consumed by humans, the glucosinolates build up in the body and can cause Lathyrism (i.e., an irreversible spastic paralysis of the legs).

The glucosinolates in rapeseed (*Brassica rapa*) can impart a bitter taste to such plants' oils and can cause goiter (goitre) when fed in large amounts to animals.

If glucosinolates from seeds of the wild mustard weed (*Sinapis arvensis*) family are mixed into canola meal (e.g., when those weeds grew in a canola field and that resultant canola is processed into canola meal), such canola meal must first be diluted (e.g., via mixing in some soybean meal) in order to reduce glucosinolate concentration (below the legal maximum allowance) before it is allowed to be fed to livestock, in Canada.

Some of the glucosinolates' enzymatic transformation products produced within certain brassica plants are beneficial, or at least benign. For example, the enzyme myrosinase that is present within cells of the wasabi plant (*Wasabi japonica*) that catalyzes the conversion of that plant's glucosinolates to the isothiocyanates that provide the spicy taste of wasabi food ingredient.

Some of the glucosinolates' metabolism (enzymatic transformation) products produced in certain brassica vegetables (e.g., broccoli) are thought to be anticarcinogenic (e.g., sulforaphane in broccoli). See also *Canola, Brassica, Isothiocyanates, Toxin, Phytotoxins, Metabolism, Hydrolysis, Digestion (within organisms), Cancer, Sulforaphane, Enzyme.*

Glufosinate

See *PAT gene, BAR gene, Herbicide-tolerant crop, Gene, Glutamine synthetase.*

Gluphosinate

See *PAT gene*, *BAR gene*, *Herbicide-tolerant crop*, *Gene*, *Glutamine synthetase*.

Glutamate Dehydrogenase

An enzyme found naturally in certain soil bacteria, which helps those bacteria to utilize soilborne nitrogen. When its gene (GDH gene) is inserted into corn plant via genetic engineering, the resultant plant production of glutamate dehydrogenase enables that corn plant to better utilize soilborne nitrogen. As a result, such genetically engineered corn (*Zea mays* L.) has a protein yield increase of approximately 10%, according to research begun in 1991 by David Lightfoot. See also *Enzyme*, *Bacteria*, *Gene*, *Corn*, *Nitrogen cycle*, *Dehydrogenases*, *Protein*, *Genetic engineering*.

Glutamic Acid

A dicarboxylic amino acid of the α-ketoglutaric acid family. See also *Amino acid*.

Glutamic Acid Decarboxylase (GAD)

Refers to a type of enzyme present in the insulin-producing cells of a mammal's pancreas. During 2003, Anthony Jevnikav discovered that feeding to mice (i.e., laboratory strain predisposed to developing Type 1 diabetes) a diet containing some GAD helped to make those laboratory mice's immune systems less likely to attack their pancreas' own insulin-producing cells (i.e., a cause of Type 1 diabetes).

The result of such feeding (of small amounts of a particular protein, to cause a mammal's immune system to tolerate and not "attack" that protein) is known as *oral tolerance*. See also *Enzyme*, *Type I diabetes*, *Insulin*, *Beta cells*, *Autoimmune disease*, *Protein*, *Strain*.

Glutamine

An amino acid; the monamide of glutamic acid. Glutamine is of fundamental importance for amino acid biosynthesis in all forms of life. See also *Glutamine synthetase*, *Amino acid*, *PAT gene*, *BAR gene*.

Glutamine Synthetase

An enzyme that catalyzes the synthesis of glutamine (which is crucial for amino acid biosynthesis). See also *Glutamine*, *Enzyme*, *PAT gene*, *BAR gene*, *Amino acid*.

Glutathione

A tripeptide that is found in all cells of higher animals, which acts to help protect against oxidative stress. Composed of the amino acids glutamic acid, cysteine, and glycine. The cysteine possesses a sulfhydryl group that makes glutathione a weak reducing agent. See also *Oxidative stress*, *Reduction (in a chemical reaction)*.

Gluten

A term that is utilized to refer to a naturally occurring mixture of two different proteins—glutenin and gliadin—in the seeds of all wheat (*Triticum aestivum*) varieties. In flour made from conventional varieties of bread wheat, glutenin proteins constitute approximately 50% of the total gluten. The relative content of those two proteins determines one of the most commercially important properties of the wheat (i.e., strength and elasticity of the flour made from that particular wheat). For example, more of the high-molecular-weight glutenin (which is "stretchy" and imparts physical strength to a dough made from such flour, so that dough holds together while rising) results in a flour that is better suited to manufacture higher-quality yeast-"raised" bread products.

Gluten (i.e., these two proteins) is also present in barley and rye grain. See also *Wheat*, *Protein*, *Glutenin*, *High-glutenin wheat*, *Yeast*, *Molecular weight*, *Polymer*.

Glutenin

A protein that is naturally present in the gluten within seeds of wheat (*Triticum aestivum*). See also *Gluten*, *Wheat*, *Protein*.

GLV

Acronym for *green leaf volatiles*. See *Green leaf volatiles*, *Volicitin*, *Lipoxygenase (LOX)*, *Jasmonates*.

Glycans

Refers to a linear or branched oligosaccharide, or polysaccharide molecule (e.g., attached to a glycoprotein or a glycolipid). Sizes of different glycan molecules are compared/expressed in terms of GU values. One GU is the molecular mass of a single glucose molecule. See *Sugar molecules*, *Oligosaccharides*, *Polysaccharides*, *Glycoprotein*, *Glycolipid*, *Glycoform*, *Glycosylation (to glycosylate)*, *Glucose (GLc)*, *Molecular weight*, *Sialic acid*, *Carbohydrate microarrays*, *Glycoinformatics*.

Glyceraldehyde (D- and L-)

One of the smallest monosaccharides, it is called an aldose because it contains an aldehyde group. Glyceraldehyde has a single asymmetric carbon atom, thus there are two stereoisomers (D-glyceraldehyde and L-glyceraldehyde). See also *Monosaccharides*, *Stereoisomers*.

Glycetein

See *Isoflavones*.

Glycine (gly)

The simplest (and smallest) of the amino acids found in proteins. It is the only amino acid that does not have an asymmetric carbon atom within its molecule. Thus, it is not optically active. See also *Amino acid*, *Protein*, *Stereoisomers*, *Optical activity*.

Glycine max

See *Soybean plant*.

Glycinin

One of the (structural) categories of proteins that are produced within seeds of legumes. In general, glycinins contain two to three

G

times more cysteine (cys) and methionine (met) per unit of protein than does β-conglycinin.

Glycinin tends to promote gelling (in water), so soybeans containing a greater proportion of glycinin would tend to enable the manufacture of a firmer tofu. See also *Protein*, *Cysteine* (*cys*), *Methionine* (*met*).

Glycitein

See *Isoflavones*.

Glycitin

The β-glycoside form (isomer in which glucose is attached to molecule at the seven position of the A ring) of the isoflavone known as glycitein (aglycone form). See also *Isoflavones*, *Isomer*, *Glycitein*.

Glycoalkaloids

See *Alkaloids*.

Glycoarrays

See *Carbohydrate microarrays*.

Glycobiology

The study of the involvement (function) of sugars in biological processes. See also *Glucose* (*GLc*), *Glucose oxidase*, *Glycogen*, *Glycolipid*, *Glycolysis*, *Glycoprotein*, *Glycosidases*, *Glycoside*, *Glycosylation*, *Sialic acid*, *Carbohydrate microarrays*.

Glycocalyx

A polysaccharide matrix that is involved (in some microorganisms) in firm attachment of the organism to a solid surface. See also *Polysaccharide*, *Microorganism*.

Glycoconjugates

Refers to various types of glycosylated molecules such as glycoproteins, glycolipids, glycosaminoglycans, and so on. See also *Glycoprotein*, *Conjugated protein*, *Glycosylation*, *Glycolipid*, *Carbohydrate microarrays*, *Glycoside*.

Glycoform

One of several molecular arrangements that a given glycoprotein can possess [variations are determined by the attachment of various oligosaccharide(s) at different attachment sites on the protein molecules, or the individual structure/arrangements of the sugar groups vary]. Some glycoforms of a given glycoprotein may exhibit greater or lesser biological activity (e.g., pharmaceutical effectiveness for biotherapeutic glycoproteins) because the oligosaccharide units of the glycoprotein molecule mediate interactions of the glycoprotein with the cells of the body. See also *Glycoprotein*, Glycosylation (*to glycosylate*), *Oligosaccharides*, *Biological activity*, *Heterogeneous glycosylation*.

Glycogen

A polymer of glucose with a branching, tree-like molecular structure. It is the chief storage form of carbohydrates in animals. In mammals, glycogen is stored mainly in the liver and muscles. Its molecular weight may be several million. See also *Glucose* (*GLc*), *Glucagon*, *Molecular weight*.

Glycoinformatics

Refers to the generation or creation, collection, storage (in databases), and efficient utilization of data or information about/resulting from all forms of analysis of sugar molecules and glycans in order to accomplish a (research) objective (e.g., to discover a new pharmaceutical). See also *Bioinformatics*, *Glycans*, *Sugar molecules*, *Molecular weight*, *Capillary electrophoresis*, *Carbohydrate microarrays*, *Glycoform*, *Glycobiology*, *Glycoprotein remodeling*.

Glycolipid

A lipid containing at least one carbohydrate group within its molecule. See also *Lipids*, *Glycoprotein*, *Glycosylation* (*to glycosylate*), *Glycolysis*.

Glycolysis

A metabolic process in which sugars are broken down into smaller compounds with the release of energy. This series of chemical reactions is found in plant and animal cells as well as in many microorganisms.

Except for the final reaction in the series, the chemical reaction pathway of glycolysis is the same as that for fermentation.

During 1926, Otto Warburg showed that cancer cells utilize glycolysis to metabolize glucose (in contrast to normal cells, which utilize catabolism to metabolize glucose). See also *Glucose* (*GLc*), *Metabolism*, *Fermentation*, *Hematopoietic stem cells*, *Cancer*, *Catabolism*.

Glycopeptide Fragments

See *Intein*.

Glycoprotein

A conjugated protein containing at least one carbohydrate (oligosaccharide) group within its molecule. A commonly occurring category of glycoproteins found in nature is called mucoproteins. These are protein–polysaccharide compounds that occur in the tissues, particularly in mucous secretions. Other glycoproteins include lymphokines (e.g., interleukins), hormones (e.g., somatotropins), receptors (e.g., GP120), enzymes (e.g., tissue plasminogen activator), and some therapeutics (e.g., CD4PE40). More than 50% of human proteins are glycoproteins. See also *Glycoform*, *Conjugated protein*, *GP120 protein*, *Conjugate*, *Protein*, *Oligosaccharides*, *Polysaccharides*, *Sialic acid*.

Glycoprotein C

A blood clot-regulating glycoprotein. See also *Protein C*, *Glycoprotein*.

G

Glycoprotein Remodeling

The use of restriction endoglycosidases to (enzymatically) remove sugar (i.e., oligosaccharide) "branches" from glycoprotein (i.e., part protein, part oligosaccharide) molecules.

When done by humans (e.g., scientists), one reason to perform such glycoprotein remodeling would be to remove one or more oligosaccharide branches so that the glycoprotein is less or no longer antigenic (i.e., triggers an immune response). This allows the glycoprotein to be injected into the body (e.g., for pharmaceutical purposes) without incurring an unwanted immune response. When done by bacteria (e.g., certain pathogenic bacteria), it can enable those bacteria to become resistant to some antibiotics. See also *Glycoprotein, Restriction endoglycosidases, Enzyme, Oligosaccharides, Antigen, Cellular immune response, Humoral immunity, Antibody, Epitope, Hapten, Bacteria, Pathogen, Pathogenic, Antibiotic, Antibiotic resistance, Intein.*

Glycosidases

Enzymes that catalyze the cleavage (hydrolysis) of glycosidic molecular bonds. For example, lysozyme (an enzyme found in human tears) lyses (cuts up) certain bacteria by cleaving the (β configuration) glycosidic linkages (bonds) between the monosaccharide units that (when linked) comprise the polysaccharide component of the bacterial cell walls. A bacterial cell devoid of a cell wall usually bursts. See also *Endoglycosidase, Exoglycosidase, Restriction endoglycosidases.*

Glycoside

Any of a group of compounds that yield sugar molecules on hydrolysis. All parts of a glycoside compound may be sugar molecules, so that sucrose, raffinose, starch, and cellulose—all of which hydrolyze into sugar molecules—may all be considered to be glycosides. However, the name (glycoside) is usually applied to a compound in which part of the molecule is not a sugar. This nonsugar component is called the aglycon. See also *Hydrolysis, Fructan.*

Glycosinolates

See *Glucosinolates.*

Glycosylation (to Glycosylate)

Addition of oligosaccharide units (e.g., to protein molecules). The oligosaccharide units are linked to either asparagine side chains by N-glycosidic bonds or to serine and threonine side chains by O-glycosidic bonds.

The particular glycosylation of a given protein molecule impacts the ability (and specificity) of antibodies and other immune system components to bind to that protein molecule. See also *Oligosaccharides, Protein, Glycosyltransferases, Golgi bodies, Plantibodies™, Baculovirus, Antibody, Sialic acid.*

Glycosyltransferases

A class of enzymes (transferases) that catalyze the attachment/addition (chemical reaction) of specific carbohydrate molecular chains/branches (molecular groups) to proteins, glycoproteins, or glycosides. Glycosyltransferases also catalyze the connection of multiple simple monosaccharides into the complex polysaccharides that are utilized in a range of plant cell structures and pathways/processes. See also *Carbohydrates (saccharides), Oligosaccharides, Polysaccharides, Monosaccharides, Protein, Pathway, Enzyme, Glycosylation, Glycoprotein, Glycoside, Carbohydrate engineering, Transferases.*

Glyphosate

An active ingredient in some herbicides, it kills plants (e.g., weeds) by inhibiting the crucial plant enzyme EPSP synthase. See also *Enzyme, EPSP synthase, CP4 EPSPS, Glyphosate oxidase, Glyphosate-trimesium, Glyphosate isopropylamine salt, GA21.*

Glyphosate Isopropylamine Salt

One of several forms of active ingredient utilized in some glyphosate-based herbicides. See also *Glyphosate, EPSP synthase, CP4 EPSPS, Glyphosate oxidase, Glyphosate-trimesium.*

Glyphosate *N*-Acetyltransferase

An enzyme that is naturally produced in the soil-dwelling bacteria *Bacillus licheniformis*. That enzyme catalyzes the *acetylation (i.e., the* "attaching" *of an acetyl group to a molecule)* of glyphosate, the active ingredient in some herbicides. Such acetylation prevents glyphosate molecules from killing plants.

If the genes that code for the production of glyphosate *N*-acetyltransferase are inserted via genetic engineering into crop plants, that could help such plants to survive postemergence applications of glyphosate-containing herbicides. See also *Enzyme, Bacteria, Gene, Genetic engineering, Coding sequence, Glyphosate.*

Glyphosate Oxidase

An enzyme that (via catalysis) chemically breaks down glyphosate (i.e., the active ingredient in some herbicides). Glyphosate oxidase is produced in nature by acclimated microorganisms.

In 1988, Michael Heitkamp discovered a strain of *Pseudomonas* bacteria that possessed a gene (GO) that caused those particular *Pseudomonas* bacteria to produce unusually large amounts of glyphosate oxidase. That GO gene can be incorporated into a variety of crop plants (e.g., soybean, cotton) in order to help enable those plants to survive postemergence applications of glyphosate-containing herbicides.

Additionally, a plant can be genetically engineered to survive postemergence applications of glyphosate-containing and/or sulfosate-containing herbicides via insertion of gene (cassette) for plant production of the enzyme CP4 EPSPS. See also *Enzyme, Acclimatization, Strain, Pseudomonas fluorescens, Gene, Genetic engineering, Bacteria, Microorganism, Soybean plant, EPSP synthase, CP4 EPSPS, Cassette, Glyphosate, Sulfosate, GA21.*

Glyphosate Oxidoreductase

An enzyme that is naturally produced in one strain of the microorganism *Ochrobactrum anthropi*. That enzyme (by catalysis) chemically breaks down glyphosate (i.e., the active ingredient in some herbicides).

If a gene (called "goxv247") that codes for the production of glyphosate oxidoreductase is inserted via genetic engineering into crop plants that would help enable such plants to survive postemergence applications of glyphosate- and/or sulfosate-containing herbicides.

Additionally, a plant can be genetically engineered to survive postemergence applications of glyphosate- and/or sulfosate-containing herbicides via insertion of gene (cassette) for plant production of the enzyme CP4 EPSPS. See also *Enzyme, Strain, Microorganism, Gene, Genetic engineering, EPSP synthase, CP4 EPSPS, Cassette, Glyphosate, Sulfosate.*

Glyphosate-Trimesium

One of several forms of active ingredient utilized in some glyphosate-based herbicides. See also *Glyphosate, EPSP synthase, CP4 EPSPS, Glyphosate oxidase, Glyphosate isopropylamine salt, GA21.*

Gm Fad2-1

A (plant) gene that codes for delta 12 desaturase (Δ 12). See also *Gene, Delta 12 desaturase, Cosuppression.*

GMAC

Acronym for the Genetic Manipulation Advisory Committee of the country of Australia, which advises the Australian government on matters pertaining to genetic engineering (e.g., new rDNA product approvals).

The GMAC is analogous to Germany's ZKBS (Central Commission on Biological Safety), Brazil's CTNBio (National Technical Biosafety Commission), and the Kenya Biosafety Council. See also *Gene Technology Regulator (GTR), ZKBS (Central Commission on Biological Safety), Recombinant DNA Advisory Committee (RAC), Genetic engineering, rDNA, Deoxyribonucleic acid (DNA), CTNBio, Kenya Biosafety Council, Gene Technology Office, Interim Office of the Gene Technology Regulator (IOGTR).*

GMO

Genetically manipulated organism, or genetically modified organism. See also *Gene, Gene splicing, Genetic engineering.*

GMP

See *Good manufacturing practices (GMP).*

GMP Guanylate

See *G-proteins.*

GMPP

See *Genetically modified pest protected (GMPP) plants.*

GMS

Genetically modified soya. See also *GMO, Soybean plant.*

GNE

Group of National Experts on Safety in Biotechnology. The group of people within the OECD that developed OECD's guidelines for nations to utilize in their safety evaluations of foods derived from biotechnology. See also *Organization for Economic Cooperation and Development (OECD), Biotechnology, Genetic engineering.*

GO Gene

See *Glyphosate oxidase.*

Gold Nanorods

See *Nanorods.*

Golden Rice

A biotechnology-derived rice (*Oryza sativa*) created in the 1990s by Ingo Potrykus and Peter Beyer, which contains large amounts of beta carotene (precursor of vitamin A) in its seeds. The human body converts beta carotene into vitamin A.

Potrykus/Beyer utilized *Agrobacterium tumefaciens* bacteria to genetically engineer rice plant (i.e., by inserting the following genes from daffodil and from the bacterium *Erwinia uredovora*):

1. Phytoene synthase—from daffodil (narcissus), which converts geranylgeranyl-diphosphate into phytoene.
2. "CRTL" gene—from *Erwinia uredovora*, which codes for phytoene desaturase, which causes the rice plant to convert phytoene (a "light harvesting" carotenoid involved in photosynthesis) into lycopene (a carotenoid that is then utilized by the rice plant in the production of beta carotene). See #3.
3. Lycopene beta-cyclase—from daffodil, which converts lycopene into beta carotene.

The United Nations (UNICEF) estimates that one to two million deaths of children age 1–4 years old could be prevented annually around the world, if they received a little more vitamin A daily in their diet (e.g., via such a rice).

Some of the diseases caused by lack of vitamin A include

- Childhood blindness (estimated to afflict 350,000–500,000 children per year)
- Coronary heart disease
- Certain cancers (e.g., cancer of the lungs, prostate)
- Macular degeneration, a leading cause of blindness in older people
- Various childhood diseases that result in death (e.g., due to a weakened immune system)

Research indicates that, when commercialized in the future, "golden rice" will also contribute more iron (bioavailable) to the human diet. That will be due to inserted genes for ferritin (an iron-rich storage protein) and phytase. Because iron deficiency anemia (IDA) is a major cause of maternal and childhood illnesses in developing countries, such a reduction in IDA via consumption of this rice could confer major health benefits to those countries' populations. See also *Biotechnology, Beta carotene, Vitamin, Phytochemicals, Nutraceuticals, Carotenoids, Gene, Genetic engineering, Bacteria, Agrobacterium tumefaciens, Photosynthesis, Lycopene, Coronary heart disease (CHD), Iron deficiency anemia (IDA), Protein, Phytase, Pathway, Metabolic pathway, Metabolic engineering.*

G

GoldenRice™

A registered trademark now owned by the company Syngenta AG. See also *Golden rice.*

Golgi Apparatus

See *Golgi bodies.*

Golgi Bodies (Also Known as Golgi Complexes)

First described by Camillo Golgi in 1898, these are each a network of interconnected sacs, located within the cytoplasm of cells.

Golgi bodies serve as the primary "sorting centers" of cells and the mechanism for glycosylation of (i.e., adding oligosaccharide and polysaccharide branches onto) proteins, thereby stabilizing them before those proteins are then transported by transfer vesicles to lysosomes, secretory vesicles, or the plasma membrane.

In plant cells, Golgi complexes are where complex polysaccharides are "sorted" and assembled in preparation for making the cell wall (located just outside the cell's plasma membrane).

Visually, a Golgi complex is a stack of flattened membranous sacs (usually 6 sacs in mammal cells and 20 sacs in plant cells).

See also *Cell, Cytoplasm, Oligosaccharides, Polysaccharides, Protein, Lysosome, Vesicles, Plasma membrane.*

Golgi Complexes

See *Golgi bodies.*

Good Laboratory Practice for Nonclinical Studies (GLPNC)

The good laboratory practice (GLPs) that is required by the U.S. Food and Drug Administration (FDA) for studies of the safety and toxicological effects of new drugs for livestock. See also *Good laboratory practices (GLPs), NADA.*

Good Laboratory Practices (GLPs)

A set of rules and regulations issued by the Food and Drug Administration (FDA) that establishes broad methodological guidelines for procedures and record keeping. They are to be followed in laboratories involved in the testing and/or preparation of pharmaceuticals. GLPs also apply to the Environmental Protection Agency (EPA) (e.g., toxicity testing of new herbicides). See also *U.S. Food and Drug Administration (FDA).*

Good Manufacturing Practices (GMP)

The set of general methodologies, practices, and procedures mandated by the U.S. Food and Drug Administration (FDA) that is to be followed in the testing and manufacture of pharmaceuticals. The purpose of GMPs is essentially to provide for record keeping and in a wider context to protect the public. GMP guidelines exist instead of specific regulations due to the newness of the technology and may later be superceded (modified) due to further advances in technology and understanding. See also *cGMP.*

Gossypol

A yellow pigment produced in glands and seeds of the cotton plant (*Gossypium* spp.) and some other plants.

When consumed by monogastric animals (e.g., swine, poultry), gossypol is somewhat toxic to those animals.

Research indicates that, when administered to human tissues (e.g., in the form of a purified pharmaceutical compound), gossypol is active against certain forms of cancer, certain bacteria, and certain fungi. See also *Cotton, Phytotoxin, Cancer, Bacteria, Fungus.*

GP120 Protein

An adhesion molecule (glycoprotein) on the surface envelope (capsid) of HIV (i.e., AIDS-causing) viruses that directly interacts with the CD4 protein on helper T cells, enabling the HIV viruses to bind to and infect helper T cells. In 1994, a group at America's Scripps Research Institute led by Dennis Burton and Carlos Barbas III announced that they had generated a recombinant human antibody to the GP120 protein, which neutralized more than 75% of HIV isolates that it was tested against. See also *Monoclonal antibodies (MAb), Human immunodeficiency virus type 1 (HIV-1), Human immunodeficiency virus type 2 (HIV-2), Acquired immune deficiency syndrome (AIDS), Soluble CD4, CD4 protein, Helper T cells (T4 cells), CD44 protein, Adhesion molecule, Conserved, Glycoprotein, Selectins, Lectins, Protein, Capsid, Viral surface proteins.*

GPA1

A gene, found in most plants, that is responsible for controlling water retention and cell division in those plants. The GPA1 gene codes for a G-protein, which transmits/regulates signals (e.g., light, temperature, phytohormones, nutrients) controlling the plant's development.

During 2001, Alan Jones and colleagues discovered that "knocking out" (i.e., silencing) the GPA1 gene caused the (then-resultant) G-protein to be insensitive to abscisic acid. Because abscisic acid is a phytohormone (i.e., plant hormone) utilized by plants to control the size of stomatal pores—that is, the openings in leaves through which plants exchange oxygen and carbon dioxide (and also water inadvertently) with the atmosphere—the "knocked-out GPA1" plants wilted due to uncontrolled water loss to the atmosphere. See also *Gene, Cell, Mitosis, G-proteins, Plant hormone, Abscisic acid, Knockout (GENE).*

GPCRs

Acronym for *G-protein-coupled receptors.* See *G-protein-coupled receptors.*

GPR120 Receptor

A specific G-protein-coupled receptor that is found in adipose (fat) tissues and on the immune cells called macrophages. When *n*-3 fatty acids (omega-3 fatty acids) dock at GPR120 receptors on macrophages, they activate applicable gene(s) within the macrophages, thereby preventing/reversing their inflammation-causing impacts. See also *Receptors, G-protein-coupled receptors, n-3 fatty acids, Cell, Adipose, Macrophage, Gene.*

G-Protein-Coupled Receptors

See *G-proteins.*

G-Proteins (Guanyl-Nucleotide-Binding Proteins)

Discovered by Rodbell and coworkers at America's National Institutes of Health and Alfred G. Gilman and coworkers at the American University of Virginia-Charlottesville, during the 1970s–1980s. G-proteins are embedded in the surface membrane of cells, analogous to the way that thread is embedded in cloth via someone using an embroidery needle (i.e., the G-protein molecule threads its way back and forth through the cell membrane seven times). G-proteins "receive chemical signals" from outside the cell (e.g., hormones) and "pass the signal" into the cell, so that cell can "respond to the signal."

For example, a hormone, drug, growth factor, neurotransmitter, photons (light), or other "signal" binds to a receptor molecule on the surface of the cell's exterior membrane. That receptor then activates the G-protein (or hundreds of G-proteins, depending on the signal), each of which causes an effector inside cell to produce a second "signal" chemical inside cell, which causes cell (nucleus) to react to the original external chemical signal. The G-proteins are called thus, because they become GTP and GDP forms alternately, as part of their reaction cycle (i.e., in "passing the signal").

G-protein-coupled receptors play crucial roles in many biological processes such as pain perception, vision, blood pressure regulation, sleep regulation, control of cancerous cell growth, allergic responses, and so on. In addition to *carrying to the cell nucleus* these signals, the cytoplasmic G-proteins also regulate certain cellular processes.

There are 24 different types of G-proteins in humans. Dysfunction of certain G-proteins in humans causes the salt and water losses inherent in cholera (the body's compromised immune defense inherent in pertussis) and is believed responsible for some symptoms of diabetes and alcoholism.

In plants (which have 1–4 types of G-proteins, depending on the plant species), G-proteins are part of the signaling process that directs the plant's response to stressors such as drought or disease. Dysfunction of G-proteins in plants can cause rapid water loss (wilting). See also *Protein, Plasma membrane, Lipid rafts, Signaling, Signal transduction, MAPK, Mitogen-activated protein kinase cascade, Hormone, Cell, Nucleus, Beta cells, GTPases, GPA1, Insulin, Receptors, Nuclear receptors, National Institutes of Health (NIH), Neurotransmitters, Transmembrane proteins, Ion channels, Cholera toxin, Prostaglandins, Ligand (in biochemistry), Growth factor, CCR5 protein, GPR120 receptor, Cancer.*

Graft-versus-Host Disease (GVHD)

The "rejection" of transplanted organs by the recipient's immune system. Also known as hyperacute rejection, it is caused by the attack of the recipient's T lymphocytes (i.e., T cells, a certain class of white blood cells) on the transplanted organ. The recipient's T cells are able to distinguish between self and foreign cells and are hence able to recognize the foreign (nonself) cells of the transplanted organ. They then naturally try to destroy the "foreign invaders" in the body. This then constitutes rejection of the transplanted organ. From this it should be understood that there is nothing wrong with the body, but that it is behaving exactly as it should.

Another source of rejection is nonmatched human leukocyte antigens (HLA, a very complex array of six proteins that cover the surface of leukocytes and the bone marrow cells that produce leukocytes). These HLA are usually different (i.e., a nonmatch) for individuals that are not genetically related to each other. Rejection of nonmatched transplanted organs then occurs because the body manufactures antibodies against the nonmatched HLA, thereby leading that HLA to work in concert with a protein molecule named integrin beta 4 to jointly stimulate cell growth and movement (e.g., overproduction of endothelial cells on the interior of applicable blood vessels, thus blocking off blood supply to the transplanted organ). See also *Cell, Cellular immune response, Humoral immunity, Protein, Xenogeneic organs, Fibroblasts, Cyclosporin A, Leukocytes, Human leukocyte antigens (HLA), Integrins, Endothelial cells.*

Gram Molecular Weight

The weight in grams of a compound that is numerically equal to its molecular weight; the weight of 1 mol ($6.022141527 \times 10^{23}$ molecules). See also *Molecular weight, Mole.*

Gram Stain

Devised by Hans Christian Joachim Gram in 1884, this is a test that illuminates the composition/makeup of the physical structure of the cell wall of bacteria being tested. It is utilized to judge the effectiveness of a given chemical compound (e.g., an antibiotic) against bacteria types.

The test consists of a differential staining procedure, which allows most bacteria to be visually separated into two groups, known as Gram-Positive (G+) and Gram-Negative (G–). An antibiotic is defined in terms of the group of (pathogenic) bacteria that it is effective against, which is known as that antibiotic's "spectrum of activity." An antibiotic is said to have a spectrum of activity against gram-positive bacteria, gram-negative bacteria, or the bacteria of *both* groups. An antibiotic that is effective against both groups of bacteria is termed "broad spectrum" or "wide spectrum." See also *Bacteria, Gram-positive (G+), Gram-negative (G–), Pathogenic, Cell, Antibiotic.*

Gram-Negative (G–)

Pertaining to one of the most important ways of classifying bacteria by means of the differences in the way they stain. The set of bacteria that are not able to be stained (blue) when treated with the gram staining procedure. Gram negativity (and gram "positivity") is conferred not by the chemical constituents of the bacteria but rather by the physical structure of the bacteria cell wall. The staining procedure involves the (attempted) staining of all cells in a sample with a blue dye. Gram-negative bacteria have a second (outer) very thin peptidoglycan cell wall known as a capsule, whose outer layer is largely comprised of lipopolysaccharide, which forms a barrier against the inflow of toxic hydrophobic compounds (e.g., an antibiotic manufactured and excreted by another nearby microbe—and this blue dye).

Hence, the washing procedure, which is an integral part of the overall staining procedure, washes out the blue dye (known as crystal violet) from the gram-negative bacteria. This leaves the gram-negative bacteria within the sample colorless. The sample cells are then stained with a red acidic counterstain (dye) such as acid fuchsin or safranin. After treatment with counterstain the gram-negative cells are red and the gram-positive cells are blue. See also *Gram-positive (G+), Bacteria, Cell, Membranes (of a cell), Gram stain, Peptidoglycan, Lipopolysaccharide (LPS).*

Gram-Positive (G+)

Pertaining to bacteria, this refers to them holding the color of the primary stain (blue) when treated with Gram's stain (a commercial staining agent) or Gentian violet solution.

In contrast to the gram-negative bacteria, the gram-positive bacteria possess a much thicker peptidoglycan cell wall (capsule). Because of this, the blue crystal violet dye (with which the bacteria were stained) does not wash out of the cell and the bacteria appear blue under the microscope.

Most gram-positive species of bacteria (e.g., *Enterococci*) utilize peptides for quorum sensing. The human pathogen *Enterococcus faecalis* utilizes the peptide *cytolysin* for both:

- Quorum sensing
- Lysing of *target cells (i.e., of the host organism)* at a distance from the bacteria cell

See also *Gram-negative* (*G*–), *Bacteria*, *Cell*, *Gram stain*, *Capsule*, *Pathogen*, *Peptide*, *Quorum sensing*, *Lyse*.

Granulation Tissue

A mixture of proteins and cells produced by the fibroblast growth that results from a wound. See also *Fibroblasts*, *Protein*.

Granulocidin

A protein produced by white blood cells, which has demonstrated (in the laboratory) an ability to kill a broad spectrum of pathogens. See also *Pathogen*, *Protein*.

Granulocyte Colony Stimulating Factor (G-CSF)

A colony stimulating factor (CSF; a protein) that stimulates production of granulocytes, particularly neutrophils. The genetically engineered analog of G-CSF (produced by genetically engineered *Escherichia coli* bacteria) is a protein molecule known as filgrastim. See also *Colony stimulating factors*, *Escherichia coliform* (*E. coli*), *Genetic engineering*, *Protein*, *Granulocytes*, *Neutrophils*.

Granulocyte-Macrophage Colony Stimulating Factor (GM-CSF) (or Granulocyte-Monocyte Colony Stimulating Factor)

A colony stimulating factor (CSF; a protein) that stimulates production of granulocytes/macrophages/monocytes.

Research indicates that injection of GM-CSF into the human body will also stimulate the growth of new blood vessels around the heart (in those people whose heart arteries are clogged, e.g., via arteriosclerosis). See also *Colony stimulating factors* (*CSFs*), *Macrophage*, *Monocytes*, *Angiogenesis*, *Arteriosclerosis*.

Granulocytes (Polymorphonuclear Granulocytes)

Phagocytic (scavenging, ingesting) cells that are part of the immune system. When their cell nucleus is segmented into lobes and they have granule-like inclusions within their cytoplasm (the neutrophils, eosinophils, and basophils) they are collectively known as polymorphonuclear granulocytes. See also *Phagocyte*.

Graphene

Refers to one-atom thick sheets of carbon. See *Nanopore*, *Nanopore sequencing*, *Polymorphism* (*chemical*).

GRAS List

A list of food additives/ingredients considered to be Generally Recognized as Safe, by the American Government's Food and Drug Administration (FDA). This list of additives is judged to be safe by a panel of FDA pharmacologists and toxicologists, who base their judgment upon data that is available for each ingredient. In practice, those additives for which extensive experience of common use in foods (without known ill effects) has been accumulated over time (e.g., common table salt) are often approved by the FDA due more to the "common use factor" than to any toxicology data, *per se*. See also *Food and Drug Administration* (*FDA*), *Delaney clause*, *Pharmacology*, *Canola*.

Grass Pea

See *Glucosinolates*.

Graves' Disease

An autoimmune disease in which the body's immune system produces an antibody that attacks the thyroid gland, resulting in inflammation and damage (especially to muscle tissue located behind the eyes). See also *Autoimmune disease*, *Antibody*, *Thyroid gland*.

Green Biotechnology

Term utilized in some countries to refer to *agricultural* applications of genetic engineering. One example would be herbicide-tolerant crops. See also *Genetic engineering*, *Herbicide-tolerant crops*.

Green Fluorescent Protein

Discovered by Osamu Shimomura in 1961, it is a protein that is naturally present within the jellyfish *Aequorea victoria*. Green fluorescent proteins (GFPs) from animals of the *Cnidarian* phylum have been utilized since 1994 by scientists to

- "Label" certain protein molecules that are of interest to scientists (e.g., in cell samples)
- Help visualize thin layers of biological tissue in fluorescence microscopy
- "Mark" certain endpoints in experiments (at which point the green light signals that endpoint was reached)

When GFP binds to double-stranded DNA, its fluorescence is greatly enhanced (i.e., also "marking" endpoint).

GFP's gene (i.e., which codes for production of the protein) was isolated in 1989 by Douglas Prasher, and it can be utilized as a "reporter gene" for monitoring gene expression (i.e., of another protein that is of interest to a researcher) in a variety of living systems, for example, inside transparent tissues of the zebrafish (*Danio rerio*), the roundworm *Caenorhabditis elegans*, in tissues that are being grown via cell culture, and so on. See also *Fluorescence*, *Protein*, *Gene*, *Transfection*, *Gene expression markers*, *Cell*, *Cell array*, *Cell culture*, *Reporter gene*, *Label* (*fluorescent*), *Deoxyribonucleic acid*

G

(DNA), Double helix, Coding sequence, Caenorhabditis elegans (C. elegans), Fluorescence-activated cell sorter (FACS), TIRF microscopy, Rapid protein folding assay.

Green Leaf Volatiles

Abbreviated GLV, this refers to the specific mixture of volatile (i.e., rapidly evaporate from liquid to vapor) chemicals known as six-carbon alcohols and aldehydes that are immediately emitted by certain plants (e.g., *Nicotiana attenuata* also known as wild tobacco) when applicable herbivorous pests (e.g., *Manduca sexta* also known as tobacco hornworm) larvae chew on their leaves.

The GLV quickly attract any nearby predator insects of the genus *Geocoris*, which come and eat the applicable pest insects (e.g., tobacco hornworm larvae in this example). See also *Volicitin, GLV, Jasmonates, Genus.*

GRF

See *Growth hormone releasing factor.*

GRH

See *Growth hormone releasing factor.*

gRNA

Acronym for **guide RNA**, which is a form of RNA that (among other things) directs Cas9 enzyme to the target DNA sequence (i.e., to remove it, repair it, or insert additional sequences). See *sgRNA, CRISPR/Cas9 gene-editing systems, Deoxyribonucleic acid (DNA), Sequence (of a DNA molecule).*

GRO

Acronym for genomically recoded organism. See *Genome editing.*

GroEL Protein

See *Chaperones.*

GroES Protein

See *Chaperones.*

Group of National Experts on Safety in Biotechnology

See *GNE.*

Growth (Microbial)

An increase in the number of cells. See also *Generation time.*

Growth Curve

The change in the number of cells in a growing culture as a function of time. See also *Generation time.*

Growth Factor

A specific substance that must be present in the organism's tissues (when *in vivo*) or growth medium (when *in vitro*) in order for the growth-factor-specific cells to grow/multiply. See also *Fibroblast growth factor (FGF), Nerve growth factor (NGF), Epidermal growth factor (EGF), Vascular endothelial growth factor (VEGF), Angiogenic growth factors, Angiogenin, Bone morphogenetic proteins (BMP), Pre-B cell colony-enhancing factor, Insulin-like growth factor-2 (IGF-2).*

Growth Hormone (GH)

A hormone produced by the anterior pituitary gland. This hormone is a protein (somatotropin) and can be obtained from the bodies of animals or produced by genetically engineered microorganisms. Its major action in humans (human growth hormone [HGH]) is a generalized stimulation of skeletal growth. However, HGH is also known to affect the growth of other tissues; to be important in fat, protein, and carbohydrate metabolism; and to enhance the effects of various other hormones. See also *Bovine somatotropin (BST), Porcine somatotropin (PST), Pituitary gland.*

Growth Hormone–Releasing Factor (GRF or GHRF)

Also termed growth hormone–releasing hormone (GRH). A factor that causes the release of growth hormone. It is 44 amino acids in length. See also *Growth hormone (GH), Growth factor, Amino acid, Hormone.*

GS

Acronym for glucosinolates. See *Glucosinolates.*

GSL

Acronym for glucosinolates. See *Glucosinolates.*

GT/PT Correlation

Abbreviation for *genotype/phenotype correlation.* See *Genotype, Phenotype.*

GT-AG Rule

Describes the presence of these constant dinucleotides at the first two and last two positions of introns of nuclear genes. See also *Intron, Gene.*

GTO

Abbreviation for *Gene Technology Office.* See *Gene Technology Office.*

GTP

See *GMP.*

GTPases

Guanosine triphosphatases. These are G-proteins (enzymes) that are crucial for growth, movement, and maintenance of the cell's shape. When active, GTPases are bound to cell membranes (surfaces) by an isoprene molecule (receptor). See also *G-proteins, Enzyme, Cell, Phosphorylation, Receptors, Protein.*

GTR

See *Gene technology regulator (GTR)*.

GTS

Acronym for glyphosate tolerant soybean. See *Herbicide-tolerant crop, Soybean plant, CP4 EPSPS, Glyphosate.*

GTS

Acronym for glufosinate-ammonium tolerant soybean. See *Herbicide-tolerant crop, Soybean plant, PAT gene, Glufosinate.*

GTs

Abbreviation for glycosyltransferases. See *Glycosyltransferases.*

Guanine

A purine base. It occurs naturally as a fundamental component of nucleic acids. See also *Purine, Nucleic acids.*

Guide RNA

See *sgRNA.*

GURTs

See *Genetic use restriction technologies.*

GUS

See *GUS gene.*

GUS Gene

A gene that codes for production of B-glucuronidase (i.e., GUS protein) in certain organisms (e.g., *Escherichia coli* bacteria).

The GUS gene is commonly utilized as a "marker gene" for genetically engineered plants. B-glucuronidase causes a color change, in the presence of the chemical 5-bromo-4-chloro-3-indoyl-beta-D-glucuronic acid, by cleaving (i.e., "cutting") a glucuronic acid molecule off of the 5-bromo-4-chloro-3-indoyl-beta-D-glucuronic acid. The (remaining) molecule is an insoluble blue dye. See also *Gene, Coding sequence, Escherichia coliform (E. coli), Marker (genetic marker), Genetic engineering, Enzyme.*

Gut Leakage

Refers to a condition during which (nondigested) bacterial endotoxins and/or bacteria pass out of the intestines into the bloodstream of an organism. Those endotoxins (and/or vigorous immune response by the organism) can result in harmful inflammation and other damage to some of the organism's tissues.

Gut leakage can result from certain diseases or from binge drinking of ethanol. See also *Bacteria, Endotoxin, Immune response, Chronic inflammation, Probiotics.*

Gut Microbiome

See *Oligosaccharides.*

Gut-Associated Lymphoid Tissues (GALT)

A variety of specialized lymph-reticular tissues that line the inside of an animal's digestive system. GALT include Peyer's patches, the appendix, and small solitary lymphoid tissues in the gut. They constitute the intestinal immune system (response to antigens).

For example, after "naive" B cells are activated (e.g., via presentation of certain pathogens' antigens by the immune system's dendritic cells), those activated B cells—called *effector B cells*)—depart the GALT and move to the intestine's ileum (exterior layer) where they secrete enteric-pathogen-fighting IgA molecules. See also *Lymphocyte, Peyer's patches, Antigen, Humoral immunity, Cellular immune response, Dendritic cells, Pathogen, Immunoglobulin, Edible vaccines, Plantigens.*

GWA

Acronym for *genome-wide association* studies. See *GWAS.*

GWAS

Acronym for *genome-wide association studies.* This refers to comparative studies of the genomes of thousands or organisms (e.g., people) using microarrays to compare the relevant DNA sequences of (sick vs. healthy individuals, susceptible vs. nonsusceptible individuals, etc.). GWAS has been utilized to find SNPs/gene variants (alleles) that increase a given person's susceptibility to certain diseases (e.g., rheumatoid arthritis, leprosy) that possess a genetic/susceptibility component. See also *Genome, Deoxyribonucleic acid (DNA), Gene, Organism, Association mapping, Gene expression analysis, Microarray (testing), Allele, Haplotype, Single-nucleotide polymorphisms (SNPs), SNP chip, Sequence (of a DNA molecule), Rheumatoid arthritis, Leprosy.*

Gyrase

See *Helicase.*

H

H. pylori

A bacteria (*Helicobacter pylori*) that has been linked (e.g., a cause) to gastric ulcers, stomach cancers, and other gastric problems in humans. That link was first announced by Barry Marshall in the early 1990s.

During 2010, research was published indicating that infection of *H. pylori* in people who have non-O blood types is associated with a threefold higher risk of pancreatic cancer. See also *Bacteria, Helicobacter pylori, Cancer, Sulforaphane.*

H. virescens

See *Heliothis virescens* (*H. virescens*).

H. zea

See *Helicoverpa zea* (*H. zea*).

HA

Abbreviation for the word *hemagglutinin*. See *Hemagglutinin.*

HAART

Acronym for *highly active antiretroviral therapy.* See *Resistin.*

Habitat

The natural environment of an organism within an ecosystem. The place, in an ecosystem, where an organism lives. See *Ecology.*

HAC

See *Human artificial chromosomes* (*HAC*).

HACCP

See *Hazard analysis and critical control points* (*HACCP*).

HAHB4 Gene

See *Drought tolerance trait.*

Hairpin Loop

A section of highly curving, single-stranded DNA or RNA formed when a long piece (string) of the DNA or RNA bends back on itself and hydrogen bonds (is able to base pair) in some regions to form double-stranded regions. The structure can be visualized by taking a human hair, bending it back on itself and holding it in such a way as to half its original length. The section where the two ends of hair lie next to each other represents the section of double-stranded DNA or RNA. At one end the hair will have to make a sharp turn and will form a loop. This loop represents the single-stranded hairpin loop.

Hairpin loops can also form in peptide (molecules). For example, during 2002, Joel P. Schneider and Darrin J. Pochan designed a 20-residue peptide that spontaneously assembles (by the millions) into a *hydrogel*, when a solution containing those peptides is caused to have a pH of 9. See also *Ribonucleic acid (RNA), Deoxyribonucleic acid (DNA), Self-assembly (of a large molecular structure), Molecular beacon.*

Halobacterium

A microbe that is able to survive and grow in water at high salt concentration, is able to withstand high levels of radioactivity (ionizing radiation), and converts incident sunlight into energy the microbe can utilize. *Halobacterium* growing in the Dead Sea and the Great Salt Lake (in Utah, USA) impart their yellow-orange color to those bodies of saltwater. See also *Microbe.*

Halophile

Microorganisms that require NaCl (salt) for growth (they are called obligate halophiles). Those that do not require it, but can grow in the presence of high NaCl concentrations, are called facultative halophiles. Natural habitats containing high salt concentrations are, for example, the Great Salt Lake in Utah, the Dead Sea in Israel, and the Caspian Sea in Russia. See also *Habitat.*

Halophytes

Refers to "salt-loving" (truly *salt-tolerant*) plants. Their primary mechanism to cope with growing in high-salt soils is to produce trichomes (i.e., outgrowths of the plants, where excess soil salt is stored after it has been taken in by halophyte plants). See also *Trichomes.*

Hanging Drop Assays

Also called 3D Hanging Drop Assays, this refers to a category of assays (e.g., utilized to test effects of drugs on certain types of living cells) in which the cells are inserted into a drop of water solution that is hanging beneath a perforated plate. The cells are added from above via the plate's holes, whereupon the cells self-aggregate into a small sphere located just above the bottom of the hanging water drop. This water-suspended sphere of cells more closely mimics the conditions under which cells live in the body of an organism (than say, cells in a Petri dish), so these cells' responses to a drug (that is subsequently added to the water drop) are more likely to be indicative of the drug's impact on cells in the body.

See also *Assay, Bioassay.*

H

HAP Gene

See *Low-phytate corn.*

Haploid

A cell with one set of chromosomes (i.e., half as many chromosomes as the normal somatic body cells contain).

A characteristic of sex cells. See also Gamete, *Doubled-haploid breeding program, Induced polyploidy.*

Haploid-Inducer Parent

See *Doubled-haploid breeding program.*

Haplophase

A phase in the life cycle of an organism in which it has only one copy of each gene. The organism is then said to be haploid. Yeast can exist as true haploids. Humans are haploid for only a few genes and cannot exist as true haploids. See also *Haploid.*

Haplotype

A subgroup (e.g., an ethnic minority, all members of a genetically related family group) of organisms (e.g., humans) whose phenotype results in their body responding in the same way to a physical agent (e.g., a certain pharmaceutical, a toxin, a food) or are predisposed to particular diseases. For example, more than 70% of black people in North America are lactose intolerant (e.g., their bodies cannot metabolize the lactose sugar in cow's milk), but fewer than 19% of Caucasian people in North America are lactose intolerant.

Analogous to that, the drugs acetaminophen, aspirin, and valium remain in the bodies of women (who constitute a haplotype) longer than in the bodies of men. Haplotypes for the β_2-adrenergic gene are predictive of asthma patients' response to the pharmaceutical *albuterol.*

Haplotypes of women possessing the BRCA 1 gene or the BRCA 2 gene have a higher-than-average chance of developing ovarian cancer or breast cancer. Haplotypes of people possessing the APOE4 gene or the CYP46 gene have a higher-than-average chance of developing Alzheimer's disease.

In terms of molecular biology, haplotypes consist of individuals whose DNA contains "grouped *SNPs*" (*single-nucleotide polymorphisms*) that *collectively* confer a particular aspect (e.g., sensitivity to certain pharmaceuticals, susceptibility to certain diseases) when they are inherited.

For example, during 2002, a haplotype consisting of people possessing *70 genes inherited together* was found to be predictive of breast cancer metastasis.

For example, during 2006, a haplotype consisting of people possessing *186 genes inherited together* was found to be predictive of breast cancer recurrence (e.g., after chemotherapy).

During 2003, Jeffrey Mogil discovered that a haplotype consisting of red-haired women possessing certain versions of the *MC1R* (*melanocortin 1 receptor*) gene was predictive of a heightened response to the opioid pharmaceutical pentazocine in those women.

See also *Pharmacogenomics, Heritability, Heredity, Trait, Gene, Genetics, Phenotype, Toxin, Insulin, Metabolism, Single-nucleotide polymorphisms (SNPs), Cancer, BRCA genes, Linkage, APOE4, Cytochrome P450 (CYP), CYP46 gene, Alzheimer's disease, Metastasis, Fluorescence in situ hybridization (FISH), Allele, Chemotherapy, GWA.*

Haplotype Map

See *Single-nucleotide polymorphisms (SNPs).*

HapMap

Acronym for *haplotype map.* See *Haplotype map.*

Hapten

A small foreign molecule that will stimulate an immune system response (e.g., antibody production) if the small molecule (now called a haptenic determinant) is attached to a macromolecule (carrier) to make it large enough to be recognized by the immune system. See also *Epitope, Cellular immune response, Humoral immunity, Carrier protein.*

Haptoglobin

A protein that is a component in human blood that can occur in one of two different molecular forms (i.e., a "large" version of that molecule or a "small" version of that molecule).

The "small" version of haptoglobin is very effective at "capturing" and removing free radicals (high-energy oxygen atoms that bear an "extra" electron) *from the bloodstream* before they damage tissues (e.g., in the eyes, kidneys, and/or arteries).

The "large" version of haptoglobin, which is the only haptoglobin molecule in the bloodstream of one particular haplotype (genetic subgroup) of people, is not effective at capture/removal of those free radicals (e.g., generated at a high rate in people with diabetes disease), so diabetics within that particular haplotype tend to suffer extreme damage to eyes, kidneys, nerves, and arteries (sometimes necessitating limb amputation).

See also *Free radical, Haplotype, Insulin, Oxidative stress, Diabetes.*

Hardening

See *Cold hardening, Hydrogenation.*

Harpin

A protein that is naturally produced by the *Erwinia amylovora* bacteria (which usually causes the plant disease known as *fire blight* in apple trees, pear trees, and some ornamental plants of the rose family).

Discovered in 1992 by Zhong-Min Wei and colleagues, harpin causes numerous species of plants to initiate a protective/defensive response (cascade) against bacteria, viruses, fungi, and some insects and nematodes. Harpin also causes plants (i.e., where it is sprayed onto) to increase their photosynthesis and to have increased root growth/proliferation, which can lead to greater crop yields. See also *Protein, Bacteria, Phytoalexins, Pathogenesis related proteins, Signaling, Signaling molecule, Signal transducers and activators of transcription (STATs), Salicylic acid (SA), Jasmonic acid, Systemic acquired resistance (SAR), Cascade, R genes, Nematodes.*

Harvesting

A term used to describe the recovery of microorganisms from a liquid culture (in which they have been grown by man). This is usually accomplished by means of filtration or centrifugation. See also *Microorganism, Culture medium, Ultracentrifuge, Dialysis.*

Harvesting Enzymes

Enzymes that are used to gently dissociate (i.e., break apart) cells in living tissues in order to produce single, separate cells that can then be established and propagated in a cell culture reactor. Harvesting enzymes are also used to dissociate cells that have been grown for some time in a cell culture reactor. See also *Cell culture, Mammalian cell culture, Enzyme, Culture medium.*

HAT

Acronym for *histone acetyl-transferase* enzymes. See also *Enzyme, Histones.*

Hazard Analysis and Critical Control Point (HACCP)

A quality control program (for food processing) to systematically prevent hazards (e.g., pathogens) from entering the production process. HACCP was initially developed in the 1950s by the Pillsbury Company to supply food products for astronauts in America's space program. Under HACCP, food processors/handlers must analyze and identify in advance the points where hazards are most likely to occur and eliminate them. For example, because melons lie in pathogen-contaminated dirt while growing, a "critical control point" for restaurants serving sliced melon is cleansing of the knife after each melon is cut (to prevent the knife carrying pathogens from one infected melon to other melons). See also *Pathogen, Rapid microbial detection (RMD).*

HB4 Gene

See *Drought tolerance trait.*

HBC

Acronym for *high beta-conglycinin.* Utilized to refer to crop varieties (e.g., of soybean) that contain higher than typical amounts of beta-conglycinin. See also *Beta-conglycinin, Soybean plant.*

H-Bonding

See *Hydrogen bonding.*

HCC

See *Angiogenesis.*

HCP

Acronym for *host cell protein* (e.g., produced in living cells that are being utilized to produce a specific biopharmaceutical). The numerous different proteins (and their amounts/concentrations) are crucial quality control parameters monitored by scientists during biopharmaceutical production (e.g., in a fermentation vat). See also *Protein, Cell, Express, Host vector (HV) system.*

HCS

Acronym for *high-content screening.* See *High-content screening.*

HD Gene

Refers to the damaged (mutant) allele that causes Huntington's disease, when present in a human's genome. See also *Gene, Dominant allele, Genome, Huntington's disease, Genetics, mutation.*

HDA

Acronym for *helicase-dependent amplification.* Refers to a DNA amplification methodology in which the enzyme helicase is utilized to denature the targeted DNA, instead of using high temperature to denature DNA (as is done in PCR). See also *Helicase-dependent amplification.*

HDAC

Acronym for *histone deacetylase* enzyme. HDACs are a family of enzymes that catalyze the removal of acetyl molecular groups from histones. See also *Enzyme, Histones, Chromatin, Sirtuins.*

HDL

See *High-density lipoproteins (HDLPs).*

HDM

Acronym for *histone demethylase* enzyme. See also *Enzyme, Histones.*

HDT

Acronym for heat and drought tolerance (i.e., traits inserted into crop plants via genetic engineering). See also *Gene, CspB gene, Genetic engineering, Drought tolerance, Drought tolerance trait.*

HDV

See *Hepatitis delta virus (HDV).*

Heat Map

Refers in general to a two-dimensional depiction of large amounts of data, where the differing values of a given variable are represented as different colors.

A common heat map utilized in molecular biology is to depict the *level of gene expression* of numerous genes across comparative samples tested via microarrays (e.g., samples from diseased/nondiseased cells, samples from cells that are treated/untreated with a drug candidate).

The *rows* of a microarray heat map usually represent specific genes, and each *column* represents a different sample. See also *Gene, Gene expression, Gene expression analysis, Gene expression profiling, Microarray (testing), DNA chip, High-throughput screening (HTS), Target (of a therapeutic agent).*

Heat Shock Protein 90

See *HSP90.*

Heat Shock Proteins

Abbreviated "Hsps." Also sometimes called stress proteins, these are special chaperone protein molecules that help other individual

H

protein molecules within a cell to properly fold into their tertiary structure (i.e., their three-dimensional structure in which they are biologically active). See also *Chaperones, Protein, Cell, Tertiary structure, Biological activity, Stress proteins.*

Heavy-Chain Variable (VH) Domains

The regions (domains) of the antibody (molecule's) "heavy chain" that vary in their amino acid sequence. The "chains" (of atoms) comprising the antibody (immunoglobulin) molecule consist of a region of variable (V) amino acid sequence and a region in which the amino acid sequence remains constant (C). An antibody molecule possesses two antigen-binding sites, and it is the variable domains of the light (VL) and heavy (VH) chains that contribute to this (antigen-binding ability). See also *Antibody, Protein, Immunoglobulin, Sequence (of a protein molecule), Antigen, Amino acid, Combining site, Domain (of a protein), Light-chain variable (VL) domains.*

Hedgehog Proteins

A "family" of related signaling molecules (consisting of "signaling protein" with cholesterol molecule attached to it), which direct/control tissue differentiation during animal and insect embryo development (into various organs, limbs, etc.). They also control left–right asymmetry of the developing body. Some of the hedgehog proteins are *sonic hedgehog (Shh), Indian hedgehog (Ihh),* and *desert hedgehog (Dhh).*

The applicable hedgehog protein (within an embryo cell) cleaves itself into two peptides, one of which then acts as a transferase (i.e., enzyme that catalyzes the addition of a functional group to a given molecule—in this case to the other "hedgehog peptide").

When the cell then secretes the cholesterol/peptide molecule, the cholesterol (functional group) "anchors" it to the cell surface, while the "signaling protein" end of the cholesterol/peptide directs differentiation of nearby cells. See also *Protein, Signaling molecules, Signaling, Cholesterol, Signal transduction, Peptide, Cell, Transferases, Enzyme, Functional group, Differentiation, Cell differentiation.*

Hedgehog Signaling Pathway

A signaling pathway that is critical to the development (i.e., of many embryonic organisms into adult organisms) and later function of cytotoxic T cells. Via this pathway, hedgehog proteins direct/control tissue differentiation (into various organs, limbs, etc.) as the embryo develops into an adult body. Hedgehog proteins also control left–right asymmetry of the developing body. When the body reaches adult form, the hedgehog signaling pathway shuts down (epigenetically), except the pathway is still utilized to facilitate cytotoxic T cell killing of tumor and virally infected cells.

Research indicates that if the hedgehog signaling pathway is (wrongly) "turned on" in an adult body, it can promote development of some cancers. See also *Pathway, Signaling, Protein, Signaling protein, Signal transduction, Hedgehog proteins, Differentiation, Cell, Cell Differentiation, Cell motility, Epigenetic, Cancer, Tumor, Cytotoxic T cells.*

HeLa Cells

A cell line (i.e., cells propagated in cell culture) utilized by researchers studying human physiology/malignancy, ever since it was donated to science (from a tumor in her body) in 1951 by *He*nrietta *La*cks. George Otto Gey was able to isolate one specific tumor cell and he discovered the progeny cells could be kept alive and keep growing indefinitely in cell culture.

They were subsequently utilized in development of the first polio vaccine, cloning, and gene mapping. See also *Cell, Cell culture, Cancer, Tumor, Monoclonal antibodies (MAb), Vaccine, Clone (an organism), Gene mapping.*

Helicase

An enzyme that "unwinds" the DNA molecule's *double-helix structure* during DNA replication. See also *Enzyme, Deoxyribonucleic acid (DNA), Double helix, Replication (of DNA), Helicase-dependent amplification.*

Helicase-Dependent Amplification

Refers to a DNA amplification methodology in which the enzyme helicase is utilized to denature the targeted DNA, instead of using high temperatures to accomplish that denaturing (i.e., as is done in the polymerase chain reaction—PCR—technique). See also *Deoxyribonucleic acid (DNA), Helicase, Replication (of DNA), Polymerase chain reaction (PCR), Polymerase chain reaction (PCCR) technique.*

Helicobacter pylori

A commensal bacteria that lives in stomachs of humans and helps to regulate the levels of stomach acids.

However if the stomach produces too much acid, some strains of *Helicobacter pylori* (i.e., those possessing the gene known as *cagA*) produce proteins that signal the stomach to reduce its acid production. Unfortunately, in certain susceptible people, those proteins can trigger stomach ulcers. *H. pylori* has also been linked (e.g., a cause) to stomach cancers and other gastric problems in humans. That link was first announced by Barry Marshall in the early 1990s.

During 2010, research was published indicating that infection of *H. pylori* in people who have non-O blood types is associated with a threefold higher risk of pancreatic cancer. See also *Bacteria, Commensal, H. pylori, Cancer, Sulforaphane.*

Helicoverpa armigera

See *Helicoverpa zea (H. zea).*

Helicoverpa zea (H. zea)

Known as the corn earworm (when it is on corn plants), known as the soybean podworm (when it is on soybean plants), and known as the tomato fruitworm (when it is on tomato plants), this is one of three insect species that is called "bollworms" (when on cotton plants). *H. zea* chews on those crop plants and is one of the insects that can act as a vector (carrier) of *Aspergillus flavus* fungus.

In the country of India, the "cotton bollworm" is *Helicoverpa armigera.*

In 1997, scientists at the U.S. Department of Agriculture created/optimized a monoclonal antibody against *"Helicoverpa zea* vitellin," which thus holds potential to be used as a means to control that insect. The fungal pathogen *Nomuraea rileyi* is a natural biological control agent for *Helicoverpa zea* in soybean fields. See also *B.t. kurstaki, Heliothis virescens (H. virescens), High-maysin corn, Fungus, Pectinophora gossypiella, Aspergillus flavus, Corn, Monoclonal antibodies (MAb), Helicoverpa zea (H. zea).*

Heliothis virescens (H. virescens)

Known as the tobacco budworm (when it is on tobacco plants), this is one of three insect species that is called "bollworms" (when they are on cotton plants). As part of integrated pest management, farmers can utilize the parasitic *Euplectrus comstockki* wasp to help control the tobacco budworm/cotton bollworm. When that wasp's venom is injected into *Heliothis* larva, it stops the larva from molting (and thus maturing). See also *B.t. kurstaki, Helicoverpa zea (H. zea), Pectinophora gossypiella, Integrated pest management* (IPM).

Helix

A spiral, staircase-like structure with a repeating pattern described by two simultaneous operations (rotation and translation). It is one of the natural conformations exhibited by biological polymers. See also *Biomimetic materials, Analogue, Spiral polypeptides.*

Helper T Cells (T4 Cells)

T cells (lymphocytes) that bind B cells (upon recognizing a foreign epitope on B cell surface). The binding stimulates B cell proliferation by secreting B cell growth factor. See also *B cells, Cytokines, T cell, T cell receptors, Suppressor T cells.*

Hemagglutinin (HA)

A special protein that some viruses (e.g., influenza) utilize to gain entry into the cells they have "targeted." The HA protein helps the virus to *adhere* to the cell it "targets." Most of the antibodies created by the body to prevent influenza infection (e.g., after person receives flu vaccine) are directed against HA.

Hemagglutinin is also utilized to refer to specific plant cell proteins (lectins) that are naturally produced by certain plants such as the soybean plant (*Glycine max* (L.) Merrill). The presence of those lectin molecules (e.g., on surfaces of root cells of the soybean plant) helps nitrogen-fixing *Rhizobium japonicum* bacteria to adhere to soybean plant roots, where they begin to "fix nitrogen" (i.e., create natural nitrate fertilizer, which improves the soil and helps plants to grow). See also *Protein, Virus, Cell, Lectins, Soybean plant, Nitrogen fixation, Bacteria, Rhizobium (bacteria), Nitrates, Nodulation.*

Hematologic Growth Factors (HGF)

A class of colony stimulating factors (proteins) that stimulates bone marrow cells to produce certain types of red and white blood cells. Some colony stimulating factors are as follows: (1) granulocyte-macrophage colony stimulating factor (GM-CSF), (2) granulocyte-monocyte colony stimulating factor, (3) granulocyte colony stimulating factor (G-CSF), (4) erythropoietin, (5) interleukin-3, and (6) macrophage colony stimulating factor. See also *Colony-stimulating factors (CSFs).*

Hematopoietic Growth Factors

Growth factors that stimulate the body to produce blood cells. See also *Growth factor, Interleukin-6 (IL-6), Hematopoietic stem cells.*

Hematopoietic Stem Cells

Certain stem cells present (e.g., in infants' bodies and in the umbilical cords of newborn infants) that can be differentiated (via chemical signals in the growing body) to give rise to red blood cells and the infection-fighting cells of the immune system.

Also present in small numbers within adult bodies, at the edge of the bone marrow (in contact with bone tissue), these are what are transplanted by doctors in "bone marrow transplant" operations.

Because they live in a low-oxygen environment (inside the bones), hematopoietic stem cells utilize glycolysis (i.e., convert glucose or other sugars into energy rather than using oxygen to release energy) as most other cells do via mitochondrial oxidative phosphorylation, to meet their energy demands. See also *Stem cells, Adult stem cell, Hematopoietic growth factors, Multipotent adult stem cells, Mesodermal adult stem cells, Cell, Organism, Signaling, Glycolysis, Glucose (GLc).*

Heme

The iron–porphyrin prosthetic group of a class of proteins called "heme proteins." See also *Prosthetic group, Chelating agent, Protein, Transferrin.*

Hemoglobin

An oxygen-transporting respiratory pigment; it is present in humans, animals, and some plants (e.g., land plants that withstand occasional immersion/flooding).

In humans, hemoglobin is carried in the red blood cells (erythrocytes) and is responsible for the red color of the blood. It is composed of two pairs of identical polypeptide chains and iron-containing heme groups, comprising the (total) hemoglobin molecule. The molecular structure of hemoglobin was determined by Max Perutz in 1959. A human disease known as sickle cell anemia is caused by (genetically induced) small change (i.e., due to SNP) in the hemoglobin molecule's structure (in victims of that disease). See also *Heme, Polypeptide (protein), Genetics, Bilirubin, Heredity, Erythrocytes, Protein structure, Single-nucleotide polymorphisms (SNPs).*

Hemoglobin AlC

See *Biomarkers.*

Hemostasis

See *Fibrin.*

Heparin

A polysaccharide sulfuric acid ester found in liver, lung, and other tissues that prolongs the clotting time of blood by preventing the formation of fibrin. Used in vascular surgery and in treatment of postoperative thrombosis and embolism. See also *Fibrin, Thrombosis.*

Hepatitis Delta Virus (HDV)

See *Farnesyl transferase inhibitors.*

H

HER-2 Gene

Abbreviation utilized for "Human Epidermal growth factor Receptor-2 gene/*neu*," which was discovered by Robert Weinberg in 1982. The HER-2 gene is an oncogene that is responsible for approximately 27% of breast cancers (i.e., in those women whose body *overexpresses* that particular oncogene), and it spreads via metastasis.

In addition to conventional treatments (e.g., mastectomy, chemotherapy), the United States' Food and Drug Administration (FDA) in 1998 approved use of a humanized monoclonal antibody (trastuzumab) to be utilized alone or in combination with certain chemotherapy agents (e.g., paclitaxel) against such HER-2 metastatic breast cancers. That monoclonal antibody attaches to the *extracellular domain* (i.e., portion of the HER-2 receptor sticking out of surface of breast tissue cells) and *downregulates* the HER-2 gene (i.e., resulting in fewer HER-2 receptors being produced on the plasma membrane surfaces of that woman's breast tissue cells).

During 2010, the United States' FDA approved use of a humanized monoclonal antibody (trastuzumab) to be utilized in combination with certain chemotherapy agents (cisplatin plus either capecitabine or 5-fluorouracil) against HER-2-positive metastatic stomach (gastric) cancer or gastroesophageal junction cancer, in patients of both gender who have not earlier received medicines for those metastatic cancers.

The latter three indications (i.e., for HER-2-positive breast cancer, metastatic stomach [gastric] cancer, or gastroesophageal junction cancer) require use of a companion diagnostic (CDx) test to determine if the patient is HER-2-positive, in advance of administration of the pharmaceuticals.

See also *Gene, Receptors, HER-2 receptor, RAS gene, EGF receptor, Oncogenes, Cancer, Express, Expressivity, Metastasis, Monoclonal antibodies (MAb), BRCA genes, Paclitaxel, Food and Drug Administration (FDA), Plasma membrane, Fluorescence in situ hybridization (FISH), Pharmacogenomics, Pharmacogenetics, Trastuzumab, Nutritional genomics, Companion diagnostic, ADO-trastuzumab emtansine.*

HER-2 Protein

See *HER-2 receptor.*

HER-2 Receptor

An epidermal growth factor receptor (protein molecule embedded in the surface of cells) that is present in abundance in the plasma membrane surface of breast tissue cells in humans possessing the "HER-2 gene." It is also present on the surface of some tumors of metastatic stomach (gastric) cancer or gastroesophageal junction cancer, in patients of both genders. See also *Receptors, Epidermal growth factor receptor, Plasma membrane, HER-2 gene, Tumor, Trastuzumab, ADO-trastuzumab emtansine.*

HER-2/neu Gene

See *HER-2 gene.*

Herbicide Resistance

See *Herbicide-tolerant crop.*

Herbicide-Resistant Crop

See *Herbicide-tolerant crop.*

Herbicide-Tolerant Crop

Crop plants, cultivated by man, which have been altered to be able to survive application(s) of one or more herbicides by the incorporation of certain gene(s), via either genetic engineering, natural mutation, or mutation breeding (i.e., soaking seeds in mutation-causing chemicals—or bombardment of seeds with ionizing radiation—to cause random genetic mutations, followed by selection of the *particular mutation in which herbicide tolerance occurs*).

Because it has been utilized for decades, most relevant national laws consider mutation breeding to be one of the so-called traditional plant breeding techniques. For example, European laws that require special labeling of food products containing genetically engineered (via rDNA) crops do not require such special labeling for food products that contain *crops that were created via mutation breeding.*

Several crops (e.g., soybean, canola, cotton) are made tolerant to glyphosate-containing or sulfosate-containing herbicides by the insertion (via genetic engineering techniques) of the aroA transgene (cassette) for CP4 EPSPS. Corn (maize) is made tolerant to glyphosate-containing herbicides by insertion (via genetic engineering techniques) of the mEPSPS or GA21 transgene (cassette).

Some soybean varieties are made tolerant to sulfonylurea-based herbicides by adding (via traditional breeding methods) the "ALS gene" (which confers the sulfonylurea-tolerance trait).

Corn (maize) and rice (*Oryza sativa*) are made tolerant to imidazolinone-containing herbicides by adding (via traditional breeding techniques) the imidazolinone-tolerance trait. That trait is imparted by the T-Gene, IT-Gene, or the IR-Gene. See also *Gene, genetic engineering, Cassette, Transgenic, Deoxyribonucleic acid (DNA), rDNA, EPSP synthase, Glyphosate oxidase, Pat gene, Bar gene, Genetics, Glyphosate, GA21, Sulfosate, ALS gene, CP4 EPSPS, Chloroplast transit peptide (CTP), Acuron™ gene, Avena gene, Transgene, Trait, Canola, Soybean plant, Corn, Mutation breeding, Traditional breeding methods, Imidazolinone-tolerant soybeans, Drought tolerance.*

Heredity

Transfer of genetic information from parent cells to progeny. See also *Informational molecules, Gene, Genetic code, Genome, Genetics, Genotype, Deoxyribonucleic acid (DNA), Heritability, Quantitative trait loci (QTL).*

Heritability

The fraction of variation (of an individual's given trait) that is due to genetics. For example, if a pig's trait (e.g., weight at birth) is 30% heritable, that means that 30% of the (birthweight) difference between that individual pig and its (statistically representative) group of contemporaries (pigs) is due to genetics. The other 70% would be due to factors such as nutrition of the mother during pregnancy, and so on. See also *Heredity, Trait, Genetics, Informational molecules, Gene, Genetic code, Genome, Genotype, Deoxyribonucleic acid (DNA), Quantitative trait loci (QTL).*

hESC

Acronym for human embryonic stem cells. See also *Human embryonic stem cells.*

Hetero-

A chemical nomenclature prefix meaning "different." For example, a *heterocyclic* compound is one with a (ring) structure made up of *more than one kind* of atom. A *heterokaryon* refers to a cell containing nuclei of *different* species. See also *Heterocyclic, Heteroduplex, Heterogeneous (catalysis), Heterogeneous (chemical reaction), Heterogeneous (mixture), Heterokaryon, Heterologous proteins, Heterologous DNA, Heterology, Heterosis, Heterotroph, Heterozygote.*

Heterochromatin

A form of DNA that is sometimes found in cells, in which the DNA is compressed (along with a protein) into a form in which the genes are not available for expression (i.e., those genes are silenced).

See also *Deoxyribonucleic acid (DNA), Cell, Protein, Gene, Gene silencing, Posttranslational modification of protein.*

Heterocyclic

See *Hetero-.*

Heteroduplex

A DNA molecule, the two strands of which come from different individuals so that there may be some base pairs or blocks of base pairs that do not match. Can arise from mutation, recombination, or by annealing DNA single strands *in vitro*. See also *Deoxyribonucleic acid (DNA).*

Heterogeneous (Catalysis)

Catalysis occurring at a phase boundary, usually a solid–fluid interface. See also *Hetero-, Heterogeneous (mixture), Catalyst.*

Heterogeneous (Chemical Reaction)

A chemical reaction in which the reactants are of different phases; for example, gas with liquid, liquid with solid, or a solid catalyst with liquid or gaseous reactants. See also *Hetero-, Heterogeneous (catalysis), Catalyst.*

Heterogeneous (Mixture)

One that consists of two or more phases such as liquid–vapor or liquid–vapor–solid. See also *Hetero-.*

Heterogeneous Glycosylation

Refers to the fact that (several) glycosyl/sugar molecular groups are attached to specific protein molecules in ways such that the attachment sites and the structures/arrangements of the sugar molecular groups are variable. See also *Glycosylation, Glycoform, Protein, Glycoprotein.*

Heterokaryon

A fused cell containing nuclei of different species. See also *Nucleoid.*

Heterologous (Chromosomes or Genes)

Chromosomes or chromosome segments that are not identical with respect to their constituent sequence, genetic loci, and/or their visible structure (in the case of chromosomes). See also *Chromosomes, Gene, Sequence (of a DNA molecule), Locus, Chromosome MAP, Caenorhabditus elegans, Model organism, Mutagenic chain reaction.*

Heterologous Proteins

Those proteins produced by an organism that is not the wild-type source of those proteins. For example, bacteria have been genetically engineered to produce human growth hormone and bovine (i.e., cow) somatotropin. See also *Protein, Wild type, Growth hormone (GH), Bovine somatotropin (BST), Homologous protein.*

Heterology

Refers to

- A sequence of amino acids in two or more proteins that are not identical to each other
- A sequence of DNA in two chromosomes/genes segments that are not identical with respect to their constituent sequence, genetic loci, and/or their visible structure (in the case of chromosomes)

See also *Amino acid, Protein, Deoxyribonucleic acid (DNA), Chromosomes, Gene, Sequence (of a DNA molecule), Locus, Chromosome Map, Homology.*

Heteroplasmy

Refers to the inheritance of DNA from both parents. Its opposite (i.e., homoplasmy—the inheritance of DNA from only one parent) is predominant for inheritance of mitochondrial DNA in eucaryotes. See also *Deoxyribonucleic acid (DNA), Homoplasmy, Mitochondrial DNA, Eucaryote.*

Heterosis

Also known as "hybrid vigor," it was first identified by George Shull in 1908. Heterosis results from the interbreeding of genetically distinct plants, which yields offspring that are more robust than either (inbred) parent plant. See also *F1 hybrids.*

Heterotroph

An organism that obtains nourishment from the ingestion and breakdown of organic matter. See also *Organism.*

Heterozygote

An individual organism with different alleles at one or more particular loci. See also *Allele.*

Hexadecyltrimethylammonium Bromide (CTAB)

A solvent that is widely utilized to dissolve plant DNA samples (e.g., when a scientist wants to sequence that sample of plant DNA). CTAB solvent helps the scientist to separate out contaminants that are commonly present in samples from plant tissues (i.e., polysaccharides, quinones) because DNA molecules are much more soluble in CTAB than are the contaminant molecules. See also *Deoxyribonucleic acid* (*DNA*), *Polysaccharides*, *Sequencing* (*of DNA molecules*), *SDS*.

Hexose

See *Glucose* (*GLc*).

HF Cleavage

A research process in which hydrofluoric acid is used to sequentially remove side-chain protective groups from peptide chains. It is also used to remove the resin support from peptides that have been prepared via solid-phase peptide synthesis. The HF cleavage reaction is a temperature-dependent process. See also *Prosthetic group*, *Synthesizing* (*of proteins*).

HFR-3 Lectin

See *Lectins*, *Chitin*.

HGPS

Acronym for *Hutchinson–Gilford progeria syndrome*. See *Hutchinson–gilford progeria syndrome*.

HGT

Acronym for *horizontal gene transfer*. See *Horizontal gene transfer*, *Introgression*.

Hh

Abbreviation for *hedgehog proteins* or *hedgehog signaling pathway*. See *Hedgehog proteins*, *Hedgehog signaling pathway*.

High-Amylose Corn

Refers to those corn (maize) hybrids that produce kernels in which the starch that is contained within those kernels is at least 50% amylose versus the average of 24%–28% amylose in traditional corn starch. See also *Corn*, *Starch*, *Amylose*.

High-Amylose Wheat

Refers to those what varieties that have been genetically engineered to produce kernels in which the starch that is contained within those kernels is at least 50% amylase versus the average of 24%–26% amylose in traditional wheat starch. Because amylose is slow to be digested (i.e., broken down to glucose) by humans, its consumption does not cause a "spike" (i.e., sudden increase) in bloodstream levels of glucose, which can be hazardous for some people (e.g., diabetics). See also *Wheat*, *Starch*, *Polymer*, *Glucose* (*GLc*), *Amylose*, *Diabetes*.

High-Content Screening

Refers to any analytical methodology/technology via which *multiple* parameters (e.g., secretion of *specific* proteins, the *amounts* of each protein secreted) of complex systems (e.g., living cells, living multicell organisms) are simultaneously analyzed. See also *Cell*, *High-throughput identification*, *High-throughput screening* (*HTS*), *Target–ligand interaction screening*, *Gene expression profiling*, *Multiplex assay*, *Confocal microscopy*, *Multiplexed* (*assay*).

High-Density Lipoproteins (HDLPs)

So-called good cholesterol, it consists of lipoproteins that can help move excess low-density lipoproteins (i.e., "bad" cholesterol, which can clog arteries) out of the human body by binding to the low-density lipoproteins (also known as LDL cholesterol) on artery walls or in the blood and then (when bound entity arrives in the liver) attaching to special LDLP receptor molecules in the liver. The liver then clears those (bound) low-density lipoproteins out of the body as a part of regular liver functions.

Studies have shown that humans having high bloodstream levels of HDLPs will offset high levels of LDLPs (e.g., the HDLPs can still help lower the risk of developing coronary heart disease).

Since cholesterol does not dissolve in water (which constitutes most of the volume of blood), the body makes HDL cholesterol into little "packages" surrounded by a hydrophilic (i.e., "water loving") protein. That protein "wrapper" is known as apolipoprotein A-1, or apo A-1, and it enables HDL cholesterol to be transported in the bloodstream because the apolipoprotein A-1 is attracted to water molecules in the blood. See also *Low-density lipoproteins* (*LDLP*), *Receptors*, *Apolipoproteins*, *Water soluble fiber*, *Cholesterol*, *Coronary heart disease* (*CHD*), *Nanoparticles*.

High-Galactomannan Soybeans

See *High-mannogalactan soybeans*.

High-Glutenin Wheat

See *Gluten*.

High-Isoflavone Soybeans

Developed in the United States in the 1990s, these are soybean varieties that contain greater content of isoflavones than do traditional soybean varieties (i.e., isoflavones constitute 0.15%–0.3% of a traditional variety soybean's dry weight).

Consumption of isoflavones helps to reduce the blood level of low-density lipoproteins (i.e., "bad cholesterol") in humans.

A human diet containing a large amount of isoflavones helps prevent osteoporosis, causes reduced risk of certain cancers (e.g., breast cancer, prostate cancer, endometrial cancer), and decreases risk of prostate enlargement. See also *Isoflavones*, *Soybean plant*, *Cholesterol*, *Cancer*, *Prostate-specific antigen* (*PSA*), *Low-density lipoproteins* (*LDLP*), *Osteoporosis*.

High-Lactoferrin Rice

Refers to rice plants (*Oryza sativa*) that have been genetically engineered to produce substantial amounts of lactoferrin in the grain they yield. Lactoferrin is a compound that is naturally produced in human breast milk. Consumption of lactoferrin by infants helps to strengthen their immune system.

Consumption of lactoferrin (e.g., from genetically engineered rice) by older humans helps their immune systems to resist some infectious diseases. Lactoferrin "binds" free iron (e.g., in body fluids), thereby *denying that iron* to pathogenic bacteria (which need free iron to grow/infect). Lactoferrin also promotes intestinal cell growth in humans. See also *Genetic engineering, Pathogen, Bacteria, Value-enhanced grains, Growth (microbial), Cell.*

High-Laurate Canola

Refers to canola (*Brassica napus/campestris*) varieties that have been genetically engineered (e.g., via insertion of gene for *lauroyl-ACP thioesterase*) to produce at least 40% laurate (lauric acid) in their oil (in seed). See also *Laurate, Canola, Genetic engineering, Fatty acid, Lauroyl-ACP thioesterase, Value-enhanced grains.*

High-Linolenic Oil Soybeans

Soybeans from soybean plants that have been genetically engineered to produce soybeans bearing oil that contains more than 40% linolenic acid, instead of the typical 8% linolenic acid content of soybean oil produced from traditional varieties of soybeans. Scientists accomplish that via up-regulation (i.e., increased expression) of the *Fad3* gene within the soybean plant's oil-synthesis pathway. See also *Soybean plant, Soybean oil, Fatty acid, Linolenic acid, Polyunsaturated fatty acids (PUFA), Gene, FAD3 gene, Genetic engineering, Express, UP-regulation, Pathway.*

Highly Available Phosphate Corn (Maize)

See *Low-phytate corn.*

Highly Available Phosphorous (HAP) Gene

See *Low-phytate corn.*

Highly Unsaturated Fatty Acids (HUFA)

Refers to a number of unsaturated fatty acids (e.g., that the human body forms from polyunsaturated fatty acids it consumes in diet) containing four or more *double* (molecular) *bonds*, such as arachidonic acid, docosahexaenoic acid, and eicosapentaenoic acid.

These HUFAs are utilized (by the human body) to make prostaglandins and other eicosanoids. See also *Polyunsaturated fatty acids (PUFA), Unsaturated fatty acids, Essential fatty acids, Coronary heart disease (CHD), N-3 fatty acids, N-6 fatty acids, Docosahexaenoic acid (DHA), Eicosapentaenoic acid (EPA), Arachidonic acid (AA), Prostaglandin endoperoxide synthase.*

High-Lysine Corn

Developed in the United States in the mid-1960s, these were initially corn (maize) varieties possessing the "opague-2" gene. The opague-2 gene causes such corn to contain 0.30%–0.55% lysine (i.e., 50%–80% more than traditional No. 2 yellow corn).

Other genes have subsequently been discovered that, when inserted into corn/maize genome (e.g., via genetic engineering techniques), cause production of larger amounts of lysine than in traditional corn/maize varieties.

High-lysine corn is particularly useful for feeding of swine, since traditional No. 2 yellow corn does not contain enough lysine for optimal swine growth. See also *Corn, Lysine (lys), Gene, Opague-2, Genetic engineering, Genome, Value-enhanced grains, "Ideal protein" Concept, MAL (multiple aleurone layer) gene.*

High-Maysin Corn

Developed in the United States during the 1980s and 1990s, these are corn (maize) varieties possessing at least 10 times the typical amount of maysin found in traditional corn (maize) varieties.

Maysin is a chemical compound that "binds up" essential amino acids within the gut of certain pest insects, so those insects starve while eating it (but humans, animals, and nonpest insects are unharmed). Maysin primarily acts against the corn earworm (*Helicoverpa zea*). See also *Corn, Maysin, Amino acid, Essential amino acids, Helicoverpa zea (H. zea).*

High-Methionine Corn

Developed in the United States in the mid-1960s, these were initially corn (maize) varieties possessing the "floury-2" gene. The floury-2 gene causes such corn to contain slightly higher levels of methionine than traditional No. 2 yellow corn.

Other genes have subsequently been discovered that, when inserted into corn/maize genome (e.g., via genetic engineering techniques), cause production of larger amounts of methionine than in traditional corn/maize varieties.

High-methionine corn is particularly useful for feeding of poultry, since traditional No. 2 yellow corn does not contain enough methionine for optimal poultry (esp. feather) growth. See also *Methionine (met), Corn, Floury-2, Gene, Genome, Genetic engineering, Value-enhanced grains, Opague-2, "Ideal protein" concept, MAL (multiple aleurone layer) gene.*

High-Mannogalactan Soybeans

Developed in the United States following the 2003 discovery by Kanwarpal S. Dhugga and coworkers that the soybean plant (*Glycine max* L.) would produce guar-gum mannogalactan (i.e., a copolymer of galactose and mannose) when the "CtManS" gene from guar plant (*Cyamopsis tetragonoloba*) is inserted via genetic engineering, these will be soybean varieties that contain significant amounts of the food thickening agent currently known as guar gum. See also *Soybean plant, Gene, Genetic engineering, Galactose (gal), Water soluble fiber.*

High-Oil Corn

Conceived in 1896 at the University of Illinois in the United States, high-oil corn (HOC) is defined to be corn (maize) possessing a kernel oil content of 5.8% or greater. Traditional No. 2 yellow corn varieties tend to contain 4.5% or less oil content. See also *Value-enhanced grains, Corn, Chemometrics.*

High–Oleic Oil Corn

Conceived in 2002 at Iowa State University in the United States, high–oleic oil corn is defined to be corn (maize) whose kernels possess oil containing more than 40% oleic acid, instead of the 20%–30% oleic acid present in kernel oil from traditional varieties of corn.

High–oleic oil corn varieties were created via the incorporation of certain genes from *eastern gamagrass* (*Tripsacum dactyloides*), which causes the higher-than-traditional amount of oleic acid in the corn oil. See also *Value-enhanced grains, Corn, Fatty acid, Oleic acid, Monounsaturated fats.*

High–Oleic Oil Safflower

Developed via mutation breeding, high–oleic oil safflower is defined to be safflower (*Carthamus tinctorius* L.) whose seeds possess oil containing more than 75% oleic acid, instead of the 12.2% oleic acid present in seed oil from traditional varieties of safflower. See also *Value-enhanced grains, Mutation breeding, Fatty acid, Oleic acid, Monounsaturated fats.*

High–Oleic Oil Soybeans

Soybeans from soybean plants that have been genetically engineered to produce soybeans bearing oil that contains more than 70% oleic acid, instead of the typical 24% oleic acid content of soybean oil produced from traditional varieties of soybeans. Cosuppression, via inserted gene for Δ 12 desaturase (i.e., enzyme that normally converts oleic acid to linoleic acid as part of the oil creation process in traditional varieties of soybean plants), causes the *higher-than-traditional* amount of oleic acid in the soybean oil.

High–oleic oil soybean oil would tend to have greater oxidative stability (especially at elevated temperatures) than soybean oil from traditional varieties of soybeans. Because of that, nuts that are fried in *high oleic oil* have been shown to possess a longer shelf life than nuts fried in traditional vegetable oils.

A human diet containing a large amount of oleic acid causes lower blood cholesterol level and thus lower risk of coronary heart disease. See also *Soybean plant, Soybean oil, Fatty acid, Oleic acid, Monounsaturated fats, Genetic engineering, Delta 12 desaturase, Cholesterol, Coronary heart disease* (*CHD*), *Palmitic acid, Cosuppression, Enzyme, Linoleic acid, FAD genes, FAD3 gene.*

High–Oleic Oil Sunflowers

Refers to sunflower (*Helianthus annuus* L.) plant varieties that have been bred so their seeds contain 80%–90% oleic acid within the oil in those seeds versus historical average of 20% oleic acid in the oil of traditional sunflower (crop) plant varieties.

To create the high–oleic oil sunflower, a mutation known as "Pervenet" was obtained via chemical mutagenesis, which acts via Δ-12 desaturase and Δ-9 desaturase enzymes. See also *Fatty acid, Oleic acid, MID-oleic sunflowers, Mutation breeding, Traditional breeding methods, Enzyme, Desaturase, Delta 12 desaturase, High–oleic oil soybeans.*

High-Phytase Corn and Soybeans

Crop plants that have been genetically engineered to contain in their grain/seed high(er) levels of the enzyme phytase (which aids digestion and absorption of phosphate in that grain/seed). High-phytase grains or oilseeds are particularly useful for the feeding of swine and poultry, since traditional No. 2 yellow corn (maize) or traditional soybean varieties do not contain phytase in amounts needed for complete digestion/absorption of phosphate naturally contained in those traditional soybeans and corn (maize) in the form of phytate.

Although some forms of the phytase enzyme are heat sensitive, the phytase produced in plants via a gene inserted from the fungus *Aspergillus fumigatus* has been shown to be stable up to a temperature of 89°C (192°F). See also *Phytase, Enzyme, Phytate, Value-enhanced grains, Low-phytate corn, Low-phytate soybeans.*

High-Protein Rice

Developed during 2002, these are varieties of rice (*Oryza sativa*) whose grain contains at least 12% protein, in contrast to traditional varieties of rice, which average 8% protein content. See also *Protein.*

High-Stearate Canola

Canola varieties that have been genetically engineered so their seed oil contains at least 15% stearate (also called stearic acid).

Cosuppression, via inserted gene for Δ-stearoyl-ACP desaturase (i.e., enzyme that normally converts stearic acid to oleic acid in the oil creation process in traditional varieties of canola), causes the *higher-than-traditional* amount of stearic acid in the canola oil. See also *Canola, Stearate, Saturated fatty acids* (*SAFA*), *Gene, Genetic engineering, Value-enhanced grains, Fatty acid, Cosuppression, Enzyme, Oleic acid, Stearoyl-ACP desaturase, Cholesterol, Coronary heart disease* (*CHD*).

High-Stearate Soybeans

Soybean plant varieties that have been bred or genetically engineered so their beans contain at least 12% stearate (also known as stearic acid) within their soybean oil (i.e., more than four times the typical 3% stearic acid content in the soybean oil produced from traditional soybean varieties). Some high-stearate soybeans contain more than 20% stearate.

Cosuppression—for example, via inserted gene for Δ-stearoyl-ACP desaturase (i.e., enzyme that normally converts stearic acid to oleic acid in the oil creation process in traditional varieties of soybeans)—is primary way to cause the *higher-than-traditional* amount of stearic acid in the resultant soybean oil.

A human diet containing stearate instead of alternative saturated fatty acids does not cause an increase in blood cholesterol levels (whereas human consumption of the *other* saturated fatty acids causes bloodstream cholesterol levels to increase, which increases risk of CHD). See also *Stearate, Value-enhanced grains, Soybean plant, Soybean oil, Gene, Genetic engineering, Fatty acid, Cosuppression, Enzyme, Oleic acid, Cholesterol, Saturated fatty acids* (*SAFA*), *Coronary heart disease* (*CHD*), *Stearoyl-ACP desaturase.*

High-Sucrose Soybeans

Another name for low-stachyose soybeans because the soybeans replace the (reduced) stachyose with (additional) sucrose. See also *Low-stachyose soybeans, Stachyose, Value-enhanced grains, Soybean plant, Sugar molecules.*

High-Throughput Identification

Determination of the identification of a given *chemical compound* (e.g., within a mixture), the desired *impact* (e.g., cell apoptosis), a specific *segment* (sequence) *of DNA* (i.e., a specific gene), a specific *ligand or receptor* (e.g., "attaching" itself to a given molecule), and so on within the overall process known as high-throughput screening. See also *High-throughput screening* (*HTS*), *Combinatorial chemistry, Biochips, Cell, Apoptosis, Gene, Deoxyribonucleic acid* (*DNA*), *Gene expression, Target–ligand interaction screening,*

Receptors, Characterization assay, Sequence (of a DNA molecule), Gene expression analysis, Caenorhabditis elegans (C. elegans), Molecular beacon, Nanosheets.

High-Throughput Screening (HTS)

A methodology utilized to quickly screen large numbers of compounds for use as pharmaceuticals or agrochemicals (e.g., herbicides).

For example, when screening chemical compounds for potential use as a pharmaceutical, the goal often is to assess differences between diseased and (treated) cells, enabling identification of a pharmaceutical candidate that favorably impacts change in protein level (i.e., gene expression) that characterizes a diseased state, or some other gene expression marker (e.g., apoptosis).

When screening compounds for potential use as herbicide active ingredients, the goal is to assess differences between normal and (treated) weed plant cells; enabling identification of a potential herbicide candidate that imparts desired (fatal) change.

Although whole living cells or whole microscopic animals such as nematodes could be utilized in HTS, it is more common to use a proxy (e.g., receptors, enzymes, or STATs from applicable cells) whose interaction with candidate compounds can be inferred to cell (and/or organism) effects. See also *Combinatorial chemistry, Biochip, Target–ligand interaction screening, Cell, Organism, Characterization assay, Protein, Gene, Gene expression, Cell array, High-throughput identification, Receptors, Gene expression analysis, Bioassay, Gene expression markers, Signal transducers and activators of transcription (STATs), Apoptosis, In silico screening, Nematodes, Caenorhabditis elegans (C. elegans), Enzyme, Northern blot analysis, Fluorescence, Molecular beacon, Fluorescence polarization (FP), Live cell array, Microarray (testing), Toxicogenomics, Label (radioactive), Whole-cell patch-clamp recording, PTEN activity, Pharmacophore searching, Nanosheets.*

Hirudin

A compound, naturally produced by leeches (e.g., *Hirudo medicinalis*), which—in humans—prolongs the clotting time of blood (i.e., hirudin acts as an anticoagulant). The U.S. Food and Drug Administration approved the use of hirudin as an anticoagulant pharmaceutical in 1998.

Used in vascular surgery and in postoperative treatment of thrombosis. See also *Thrombosis, Food and Drug Administration (FDA).*

Histamine

A base that is naturally present in ergot (a fungus) and plants; it is also naturally produced by basophils (basophilic leukocytes) in the human body. It is formed from histidine by decarboxylation and is held to be responsible for the dilation and increased permeability of blood vessels that play a major role in allergic reactions. See also *Base, Histidine (HIS), Basophils.*

Histidine (his)

A basic amino acid that is essential in the nutrition of the rat. It is formed by the decomposition of most proteins (as globin). See also *Protein.*

Histiocyte

See *Macrophage.*

Histoblasts

See *B lymphocytes.*

Histocompatibility Leukocyte Antigens (HLA)

See *Histocompatibility leukocyte antigens (HLA).*

Histone Deacetylase Inhibitors

Refers to chemicals (e.g., certain pharmaceuticals) that inhibit the activity of histone deacetylases (HDACs). HDAC inhibitors can reactivate certain (inappropriately silenced via epigenetics) genes that normally suppress tumor growth. See also *Histone deacetylases, Gene, Gene silencing, Epigenetics, Tumor.*

Histone Deacetylases

Histone deacetylases (HDACs) are enzymes that can catalyze the removal of acetyl molecular groups from histones. Some food compounds such as sulforaphane are inhibitors of HDAC. See *Histones, Acetylation, Chromatin, Posttranslational modification of protein, Sulforaphane.*

Histone Modification

See *Chromatin remodeling, Posttranslational modification of protein.*

Histones

Proteins rich in basic amino acids (e.g., lysine) that are found complexed with DNA in the chromosomes of all eucaryotic cells except sperm.

Histones play a significant role in the regulation of gene expression. Examples include the following:

- Acetylation (i.e., addition of acetyl molecular group) of histones results in some genes within the DNA looped around that histone to become (more) accessible to the cell's transcriptional "machinery," thereby turning on those genes.
- Methylation (i.e., addition of methyl molecular group) of a protruding amino acid (e.g., lysine) in a histone, thereby "turning on"/upregulating those genes.
- Sumoylation (i.e., addition of SUMO protein group) of histones results in repression/"turning off" those genes.

See also *Chromosomes, Chromatids, Chromatin, Cell, Protein, Deoxyribonucleic acid (DNA), Gene, Gene expression, Transcription, Expressivity, Methylated, Amino acid, Lysine (lys), UP regulating, Chromatin remodeling, Small ubiquitin-related modifier, Sumoylation, Repression (of gene transcription or translation), Differentiation pathways, Epigenetic, Histone deacetylases, Epigenetic, Tetrasomes.*

Histopathologic

Refers to changes in tissue caused by a disease. For example, certain diseases (e.g., jaundice) cause the skin to turn yellow. See also *Pathogenic, Virus, Cancer, Adhesion molecule.*

H

HIV-1 and HIV-2

See *Human immunodeficiency virus type 1 (HIV-1)*, *Human immunodeficiency virus type 2 (HIV-2)*.

HLA

See *Human leukocyte antigens (HLA)*, *Human leukocyte antigen gene*.

HMG-CoA Reductase

See *Cholesterol* and *Statins*.

HMO

Acronym for *human milk oligosaccharides*. See also *Oligosaccharides*.

HMT

Acronym for *histone methyltransferase enzymes*. See *Enzyme*, *Histones*.

HNE

The common chemical (by) product of lipid oxidation, known as 4-hydroxy-2-nonenal, which is an aldehyde. See *Oxidative stress*, *Oxidation*, *Plasma membrane*, *Lipids*.

HNGF

Human nerve growth factor. See *Nerve growth factor (NGF)*.

HOC

See *High-oil corn*.

Holins

Small proteins that are produced by bacteriophages during the infection (of bacteria) by bacteriophages. Holins "punch" holes into the bacterial cell membranes, thereby allowing the cell contents to leak out, and the bacteria thus die. See also *Bacteria*, *Bacteriophage*, *Protein*, *Lytic infection*.

Hollow Fiber Separation (of Proteins)

The separation of proteins from a mixture by means of "straining" the mixture through hollow, semipermeable fibers (e.g., polysulfone fibers) under pressure. The hollow fibers are constructed in such a way that they have very tiny (molecular size) holes in them. In this way large molecules are retained in the original liquid while smaller molecules, which are able to pass through the holes, are filtered out. See also *Dialysis*, *Protein*, *Ultrafiltration*.

Holoenzyme

The entire, functionally complete enzyme. The term is used to designate an enzyme that requires a coenzyme in order for it to function (possess catalytic abilities). The holoenzyme consists of the protein part (apoenzyme) plus a dialyzable, nonprotein coenzyme part that is bound to the apoenzyme protein. See also *Coenzyme*, *Apoenzyme*, *Dialysis*.

Homeobox

A short sequence of DNA that is 180 base pairs long and located in the 3′ exon of certain genes of the *Drosophila* fly (where they were discovered by Walter Gehring during the 1970s). In the 1980s, Jani Christian Nusslein-Volhard discovered that one homeobox was attached (in adjacent exon) to each of the genes that are responsible for embryonic development (i.e., "switched on" only in an embryo that is developing into an adult), in a wide variety of species including invertebrates, birds, and mammals. Thus, it is now possible to locate many embryonic development genes in many species by using a DNA probe (made via a *Drosophila* homeobox DNA sequence) to find homeobox sequences attached to those embryonic development genes. In such a role, the respective homeobox sequences attached to each gene are known as DNA markers. See also *Gene*, *Deoxyribonucleic acid (DNA)*, *DNA probe*, *DNA marker*, *Sequence (of a DNA molecule)*, *Base pair (bp)*, *Drosophila*, *Exon*, *Species*.

Homeostasis

A tendency toward maintenance of a relatively stable internal environment in the bodies of higher animals through a series of interacting physiological processes. An example is the mammal's maintenance of a constant body temperature despite extremes in weather temperature. See also *Selectins*, *Lectins*, *Adhesion molecule*, *Cortisol*, *AMPK*.

Homing Receptor

Also known as L-selectin. See *Selectins*, *Lectins*, *Adhesion molecules*.

Homocysteine

A metabolite compound (i.e., amino acid derived via metabolism from methionine) that, when present in the bloodstream in elevated amount, increases the risk of stroke and increases the likelihood for a person to develop arteriosclerosis and/or coronary heart disease (CHD).

Folic acid, vitamin B_{12}, and vitamin B_6 act as cofactors in the conversion of homocysteine back to methionine or cysteine.

Consumption of choline or folic acid has been shown to reduce bloodstream levels of homocysteine. Research also has shown that moderate consumption of beer and B vitamins will reduce bloodstream levels of homocysteine. See also *Metabolism*, *Arteriosclerosis*, *Coronary heart disease (CHD)*, *Methionine*, *Amino acid*, *Cofactor*, *Cysteine (cys)*, *Choline*, *Vitamin*.

Homologous (Chemically)

See *Homology*.

Homologous (Chromosomes or Genes)

Chromosomes or chromosome segments that are identical with respect to their constituent sequence, genetic loci, and/or their visible structure (in the case of chromosomes).

So, for example, a gene of "unknown" function in humans could be compared (in a database) with genes of a simpler organization (e.g., *Caenorhabditus elegans*). If the human gene is homologous, and the function of the *Caenorhabditus elegans* gene is known, the function of the human gene could be inferred by comparison. See also *Chromosomes, Gene, Sequence (of a DNA molecule), Locus, Caenorhabditus elegans, Model organism.*

Homologous Protein

A protein having identical functions and similar properties in different species. For example, the hemoglobins that perform identical functions in the blood of different species. See also *Protein, Species.*

Homologous Recombination

Refers to the fact that insertion (into living cells/organism) of the DNA sequence of a given gene can (under certain conditions) "knock out"/silence that particular gene in that cell. See also *Gene, Cell, Organism, Deoxyribonucleic acid (DNA), Knockout, Gene silencing, Cosuppression.*

Homology

A sequence of amino acids in two or more proteins that are identical to each other. *Nucleic acid homology* refers to complementary strands that can hybridize with each other. See also *Tata homology, Protein, Hybridization (molecular genetics).*

Homology Modeling

Refers to the use (e.g., in computerized molecule models) of *known* proteins' structural and functional properties as a "predictive template" for computer-generated *hypothetical proteins* (whose structure is not known).

Such predictive structural modeling of hypothetical proteins becomes more accurate as more and more of the known structures (i.e., *parts* comprising the large protein molecule) are added to the computer model. See also *Protein, Conformation, Protein folding, Protein structure, Protein engineering, Absolute configuration.*

Homoplasmy

Refers to the inheritance of mitochondrial DNA from only one parent (usually the mother). Homoplasmy occurs in >99% of eucaryotes. See also *Mitochondrial DNA, Eucaryote.*

Homotropic Enzyme

An allosteric enzyme whose own substrate functions as an activity modulator. See also *Enzyme.*

Homozygote

An organism in which the corresponding genes (alleles) on the two genomes are identical. An organism that possesses an identical pair of alleles in regard to a given (genetic) characteristic. See also *Gene, Allele, Genome, Genotype, Phenotype, Homozygous, Heterozygote.*

Homozygous

In a diploid organism, a state where both alleles of a given gene are the same. See also *Heterozygote, Allele, Diploid, Diplophase, Homozygote.*

Horizontal Gene Transfer

Refers to the exchange of genes between species that are unable to mate with each other. Horizontal gene transfer is common among many bacteria, but it can also occur between some plant species such as between the root-parasite weed known as purple witchweed (*Striga hermonthica*) and its host plants such as sorghum (*Sorghum bicolor*) and rice (*Oryza sativa*). See also *Gene, Species, Rice, Bacteria.*

Hormesis

Refers to a typically mild, health-promoting activation of a natural stress response. For example, because consumption of genistein by humans causes a reduction in the production of stress proteins, genistein thereby helps the human immune system to destroy cancerous cells. A human diet containing a large amount of genistein has been shown to increase bone density and to decrease total serum (blood) cholesterol, thereby lowering risk of osteoporosis and coronary heart disease. Research indicates that human consumption of genistein can help to prevent breast cancer, help prevent prostate cancer/metastasis, prevent adverse increases in blood platelet aggregation, and inhibit the proliferation of smooth-muscle cells in plaque deposits (inside blood vessels). See also *Stress response, Stress proteins, Stress hormones, Genistein (Gen), Cell, Cancer, Metastasis, Osteoporosis, Coronary heart disease (CHD), Cholesterol.*

Hormone

Coined in 1905, the term hormone refers to a type of chemical messenger (peptide), occurring both in plants and animals, that acts to inhibit or excite metabolic activities (in that plant or animal) by binding to receptors on specific cells to deliver its "message." A hormone's site of production is distant from the site of biological activity (i.e., where the message is delivered). See also *Peptide, Minimized proteins, Signaling, Signaling molecule, Nuclear hormone receptors, Albumin, Stress hormones, Indole-3-acetic acid, Receptor-binding mapping.*

Hormone Response Elements

See *Nuclear receptors.*

Hormone-Sensitive Lipase (HSL)

See *Lipase.*

Host Cell

A cell whose metabolism is used for growth and reproduction by a virus. Also the cell into which a plasmid is introduced (in recombinant DNA experiments). See also *Cell, Plasmid.*

Host Vector (HV) System

The host is the organism into which a gene from another organism is transplanted. The guest gene is carried by a vector (i.e., a larger

H

DNA molecule, such as a plasmid, or a virus into which that gene is inserted) that then propagates in the host. See also *Organism*, *Gene*, *Vector*.

Hot Spots

Sites in genes at which events, such as mutations, occur with unusually high frequency. See also *Gene*, *Jumping genes*, *Mutation*, *Translocation*.

Housekeeping Gene

Refers to internal control genes utilized in RT-PCR to normalize the mRNA fraction. See also *Gene*, *RT-PCR*.

HPLC

Initially known as *high-performance liquid chromatography* when developed during the 1970s, this separation/analysis technology was later renamed *high-pressure liquid chromatography*. See *Chromatography*.

HPV

Acronym for human papilloma virus. See *Tumor-suppressor proteins*.

HR

Acronym for *hypersensitive response* or *hypersensitive defense response* in some plants. See *Hypersensitive response*.

HSE

Acronym for *human skin equivalent*, a three-dimensional "model" of human skin tissue (created via tissue engineering).

One use of HSEs is to determine if a particular disease is an autoimmune disease. For example, by extracting from a patient some apparent autoantibodies (i.e., antibodies produced via an immune response to the body's own tissue) and applying them within an applicable HSE, Lynn Solomon in 2011 was able to show that chronic ulcerative stomatitis is an autoimmune disease because those antibodies "attacked" the adhesion molecules that hold the skin surface layer (epithelium) to subsurface cells in skin tissue. See also *Tissue engineering*, *Epithelium*, *Adhesion molecule*, *Autoimmune disease*, *Antibody*.

HSOD

See *Human superoxide dismutase* (*hSOD*).

HSP

Acronym for *heat shock protein*. See *Heat shock proteins*.

HSP90

Acronym for *heat shock protein 90*. It is a heat shock protein (also known as stress proteins) that helps individual protein molecules within a cell to properly fold into their tertiary structure (i.e., their three-dimensional structure in which they are biologically active). See also *Heat shock proteins*, *Protein*, *Cell*, *Tertiary structure*, *Biological activity*.

HTC

See *Herbicide-tolerant crop*, *STS*, *Pat gene*, *EPSP synthase*, *ALS gene*, *BAR gene*, *CP4 EPSPS*, *Glyphosate oxidase*.

HTMS

Acronym for *high-throughput mass spectrometry*. See *High-throughput screening* (*HTS*), *Mass spectrometer*, *MALDI-TOF-MS*.

HTS

Herbicide-tolerant soybeans. See *Soybean plant*, *Glyphosate*, *CP4 EPSPS*, *EPSP synthase*, *Glyphosate oxidase*, *Herbicide-tolerant crop*, *STS*, *Glufosinate*, *PAT gene*, *Bar gene*.

HTS

See *High-throughput screening* (*HTS*).

Human Artificial Chromosomes (HAC)

Chromosomes that have been synthesized (made) from chemicals that are identical to chromosomes within human cells. See also *Yeast artificial chromosomes* (*YAC*), *Bacterial artificial chromosomes* (*BAC*), *Chromosomes*, *Arabidopsis thaliana*, *Synthesizing* (*of DNA molecules*).

Human Chorionic Gonadotropin

A human hormone. In 1986, Mark Bogart discovered that elevated levels of human chorionic gonadotropin in pregnant women are correlated with babies (later) born with Down Syndrome. See also *Hormone*.

Human Colon Fibroblast Tissue Plasminogen Activator

A second generation tissue plasminogen activator (tPA), which has the clot-sensitive activation of plasminogen with potentially great selectivity and (clot) specificity. See also *Tissue plasminogen activator* (*tPA*).

Human EGF-Receptor-Related Receptor (HER-2)

A gene that appears to be directly related to human breast cancer mortality. The more copies of the HER-2 gene (in a patient's breast tumor cells) the more dismal that patient's prospects for survival. See also *Gene*, *Cancer*.

Human Embryonic Stem Cells

Those cells (in the early embryo's inner cell mass) from which each of the human body's 210 different types of tissues arise via differentiation, proliferation, and growth processes. See also *Stem cells*, *Pluripotent*, *Stem cell growth factor* (*SCF*), *Differentiation*, *Adult stem cell*.

Human Gamma-Glutamyl Transpeptidase

A glycoprotein that is thought to possess a different oligosaccharide when it is produced by a (liver) tumor cell instead of a healthy cell. Thus, it is a possible early warning marker for liver cancer. See also *Glycoprotein*, *Oligosaccharides*.

Human Growth Hormone (HGH)

See *Growth hormone* (GH).

Human Immunodeficiency Virus Type 1 (HIV-1)

One of the two "families" of the viruses identified (so far) that cause acquired immune deficiency syndrome (AIDS), although not all strains of HIV-2 cause AIDS. HIV-1 and HIV-2 show a preferential tropism (affinity) toward the helper T cells, although other immune system (and central nervous system) cells are also infected. Within 2 years of the initial HIV infection, a genetically distinct version of HIV is established and replicating within the brains of as many as one in four patients, which can result in dementia.

The GP120 envelope (surface) protein of HIV-1 and HIV-2 directly interacts (binds) with the CD4 proteins (receptors) on the surface of helper T cells, enabling the viruses to bind (attach to) and infect the helper T cells. In order to successfully enter and infect cells, the HIV must also bind with CKR-5 proteins (receptors) located on the surface of cells of most humans. In 1996, Nathaniel Landau and Richard Koup discovered that approximately 1% of humans carry a gene for a version of CKR-5 receptor that resists entry to cells by HIV. As of 1996, a total of nine separate strains (serotypes) of Human Immunodeficiency Virus were known; identified by the letters A, B, C, D, E, F, G, H, I. See also *CD4 protein*, *TAT*, *TATA homology*, *Adhesion molecule*, *GP120 protein*, *Acquired immune deficiency syndrome* (*AIDS*), *Receptors*, *Tropism*, *Helper T cells* (*T4 cells*), *Strain*, *T cell receptors*, *Virus*, *Serotypes*, *Human immunodeficiency virus type 2* (*HIV-2*).

Human Immunodeficiency Virus Type 2 (HIV-2)

See *Human immunodeficiency virus type* 1 (HIV-1).

Human Leukocyte Antigen Gene

Refers to a gene that codes for a human leukocyte antigen. See also *Human leukocyte antigens* (*HLA*).

Human Leukocyte Antigens (HLA)

A very complex array of six proteins that cover the surface of leukocytes (and the bone marrow cells that produce leukocytes). These HLA are usually different (i.e., a nonmatch) for individuals that are not genetically related to each other (e.g., a father–son or a father–daughter), so they have been used in the past to prove paternity.

HLA must also be matched (as nearly as possible) for successful bone marrow and organ transplants to prevent the donated organ or bone marrow (and the recipients') from "rejecting" each other. Such "rejection" of transplanted organs usually occurs because the body manufactures antibodies against nonmatched HLA, thereby leading that HLA to work in concert with a protein molecule named integrin beta 4 to jointly stimulate cell growth and movement (e.g., overproduction of endothelial cells on the interior of applicable blood vessels, thus blocking off blood supply to the transplanted organ). See also *Leukocytes*, *Antigen*, *Major histocompatibility complex* (*MHC*), *Protein*, *Graft-versus-host disease* (*GVHD*), *Antibody*, *Integrins*, *Endothelial cells*.

Human Papilloma Virus

See *Tumor-suppressor proteins*.

Human Protein Kinase C

An enzyme that is involved in the control of blood coagulation and fibrinolysis. See also *Fibrin*.

Human Superoxide Dismutase (hSOD)

An enzyme that "captures" *oxygen free radicals* (*oxygen atoms bearing an extra electron, thus high in energy*—e.g., which are sometimes generated in a biological system such as within the body of an organism). Oxygen free radicals are generated within occluded blood vessels when a blood clot blocks arteries in the heart, causing a heart attack. These oxygen free radicals are highly energized and can cause damage to blood vessel walls after the clot is dissolved (e.g., with tissue plasminogen activator), so hSOD may profitably be administered in conjunction with clot-dissolving pharmaceuticals to minimize damage when occluded arteries are reopened.

Research indicates that hSOD may help protect elderly patients from the lethal effects of influenza (i.e., the flu), because influenza often causes overproduction of free radicals in the victim's body.

Research indicates that administration of hSOD can help to relieve some pain and inflammation caused by certain clinical procedures (e.g., dental surgery), because overproduction of free radicals can result from those particular procedures. See also *Free radical*, *PEG-SOD* (*Polyethylene glycol superoxide dismutase*), *Catalase*, *Xanthine oxidase*, *Tissue plasminogen activator* (*tPA*), *Antioxidants*, *Sitosterol*.

Human Thyroid-Stimulating Hormone (hTSH)

A naturally occurring hormone that causes the thyroid gland to develop. See also *Hormone*, *Thyroid gland*.

Humanized Antibody

Refers to a (genetically engineered) antibody in which the complementarity-determining (i.e., antigen-binding) portion of an (animal-source) antibody is imparted to a human antibody molecule via splicing the (sequence of) DNA responsible for that animal antibody's *complementarity* (*to a specific antigen*) into a cell line producing human monoclonal antibodies.

The pharmaceutical panitumumab (Vectibix™) is a ("humanized") monoclonal antibody used to treat certain metastatic colorectal cancers and head and neck cancer. Its complementarity-determining portion binds to EGF receptor, a receptor found in abundance on the surface of those tumors' cells. That binding to EGF receptors induces tumor cell death via apoptosis or humoral immune response.

The pharmaceutical trastuzumab is a ("humanized") monoclonal antibody against the HER-2 gene that was approved by the U.S. Food and Drug Administration (FDA) during 2002 for use as a pharmaceutical in conjunction with chemotherapy against metastatic breast cancer.

H

The pharmaceutical obinutuzumab (Gazyva®) is a ("humanized") monoclonal antibody-against the CD20 B-cell-specific protein (on the surface membrane of malignant B-cells) that was approved by the U.S. FDA during 2013 for use as a pharmaceutical in conjunction with chlorambucil chemotherapy in people with previously untreated chronic lymphocytic leukemia.

See also *Antibody*, *Chimeric antibody*, *Antigen*, *Avidity*, *Chimeric proteins*, *Sequence (of a DNA molecule)*, *Monoclonal antibodies (MAb)*, *Genetic engineering*, *Trastuzumab*, *Fluorescence in situ hybridization (FISH)*, *Antiepidermal growth factor receptor*, *Chemotherapy*, *CD20 protein*, *Obinutuzumab*, *Food and Drug Administration (FDA)*, *PCSK9 inhibitors*.

Humanized Monoclonal Antibody

See *Humanized antibody*.

Humic Acids

A class of oligomeric molecules created from lignins and tannins via enzymes within the guts of earthworms as the earthworms digest lignins and tannins present in dead plant material (e.g., on the surface of a farm field). Humic acids act to beneficially buffer soil pH, and they function as plant growth promoters. See also *Oligomer*, *Lignins*, *Tannins*, *Enzyme*, *Earthworms*.

Humoral Immune Response

Refers to the rapid manufacture and secretion by the body of the soluble blood serum components—examples include the following:

- Antibodies (by B cells)
- Complement proteins
- Lymphokines
- Cecrophins

in response to an infection. See also *Antibody*, *Complement*, *Complement cascade*, *Cecrophins*, *Humoral immunity*, *Lymphokines*, *Gamma interferon*.

Humoral Immunity

The immune system response consisting of the soluble blood serum components that fight an infection (e.g., antibodies, complement proteins, cecrophins). See also *Antibody*, *Complement*, *Complement cascade*, *Cecrophins*, *Cellular immune response*, *Immunoglobulin*.

Huntington's Disease

A neurodegenerative disease that is dominantly inherited (i.e., disease results even if there is only one copy of the damaged gene in the genome). The gene for predisposition to Huntington's disease was discovered by Nancy Wexler. Huntington's disease is a progressive neurological condition that results in involuntary bodily movements, emotional disturbance, and eventual cognitive impairment. See also *Gene*, *Dominant allele*, *HD gene*, *Genome*, *Genetics*, *Mutation*.

HuSNPs

Abbreviation for human single-nucleotide polymorphisms (SNPs). See *Single-nucleotide polymorphisms (SNPs)*.

Hutchinson–Gilford Progeria Syndrome

A disease characterized by a mutation in the gene that codes for production of lamin A protein (which thereby results in a buildup of *prelamin A* at cells' nuclear envelopes) whose symptoms include accelerated osteoporosis, slow growth, loss of muscle strength, cardiovascular disease in children having the disease, and so on.

Farnesyl transferase inhibitors administered as a pharmaceutical have been shown in very preliminary research to alleviate the buildup of prelamin A at the cells' nuclear envelopes and to alleviate some of the symptoms of the disease. See also *Gene*, *Protein*, *Mutation*, *Coding sequence*, *Cell*, *Nuclear envelope*, *Osteoporosis*, *Farnesyl transferase inhibitors*, *Farnesyl transferase*.

Hybrid Vigor

See *Heterosis*, *F1 hybrids*, *Hybridization (plant genetics)*.

Hybridization (Molecular Genetics)

The pairing (tight physical bonding) of two complementary single strands of RNA and/or DNA to give a double-stranded molecule. See also *Anneal*, *Sticky ends*, *Ribonucleic acid (RNA)*, *Messenger RNA (mRNA)*, *Base pairing*, *Biosensors (electronic)*, *Biosensors (chemical)*, *Hybridization surfaces*, *DNA probe*, *Deoxyribonucleic acid (DNA)*, *Antisense (DNA sequence)*, *Biomotors*.

Hybridization (Plant Genetics)

The mating of two plants from different species or *genetically very different* members of the same species to yield hybrids (first filial hybrids) possessing some of the characteristics of each parent. Those (hybrid) offspring tend to be more healthy, productive, and uniform than their parents—a phenomenon known as "hybrid vigor." Hybrids can also arise from more than two ("parent") species.

Hybrid corn/maize seed was first commercialized (in the United States) in 1922. Other recently created crop hybrids include tangelos (produced by crossing grapefruit with tangerines), nectarines (bred from peaches), broccoflower (produced by crossing broccoli with cauliflower), and so on.

Some hybrids have occurred spontaneously in nature. For example, wheat (*Triticum aestivum*) arose centuries ago from a naturally occurring interbreeding of three Middle East grasses. In the 1980s, sugar beet (*Beta vulgaris*, subspecies *vulgaris*) naturally interbred with the wild native weed known as sea beet (*Beta vulgaris*, subsp. *maritima*) in Europe resulting in an annual weed (in contrast to sugar beet, which is a biannual). Because that (new hybrid weed) is closely related to sugar beet, any herbicide that kills the (new hybrid weed) is likely to harm the sugar beet crop (unless the sugar beet crop is made herbicide tolerant). See also *F1 hybrids*, *Species*, *Transgressive segregation*, *Segregant*, *Genetics*, *Corn*, *Wheat*, *GEM*, *Exotic germplasm*, *Barnase*, *Herbicide-tolerant crop*.

Hybridization Surfaces

Various physical substrates (surfaces) onto which have been "attached" genetic materials (DNA, RNA, oligonucleotides, etc.). Relevant complementary genetic materials (e.g., DNA, RNA, oligonucleotides) then are hybridized onto those attached-to-surface genetic materials for various specific purposes (e.g., detection of the presence of those unattached genetic materials, in the case of biosensor's hybridization surface). One of the technologies that can be utilized to assay (evaluate) DNA from hybridization surfaces is matrix-assisted laser desorption ionization time of flight mass spectrometry. See also *Substrate (structural), Hybridization (molecular genetics), Complementary DNA (c-DNA), Deoxyribonucleic acid (DNA), Ribonucleic acid (RNA), Nanocrystal molecules, Double helix, Biosensors (electronic), Biosensors (chemical), Biochips, Oligonucleotide, Oligonucleotide probes, MALDI-TOF-MS, Assay, Microarray (testing), Directed self-assembly, Massively parallel signature sequencing.*

Hybridoma

The cell line produced by fusing a myeloma (tumor cell) with a lymphocyte (which makes antibodies); it continues indefinitely to express the immunoglobulins (antibodies) of both parent cells. See also *Monoclonal antibodies (MAb), Aging.*

Hydrazine

A chemical with formula N_2H_4. Used as a rocket fuel and in the hydrazinolysis of glycoproteins.

Some hydrazine compounds are also naturally produced in certain mushrooms (*Agaricus bisporus, Gyromitra esculenta*, etc.). See also *Hydrazinolysis (of glycoproteins, to isolate unreduced oglio-saccharide side chains), Glycoprotein, Reduction (in a chemical reaction).*

Hydrazinolysis (of Glycoproteins to Isolate Unreduced Oligosaccharide Side Chains)

A technique that used the chemical hydrazine to separate and isolate the oligosaccharide portion from the protein portion of a glycoprotein. The hydrazine chemically "chews up" the polypeptide (i.e., protein) portion of a glycoprotein molecule, leaving the intact oligosaccharides behind. It can subsequently be analyzed (after chromatographic separation from the peptide pieces and other chemical components). See also *Reduction (in a chemical reaction), HF cleavage, Polypeptide (protein), Glycoprotein, Sequencing (of oligosaccharides), Hydrazine, Chromatography.*

Hydrilla verticillata

An aquatic plant that contains a very basic/"primitive" C4 photosynthesis system, so it thus is utilized as a model organism for research on C4 photosynthesis applicable to all plants. See also *C4 photosynthesis, Model organism.*

Hydrofluoric Acid Cleavage

See *HF cleavage.*

Hydrogels

Water-saturated gelatin-like materials consisting of hydrated "networks" of polymers between and among (entrained) water molecules. Because hydrogels possess many of the chemical properties of the extracellular matrix, they can be used to construct "scaffolds" for use in tissue engineering.

During 2012, Dan Luo created a hydrogel in which the polymers consisted of long molecular chains of synthetic DNA. This hydrogel can be manufactured into specific shapes, reversibly becomes a free-flowing liquid in air, and then reverts to its original three-dimensional shape when immersed in water. See also *Tissue engineering, Extracellular matrix, Deoxyribonucleic acid (DNA), Metamaterials.*

Hydrogen Bonding

Refers to the electrostatic attraction (pseudo-chemical bond) that occurs between a hydrogen atom on one molecule and the electron cloud of another molecule (or one atom located some distance away on the *same* molecule). See also *Weak interactions, Van der Waals forces, Deoxyribonucleic acid (DNA).*

Hydrogen Sulfide (H_2S)

A toxic gas that is synthesized by bacteria to protect them against oxidative stress; it also protects H_2S-emitting bacteria against many antibiotics (i.e., those that kill bacteria via oxidative stress). See also *Oxidative stress, Reactive oxygen species, Bacteria, Antibiotics.*

Hydrogenation

Invented by Wilhelm Normann in 1901, it is a chemical reaction/process in which hydrogen atoms are added to molecules (e.g., of unsaturated fatty acids) in edible oils. In the case of fatty acids, the fraction of each isomeric form (*trans* vs. *cis* fatty acids) and the molecular chain length (of the fatty acids present) have a large impact on the melting characteristics of each (fat or oil), with shorter-chain fats melting at lower temperature.

Hydrogenation is the most common chemical reaction utilized in the edible oils (processing) industry. Hydrogenation increases the solids (i.e., crystalline fat) content of edible fats/oils and improves their resistance to thermal and atmospheric oxidation (e.g., for frying of foods). Those increases in solids and resistance to oxidation result from the reduction in the fat/oil relative unsaturation, plus increased geometric and positional isomerization of the fat/oil molecules. The edible oil/fat hydrogenation reaction is accomplished by treating fats/oils with pressurized hydrogen gas in the presence of a catalyst. As a result, the (usually) liquid oils are converted to more saturated fats, which are semisolid at an ambient temperature of 72°F (22°C). The presence of *trans* fatty acids in hydrogenated edible oils can be reduced significantly via changes in catalyst, temperature, pressure, and so on, utilized in the hydrogenation reaction. In general, natural oils and fats possessing melting points lower than 121°F (50°C) are nearly completely absorbed in the digestive system of typical humans. See also *Fatty acid, Monounsaturated fats, Saturated fatty acids (SAFA), Dehydrogenation, Essential fatty acids, Laurate, Lecithin, Triglycerides, Unsaturated fatty acid, Soybean oil, Conjugated linoleic acid (CLA), Oxidation, Isomer, Stereoisomers, Catalyst, Substrate (chemical), Trans fatty acids.*

Hydrolysis

Literally, means "cleaved by water." It is used for a chemical reaction in which the chemical bond attaching an atom, or group of atoms to the (rest of the) molecule is cleaved, followed by attachment of a hydrogen atom at the same chemical bond. See also *Digestion (within organisms).*

Hydrolytic Cleavage

A chemical reaction in which a portion (e.g., an atom or a group of atoms) of a molecule is "cut" off the molecule via hydrolysis. See also *Hydrolysis.*

H

Hydrolyze

To "cut" a chemical bond (i.e., with a molecule) via hydrolysis. See also *Hydrolysis*.

Hydrophilic

This term means water loving or having a great affinity for water. It is used to describe molecules or portions of molecules that have an affinity for water. The property of having an affinity for water at an oil–water interface. For example, ordinary sugar that dissolves readily in water is said to be hydrophilic (i.e., a molecule that is "water loving"). See also *Amphiphilic molecules*.

Hydrophobic

This term means "water hating" or having a great dislike for water. It is used to describe molecules or portions of molecules that have very little or no affinity for water. The property of having an affinity for oil (nonpolar environments) at an oil–water interface. For example, a nonpolar hydrocarbon such as butane (as used in lighters) that will not dissolve in water, but which will dissolve (be miscible) in oil is said to be hydrophobic (i.e., a molecule that is "water hating"). See also *Amphiphilic molecules*, *Phytol*.

Hydroxylation Reaction

A chemical reaction in which one or more hydroxyl groups (i.e., the –OH group) is introduced (i.e., is chemically attached) to a molecule.

Hyperacute Rejection

See *Graft-versus-host disease* (*GVHD*).

Hyperchromicity

The increase in optical density that occurs when DNA is denatured. See also *Deoxyribonucleic acid* (*DNA*), *Denatured DNA*, *Optical density* (OD).

Hypersensitive Response

A protective/defensive response by certain plants to "infection" by plant pathogens (e.g., bacteria, fungi, viruses), in which those plant cells that are immediately adjacent (to the infected area of plant) are "instructed" to self-destruct via apoptosis, in order to cordon off the infected area (to prevent further spread of the infection).

The initiation of the hypersensitive response is often triggered by *signaling molecules* that are produced by the pathogens themselves. For example, one particular protein produced by the soil fungus *Fusarium oxysporum* triggers a hypersensitive response that often is so severe that the entire plant dies. See also *Pathogenesis related proteins*, *Protein*, *Pathogen*, *Bacteria*, *Fungus*, *Virus*, *Cell*, *Apoptosis*, *Signaling*, *Signaling molecule*.

Hyperthermophilic (Organisms)

See *Thermophile*, *Thermophilic bacteria*.

Hypostasis

Interaction between nonallelic genes in which one gene will not be expressed in the presence of a second. See also *Epistasis*, *Gene*, *Express*, *Allele*.

Hypothalamus

A part of the brain structure, lying near base of brain, it regulates a number of hormones. As a part of the brain, it constantly receives (neurochemical) signals from nerve cells (neurons). The hypothalamus monitors those signals and converts them into hormonal "signals" (e.g., it generates a "burst" of hormones in response to certain visual stimuli, certain physical [e.g., sexual] stimuli, etc.). Also, the hypothalamus is able to monitor and detect changes in the blood levels of hormones coming from endocrine glands. For example, the metabolic hormone insulin (from the pancreas) and the reproductive hormone estrogen (from the ovaries) both trigger changes in function in the hypothalamus. The hypothalamus regulates biological processes (e.g., metabolic rate, appetite). A major function of the hypothalamus is to control reproduction via secretion of gonadotropin-releasing hormone from the tips of hypothalamic nerve fibers that extend downward toward (into) the pituitary gland. Similarly, the hypothalamus also helps to control the body's growth (from birth until the end of puberty) via secretion of growth hormone–releasing factor to the pituitary gland. See also *Hormone*, *Endocrine hormones*, *Endocrine glands*, *Endocrinology*, *Pituitary gland*, *Growth hormone* (*GH*), *Neurotransmitter*, *Growth hormone–releasing factor* (*GHRF*).

Hypoxia

Used to refer to a state (e.g., of cells within a specific tissue in an organism) in which the media lacks enough oxygen (e.g., to sustain growth). Hypoxia can lead to epigenetic events in some organisms. See also *Cell*, *Organism*, *Epigenetic*.

Hypoxia-Inducible Factors

Refers to master control proteins that "turn on" numerous particular genes that help cells adapt to a scarcity of oxygen (e.g., cells within a specific tissue in an organism). Although these genetic responses are essential for continuation of the cell's life, the hypoxia-inducible factors also can turn on additional genes that help certain cancerous cells (within oxygen-starved tumors) to metastasize (i.e., escape the tumor) via moving on their own to invade the body's blood vessels, through which they spread to other tissues of the body. See also *Protein*, *Gene*, *Cell*, *Cancer*, *Tumor*, *Metastasis*.

I

IAA

Acronym for indole-3-acetic acid. See *Indole-3-acetic acid, Auxins.*

IBA

See *Industrial Biotechnology Association.*

IBD

Acronym for inflammatory bowel disease. See *Inflammatory bowel disease.*

IBG

See *International Biotechnology Group.*

Ibrutinib

A tyrosine kinase inhibitor that is approved by the U.S. Food and Drug Administration (FDA) as the pharmaceutical Imbruvica® for the treatment of

- Chronic lymphocytic leukemia (CLL) in patients who have received at least one prior therapy
- CLL in patients with 17p deletion within their DNA
- Mantle cell lymphoma (MCL) in patients who have received at least one prior therapy

See also *Enzyme, Protein tyrosine kinase inhibitor, Tyrosine kinase inhibitors (TKI).*

ICAM

Acronym for *intercellular adhesion molecule.* See *Adhesion molecule.*

ICM

Acronym for *intact-cell MALDI-TOF-MS.* Beginning in 1975, Catherine Fenselau and John Anhalt extended the use of MALDI-TOF-MS (previously utilized to identify only *molecules*) to encompass identification of certain intact cells (e.g., gram-positive bacteria, after they were gently heated and dislodged via laser from a "soft" matrix/substrate that Fenselau/Anhalt had adhered them to). See also *MALDI-TOF-MS, Cell, Bacteria, Gram-positive.*

ICS Gene

A gene in plants that codes for the production of the enzyme isochorismate synthase. See also *Gene, Enzyme, Alternative splicing.*

IDA

Acronym for *iron deficiency anemia.* See *Iron deficiency anemia (IDA).*

IDE

"Investigational Device Exemption" application to the Food and Drug Administration seeking approval to begin clinical studies of a new medical device. See also *Food and Drug Administration (FDA).*

Ideal Protein Concept

Refers to the protein content in the feed ration (food) eaten by livestock, poultry, and humans. Feed that contains *ideal protein* contains protein(s) that—when digested by animal—yields all of the essential amino acids, in proper proportions, for the growth and/or maintenance needs of that animal.

"Ideal protein" varies for different species (e.g., pigs require different amino acids/rations than chickens do). "Ideal protein" varies for different stages in the life of a given animal (e.g., poultry require more sulfur-containing amino acids, such as methionine, during life stages when feather growth is at a comparatively high rate).

The animal's requirement for one essential amino acid is proportionally linked to the animal's requirements for another. Increasing the supply (when deficient) of one essential amino acid in the animal's diet would improve that animal's (growth) performance if no other amino acids were limiting.

Feed rations formulated to contain "ideal protein" have been shown to reduce the amount of nitrogen (nitrates) excreted by livestock and poultry, by as much as 50%. See also *Amino acid, Protein, Essential amino acids, Essential nutrients, Methionine (met), Digestion (within organisms), Soy protein, High-lysine corn, High-methionine corn.*

Idiotype

The region of the antibody molecule (i.e., antigen combining site or antigenic determinant) that enables each antibody to recognize a specific foreign structure (i.e., epitope or hapten) is said to have an idiotype (for that epitope or hapten). An identifying characteristic (or property) of the epitope or hapten that one is talking about. See also *Epitope, Hapten, Antigen, Antibody, Catalytic antibody, Antigenic determinant.*

IDM

See *Integrated disease management.*

IFBC

See *International Food Biotechnology Council.*

IFC

Acronym for integrated fluidic circuit. See *Integrated fluidic circuit.*

IFN-Alpha

Alpha interferon. See *Interferons.*

IFN-Beta

Beta interferon. See *Interferons.*

IGF-1

See *Insulin-like growth factor-1.*

IGF-2

See *Insulin-like growth factor-2.*

IGF-I

See *Insulin-like growth factor-1.*

IGF-II

See *Insulin-like growth factor-2.*

IGR

Acronym for *intergenic region* (of an organism's DNA). See *Gene, Deoxyribonucleic acid (DNA), Intron.*

IL-22

See *Neutrophils.*

IL-lra

See *Interleukin-1 receptor antagonist.*

Imidazilinone-Tolerant Soybeans

See *Imidazolinone-tolerant soybeans.*

Imidazolinone-Tolerant Soybeans

Refers to soybeans (*Glycine max* (L.) Merrill) that are able to resist the (weed killing) effects of imidazolinone-based herbicides (inc. imazethapyr and imazaquin). During 2003, Brazilian researchers developed such soybeans via genetic engineering. See also *Herbicide-tolerant crop, Soybean plant, STS sulfonylurea (herbicide)-tolerant soybeans, Genetic engineering.*

Imiglucerase

A commercially produced (via biotechnology) enzyme that is utilized in the treatment of Gaucher disease. See also *Orphan drug.*

Immobilization

Refers to the process of "attaching" the molecular capture agents, biosensors/probes (e.g., fluorophore-labeled DNA segment or antibody) to the glass/silicon/plastic/gold surface of a *microarray* (e.g., DNA chip, SNP chip, protein microarray, proteome chip, cell array), *magnetic particle, surface plasmon resonance chip*, or other hybridization surface.

The particular immobilization that is utilized is dependent upon the physical properties of the (chip) surface and the capture agent/probe molecule. Immobilization can be accomplished via a (covalent) chemical reaction between capture agent/probe and surface or via a noncovalent means such as physical adsorption onto surface, Van der Waals forces, hydrogen bonding, electrostatic forces, and so on. See also *Biosensors (chemical), Probe, DNA probe, Deoxyribonucleic acid (DNA), Label (fluorescent), Fluorophore, Microarray (testing), DNA chip, Biochip, Cell array, Protein microarrays, Proteome chip, Target–ligand interaction screening, Multiplexed assay, Immunosensor, SNP chip, Surface plasmon resonance (SPR), Magnetic particles, Hybridization surfaces, Capture agent.*

Immune Effector Sites

See *Peyer's patches.*

Immune Profiling

Refers to the process of

A. Sequencing of the DNA that makes up each of an individual's relevant subpopulation of immune system's B cells and T cells (by doctors who want to treat a disease in that individual). In such B and T cells, the DNA has been recombined (shuffled around) in a vast array of new combinations as part of the immune response, thereby allowing T cells to recognize specific pathogens such as influenza viruses, and allowing the B cells to generate antibodies against those pathogens.
B. Tailoring the disease treatment(s) for the individual based upon the information gleaned via #A.

See also *Immune response, Deoxyribonucleic acid (DNA), Pathogen, Virus.*

Immune Response

See *Cellular immune response, Adaptive immune response, Antibody, Humoral immunity, Innate immune response, Immunomodulating agent, Neutrophil extracellular trap, Large intervening noncoding RNA, Systemic acquired resistance (SAR), Tomatidine.*

Immunoadhesins

See *Adhesion molecule.*

Immunoassay

The use of antibodies to identify and quantify (measure) substances by a variety of methods. The binding of antibodies to antigen (substance being measured) is often followed by tracers, such as fluorescence or (radioactive) radioisotopes, to enable measurement of the substance. See also *Antibody, Tracer (radioactive isotopic method), Antigen, ELISA, Radioimmunoassay, Assay, EIA, Fluorescence, Near-infrared spectroscopy (NIR), Chemiluminescent immunoassay (CLIA).*

Immunoconjugate

A molecule that has been formed by attachment to each of two originally different molecules. One of these is generally an antibody and, hence, the word "immunoconjugate." Classic organic drug molecules such as methotrexate, adriamycin, and chlorambucil; radionuclides; enzymes; cytotoxins; and ribosome-inhibiting proteins may be conjugated to antibodies. The salient point is that the antibody portion of the conjugate is there to "steer" the biologically active molecule to its target (e.g., receptor, tumor).

For example, during 2013, the European Commission approved for use in Europe the Roche immunoconjugate Kadcyla (trastuzumab emtansine or T-DM1) for people with previously treated HER2-positive advanced breast cancer. Kadcyla is indicated as a single agent for the treatment of adults with HER2-positive, unresectable locally advanced or metastatic breast cancer who previously received Herceptin (trastuzumab) and a taxane, separately or in combination. See also *Conjugate, Magic bullet, Antibody, Trastuzumab, Herceptin, Monoclonal antibodies (MAb), Radioimmunotherapy, Magnetic particles, Diphtheria toxin, Receptor, Tumor.*

Immunocontraception

Any process or procedure in which an organism's immune system is utilized to attack or inactivate the reproductive cells (e.g., sperm) within the organism. See also *Cellular immune response, Antibody, Humoral immunity, Germ cell.*

Immunodominant

Term utilized to refer to a compound (e.g., a food allergen) that causes an organism's immune system to respond so strongly that it causes harm to the organism. See also *Allergies (foodborne), Antigen, Immune response.*

Immunogen

A molecule or an organism (e.g., pathogenic bacteria) that is specifically "recognized" by the immune system (e.g., of humans it has entered) and triggers an immune response. See also *Antigen, Pathogenic, Humoral immunity, Cellular immune response.*

Immunoglobulin (IgA, IgE, IgG, and IgM)

A class of (blood) serum proteins representing antibodies. Often used, along with the more specific monoclonal antibodies, in health diagnostic reagents. In certain people who are genetically predisposed to foodborne allergies, immunoglobulin-E (IgE) initiates an immune system response to antigen(s) present on protein molecule(s) in the particular food that person is allergic to. Severe allergic reactions to foods may lead to death. See also *Protein, Antigen, Allergies (foodborne), Antibody, Immunoassay, B lymphocytes, Gut-associated lymphoid tissues (GALT), Joining segment.*

Immunomagnetic

Refers to the usage of antibody molecules linked to magnetic particles (e.g., as part of an immunoassay).

See also *Antibody, Magnetic particles, Immunoconjugate, Immunoassay, Cell sorting.*

Immunomodulating Agent

Refers to any agent (e.g., chemical compound) that increases or decreases the immune response. For example, the presence of the microscopic eggs of the porcine whipworm (*Trichuris suis* ova) in human bloodstream causes the human immune system to downregulate (overactive) T cells and proinflammatory cytokines, thereby relieving Crohn's disease (i.e., an autoimmune intestinal disease of humans that can cause inflammation of the colon, abdominal pain, diarrhea, and weight loss and decrease the body's ability to absorb dietary source vitamin D). See also *Autoimmune disease, Crohn's disease, T cells, T cell modulating peptide (TCMP), Cytokines.*

Immunosensor

A biosensor with a selected antibody attached, which can sense when a given molecule (from sample) binds (i.e., "attaches to") that antibody.

For example, the selected antibody "binding" can be made to cause (simultaneous) fluorescence. If the biosensor (which the antibody is attached to) incorporates a fiber optic and light detector (e.g., CCD detector), the "binding" can be detected automatically and at a distance (e.g., from outside a reactor, or outside the body—in the case of an implanted-in-body sensor). See also *Biosensors (electronic), Antibody, Fluorescence, Biosensors (chemical), Catalytic antibody.*

Immunosuppressive

That which suppresses the immune system response (e.g., certain chemicals). See also *Cellular immune response, Humoral immunity, Cyclosporin A.*

Immunotherapy

See *Modulatory nanotechnologies.*

Immunotoxin

A conjugate formed by attaching a toxic molecule (e.g., ricin) to an agent of the immune system (e.g., a monoclonal antibody) that is specific for the pathogen or tumor to be killed. The immune system agent portion (of the conjugate) delivers the toxic chemical directly to the specified (disease) site, thus sparing other healthy tissues from the effect of the toxin. See also *Ricin, Monoclonal antibodies (MAb), Magic bullet, Diphtheria toxin.*

Importins

See *Nuclear proteins.*

Imprinting

An epigenetic process in which certain genes within an organism's cells are "disabled" (e.g., via methylation) during the earliest stage(s) of the organism's development. For example, the embryo of a female mammal (which receives two copies of the X chromosome—one from each parent) disables one of those copies, at random, in each of its cells, so the female becomes a *genetic mixture* of its two parents.

Loss of imprinting (LOI) can sometimes occur in an adult organism. For example:

- In mice, LOI of the gene that codes for insulin-like growth factor-2 (IGF-2) results in the intestine's epithelial cells reverting to a less developed state and also development of significantly more intestinal tumors.
- In humans, LOI of the gene that codes for IGF-2 is correlated with development of colorectal cancer.

See also *Cell, Epigenetic, Gene, Genetic code, Chromosomes, X chromosome, Long noncoding RNAs, Methylated, DNA methylation, Embryology, Tumor, Cancer, Insulin-like growth factor-2.*

Inbreeding Depression

Refers to the fact that offspring resulting from the mating of two closely related individuals are less fit and less fertile than offspring from mating of individuals who are not related). See *DNA methylation.*

In Silico

See *In silico biology.*

In Silico Biology

A set of computer modeling technologies via which researchers can

- Create computer models of specific cells, how a given disease impacts that cell, how a given pharmaceutical impacts that cell (e.g., by "docking" to it) or fails to impact that cell, and so on
- Create computer models of specific organs, how a given disease impacts that organ, how a given pharmaceutical impacts that organ, and so on
- Create computer models of specific organisms, how a given disease impacts that organism, how a given pharmaceutical then impacts that disease within that organism, and so on
- Create computer models of specific organisms that possess a given genome, how a given disease impacts that specific organism/phenotype, how a given pharmaceutical then impacts that disease within that organism/phenotype, and so on
- Create computer models of protein "digestion" (i.e., breaking apart into constituent peptides), for comparison with the *actual* peptides (fragments) that are determined (e.g., via MALDI-TOF-MS) to have resulted from *chemical digestion* of those protein molecules (e.g., via immersion in trypsin)

See also *Rational drug design, Receptor mapping, Cell, Biochips, Genome, Genomics, Pharmacogenomics, Protein, Proteomics, Phenotype, MALDI-TOF-MS, Peptide, Trypsin, Docking (in computational biology), Synthetic biology.*

In Silico Screening

A set of computer modeling technologies via which researchers can (vicariously) screen chemical compounds for their potential as pharmaceutical candidate compounds, pesticide candidate compounds, and so on.

The chemical compounds are "generated" (e.g., from data available about compounds actually *created* in a laboratory in the past) and then computer modeling is utilized to

- Assess their impact on "generated" specific cells, tissues, and so on, via "docking" (e.g., from data available about that *chemical type* of molecule's impact on that *type of cell/tissue* when actually tested on it in a laboratory/clinic in the past)
- Generate an analogous chemical compound that is likely to be more efficacious or have fewer undesirable side effects
- Repeat the process

For example, when *screening compounds* for potential usefulness as a pharmaceutical, the goal is to assess (modeled/predicted) differences between *diseased* (untreated) and *treated* cells, thus enabling prediction of (better) pharmaceutical candidate compounds for eventual actual testing on *real* cells/tissues.

Some of the more sophisticated *in silico* screening software can even "model" ADME properties for selected pharmaceutical candidate compounds. See also *Rational drug design, In silico biology, Receptor mapping, Cell, Biochips, High-throughput screening (HTS), Combinatorial chemistry, Pharmacogenomics, Proteomics, Quantitative structure–activity relationship (QSAR), ADME tests, Target (of a therapeutic agent), Target (of a herbicide or insecticide), Docking (in computational biology), Pharmacophore searching.*

In Situ

In the natural or original position (e.g., inside the body).

In Vitro

In an unnatural position (e.g., outside the body, in the test tube). "*In vitro*" is Latin for "in glass." For example, the testing of a substance or the experimentation in (using) a "dead" cell-free system. See also *In vitro selection.*

In Vitro Selection

A search process (e.g., for a new pharmaceutical) that first involves the construction of a large "pool" of polynucleotide sequences (at least some of which are likely to possess the desired pharmaceutical properties) synthesized by a totally random process. This is followed by repeated cycles of screening (for those sequences possessing desired properties) and/or enriching, and amplification (of the screened/enriched sequences). Common amplification techniques include polymerase chain reaction, ligase chain reaction, self-sustained sequence replication, Q-beta replicase technique, and strand displacement amplification. See also *In vitro, Amplification, Gene amplification, Polymerase chain reaction (PCR), Q-beta replicase technique, Nucleotide, Deoxyribonucleic acid (DNA), Synthesizing (of DNA molecules), Oligonucleotide, DNA probe, Gene machine, Combinatorial chemistry, Pharmacophore searching.*

In Vivo

Latin for "*in living*" (e.g., the testing of a new pharmaceutical substance or experimentation in (using) a living, whole organism. An *in vivo* test is one in which an experimental substance is injected into an animal such as a rat in order to ascertain its effect on the organism. See also *Model organism.*

In/Dels

Abbreviation for *insertions/deletions*. Refers to insertions and/or deletions (e.g., of alleles within one organism's DNA versus another organism of that same species). These can occur either spontaneously or as a result of a scientist's use of CRISPR/Cas9 Gene-editing System. See also *Deoxyribonucleic acid (DNA), Gene, Allele, Organism, Species, Insertional knockout systems, Colinearity, CRISPR/Cas9 gene-editing systems.*

Inclusion Bodies

See *Refractile bodies (RB).*

IND

"Investigational New Drug" application to the Food and Drug Administration seeking approval to begin human clinical studies of a new pharmaceutical compound. See also *"Treatment" IND, IND exemption, Phase I clinical testing, Food and Drug Administration (FDA).*

IND Exemption

A permit by the Food and Drug Administration (FDA) to begin clinical trials on humans (of a new pharmaceutical) after toxicity data have been reviewed and approved by the FDA. See also *Kefauver rule, IND, Phase I clinical testing.*

INDA

Acronym for *investigational new drug application*. See *IND.*

Indel Mutations

See *IN/Dels.*

Indian Department of Biotechnology

The governmental body in India that regulates all recombinant DNA research. It is the Indian counterpart of the American Government's Recombinant DNA Advisory Committee, the Australian government's Gene Technology Regulator (GTR), and the French government's Commission of Biomolecular Engineering. See also *Recombinant DNA Advisory Committee (RAC), ZKBS (Central Commission on Biological Safety), Genetic engineering, Recombinant DNA (rDNA), Recombination, Biotechnology, Gene Technology Office, Commission of Biomolecular Engineering, Gene technology regulator (GTR).*

Indian Hedgehog Protein (Ihh)

See *Hedgehog proteins.*

Indole-3-Acetic Acid

A plant hormone (abbreviated IAA) that regulates how plants grow, causing them to extend their shoots toward sunlight.

During 2010, Reeta Prusty Rao and Jennifer Normanly discovered that yeasts also produce IAA as a signal they utilize for quorum sensing (i.e., to trigger the yeasts to produce filaments as part of "yeast infection" attack). See also *Hormone, Quorum sensing, Auxins.*

Induced Fit

A substrate-induced change in the shape of an enzyme molecule that causes the catalytically functional groups of the enzyme to assume positions that are optimal for catalytic activity to occur. See also *Enzyme.*

Induced Pluripotent Stem Cells

Abbreviated *iPS cells or iPSC,* this term was coined by Shinya Yamanaka in 2007 when he discovered how to "reprogram" (i.e., induce) adult mammalian skin cells so they would return to an embryonic-stem-cell-type state. This "reprogramming" was initially accomplished via insertion of four specific genes (later reduced to three genes).

During 2009, researchers discovered how to similarly turn fat (adipose tissue) cells into iPS cells. iPS cells, like all pluripotent cells, can differentiate into the numerous different types of tissues comprising the body of an organism. When iPS cells are created from the cells of people who have certain diseases (e.g., amyotrophic lateral sclerosis), scientists can utilize them to establish stable, growing populations of cells and sometimes tissues *that evidence the particular disease* (*i.e., known as a* "disease in a dish") for experiments to try to find a treatment.

During 2014, Masayo Takahashi of Riken Institute turned a patient's skin cells into iPSC that subsequently became retinal epithelial cells and then injected them into that patient's eye, as part of efforts to treat that patient's macular degeneration disease.

Research indicates that neurons derived from human-induced pluripotent stem cells might someday be useful for the treatment of some spinal cord injuries. See also *Cell, Stem cells, Adult stem cell, Embryonic stem cells, Gene, Mammalian cell culture, Pluripotent stem cells, Differentiation, Macular degeneration, Adipose, RIKEN, Neuron.*

Induced Polyploidy

Refers to a technique utilized within certain commercial crop breeding programs (e.g., to produce cereal grains, certain forage crops, etc.) in which the "parents" (i.e., the two that will be bred together to produce the commercial seed that is subsequently sold to farmers) are created as follows:

1. First separately treating the elite crop germplasm "grandparent" seeds with colchicine (a chemical that doubles the number of chromosomes in cells by interfering with the cell division) and then growing and cross-pollinating those elite crop germplasm (e.g., optimized to the applicable growing climate/latitude).
2. When seeds resultant from that pollinating are harvested, their genome contains four times the usual number of chromosomes as normal plants of that species.
3. Those tetraploid plants (i.e., the "parent" plants) are then bred together to produce the commercial seed that is subsequently sold to farmers.

The net result of a crop breeding program's use of induced polyploidy is creation of commercial seed in less time than required by a conventional crop breeding program. See also *Colchicine, Tetraploid, Haploid, Doubled-haploid breeding program, Elite germplasm, Gene, Genome, Chromosomes.*

Inducer Line

See *Doubled-haploid breeding program.*

Inducer Parent

See *Doubled-haploid breeding program.*

Inducers

Molecules that cause the production of larger amounts of the enzymes that are involved in the uptake and metabolism of the inducer (such as galactose). Inducers may be enzyme substrates. See also *Enzyme, Inducible enzymes, Substrate (chemical).*

Inducible Enzymes

Enzymes whose rate of production can be increased by the presence of certain chemical molecules. For example, *Paneth cells* that line the human small intestine are induced by the presence of plant "natural pesticidal compounds" to excrete into passing food/plant materials large amounts of nucleases that degrade those plant natural pesticidal compounds (e.g., psoralene, caffeine), thereby protecting the human body.

Other inducible enzymes include the *Phase I and Phase II detoxification enzymes* that work in tandem to eliminate some toxins from the body. The phase I enzymes metabolize certain food compounds (sometimes into chemicals that happen to themselves be carcinogens), which are then transformed into harmless compounds by Phase II enzymes.

Research published during 2004 indicates that the presence of lycopene or sulforaphane in the human digestive tract induces excretion of some cancer-inhibiting phase II detoxification enzymes.

Some diseases result in the production of certain chemicals that also thereby induce Phase I/II enzymes. See also *Enzyme, Nuclease, Caffeine, Toxin, Psoralene, Lycopene, Sulforaphane, Cancer, Carcinogen.*

Inducible Promoter

Refers to a particular promoter, in which start/increase of promotion is caused (to initiate "defense" of the organism) by the presence of disease/pathogen or a toxin. See also *Promoter.*

Industrial Biotechnology Association (IBA)

An American trade association of companies involved in biotechnology. Formed in 1981, the IBA tended to consist of the larger firms involved in biotechnology. In 1993, the Industrial Biotechnology Association (IBA) was merged with the Association of Biotechnology Companies (ABC) to form the Biotechnology Industry Organization (BIO). See also *Association of Biotechnology Companies (ABC), Biotechnology Industry Organization (BIO), Biotechnology.*

Infant Gut Microbiome

See *Oligosaccharides.*

Inflammatory Bowel Disease

Refers to conditions including Crohn's disease and ulcerative colitis. See *Crohn's disease.*

Inflammatory Response

See *Chronic inflammation, Cilia.*

Infliximab

A chimeric monoclonal antibody against tumor necrosis factor that was approved by the U.S. Food and Drug Administration (FDA) as the pharmaceutical Remicade™ for the treatment of several types of arthritis, colitis, and psoriasis. See *Monoclonal antibody, Food and Drug Administration (FDA), Tumor necrosis factor (TNF), Arthritis.*

Information RNA (iRNA)

Refers to an RNA molecule (within cell) that does not code for the production of a protein but only provides some "information" to *regulate* one or more cell functions (e.g., protein synthesis). See also *Ribonucleic acid (RNA), Cell, Genetic code, Protein, Gene, Translation, Synthesizing (of proteins).*

Informational Molecules

Molecules containing information in the form of specific sequences of different building blocks. They include proteins and nucleic acids. See also *Heredity, Gene, Genetic code, Genome, Genotype, Nucleic acids, Messenger RNA (mRNA), Deoxyribonucleic acid (DNA), Ribonucleic acid (RNA), Editosome.*

Ingestion

Taking a substance into the body. For example, the amoeba surrounds a food particle and then ingests the particle.

Inhibition

The suppression of the biological function of an enzyme or system by chemical, physical, or epigenetic means.

For example, bone consists of nanocrystals of carbonated apatite (one form of calcium phosphate) within a tight matrix of collagen protein. Those nanocrystals in bones do not grow larger than 3 nm in size because citrate molecules (in the bone) tightly bind to the surface of the nanocrystals and inhibit formation of more phosphate atom layers (beyond 3 nm nanocrystal size). See also *Aptamers, Enzyme, Protein, Protein tyrosine kinase inhibitor, Solanine, Epigenetic, Micro-RNAs, Nanocrystals, Collagen, Nanometers (nm).*

Initiation Factors

Refers to either of the following:

- Specific proteins required to initiate synthesis of a polypeptide on ribosomes
- Specific proteins (e.g., C-reactive protein) that initiate an immune system response

See also *Ribosomes, Protein, Polypeptide (protein), C-reactive protein (CRP), Immune response, Complement factor H gene.*

Initiator Codon

See *Start codon.*

Innate Immune Response

Refers collectively to the inherent "first lines of immune defense" in the organism (e.g., complement cascade), which are initiated, for example

- In humans and some animals by TLR (i.e., *toll-like receptors*), a category of cellular transmembrane proteins that "recognize" certain features (e.g., antigens) present on or in certain invading pathogens
- In plants by pattern recognition receptors and pathogen-associated molecular patterns that "recognize" certain features on surfaces of pathogens (e.g., a 22-amino acid peptide on the exterior of flagella of certain invading pathogenic bacteria)

For example, the *TLR 11* class of TLRs specifically senses the presence of pathogenic bacteria that infect the urinary tract. The *TLR 7* and *TLR 8* collectively specifically sense the single-stranded RNAs that are present within some pathogenic viruses. The *TLR 9* senses unmethylated CG motifs in DNA (which are typical for bacterial DNA but not for human DNA).

When thus activated, one of the actions of the innate immune response is production of certain reactive molecules designed to neutralize invading pathogen, including hydrogen peroxide, nitric oxide, and hypochlorous acid. See also *Innate immune system, Humoral immune response, Complement, Complement cascade, Receptors, Cell, Transmembrane proteins, Antigen, Pathogen, Pathogenic, Bacteria, Flagella, Virus, Peptide, Ribonucleic acid (RNA), Neutrophil extracellular trap, Pathogenesis related proteins, Pattern recognition receptor, Bacteria, Deoxyribonucleic acid (DNA), Ribonucleic acid (RNA), Methylated, Long non-protein-coding RNA (lncRNA), CD8⁺ T cells, Nitric oxide.*

Innate Immune System

Refers to an organism's "first line of defense" against pathogens. Typically consists of

- Physical barriers (e.g., skin, epithelium)
- Chemical barriers (e.g., digestive enzymes and acids)
- Receptors (located on the surface of certain cells) that initiate the *innate immune response*
- Certain cells (e.g., neutrophils) that ingest/envelope pathogens, as a part of the immune response
- Cytokines and other relevant signaling molecules, which help regulate immunological and inflammatory processes (e.g., certain $CD8^+$ T cells)

The active (i.e., nonbarrier ones listed earlier) are generally signaled to become active by MyD88, a signaling protein.

See also *Organism, Cell, Pathogen, Epithelium, Enzyme, Digestion (within organisms), Receptors, Innate immune response, Cytokines, Protein, Signaling molecule, Neutrophil extracellular trap, Long noncoding RNAs, $CD8^+$ T cells.*

Inositol

A cyclic (i.e., ring-shaped molecule) alcohol, initially characterized as a vitamin in 1941, which imparts certain (nutritional and other) benefits to animals and humans that consume it. Because it is critically important (nutritionally) during periods of rapid growth, the U.S. Food and Drug Administration (FDA) has mandated the inclusion of inositol in nonmilk infant formula products.

Research indicates that inositol and inositol-containing metabolites may also help to prevent Type II diabetes, may help reduce/avoid several mental illnesses, and certain cancers. See also *Vitamin, Metabolite, Type II diabetes, Cancer, Phytate, Food and Drug Administration (FDA).*

Inositol Hexaphosphate (IP-6)

See *Phytate.*

Insect Cell Culture

The propagation *in vitro* (e.g., in a vat or other container) of a population of living cells isolated from insects. Two insect species commonly utilized are fall armyworm (*Spodoptera frugiperda*) and cabbage looper (*Trichoplusia ni*).

Glycosylation of protein molecules produced in these (insect-source) cells is not identical to the glycosylation of rel. protein molecules produced by mammalian cells. That is because insect cells cannot put the sialic acid or galactose units onto the "ends" of the glycosylation molecular chains/branches. However, the glycoproteins (i.e., glycosylated protein molecules) produced by insect cells are *similar enough to mammalian-source glycoproteins* to possess a similar biological activity. See also *Cell, Cell culture, In vitro, Protein, Baculovirus expression vector system (BEVS), Glycosylation, Sialic acid, Galactose (gal), Biological activity, Mammalian cell culture, Fall armyworm.*

Insertional Knockout Systems

See *Gene silencing.*

Insertional Mutagenesis

A mutation that can arise via:

- A mobile genetic element (e.g., transposon) naturally inserting itself at a particular (new) point within an organism's DNA.
- A genetic cassette inserted by man (i.e., genetic engineering) at a particular (new) point within an organism's DNA.

See also *Mutation, Organism, Transposon, Deoxyribonucleic acid (DNA), Genetic cassette, TALENs.*

Insitu

See *In situ.*

Insulin

A protein hormone normally secreted by the beta (β) cells of the pancreas (when stimulated by glucose and the parasympathetic nervous system). Insulin and glucagon are the most important regulators of fuel (food) metabolism. In essence, insulin signals the "fed" state to the body's cells, which stimulates the storage of energy (fuel) in the form of fat and the synthesis of proteins (i.e., tissue building/repair) in a variety of ways.

Other impacts of insulin are to stimulate the uptake of amino acids by tissues, increase the permeability of cells to some ions (e.g., potassium), cause secretion of the hormone *angiotensin II* that constricts arteries, promote synthesis of free fatty acids in the liver, inhibit the breakdown of fat in adipose tissue, and so on. The disease known as *diabetes* results from a body's inability to produce insulin or its insensitivity to the insulin that is produced. That inability/insensitivity, and thus the disease, can result from several different causes:

- Type I (also known as *childhood* or *juvenile* or *early-onset*) diabetes results when the body's insulin-making tissue is destroyed by autoimmune disease. See also the entry for *Insulin-dependent diabetes mellitus* (*IDDM*).
- Type II diabetes results when the body's insulin-utilizing tissues become insensitive to insulin. This can occur when insulin causes the liver to synthesize an overabundance of free fatty acids, which get stored in adipose tissue in the form of triglycerides, which can subsequently result in those triglyceride-laden tissues producing far fewer insulin receptors (i.e., thereby becoming insensitive to insulin).

The too-high sugar content in the bloodstream that results from diabetes causes creation of *free radicals* (high-energy oxygen atoms bearing an "extra" electron) that can damage the eyes, kidneys, and extremity arteries (sometimes necessitating limb amputation) in one **haplotype (i.e., genetic subgroup) of people (i.e., those possessing the larger-size molecules of haptoglobin—a blood protein)**.

Some research indicates that consumption of amylose (starch only) or inulin (fructose oligosaccharide) in human diet as the primary carbohydrate source, instead of glucose (or other sugars that the human body converts to glucose), can help the human body to avoid Type II diabetes, by avoiding gluconeogenesis.

In 1922, Canadian scientists Frederick Banting, Charles Best, J. J. R. MacLeod, and J. B. Collip succeeded in extracting insulin from the pancreas of slaughtered livestock (cows, pigs) in a form that could be injected into diabetes patients as a substitute for human insulin. The English biochemist, Fred Sanger, was first to determine the complete amino acid sequence of the insulin molecule. In 1977, the American scientist Howard Goodman, collaborating with William Rutter, announced the first cloning of insulin genes. This led to human insulin production by genetically engineered microorganisms (approved by FDA in 1982). See also *Beta cells, Islets of Langerhans, Hormone, Protein, Receptors, Glucose* (*GLc*), *Amino acid, Polypeptide* (*protein*), *Sequence* (*of protein molecule*), *Genetic engineering, Glucagon, Insulin-dependent diabetes mellitus* (*IDDM*), *G-proteins, Carbohydrates, Pancreas, Autoimmune disease, Inulin, Free radical, Haplotype, Oxidative stress, Haptoglobin, Type I diabetes, Type II diabetes, Adipose, Triglycerides, Resistin, Serotonin, Adipokines.*

Insulin-Dependent Diabetes Mellitus (IDDM)

An autoimmune disease in which the insulin-producing cells of the pancreas (i.e., *beta cells*, also known as *islets of Langerhans*) are attacked and destroyed by the cytotoxic T cells of the body's immune system. See also *Autoimmune disease, Insulin, Islets of Langerhans, Beta cells, Cytotoxic T cells, Haptoglobin, Diabetes, Type I diabetes, Glutamic acid decarboxylase* (*GAD*).

Insulin-Like Growth Factor-1 (IGF-1)

A protein hormone that is produced by the body's liver (when those cells have been stimulated by human growth hormone) and bone cells (when those bone cells have been stimulated by parathyroid hormone and/or estrogen), which is a promoter of bone formation and follicle development (in ovaries). When muscle tissue is damaged via injury, macrophages enter the muscle tissue and also produce IGF-1.

Another function of IGF-1 is to facilitate the transport of amino acids into cells, and further inhibit protein breakdown in cells. If the body is injured, IGF-1 works with platelet-derived growth factor to stimulate fibroblast and collagen cell division/metabolism to cause healing of wounds and bones. IGF-1 also occurs naturally in cow's milk. See also *Hormone, Human growth hormone* (*HGH*), *Cell, Estrogen, Fibroblasts, Amino acid, Collagen, Macrophage, Essential amino acids, Digestion* (*within organisms*), *Metabolism, Protein, Messenger RNA* (*mRNA*), *Ubiquitin, Platelet-derived growth factor* (*PDGF*).

Insulin-Like Growth Factor-2 (IGF-2)

A protein hormone that is produced by the body's brain, kidney, pancreas, and muscle tissues. IGF-2 is a primary growth factor important for early mammal development and especially for the development of the liver and kidneys. See also *Hormone, Protein, Embryology, Growth factor, Insulin-like growth factor-1* (*IGF-1*).

Intact-Cell MALDI-TOF.MS

See *ICM.*

Integrated Crop Management

See *Integrated pest management* (*IPM*).

Integrated Disease Management

See *Integrated pest management* (*IPM*).

Integrated Fluidic Circuits

See *Microfluidics, Lab-on-a-chip.*

Integrated Pest Management (IPM)

A holistic (system) approach utilized by some farmers to try to control agricultural pests (e.g., tobacco budworm, European corn borer [ECB], soybean cyst nematode, root-knot nematode, weevils) that was initially developed as a formal methodology by Ray Smith and Perry Adkisson.

For example, farmers can minimize a field's root-knot nematode populations in soil by planting *trap crops* (i.e., host plant species/strains that the nematodes cannot reproduce in, but that "trick" the nematodes into starting their life cycle) instead of letting the field lie fallow between nematode-susceptible crops. For example, California tomato growers can reduce their tomato losses to root-knot nematodes by planting a strain of wheat known Lassik in the field between tomato crops.

IPM also helps to control plant diseases. For example, farmers can plant buckwheat near their cornfields in order to help control ECB, a serious pest of corn (maize) *Zea mays* L. plants. Green lacewing beetles (*Chrysoperla carnea*), which prey on ECBs, are attracted by the buckwheat and consume ECB in the corn while they live in the buckwheat areas. Because ECB is a vector (carrier) of *disease and/or mycotoxin-producing microorganisms* such as the fungi

Aspergillus flavus, *Aspergillus parasiticus*, and *Fusarium* spp., this lacewing beetle (IPM) control of ECB also helps reduce those plant diseases and mycotoxins.

IPM is often utilized in conjunction with no-tillage crop production. See also *Weevils*, *Heliothis virescens* (*H. virescens*), *European corn borer* (*ECB*), *Fungus*, *Mycotoxins*, *Aflatoxin*, *Low-tillage crop production*, *No-tillage crop production*, *Soybean cyst nematodes* (*SCN*), *Corn*, *Soybean plant*, *Bacillus thuringiensis* (*B.t.*), *Root-knot nematode*, *Trap crop*.

Integrin Beta 4

See *Integrins*, *Graft-versus-host disease* (*GVHD*), *Human leukocyte antigens* (*HLA*).

Integrin Receptors

Also known as *cellular adhesion receptors*. See *Integrins*.

Integrins

A class of proteins that is found on the surface (membranes) of cells and that function as cellular adhesion receptors, thereby forming the structural architecture of organs and multicellular organisms. For example, integrin $\alpha_v\beta_3$ is a receptor on the surface of endothelial cells in growing blood vessels (e.g., the new blood vessels forming in a body with cancer to supply blood to growing tumors). It binds angiogenic endothelial cells, enabling them to form new blood vessels.

This is usually preceded by the body manufacturing antibodies against a tumor cells' human leukocyte antigens (HLA), thereby leading that HLA to work in concert with a protein molecule named integrin beta 4 to jointly stimulate cell growth and movement (e.g., creation of endothelial cell–lined new blood vessels to provide a blood supply to growing/metastasizing tumor). See also *Adhesion molecule*, *Protein*, *Glycoproteins*, *Cell*, *Receptors*, *Lectins*, *Selectins*, *Signal transduction*, *Angiogenesis*, *Tumor*, *Metastasis*, *Endothelial cells*, *Plasma membrane*, *Invasin*, *Human leukocyte antigens* (*HLA*), *Graft-versus-host disease* (*GVHD*).

Intein

Abbreviation for *intervening domain*; it is an internal-within-protein molecular sequence that is excised (i.e., "popped out") during a self-splicing process. An intein is a protein domain in the "center" of a protein molecule. The two halves of the protein molecule that remain after intein has been "popped out" are called exteins. The first intein was discovered in 1990.

Following translation of an intein-containing protein molecule, as soon as that protein molecule folds up:

- The end of the protein segment (C-extein) that is attached to the carboxyl end of the intein section attaches itself to the protein segment (N-extein) at the opposite end of the intein section.
- The new (and shorter) resultant protein molecule is thus freed from the intein section, which then departs.

For example, scientists working with glycoproteins (i.e., protein molecules on whose surfaces oligosaccharides are attached) can sometimes cause two such adjacent oligosaccharides (sometimes called *glycopeptide fragments*) to link chemically, by removing the intein located between their respective points of attachment to the protein molecule.

For example, during 2002, Tom Muir discovered that addition of the immunosuppressant drug rapamycin to cells being studied would cause some inteins to splice out of specific protein molecules within those cells (thereby activating or inactivating those specific proteins in the cell—allowing Dr. Muir to determine the function of those protein molecules in the cells). See also *Splicing* (*of protein molecule*), *Extein*, *Chemical genetics*, *Protein*, *Transcription*, *Translation*, *Excision* (*of protein molecule*), *Domain* (*of a protein*), *Sequence* (*of a protein molecule*), *Glycoprotein*, *Oligosaccharides*, *Cell*, *Immunosuppressive*, *Functional genomics*, *Carboxyl terminus* (*of a protein molecule*).

Intein-Based Coupling

See *Intein*.

Intercellular Adhesion Molecule (ICAM)

See *Adhesion molecule*.

Interfering RNAs

See *Short interfering RNA* (*siRNA*).

Interferons

Discovered in 1957 by Alick Isaacs and Jean Lindenman, they are a family of small (cytokines) proteins (produced by vertebrate cells following a virus infection) that *interfere* with (i.e., block) the translation of viral DNA.

Via that blocking, interferons prevent synthesis of proteins needed for viral reproduction, so interferons possess potent antiviral effects. Secreted interferons bind to the plasma membrane of other cells in the organism and induce an antiviral state in them (conferring resistance to a broad spectrum of viruses). Three classes of interferons have been isolated and purified, so far: α-interferon (originally called leukocyte interferon), β-interferon (beta interferon or fibroblast interferon), and γ-interferon (gamma interferon or immune interferon, a lymphokine). These proteins have been cloned and expressed in *Escherichia coli* (*E. coli*), which has enabled large quantities to be produced for evaluation of the interferons as possible antiviral and anticancer agents. To date, interferons have been used to treat Kaposi's sarcoma, hairy cell leukemia, venereal warts, multiple sclerosis, and hepatitis. See also *Alpha interferon*, *Beta interferon*, *Gamma interferon*, *Cytokines*, *Protein*, *Lymphokines*, *Escherichia coliform* (*E. coli*).

Interim Office of the Gene Technology Regulator (IOGTR)

The regulatory body of Australia's government that was responsible for approvals of new rDNA products (e.g., new genetically engineered crops) before they could be introduced into Australia, during 1999–2001. IOGTR replaced/superseded Australia's Gene Technology Office (in this role) 1999 and was itself replaced by the GTR in 2001. See also *Gene technology regulator* (*GTR*), *Gene Technology Office*, *Genetic Manipulation Advisory Committee* (*GMAC*), *rDNA*, *Deoxyribonucleic acid* (*DNA*), *Genetic engineering*, *Recombinant DNA Advisory Committee* (*RAC*), *Commission of Biomolecular Engineering*, *Indian Department of Biotechnology*.

I

Interleukin-1 (IL-1)

A cytokine (glycoprotein) released by activated macrophages, during the inflammatory stage of immune system response to an infection, which promotes the growth of epithelial (skin) cells and white blood cells. Research has indicated that too much IL-1 is linked to the development of rheumatoid arthritis, diabetes, inflammatory bowel disease, and other autoimmune diseases. See also *Macrophage, Autoimmune disease, Adhesion molecule, Tumor necrosis factor (TNF), Cytokines, Glycoprotein, White blood cells, Islets of Langerhans, Epithelium, Interleukin-1 receptor antagonist (IL-Ira), Interleukins.*

Interleukin-1 Receptor Antagonist (IL-1ra)

A glycoprotein (produced by macrophages in response to presence of interleukin-1 (IL-1) and endotoxin in tissues) that preferentially binds to those cell receptors in the body that typically bind the lymphokine, IL-1. When manufactured by man (e.g., via genetic engineering) and injected into the body in large quantities. IL-Ira can block the deleterious effects of (too much) IL-1. See also *Interleukin-1 (IL-1), Receptors, Receptor fitting, Glycoprotein, Macrophage, Endotoxin, Adhesion molecule, Cellular immune response, Protein, Lymphokines, Antagonists.*

Interleukin-12 (IL-12)

A cytokine (glycoprotein) produced by the body, which serves to activate the immune system against certain tumors and pathogens. See also *Cytokines, Glycoprotein, Tumor, Tumor-associated antigens, Major histocompatibility complex (MHC), T cell receptors, Cytotoxic T cells, Pathogen, Interleukins, Electroporation.*

Interleukin-18 (IL-18)

An inflammation-promoting cytokine that gets deposited in the retinas of patients with the "dry" form of age-related macular degeneration (AMD) disease. Because it is antiangiogenic (i.e., inhibits the formation/growth of new blood vessels), IL-18 helps to prevent or at least slow the progression to the "wet" form of AMD disease. See also *Cytokines, Immune response, Chronic inflammation, Age-related macular degeneration (AMD), Antiangiogenesis.*

Interleukin-2 (IL-2)

Also known as *T cell growth factor.* A cytokine (glycoprotein) secreted by (immune system response) stimulated helper T cells that promotes the proliferation/differentiation of more helper T cells and promotes the growth of lymphocytes to combat an infection. Interleukin-2 also stimulates the lymphocytes to produce gamma interferon. It is gamma interferon that prompts the cytotoxic T cells to attack virus-infected cells and kill the virus within them. The structure of the gene that codes for synthesis of IL-2 (by immune system cells) was determined by Tadatsugu Taniguchi in 1983. See also *Immune response, Humoral immunity, Cytokines, Glycoprotein, Cytotoxic T cells, T cells, Helper T cells, T cell receptors, Interferons, Interleukins, Gene.*

Interleukin-3 (IL-3)

A hematologic growth factor (glycoprotein) cytokine that stimulates the proliferation of a wide range of white blood cells (to combat an infection). See also *Hematologic growth factors (HGF), Glycoprotein, Cytokines, White blood cells, Interleukins.*

Interleukin-4 (IL-4)

A cytokine (glycoprotein) that stimulates production of antibody-producing B cells, immunoglobulin-E (I_gE), and promotes cytotoxic T cell (i.e., killer T cells) growth. See also *Antibody, Cytotoxic T cells, B cells, Glycoprotein, Cytokines, Immunoglobulin, Interleukins.*

Interleukin-5 (IL-5)

A cytokine (glycoprotein) that stimulates eosinophil growth. See also *Eosinophils, Protein, Glycoprotein, Cytokines, Cellular immune response, Interleukins.*

Interleukin-6 (IL-6)

A cytokine (glycoprotein) that is pleiotropic (i.e., stimulates several different types of immune system cells) and is a hematopoietic growth factor.

For example, infections and certain physical trauma can cause the body to produce IL-6, which subsequently causes the liver to synthesize (manufacture) *C-reactive protein.* See also *Hematopoietic growth factors (HGF), Growth factor, Glycoprotein, Pleiotropic, Macrophage, Cytokines, C-reactive protein (CRP), Interleukins, Chronic inflammation.*

Interleukin-7 (IL-7)

A cytokine (glycoprotein) synthesized in the bone marrow that stimulates early (fetal) proliferation and differentiation of B cells and T cells. May be useful in regenerating lymphoid cells in patients whose immune systems have been devastated by cancer chemotherapy. See also *Cytokines, Glycoprotein, Stem cell one, T cells, Cancer, Interleukins.*

Interleukin-8 (IL-8)

A basic polypeptide (glycoprotein) with heparin-binding activity that, as part of the body's response to some diseases, attracts white blood vessels to the applicable tissues and activates those white blood vessels. Endogenous endothelial IL-8 appears to regulate transvenular traffic during acute inflammatory responses. See also *Polypeptide (protein), Glycoprotein, Heparin, Endothelial cells, Endothelium, Polymorphonuclear leukocytes (PMN), Cellular immune response, Interleukins.*

Interleukin-9 (IL-9)

A cytokine (glycoprotein) that is released at sites in the body where inflammation has occurred. See also *Cytokines, Glycoprotein, Cellular immune response, Interleukins.*

Interleukins

A class of 24 different cytokines that "carry a signal" *between different leukocyte populations* within the immune system of an organism. See also *Cytokines, Leukocytes, Interleukin-1 (IL-1), Interleukin-2 (IL-2), Interleukin-3 (IL-3), Interleukin-4 (IL-4), Interleukin-5 (IL-5), Interleukin-6 (IL-6), Interleukin-7 (IL-7), Interleukin-8 (IL-8), Interleukin-9 (IL-9), Interleukin-12 (IL-12).*

Intermediary Metabolism

The chemical reactions that take place in the cell that transform the complex molecules derived from food into the small molecules needed for the growth and maintenance of the cell. See also *Metabolism*, *Cell*, *Digestion* (*within organisms*), *Metabolic pathway*.

International Food Biotechnology Council (IFBC)

An organization that was established in 1988 by the Industrial Biotechnology Association (IBA) and the International Life Sciences Institute (ILSI), in order to "produce a (recommended) set of guidelines that could be used to assess the safety of genetically altered foods." See also *GNE*, *Industrial Biotechnology Association* (*IBA*), *International Life Sciences Institute* (*ILSI*), *Senior Advisory Group on Biotechnology*, *Biotechnology Industry Organization* (*BIO*), *Genetic engineering*, *Polygalacturonase*, *Antisense* (*DNA sequence*), *Biotechnology*, *Bacteriocins*.

International Life Sciences Institute (ILSI)

A nonprofit foundation that was established in 1978 to advance the understanding of scientific issues relating to nutrition, food safety, toxicology, risk assessment, and the environment. ILSI is headquartered in Washington, DC, and has branches in Argentina, Brazil, Europe, India, Japan, Korea, Mexico, Africa, Thailand, Singapore, China, and other nations.

International Office of Epizootics (OIE)

One of the three international SPS standard-setting organizations that is recognized by the World Trade Organization (WTO), the OIE is an international veterinary organization headquartered in Paris. Also known as the World Organization for Animal Health, the OIE was established in 1924, originally as part of the League of Nations, and is the worldwide authority for development of animal health and zoonoses standards, guidelines, and recommendations. See also *SPS*, *International Plant Protection Convention* (*IPPC*), *Zoonoses*, *World Trade Organization* (*WTO*).

International Plant Protection Convention (IPPC)

One of the three international SPS standard-setting organizations that is recognized by the World Trade Organization, the IPPC is the worldwide authority for development of plant health standards, guidelines, and recommendations (e.g., to prevent transfer of a plant disease or plant pest from one country to another). The treaty establishing the IPPC was signed in 1952 (amended in 1979 and 1997) and currently has 107 member countries (i.e., signatories to the 1979 text).

The IPPC Secretariat is within the United Nations' Food and Agriculture Organization. IPPC standards are set (and enforced) via regional SPS institutions such as the North American Plant Protection Organization, European Plant Protection Organization, Southern Cone Plant Protection Organization, and so on. There are currently nine RPPOs (i.e., regional plant protection organizations) under Article VIII of the 1979 IPPC text. See also *SPS*, *European Plant Protection Organization* (*EPPO*), *International Office of Epizootics* (*OIE*), *World Trade Organization* (*WTO*), *North American Plant Protection Organization* (*NAPPO*), *Southern Cone Plant Protection Organization* (*COSAVE*), *National Plant Protection Organization* (*NPPO*), *Quarantine pest*, *Introduction*, *Establishment potential*.

International Society for the Advancement of Biotechnology (ISAB)

A nonprofit organization of individuals that was started in 1994 "to advance and promote the general welfare of the science and commercialization of genetic engineering and industrial biotechnology." See also *Genetic engineering*, *Biotechnology*, *Biotechnology Industry Organization* (*BIO*).

International Union for Protection of New Varieties of Plants (UPOV)

See *Union for Protection of New Varieties of Plants* (*UPOV*).

Internaulin

See *Cadherins*.

Intervening Domain

See *Intein*.

Intracellular Transport

See *Cell*, *Gated transport*, *Lipids*, *Membrane transport*, *Transport proteins*.

Intragenesis

A form of genome editing that results in the organism's resultant DNA containing a combination of different genes and/or expression elements from donor organisms of the same or sexually compatible specie(s)—in either a *sense* or an *antisense* orientation. See also *Deoxyribonucleic acid* (*DNA*), *Gene*, *Organism*, *Genome*, *Genome editing*, *Sense*, *Antisense* (*DNA sequence*), *Cisgenesis*.

Intrinsic Protein

Refers to a protein molecule that is embedded within a cell membrane (and protrudes from each side of the membrane). See also *Protein*, *Cell*, *Plasma membrane*, *Membranes* (*of a cell*), *Transmembrane proteins*, *Ion channels*, *Ionotropic*.

Intrinsically Unstructured Proteins

See *IUP*.

Introduction

Term utilized (e.g., by the IPPC) to refer to the *entry* and *successful establishment* of a given pest (e.g., weed, insect, disease) into a (formerly) "pest-free area" (i.e., country or region where *that pest* is not yet present, or is present but not widely distributed and thus officially controlled). See also *International Plant Protection Convention* (*IPPC*), *Establishment potential*, *Quarantine pest*, *National Plant Protection Organization* (*NPPO*).

Introgression

The incorporation of exotic (i.e., wild type) genes into elite germplasm (i.e., domesticated breeding lines) or of transgenes (i.e., genes from transgenic organisms) or cisgenes into a wild type's genome. See also *Transgenic, Outcrossing, Wild type, Genome, Gene, Cisgenics, Variety (e.g., of crop plant), Translocation.*

Intron

Discovered in 1977, an intron is an (intervening sequence) segment of deoxyribonucleic acid (DNA) within a gene that is transcribed but is removed from within the mRNA transcript by splicing together the sequences (exons) on either side of it (in the molecule) by snRNP during the final step of the transcription process. In the past, it was generally considered to be a "nonfunctioning" portion of the DNA molecule.

However, during the 1990s, Malcolm Simon showed that some introns contain the "markers" that scientists utilize to identify where a given gene (within DNA strand) begins and ends. For example, the genetic test (conducted on women) for the presence of a "BRCA 1" gene actually detects a *DNA* "marker" in the intron sequence *near* the "BRCA 1" gene, not the "BRCA 1" gene itself *per se.*

Some small RNAs (short interfering RNAs) are coded for by specific DNA segments within certain introns. There also exist within many introns, enhancers (i.e., particular short DNA segments that act to either increase/activate/"turn on" a given gene, or that act to decrease/silence/"turn off" a given gene). The genes that enhancers thereby regulate may be located up to several thousand base pairs distant from that enhancer.

Certain DNA segment(s) within some introns in an organism's DNA can interact with (e.g., "turn on," "turn off," etc.) a specific gene that is located a long distance away (within same DNA molecule) from that initial DNA segment. This physical interaction (i.e., creation of a large loop in the organism's DNA molecule, to cause the applicable intron to "touch" the relevant gene) results in an apparent genetic effect. For example, via such DNA looping, aberrant DNA segments in 14 different introns result in increased risk of bowel cancer for those people whose DNA contains one or more of those aberrant DNA segments.

Also, sometimes a given intron remains in the transcript (e.g., via alternative splicing), resulting in a *different* protein expressed by the same gene. For example, the COX-3 enzyme and the COX-1 enzyme are both produced from the COX-1 gene. The COX-3 enzyme results when *intron 1* is retained in the mRNA transcript. See also *Transcription, Deoxyribonucleic acid (DNA), DNA looping, Messenger RNA (mRNA), Exon, Gene, Editing, Splicing, Alternative splicing, Splicing junctions, Marker (DNA sequence), BRCA genes, Short interfering RNA (siRNA), Enzyme, COX-1, COX-2, COX-3, Cyclooxygenase, Enhancer.*

Inulin

A fructose oligosaccharide (FOS) that is naturally produced in more than 30,000 plants. Like many other FOS, consumption of inulin by humans results in several health benefits (e.g., help prevent coronary heart disease, promote growth of bifidobacteria in the intestines, reduce likelihood of developing diabetes, promote absorption of calcium from foods). During 2000, the European Union's government regulatory agencies agreed to classify inulin as a water soluble fiber (because humans cannot digest inulin). See also *Fructose oligosaccharides, Water soluble fiber, Bifidobacteria, Coronary heart disease (CHD), Diabetes.*

Invadosomes

See *Actin.*

Invasin

A transmembrane (i.e., through the membrane of the cell) protein present on the surface of some bacteria cells that enables those bacterial cells to attach themselves to 1-integrins (a protein present in the plasma membrane of certain mammal cells) and thereby enter a mammal's normal (body) cells to cause infection. See also *CD4 protein, Receptors, Cell, T cell receptors, Endocytosis, Plasma membrane, Integrins.*

Inverted Micelle

See *Reverse micelle (RM), Micelle.*

Investigational New Drug

See *IND.*

Invitro

See *In vitro.*

In Vitro Evolution

See *In vitro selection.*

In Vitro Selection

See *In vitro selection.*

Invivo

See *In vivo.*

IOGTR

See *Interim Office of the Gene Technology Regulator (IOGTR).*

Ion

From the Greek *ion* = "something that goes." An ion is an atom or molecule possessing a positive or a negative electrical charge. Ions are produced by the dissociation (coming apart) of an (electrolyte) molecule resulting from the electrolyte dissolving in solution. One example is the dissociation of common table salt (i.e., sodium chloride) in water, which results in positively charged sodium ions (called cations) and negatively charged chloride ions (called anions). Ions play critically important roles in many biological processes such as nerve activity. See also *Chelation, Chelating agent, Ion channels, Citric acid, Citrate synthase (CSb) gene.*

Ion Channels

Refers to specialized proteins that act as "pores" (e.g., through the plasma membrane of a cell) through which certain ions (i.e., atoms or molecules bearing an electrical charge) are *selectively* allowed to pass. Examples include calcium channels, sodium channels, and potassium channels. The selectivity of ion channels can be altered when specific molecules (e.g., in the blood or digestive fluids) come in contact with the plasma membrane (i.e., G-protein receptors coupled to the ion channel).

For example, the group of pharmaceuticals known as *calcium channel blockers* (e.g., verapamil, amlodipine, diltiazem, nifedipine) acts to "block"/hinder the movement of calcium ions through *calcium ion channels* (i.e., "pores" that had previously allowed calcium ions to enter relevant cells [i.e., in blood vessel walls] easily).

Another example is the mode of action of the "cry" (crystal-like) *proteins* that are naturally present within *Bacillus thuringiensis* (*B.t.*) bacteria. When eaten by certain insects (possessing alkaline digestive fluids in their stomach/gut), cry proteins are hydrolyzed (i.e., chemically "cut") into fragments. One of those fragments—60 kDa in size—attaches to specific receptors located on the surface (membrane) of certain cells that line the inside (i.e., epithelium) of the insect's midgut. That *attachment to those receptors* triggers ion channels in the (epithelium) cell's membrane to suddenly allow cations (i.e., atoms or molecules with positive electrical charge) to quickly flow out of the cell (which leads to death of all insect gut cells that the *cry protein piece* attached to).

The ion channel known as SLAC1 is utilized by plants to control the opening and closing (e.g., in response to drought conditions) of the stomatal pores located on surface of plant leaves. Because those stomatal pores must be open *enough* to allow carbon dioxide to enter the plant leaves (i.e., it is needed for photosynthesis) and for oxygen to enter/exit the leaves—but must not allow *too much* water vapor to exit the leaves (especially during drought conditions)—the survival of a plant depends on the precise regulation of stomatal pore openings in response to environmental stimuli that it achieves via SLAC1. See also *Cell, Plasma membrane, Ion, Calcium channel-blockers, Membrane transport, Protein, Cry proteins, G-proteins, Bacillus thuringiensis* (*B.t.*), *Bacteria, Protoxin, Hydrolyze, Kilodalton* (*Kd*), *Receptors, Epithelium, Ionotropic, Gated channel, Intrinsic protein, Stomatal pores, Photosynthesis.*

Ion Trap

Invented by Wolfgang Paul in 1954, it is a device that is utilized to confine ions (e.g., from a sample entering a mass spectrometer) within a small volume of space, without the use of physical walls. Instead, it utilizes three carefully placed hyperbolic electrodes to which applicable radio-frequency voltage potential is applied. The ions are thereby confined within the desired volume of space by high-frequency electrical fields.

In *ion trap–based mass spectrometers*, the voltages of the electrodes are selectively changed to cause specific ions (i.e., pieces of the original sample molecules) to be ejected from the ion trap into the spectrometer's detector. As with all mass spectrometers, those "pieces of sample" are processed as follows:

- They are separated by the differences in their mass-to-charge ratios.
- Their exact mass is determined based on the measurement of their mass-to-charge ratios while *those* "pieces" are passing through precisely known strength electromagnetic fields.
- The identity of the "pieces" is determined by comparison of their mass/charge (m/e) spectra to those within a database of known "pieces" (ions).

See also *Ion, Mass spectrometer, Molecular weight.*

Ion-Exchange Chromatography

Separation of ionic compounds (which include nucleic acids and proteins) in a chromatographic column containing a polymeric resin (i.e., the stationary phase) having fixed charge groups. The process works in that the charges of the column (stationary phase) interact with the opposite charges of the material dissolved in the solution that is flowing through the column (mobile phase). The charge interaction between the column material and, say, the protein has the effect of slowing down the rate of movement of the protein through the column. The other molecules, meanwhile, which do not interact with the column, flow right on through. This then constitutes the separation process. See also *Chromatography.*

Ionic Liquids

Refers to a category of organic salts that possess melting points of lower than 100°C, dissolve both polar and nonpolar molecules, and have very little or no vapor pressure in the temperature range typically utilized for bioconversions done by man (e.g., use of an enzyme to convert plant starch to sugars, prior to fermentation production of alcohol in a vat).

Because relevant ionic liquids are *protective of such enzymes* (*e.g., prevent degradation of the enzyme molecules*), those ionic liquids are sometimes the solvent of choice for bioconversions done by man. See also *White biotechnology, Enzyme, Fermentation, Feedstock, Substrate* (*chemical*).

Ionotropic

Refers to a cellular receptor that impacts (mediates) that cell's processes/states, etc. via regulation of the cell's ion channels. See also *Ion channels, Cell, Receptors.*

IP-6

Inositol hexaphosphate. See *Phytate.*

IPM

See *Integrated pest management* (*IPM*).

IPPC

See *International Plant Protection Convention.*

iPS Cells

See *Induced pluripotent stem cells.*

iPSC

See *Induced pluripotent stem cells.*

iRNA

Acronym for information RNA. See *Information RNA* (*iRNA*).

Iron Bacteria

See *Ferrobacteria.*

Iron Deficiency Anemia (IDA)

A disease caused by lack of iron in an organism's body, due to shortfall in diet or due to dietary iron not being bioavailable (digestible) to that organism's body. For example, the phytate that is naturally present in traditional varieties of corn (maize) inhibits absorption of the iron in that corn (maize) by humans, swine, and poultry.

IDA is a major cause of childhood diseases and maternal death (i.e., death of the mother following childbirth) in many developing countries. IDA also makes people more susceptible to diphtheria.

Ascorbic acid (vitamin C) is important in the human diet because it enables more iron, which carries oxygen to all cells, to be taken up and absorbed. See also *Golden rice, Phytate, Low-phytate corn, Low-phytate soybeans, Organism, Ascorbic acid.*

Islets of Langerhans (Also Called Beta Cells)

Cells in the pancreas that produce insulin in response to the presence of glucose (sugar) in the bloodstream. The failure of insulin production results in the disease called diabetes. See also *Glucose (GLc), Glycolysis, Autoimmune disease, Insulin, Insulin-dependent diabetes mellitus (IDDM), Serotonin.*

Isobaric

Refers to two items (e.g., two different chemical reagents) possessing identical mass. See also *Dalton, Molecular weight.*

Isoelectric Focusing (IEF)

An electrophoresis methodology in which protein molecules are moved (via application of an electrical charge/potential) through a pH gradient (e.g., in a 2D gel until they reach their individual isoelectric points).

IEF is the first step in many *gene expression studies*, followed by extraction of the individual (separated) proteins for identification and quantitation (i.e., *how much* of each protein was produced by the cell/tissue/organism being evaluated). See also *Two-dimensional (2D) gel electrophoresis, Gene expression analysis, Protein, Gene expression profiling, Cell, Gene function analysis, Organism, Isoelectric point, Capillary electrophoresis.*

Isoelectric Point

Abbreviated as *pI*, this refers to the point

- In an ionic solution, at which the pH of the solution results in the (solute) molecule possessing no net charge.
- In a 2D gel, at which the charge/mass of a given protein is exactly matched by the electrical charge/potential applied to that 2D gel. Because the isoelectric point is different for virtually every protein (e.g., in a sample applied to the 2D gel), this enables separation of individual proteins from a (mixed) sample.

See also *Ion, Two-dimensional (2D) gel electrophoresis, Protein, Isoelectric focusing (IEF).*

Isoenzymes

See *Isozymes.*

Isoflavins

See *Isoflavones.*

Isoflavones

A group of phytochemicals (including genistein, glycitein, and daidzein) that are produced within the seeds of the soybean plant (*Glycine max* (L.) Merrill) at a typical concentration of approximately 0.04%–0.24%. Isoflavones are also produced within other types of tissues of the soybean *plant* (e.g., to ward off infection by plant diseases such as *Phytophthora* ones) and the soybean plant's *roots* (e.g., to signal and attract the *Rhizobium japonicum* bacteria that live symbiotically among the soybean plant's roots and "fix" nitrogen from the air, thereby providing natural fertilizer for the plant). Much smaller amounts of isoflavones are produced in some wheat, lentils, chickpeas, and edible bean plants.

Evidence shows that consumption of soybean isoflavones by humans can help lower the blood content of low-density lipoproteins, help prevent osteoporosis, help prevent prostate enlargement, and help reduce the risk of certain types of cancer (e.g., breast cancer, colon cancer, lung cancer, prostate cancer, uterine cancer).

A human diet containing a large amount of isoflavones has been shown to increase bone density and to decrease total serum cholesterol, thereby lowering the risk of osteoporosis and coronary heart disease.

Isoflavones also exhibit antioxidant properties. See also *Genistein (GEN), Soybean plant, Bradyrhizobium japonicum, Phytoalexins, Phytochemicals, Low-density lipoproteins (LDLP), Osteoporosis, Prostate-specific antigen (PSA), Cancer, Selective estrogen effect, Stress proteins, Cholesterol, Nitrogen fixation, Nodulation, Coronary heart disease (CHD), Osteoporosis, Rhizobium (bacteria), Phytophthora megasperma* f. sp. *glycinea, Phytophthora root rot, Signaling, Signaling molecules, High-isoflavone soybeans, Antioxidants, Oxidative stress.*

Isoflavonoids

See *Isoflavones.*

Isolated Soy Proteins

See *Soy protein.*

Isoleucine (ile)

A monocarboxylic amino acid occurring within most dietary proteins. See also *Amino acid, Protein, ALS gene.*

Isomer

One of the two or more chemical substances having the same elementary percentage composition (i.e., same atoms) and molecular weight, but differing in structure and therefore in properties. There are many ways in which such structural differences (between the two or more isomeric molecules) occur. One example is *n*-butane $[CH_3(CH_2)_2CH_3]$ and isobutane $[CH_3CH(CH_3)_2]$. See also *Stereoisomers.*

Isomerase

A category of enzymes that can catalyze transformation of a given compound into its positional isomer.

One of the isomerases (i.e., topoisomerase) can either cause, or reduce, supercoiling in DNA molecules. See also *Enzyme, Isomer, Deoxyribonucleic acid (DNA), Supercoiling, Topo-isomerase.*

Isoprene

The five-carbon hydrocarbon molecule: 2-methyl-1,3 butadiene. It is a recurring structural unit of the terpenoid molecules, which are either linear or cyclic. There exists a very large number of terpenes and many are major components of essential plant oils. See also *GTPases*.

Isotachophoresis

Refers to one of the capillary electrophoresis technologies, in which the sample's components are (additionally) separated between the *leading electrolyte* (i.e., injected into the capillary tube first) and the *terminating electrolyte* (i.e., injected into the capillary tube last). See also *Capillary electrophoresis*, *Electrolyte*.

Isothiocyanates

A category of nutritionally beneficial chemicals naturally derived via enzymatic transformation from glucosinolates. For example, the enzyme myrosinase that is present within cells of the wasabi plant (*Wasabia japonica*) catalyzes the conversion of that plant's glucosinolates to the isothiocyanates that provide the spicy taste of Wasabi food ingredient. See also *Enzyme*, *Cell*, *Sulforaphane*, *Glucosinolates*.

Isotope

A term coined in 1913 by Frederick Soddy, it refers to one of the several "varieties" of atoms that exist, of the same element, that differ from each other in the number of neutrons in the atom's nucleus. For example, the element chlorine exists primarily in two forms (isotopes) in nature—with 18 neutrons (76% of the time) and with 20 neutrons (24% of the time).

From the Greek *isos* ("same") and *topos* ("place"), because different isotopes of a given element occupy the same place in the periodic table. The chemical properties of isotopes of a given element are virtually identical. See also *Atomic weight*.

Isozymes (Isoenzymes)

Multiple forms of an enzyme that differ from each other in their substrate (substance acted upon) affinity, in their maximum activity, or in their regulatory properties. See also *Enzyme*, *Substrate (chemical)*, *Ribozymes*, *PGHS*.

ISPM

Acronym for International Standards for Pest Management. See also *International Plant Protection Convention (IPPC)*.

ITP

Acronym for Isotachophoresis. See *Isotachophoresis*.

IUP

Acronym for *intrinsically unstructured proteins*. See *Structural biology primary structure*, *Protein folding*, *Conformation*, *Tertiary structure*.

IκB Kinase

A kinase that helps regulate the NFκB pathway. See *NFκB*, *Kinases*.

J

JAK

Abbreviation for Janus kinases. See *Janus kinases.*

Janus Kinases

Refer to a "family" of four different tyrosine kinase enzymes (abbreviated JAK1, JAK2, JAK3, and TYK2) that play crucial roles in numerous signaling pathways within the body. Such signaling pathways (e.g., of certain hormones, growth factors, cytokines, etc.) serve to regulate specific body responses such as the immune response(s) to pathogens, the process of erythropoiesis, and so on.

The inhibition of certain Janus kinases (e.g., those acting aberrantly in the body) may help in the treatment of certain diseases caused by those aberrant Janus kinases (e.g., rheumatoid arthritis, allergies, asthma).

When certain Janus Kinases get "switched on," that can lead to (muscle-like) contractions within tumor cells that generate a force that causes the tumor's cells to move (e.g., through narrow spaces) and metastasize. See also *Enzyme, Kinases, Pathway, Signaling, Pathogen, Erythropoiesis, Tyrosine kinase inhibitors (TKI), Cancer, Tumor, Metastasis.*

Japan Bio-Industry Association

An association of the largest Japanese companies that are engaged in at least some form of genetic engineering research or production. Similar to America's Biotechnology Industry Organization, it is headquartered in Tokyo. See also *Biotechnology Industry Organization (BIO), Biotechnology, Genetic engineering, Recombinant DNA (rDNA), Senior Advisory Group on Biotechnology (SAGB), International Food Biotechnology Council.*

Jasmonate Cascade

Refers to the cascade of different (signaling, etc.) natural chemicals that are produced in response to certain pest insects chewing on some plant species.

For example, in response to such insects chewing on the *Nicotiana attenuata* plant, that plant expresses *lipoxygenase 3* and certain other enzymes that cause production (via oxylipin pathways) from linolenic acid of jasmonic acid, which triggers specific plant defenses (e.g., systemic acquired resistance). See also *Cascade, Signaling molecule, Enzyme, Pathway, Jasmonic acid, Systemic acquired resistance (SAR), Lipoxygenase (LOX), Polyunsaturated fatty acids (PUFA), Linolenic acid, Oxylipins, Green leaf volatiles.*

Jasmonates

A category of plant hormone that plants use to regulate the production of metabolites that interfere with insect digestion. See *Jasmonate cascade, Hormone, Metabolite.*

Jasmonic Acid

Jasmonic acid is a signaling molecule produced by the soybean plant (*Glycine max* L. Merrill) and several other plants in response to insects chewing on their leaves. The presence of that jasmonic acid signal causes the plant to increase its defenses, such as to increase its production of protease inhibitor(s) that decrease the insects' ability to digest food.

Jasmonic acid is also produced by plants as part of the systemic acquired resistance (SAR) when SAR is triggered in plants (e.g., via spray application of harpin protein to various plants, via chewing of insects on the leaves of certain plants, and/or via the entry into plant of certain pathogenic bacteria/fungi).

In some plants (e.g., corn *Zea mays* L.), jasmonic acid plays a part in the development of the male flower. See also *Systemic acquired resistance (SAR), Signaling molecule, Soybean plant, Fungus, Pathogen, Protein, Pathogenesis related proteins, Harpin, Phytoalexins, Jasmonate cascade, Oxylipins, Protease, Corn.*

Joining Segment

Refers to a certain short segment of DNA that physically links two genes, resulting in a large *functional gene* that codes for an immunoglobulin molecule. See also *Deoxyribonucleic acid (DNA), Gene, Immunoglobulin.*

Jumping Genes

Genes that move (change positions) within the genome. Genes associated with transposable elements. A segment fragment of deoxyribonucleic acid (DNA) that can move from one position in the genome to another. See also *Gene, Genome, Deoxyribonucleic acid (DNA), Genetic code, Transposition, Transposon, Translocation, Introgression, Hot spots.*

Juncea

Refers to a group of related plants; often commonly called "wild mustard." See *Brassica.*

J

Junk DNA

A term historically utilized by some to refer to portions of an organism's DNA that were not *obviously* genes (i.e., not transcribed into the mRNA, thus not part of the DNA "tagged"/labeled with ESTs). However, it was subsequently discovered that at least some of what was formerly called "junk DNA" (e.g., introns) helps enable more than one specific protein molecule to be expressed from certain genes.

In the human genome (DNA), approximately 2% of the total DNA is obviously genes (i.e., transcribed to mRNA from which proteins are subsequently synthesized). The other 99% of human DNA is transcribed to so-called noncoding RNA. See also *Deoxyribonucleic acid (DNA), Gene, Intron, Protein, Express, Messenger RNA (mRNA), Ribosomes, Expressed sequence tag (EST), Central dogma (new).*

K

KARI

Acronym for either the *Kenya Agricultural Research Institute* or the *Kawanda Agricultural Research Institute in Uganda.*

Karnal Bunt

A plant disease that can be caused by the smut fungus *Tilletia indica* in wheat. See also *Fungus, Wheat.*

Karyopherins

See *Nuclear proteins.*

Karyotype

A size-order alignment of an organism's chromosome pairs in the format of a (photomicrograph) chart. It enables the connecting of chromosomes to symptoms (e.g., of genetic diseases in the organism) and traits. See also *Chromosomes, Gene, Genotype, Trait, Linkage, Linkage group, Muscular dystrophy (MD), Chromatids, Chromatin, Aneuploid.*

Karyotyper

A scientist (or more frequently an automated analytical machine) that

- Takes a video picture of a given cell under a microscope
- Digitizes that picture within a computer
- "Cuts out" the individual chromosomes contained within that cell's genome
- Arranges the cell's chromosomes in pairs by size order into a chart (called a karyotype)

See also *Chromosomes, Genome, Karyotype, Aneuploid.*

kb

An abbreviation for 1000 (kilo) base pairs of deoxyribonucleic acid (DNA). See also *Deoxyribonucleic acid (DNA), Kilobase pairs (kbp).*

Kd

An abbreviation for kilodalton. See *Kilodalton (kDa).*

Kefauver Rule

A 1962 United States' law that mandates that the Food and Drug Administration (FDA) requires proof of pharmaceutical *efficacy* for drugs to be sold in the United States. See also *Food and Drug Administration (FDA).*

Kenya Biosafety Council

The country of Kenya's national regulatory body for granting approval to a new genetically engineered plant (e.g., a new genetically engineered crop to be planted).

The Kenya Biosafety Council is analogous to Germany's ZKBS (Central Commission on Biological Safety), Australia's GMAC (Genetic Manipulation Advisory Committee), or Brazil's CTNBio (National Biosafety Commission). See also *GMAC, Recombinant DNA Advisory Committee (RAC), ZKBS (Central Commission on Biological Safety), Genetic Engineering, CTNBio.*

Keratins

Insoluble protective or structural proteins consisting of parallel polypeptide chains arranged in an α-helical or β conformation. See also *Protein, Conformation.*

Ketose

A simple monosaccharide having its carbonyl groups at other than a terminal position. See also *Monosaccharides.*

Killer T Cell

See *Cytotoxic T cells.*

Kilobase Pairs (kbp)

A unit of DNA equals to 1000 bp. See also *Base pair (bp), Deoxyribonucleic acid (DNA).*

Kilodalton (kDa)

A unit of mass equal to 1000 Da. See also *Dalton.*

Kinase Assays

Refer to a variety of assays (e.g., radiolabeling, antibody-binding assays) that are utilized to assess the biological activity of kinase inhibitor compounds (e.g., certain pharmaceuticals) against kinases.

For example, the pharmaceutical Gleevec™ (imatinib mesylate) inhibits the kinase known as *Bcr-Abl tyrosine kinase*, which can cause excessive production of white blood cells (leukemia), if unchecked. See also *Assay, Bioassay, Radiolabeled, Label (radioactive), Kinase inhibitors, Antibody, Radioimmunoassay, Gene, Gene expression analysis, Enzyme, Kinases, Gleevec™, White blood cells, Kinome.*

Kinase Cascades

Discovered by Edwin G. Krebs and Edmond Fischer in 1955. See *Kinases.*

K

Kinase Inhibitors

Refers to compounds that inhibit the action of kinases (i.e., a category of enzymes that facilitate the transfer of "phosphoryl groups" from one molecule to another molecule).

Because some kinases regulate the signal transduction inherent in the diseases of diabetes, cancer, and Alzheimer's, certain kinase inhibitors may be useful in treating those diseases.

For example, the pharmaceutical Gleevec™ (imatinib mesylate) inhibits the kinase known as *Bcr-Abl tyrosine kinase*, which can cause excessive production of white blood cells (leukemia), if unchecked. For example, the pharmaceutical bafetinib™ inhibits the kinases known as *Bcr-Abl*, *Lyn*, and *Fyn tyrosine kinases*.

See also *Kinases, Enzyme, Phosphorylation, Signal transduction, Diabetes, Cancer, Alzheimer's disease, Kinome.*

Kinases

A category of enzymes that (assist/facilitate) transfer of "phosphoryl groups" (from one molecule to another molecule that is "targeted" by that kinase).

The subcategory known as *MAP kinases* (*MAPK*) helps transfer certain "signals" from the cell's exterior (receptors) to its nucleus (i.e., thereby causing phosphorylation of certain protein molecules in the nucleus), resulting in changes to the cell's protein-synthesizing processes. That "transfer" occurs via a *cascade* in which each of a *series* of kinases transfers a phosphoryl group to another molecule that is itself a kinase (and that kinase then passes it to another kinase, etc.).

Such kinases signaling cascades are involved in cell apoptosis, differentiation, transcription, growth regulation pathway(s), and other cellular processes.

Some kinases regulate the signal transduction inherent in the diseases of diabetes, cancer, and Alzheimer's. Thus, certain kinase inhibitors may be useful in treating those diseases. See also *Enzyme, Kinome, Phosphorylation, Protein, Protein kinases, Cell, Receptors, Signal transduction, Cascade, Mitogen-activated protein kinase cascade, MAPK, Tyrosine kinase, Gleevec™, Nucleus, Tyrosine kinase inhibitors (TKI), Apoptosis, Transcription, Knockin, Cell differentiation, Kinome, Amyloid β protein (AβP), Pathway, Alzheimer's disease, Cancer, Diabetes, Kinase inhibitors.*

Kinesin

A contractile (i.e., periodically contracting) protein—also called a "motor protein"—within cells, which transports cellular "cargo" such as vesicles or proteins complexed with chaperones along microtubules (string-like structures) within the cell.

That transportation is accomplished via "walking" of the kinesin molecule along the microtubule, powered by hydrolysis of ATP molecules (one of which binds to the kinesin molecule between each "step" taken).

Some viruses (e.g., *Vaccinia*) also utilize kinesin to transport their viral core particle (i.e., following replication of the viral DNA in cell's nucleus) to the surface of the cell, where it is released to go infect new cells. See also *Protein, Cell, Vesicle, Chaperones, Microtubules, Virus, Nucleus, Deoxyribonucleic acid (DNA), Hydrolysis, Adenosine triphosphate (ATP).*

Kinome

Refers to the set of all kinases and their products (i.e., phosphorylated proteins) present within the cells of a given organism, sometime in its lifetime. For example, the human kinome is currently known to contain approximately 520 kinases. Plus, knowledge of each kinase's function, its gene express/activation pattern in different types of tissue (diseased and normal), and each kinase's substrate (i.e., what it chemically acts upon). See also *Kinases, Kinase inhibitors, Kinase assays, Enzyme, Protein, Phosphorylation, Cell, Gene, Genetic map, Genomics, Organism, Functional genomics, Gene expression analysis, Protein interaction analysis, Substrate (chemical).*

Knockdown

Refers to (a scientist's) alteration of a particular gene within an organism, so that a specific gene may subsequently *not be expressed*, or be expressed only under (controlled) condition(s) selected by that scientist. See also *Gene, Organism, Express, Expressivity, RNA interference (RNAi), Short interfering RNA (siRNA), Short hairpin RNA, Knockout, Homologous recombination, Transfection, CRISPR/Cas9 gene-editing systems.*

Knockin

Refers to (a scientist's) alteration of a particular gene within an organism, so that specific organism gains a desired function (e.g., to be able to produce a therapeutic protein in its mammary gland, etc.).

For example, an "ASKA gene" can be "knocked in" to laboratory mice, in which the ASKA (i.e., analog-sensitive kinase allele) gene codes for a kinase (in the mouse's cells) that is susceptible to modulation by certain compounds that act as chemical analogues of kinases. See also *Gene, Organism, Protein, Kinases, Genetic code, Express, Analogue, CRISPR/Cas9 gene-editing systems.*

Knock-in

See *Knockin.*

Knockout

Refers to one of the following:

- (A scientist's) Alteration of a *particular gene* within an organism, so that the organism loses a (specific) function (e.g., the ability to produce a given needed clotting factor in its blood, the ability to produce a given allergen in its seeds).
- The *altered organism* itself (i.e., in which the particular gene has been inactivated as detailed earlier).
- (A scientist's) Alteration of a *particular protein* (e.g., within an organism's cell) so that protein loses its biological activity (e.g., the ability to cause blood clotting). That can enable a detailed study of which proteins within a cell are responsible for particular diseases, and so on.

Such "gene knockout" can be accomplished via any one of several different methods/technologies, such as gene silencing, cosuppression, site-directed mutagenesis, short interfering RNA (siRNA), zinc finger nuclease, TALENs, CRISPR/Cas9 gene-editing systems, and so on.

Such "protein knockout" can be accomplished via any one of several different methods/technologies, such as laser inactivation and so on. See also *Gene, Organism, Protein, Biological activity, Cell, Gene silencing, Cosuppression, GPA1, Site-directed mutagenesis (SDM), Laser inactivation, RNA interference (RNAi),*

Reduced-allergen soybeans, Short interfering RNA (siRNA), Zinc finger nuclease, Proteomics, Deletions, Cre-Lox system, TALENs, CRISPR/Cas9 gene-editing systems.

Knockout (Gene)

See *Knockout, Zinc finger nuclease, Gene silencing, GPA1, Nuclear transfer, Cre-Lox system, Deletions, RNA interference (RNAi), TALENs, CRISPR/Cas9 gene-editing systems.*

Knottins

Refers to a structural category of molecules, whose (molecule) shape visually "looks like" a knot in a rope. First discovered in 1982.

Examples of knottins include *Ecballium elaterium* trypsin inhibitors. See *EETI.*

KO

Acronym for Kusabira Orange. See *Kusabira Orange.*

Konzo

A term used in some countries to refer to *lathyrism*. See *Lathyrism, Glucosinolates.*

Koseisho

The Japanese government agency that must approve new pharmaceutical products for sale with Japan. It is the equivalent of the U.S. Food and Drug Administration. See also *NDA (to Koseisho), Food and Drug Administration (FDA), Committee for Proprietary Medicinal Products (CPMP), Committee on Safety in Medicines, Medicines Control Agency (MCA), European Medicines Evaluation Agency (EMEA), Bundesgesundheitsamt (BGA).*

Kozak Sequence

Refers to the DNA sequence that "surrounds" (both ends of) the ATG start signal (for translation of mRNA). See also *Sequence (of a DNA molecule), Startpoint, Messenger RNA (mRNA), Deoxyribonucleic acid (DNA).*

Krebs Cycle

See *Citric acid cycle.*

Kunitz Trypsin Inhibitor (TI)

See *Trypsin inhibitors.*

Kusabira Orange

A protein that is naturally present within the stony coral *Fungia concinna.* Kusabira Orange is utilized by scientists to

- Help visualize thin layers of biological tissue in fluorescence microscopy
- "Mark" certain endpoints in experiments (at which the orange light signals that endpoint was reached)

See also *Fluorescence, Protein, Transfection, Gene expression markers, Reporter gene, TIRF microscopy.*

L

LAAM

Acronym for *light active antimicrobials*, a coating consisting of nanometer-scale particles that render the air above them (e.g., when applied as a coating to a hospital room surface) to be antimicrobial (i.e., toxic to microbes). Invented in 2006 by Stephen Michielsen, Igor Stojiljkovic, and Gordon Churchward. See also *Nanometers (nm), Nanotechnology, Microbe, Antibiotic, Antibiosis, Nanostructured material.*

Label (Fluorescent)

Refers to the practice of "attaching" a fluorophore (i.e., atom/molecule that emits fluorescent light when a light of a specific wavelength is shined onto it) by a scientist, thereby enabling that molecule to later be tracked (e.g., when inside living cells). See also *Fluorescence, Cell, Fluorophore, Green fluorescent protein, Fluorescence resonance energy transfer (FRET), Luciferase, Fluorescence-activated cell sorter (FACS), Rapid protein folding assay, Streptavidin, Quantum dot, Metamaterials, Click chemistry.*

Label (Radioactive)

A radioactive atom, introduced into molecule(s) in order to

- Enable observation of that molecule's metabolic transformation (within an organism). For example, if radioactive hydrogen in the form of water (known as deuterium) is supplied to a living cell, a series of "photographs" (e.g., taken via an electron microscope, which has photographic film in it that is sensitive to radiation) will reveal how rapidly that deuterium enters the cell, and into what structures within the cell that water is incorporated.
- Enable observation of which specific substrate within a living organism/cell gets acted upon by a given compound. For example, the radiolabel (isotope) known as *phosphorous-33* can be utilized to determine which substrate gets phosphorylated via a specific kinase (e.g., as a target in an assay).
- Quantify the rate at which certain (non)radioactive atoms are being introduced into a polymer (e.g., DNA) that is being polymerized (i.e., "manufactured") as part of a biological test or testing process (e.g., quantitative PCR, reverse transcriptase PCR).

See also *Autoradiography, Cell, Deoxyribonucleic acid (DNA), Organism, Substrate (chemical), Kinases, Target (of a therapeutic agent), High-throughput screening (HTS), Gene expression analysis, QPCR, RT-PCR, Radioimmunoassay, Radioimmunotechnique.*

Labeled (Molecules or Cells)

Also sometimes referred to as *tagged (molecules or cells)*. See *Label (fluorescent), Label (radioactive), Molecular beacon, Quantum dot, Nanoparticles, Microarray (testing), DNA microarray, Cell, Cell surface engineering, Bio-bar codes, Affinity tag, Affinity chromatography, Expressed sequence tags (EST), Bacterial expressed sequence tags (BEST), Streptavidin.*

Label-Free Detection

Refers to devices/methods utilized for identifying molecules or detecting changes in DNA hybridization, mass, concentration, or number of molecules present within a sample. These include optical methods (e.g., imaging ellipsometry), piezoelectric methods, potentiometric methods (i.e., detect changes in potential at constant current), conductive methods (i.e., detect changes in conductivity within sample), amperometric (i.e., detect changes in current at constant potential), thermal (i.e., measure changes in the sample temperature), and capacitive (i.e., when the biorecognition reaction causes a change in the dielectric constant). See also *Labeled (molecules or cells), Deoxyribonucleic acid (DNA), Hybridization (molecular genetics), Hybridization surfaces, Piezoelectric effect.*

Labile

Refers to a compound/molecule that is unstable at elevated temperatures, mechanical stress/shear, and so on (e.g., it disintegrates).

For example, some forms of phytase enzyme disintegrate when they encounter mild mechanical stress/shear. See also *Phytase, Thermolabile.*

Lab on a Chip

Term utilized to refer to microfluidic devices that perform applications such as nucleic acid separations, protein analysis, small-molecule organic synthesis, detection and hybridization of DNA, and so on.

To move fluid (samples), *microfluidic chips* utilize either capillary action or else they "pump" fluid (through microchannels in those chips) electrokinetically (i.e., cause the flow to occur by applying a controlled electrical field, so liquid is attracted to electrical charge and thereby flows).

Such "pumping" can be used to perform multiple chemical analyses (e.g., of body fluids within diseased tissues). For example, in 2006, Richard N. Zare and colleagues created a lab on a chip that

- Will lyse a single cell (in an individual cavity on the chip)
- Will separate all the individual protein molecules thus extracted from that cell
- Will subsequently identify via fluorescence mapping each of the protein molecules from that cell

In 2006, Michael J. Lochhead and colleagues created a lab on a chip that

- Will capture (within a cavity on the chip) pathogenic bacteria from a sample taken from a sick patient

- Will hold the bacteria cells in the cavity via a hydrogel augmented by an electrical field
- Will identify the bacteria by subsequently passing over them a mixture of fluorescent-labeled *antibodies that are specific to numerous different strains* of a variety of human-pathogenic bacteria, followed by strain identification via fluorescence mapping of each of the now antibody-laden bacteria

See also *Biochip, Nanotechnology, Microfluidics, Genosensors, Gene expression, Biosensors* (*electronic*), *Biosensors* (*chemical*), *Cell, Gene expression analysis, Nucleic acids, Protein, Deoxyribonucleic acid* (*DNA*), *Hybridization* (*molecular genetics*), *Lyse, Lysis, Fluorescence, Label* (*fluorescent*), *Fluorescence mapping, Pathogen, Bacteria, Antibody, Strain.*

Lac Operon

An operon in *Escherichia coli* that codes for three enzymes involved in the metabolism of lactose. See also *Operon, Coding sequence, Escherichia coliform* (*E. coli*).

Laccase

An oxidase enzyme that can

- Break down indigo dye (used in some manufacturing processes for blue jeans).
- Catalyze certain resins (e.g., containing lignin molecules that bear phenolic hydroxyl groups) to "cure" and cause those resins to harden in place, acting as an adhesive. For example, the numerous small pieces of so-called waste wood from lumber companies can be heated to approximately 200°C (392°F), whereupon the lignin within the wood fibers breaks into smaller molecules that bear some phenolic hydroxyl groups. Addition of laccase under appropriate conditions then results in phenoxy radical molecules, which cause the lignin-and-wood-fiber combination to harden (e.g., into a useful wood product such as a sheet of siding, within a pressurized mold).

See also *Enzyme, Catalyst, Lignins, Oxidation* (*chemical reaction*).

Lachrymal Fluid (Tears)

A salty solution produced by the tear glands to bathe and lubricate the eye. Possesses antimicrobial properties.

Lactoferricin

A protein compound that acts to inhibit pathogenic (i.e., disease-causing) bacteria and yeasts (e.g., in the human body). See also *Protein, Pathogen, Bacteria, Yeast, Lactoferrin.*

Lactoferrin

A transferrin protein compound that is naturally produced in human breast milk. Also found within specific granules inside neutrophils/leukocytes and produced in cow's milk. Lactoferrin assists transport of iron in the body, and it supports cell growth.

Consumption of lactoferrin by infants (e.g., via nursing) helps to strengthen their immune system. Consumption of lactoferrin by older humans helps their immune system to resist infectious diseases. Lactoferrin binds free iron (e.g., in body fluids), thereby denying that iron to pathogenic bacteria (which need that iron to grow/infect).

Pepsin and some other proteases (enzymes) can convert lactoferrin to lactoferricin. See also *Protein, Cell, Pathogen, Bacteria, Growth* (*microbial*), *Lactoferricin, Pepsin, Protease, High-lactoferrin rice, Lactoperoxidase.*

Lactonase

An enzyme that "breaks open" the lactone ring in (molecular structure of) the mycotoxin *zearalenone*. See also *Enzyme, Mycotoxin, Zearalenone, Toxin.*

Lactoperoxidase

A protein compound (enzyme) that acts to inhibit pathogenic bacteria (e.g., in the human body). See also *Protein, Enzyme, Pathogen, Bacteria.*

Lagging Strand

The one strand (of the two) in a DNA molecule during the DNA replication process that is *not* continuously being synthesized. That is because DNA synthesis can only proceed in one direction, so the geometry of DNA's double helix shape forces the lagging strand to be synthesized only discontinuously. See also *Deoxyribonucleic acid* (*DNA*), *Replication* (*of DNA*).

Lambda Bacteriophage

See *Lambda phage.*

Lambda Phage

A bacteriophage that infects *Escherichia coli* (*E. coli*). It is commonly used as a vector in recombinant deoxyribonucleic acid (DNA) research. See also *Phage, Escherichia coliform* (*E. coli*).

Langerhans Cells

See *Dendritic Langerhans cells, Islets of Langerhans.*

Lantibiotics

A class of posttranslationally modified peptides, which have had thioether rings (molecular substructure) attached to those peptide molecules. See also *Nisin, Peptide.*

Large Intervening Noncoding RNA

Discovered in 2003 by John Rinn, *large intervening noncoding ribonucleic acids* are responsible for directing a number of transcription factors' interactions with applicable genes, responsible for assisting in regulation of some immune responses, production of some stem cells, and so on. See also *Ribonucleic acid* (*RNA*), *Gene, Transcription factors, Immune response, Stem cells.*

Laser Capture Microdissection

Abbreviated LCM, it refers to a methodology in which a scientist is able to extract (e.g., from living tissue) a very specific type of cell.

In the LCM procedure, the scientist covers the relevant area of tissue with a special thin thermoplastic film. Using a microscope, the scientist then shines a pulse of applicable-wavelength laser beam onto the desired cell, which causes the plastic film to fuse onto that cell. When the plastic film is subsequently lifted, the desired/fused cells are lifted out of the tissue.

LCM enables scientists to biopsy/analyze rare cells (e.g., certain malignant cells), for comparison with others (e.g., nonmalignant cells). LCM enables scientists to preserve the cell's original structure and the cell's intact molecular composition, for analysis (e.g., of transcription and translation products). See also *Cell*, *Transcription*, *Translation*.

Laser Inactivation

Refers to a "protein knockout" technique in which a chromophore (i.e., chemical that is "triggered" to react by light shown onto that chemical) is first chemically bound to a certain protein molecule than a specific-wavelength laser beam is shined onto that protein–chromophore complex in order to inactivate that protein. Such inactivation results in loss of biological activity of that protein. See also *Protein*, *Knockout*, *Denaturation*, *Conformation*, *Protein folding*, *Protein structure*, *Biological activity*.

Laser Tweezer

See *Optical tweezer*.

Lathyrism

See *Glucosinolates*.

Laurate

A medium chain length (i.e., C12) fatty acid that is naturally produced by coconut trees, oil palm trees, and certain species of wild plants. In 1992, some canola varieties were genetically engineered so that they could also produce (desirable) laurate in their seeds. See also *Fatty acid*, *Fats*, *Canola*, *Genetic engineering*, *Genetic code*, *LPAAT protein*, *ACP*, *Lauroyl-ACP thioesterase*, *High-laurate canola*.

Lauric Acid

See *Laurate*.

Lauroyl-ACP Thioesterase

The enzyme that is required for the synthesis ("manufacturing") of laurate in plants. For example, the presence of this enzyme in the California bay tree (*Umbellularia californica*) causes its seed oil to contain as much as 45% laurate. See also *Laurate*, *Enzyme*, *LPAAT protein*, *High-laurate canola*.

Lazaroids

A class of drugs being developed to "bring back from the dead" tissues that have been (almost) killed due to a lack of oxygen (e.g., caused by a clot blocking a vital artery). See also *Human superoxide dismutase* (*hSOD*), *Fibrin*, *Reperfusion*.

L-Carnitine

See *Carnitine*.

LCM

Acronym for *laser capture microdissection*. See *Laser capture microdissection*.

LCN

Acronym for *low copy number*. See *Copy number*.

LCPUFA

Acronym for *long-chain polyunsaturated fatty acids* such as the essential fatty acids eicosapentaenoic acid (EPA) and docosahexaenoic acid (DHA). See also *Essential fatty acids*, *Eicosapentaenoic acid* (*EPA*), *Docosahexaenoic acid* (*DHA*).

LD

Acronym for *linkage disequilibrium*. See *Linkage disequilibrium*.

LD50

Acronym for *lethal dose 50%*. It was developed by J. W. Trevan in 1927 and is the amount of a given substance (or radiation) required to kill half of a sample population (e.g., of laboratory mice). For example, the LD50 of arsenic is 13 mg per kg (of mice).

LDL

See *Low-density lipoproteins* (*LDLPs*).

LDL-c

Acronym for *low-density lipoprotein cholesterol*. See also *Low-density lipoproteins* (*LDLPs*).

LDLP

See *Low-density lipoproteins* (*LDLPs*).

LDLP Receptors

See *Low-density lipoproteins* (*LDLPs*).

LDT

Acronym for *lab-developed test*. See also *Companion diagnostic*.

Leader

See *Leader sequence*.

Leader Peptide

See *Signal sequence*.

L

Leader Sequence (mRNA)

The nontranslated sequence at the 5′ end of mRNA that precedes the initiation codon. See also *Messenger RNA (mRNA)*, *Codon*.

Leader Sequence (Protein Molecule)

A (short) sequence of amino acids within a given protein molecule that determines *where* within a living cell that particular protein molecule will "reside." See also *Protein*, *Amino acid*, *Chaperones*, *Sequence (of a protein molecule)*.

Leading Strand

The one strand (of the two) in a DNA molecule during the DNA replication process that is continuously being synthesized. See also *Deoxyribonucleic acid (DNA)*, *Replication (of DNA)*.

LEAFY Gene

Refers to one plant gene, which governs leaf growth in some plants. See also *FT gene*, *FT protein*.

Leaky Gut Syndrome

See *Gut leakage*, *Probiotics*.

Leaky Mutants

A mutant in which the mutated gene product, such as an enzyme, still possesses a fraction of its normal biological activity. See also *Mutation*, *Gene*, *Protein*, *Biological activity*, *Enzyme*.

Lear

See *Canola*.

Lecithin

From the Greek *lekithos*, which meant *egg yolk*. See *Lecithin (crude, mixture)*, *Lecithin (refined, specific)*.

Lecithin (Crude, Mixture)

Also called *lecithin complex*, it is a mixture of phospholipids (i.e., lecithin–phosphatidylcholine, cephalin, inositol phosphatides, glycerides, tocopherols, glucosides, and certain pigments) that constitutes approximately 2% of soybeans by weight.

Historically, lecithin complex has often been utilized commercially in food processing as an emulsifier, dough conditioner, instantizing agent, and lubricating agent. Lecithin added to wheat-based flour dough at a rate of 0.25%–0.6% acts as a dough conditioner. It disperses fat evenly throughout the dough, enabling that conditioned dough to trap more of the carbon dioxide produced by yeast during fermentation (rising) process. Such conditioned doughs tend to produce a bread of fine grain, larger baked volume, and improved slicing properties.

Because lecithin–phosphatidylcholine naturally contains a high content of linoleic acid, consumption by humans of lecithin–phosphatidylcholine results in similar impact (e.g., lowered cholesterol levels in blood) as consumption of linoleic acid.

Because dietary fats are generally not absorbed directly through the intestinal wall (when eaten), they must first be emulsified, to form micelles that can pass through the intestinal wall and thus be absorbed by the body. That emulsification/micelle formation is aided by lecithin, since it is an emulsifier. Lecithin contains 1-palmitoyl-2-oleoyl-*sn*-glycerol-3-phosphocholine, which binds to a particular protein within the liver and thereby aids the liver's metabolism of fats and glucose.

Lecithin itself (also known as phosphatidylcholine) is a source of choline when digested and is a critical component of the lipoproteins that transport fat and cholesterol molecules in the bloodstream (e.g., from the digestive system, to body cells, to and from the liver).

Lecithin (phosphatidylcholine) promotes synthesis of high-density lipoproteins (HDLPs, also known as "good" cholesterol) by the liver, when lecithin is consumed by humans, thereby helping to lower blood levels of low-density lipoproteins (LDLPs, also known as "bad" cholesterol). See also *Lecithin (refined, specific)*, *Protein*, *Lipoprotein*, *Lipids*, *Conjugated protein*, *High-density lipoproteins (HDLPs)*, *Low-density lipoproteins (LDLPs)*, *Soybean plant*, *Soybean oil*, *Choline*, *Signal transduction*, *Linoleic acid*, *Acetylcholine*, *Fats*, *Micelle*, *Digestion (within organisms)*, *Cholesterol*, *Bile acids*, *Choline*, *Metabolism*, *Glucose*.

Lecithin (Refined, Specific)

A by-product of the refining process for soybean oil (deoiled lecithin from processed soybeans is composed of approximately 20%–25% phosphatidyl choline [PC] by weight). The lecithin molecule (i.e., PC) naturally contains a high content of linoleic acid, so consumption of lecithin by humans results in a similar impact (e.g., lowered cholesterol levels in blood) as consumption of linoleic acid.

Because dietary fats are generally not absorbed directly through the intestinal wall (when eaten), they must first be emulsified to form micelles that can pass through the intestinal wall and be absorbed by the body. That emulsification/micelle formation is aided by lecithin, since it is an emulsifier. Lecithin contains 1-palmitoyl-2-oleoyl-*sn*-glycerol-3-phosphocholine, which binds to a particular protein within the liver and thereby aids the liver's metabolism of fats and glucose.

Lecithin (also known as phosphatidylcholine) is a source of choline when digested and is a critical component of the lipoproteins that transport fat and cholesterol molecules in the bloodstream (e.g., from the digestive system, to body cells, to and from the liver). Lecithin (phosphatidylcholine) promotes synthesis of high-density lipoproteins (i.e., HDLPs, also known as "good" cholesterol) by the liver, when it is consumed by humans, thereby helping to lower blood levels of low-density lipoproteins (LDLPs, also known as "bad" cholesterol).

PC is involved in cell signal transduction (e.g., via which a cell reacts to an external chemical "signal").

Some other common dietary sources of lecithin include eggs, red meats, spinach, and nuts. See also *Lipoprotein*, *Lipids*, *Protein*, *Conjugated protein*, *High-density lipoproteins (HDLPs)*, *Low-density lipoproteins (LDLPs)*, *Soybean plant*, *Soybean oil*, *Choline*, *Signal transduction*, *Linoleic acid*, *Acetylcholine*, *Lecithin (crude, mixture)*, *Fats*, *Micelle*, *Digestion (within organisms)*, *Cholesterol*, *Metabolism*, *Glucose*.

Lecithin Complex

See *Lecithin (crude, mixture)*.

L

Lectin Pathway

See *Reperfusion.*

Lectins

A class of glycoproteins that have the capability to rapidly (and reversibly) combine with *specific* sugar molecules (e.g., those sugar molecules or glycoproteins on the surface of adjacent cells, within an organism). Lectins are a common component of the surface (membranes) of plant and animal cells and are so specific (regarding sugar molecules that they will or won't combine with/(attach to) that they discriminate between different monosaccharides *and* different oligosaccharides (i.e., on the surfaces of adjacent cells within an organism).

This capability to reversibly combine with sugar (i.e., carbohydrate) molecules (on the surface of adjacent cells) is utilized by

- Bacteria and other microorganisms, to adhere to (sugar molecules on surface of) host cells, as the first step in the process of infecting those host cells.
- White blood cells (e.g., lymphocytes), to adhere to the walls of blood vessels (endothelium), as the first step to leaving the bloodstream to go fight infection (pathogens, trauma) in tissue adjacent to that blood vessel. The lectin (glycoprotein) that adheres to the (endothelial sugar molecule on) blood vessel wall is called L-selectin, or the homing receptor. The two sugar molecules (glycoproteins) on the blood vessel wall (endothelium) are called P-selectin and E-selectin (also known as ELAM-1).
- Cancerous tumor cells, to adhere to the walls of blood vessels (endothelium) as part of the tumor-proliferation process known as metastasis (i.e., new tumors are "seeded" throughout the body via this process).

Separate and apart from the above impacts, some plant lectins (e.g., in the seeds of certain plants) are toxic to some of the animals that consume those seeds.

Because the lining of the midgut ("stomach") of certain insect pests is composed at least partially of chitin, genetically engineering a crop plant to produce within its applicable tissues the lectin known as HFR-3 (which tightly latches onto chitin molecules) can help such crop plants to resist being attacked by that particular insect pest. See also *Protein, Sugar molecules, Glycoprotein, Leukocytes, Cell, Selectins, Lymphocytes, Monocytes, Neutrophils, Endothelial cells, Endothelium, Cancer, Metastasis, Signal transduction, Ricin, Paneth cells, Chitin, Genetic engineering.*

Lentivirus

From the Latin *lenti* meaning "slow," it is a genus of retroviruses possessing slow life cycles. See also *Retroviruses.*

Leprosy

An infection caused by *Mycobacterium leprae* bacteria. See *GWAS, Bacteria.*

Leptin

Identified by Jeffrey M. Friedman in 1994, it is a protein hormone that is produced by fat cells (adipose tissue) in the body (e.g., following consumption of food). When leptin is produced and travels to neuron cells whose surface bears leptin receptors (e.g., in the brain), those brain cells receive signal (transduction) indicating fullness/satiety. The brain cells are also informed about the body's current metabolic state.

For example, Canadian scientists devised a test to detect which variant (SNP) of the "leptin gene" is possessed by dairy cattle breeding stock. By selecting only cattle possessing the "leptin-tt" SNP within their DNA (i.e., SNP with the lowest level of leptin production), it is expected that such future cattle herds will have inherently larger appetites and thus larger milk production potential.

Leptin has been found to be present in the bloodstream of obese humans at a concentration of approximately four times the concentration found in bloodstreams of lean humans. Obese humans tend to be less insensitive to leptin. High levels of leptin present in the bloodstream disrupt some of the activities of insulin (hormone that regulates blood sugar levels) and may possibly lead to diabetes.

Research indicates that sleep deprivation (e.g., less than 8 hours per night, for most people) leads to decreased leptin levels.

See also *Hormone, Adipokines, Protein, Biological activity, Gene, Single-nucleotide polymorphisms (SNPs), Metabolism, Insulin, Adipose.*

Leptin Receptors

Cellular receptors that are specific to leptin. In 1996, H. Ralph Snodgrass discovered that leptin receptors are involved in the "sorting" of immature blood cells (from bone marrow) to create subpopulations. See also *Leptin, Receptors.*

Lesion

Refers to the site (locus) within a DNA molecule where either the DNA molecule's structure is broken (e.g., double-strand break, single-strand break), or where one of the bases within the molecule is missing, mismatched, and so on. See *Deoxyribonucleic acid (DNA), Double-strand breaks (in DNA), Locus, Base (nucleotide), Base pair (bp), Base substitution, Mismatch repair.*

Lethal Mutation

Mutation of a gene to yield no or a totally defective gene product (protein), thereby making it unable to function and hence unable to sustain the life of the organism. See also *Gene, Protein, Mutation.*

Leucine (leu)

A monocarboxylic essential amino acid. See also *Amino acid, Essential amino acids, ALS gene.*

Leukocytes

A diverse "family" of nucleated white blood cells (including mast cells) that has many immunological functions. See also *Neutrophils, Interleukins, Eosinophils, Basophilis, Lymphocyte, B lymphocytes, Monocytes, Granulocytes, Mast cells.*

Leukotrienes

Lipid mediator molecules (synthesized from arachidonic acid via 5-lipoxygenase enzyme) that are synthesized and released by certain

L

inflammatory cells (i.e., macrophages, polymorphonuclear leukocytes, mast cells, T cells), which "signal" leukocytes (white blood cells) during the initial stages of an infection or an allergic reaction. Their primary mode of action is through certain G-protein-coupled receptors.

When thus activated, the leukocytes migrate to the site of infection to combat the pathogens (or allergens) and mediate the inflammation. See also *Eicosanoids, Lipids, Macrophages, Leukocytes, Receptors, Polymorphonuclear leukocytes (PMN), Mast cells, Signaling, Signal transduction, T cells, Pathogen, Arachidonic acid, Allergies (foodborne), Allergies (airborne), Signaling molecule, G-protein-coupled receptors, Oxylipins, Polyunsaturated fatty acids (PUFA).*

Levorotary (L) Isomer

An isomer of an optically active compound; rotates (when illuminated) the plane of plane-polarized light to the left. See also *Stereoisomers, Dextrorotary (D) isomer.*

Lfng Gene

See *MicroRNAs.*

LH

See *Luteinizing hormone.*

Library

A set of cloned DNA fragments together representing the entire genome of an organism. See also *Deoxyribonucleic acid (DNA), Genome, Clone (a molecule).*

LIF

Acronym for *laser-induced fluorescence.* See *Fluorescence, Fluorescence-activated cell sorter (FACS), Fluorescence in situ hybridization (FISH), Capillary electrophoresis.*

Ligand (in Biochemistry)

In general, a molecule or ion that can bind to (interact with) a protein molecule. For example, a pharmaceutical that binds to a receptor protein molecule on the surface of a cell may be called a ligand.

For example, the effector (inside the cell) that binds to a G-protein (inside cell) after that G-protein has received (outside the cell) a chemical signal (e.g., via a hormone molecule binding onto the exterior end of the G-protein molecule) may be called a ligand. See also *Protein, Receptors, T cell receptors, Endocytosis, CD4 protein, Invasin, Ligand (in chromatography), Chelation, Structure–activity models, G-proteins.*

Ligand (in Chromatography)

A term used to describe a substance (the ligand) that has the capacity for specific and noncovalent (reversible) binding to some protein. A ligand may be a coenzyme for a specific enzyme. The ligand can be covalently attached (immobilized) by means of the appropriate chemical reaction to the surface of certain porous column material. When a mixture of proteins containing the enzyme to be isolated is passed through the column, the enzyme, which is capable of tightly binding to the ligand, does so, and is in this manner held to the column. The other proteins present, which have no specific affinity for the ligand, pass on through the column. The protein/ligand complex is then dissociated and the enzyme eluted from the column, which may be accomplished by passing more free (unbound) coenzyme through the column. The ligand may be hormones (i.e., used to isolate receptor molecules) or any other type of molecule that is capable of binding specifically and reversibly to the desired protein or protein complex. See also *Affinity chromatography, Substrate (in chromatography), Chromatography, Protein, Peptide, Antibody, Monoclonal antibodies (MAb).*

Ligand-Activated Transcription Factors

See *Nuclear receptors.*

Ligase

An enzyme used to catalyze the joining together (i.e., "ligating") of two separate molecules, in an energy-requiring process. For example, the joining together of two single-stranded DNA segments. See also *Deoxyribonucleic acid (DNA), Enzyme.*

Ligation

The formation of a phosphodiester bond to link two adjacent bases separated by a nick in one strand of a double helix of DNA.

This term can also be applied to

- Blunt-end ligation
- The joining of RNA (ribonucleic acid) strands
- The joining of two adjacent oligosaccharides on a glycoprotein molecule

See also *Deoxyribonucleic acid (DNA), Ligase, Editing, Spliceosomes, Intein.*

Light-Chain Variable (VL) Domains

The regions (domains) of the antibody (molecule's) "light chain" that vary in their amino acid sequence. The "chains" (of atoms) comprising the antibody (immunoglobulin) molecule consist of a region of variable (V) amino acid sequence and a region in which the amino acid sequence remains constant (C). An antibody molecule possesses two antigen-binding sites, and it is the variable domains of the light (VL) and heavy (VH) chains that contribute to this (antigen-binding ability). See also *Antibody, Immunoglobulin, Protein, Sequence (of a protein molecule), Antigen, Amino acid, Combining site, Domain (of a protein), Heavy-chain variable (VH) domains.*

Lignans

A category of phytochemicals that play defensive roles (e.g., against infections by bacteria, fungi, etc.) within land plants (e.g., those grown by man for crops).

Lignans are also sometimes referred to by some people as "phytoestrogens" and are typically beneficial to the health of humans that consume them. Lignans are found in virtually all fruits, vegetables, and cereals (grains); generally within the seed coats, stems, leaves, or flowers. Some of the beneficial lignans commonly consumed by humans include the following:

- *Sesamin*, found in seeds of the Sesame plant (*Sesamum indicum*), which acts as an antioxidant
- The lignans that are found in seeds of the flax plant (*Linum usitatissimum*) and the rye plant

See also *Phytochemicals, Phytoestrogens, Isoflavones, Antioxidants, Oxidative stress.*

Lignin Nanotubes

Refers to nanotubes made up of lignin (residue) from plant material after processing within certain biofuel manufacturing facilities. Lignin nanotubes are less toxic to living cells than carbon nanotubes and can be utilized to ferry plasmid DNA (e.g., to correct certain single mutation impacts) into cells.

See also *Nanotube, Carbon nanotubes, Nanoscience, Nanotechnology.*

Lignins

From the Greek *lignum* meaning wood. A category of *phenolic* ("ring-shaped" molecules) *polymeric* (i.e., composed of more than one molecular unit) *compounds* produced by land plants within the cell walls (i.e., exterior of cell's plasma membrane) of those plants, to reinforce/strengthen those cell walls. See also *Cell, Polymer, Plasma membrane.*

Lignocellulose

A complex biopolymer comprising the bulk of woody plants. It consists of polysaccharides and polymer phenols. See also *Polysaccharides, Lignins.*

Limonene

See *Phytochemicals.*

LINC RNAs

Acronym for *large intervening noncoding ribonucleic acids.* See *Large intervening noncoding RNA.*

Linkage

A phenomenon discovered by Thomas Hunt Morgan in the early 1900s via his experiments with fruit flies. This term describes the tendency of genes to be inherited together as a result of their locations being physically close to each other on the same chromosome, measured by percent recombination between loci. Because the locus (i.e., location of gene on the chromosome) determines the likelihood that two genes will go together into offspring, "marker genes" that are linked to a gene (e.g., for a given trait or disease) of interest can be utilized to predict the presence of that (trait or disease-causing) gene. See also *Gene, Locus, Chromosomes, Linkage group, Marker (genetic marker), Map distance, Linkage map, Haplotype, "Nude" mouse.*

Linkage Disequilibrium

See *Linkage map.*

Linkage Group

Includes all loci (in DNA molecule) that can be connected (directly or indirectly) by linkage relationships; equivalent to a chromosome. See also *Locus, Chromosomes, Linkage, Chromatids, Chromatin, Linkage map, Deoxyribonucleic acid (DNA).*

Linkage Map

A depiction of gene loci (on chromosomes) based on the frequency of recombination (of linked genes) in the offspring's genome. Close (linked) genes tend to be inherited *together.*

When that does not happen (i.e., fewer linked-together genes inherited across generations than would be expected from mathematical prediction), the phenomenon is known as *linkage disequilibrium.* See also *Linkage, Linkage group, Gene, Locus, Marker (genetic marker), Genome.*

Linker

A short synthetic duplex oligonucleotide containing the target site for some restriction enzyme. It may be added to the ends of a DNA fragment prepared by cleavage with some other enzyme reconstructions of recombinant DNA.

Linking

The process of "attaching" a drug or a toxin to a monoclonal antibody, or another homing molecule of the immune system. Because this attachment must be reversible, so that the homing molecule can release the drug or toxin after delivering that drug or toxin to the desired site in the body (e.g., delivery of a toxin to a tumor, to kill the tumor), linking is a difficult process to reliably achieve. See also *Immunotoxin, Conjugate, Monoclonal antibodies (MAb), Toxin.*

Linoleic Acid

One of the so-called omega-6 (*n*-6) polyunsaturated fatty acids (PUFAs), it has historically comprised approximately 53% of the total fatty acid content of soybean oil. It is an essential fatty acid for humans. When consumed by humans, linoleic acid causes *LDLP cholesterol* levels in the blood to decrease, which reduces the risk of coronary heart disease (CHD). The human body converts linoleic acid to the n-6 highly unsaturated fatty acid (HUFA) *arachidonic acid.* See also *Polyunsaturated fatty acids (PUFA), N-6 fatty acids, Fats, Unsaturated fatty acids, Essential fatty acids, Low density lipoproteins (LDLPs), Cholesterol, Lecithin, Conjugated linoleic acid (CLA), Coronary heart disease (CHD), Volicitin, Soybean oil, Arachidonic acid, Cosuppression.*

Linolenic Acid

The nutritionally relevant form (i.e., an essential fatty acid) is known as α-linolenic acid (ALA). One of the so-called omega-3 (*n*-3) polyunsaturated fatty acids (PUFAs), it has historically comprised approximately 8% of the total fatty acid content of soybean oil. It is an essential fatty acid for humans (i.e., required by the human body).

L

One of the many human health benefits of linolenic acid consumption is that of helping to decrease unnecessary platelet aggregation that can lead to thrombosis or stroke. ALA consumption also dampens inflammatory reactions within the human body via blocking the formation of certain compounds that promote inflammation such as omega-6 (*n*-6)-derived eicosanoids, cytokines, platelet-activating factor, and C-reactive protein.

Humans and animals convert linolenic acid to the *n*-3 highly unsaturated fatty acids (HUFAs) *docosahexaenoic acid* (*DHA*) and *eicosapentaenoic acid* (*EPA*). When thus produced, or consumed by humans, both DHA and EPA each confer various additional health benefits to the human body.

Some plants convert linolenic acid to jasmonates (e.g., as part of their response to an insect pest attack). See also *N-3 fatty acids*, *Polyunsaturated fatty acids* (*PUFA*), *Unsaturated fatty acids*, *Essential fatty acids*, *Coronary heart disease* (*CHD*), *Thrombosis*, *Inflammation*, *C-reactive protein* (*CRP*), *Chronic inflammation*, *Cancer*, *Highly unsaturated fatty acids* (*HUFA*), *Docosahexaenoic acid* (*DHA*), *Eicosapentaenoic acid* (*EPA*), *Fats*, *Gene*, *Fad3 gene*, *High–linolenic oil soybeans*, *Jasmonate cascade*.

Lipase

An enzyme (one of a class of enzymes) that catalyzes the hydrolytic cleavage of lipid molecules (triglycerides) to yield free fatty acids. A lipase was the first enzyme to be produced via genetic engineering and marketed. Lipase also occurs naturally in cow's milk and in the intestines of many animals (where it aids/assists digestion of fats that the animal consumes).

For example, the two lipase enzymes known as *hormone-sensitive lipase* (*HSL*) and *adipose triglyceride lipase* (*ATGL*) are utilized by the human body to metabolize fats. See also *Enzyme*, *Hydrolytic cleavage*, *Triglycerides*, *Fats*, *Adipose*, *Fatty acid*, *Free fatty acids*, *Digestion* (*within organisms*), *Metabolism*.

Lipid Bilayer

A membrane (i.e., thin sheet–type) structure composed of relatively small lipid molecules that possess both a *hydrophilic* (i.e., "water loving") and a *hydrophobic* (i.e., "water hating") moiety. These (membrane) lipids thus spontaneously form closed bimolecular sheets in aqueous (water-containing) media, in which the hydrophobic ends of each lipid molecule are in the center of the bimolecular membrane and the hydrophilic ends of the lipid molecules are on the outside (i.e., touching the water molecules). See also *Lipids*, *Plasma membrane*, *Moiety*, *Giant vesicles*.

Lipid Rafts

Specific domains ("islands") within a mammal cell's plasma membrane in which are embedded certain receptors and/or whole functional systems (e.g., signaling systems, amino acid transport systems).

Some specific protein molecules located on the surface of certain pathogens are able to gain entry into cell of their pathogen via interaction with receptors on lipid rafts. See also *Cell*, *Lipids*, *Plasma membrane*, *Membrane transport*, *Amino acid*, *Protein*, *Transmembrane proteins*, *G-proteins*, *Receptors*, *Nuclear receptors*, *Liver X receptors* (*LXR*), *Farnesoid X receptors* (*FXR*), *Retinoid X receptors* (*RXR*), *Signaling*, *Signal transduction*, *Pathogen*, *Effectors* (*fungal*).

Lipid Sensors

See *Orphan receptors*.

Lipid Vesicles

See *Liposomes*, *Giant vesicles*.

Lipidation

See *Prenylation*, *Posttranslational modification of protein*.

Lipidoids

Refers to nanometer-scale particles in which a compound (e.g., a pharmaceutical, dsRNA) is encased inside a lipid layer in order to facilitate its entry into living cells. See *Nanoparticles*, *Lipids*, *Cell*, *Ribonucleic acid* (*RNA*), *RNA interference* (*RNAi*).

Lipidomics

The scientific study of an organism's lipids and their role in an organism's structure, metabolism, growth, health, disease (and/or the organism's resistance to disease, etc.). Some methods utilized to determine which impact results from which lipid, are as follows:

- *Lipid profiling*: Determination of the identities of each lipid present within a cell/tissue/organism (e.g., via mass spectrometry techniques such as MALDI-TOF-MS, etc.) and the function of each lipid
- *Metabolite profiling*: Determination of specifically which metabolic pathways (and/or related genes) are "switched on," inhibited, and so on, within a cell/tissue/organism (e.g., by the presence of a particular lipid)

See also *Lipids*, *Cell*, *Lipid bilayer*, *Lipase*, *Plasma membrane*, *Fats*, *Lipid sensors*, *Lipoprotein*, *Metabolism*, *Mass spectrometer*, *MALDI-TOF-MS*, *Pathway*, *Metabolic pathway*, *Metabolite profiling*, *Gene*, *Organism*.

Lipids

From the Greek word *Lipos* (fat), lipids are water-insoluble biomolecules (e.g., fats, oils, waxes, phospholipids, steroids) that are highly soluble in organic solvents such as chloroform, acetone, and so on. Lipids serve as "fuel" molecules in organisms, highly concentrated energy stores, "signaling" molecules (e.g., hormones, secondary messengers), and are fundamental components of cell membranes and enzymes. Lipids also play important roles in signal transduction, gene transcription, and intracellular transport (e.g., movement of certain protein molecules from one part of a cell to another part, ensuring that protein molecules adopt optimum conformations for transport or to react with other molecules).

Membrane lipids are relatively small molecules that have both a hydrophilic (i.e., "water loving") and a hydrophobic (i.e., "water hating") moiety. These (membrane) lipids spontaneously form closed bimolecular sheets in aqueous media (water) that are barriers to the free movement (flow) of polar molecules. See also *Fats*, *Moiety*, *Lipoprotein*, *Cholesterol*, *Cell*, *Signaling*, *Signaling molecule*, *Signal transduction*, *Plasma membrane*, *Membrane transport*, *Protein*, *Conformation*, *Phosphatidyl serine*, *Antioxidants*, *Oxidative stress*, *Phospholipids*, *Lipid bilayer*, *Cationic lipids*,

Prenylation, Leukotrienes, Oleosomes, Gene, Transcription, Lipid rafts, Medium chain triacylglycerides, Lipidomics, Sphingolipids, Sphingosine-1-phosphate, Giant vesicles.

Lipolytic Enzymes

See *Lipase.*

Lipophilic

A "fat-loving" molecule, or portion of a molecule. Relating to, or having strong affinity for fats or other lipids. See *Lipids, Fats.*

Lipopolysaccharide (LPS)

See *Endotoxin.*

Lipoprotein

A conjugated protein containing a lipid or a group of lipids. For example, *low-density lipoproteins* (LDLPs, also known as "bad" cholesterol) are a "package" of cholesterol (lipid) surrounded by a hydrophilic protein.

LDLPs and very-low-density lipoproteins (VLDLs) are the specific lipoproteins that are most likely to deposit cholesterol (plaque) on artery walls, which increases the risk of coronary heart disease (CHD). See also *Protein, Low-density lipoproteins (LDLPs), Very low-density lipoproteins (VLDL), Conjugated protein, Hydrophilic, Lipids, Prenylation, Cholesterol, Apolipoproteins.*

Lipoprotein-Associated Coagulation (Clot) Inhibitor (LACI)

A protein that prevents formation of blood clots. This occurs because LACI inhibits the controlled series of zymogen activations (enzymatic cascade) that cause the formation of fibrinogen (precursor to fibrin), leading subsequently to clot formation. See also *Fibrin, Fibronectin, Zymogens.*

Liposomal Nanoparticles

Refers to nanoparticles encapsulated within a shell composed of a combination of liposomes and polyethylene glycol. Certain pharmaceutical compounds (e.g., doxorubicin) can thereby be delivered to certain sites within a patient's body (e.g., the immature/growing blood vessels that supply nutrients to certain cancerous tumors) without the pharmaceutical compound dispersing to tissues throughout the body. That is because such pharmaceutical compounds that tend to be rapidly degraded in the bloodstream are protected from degradation/dispersion in the bloodstream via being enclosed within the liposomes and polyethylene glycol, so that more of the nondegraded/nondispersed pharmaceutical would remain by the time it reached the targeted tissue (i.e., the tumor). For example, in 1995, U.S. regulators approved pharmaceutical-in-liposome Doxil for treatment of Kaposi's sarcoma tumors. See also *Nanoparticles, Liposomes, Lipids, Cancer, Tumor.*

Liposomes

Also called lipid vesicles or vesicle. Aqueous (i.e., watery) compartments enclosed by a lipid bilayer. They can be formed by suspending a suitable lipid, such as phosphatidyl choline, in an aqueous medium. This mixture is then sonicated (i.e., agitated by high-frequency sound waves) to give a dispersion of closed vesicles (i.e. compartments) that are quite uniform in size. Alternatively, liposomes can be prepared by rapidly mixing a solution of lipid in ethanol with water, which yields vesicles that are nearly spherical in shape and have a diameter of 500 Å (Angstroms). Larger vesicles (10,000 Å or 1 μm, or 0.00003937 in. in diameter) can be prepared by slowly evaporating the organic solvent from a suspension of phospholipid in mixed solvent system.

Liposomes can be made to contain certain drugs for protective, controlled release delivery to targeted tissues. Pharmaceuticals that tend to be rapidly degraded in the bloodstream can be enclosed within liposomes so that more of the nondegraded pharmaceutical would remain by the time it reached the targeted tissue. For example, in 1995, U.S. regulators approved pharmaceutical-in-liposome Doxil for treatment of Kaposi's sarcoma.

For example, in 2005, Tracy S. Zimmermann injected *siRNA that acts to silence the ApoB gene* encased within liposomes into monkeys. Thus, protected from degradation in the bloodstream, that siRNA decreased expression of the ApoB gene in monkey's liver by approximately 90%, resulting in greatly reduced levels of cholesterol and low-density lipoproteins (LDLPs) in the bloodstream.

The controlled release property also could enable larger doses (e.g., of drugs possessing toxic side effects) to be prescribed, knowing that the drug will be released in the body over an extended period of time. In 2011, William T. Phillips, Beth A. Goins, and Ande Bao encapsulated tiny particles of rhenium-186 (which emits radiation that only travels out a few millimeters) in liposomes of approximately 100 nm diameter, which effectively delivered the rhenium-186 to brain tumors called glioblastomas. See also *Lipids, Cationic lipids, Micron, Angstrom (Å), Phosphatidyl choline, siRNA, Gene, Gene expression, Apolipoprotein B, Cholesterol, Low-density lipoproteins (LDLPs), Gene delivery, Gene therapy, Nanometers (nm), Nanotechnology.*

Lipoxidase

See *Lipoxygenase (LOX).*

Lipoxygenase (LOX)

A "family" of enzymes that include the following:

- At least one (i.e., 5-lipoxygenase) that is naturally produced within humans and some other animals, which is the primary enzyme utilized for their synthesis of leukotrienes (inflammation-promoting compounds). See *LEUKOTRIENES.*
- At least three (LOX-1, LOX-2, LOX-3) that are naturally produced within some plants. For example, lipoxygenase enzymes are produced within the seeds (soybeans) of the soybean plant (*Glycine max* (L.) Merrill). Among other purposes, some lipoxygenase enzymes are utilized by some plants in their defense against certain pest insects. When "enzyme active" (i.e., not heated during processing) soy flour is added to wheat-based flours, the lipoxygenases present within that soyflour brighten (make whiter) the wheat-based flour.

For example, in response to such insects chewing on the *Nicotiana attenuata* plant, that plant expresses *lipoxygenase 3* and certain other enzymes that cause production (via oxylipin pathways) of jasmonic acid, which triggers specific plant defenses (e.g., systemic acquired resistance).

Lipoxygenase enzymes also catalyze a reaction within some plants in which certain volatile chemicals (known as green leaf volatiles or GLV) are produced that inhibit growth of any *Aspergillus flavus* fungus.

In the presence of moisture and certain other conditions, lipoxygenase enzymes catalyze a chemical reaction in which objectionable "beany" flavor can be produced from certain components of the soybean. That "beany" flavor decreases the suitability of resultant soybean raw materials for manufacture of human foods in some countries.

Prevention of the reactions that create the "beany" flavor can be accomplished via heat denaturation (of lipoxygenases present in the soybeans) or via creation of soybeans that do not contain any lipoxygenase enzymes (known as "LOX null" soybeans). See also *Enzyme, Leukotrienes, Soybean plant, Express, Jasmonic acid, Cascade, Jasmonate cascade, Pathway, Systemic acquired resistance* (*SAR*), *LOX null soybeans, LOX-1, LOX-2, LOX-3, Chronic inflammation, Oxylipins.*

Lipoxygenase Null

See *LOX null soybeans, Lipoxygenase* (*LOX*).

Listeria monocytogenes

Refers to the "family" (numerous strains) of *Listeria monocytogenes* bacteria that can grow in many different foodstuffs (e.g., meats, cheese, meat products such as sausage) under specific conditions and can cause food poisoning (Listeriosis) in humans that subsequently consume those foodstuffs. When consumed by humans, certain strains/serotypes of *L. monocytogenes* can cause fever, severe headaches, stiffness, nausea, diarrhea, and possibly miscarriages in pregnant women. *Listeria* is particularly dangerous for pregnant women because it can be passed to the unborn baby within her even if the mother is not showing outward signs of the illness.

Following infection of human cells by *L. monocytogenes*, that bacteria are able to "commandeer" actin in those cells, to be transported quickly within those cells (to multiply and further infect).

As of January 19, 2001, all meat processed in the United States is required to be tested for the presence of *L. monocytogenes*.

Recent research indicates that growth of *L. monocytogenes* in dairy products can be inhibited by the presence of the compound *pediocin*, which is produced by some bacteria (e.g., *Lactobacillus plantarum*). See also *Bacteria, Strain, Serotypes, Cell, Actin, Enterotoxin, Bacteriocins, Cadherins.*

Live Cell Array

Refers to a microarray (e.g., a piece of glass, plastic, or silicon) onto which has been attached a number of living cells that are subsequently utilized to bioassay (e.g., a pharmaceutical, a toxin). See also *Cell, Bioassay, Cell array, Microarray* (*testing*), *Biosensors* (*chemical*), *Toxicogenomics, High-throughput screening* (*HTS*).

Liver X Receptors (LXR)

Refers to nuclear receptors that are primarily present within the body's tissues engaged in lipid metabolism (i.e., liver, kidney, lung, intestine, adrenals, macrophage, and adipose tissues).

Elevated levels (in the body) of certain sterols (e.g., 24(*S*),25-epoxycholesterol or 27-hydroxycholesterol) activate the LXRs. That causes the LXRs to act as *cholesterol sensors*, transactivating a "family" of genes that collectively control the catabolism (i.e., breakdown to yield energy), transport, and elimination of cholesterol. See also *Nuclear receptors, Lipids, Cholesterol, Metabolism, Macrophages, Adipose, Sterols, Protein, Transactivation, Transactivating protein, Gene, Catabolism.*

Living Modified Organism (LMO)

See *GMO.*

lncRNAs

See *Long noncoding RNAs.*

LOC

Acronym for *lab on a chip.* See *Lab on a chip.*

Loci

The plural of locus. See *Locus.*

Locus

The position (location) of a gene on a chromosome, or the position of a base pair on a DNA molecule. See also *Deoxyribonucleic acid* (*DNA*), *Gene, Chromosomes, Base pair* (*bp*).

LOI

Acronym for *loss of imprinting.* See *Imprinting.*

Lonafarnib

See *Farnesyl transferase inhibitors.*

Long Noncoding RNAs

Abbreviated "lncRNAs," these molecules are one form of RNA (ribonucleic acid) with a length of >100 nucleotides that are involved in the following:

- Epigenetic regulation of the organism's genome (e.g., control of when, how, and how much a gene is expressed).
- Genomic imprinting (i.e., the epigenetic process in which an offspring's X-chromosome gene expression is ensured to be a *mixture* of the genes contributed by *each* of its parents via silencing some of the genes contributed by each parent). In 2012, Paulina Latos and colleagues showed that the lncRNA known as *Airn* accomplishes that via *Airn* transcriptional overlap of the lncRNA with the applicable gene's promoter, which interferes with RNA polymerase II *recruitment (i.e., attracting the RNA polymerase to come to that site on the DNA).*
- The organism's innate immune response to viral infection.
- Synthesis of some peptides.

See also *Ribonucleic acid (RNA), Nucleotide, Deoxyribonucleic acid (DNA), Gene, Junk DNA, Promoter, Expression, Transcription, Transcription activators, RNA polymerase, Organism, Epigenetic, Imprinting, Gene silencing, Differentiation, Transcriptome, Virus, Innate immune response, Peptide.*

Long Non-Protein-Coding RNAs

See *Long noncoding RNAs.*

Long Terminal Repeat

Refers to a particular sequence of (repeated) nucleotides that appears within the end portion/segment of a retrovirus element that was incorporated into the DNA of a (host) organism. See also *Sequence (of a DNA molecule), Deoxyribonucleic acid (DNA), Nucleotide, Retroviruses, Organism.*

Loop

A single-stranded region at the end of a hairpin in RNA (or single-stranded DNA). It corresponds to the sequence between inverted repeats in duplex DNA. See also *Ribonucleic acid (RNA), Deoxyribonucleic acid (DNA), Sequence (of a DNA molecule).*

LOSBM

Low-oligosaccharide soybean meal. See also *Low-stachyose soybeans, Soybean plant.*

Loss of Imprinting

See *Imprinting.*

Loss-of-Function Mutations

See *Mutation, Knockout, Gene silencing, Functional genomics, RNA interference (RNAi).*

Low-Density Lipoproteins (LDLPs)

So-called bad cholesterol (i.e., LDL cholesterol), which carries cholesterol molecules from the digestive system (e.g., intestine) to body cells and can sometimes clog arteries over time (a disease called atherosclerosis, or coronary heart disease). Since cholesterol does not dissolve in water (which constitutes most of the volume of blood), the body makes LDL cholesterol (derived from the digestion of fatty foods) into little "packages" surrounded by a hydrophilic (i.e., "water loving") protein. That protein "wrapper" is known as apolipoprotein B-100, or apo B-100, and it enables LDL cholesterol to be transported in the bloodstream because the apolipoprotein B-100 is attracted to water molecules in the blood. Part of the apolipoprotein B-100 molecule also will bind to special LDLP receptor molecules in the liver, which then clears those (bound) cholesterol packages out of the body as part of regular liver functions. See also *High-density lipoproteins (HDLPs), Hydrophilic, Receptors, Protein, Sitostanol, Isoflavones, Water soluble fiber, Cholesterol, Coronary heart disease (CHD), Apolipoproteins, Very low-density lipoproteins (VLDL).*

Low–Linolenic Oil Soybeans

Soybeans from soybean (*Glycine max*) plant varieties that have been bred specifically to produce soybeans bearing oil that contains less than 3% linolenic acid, instead of the typical 8% linolenic acid content of soybean oil produced from traditional varieties of soybeans.

Low–linolenic soybean oil would tend to have greater flavor stability (especially at elevated temperatures utilized in frying foods) than soybean oil from traditional varieties of soybeans. See also *Soybean plant, Soybean oil, Fatty acid, Linolenic acid, Polyunsaturated fatty acids (PUFA).*

Low-Lipoxygenase Soybeans

See *LOX-null soybeans.*

Low-Phytate Corn

Developed in the United States during the 1990s, these are corn (maize) hybrids possessing the Lpa1 gene, the Lpa2 gene, or the highly available phosphorous (HAP) gene (which was discovered by Victor Raboy). That gene causes corn (maize) hybrids possessing it to produce much less phytate than the 0.15% typically present in traditional varieties of corn (maize). Because phytate is not digestible in humans and other monogastric animals (e.g., swine, poultry), substituting low-phytate corn in place of traditional corn varieties in those animals' diets helps to lessen the adverse environmental impact of animal feeding (e.g., phosphorous emissions in excess of annual cropland requirements).

Swine fed a diet in which traditional corn (maize) varieties have been replaced by low-phytate corn (maize) produce up to 30% less phosphorous in their manure, thereby lessening the phosphorous impact of those swine on the environment.

Humans consuming a diet based heavily on corn/maize (e.g., tortillas) absorb 50% more iron when traditional corn varieties are replaced by low-phytate corn varieties. That is because the phytate (inositol hexaphosphate) molecule "binds"/chelates iron (and some other metals) within the digestive system and prevents their absorption into the body. See also *Corn, Phytate, High-phytase corn, Phytase, Value-enhanced grains, Highly available phosphorous (HAP) gene, Chelation, Chelating agent, Iron deficiency anemia (IDA).*

Low-Phytate Soybeans

Developed in the United States during the 1990s, these are soybean varieties possessing less than 0.3% (of total soybean weight) phytate versus the typical 0.6% phytate content of soybeans from traditional soybean varieties.

One type of low-phytate soybean is derived via a single recessive mutation (i.e., an SNP) in the gene that codes for seed-expressed *myoinositol L-phosphate synthase.*

Because phytate (myoinositol hexaphosphate) is not digestible in humans and other monogastric animals (e.g., swine, poultry), substituting low-phytate soybeans in place of traditional soybean varieties in those animals' diets helps to lessen adverse environmental impact of animal feeding (e.g., manure phosphorous emissions in excess of cropland requirements).

Swine fed a diet in which traditional soybean varieties have been replaced by low-phytate soybeans produce up to 20% less phosphorous in their manure, thereby lessening the phosphorous impact of those swine on the environment. Due to the fact that the amino acids lysine, methionine, cysteine, arginine, and threonine all become more "bioavailable" (i.e., available for the animal to build its protein-containing body tissues, or otherwise utilize) in a low-phytate diet, low-phytate diets also help reduce excess nitrogen emissions. See also *Soybean plant*, *Phytate*, *Mutation*, *Gene*, *Single-nucleotide polymorphisms (SNPs)*, *Recessive allele*, *Low-phytate corn*, *High-phytase corn/soybeans*, *Lysine*, *Cysteine*, *Methionine*, *Arginine*, *Threonine*, *Deamination*.

Low-Stachyose Soybeans

Those soybean varieties that contain lower-than-1% levels of the relatively indigestible stachyose carbohydrate (and thus higher levels of easily digestible other nutrients) than traditional varieties of soybeans (which typically contain 1.4%–4.1% stachyose in traditional soybean varieties). Compared to traditional varieties of soybeans, low-stachyose soybeans have approximately 10% more metabolizable (i.e., useable by animals) energy content and a 3% increase in amino acid digestibility.

Low-stachyose soybeans are particularly useful for feeding of monogastric animals (e.g., swine, poultry), since their single stomach cannot digest stachyose. Thus, stachyose tends to "ferment" (promote excess bacterial growth) in their intestines, causing them to feel prematurely full. See also *Stachyose*, *Carbohydrates (saccharides)*, *Value-enhanced grains*, *Soybean plant*, *High-sucrose soybeans*, *Digestion (within organisms)*, *Metabolism*.

Low-Tillage Crop Production

A methodology of crop production in which the farmer utilizes a minimum of mechanical cultivation (i.e., only two to four passes over the field with tillage equipment instead of the conventional five passes per year utilized for traditional crop production). This reduced mechanical tillage leaves more carbon in the (less-disturbed) soil, leaves more earthworms (*Aporrectodea caliginosa*, *Eisenia fetida*, etc.) per cubic foot or per cubic meter living in the topsoil, and reduces soil compaction (i.e., the reduction in interstitial spaces between individual soil particles), thereby increasing the fertility of "low-till" farm fields.

The plant residue remaining on field's surface helps to control weeds and reduce soil erosion; it also provides sites for insects to shelter and reproduce, leading to a need for increased pest insect control via methods such as inserting a *Bacillus thuringiensis* (*B.t.*) gene into certain crop plants. But if a farmer needs to apply synthetic chemical pesticides, the plant residue remaining on the field's surface helps to cause breakdown (into substances such as carbon dioxide and water) of those pesticides. That is because that plant residue helps to retain moisture in the field-surface environment, thereby enhancing growth of the types of microorganisms that help to break down pesticides. See also *No-tillage crop production*, *Glomalin*, *Earthworms*, *Microorganisms*, *Integrated pest management (IPM)*, *Corn*, *Soybean plant*, *Bacillus thuringiensis (B.t.)*, *Gene*, *Genetic engineering*, *European corn borer (ECB)*, *Helicoverpa zea (H. zea)*, *Corn rootworm*, *Cold hardening*.

LOX Null Soybeans

Refers to soybeans that do not contain any of the three lipoxygenase enzymes (thus, they result in a "null" test reading). See also *Lipoxygenase (LOX)*, *LOX-1*, *LOX-2*, *LOX-3*, *Soybean plant*, *Enzyme*.

LOX-1

One of the isozymes (enzyme molecule variations) of the lipoxygenase (LOX) enzyme "family." See also *Lipoxygenase (LOX)*, *Isozymes (isoenzymes)*.

LOX-2

One of the isozymes (enzyme molecule variations) of the lipoxygenase (LOX) enzyme "family." See also *Lipoxygenase (LOX)*, *Isozymes (isoenzymes)*.

LOX-3

One of the isozymes (enzyme molecule variations) of the lipoxygenase (LOX) enzyme "family." See also *Lipoxygenase (LOX)*, *Isozymes (isoenzymes)*.

LPAAT Protein

A protein consisting of lysophosphatidic acid acyl transferase (enzyme), which (when present in a plant) causes production of triglycerides (in the seeds) possessing saturated fatty acids in the "middle position" of the triglycerides' molecular (glycerol) "backbone." For example, canola (rapeseed) plants genetically engineered to contain LPAAT protein are able to produce high levels of saturated fatty acids (including laurate) in their oil. See also *Protein*, *Laurate*, *Enzyme*, *Triglycerides*, *Saturated fatty acids (SAFA)*, *Monounsaturated fats*, *Canola*, *Genetic engineering*.

LPE

See *Lysophosphatidylethanolamine*.

LPS

See *Endotoxin*.

LR11

See *Docosahexaenoic acid (DHA)*.

LSD1

See *Lysine specific demethylase 1 (LSD1)*.

L-Selectin

Also known as the homing receptor. See *Selectins*, *Lectins*, *Adhesion molecules*.

LTR

Abbreviation for *long terminal repeat*. See *Long terminal repeat*.

Luciferase

Refers to a group of enzymes that can catalyze a chemical reaction that results in the *production of light* (*i.e., bioluminescence*) within certain living organisms.

For example, the common firefly (*Photinus pyralis*) is able to emit light from its tail (photophores) via luciferase-catalyzed bioluminescence. The ocean jellyfish known as the *sea pansy* (*Renilla reniformis*) is able to emit light via similar use of a slightly different luciferase-type molecule. See also *Bioluminescence, Luciferin, Enzyme, Catalyst, Organism, Nitric oxide, Luminophore.*

Luciferin

Broadly speaking, it is any chemical substrate that becomes luminescent when catalyzed by the enzyme known as luciferase. See also *Bioluminescence, Luminophore, Luminescence, Enzyme, Catalyst, Luciferase, Substrate (chemical).*

Lumen

The interior (opening through which blood flows), for example, within a blood vessel. See also *Endothelium.*

Luminase

An enzyme that could potentially be utilized to help bleach wood pulp during the papermaking process; resulting in less adverse impact on the environment. See also *Enzyme.*

Luminesce

See *Bioluminescence.*

Luminescence

See *Bioluminescence.*

Luminescent Assays

Refers to assays (i.e., tests/test techniques) that detect or measure

- The presence of a specific substance (e.g., bacteria ATP on surfaces in a slaughterhouse). Utilization of firefly luciferase in combination with luciferin can result in assays that can (visually) detect the presence of ATP
- The efficacy (i.e., effectiveness) of a specific substance

via the *enzyme (e.g., luciferase)-catalyzed production of light.*

For example, one (rapid) luminescent assay utilizes two chemical reagents that first break down bacteria cell membranes and then cause ATP from those broken-open cells to luminesce. Subsequent measurement of that light is the assay's proof (e.g., that bacteria had been present on the tested surface in a slaughterhouse). See also *Assay, Bioluminescence, Enzyme, Bacteria, Plasma membrane, Adenosine triphosphate (ATP), Luciferase, Luciferin.*

Luminophore

Refers to any substance that becomes luminescent. See also *Luminescence, Bioluminescence, Luciferin, Luciferase, Lux proteins.*

Lunasin

A 43 amino acid soy peptide, whose consumption has been reported to reduce inflammation. See *Peptide, Soy protein.*

Lupus

An autoimmune disease of the body, in which anti-DNA antibodies bind to DNA. The resulting complexes (of DNA and antibodies) travel to the kidneys via the bloodstream, and become lodged in kidneys, where they cause inflammatory reactions (that can lead to kidney failure).

Sometimes the skin, joints, blood vessels, bone marrow, the liver, and fibrous tissue around the heart are also damaged by this disease.

Children who contract lupus have been shown to be at higher risk for coronary heart disease when they become adults.

Women are far more likely to contract lupus than are men. U.S. pharmacogenetic research has shown that African-Americans are three times more likely to contract lupus than Caucasians and that at least one of the drugs utilized to treat lupus (belimumab, trade name Benlysta™) does not work when administered to most African-American Lupus patients.

In 2012, some researchers utilized receptor-binding mapping to link vitamin D deficiency to an increased risk for cancer and the autoimmune diseases rheumatoid arthritis, multiple sclerosis, and lupus. See also *Antibody, Deoxyribonucleic acid (DNA), Autoimmune disease, Superantigens, Coronary heart disease (CHD), Pharmacogenetics.*

Lupus Erythematosus

See *Lupus.*

Lutein

A carotenoid (i.e., "light-harvesting" compound utilized in photosynthesis) that is naturally produced in soybeans, carrots, summer squash, corn (maize), broccoli, spinach, dark lettuce, and green peas. Lutein is also naturally present within the retina of the human eye.

Lutein is a phytochemical/nutraceutical conducive to good eye health and regular consumption of large amounts of lutein has been shown to reduce the risk of the disease *age-related macular degeneration*, a leading cause of blindness in old people.

Research indicates that consumption of lutein by humans also reduces risk of prostate cancer and breast cancer. Recent research indicates that consumption of lutein by chickens increases their body's response (i.e., antibody production) to a vaccine for infectious bronchitis virus. See also *Phytochemicals, Nutraceuticals, Carotenoids, Soybean plant, Cancer, Photosynthesis, Age-related macular degeneration (AMD), Antibody, Vaccine.*

Luteinizing Hormone (LH)

A reproductive hormone that acts upon the ovaries to stimulate ovulation. It is secreted by the pituitary gland. See also *Hormone, Pituitary gland, Endocrine hormones, Estrogen.*

Luteolin

See *Nodulation.*

LUX

See *Docking proteins.*

L

Lux Gene

A gene within the DNA of *Vibrio fischeri*, a bacterium that lives in light-producing organs ("spotlights") of certain deep sea fish. The lux gene codes for *lux proteins* (luminophores), which causes those bacteria (and thus the fish's light-producing organs) to emit light.

The lux gene can be utilized as a "reporter gene" by inserting it into the DNA of (e.g., certain bacteria species that can be genetically engineered to biodegrade diesel fuel spilled in soil). Then, when those engineered bacteria encounter diesel fuel and begin "eating" it (i.e., breaking it down), those engineered bacteria will glow (bioluminesce) to "report" that they are biodegrading the spilled diesel fuel. See also *Gene, Reporter gene, Deoxyribonucleic acid (DNA), Bacteria, Bioluminescence, Luminophore, Protein, Lux proteins, Genetic engineering, Bioremediation.*

Lux Proteins

Refers to bioluminescent proteins found in some species of (usually deep ocean) marine organisms. Also utilized by man to make some luminescent assays. See also *Protein, Bioluminescence, Luminophore, Luminescent assay, Lux gene.*

LXR

Acronym for *liver X receptors*. See *Liver X receptors (LXR).*

Lycopene

An antioxidant carotenoid ("light-harvesting" pigment utilized by plants in the photosynthesis process) that is a naturally occurring phytochemical in tomatoes, watermelon, guava, and pink grapefruit (and some other fruits).

Consumption of significant amounts of lycopene by humans causes an increase in the concentration of lycopene in the blood plasma. Lycopene is a natural constituent of blood plasma and certain tissues in the human body, but it must be consumed in the diet, because the human body does not synthesize ("manufacture") lycopene. Consumption of lycopene by humans has been linked to a reduction in atherosclerosis, coronary heart disease, some cancers (e.g., prostate cancer, colorectal cancer), and inhibition of *oxidation of low-density lipoproteins (LDLPs).*

Lycopene is also converted (in some instances) into alpha-carotene and/or beta-carotene. Because beta-carotene is processed into vitamin A by the human body, consumption of this phytochemical can help prevent human diseases (e.g., in developing countries) that result from deficiency of vitamin A, for example:

- CHD
- Certain cancers (e.g., cancer of the prostate and lungs)
- Childhood blindness
- Age-related macular degeneration, a leading cause of blindness in older people
- Various childhood diseases that can cause death, due to weakened immune system

Research published in 2004 indicates that the presence of lycopene in the human digestive tract induces the excretion of some cancer-inhibiting enzymes known as *phase II detoxification enzymes*. See also *Phytochemicals, Nutraceuticals, Cancer, Antioxidants, Carotenoids, Coronary heart disease (CHD), Plasma, Atherosclerosis, Prostate-specific antigen (PSA), Tomato, Beta carotene, Vitamin, Lutein, Photosynthesis, Low-density lipoproteins (LDLPs), Inducible enzymes.*

Lymphocyte

A type of cell found in the blood, spleen, lymph nodes, etc. of higher animals. They are formed very early in fetal life, arising in the liver by the sixth week of human gestation. There exist two subclasses of lymphocytes: B lymphocytes and T lymphocytes.

B lymphocytes make antibodies (immunoglobulins) of which there are five classes: IgM, IgA, IgG, IgD, and IgE. The antibodies circulate in the bloodstream.

T lymphocytes recognize and reject foreign tissue, modulate B cell activity, kill tumor cells, and kill host cells infected with virus. T-lymphocytes are also called T cells.

The bone marrow of humans continues to make more lymphocytes throughout its lifetime. The lipid known as sphingosine-1-phosphate determines what fraction of total lymphocytes are present within lymph nodes versus the bloodstream. See also *B lymphocytes, T cells, Antibody, Helper T cells (T4 cells), Blast cell, Cytotoxic T cells, Antigen, Dendritic cells, Sphingosine-1-phosphate.*

Lymphokines

Peptides and proteins secreted by (immune system response) stimulated T cells. These hormone-like (peptide and protein) molecules direct the movements and activities of other cells in the immune system. Some examples of lymphokines are interleukin-1, interleukin-2, tumor necrosis factor, gamma interferon, colony-stimulating factors, macrophage chemotactic factor, and lymphocyte growth factor. The suffix "-kine" comes from the Greek word kinesis, meaning movement. See also *Protein, Peptide, T cells.*

Lynparza®

See *Olaparib.*

Lyochrome

See *Flavin.*

Lyophilization

The process of removing water from a frozen biomaterial (e.g., a microbial culture or an aqueous protein solution) via application of a vacuum. It is a drying method for long-term preservation of proteins in the solid state and for long-term storage of live microbial cultures. See also *Culture, Protein.*

Lyse

To rupture a membrane (cell). The act of lysis (rupturing a membrane). See also *Lysis.*

Lysine (lys)

An essential amino acid that can be obtained from many proteins by hydrolysis (i.e., cutting apart the protein molecule). See also

Essential amino acids, Protein, Opague-2, Photorhabdus luminescens, Hydrolysis.

Lysine Specific Demethylase 1 (LSD1)

See *Methylated.*

Lysis

The process of cell disintegration; membrane rupturing; breaking up of the cell wall. See also *Cytolysis, Cell, Lysozyme, Membrane transport, Biocide, Gram-positive* (*G*+).

Lysogeny

Refers to the ability of a bacteriophage to be able to itself become a part of a bacteria's DNA. See also *Bacteriophage, Deoxyribonucleic acid* (*DNA*).

Lysophosphatidylethanolamine

Also known by the abbreviation LPE; also known as phosphatidyl ethanolamine. It is one of the lipids (phospholipids) naturally found in soybean oil. In plants, it functions as a *signaling molecule* (e.g., speeding the ripening process). See also *Lipids, Soybean oil, Signaling molecule.*

Lysosome

A membrane-surrounded organelle within the cytoplasm of eucaryotic cells that contains many hydrolytic enzymes. Discovered by Christian de Duve.

The lysosome internalizes and digests foreign proteins as well as cellular debris (e.g., parts of worn-out cellular protein molecules). The foreign protein fragments (epitopes) are "presented" to T cells by the major histocompatibility complex proteins on the surface of the eucaryotic cell. The cellular debris is carried to the lysosomes by phagophores (i.e., open-ended globules formed by sheets of proteins and lipids).

When a cell's exterior is mechanically wounded (e.g., by scraping), the release of calcium ions triggers fusing of lysosomes with the cell's plasma membrane to quickly reseal any holes in the plasma membrane, to prevent leakage of the cell's contents. See also *Cell, Antigen, Major histocompatibility complex* (*MHC*), *T cells, Plasma membrane, Ion, Shigellosis, Autophagy.*

Lysozome

See *Lysosome.*

Lysozyme

An enzyme, naturally produced by some animals, which possesses antibacterial (i.e., bacteria killing) properties. Discovered in 1922 by Alexander Fleming, in his nasal mucus, Mr. Fleming named it (from the Greek) *lyso-* due to its ability to lyse (cut) bacteria and *zyme-* due to its being an enzyme. Lysozyme lyses certain kinds of bacteria, by dissolving the polysaccharide components of the bacteria's cell wall. When that cell wall is weakened, the bacteria cell then bursts because osmotic pressure (inside that bacteria cell) is greater than the weakened cell wall can contain. Tears and egg whites both contain significant amounts of lysozyme, as agents to prevent bacterial infections (e.g., against bacteria entering body via eye openings, against bacteria entering chicken embryo through the eggshell). See also *Enzyme, Lysis, Cell, Cytolysis, Polysaccharides, Bacteria, Paneth cells.*

M

M

M Cells

The immune system cells that constitute the surface of Peyer's patches. M cells preferentially sample/evaluate certain particles (e.g., viruses) passing over the Peyer's patch—embedded in the wall of intestine—during the digestive process. Those particles that meet certain criteria (and also specific toxin molecules that adhere to the M cells) are absorbed by the M cells and their antigens "presented" to adjacent lymphoid tissue underlying the Peter's patch.

This activates the lymphocytes present in the patches, which then migrate into the blood where they float in the tissue spaces just inside the intestinal lining. There, they secrete antibodies (primarily IgA), which are then transported into the lumen (contents) of the gut and subsequently attack (bind) the antigens. See also *Peyer's patches, Virus, Toxin, Antigen, Lymphocyte, Antibody, Immunoglobulin.*

MAA (Marketing Authorization Application)

It is the European Union (EU) equivalent to a U.S. New Drug Application (NDA). An MAA is an application to the EU's Committee for Proprietary Medicinal Products (CPMP) seeking approval of a new drug that has undergone Phase 2 and Phase 3 clinical trials. See also *NDA (to FDA), CANDA, Food and Drug Administration (FDA), MAA, NDA (to Koseisho), CPMP, Phase I clinical testing, Phase II clinical tests, Phase III clinical tests.*

MAb

See *Monoclonal antibodies (MAb).*

MAB

See *Marker-assisted breeding.*

Macromolecules

Large molecules with molecular weights ranging from approximately 10,000 to hundreds of millions. See also *Molecular weight.*

Macrophage

A phagocytic cell that is the counterpart of the monocyte. A monocyte that has left the bloodstream and has moved into the tissues. Macrophages have basically the same functions as monocytes, but they carry these out in the tissues. In summary, they engulf and kill microorganisms, present antigen to the lymphocytes, kill certain tumor cells, and their secretions (e.g., leukotrienes) regulate inflammation. Macrophages utilize nitric oxide and hydrogen peroxide (which they synthesize) to kill the microorganisms they engulf (via oxidation), and the nitric oxide also helps to regulate the immune system.

In the spleen, macrophages engulf and destroy old red blood cells. They are required to initiate production of large amounts of new red blood cells after massive injury (stress erythropoiesis) by physically touching erythroblasts, the "factories" located primarily within the spleen, that make red blood cells.

When macrophages reside in the bone marrow they store iron and then transfer it to red blood cells. In the lungs and GI tract they are scavengers and keep tissues clean. They also serve as a reservoir for the AIDS virus, in infected people. They (and other phagocytic cells) are largely responsible for the localization and degradation of foreign materials at inflammatory sites.

Macrophages display chemotaxis (i.e., the sensing of, and movement toward or away from a specific chemical). For example, consumption (in food/feed) of mannanoligosaccharides by mammals causes macrophages (within that mammal's bloodstream) to depart from the bloodstream and move toward the gastrointestinal tract (tissues) where those macrophages eliminate some pathogens (i.e., those growing/reproducing in the gastrointestinal tract).

When *n*-3 fatty acids ("omega-3" fatty acids) are present within macrophages, the enzyme known as cyclooxygenase-2 (COX-2) changes those *n*-3 fatty acids into metabolites known as electrophilic fatty acid oxidation (EFOX) derivatives that alter expression patterns of some macrophage genes, thereby reducing the inflammation impact noted within the first paragraph. See also *Cell, Innate immune response, Cellular immune response, Chemotaxis, Monocytes, Phagocyte, Adhesion molecule, lysosome, Nitric oxide, Nitric oxide synthase, Mannanoligosaccharides (MOS), Pathogen, Leukotrienes, Interleukin-1 (IL-1), Phosphatidyl serine, GPR120 receptor, n-3 fatty acids, Enzyme, COX-2, Cyclooxygenase, Metabolite, Insulin-like growth factor-1 (IGF-1), Tumor necrosis factor (TNF), Parkin.*

Macrophage Colony-Stimulating Factor (M-CSF)

A colony-stimulating factor (CSF) that stimulates production of macrophages in the body. See also *Colony-stimulating factors (CSFs), Macrophage.*

macroRNA

Coined by Jörg Hackermüller in 2014, this term refers to the very large RNA molecules (50–200 times the size of regular, protein-coding RNA) that are coded for by some of a cell's so-called noncoding RNA (i.e., the RNA that is transcribed from the cell's so-called junk DNA). See also *Deoxyribonucleic acid (DNA), Ribonucleic acid (RNA), Junk DNA, Long noncoding RNAs, Protein, Cell.*

MACS

Acronym for *magnetic cell sorting*. See *Magnetic particles.*

Macular Degeneration

See *AMD.*

M

Magainins

Discovered within frog skin tissues by Michael Zasloff in 1987, magainins are antimicrobial, amphopathic peptides that lyse (i.e., burst) certain cells upon contact by "worming" their hydrophobic portion into the cell's membrane, which creates a transmembrane (i.e., through the surface) pore (allowing ions to flow into the cell, causing osmotic bursting). Magainins are selective against bacteria, fungi, and protozoa cells (the word magainin comes from the Hebrew word for "shield"). See also *Amphiphilic molecules, Cell, Peptide, Bacteria, Fungus, Antibiotics, Plasma membrane.*

Magic Bullet

When this term was first coined by Paul Ehrlich in 1905, it initially referred only to antibodies (e.g., because antibodies seek their own target, without damaging other nearby tissues).

However, over time, this term has come to be applied to immunotoxins and other immunoconjugates (i.e., toxic or pharmacological molecules that are "attached" to an antibody that "steers/guides" the toxic or pharmacological molecule to the intended "target" in the body such as a tumor).

In 2005, Stephen Russell was able to modify a measles virus so that

- It expressed an antibody that targeted an antigen on the surface of a cancer tumor's cells in a mouse, thereby causing the (injected) virus to accumulate on the cancerous cells
- It "infected" those cancerous cells and killed them without harming adjacent healthy tissue

In 2014, Samir Mitragotri and Aaron Anselmo created and attached to monocytes a tiny disc-shaped polymer "nano-backpack" that could contain pharmaceutical(s) that can be released by the backpack when the monocyte reaches the site of (chronic) inflammation within body tissues. These polymer backpacks are coated on one of their sides with an antibody that can bind to receptors on the monocyte's surface. At the site of the chronic inflammation, the other side of the backpack degrades to release the pharmaceutical. See also *Antibody, Immunoconjugate, Immunotoxin, Genistein, Ricin, Monoclonal antibodies (MAb), HER2gene, Virotherapy, Cell, Antigen, Cancer, Tumor, Diphtheria toxin, Monocytes, Polymer, Nanobackpack, Chronic inflammation.*

Magnetic Antibodies

See *Magnetic particles.*

Magnetic Beads

See *Magnetic particles.*

Magnetic Cell Sorting

See *Magnetic particles.*

Magnetic Fluid Hyperthermia

See *Nanoshells.*

Magnetic Labeling

See *Magnetic particles.*

Magnetic Particles

Refer to various tiny pieces of natural magnetic materials, which are bonded (attached) to *capture molecules* such as specific molecular ligands, receptors, aptamers, antigens, antibodies (e.g., monoclonal antibodies that are specific to a particular type of cell), and so on.

These can then be mixed with a large population of many cell types (e.g., crude tissue samples, cells grown in a vat/reactor), where the *now-magnetic capture molecules* will attach themselves to *only the desired cells*, and then the desired cells are separated out using a magnetic field (and the magnetic particles/antibodies are subsequently removed from those cells). For example, magnetic nanoparticles (100 nm diameter) attached to *antibodies against epithelial cells* can be utilized to detect metastasis of cancer in a human. The magnetized antibodies attach themselves to epithelial cells (a biomarker of metastasis) in a blood sample, enabling the epithelial cells to be detected/counted by doctors.

In similar fashion, specific nucleic acids/DNA can be attached to magnetic particles. These can then be mixed with a mixture of nucleic acids/DNA whereby the magnetic particles will attach themselves (via hybridization) to only the desired nucleic acids/DNA; then they are separated out using a magnetic field. See also *Capture molecule, Antibody, Monoclonal antibodies (MAb), Cell, Immunoconjugate, Cell sorting, Nucleic acids, Deoxyribonucleic acid (DNA), Hybridization (molecular genetics), Hybridization surfaces, Bio-bar codes, Nanoparticles, Ligand (in biochemistry), Receptors, Aptamers, Antigen, Cancer, Metastasis, Biomarkers, Nanobackpack.*

Maillard Reaction

Refers to a set of chemical reactions discovered in 1912 by Louis-Camille Maillard, which occur when certain foodstuffs are cooked at temperatures exceeding 121°C (250°F). Certain amino acids within the foodstuffs react with sugars to produce flavorful chemical compounds (e.g., melanoidins).

Carefully controlled Maillard reactions of soybean meal can also be utilized to produce a *ruminal-bypass feed* ingredient. Such feed (protein) would be protected from breakdown in the rumen (e.g., of a dairy cow), so that optimal digestion would occur in the cow's intestine (abomasum). See also *Amino acid, Melanoidins, Protein, Soybean plant, Soy protein, Rumen (of cattle).*

Maize

See *Corn.*

Major Histocompatibility Antigen: Class I

A "family" of glycoproteins that appear on the surfaces of most cells of an organism, which help enable that organism's immune system to distinguish "self" (cells) from "nonself" (e.g., invading pathogens). See also *Glycoprotein, Cell, Organism, Pathogen, Major histocompatibility complex (MHC), Major histocompatibility antigen—class II.*

Major Histocompatibility Antigen: Class II

A "family" of glycoproteins that appear only on the surface of specific lymphocyte cells (dendritic cells) and on the surface of certain macrophages, within an organism. See also *Glycoprotein, Cell, Lymphocyte, Dendritic cells, Macrophage, Organism, Major histocompatibility complex (MHC), Major histocompatibility antigen—class II.*

Major Histocompatibility Complex (MHC)

A genetic loci or chromosomal region (approximately 3000 kb) that encodes for three classes of transmembrane (cell) proteins. MHC I proteins (located on the surface of nearly all cells) present foreign epitopes (i.e., fragments of antigens that have been ingested; peptides) to cytotoxic T cells (killer T cells). MHC II proteins (located on the surface of immune system lymphocyte/dendritic cells and phagocytes) present foreign epitopes to helper T cells. That *presenting* of epitopes induces the organism's immune response.

MHC III proteins are components of the complement cascade. Genes in the MHC must be matched (between an organ donor and organ recipient) to prevent rejection of organ transplants. See also *Complement cascade, Loci, Locus, Chromosomes, Graft-versus-host disease (GVHD), kb, Lymphocyte, Dendritic cells, Macrophage, Protein, Cell, T cell receptors, Antigen, T cells, Cytotoxic T cells, Epitope, Humoral immunity, Gene, Tumor-associated antigens, Human leukocyte antigens (HLA), Cellular immune response.*

MAL (Multiple Aleurone Layer) Gene

A gene in corn (maize) that (when present in the DNA of a given plant) causes that plant to produce seed that contains higher-than-normal levels of calcium, magnesium, iron, zinc, and manganese. These higher mineral levels are particularly useful for feeding of swine, since traditional No. 2 yellow (dent) corn does not contain enough for optimal pig growth. See also *Gene, Deoxyribonucleic acid (DNA), High-methionine corn, High-lysine corn, Floury-2, Opague-2.*

MALDI-TOF-MS

Acronym for matrix-associated laser desorption ionization time of flight mass spectrometry. A mass spectrometry methodology/technology that was initially developed by Franz Hillenkamp for the analysis of biological molecules.

MALDI-TOF-MS can establish, in seconds, the identity, purity, etc. of a sample of proteins, oligonucleotide, or (poly)peptides. Also the identification of gram-positive microorganisms or characterization of genetic materials (e.g., DNA, RNA) on hybridization surfaces.

MALDI-TOF utilizes measurement of the *time for particles (e.g., proteins) to transit a specific distance* after being "dislodged" from within a specific point on an anode where each was placed (e.g., by a robot arm, which picks proteins, for instance, out of the gel after running them through two-dimensional gel electrophoresis to separate from others in a sample).

After being placed by the robot arm onto the anode and then dried into a crystalline matrix adhered to its surface, MALDI-TOF-MS dislodges (molecules) from the ("adhered") surface by vaporization with a specific amount of laser energy to precisely determine the molecular weight (e.g., of proteins). See also *Mass spectrometer, Microorganism, Oligonucleotide, Gram-positive, Ribonucleic acid (RNA), Hybridization surfaces, Deoxyribonucleic acid (DNA), In silico biology, Protein, Peptide, Two-dimensional (2D) gel electrophoresis, ICM.*

Male-Sterile

See *Barnase.*

Malonyl CoA

See *Fats.*

Mammalian Cell Culture

Technology to artificially cultivate cells, of mammal origin, in a laboratory or production-scale device (i.e., *in vitro*). Can be either a batch or continuous process device. The first mammalian cell culture was performed by a neurobiologist named R. G. Harrison in 1907, when he added chopped-up spinal cord tissue to clotted (blood) plasma in a humidified growth chamber. The nerve cells from this spinal cord tissue successfully grew, divided, and extended long fibers into the clot. Many improvements to cell culture process have been made over the years, including special growth media (fluids that bathe the cultured cells with the right amounts of amino acids, salts, and other minerals). See also *Continuous perfusion, Dissociating enzymes, CHO cells, Harvesting enzymes, In vitro, Plasma, Cell, Medium, Amino acid, Induced pluripotent stem cells.*

Management Zones

Refer to precision agriculture software products that utilize a combination of public soil databases (including soil depth, texture, soil organic matter, and water-holding capacity), company proprietary (analytics, high-resolution field elevation data, field topography, watersheds/hydrogeology), and electrical conductivity soil testing (basis: topsoil depth, pH, salt concentrations, and available water-holding capacity) to divide farm fields into contiguous *management zones* (subportions of field, also sometimes known as *yield environments* or *ERU maps*) where crops planted there will respond positively to farmer management decisions regarding crop inputs such as amounts and timing of fertilizer applied to field, amounts and timing of irrigation water applied to field, number of crop seeds planted per hectare, and so on.

For example, such precision agriculture software might thereby recommend far less irrigation water be applied by the farmer to a *management zone* located in a low-elevation, heavy-soil-type area of farm field that is naturally wet throughout the growing season due to the field's hydrogeology, plus a higher number of crop seeds planted per hectare (of a wet-environment-tolerant crop variety). The precision agriculture software would recommend different inputs/rates for a *management zone* located in a sandy soil, higher-elevation area of the farm field (e.g., more irrigation water applied, lower number of crop seeds planted per hectare, of a drought-tolerant crop variety). Such properly integrated "precision management of management zones" would maximize the field's crop yield while minimizing its consumption of inputs such as fertilizer, irrigation water, crop seeds, and so on. See also *Drought-tolerance trait, ERU maps, pH.*

Mannan-Binding Lectin-Associated Serine Protease-2 (MASP-2)

See *Reperfusion.*

Mannan Oligosaccharides

See *Mannanoligosaccharides* (MOS).

Mannanoligosaccharides (MOS)

A "family" of oligosaccharides that can be produced by man in commercial quantities via certain yeast cells. When consumed (e.g., by humans or monogastric livestock such as swine or poultry), mannose sugars in the MOS stimulate the liver to secrete the mannose-binding protein. Mannose-binding protein enters the digestive system and binds to the (mannose-containing) capsule (surface membrane) of pathogenic bacteria. That binding to pathogens triggers the immune system's complement cascade to combat those pathogenic bacteria.

Consumption of mannanoligosaccharides by mammals also causes macrophages to move toward the gastrointestinal tract (in body's tissues), where those macrophages eliminate some pathogens (i.e., growing/reproducing in the gastrointestinal tract). See also *Oligosaccharides, Fructose oligosaccharides, Sugar molecules, Yeast, Complement cascade, Pathogenic, Bacteria, Immune response, Complement, Capsule, Macrophage, FOSHU, Nutraceuticals.*

Mannogalactan

See *High-mannogalactan soybeans.*

Map Distance

A number proportional to the frequency of recombination between two genes. One map unit corresponds to a recombination frequency of 1%. See also *Genetics, Genetic code, Genetic map, Gene, Linkage, Quantitative trait loci (QTL).*

MAP Kinase Pathway

See *Stem cells, Mitogen-activated protein kinase cascade.*

MAPK

Acronym for *mitogen-activated protein kinase.* See *Mitogen-activated protein kinase cascade.*

MAPK System

See *Mitogen-activated protein kinase cascade.*

Mapping (of Genome)

See *Genetics, Genetic code, Genetic map, Quantitative trait loci, Position effect.*

Marker (DNA Marker)

A DNA fragment of known size used to calibrate an electrophoretic gel. See also *Electrophoresis, Two-dimensional (2D) gel electrophoresis, Deoxyribonucleic acid (DNA).*

Marker (DNA Sequence)

A specific sequence of DNA that is virtually always associated with a specified trait, because of "linkage" between that DNA sequence (the "marker") and the gene(s) that cause that particular trait. Such markers have been utilized to aid/speed up the process of plant (e.g., crop) breeding since the mid-1970s via *marker-assisted selection.* See also *Deoxyribonucleic acid (DNA), Trait, Linkage, Linkage group, Linkage map, Gene, Sequence (of a DNA molecule), Marker-assisted selection, YSTR DNA.*

Marker (Genetic Marker)

A trait that can be observed to occur or not to occur in an organism such as, for example, bacteria or plant(s). Genetic markers include such traits as

- Expression of luciferase-catalyzed bioluminescence in leaf cells (causing leaves to glow when illuminated by certain light sources)
- Resistance to specific antibiotics
- The nature of the cell wall and capsule characteristics
- Requirements for a particular growth factor, and carbohydrate utilization, to mention a few

For example, if a culture of dividing (growing) bacteria that is not resistant to a particular antibiotic (i.e., lacks the trait of antibiotic resistance) is exposed to only the DNA isolated from bacteria that are resistant to the antibiotic, then a fraction of the cells exposed will directly incorporate this trait (some DNA) into their genome, hence acquiring the trait. The first genetically engineered plants bearing a marker gene were field tested in 1986. See also *Allele, Genetic engineering, Positive and negative selection (PNS), Transformation, Transfection, NPTII gene, Bioluminescence, Marker-assisted selection, GUS gene, bla gene, Recombinase.*

Marker-Assisted Breeding

See *Marker-assisted selection.*

Marker-Assisted Selection

The utilization of DNA sequence "markers" (molecular markers) by commercial breeders to select the organisms (e.g., crops, livestock) that possess gene(s) for a particular performance trait (e.g., rapid growth, high yield) desired for subsequent breeding/propagation. Marker-assisted selection has been utilized in many plant (e.g., crop) breeding programs since the mid-1970s.

In 2014, Richard Oliver reported finding molecular markers that are useful for faster/more efficient breeding of wheat (*Triticum aestivum*) for resistance to the fungal diseases known as yellow (tan) spot and *Septoria nodorum* blotch. These molecular markers denote specific *disease-sensitivity genes* that code for protein molecule(s) in wheat plant that the relevant fungi latch on to, in order to enter wheat plant cells to cause those diseases. A wheat breeder can utilize those markers to avoid including the disease-susceptible wheat varieties in his breeding efforts. See also *Deoxyribonucleic acid (DNA), Sequence (of a DNA molecule), Marker (DNA sequence), Gene, Trait, Genetic map, Linkage, Linkage group, Molecular breeding™, Linkage map, Quantitative trait loci (QTL).*

MARS

Acronym for *marker-assisted recurrent selection.* Refers to the cycle utilized within MAS (marker-assisted selection) formal breeding programs (e.g., conducted by modern crop seed companies) to help the plant breeder to more rapidly increase the frequency of favorable-trait (desired) genes in the DNA of the breeding population. See also *Marker-assisted selection, Gene, Trait, Deoxyribonucleic acid (DNA).*

MAS

See *Marker-assisted selection.*

MASP-2

Acronym for *mannan-binding lectin-associated serine protease-2.* See *Reperfusion.*

Mass Applied Genomics

See *Genomics, Biochips, Microarrays (testing), Bioinformatics.*

Mass Spectrometer

An analytical device that can be used to determine the molecular weights (mass) of proteins and nucleic acids, the sequence of (composition and order of amino acids comprising) protein molecules, the chemical composition of virtually any biomaterial (e.g., lipids), and the rapid identification of intact gram-negative and gram-positive microorganisms (the latter, using matrix-assisted laser desorption ionization time of flight mass spectrometry).

The exact mass of such charged particles (e.g., ions) is based on measurement of their mass-to-charge ratios (while the particles are passing through precisely known strength magnetic and electrical fields).

To utilize a mass spectrometer to identify each protein present within a given sample, the protein molecules are first separated (e.g., via two-dimensional gel electrophoresis or via liquid chromatography). Next, those protein molecules are *very specifically* reduced, alkylated, and broken (in specifically known ways via enzymes) into peptides. When passed through the mass spectrometer, the peptides (and by derivation, the initial proteins) are identified by comparing their mass/charge spectra to those within a database of known proteins (i.e., which were earlier passed through the mass spectrometer). See also *Gram-negative, Gram-positive, Molecular weight, Ion, Sequencing (of DNA molecules), Protein, Amino acid, Nucleic acids, Gene machine, MALDI-TOF-MS, Two-dimensional (2D) gel electrophoresis, Chromatography, Reduction (in a chemical reaction), Peptide, Enzyme, Lipids, Ion trap.*

Massively Parallel Signature Sequencing

Refers to a form of gene expression analysis, in which a given cell's small RNA molecules (21–24 ribonucleotides in length) can be thoroughly profiled. That profiling is accomplished by

1. Cloning each (known) small RNA molecule of that organism/cell and attaching it to a single bead of approximately 5 μ size
2. Placing such RNA-attached beads into a suitable container and flowing the relevant cellular (fluid) sample around those beads
3. Determining the amounts of each small RNA molecule that hybridizes to each bead

Because the *known* small RNA molecule depicted in #1 acts as a *sequence tag* for the (agglomeration on each bead), the respective amounts of each of the cell's small RNA molecules can be subsequently determined. See also *Gene, Gene expression analysis, Cell, Ribonucleic acid (RNA), Nucleotide, Clone (a molecule), Micron, Hybridization (molecular genetics), Hybridization surfaces.*

Mast Cells

Fixed (noncirculating) leukocyte cells that are present in many different kinds of body tissues. When two IgE molecules of the same antibody "dock" at adjacent receptor sites on a mast cell, then (the two IgE molecules) capture an allergen (e.g., a particle of pollen) between them, a chemical-energetic signal is sent to the interior (inside mast cell) portion of receptor molecules, which causes that interior portion of molecule to change (i.e., transduction). That signal transduction causes a protein named "syk" to set off a chemical chain reaction inside the mast cell, thereby causing that mast cell to release leukotrienes, histamine, serotonin, bradykinin, and "slow reacting substance." Release of these chemicals into the body causes the blood vessels to become more permeable (leaky) and causes the nose to run, itchy, and watery eyes. These chemicals also cause smooth muscle contraction—causing sneezing, breath constriction coughing, wheezing, and so on. See also *Basophils, Antigen, Antibody, Receptors, Signal transduction, Histamine, Allergies (foodborne), Signaling, Leukotrienes, Leukocytes.*

Matrix Metalloproteinases (MMP)

A "family" of enzymes that contain the zinc metal ion (Zn^{2+}) at their active sites. MMPs can change the molecular conformations/configurations of the protein molecules they interact with. Because certain cancerous tissues produce high levels of matrix metalloproteinases, the resultant changes in protein molecular conformations/configurations can lead to metastasis (i.e., spread of the cancer). Applicable matrix metalloproteinases exuded by cancerous cells enable those cancer cells to spread from their original locations (within body tissue) by "cutting" through proteins of the extracellular matrix that normally holds cells in place within tissues.

Also among the MMP family are the collagenases. See also *Enzyme, Ion, Active site, Catalytic site, Stromelysin (MMP-3), Cancer, Metastasis, Collagenase, Extracellular matrix (ECM).*

Matrix-Assisted Laser Desorption/Ionization Time of Flight Mass Spectrometry

See *MALDI-TOF-MS.*

Maximum Residue Level (MRL)

Term used for an officially established upper allowable limit, of a given compound (e.g., a synthetic hormone) in a particular product, such as meat. For example, in 1994, the Codex Alimentarius Commission in Rome, Italy, decided to establish maximum residue levels for each of five growth promotants that are commonly utilized by the U.S. beef industry. Because the World Trade Organization (WTO) subsequently stated that it would respect MRLs, a WTO member nation cannot legally refuse to allow import of meat products on growth promotant-content basis if the content of the promotant contained in the meat is less than its maximum residue level. See also *Growth hormone, Growth factor, Codex Alimentarius Commission, World Trade Organization (WTO).*

Maysin

A chemical that is naturally produced at low amount within most varieties of corn/maize (*Zea mays* L.) plants. During the 1980s and 1990s, certain corn/maize varieties with very high maysin content were developed in the United States.

Because maysin "binds up" certain (essential) amino acids in the gut of the corn earworm (*Helicoverpa zea*) caterpillar (larvae), those insect pest caterpillars stop growing when they are consuming a large amount of maysin. See also *Corn, Helicoverpa zea (H. zea), Amino acids, Essential amino acids, High-maysin corn.*

MC

Acronym for *mast cells.* See *Mast cells.*

M

MCA

See *Medicines control agency (MCA).*

MCL

Acronym for mantle cell lymphoma, a rare and aggressive type of B-cell non-Hodgkin lymphoma that typically occurs in older adults. See also *B-cells, Bortezomib.*

mCNVs

Abbreviation for multiallelic copy number variation loci. See *Multiallelic copy number variation loci.*

MCR

Acronym for mutagenic chain reaction. See *Mutagenic chain reaction.*

mCry3Aa Protein

One of the "Cry" (i.e., "crystal-like") proteins, it is a protoxin that when eaten by larvae of the northern, western, and Mexican corn rootworms (*Diabrotica virgifera virgifera*) is toxic to them. See also *Corn rootworm, Cry proteins, Bacillus thuringiensis (B.t.), Protoxin, Protein, Ion channels.*

MCT

Acronym for *medium-chain triacylglycerides*. See *Medium-chain triacylglycerides.*

MDCK

Acronym for *Madin–Darby canine kidney* epithelial cell lines. These are cell lines that are widely utilized in vaccine research, which are "descended" from kidney epithelial cells that were taken from a female cocker spaniel dog in 1958. See also *Cell, Cell culture, Epithelium.*

MDR-TB

Acronym for multidrug-resistant tuberculosis. See *ATP synthase.*

MEA

Acronym for *multilateral environmental agreement*; an agreement (e.g., treaty) between a number of nations that is intended to protect/benefit the environment. See also *Convention on biological diversity (CBD).*

Mechanobiology

A term coined in 1997 by Dennis Carter, it refers to how mechanical (force) conditions help regulate certain biological processes (e.g., the signaling cascade that occurs at the cellular level when applicable mechanical forces are applied to certain cells/tissues). Examples are as follows:

- Osteoblasts (bone-building specialist cells) and osteoclasts (bone-consuming specialist cells) work in response to such signals, to actually reshape support bones (e.g., in legs) as a consequence of the particular walking gait adopted by each human.
- Certain cells within tissues that are under different types or amount of mechanical force will have *differential display* (i.e., different sets of proteins expressed, or different expression levels of said proteins, out of a suite of the tissue-applicable proteins).

See also *Cell, Protein, Signaling, Signal transduction, Cascade, Osteoinductive factor (OIF), Differential display.*

Mediator

Refers to a protein molecule that, together with another protein known as *cohesin*, forms a protein complex (structure) that helps a cell's DNA form into the specific loop that is necessary for the applicable gene(s) in the DNA to be activated that control that particular cell's state (e.g., the tissue it has differentiated into, if the cell is no longer in its embryonic state). See also *Protein, Cell, Deoxyribonucleic acid (DNA), Cohesin, Loop, Gene, Activator (of gene), Expressivity, Cell differentiation, Embryonic stem cells, Pluripotent stem cells, Differentiation.*

Medicines Control Agency (MCA)

The British Government agency that, in concert with the Committee on Safety in Medicines, regulates the approval and sale of pharmaceutical products in the United Kingdom. See also *Committee on Safety in Medicines, Food and Drug Administration (FDA), Committee for Proprietary Medicinal Products (CPMP), Koseisho, NDA (to Koseisho), IND, Bundesgesundheitsamt (BGA).*

Medifoods

See *Nutraceuticals, Phytochemicals.*

Medium

A substance used to provide nutrients for cell growth. It may be liquid (e.g., broth) or solid (e.g., agar). See also *Culture medium, AGAR, Mammalian cell culture.*

Medium-Chain Saturated Fats

See *Medium-chain triacylglycerides.*

Medium-Chain Triacyglycerides

Refer to a category of saturated fatty acid molecule (fragments or derivatives). When consumed by humans, *medium-chain triacylglycerols (MCTs)* are more rapidly digested, metabolized, and absorbed than the corresponding full-length fatty acid molecules (triacylglycerols). For that reason, MCTs are utilized to deliver lipid nourishment to people (e.g., certain hospital patients) whose bodies suffer from lipid nutrient maladsorption.

Research indicates that consumption of MCTs increases the body's caloric consumption more than long-chain triacylglycerols. See also *Fats, Fatty acid, Lipids, Saturated fatty acids (SAFA), Triacylglycerides, Triglycerides, Digestion (within organisms).*

Medium-Chain Triglycerides

See *Medium-chain triacylglycerides.*

Megabase

A unit of length for DNA, equal to 1,000,000 bp. See also *Deoxyribonucleic acid (DNA), Base pair (bp).*

Megakaryocyte-Stimulating Factor (MSF)

A colony-stimulating factor (protein) involved in the regulation of platelet production, white blood cell production, and red blood cell production from stem cells in bone marrow. See also *Colony-stimulating factors (CSFs), Platelets, Stem cells.*

Megakaryocytes

See *Platelets.*

Meganuclease

Refers to a type of much-larger-than-usual endonuclease (DNA-cutting enzyme) that is coded for by certain mobile elements (i.e., DNA segments that can move from one chromosomal locus to another). Initially developed commercially by the Cellectis SA company, these meganucleases possess very large recognition sites (i.e., portion of the endonuclease molecule that "latches on" to the DNA or RNA molecule at a specific location, so the active site of the endonuclease will cleave/cut the DNA or RNA at a precise spot). Meganucleases can be utilized by man to change the sequence of a selected gene (thereby changing its function) or knock out that gene. See also *Endonucleases, Enzyme, Active site, Deoxyribonucleic acid (DNA), Gene, Coding sequence, Mobile element, Chromosomes, Locus, Ribonucleic acid (RNA), Genetic engineering, Knockout.*

Mega-Yeast Artificial Chromosomes (mega YAC)

A large (i.e., greater than 500 bp in length) piece of DNA that has been cloned (made) inside a living yeast cell. While most bacterial vectors cannot carry DNA pieces that are larger than 50 bp, and "standard" YACs typically cannot carry DNA pieces that are larger than 500 bp, mega YACs can carry DNA pieces (chromosomes) as large as one million base pairs in length. See also *Yeast, Chromosomes, Human artificial chromosomes (HAC), Arabidopsis thaliana, Deoxyribonucleic acid (DNA), Clone (a molecule), Vector, Base pair (bp), Yeast artificial chromosomes (YAC).*

Meiosis

From the Greek *meioun*, meaning *to make smaller.* Discovered by Edouard Van Beneden in the 1870s, meiosis is the sequence of complex cell nucleus changes resulting in the production of cells (as gametes) with half the number of chromosomes present in the original cell and typically involving an actual *reduction division* in which the chromosomes without undergoing prior splitting join in pairs with homologous chromosomes (of maternal and paternal origin) and then separate (i.e., pulled apart by actin and microtubules within the cell) so that one member of each pair enters each product cell nucleus and undergoes a second division not involving reduction.

Occurs by two successive divisions (meiosis I and II) that reduce the starting number of $4n$ chromosomes to $1n$ in each of four product cells. Product cells may mature to germ cells (sperm or eggs). See also *Oocytes, Cell, Chromosome, Nucleus, Microtubules, Actin, Motor proteins, Centromere.*

Melanin

A pigment found in most fungi that enables applicable fungi to harvest ionizing radiation (e.g., emitted by nuclear reactors) as a source of energy for manufacturing "food" for those fungi to live on. See also *Fungus.*

Melanoidins

A class of flavorful chemicals that are formed as reaction products during the *Maillard reaction.* Melanoidins act as strong antioxidants in the human body. See also *Antioxidants, Maillard reaction.*

Melanoma

A potentially fatal skin cancer, in which melanocytes (i.e., pigment-manufacturing cells in the skin) grow and proliferate rapidly to form cancerous tumors. See also *Cancer, Tumor, Electroporation.*

Melatonin

A hormone that is present in animals, plants, and some microorganisms. Melatonin plays a role in regulating animals' circadian (e.g., "biological clock") rhythms and sleep cycles. Significant consumption of melatonin (e.g., via dietary supplement) may help to strengthen bones of the elderly, because activity of osteoclasts (the body's agents of bone breakdown) is greatest during evening hours, and increased sleep amounts slow down the process of bone breakdown, potentially allowing the osteoblasts (the body's bone-building agents) to exceed it.

Because high melatonin levels at night put breast cancer cells to "sleep" by turning off key cellular growth mechanisms, the breast cancer cells are made more vulnerable to tamoxifen (a pharmaceutical that acts against breast cancer). See also *Hormone, Microorganism, Cell, Cancer, Tamoxifen.*

Melittin

A peptide that is naturally present within the venom of bees. When injected via stinging, melittin causes cells to break open and to release the contents of those cells. See also *Peptide.*

Melting (of DNA)

Melting DNA means to heat-denature it. When this happens the hydrogen bonds holding the DNA molecule together in the normal way are disrupted, allowing a more random polymer structure to exist. See also *Denatured DNA.*

Melting (of Substance Other Than DNA)

To change from a solid to a nonsolid (e.g., liquid) state by the addition of heat (to the solid substance).

M

Melting Temperature (of DNA) (Tm)

The midpoint of the temperature range over which DNA is denatured. See also *Melting (of DNA)*.

Membrane Channels

See the links. See also *Plasma membrane, Ion, Aquaporins, Membrane transporter protein, Calcium ion channels, Potassium ion channels.*

Membrane Transport

The facilitated transport of a solute across a membrane, usually by a specific membrane protein (e.g., adhesion molecule, receptor). See also *Endocytosis, Exocytosis, Signal transduction, G-proteins, Vaginosis, Receptors, Adhesion molecule, Vesicular transport, Gated transport, Translocon, Calcium channel-blockers, Lipids, Cage carrier, Gated channel, Translocation (of protein molecules), Aquaporins, Effectors (fungal).*

Membrane Transporter Protein

A class of transmembrane proteins (i.e., protein molecules embedded in a cell's membrane, extending through both sides of the membrane) that function to *transport certain molecules through the cell's membrane*. Such molecules that are thus "transported" include

- Sugar molecules (utilized by the cell as "fuel")
- Inorganic ions (which catalyze certain cellular processes)
- Polypeptides (e.g., "manufactured" in the cell's ribosome(s) and then secreted from cell to perform some function elsewhere in the body of the organism)
- Anticancer drugs
- Antibiotics

See also *Protein, Cell, Plasma membrane, Membrane transport, Ribosomes, Polypeptide (protein), ABC transporters, Translocation (of protein molecules).*

Membranes (of a Cell)

Refer to the thin "*skin-like*" structures that surround the exterior of a cell (i.e., plasma membrane) and also surround various specialized bodies (e.g., nucleus, mitochondria) *within* the cell itself (e.g., the membrane that surrounds the cell's nucleus is called the "nuclear envelope").

Membranes are lipoidal, that is, "made of lipoidal (fatlike) material," in which proteins and protein complexes are embedded. For example, protein molecules known as receptors are embedded in the plasma membrane (i.e., the outermost membrane of the cell) and in the nuclear envelope. See also *Cell, Cecrophins (lytic proteins), Magainins, Plasma membrane, Transmembrane proteins, Ion channels, Receptors, Nuclear receptors.*

Memory T Cells

See *CD8$^+$ T cells.*

MEMS (Nanotechnology)

Acronym utilized by Americans to refer to "microelectromechanical systems" (which Europeans tend to refer to as "microsystems technology"—MST). See also *Nanotechnology, Biochip, Genosensors, Biosensors (electronic), Biosensors (chemical), Nanocrystal molecules microfluidics, Quantum wire, Quantum dot, Molecular machines, Biomotors, Biomems.*

mEPSPS

The "m" *variant* (of the many forms of) the enzyme *5-enolpyruvylshikimate-3-phosphate synthase*. mEPSPS is unaffected by glyphosate-containing or sulfosate-containing herbicides, so introduction of the gene (coding for mEPSPS) into crop plants (e.g., corn/maize) makes those crop plants essentially impervious to glyphosate-containing or sulfosate-containing herbicides. See also *Enzyme, Gene, Genetic engineering, EPSP synthase, Glyphosate, Sulfosate, Corn, Herbicide-tolerant crop, AroA.*

Meristem

From the Greek *meristos* meaning "divisible," it refers to certain plant cells that are able to actively divide and differentiate into plant parts such as roots, shoots, and so on. See also *Cell, Differentiation.*

Mesenchymal Adult Stem Cells

See *Mesodermal adult stem cells.*

Mesenchymal Stem Cell (MSC)

Refers to a pluripotent stem cell found in bone marrow and in muscle tissue, which differentiates (and migrates where needed) to become primarily bone and connective tissue within the body of an organism. Their differentiation is impacted by whether they were initially grown on a stiff (e.g., bone, or lactic acid–glycolic acid polymer scaffold in the case of tissue engineering) substrate or on a soft substrate. Those that were initially grown on a stiff substrate tend to differentiate into bone cells, whereas those initially grown on soft substrates differentiate into bone or adipose (fat) cells. See also *Cell, Stem cells, Pluripotent stem cells, Cell differentiation, Cell motility, Adipose, Tissue engineering.*

Mesodermal Adult Stem Cells

Certain stem cells present within (adult) bodies of organisms that can be differentiated (via chemical signals) to give rise to bone, muscle, and/or fat cells. See also *Stem cells, Multipotent adult stem cells, Cell, Organism, Signaling.*

Mesophile

An organism that grows best in the temperature range of 25°C (77°F) to 40°C (104°F). See also *Thermophile, Psychrophile.*

Mesoscale

See *Nanoscience.*

Messenger RNA (mRNA)

Messenger ribonucleic acid, first identified by Francis Crick, Sydney Brenner, and Matthew Meselson. mRNA is the intermediary molecule that moves between DNA and ribosomes in same cell (or in a distant cell of that organism, when the organism is under certain types of stress) that synthesize (i.e., manufacture) those proteins coded for by the cell's DNA. Upon receiving the "message" that was encoded in the DNA, the messenger RNA passes through the ribosomes like a reel of punched paper passes through an old player piano (pianola) giving the ribosomes the specifications for making the coded-for proteins (akin to how the punched paper causes a pianola to play the tune that is coded in the paper).

This process is aided by transfer RNA (tRNA) molecules, which forage for amino acids that float around in the cell (outside of the cell's nucleus and ribosomes). The transfer RNA (tRNA) molecules attach to, and escort individual amino acids to the ribosome, as and when the messenger RNA (mRNA) directs. Each of the 20 different amino acids has at least one of its own purpose-built tRNA molecules, which possess a three-letter code of nucleotides at the stem of the cloverleaf-shaped rRNA molecule.

The ribosome has room for only two tRNA molecules at a time. The messenger RNA (mRNA) molecule (which itself is passing through the ribosome) calls over the first tRNA molecule, which brings with it the specified amino acid. Short sections of the messenger RNA (mRNA) and transfer RNA (tRNA) molecules lock together inside the ribosome (because where these two molecules meet, their three nucleotides are complementary), the whole (locked together) apparatus shifts along by three notches (i.e., nucleotides), and a second tRNA molecule (bearing another amino acid) slips in next to the first tRNA molecule.

Next, the first amino acid (brought in by the first tRNA molecule) jumps over to the second tRNA molecule, joining to the amino acid that was brought in by the second tRNA molecule and thus making the start of a protein (i.e., a poly-amino acid molecule, also known as polypeptide or protein molecule). The empty (first) tRNA molecule falls out of the ribosome, and the whole (locked together) apparatus (i.e., mRNA plus second tRNA molecule) moves three more notches (i.e., nucleotides) along the mRNA molecule to make room for a third tRNA molecule bearing another amino acid, and so on.

This process of creating ever-longer chains of amino acids continues to repeat itself inside the ribosome until the protein (coded for by the DNA, which code was transferred to mRNA, which transferred it to the ribosome) is completed. See also *Transcription, Complementary DNA (c-DNA), Central dogma, Deoxyribonucleic acid (DNA), Ribonucleic acid (RNA), Nucleic acids, Coding sequence, Genetic code, Cell, Informational molecules, Codon, Ribosomes, Polyribosome (polysome), rRNA (ribosomal RNA), Nucleotide, Polymer, Transfer RNA (tRNA), Protein, Amino acid, Polypeptide (protein), Antisense (DNA sequence), Transcriptome, Abiotic stresses, Drought tolerance.*

Messenger™

See *Harpin.*

Metabolic Cell

Refers to a cell that is only metabolizing (i.e., is not dividing). See also *Cell, Metabolism.*

Metabolic Engineering

The selective, deliberate alteration of an organism's metabolic pathway(s) via genetic engineering of the genes that define/control the organism's metabolism. Some reasons to do metabolic engineering of an organism include the following:

- Altering cell "behavior" and organism metabolic patterns to induce production of proteins/polypeptides and/or metabolites that are desired by mankind (e.g., "golden rice").
- Altering cell "behavior" and organism metabolic patterns to induce a given organism to consume or accumulate toxic wastes or valuable materials (e.g., gold) that are present at a site in low concentration or highly dispersed.
- Altering cell "behavior" and organism metabolic patterns to cure disease. For example, in 2004, Koji Yanai et al. utilized metabolic engineering of *Rosellinia* sp. filamentous fungi to make a more potent version of the naturally occurring compound known as "PF1022A," which can be used to kill parasitic nematodes.

See also *Metabolism, Intermediary metabolism, Cell, Pathway, Metabolic pathway, Flux, Genetic engineering, Organism, Gene, Gene splicing, Protein, Fungus, Phyto-manufacturing, Polypeptides, Bioleaching, Biodesulfurization, Biorecovery, Bioremedition, Golden rice, Phytoremediation, Cell surface engineering, Metabonomics, Metabolite profiling, Nematodes.*

Metabolic Flux Analysis

See *Flux.*

Metabolic Pathway

Refers to a particular pathway (i.e., series of chemical reactions, each of which is dependent on previous one(s)) within the overall process of metabolism in an organism. For example, when humans consume the herb known as Saint John's wort (*Hypericum perforatum*), certain components in that herb induce a (new) metabolic pathway—catalyzed by cytochrome P450 enzymes—that (more) rapidly metabolizes (i.e., breaks down) a number of commercial pharmaceuticals (thereby lowering the effectiveness of a given dose of that particular pharmaceutical). See also *Metabolism, Pathway, Organism, Intermediary metabolism, Cytochrome P450, Cytochrome P4503A4, Catalyst, Golden rice, Flux, Metabonomics, Metabolite profiling.*

Metabolism

From the Greek *metobolos* meaning to change, it is the entire set of enzyme-catalyzed transformations of organic nutrient molecules (to sustain life) in living cells. Conversion of food and water into nutrients that can be used by the body's cells, *and* the use of those nutrients by those cells (to sustain life, grow, etc.). See also *Enzyme, Cell, Anabolic pathway, Anabolism, Intermediary metabolism, Metabolite, Combinatorial biology, Citric acid, Aflatoxin, Fusarium, Cytochrome P4503A4, Pathway, Metabolic pathway, Flux, Metabonomics, Metabolite profiling.*

Metabolite

A chemical intermediate in the enzyme-catalyzed chemical reactions of metabolism. See also *Metabolism, Enzyme, Cell, Intermediary metabolism, Aflatoxin, Fusarium, Metabonomics, Metabolite profiling, Anabolic pathway.*

Metabolite Profiling

Determination of specifically which metabolic pathways (and/or related genes) are "switched on" (e.g., within a cell, tissue, or organism), thereby enabling the precise definition of the *metabonomic condition* of that cell/tissue/organism at that moment in time (e.g., cellular response to an environmental stimulus or a genetic modification). See also *Metabolism, Metabolite, Gene, Metabolome, Metabolic pathway, Metabolic engineering, Cell, Organism, Metabonomics.*

Metabolome

The complete set/complement of all metabolite and other molecules (e.g., metabolon) involved in, or produced during, a cell's metabolism. See also *Metabolism, Metabolites, Metabolon, Cell.*

Metabolomics

Refers to metabonomics of a given *single cell* or within a given *single cell type.* See also *Metabonomics, Cell, Metabolite profiling, Metabolome.*

Metabolon

A large agglomeration (within cell) of the enzymes involved (i.e., sequentially) in a metabolic pathway, in an organism. A metabolon enables the product of one enzymatic reaction to be passed directly to the next enzyme in the pathway, and so on. See also *Enzyme, Metabolism, Pathway, Metabolic pathway, Organism, Flux.*

Metabonomic Signature

Refers to the complete set of metabolites (both *quantitative* compounds—metabolite amounts—and *qualitative* metabolic pathways) of a cell/tissue/organism at a specific moment in time. See also *Metabolism, Metabolite, Metabolic pathway, Metabolic engineering, Metabonomics, Cell, Organism.*

Metabonomics

The scientific study (e.g., delineation, measurement) of an organism's *metabolic response* (e.g., delineation of its metabolic pathways, measurement of all metabolites) to an environmental stimulus or a genetic modification. See also *Metabolism, Metabolite, Metabolic pathway, Organism, Genetic engineering, Metabolic engineering, Activator (of gene), Metabonomic signature, Biomarkers.*

Metagenomics

A term coined by Jo Handelsman in 1998, it refers to the sequencing and analysis of all (pieces of) microorganisms' genomes (DNA) that are found within a given environmental sample (e.g., a milliliter of seawater) or series of samples from that environment, plus postanalysis reassembly of individual microorganisms' genomes. After sequencing of the sample's DNA en masse yields its (mixed) information, binning refers to how computational tools (bioinformatics) are utilized to do data assembly and the subsequent assignment of DNA fragments to each of the respective microorganisms present.

In this particular context, the Greek word *meta* means "being beyond" (i.e., beyond the genomics of a single organism). See also *Deoxyribonucleic acid (DNA), Metagenomics, Microorganism, Sequence (of a DNA molecule), Sequencing (of DNA molecules), Sequence map, Binning, Gene, Genetic code, Genome, Genomics.*

Metalloenzyme

An enzyme having a metal ion as its prosthetic group. See also *Enzyme, Prosthetic group, Metalloproteins.*

Metalloproteins

A term that is utilized to refer to any protein molecule that contains within it (i.e., in "peptide chain") a metal atom (e.g., zinc, iron, copper).

Approximately one-third of all proteins are metalloproteins. Those that contain a zinc atom (Zn^{2+}) are generally enzymes (thus called *metalloenzymes*), because that metal acts as a catalyst. For example, botulinum toxin is a zinc proteinase that cleaves the fusion proteins via which the vesicles (at neuromuscular junctions) release acetylcholine. Without that acetylcholine, contraction of relevant muscle (e.g., diaphragm muscle needed for breathing) is prevented.

In swine, a metalloprotein named metallothionein protects young piglets from overconsumption of zinc. It binds up excess zinc (i.e., zinc amounts in excess of what is needed by piglet's body) and causes it to be discarded via the animal's waste. See also *Protein, Peptide, Enzyme, Catalyst, Metalloenzyme, Fusion protein, Acetylcholine, ET, Tryptophan (trp).*

Metallothionein

See *Metalloproteins.*

Metamaterials

Refer to certain materials that are man-synthesized media that are structured on a size scale that is smaller than the wavelength of external stimuli (acting on those materials). Metamaterials exhibit properties not found in nature, such as negative index of refraction; particularly with respect to materials interacting with electromagnetic radiation/fields and can be "tuned" via their design and manufacture/dimensions. For example, several categories of specifically sized and shaped nanoparticles respond to the following:

- Certain wavelengths of infrared light by getting very hot (e.g., nanoshells within a tumor, thereby killing the tumor)
- Certain wavelengths of ultraviolet light by fluorescing in different colors (e.g., quantum dots within tissue/cells, the specific color of which depends on the precise dimensions of the quantum dot)
- Oscillating magnetic fields by getting very hot (e.g., nanoshells within a tumor, thereby killing the tumor)

Some metamaterials interact with visible light in a manner that causes *negative* refraction and those have been utilized for the following:

- *Superlensing* (e.g., make a "microscope" that can focus on objects that are smaller than the wavelength of the light utilized to illuminate them).
- *Cloaking* (e.g., make a device that conducts the electromagnetic radiation *around* an object), so that the object does not reflect/absorb the light. One example of such a metamaterial manufactured in 2008 by Xiang Zhang consisted of silver nanowire arrays in which the nanowires' separation distances within the array are significantly smaller than the wavelength of the visible light.
- *Biosensing* by forming multiple cloaks into a device that can positively identify biological materials based on the amount of light they absorb and then subsequently emit (i.e., fluorescence spectroscopy), because the cloaks slow down that light, and slowed-down light has a more pronounced interaction with molecules than does light travelling at normal speed, thereby enabling a more thorough analysis of the biological materials.

See also *Nanoparticles, Nanoshells, Quantum dot, Label (fluorescent), Nanometers (nm), Tumor, Nanotechnology, Nanoscience, Nanostructured material, Nanowire, Optical activity, Biosensors (light-based), Hydrogels.*

Metamodel Methods (of Bioinformatics)

These refer to methods utilized to integrate data that have been independently generated/created (and generally stored in separate database models) via independent *genomics* research projects, *combinatorial chemistry* projects, *high-throughput screening* projects (e.g., via biochip use), and so on.

Metamodel methods sometimes reveal important interrelationships that were not apparent in the *individual* models (i.e., created solely for the *genomics* project data, or created solely for the *combinatorial chemistry* project data, or created solely for the *high-throughput screening* project data). See also *Bioinformatics, Genomics, Functional genomics, Structural genomics, Combinatorial chemistry, High-throughput screening, Biochip.*

Metanomics

See *Metabonomics, Metabolomics.*

Metaphase

The second of the four phases of eucaryotic mitosis (i.e., cell replication via division) during which the cell's now-doubled chromosomes move to the cell's "equator" (i.e., line where the cell will soon split) and subsequently align themselves in pairs along that equator. See also *Mitosis, Eukaryotes, Cell, Chromosomes, S-phase.*

Metastasis

The process via which cells of a given cancer (e.g., initial tumor) spread from the site of its initial formation (in body) to other parts of the body.

One symptom of one particular metastasis beginning/occurring in humans is the presence of epithelial cells in the bloodstream. That is because a transition of *epithelial cells becoming mesenchymal* (thereby enabling cell motility) is the typical first step of cancer metastasis. Transforming growth factor-beta (TGF-beta), synthesized/exuded by platelets and by the cancerous tumor, leads epithelial tissue to dissolve the tight junctions between cells (e.g., in epithelium), and then those cells undergo morphogenesis to mesenchymal phenotype, as prelude to cancer "invasion" of adjacent tissues.

One symptom of prostate cancer metastasis beginning/occurring in humans is expression on those cancer cells of the RANKL signaling protein (which activates osteoclasts), thereby making the bones more receptive to metastatic tumor entry.

In some cancers hypoxia-inducible factors, which are master control proteins, "turn on" numerous particular genes that help cells adapt to a scarcity of oxygen (e.g., in cells within a specific tissue in an organism). Although these genetic responses are essential for continuation of a normal cell's life, the hypoxia-inducible factors also can turn on additional genes that help certain cancerous cells (within oxygen-starved tumors) to metastasize (i.e., escape the tumor) via moving on their own to invade the body's blood vessels, through which they spread to other tissues of the body.

When certain Janus kinases get "switched on," it can lead to (muscle-like) contractions within tumor cells that generate a force that causes the tumor's cells to move (e.g., through narrow spaces) and thereby metastasize. See also *Cancer, Tumor, Cell, Epithelium, Oligosaccharides, Lectins, Protein, Angiogenesis, Antiangiogenesis, Integrins, Genistein (Gen), Isoflavones, Phenotype, Haplotype, Transforming growth factor-beta (TGF-beta), Oncolytics, CTC, HER-2 gene, HER-2 receptor, Prostate cancer, Signaling, Osteoclasts, Janus kinases, Matrix metalloproteinases (MMP), Artificial interfering RNA (aiRNA).*

Meter

A unit of measurement that was contrived by French scientists during the 1670s. It was initially defined to be one ten-millionth of the distance from the Earth's equator to its poles.

In 1983, the world's nations defined the meter to be the distance light travels in 1/299,792,458 of a second. See also *Nanometers* (nm).

Methionine (met)

An essential amino acid; furnishes (to organism) both labile methyl groups and sulfur necessary for normal metabolism. See also *Essential amino acids, Metabolism, Cystine, High-methionine corn.*

Methyl Jasmonate

The volatile chemical compound that results when methyl groups (CH_3) are chemically added to a molecule of jasmonic acid. See also *Jasmonic acid.*

Methyl Salicylate

The volatile chemical compound that results when methyl groups (CH_3) are added to a molecule of salicylic acid. In 1997, Ilya Raskin showed that methyl salicylate emitted by one tobacco plant (e.g., under "attack" by insects, fungi, bacteria, or viruses) could cause other nearby tobacco plants to "turn on" their self-defense

M

mechanism (systemic acquired resistance). See also *Salicylic acid (SA)*, *Bacteria*, *Systemic acquired resistance (SAR)*, *Fungus*.

Methylated

Refers to either:

- A DNA molecule or a motif (e.g., a "CG" segment) within DNA molecule that is saturated with methyl groups (i.e., methyl submolecule groups, $-CH_3$, have attached themselves to the DNA molecule or motif at all possible locations such as cytosines, thereby resulting in 5-methylcytosine [5-mC] at cytosine loci). Generally, when a DNA molecule is methylated, the genes comprising that DNA molecule are silenced/"turned off" (i.e., inactivated).
- A histone (i.e., the protein portion of chromosome that the DNA condenses around) in which a protruding amino acid (e.g., lysine) has a methyl group attached to it. Generally, when a histone is methylated, the gene(s) regulated by that histone are upregulated/"turned on."

In 2004, Yang Shi discovered an enzyme named LSD1 that demethylates (i.e., "cuts" methyl group off) histone's protruding lysine amino acid. That demethylation results in repression of the gene(s) regulated by that histone.

Even though it is not an inherent part of a gene's DNA sequence, methylation of a gene (or DNA motif) is often copied as part of a cell's division process, so its effect (e.g., gene silencing) is perpetuated. In the case of domesticated chickens, methylation is passed down through many generations.

Along with many other causes of methylation, one cause of (more) methylation in plants is when a plant is grown under stress (either biotic or abiotic stress). See also *DNA methylation*, *Deoxyribonucleic acid (DNA)*, *Alkylating agents*, *Transcription*, *Messenger RNA (mRNA)*, *Gene*, *Genetic code*, *Gene expression*, *Expressivity*, *P53 gene*, *Tumor-suppressor genes*, *Imprinting*, *Protein*, *Histones*, *Chromosomes*, *Amino acid*, *Lysine (lys)*, *Up-regulation*, *Positive control*, *Transcriptional activator*, *Enzyme*, *Repression (of gene transcription/translation)*, *Demethylation*, *Lysine specific demethylase 1 (LSD1)*, *Differentiation pathways*, *Biotic stresses*, *Abiotic stresses*, *Cell*, *Promoter*, *Demethylation*, *Oxidative demethylation*, *Epigenetic*.

Methylation

See *Methylated*, *DNA methylation*.

MFA

Acronym for *metabolic flux analysis*. See *Flux*.

MFH

Acronym for magnetic fluid hyperthermia. See *Magnetic fluid hyperthermia*.

MGED

Acronym for Microarray Gene Expression Data Society. It consists of scientists who are attempting to jointly create a set guidelines (known as MIAME standards) governing the types of information to record/publish concerning experiments in which DNA microarrays are utilized. The goal is to make it easier to analyze and compare the results achieved by differing researchers utilizing different microarrays. See also *Microarray (testing)*, *Deoxyribonucleic acid (DNA)*, *DNA microarray*, *Gene*, *Express*, *Gene expression analysis*, *Bioinformatics*.

MHC

See *Major histocompatibility complex (MHC)*.

MHC I

See *Major histocompatibility complex (MHC)*, *Major histocompatibility antigen—class I*.

MHC II

See *Major histocompatibility complex (MHC)*, *Major histocompatibility antigen—class II*.

MIAME

Acronym for *minimum information about a microarray experiment*. See *MGED*.

Micelle

The spherical structure formed by the association of a number of amphiphilic molecules dissolved in water. Structurally, the outer surface of the micelle (sphere) is covered with the polar domains (head groups) that are directed toward (stick into) the water while the interior of the micelle contains the nonpolar domains (tails) that self-associate to create an "oil droplet" microenvironment. Micelles may be used to solubilize non-water (oil)-soluble or sparingly water-soluble molecules in water. They may be formed by ionic or nonionic surfactants. See also *Amphiphilic molecules*, *Supercritical carbon dioxide*, *Critical micelle concentration*, *Reverse micelle (RM)*, *Surfactant*, *Fats*, *Self-assembly*.

Micro Sensors

See *Biochip*, *Microarray (testing)*, *Biosensor*, *Charge coupled device*.

Micro Total Analysis Systems

Abbreviated µTAS or mTAS. See *Gene expression analysis*, *Biochip*, *Genosensors*, *Nanotechnology*, *Biosensors (electronic)*, *Biosensors (chemical)*, *Lab on a chip*.

Micro Total Analytical Systems

See *Micro total analysis systems*.

Microaerophile

An organism that grows best in the presence of a small amount of oxygen. See also *Organism*, *Microorganism*, *Facultative anaerobe*.

Microarray (Testing)

Refers to a piece of glass, plastic, silicon, metallic film, etc. onto which have been placed a large number of biosensors at known, specific locations. These microarrays (sometimes called "biochips" or "DNA chips") can then be utilized to test a single biological sample for a variety of attributes or effects.

For example, by placing protein-detection molecules (e.g., ligands, dyes that change color, fluoresce, or cause electronic signal upon contact with specific protein molecules) onto a microarray, a scientist can perform *gene expression analysis* (i.e., evaluation of the protein expression and *expression levels* of genes in a biological sample).

Another application would be to place (cellular) receptors, nucleic acids/probes, oligonucleotides, adhesion molecules, messenger RNA (specific to which gene is "turned on" in a given disease state), cDNA (complementary to mRNA coded for by each gene that is "turned on"), oligosaccharides and other relevant carbohydrate molecules, or cells (indicating which cellular pathway is "turned on," etc.) onto a microarray, to utilize that microarray to screen for proteins or other chemical compounds that act against a disease (i.e., therapeutic target), as indicated by (the relevant component from biological sample) adhesion or hybridization to a specific spot (location) on the microarray where a specific (target) molecule was earlier placed/attached.

The "detection event" (e.g., hybridization of sample molecules to cDNA) is indicated to the scientist via "tagging"/"labeling" of the sample molecules prior to the testing. Examples include

- *Fluorescent tags*: Reveal which location (and thus which cDNA molecule) on the biochip that the sample molecule hybridized to, when the biochip is scanned with a laser of appropriate wavelength
- *Radioactive labels*: Reveal which location (and thus which cDNA molecule) on the biochip that the sample molecule hybridized to, when the biochip is utilized to expose/develop relevant photographic film
- *Enzymatic*: Reveal which location (and thus which target molecule) on the biochip that the sample molecule hybridized to, adhered to, and so on, when the biochip is scanned (e.g., to detect location of the product of the enzyme "tag" that is released when the sample molecule hybridizes to or adheres to the specific target molecule on the biochip)

"Quantum dots" could potentially be used on microarrays in place of cellular receptors. See also *DNA chip, Multiplexed (assay), Biochips, Gene, Coding sequence, Gene expression, Gene expression analysis, Genosensors, Nanotechnology, Genomics, Functional genomics, Biosensors (electronic), Biosensors (chemical), High-throughput screening (HTS), Target-ligand interaction screening, oligosaccharides, Oligosaccharide microarrays, Fluorescence, ADME/Tox, Receptors, Bioreceptors, Combinatorial chemistry, Target (of a therapeutic agent), Immunosensor, Target (of a herbicide or insecticide), Adhesion molecule, Microfluidics, Bioelectronics, Assay, Bioassay, Messenger RNA (mRNA), Characterization assay, Probe, Hybridization (molecular biology), Bioinformatics, Cell array, Hybridization surfaces, Pathway, Deoxyribonucleic acid (DNA), Quantum dot, Nanoparticles, Protein microarrays, Proteome chip, Dip-pen nanolithography, SNP chip, Immobilization, Carbohydrate microarrays, Tiling arrays, ChIP, Expressed sequence tags (EST), Peptide nucleic acid, Nanosheets, Plasmonic nanohole arrays.*

Microarray Heat Map

See *Heat map.*

Microbe

A microscopic organism; applied particularly to bacteria. The word "microbe" was coined by Monsieur Sedillot, a colleague of Louis Pasteur. See also *Bacteria, Genetically engineered microbial pesticides (GEMP), Phytoalexins.*

Microbial Pesticide

See *Crop biologicals.*

Microbial Physiology

The cell structure, growth factors, metabolism, and genetics of microorganisms. See also *Microorganism, Cell, Metabolism, Genetics, Microbiology.*

Microbial Source Tracking (MST)

The process of systematically determining the *original source* (in a specific environment) *of a microbe* (e.g., the one that has caused a given disease outbreak). Some of the technologies utilized in MST include genetic fingerprinting, polymerase chain reaction (PCR), serotyping, lab on a chip, and so on. See also *Microbe, Pathogen, Polymerase chain reaction (PCR) Technique, Serotypes, Lab on a chip.*

Microbicide

Any chemical that will kill microorganisms. Used synonymously with the terms biocide and bactericide. See also *Microorganism, Biocide, Tannins.*

Microbiology

The science dealing with the structure, classification, physiology, and distribution of microorganisms, and with their technical and medical significance. The term microorganism is applied to the simple unicellular and structurally similar representatives of the plant and animal kingdoms. With few exceptions, the unicellular organisms are invisible to the naked eye and generally have dimensions of between a fraction of a micron and 200 μ. See also *Micron.*

Microbiomes

Refer to the entire population of all microorganisms living within a given specified environment (e.g., in the topsoil of a particular farm field, in the gut of a human infant, within the root cluster of a specific plant). See *Microorganism, Oligosaccharides, Salicylic acid (SA).*

Microcallus

See *Regeneration.*

Microchannel Fluidic Devices

See *Microfluidics.*

M

Microelectromechanical Systems

See *MEMS (nanotechnology)*.

Microfilaments

Very thin filaments found in the cytoplasm of cells. See also *Cell, Cytoplasm, Microtubules*.

Microfluidic Chips

See *Biochip, Microfluidics, Nanotechnology, Nanofibers*.

Microfluidics

Refers to the science and properties of fluids when flowing through *very small passages* (e.g., micron or nanometer dimensions) and/or in *very small amounts* (e.g., femtogram quantities).

For example, to mix together fluid (samples), *microfluidic chips* can utilize a nanomotor that comprises a twisted yarn made by twist-spinning carbon nanotubes together, which subsequently untwist/twist when different charges are injected into the electrolyte it is immersed in.

For example, to move fluid (samples), *microfluidic chips* utilize either capillary action or else they "pump" fluid (through micro-channels in those chips) electrokinetically (i.e., cause the flow to occur by applying a controlled electrical field, so liquid is attracted to electrical charge, and thereby flows).

Such "pumping" could also be utilized to deliver certain medicines in very small, precisely timed and metered doses (e.g., if the microfluidic chip is embedded into diseased tissue within the body).

Another potential application of such "pumping" could be to perform multiple chemical analyses (e.g., of body fluids within diseased tissues), in which case such microfluidic chips are known as "lab-on-a-chip"/laboratory-on-a-chip analytical devices. For example, in 2006 Richard N. Zare and colleagues created a lab on a chip that

- Will lyse a single cell (in an individual cavity on the chip)
- Will separate all the individual protein molecules thus extracted from that cell
- Will subsequently identify via fluorescence mapping each of the protein molecules from that cell

See also *Biochip, Nanotechnology, Nanofibers, Microarray (testing), Cell, Protein, Nanoscience, Micron, Lyse, Lysis, Fluorescence, Fluorescence mapping, Lab on a chip*.

Microgram

10^{-6} g, or 2.527×10^{-8} oz (avoirdupoir).

Micromachines

See *Nanoelectromechcanical system (NEMS), Self-assembling molecular machines, Nanopiezoelectronics, Casimir force, Nanobots*.

Micromachining

Refers to the technology and tools/methods utilized to create the very small parts, grooves (in chips/arrays), and so on, in nanoelectromechanical systems (NEMS), biochips, microarrays, and other devices of the field of nanotechnology. See also *Nanotechnology, Nanoelectromechanical systems (NEMS), Biochip, Microarray (testing), Dip-pen nanolithography*.

Micromodification

Term utilized by some people to refer to posttranslational modification of protein molecules. See also *Posttranslational modification of protein, Translation*.

Micron

From the Greek *mikros* meaning small. Also called micrometer. A unit of length convenient for describing cellular dimensions; the Greek letter μ is used as its symbol. For example, animal and human cells range in size from 10 to 25 μ. Plant cells range in size from 10 to 100 μ. Bacteria range in size from 1 to 3μ. A micron is equal to 10^{-3} mm (millimeter) or 10^4 Å or 0.00003937 in. See also *Microbiology, Cell, Microfluidics*.

Microorganism

Any organism of microscopic size (i.e., requires a microscope to be seen by man). First viewed by Antoni van Leeuwenhoek in 1676.

Some microorganisms are pathogenic (i.e., disease causing) and some are not. See also *Microbiology, Bacteria, Pathogenic, Nematodes, Capsule*.

Microparticles

Refer to the metal particles (<1 μ in diameter) that are coated with gene(s) and "shot" into cells with the Biolistic® gene gun. See also *Biolistic® gene gun, Vectors, Micron, Gene*.

Microphage

See *Polymorphonuclear leukocytes*.

Micropropagation

From the Latin word *propagare* meaning "to propagate" and the word micro meaning "tiny." A technique used by man to replicate (mass produce) a given (e.g., valuable) plant by making genetic clones ("copies") of that original plant via *in vitro* tissue culture of tiny pieces of the original plant. See also *Clone (an organism), Genetics, Tissue culture*.

microRNA Genes

See *miRNA genes*.

MicroRNA Pathway

See *Pathway, MicroRNAs*.

MicroRNAs

Refer to either:

- Naturally occurring small segments of RNA (approximately 21–23 nucleotides in length), which play an important role in gene regulation (i.e., "turning on" or "turning off"/silencing genes in DNA) via binding to and impacting the translation of specific mRNAs. MicroRNAs thereby help regulate an organism's early development, cell differentiation, apoptosis, immune system, and so on, including via RNA interference (RNAi). MicroRNAs were proven by Ji-Young Lee in 2010 to routinely move between cells as a regulatory signaling molecule. The first microRNA was discovered by Victor Ambros in 1993.

MicroRNAs are directly coded for by an organism's genes. For example, in animals the enzyme known as *Drosha* subsequently recognizes the applicable microRNA (i.e., within an RNA hairpin loop) and cuts it out of applicable RNA transcripts. That cutout RNA segment is then transported within the cell (to relevant DICER enzyme) where it is cleaved into the same length (21–24 nucleotides long) as siRNA.

For example, during the embryo development of vertebrate animals, the gene known as "Lfng" recognizes the microRNA known as "mir-125a-5p" in the "mir-125a-5p"-controlled "Lfng" cyclical gene activity that defines the timing of the formation of somites (i.e., tissue segments that later become muscle and vertebrae).

It is estimated that approximately 60% of human genes are regulated by microRNAs.

- Small chemically synthesized RNA segments, which are designed to mimic the products (i.e., siRNA segments) of Dicer enzymes, in terms of causing RNA interference.

Some malfunctions of microRNAs can lead to some cancers, heart disorders, immune system disorders, or dementia diseases.

Research indicates that microRNAs can also cause

- Some genes within certain invading (e.g., pathogenic) viruses to shut down (via RNA silencing), thereby preventing infection of the host.
- Some genes within a plant's DNA to revert to (grandparent's) wild-type version of the gene, when both parents' DNA contained a mutated version of the gene.
- Inhibition of expression of some cancer-causing enzymes. For example, the microRNA known as "miR-101" inhibits the histone methyltransferase known as "EZH2" (enhancer of zeste homolog 2). When certain tumors (e.g., prostate cancer) lose miR-101 via genomic deletion, then EZH2 can be overexpressed (resulting in an epigenetic change that leads to cancer metastasis).

See also *Ribonucleic acid (RNA), Messenger RNA (mRNA), Translation, Deoxyribonucleic acid (DNA), Gene, Gene expression, Regulatory sequence, Regulatory genes, RNA interference (RNAi), Epigenetic, Organism, Embryology, Differentiation, Cell differentiation, Wild type, Mutation, Coding sequence, Apoptosis, Dicer enzymes, Short interfering RNA (siRNA), Pathogen, Virus, Paramutation, Hairpin loop, Somites, Enzyme, Inhibition, Histones, Methylated, Homologous (chromosomes or genes), Cancer, Cancer epigenetics, Metastasis, miRNA genes, Gene silencing, Signaling molecule, Transcriptome, Artificial interfering RNA (aiRNA).*

Microsatellite DNA

Pieces of the same small DNA segment (e.g., a DNA sequence such as CACACACA) that are "repeated" (appear repeatedly in sequence within the DNA molecule) adjacent to a specific gene within the DNA molecule. Thus, these "microsatellites" are linked to that specific gene. Also known as variable number of tandem repeats.

Each small *repeat unit* tends to be five bases in length, or less. See also *Deoxyribonucleic acid (DNA), Linkage, sequence (of a DNA molecule), Satellite DNA, Gene, Linkage group, Anonymous DNA marker.*

Microsomes

Refer to small vesicles that are derived from broken-up smooth endoplasmic reticulum that result when desired tissues such as liver are homogenized (finely chopped) by scientists. Researchers want liver microsomes because they contain the liver cell's cytochrome P450 (CYP) enzymes, which catalyze oxidative metabolism (e.g., of pharmaceuticals, toxins).

Microsomes are thus potentially useful for ADME tests. See also *Vesicle, Endoplasmic reticulum (ER), Enzyme, Cytochrome P450, Cytochrome P450 (CYP), ADME tests, Absorption, Caco-2, Plasma, Cell, Cell culture, Receptors, Active transport, Membrane transport, Protein, Passive transport, Microarray (testing), Metabolism, Metabolite, Haplotype, Cell array, Live cell array, Toxicogenomics.*

Microsystems Technology

See *MST (nanotechnology).*

Microtubules

Tiny hollow filaments (i.e., stringlike structures) within eukaryotic cells that are made of tubulin (α and β proteins). Some microtubules give the cell its shape (e.g., act as structural components of cell). Other microtubules are the "towropes" utilized to move proteins within cells via vesicular transport (vesicles are small hollow structures that contain those protein molecules). Microtubules also "tow" apart the paired chromosomes within the nuclear DNA of cells undergoing meiosis.

Microtubule proteins present in bacterial cells include "FtsZ." Microtubules also "power" flagella (i.e., the whiplike structures used by sperm and some bacteria to "swim") and cilia (i.e., the tiny hairlike projections that line many mucosal surfaces in humans and that "sweep" dust particles, etc., to clean those mucosal surfaces).

Within neurons (cells of the mammal nervous system), microtubules transport messenger RNAs (mRNA) from the nucleus (where they are "manufactured") to the ribosomes in the dendrites (i.e., long extensions of the neuron cell), where the mRNAs are "translated" into protein molecules (i.e., proteins are "manufactured" by ribosome). See also *Cell, Meiosis, Nuclear DNA, Deoxyribonucleic acid (DNA), Bacteria, Flagella, Cilia, Neuron, Messenger RNA (mRNA), Nucleus, Ribosomes, Dendrites (in brain), Protein, Vesicular transport (of a protein), Mreb, ParM, Plasmid, Eukaryote, Analogue, Cytoskeleton.*

Mid-Oleic Sunflowers

Refer to sunflower (*Helianthus annus* L.) plant varieties that have been bred so their seeds contain 50%–75% oleic acid within the oil in those seeds versus historical average of 20% oleic acid in the oil of traditional sunflower (crop) plant varieties. See also *Fatty acid, Oleic acid, High-oleic sunflowers, High-oleic oil soybeans.*

M

Mid-Oleic Vegetable Oils

Refer to any vegetable oils (other than sunflower oil) that contain 50%–70% oleic acid. The range of *oleic acid content* is slightly different for mid-oleic sunflower oil definition. See also *Mid-oleic sunflowers, Fatty acid, Oleic acid.*

Mimetics

See *Biomimetic materials.*

Minimized Domains

See *Minimized proteins.*

Minimized Proteins

The domain/active site of a (former) native protein after all or most of its extraneous (unneeded) portions (peptides) have been removed. In 1995, Brian Cunningham and James A. Wells reduced the 28-residue (peptide) protein (hormone) *atrial natriuretic factor* to 15-residue (peptide) size without reducing its potency (biological activity). Minimized proteins—which retain their potency—hold the potential for medicines possessing a greater serum lifetime (when injected into a patient's body) and as "models" for the creation of organic-chemical-synthesized mimetic drugs possessing the same therapeutic effect as the native protein did. See also *Protein, Peptide, Active site, Enzyme, Catalytic site, Domain (of a protein), Hormone, Atrial natriuretic factor, Biomimetic materials, Serum lifetime, Biological activity, Nanobodies, Ranibizumab.*

Minimum Tillage

See *Low-tillage crop production, No-tillage crop production.*

Miniprotein Domains

See *Minimized domains.*

Miniproteins

See *Minimized proteins.*

MIQE2

Acronym for Minimum Information for Publication of Quantitative Real-Time PCR Experiments, which is a set of detailed guidelines intended to standardize the methodologies utilized by different scientists for conducting Real-Time PCR tests. MIQE was published in 2009 because the qPCR (real-time PCR) test is inherently so prone to being misused or miscalibrated (and thus misled) that numerous published scientific papers in which a qPCR test was performed as part of the author's research have recently been retracted. See also *Real-time PCR, QPCR, Polymerase chain reaction (PCR), Polymerase chain reaction (PCR) technique.*

miR-101

See *Micro-RNAs.*

Mir1-CP

Acronym for the enzyme known as *maize insect resistance cysteine protease*, which is naturally present in some strains of maize (corn). In 2005, Dawn Luthe and colleagues found that in those maize strains, Mirl-CP accumulates at sites on the plant where insects have chewed into the plant.

When ingested by relevant pest insect caterpillars (i.e., pupae), the Mirl-CP degrades these insects' peritrophic matrix (i.e., the membrane that lines the insect gut and aids digestion of what those insects eat). Subsequently, those insects have difficulty digesting their food and they typically die. See also *Enzyme, Protease, Cysteine (cys), Corn, Strain.*

miR-21 Gene

See *miRNA genes.*

miRNA Genes

Refer to particular genes that code for the production of specific microRNAs (miRNAs) instead of coding for the production of specific proteins as most genes do.

For example, the miR-21 gene (an miRNA gene) is not expressed correctly (e.g., due to mutations at miRNA-coding loci, epigenetic silencing, deletions, or amplifications, etc.) in many different cancers. See also *Gene, Coding sequence, microRNAs, Protein, Express, Mutation, Loci, Gene silencing, Epigenetic, Deletions, Amplification, Cancer, Transcriptome.*

miRNAs

Acronym for *microRNAs*. See *MicroRNAs.*

Mismatch Repair

Refers to cellular enzyme systems that repair any nucleotide-insertion errors (i.e., resulting from the DNA replication process) via excising (i.e., cutting out) the incorrect DNA sequence and replacing it with correct DNA sequence. Thus, it repairs mismatched base pairs. See also *Cell, Enzyme, Nucleotide, Deoxyribonucleic acid (DNA), DNA repair, Replication (of DNA), Sequence (of a DNA molecule), Unwinding protein, Base pair (bp), Base pairing.*

Mitochondria

Granular or rod-shaped bodies (organelles) in a cell's cytoplasm that contain the zyme systems required in the citric acid cycle, electron transport, beta oxidation of fatty acids, and synthesis of ATP via oxidative phosphorylation. They generate the majority of the cell's energy. See also *Zyme systems, Cell, Mitochondrial DNA, Siderophore, Carnitine, Ubiquinone, Adenosine triphosphate, Fatty acids, Fats, Phospholipids, Cytochrome, Cytoplasm, Citric acid cycle, ATP, Ac-CoA, Tumor necrosis factor (TNF).*

Mitochondrial DNA

The DNA within an organism's (e.g., human) cells that is located inside the mitochondria (organelles), not inside the cell nucleus. In virtually all eukaryotes, the mitochondrial DNA is only passed down from *mother* to offspring, not from father to offspring, as *nuclear DNA* is. See also *Deoxyribonucleic acid (DNA), Cell, Mitochondria, Homoplasmy, Nucleus, Nuclear DNA, Cytoplasmic DNA.*

Mitogen

A substance (e.g., growth factor, hormone) that initiates cell division within the body. For example, most angiogenic growth factors (e.g., fibroblast growth factor) stimulate cell division of the endothelial cells that line blood vessel walls. See also *Mitosis, Growth factor, Hormone, Angiogenic growth factors, Endothelial cells, MAPK, Mitogen-activated protein kinase cascade.*

Mitogen-Activated Protein Kinase Cascade

A cellular signaling pathway via which many fundamental cell processes such as differentiation, transcription, proliferation, apoptosis, etc. are controlled. Components of the pathway consist of serine/threonine kinases that respond to certain stimuli impacting the outside of the cell. For example, cell proliferation is promoted by growth factor(s) activation of receptor tyrosine kinases, which then "recruit" *ras-family small G proteins* (to cell's internal plasma membrane surface), followed by activation of other cellular kinases (thereby causing phosphorylation of certain proteins in the cell's nucleus), resulting in changes to the cell's protein-synthesizing processes.

In certain plants, exposure to cold temperatures can cause oxidative stress. That oxidative stress then can initiate activation of the mitogen-activated protein kinase (MAPK) cascade, resulting in the production of several *stress-responsive proteins* (e.g., heat shock proteins). Those "stress proteins" help protect such plants from cold temperatures. However, infection/infestation of a plant by the fungi *Penicillium syringiae* can result in an inhibition of the plant's mitogen-activated protein cascade defensive response.

In certain fungi (e.g., *Eurotium herbariorum*), exposure to high concentrations of salt can cause stress. That saline stress then can initiate activation of a MAPK cascade, to help protect those fungi from high salt concentration environments. See also *Cascade, Kinases, Serine (ser), Threonine (thr), Receptors, Mitogen, Protein, Cell, Differentiation, Apoptosis, Transcription, Growth factors, ras protein, G-proteins, Phosphorylation, Nucleus, Cell differentiation, Translation, Cold hardening, Oxidative stress, Stress proteins, Fungus, Penicillium syringiae.*

Mitosis

From the Greek *mitos* meaning "a thread." First described in 1882 by Walther Flemming (who also named it), mitosis is a four-phase process of cell duplication, or reproduction, during which one cell gives rise to two identical daughter cells.

During mitosis, the cell's chromosomes become visible under a microscope, looking like a seamstress's thread. See also *Mitogen, S-phase, Prophase, Metaphase, Anaphase, Telophase, Tubulin, Actin, Microtubules, Dynein, Centromere.*

Mixed-Function Oxygenases

Enzymes catalyzing simultaneous oxidation of two substances by oxygen, one of which is usually NADPH or NADH. See also *NADPH, NADH, Oxidation, Enzyme.*

MMR

Acronym for *mismatch repair.*

Mobile Element

Refers to certain short DNA segments (e.g., transposons) that can move from one chromosomal locus to another.

See also *Deoxyribonucleic acid (DNA), Gene, Chromosomes, Locus, Transposon, Jumping genes, Meganuclease.*

Model Organism

Refers to an organism that is utilized (e.g., in scientific experiments) to conduct tests, etc. in an attempt to infer results applicable to larger, more complex organisms. For example:

- The use of the microscopic roundworm *Caenorhabditis elegans* in high-throughput screening to attempt to find pharmaceuticals that will be useful for humans
- The use of the zebrafish (*Danio rerio*) for research on embryonic development applicable to all vertebrates
- The use of the aquatic plant (*Hydrilla verticillata*) for research on C4 photosynthesis applicable to all land plants

See also *Organism, Drosophila, Caenorhabditis elegans (C. elegans), High-throughput screening (HTS), Arabidopsis thaliana, Homologous (chromosomes or genes), Ortholog, Phylogenetic profiling, Hydrilla verticillata.*

Modulatory Nanotechnologies

Refer to the utilization of one or more forms of nanotechnology to control/improve an existing process (e.g., immunotherapy, in which one or more components of a patient's immune system are removed from the body and strengthened or "trained" to better combat a particular disease, then reinjected back into the body). For example, in 2014, Tarek Fadel improved cytotoxic T cell immunotherapy (e.g., against cancer tumors) by culturing the T cells outside the body on surfaces composed of carbon nanotube–polymer nanoparticle composite in which

- The polymer nanoparticle portion of the composite was impregnated with interleukin-2, a cell signaling protein that encourages T cell growth and proliferation
- The surface of the nanotube portion of the composite was seeded with specific antigen-presenting molecules to signal to the T cells that of the cells (e.g., tumor) within the patient were foreign

See also *Nanotechnology, Carbon nanotubes, Immune response, Cell, Cytotoxic T cells, Cancer, Tumor, Tissue culture, Interleukin-2 (IL-2), Antigen, Dendritic cells, Polymer.*

Moiety

Referring to a part or portion of a molecule, generally complex, having a characteristic chemical or pharmacological property. See also *Analogue, Pharmacophore.*

Mold

See *Fungus.*

Mole

An Avogadro's number ($6.02214129 \times 10^{23}$) of whatever units are being considered. One gram molecular weight of an element or a compound (i.e., same number of grams of an element or a compound

M

as that substance's molecular weight, equal to $6.02214129 \times 10^{23}$ molecules). See also *Molecular weight.*

Molecular Beacon

Term that refers to specific stem loop DNA segments—*oligonucleotides possessing a* "hairpin loop" *and bearing a fluorescent dye.* A "quencher dye" located on a *nearby portion of the hairpin loop* prevents fluorescence until the hairpin loop is opened up.

Molecular beacons (sometimes called *fluorogenic probes*) are utilized (e.g., in *high-throughput screening* or *high-throughput identification*) to detect the presence of a desired "target" molecule. When the "target" (i.e., a specific DNA sequence or a molecule possessing the desired functional group or desired property) is present within a given sample being evaluated, the "hairpin loop" opens up because a portion of it forms a stronger bond to the "target" (than to the rest of the loop), thereby allowing the fluorescent dye to emit light.

In 2005, Weihong Tan showed that molecular beacons can be utilized to monitor the expression of multiple genes (simultaneously) in a living cell. See also *Deoxyribonucleic acid (DNA), Oligonucleotide, Hairpin loop, Fluorescence, Target (of a therapeutic agent), Target (of a herbicide or insecticide), High-throughput identification, High-throughput screening (HTS), Cell-based assays.*

Molecular Biology

A term coined by Vannevar Bush during the 1940s that eventually came to mean the study and manipulation of molecules that constitute, or interact with, cells. Molecular biology as a distinct scientific discipline originated largely as a result of a decision to provide "support for the application of new physical and chemical techniques to biology" during the 1930s by Warren Weaver, director of the biology (funding) program at America's Rockefeller Foundation (a philanthropic organization). See also *Molecular genetics, Genetics, Genetic engineering, Biological activity, Biopolymer, Biogenesis, Biochemistry, Deoxyribonucleic acid (DNA), Mitosis, Meiosis.*

Molecular Breeding™

A trademarked term that refers to certain "molecular evolution" technologies developed by Maxygen Company.

This term is also sometimes used to refer to the utilization of molecular genetics and/or *marker-assisted selection* in a breeding program (e.g., within a seed company or within a university) to select the organisms (e.g., crop varieties) that possess gene(s) for a particular trait (e.g., higher yield, disease resistance). See also *Marker-assisted selection, Molecular evolution, Gene, Trait, Marker (DNA sequence), Quantitative trait loci (QTL).*

Molecular Bridge

Refers to the use of high-affinity molecules such as biotin and streptavidin, or other surface treatments, in order to "attach" something (e.g., quantum dot, nanoshell, enzyme, probe) to specific molecules (e.g., antibody targeted to tumor) or to a specific surface (e.g., the surface of a biosensor, biochip/microarray). See also *Antibody, Streptavidin, Quantum dot, Nanoshells, Ligand (in biochemistry), Enzyme, Probe, Tumor, Biosensors (chemical), Biosensors (electronic), Biochips, Microarray (testing), Hybridization surfaces, Proteome chip, Biotin.*

Molecular Chaperones

See *Chaperones, Protein folding.*

Molecular Diversity

Sometimes referred to as "irrational drug design," this refers to the drug design technique of generating large numbers of diverse candidate molecules (e.g., pieces of DNA, RNA, proteins, or other organic moieties) at random (via a variety of methods). These diverse candidate molecules are then tested to see which is best at working against a disease/condition (e.g., fitting a cell receptor or category of receptors relevant to the disease in question). Molecular candidates that show promise (e.g., via a "pretty good fit" to receptor) are then produced in larger quantities (e.g., via polymerase chain reaction techniques) along with additional molecules that are similar though slightly different in structure (e.g., via site-directed mutagenesis) in an attempt to create a molecule that is a "perfect fit" (e.g., to receptor). See also *Rational drug design, Deoxyribonucleic acid (DNA), Ribonucleic acid (RNA), Receptors, Receptor fitting (RF), Receptor mapping (RM), Moiety, Polymerase chain reaction (PCR), Site-directed mutagenesis, Diversity Biotechnology Consortium, Combinatorial chemistry, Combinatorial biology.*

Molecular Evolution

See *Combinatorial chemistry.*

Molecular Fingerprinting

See *Combinatorial chemistry.*

Molecular Genetics

The science dealing with the study of the nature and biochemistry of the genetic material. Includes the technologies of genetic engineering. See also *Genetics, Genetic engineering, Molecular biology, Biological activity, Biopolymer, Biogenesis, Biochemistry, Deoxyribonucleic acid (DNA), Mitosis, Meiosis, Molecular diversity, Central dogma.*

Molecular Imprinting

Refers to the process of creating nanoparticles (e.g., from acrylamide polymers) onto the surface of which are made (i.e., imprinted) binding/interaction sites for a specific molecule.

For example, in 2010, Kenneth J. Shea and Yu Hoshino manufactured (i.e., polymerized) acrylamide polymeric nanoparticles in the presence of molecules of melittin (a peptide that is a component of bee venom). Afterward, the melittin was removed, leaving plastic nanoparticles that would bind to any melittin (e.g., when in the human bloodstream) they encountered, much like an antibody would. Thus, these molecularly imprinted polymeric nanoparticles were called *plastic antibodies.* See also *Nanoparticles, Polymer, Melittin.*

Molecular Lithography

See *Bioelectronics, Nanolithography.*

Molecular Machines

Refer to nanometer-dimension "machines" capable of doing various tasks. See also *Nanotechnology, Nanometers (nm), Biomotors, Nanobots, Nanoelectromechanical system (NEMS), Nanoscience, Dynein.*

Molecular Marker

See *Marker-assisted selection.*

Molecular Motors

See *Motor proteins.*

Molecular Pharming™

A trademark of the Groupe Limagrain company, it refers to the production of pharmaceuticals and certain other chemicals (e.g., intermediate chemicals utilized to manufacture pharmaceuticals) in agronomic plants (which have been genetically engineered). See also *Antibiotic, Genetic engineering, Phytochemicals, Edible vaccines, Corn, Plantibodies™.*

Molecular Profiling

See *Gene expression profiling, Metabolite profiling, Gene expression analysis.*

Molecular Sieves

Refer to any structure bearing pores/channels whose internal diameters are smaller than 0.5 nm and thus can be utilized to separate small molecules (in a solution) from "large" molecules in the solution. See also *Nanometers (nm), Nanotube, Carbon nanotubes, Nanotube membranes.*

Molecular Stacking (of Multiple Traits in a Single Transgene Locus)

Refers to the delivery of multiple trait genes (cassettes) in one or more pieces of recombinant DNA (rDNA) simultaneously or consecutively into a single locus within the DNA of an organism (e.g., crop plant). Trait molecular stacks are tightly linked and have an extraordinarily low rate of segregation (i.e., they behave as a single gene), thereby making trait introgression/variety creation easier.

See *Gene, Trait, Cassette, Deoxyribonucleic acid (DNA), Introgression, Locus, Variety (e.g., of a crop plant), Organism.*

Molecular Tweezers

Refer to certain complex molecular compounds that are capable of binding (reversibly) to other proteins. Shaped like the letter "C," these compounds can encircle lysine, a basic amino acid that is a constituent of most proteins.

In 2011, Jeff Bronstein and Gal Bitan showed that a molecular tweezer named CLR01 was able to prevent α-synuclein from forming the aggregates that cause Parkinson's disease, prevent their toxicity to brain cells, and even break up existing aggregates.

See also *Parkinson's disease, Alpha-synuclein.*

Molecular Weight

The sum of the atomic weights of the constituent atoms in a molecule. See also *Atomic weight, Molecular-weight-size marker.*

Molecularly Imprinted Polymeric Nanoparticles

See *Molecular imprinting.*

Molecular-Weight-Size Marker

Also called a DNA ladder, protein ladder, or RNA ladder, it is a set of standards (i.e., DNA, RNA, or protein molecular fragments of known sizes/concentrations) that are placed in preidentified "lanes" of the agarose or polyacrylamide gels that are used in gel electrophoresis (prior to start of the run). The sizes of the sample (i.e., unknown) molecular fragments can be estimated by comparison of how far those unknown fragments travel within the gel, compared to the distance traveled by the standards. See also *Molecular weight, Protein, Deoxyribonucleic acid (DNA), Ribonucleic acid (RNA), Two-dimensional (2D) gel electrophoresis, Electrophoresis, Polyacrylamide gel electrophoresis.*

Monarch Butterfly

Refers to the insect (Lepidoptera: *Danaidae* or *Danaus plexippus*) whose pupae (caterpillars) feed exclusively on tissue of the plant known as *common milkweed* (*Asclepias syriaca*) and whose territory extends from northern Mexico to approximately Canada's southern border. See also *Bacillus thuringiensis (B.t.), B.t. kurstaki, B.t. tolworthi, CryIA(b) protein.*

Monoclonal Antibodies (MAb)

Discovered and developed in the 1970s by Cesar Milstein and Georges Kohler, monoclonal antibodies are the name for antibodies derived from a single source or clone of cells that recognize only one kind of antigen. Made by fusing myeloma cancer cells (which multiply very fast) with antibody-producing cells, then spreading the resulting conjugate colony so thin that each cell can be grown into a whole, separate colony (i.e., cloning). In this way, one gets whole batches of the same (monoclonal) antibody, which are all specific to the same antigen. Monoclonal antibodies have found markets in diagnostic kits, pharmaceuticals (e.g., trastuzumab, rituximab, bevacizumab, adalimumab, infliximab, cetuximab, blinatumomab), imaging agents, and purification processes.

One example of a diagnostic use is the invention in 1997 by Bruno Oesch of a monoclonal antibody-based rapid test to detect the prion (PrP 5c) that causes bovine spongiform encephalopathy (BSE) in cattle.

During the 1990s, scientists were able to insert gene(s) cassette into plants to cause those plants to produce such antibodies in plant tissue (e.g., seeds). See also *Ascites, Myeloma, Tumor, Corn, Immunocojugate, Immunotoxin, Radioimmunotherapy, Blast cell, Antigen, Antibody, Single-domain antibodies (dAbs), Murine, Catalytic antibody, Semisynthetic catalytic antibody, Angiogenesis inhibitors, BSE, Prion, Gene, HER-2 gene, Rheumatoid arthritis, Adalimumab, Infliximab, Rituximab, Humanized antibody, Trastuzumab, Bevacizumab, Natalizumab, cetuximab, Blinatumomab, Ranibizumab, Plantibodies, Cytoskeleton, Cassette, EGF receptor, Antiepidermal growth factor receptor monoclonal antibodies, PCSK9 inhibitors.*

M

Monocot

See *Monocotyledon.*

Monocotyledon

From the Greek *monos* meaning "solitary" and *kotyledon* meaning "a saucer-shaped hollow," it refers to a plant whose seed bears one seed leaf (known as a cotyledon). Examples of monocotyledons include wheat (*Triticum aestivum*), rice (*Oryza sativa* or *Oryza glaberrima*), and corn/maize (*Zea mays* L.). See also *Wheat, Rice, Corn.*

Monocytes

Also called monocyte macrophages. The round-nucleated cells that circulate in the blood. In summary they engulf and kill microorganisms, present antigen to the lymphocytes, kill certain tumor cells, and are involved in the regulation of inflammation. These cells are often the first to encounter a foreign substance or pathogen or normal cell debris in the body. When they do, the material is taken up (engulfed) and degraded by means of oxidative and hydrolytic enzymatic attack. Peptides that result from the degradation of foreign protein are then bound to a monocyte protein called class II MHC (major histocompatibility complex) and this self-foreign complex then migrates to the surface of the cell where it is embedded into the cell membrane in such a way as to present the peptide to the outside of the cell. This positioning allows T lymphocytes to recognize (inspect) the peptide. Whereas self-peptides derived from normal cellular debris are ignored, foreign peptides activate precursors of helper T cells to further mature into active, lymphokine-secreting helper T lymphocytes, also known as TH cells. When monocytes move out of the bloodstream and into the tissues they are then called macrophages. See also *Macrophage, Cellular immune response, Pathogen, MHC.*

Monoecious

A category of plants (e.g., the soybean plant is one) that possess both male and female reproductive structures on the same plant. Thus, such plants are capable of self-pollination.

For example, 95% of the pollen from a soybean plant (*Glycine max*) does not leave the flower that it was produced in. Virtually none of a given soybean plant's pollen leaves the plant that it was produced in. See also *Soybean plant, Barnase.*

Monolithic Chromatography Substrates

Refer to chromatographic stationary phases (e.g., resins inside chromatography columns) that are polymerized directly into the chromatography column in the form of a single unit (versus older resin/ceramic beads stacked within a chromatography column).

Such monolithic substrates offer the potential for higher throughput of the mixtures being separated and higher resolution (i.e., degree of separation of desired compound from mixture). See also *Chromatography, Substrate (chromatography).*

Monomer

The basic molecular subunit from which, by repetition of a single reaction, polymers are made. For example, amino acids (monomers) link together via condensation reactions to yield polypeptides or proteins (polymers). A monomer is analogous to a link (monomer) in a metal chain (polymer). See also *Polymer.*

Monosaccharides

The chemical building blocks of carbohydrates, hence known as "simple sugars." They are classified by the number of carbon atoms in the (monosaccharide) molecule. For example, pentoses have five and hexoses have six carbon atoms. They normally form ring structures. The empirical formula for monosaccharides is $(CH_2O)_n$. See also *Oligosaccharides, Carbohydrates, Sugar molecules.*

Monounsaturated Fats

Fat molecules possessing one less than the maximum possible number of hydrogen atoms (on that given fat molecule). Diets that are high in monounsaturated fat content have been shown to reduce low-density lipoproteins ("bad" cholesterol) blood content, while leaving blood levels of high-density lipoproteins ("good" cholesterol) essentially unchanged. See also *Fatty acid, Saturated fatty acids, Dehydrogenation, Unsaturated fatty acid, Low-density lipoproteins (LDLP), High-density lipoproteins (HDLPs), Oleic acid, Fats.*

Monounsaturated Fatty Acids (MUFAs)

Refer to category of those fatty acids (e.g., oleic acid) that possess one less than the maximum possible number of hydrogen atoms (e.g., possible to be attached to the molecular structure of oleic acid, in this example).

Enzymes (e.g., Δ12 desaturase) present in some oilseed plants (e.g., soybean, corn/maize, canola) convert some MUFAs to polyunsaturated fatty acids (PUFAs), within their developing seeds.

Diets that are high in monounsaturated fatty acid content have been shown to reduce low-density lipoproteins' (i.e., so-called bad cholesterol) blood content while simultaneously leaving blood levels of high-density lipoproteins (i.e., "good cholesterol") essentially unchanged. Soybean oil has historically averaged approximately 24.5% monounsaturated fatty acids content by weight. See also *Monounsaturated fats, Fatty acid, Unsaturated fatty acid, Soybean oil, Oleic acid, Low-density lipoproteins (LDLP), Delta 12 desaturase, Polyunsaturated fatty acids (PUFA).*

Morphogenetic

An adjective referring to formation and differentiation of tissues and organs in an organism. See also *Differentiation, Morphogens, Morphology, Stem cells, Totipotent stem cells, Organism.*

Morphogens

Refer to specific protein molecules (e.g., distributed to needed locations during the development of an embryo) that subsequently signal relevant cells (within the embryo) to create differentiated tissues (organs), networks of new blood vessels/capillaries (to support those growing organs/tissues), and so on. See also *Differentiation, Cell differentiation, Morphogenetic, Cell, Signaling, Protein, Protein signaling.*

Morpholino

Refers to one methodology utilized for gene silencing. See *Gene silencing.*

Morphology

First used in print by the poet Johann Wolfgang von Goethe, this word is utilized to refer to the form/structure of an organism or any of its parts (e.g., cells). See also *Trait*, *Phenotype*, *Actin*.

MOS

See *Mannanoligosaccharides*.

Motif

See *Methylated*.

Motility

Refers to *movement* (e.g., of cells within the body, of certain proteins within a cell or out of the cell). See also *Cell*, *Cell motility*, *Cellular affinity*, *Protein*, *Chaperones*, *Sphingolipids*.

Motor Proteins

Refer to specialist protein molecules (e.g., kinesin, dynein, myosin, actin) within cells, which transport various items (e.g., newly made protein molecules, vesicles, organelles) from one part of the cell to another. Some of them sometimes also move the entire cell to another location inside the organism (e.g., via "towropes" called *filopodia*). See also *Protein*, *Kinesin*, *Dynein*, *Myosin*, *Actin*, *Cell*, *Organelles*, *Vesicle*, *Organism*.

Mouse-Ear Cress

A common name for *Arabidopsis thaliana* utilized in some countries. See *Arabidopsis thaliana*.

Movable Genetic Element

See *Transposon*, *Jumping genes*.

MPSS

Acronym for *massively parallel signature sequencing*. See *Massively parallel signature sequencing*.

MRA

See *Mutual recognition agreements*, *Mutual recognition arrangements*.

MreB

A contractile (i.e., periodically contracting) stringlike protein that is present in at least the rod-shaped bacteria (probably some other bacterial types too), just beneath the surface (membrane) of those bacteria. MreB filaments thus give rod-shaped bacteria their shape.

MreB is a homologous protein to the protein actin in eukaryotic cells. See also *Protein*, *Homologous protein*, *Motor proteins*, *Actin*, *Cell*, *Bacteria*, *Eukaryote*.

MRL

See *Maximum residue level*.

mRNA

See *Messenger RNA*.

MRSA

See *Nanosponges*.

MS

Acronym for Mass Spectrometer. See *Mass spectrometer*.

MSA

Acronym for *molecular self-assembly*. See *Self-assembly (of a large molecular structure)*, *Self-assembling molecular machines*.

MSF

See *Megakaryocyte-stimulating factor*.

MST (Microbes)

See *Microbial source tracking*.

MST (Nanotechnology)

Acronym utilized by Europeans to refer to "microsystems technology" (i.e., their common term for "microelectromechanical systems"). See also *Nanotechnology*, *Biochip*, *Genosensors*, *Biosensors (electronic)*, *Biosensors (chemical)*, *Quantum wire*, *Quantum dot*, *Nanocrystal molecules*, *Microfluidics*, *Biomotors*, *Molecular machines*.

MSX Genes

See *Bone morphogenetic proteins (BMP)*.

mTAS

Acronym for *Micro Total Analysis Systems*. See the link *Micro total analysis systems*.

MTAS

See *Micro total analysis systems*.

mtDNA

Acronym for *mitochondrial DNA*. See *Mitochondrial DNA*.

MUDPIT

Abbreviation for *mu*lti*d*imensional *p*rotein *i*dentification *t*echnology, a technology utilized in proteomics that combines liquid chromatography with mass spectrometry. See also *Proteomics*, *Chromatography*, *HPLC*, *Mass spectrometer*, *MALDI-TOF-MS*.

MUFA

See *Monounsaturated fatty acids (MUFA)*.

Multiallelic Copy Number Variation Loci

Abbreviated mCNVs, these are the loci (i.e., the "locations" on a given organism's DNA) where multiple copy number mutations (i.e., those in which a particular DNA sequence occurs more than once within the organism's DNA) occur. Multiple copy variants can perturb (affect the function of) many genes within an organism's DNA simultaneously. See also *Locus, Deoxyribonucleic acid (DNA), Sequence (of a DNA molecule), Gene, Mutation, Organism, Copy number variation, Copy number (protein molecules), Synthesizing (of proteins), Copy number variant.*

Multicopy Plasmids

Plasmids that are present inside bacteria in quantities greater than one plasmid per (host) cell. This results in a larger than one number of the applicable protein molecules being produced during transcription/expression. That larger number is known as the *copy number.* See also *Plasmid, Vector, Copy number, Transcription, Expression.*

Multidrug Resistance

See *P-glycoprotein, Flux, Antibiotic resistance.*

Multiphoton Microscopy

Refers to microscopy/imaging in which photons of light (i.e., from a longer-wavelength laser) are fed into a sample (e.g., a living cell deep inside tissue) at a high enough density that two or more photons are absorbed *simultaneously*, by the relevant fluorophore. This results in that fluorophore being raised to the "excited" state, whereupon it emits light (to return to lower energy state), which enables the scientist to see desired image of that cell.

Because longer-wavelength light is less harmful to living cells, multiphoton microscopy enables the imaging of such cells without harming them. For example, instead of utilizing harmful 380 nm ultraviolet light to cause a given cellular fluorophore to fluoresce, a scientist can utilize two 760 nm infrared photons *simultaneously absorbed* via multiphoton microscopy. See also *Cell, Fluorophore, Nanometers (nm).*

Multienzyme System

A sequence of related enzymes participating in a given metabolic (chemical reaction) pathway. See also *Enzyme.*

Multigenic

See *Polygenic.*

Multiple Myeloma

A blood cancer that arises from plasma cells, an antibody-producing B lymphocyte found in bone marrow. Multiple myeloma primarily afflicts older adults. It causes their plasma cells to multiply rapidly and thereby crowd out the other types of blood cells from the affected bone marrow, which results in those cells moving to other tissues of the body—which weakens the body's immune system and can lead to anemia, kidney problems, and other bone problems.

Panobinostat (an HDAC inhibitor) slows the progression of multiple myeloma by inhibiting the activity of enzymes known as histone deacetylases, which slows the overly rapid multiplication of plasma cells and/or causes those overactive plasma cells to die. See also *Cancer, Plasma cell, B lymphocytes, Antibody, Panobinostat.*

Multiple Sclerosis

A disease in which the human body's immune cells attack myelin (i.e., the "insulation" that surrounds nerve fibers in the spinal cord and brain) and the body's acetyl choline receptors. That leads to recurrent muscle weakness, loss of muscle control, and (potentially) eventual paralysis. With time, the axons (nerve fibers in spinal cord and brain) also begin to deteriorate.

In 2012, some researchers utilized receptor-binding mapping to link vitamin D deficiency to an increased risk for cancer and the autoimmune diseases rheumatoid arthritis, multiple sclerosis, and lupus. Research published in 2001 indicated that women whose blood contained significant levels of antibodies to Epstein–Barr virus were four times more likely to develop multiple sclerosis than women without significant blood levels of those antibodies.

Some research has implicated degradation of myelin (during midlife) with Alzheimer's disease. See also *Autoimmune disease, Thymus, Acetylcholine, Receptors, Immune response, Neurotransmitter, Excitatory amino acids (EAAs), Axon, Oligodendrocytes, Antibody, Virus, Alzheimer's disease, Natalizumab, Receptor-binding mapping.*

Multiplex Assay

Refers to an assay that generates *more than one data point* in each (assay) evaluation that is performed. For example, fluorescence mapping (i.e., scanning an *x–y* plane within tissue at varying depths with a microscope/light of selected wavelength) designed to cause/detect any fluorescence resulting from a biological event of interest (e.g., the binding of a particular protein to a given cell receptor, the expression of particular gene(s)). See also *Assay, Bioassay, Cell, Fluorescence, Fluorescence mapping, Receptors, High-content screening, High-throughput screening (HTS), Target–ligand interaction screening, Gene, Gene expression profiling, Cellular pathway mapping, Metabolomics, Multiplexed (assay).*

Multiplexed (Assay)

Refers to an assay that simultaneously measures several different aspects (e.g., several different proteins produced within a cell, several different products produced in the same chemical reaction). See also *Assay, Protein, Cell, Microarray (testing), High-content screening, Multiplex assay.*

Multipotent

Refers to the property of certain cells (e.g., stromal cells, some adult stem cells) to be differentiated (i.e., via chemical signals) so those cells then give rise to one of several different cell/tissue types.

For example, *stromal cells* within mammary tissue can be influenced by the presence of conjugated linoleic acid to become fat cells (adipose tissue). Otherwise, stromal cells tend to become *blood vessel lining* cells (endothelial cells), or *connective tissue–producing* cells (fibroblasts). See also *Cell, Adult stem cell, Multipotent adult stem cell, Signaling, Differentiation, Conjugated linoleic acid (CLA), Adipose, Endothelial cells, Fibroblasts.*

Multipotent Adult Stem Cell

Certain stem cells present within (adult) bodies of organisms that can be differentiated (via chemical signals) to give rise to a variety of different cell/tissue types (e.g., bone, cartilage, fat, muscle, red blood cells, B cells, T cells).

For example, adipose (body fat) cells have been removed from the bodies of some mammals by researchers and subsequently coaxed into expressing genes and otherwise exhibiting characteristics of (differentiated) bone cells, cartilage cells, muscle cells, and so on. See also *Stem cells, Cell, Multipotent, Differentiation, Organism, Signaling, Red blood cells, B cells, T cells, Mesodermal adult stem cells, Adipose, Gene, Express, Gene expression analysis.*

Multipotent Progenitors

One type of adult stem cell that is present in human breast tissue. It typically differentiates into luminal (i.e., milk-producing) cells. When a breast cancer tumor is growing adjacent to a multipotent progenitor cell, the multipotent progenitor cell produces tumor-fighting myoepithelial cells. See also *Cell, Stem cells, Multipotent, Tumor, Cancer.*

Murine

Of, or pertaining to, mice. For example, the first monoclonal antibodies were produced using cells from mice. This frequently caused adverse immune responses to monoclonal antibodies when they were injected into the human body (e.g., thus limiting their use in therapeutic purposes). However, researchers have recently discovered how to make monoclonal antibodies in human cells. See also *Monoclonal antibodies (MAb).*

Muscular Dystrophy (MD)

A genetic disease caused by a defect in the X chromosome (resulting in nonexpression of the Duchenne muscular dystrophy [DMD] gene); first recognized by G. A. B. Duchenne in 1858. The disease afflicts males almost exclusively because males have only one X chromosome, whereas females inherit two copies of the X chromosome and have a "backup" in case one X chromosome is damaged (as is the case for MD victims). In 1981, Kay E. Davies used DNA probes (genetic probes) to discover that the DMD gene must lie somewhere between two unique (to MD victims) segments on the upper, shorter arm of the X chromosome. See also *DNA probe, Chromosomes, Karyotype, Chromatids, Chromatin, Single-nucleotide polymorphisms (SNPs).*

Mutagen

A chemical substance (or physical phenomena, such as ultraviolet light) capable of producing a genetic mutation (change), by causing changes in the DNA of living organisms. For example, Dr. Gary Shaw discovered in 1996 that women who smoke cigarettes during their pregnancies are twice as likely to have babies with the genetic deformity known as cleft lip and palate. If those women have a particularly susceptible (to smoke) gene variant (allele) within their DNA, they are as much as eight times as likely to have babies with cleft lip and palate.

According to the World Health Organization, 60%–80% of all known mutagens are also carcinogens (i.e., cancer causing). See also *Mutation, Gene, Genetics, Heredity, Genetic code, Cancer, Carcinogen, Allele, Deoxyribonucleic acid (DNA), Oncogenes, Mutant, Toxicogenomics, Antioxidants, Alkylating agents.*

Mutagenesis Breeding

See *Mutation breeding.*

Mutagenic Chain Reaction

Abbreviated MCR, it refers to the use of the CRISPR/Cas9 genome editing system for creating heterozygous autocatalytic (self-catalyzed/powered) mutations to generate homozygous loss-of-function mutations in most somatic and germline cells. See also *CRISPR/Cas9 gene-editing systems, Mutation, Heterologous (chromosomes or genes), Heterozygous (chromosomes or genes), Gene drive.*

Mutant

An altered cell or organism resulting from mutation (an alteration) of the original wild (normal) type. A change from the normal to the unique or abnormal.

For example, in 1970, an *orange*-colored wild cauliflower was discovered growing in Bradford Marsh in Canada. Its orange color resulted from a mutation that caused it to produce approximately 100 times more beta-carotene than original (wild type) cauliflower. If that particular mutation was of a *single nucleotide* in the cauliflower's DNA (e.g., caused by ultraviolet radiation striking it), then it is a *point mutation.* See also *Cell, Organism, Mutagen, Mutation, Heredity, Wild type, Beta carotene, Point mutation, Deoxyribonucleic acid (DNA), Nucleotide.*

Mutase

An enzyme catalyzing transposition of a functional group in the substrate (substance acted upon by the enzyme). Intramolecular transfer of a chemical group from one position (i.e., carbon atom) to another within the same molecule. An example of a mutase is phosphoglucomutase. It has a molecular weight of about 60,000 Da with about 600 amino acid residues (monomers). The mutase can interchange (move) a phosphate unit between the 1 and 6 position. The 1 refers to a carbon atom designated as "#1" and the 6 refers to a different carbon atom designated as "#6." See also *Enzyme.*

Mutation

From the Latin term *mutare*, meaning "to change." Any change that alters the sequence of the nucleotide bases in the genetic material (DNA) of an organism or cell; with alteration occurring either by displacement, addition, copy number variation, deletion, cross-linking, or other destruction.

The mutation alteration to the DNA sequence would alter its meaning, that is, its ability to produce the normal amount or normal kind of protein, or normal miRNA gene; so the (organism or cell) is itself altered. Such an altered organism is called a mutant.

For example, in 1820, one orange tree in Brazil was discovered to be producing *seedless* oranges (navel oranges), as a result of a naturally occurring mutation in its DNA. If that particular mutation was of a *single nucleotide* in that orange tree's DNA (e.g., caused by ultraviolet radiation striking it), then it was a *point mutation.* Since all commercial seedless orange trees existing today are direct descendants *via cloning* of that one Brazilian tree, that mutation was a stable mutation. See also *Mutant, Informational molecules, Heredity, Gene, Genetic code, Genetic map, Protein, Deoxyribonucleic acid (DNA), Sequence (of a DNA molecule), Point mutation, Copy number (protein molecules), miRNA gene, Organism, Multiallelic copy number variation loci, Cell, Cystic fibrosis transmembrane regulator protein (CFTR).*

Mutation Breeding

Refers to several techniques, involving induced mutations, that were utilized by some crop plant breeders (e.g., primarily in the 1960s and 1970s) to introduce desirable genes into the plants they were working with. For example, gene(s) to confer

- Resistance to plant diseases
- Increased yield per acre/hectare
- Improvements in composition

that were not present within the historic/natural germplasm of that plant species.

These *new to that species* genes were "created" via

- Soaking its seeds or pollen in mutation-causing chemicals (i.e., mutagens) such as colchicine or sodium azide
- Bombardment of seeds with X-rays
- Bombardment of seed with *fast neutron* radiation (i.e., causes deletion of approximately 1000 bp of DNA each time)
- Bombardment of seed with gamma radiation (i.e., causes deletion of several hundred bp of DNA each time). For example, cauliflower was the result of such a naturally occurring mutation (to wild cabbage)

followed by grow out of the resultant plants and selection of the particular mutation (i.e., beneficial trait) desired by the plant breeder. That plant was then propagated via straightforward breeding to yield seeds that are still sown today. See also *Traditional breeding methods, Mutation, Mutagen, Gene, Trait, Wheat, Barley, Point mutation, Colchicine, Base pair (bp), Deletions, High-oleic safflower oil.*

Mutual Recognition Agreements (MRAs)

Legal agreements (e.g., treaties) between two or more nations to recognize and respect each other's approval process (e.g., for new crops derived via biotechnology). See also *GMO, Committee for Veterinary Medicinal Products (CVMP), Organization for Economic Cooperation and Development (OECD), Event, European Medicines Evaluation Agency (EMEA), Committee for Proprietary Medicinal Products (CPMP), Union for Protection of New Varieties of Plants (UPOV).*

Mutual Recognition Arrangements

See *Mutual recognition agreements (MRAs).*

Mycelia

The fine, threadlike filaments that are sent out into the surrounding soil by mycorrhizal species of fungi. See also *Mycorrhizae.*

Mycelium

See *Mycelia.*

Mycobacterium tuberculosis

The pathogen that causes tuberculosis, a human disease in which the lungs are destroyed as these bacteria grow (within lung tissue).

In 1998, scientists completed sequencing of the genome of *Mycobacterium tuberculosis.*

Recently, a new strain of *Mycobacterium tuberculosis* has begun to infect some people, which is resistant to virtually all commercial antibiotics. See also *Bacteria, Pathogen, Sequencing (of DNA molecules), Antibiotic, Antibiotic resistance, Genome, Strain, Parkin.*

Mycorrhizae

Refers to the symbiotic (i.e., mutually beneficial) relationship that exists between certain plants and the specific soil-dwelling species of fungi dwelling among the roots of those particular plants. The fungi provide certain minerals (e.g., phosphorous) to the plant roots (which the fungi's mycelia are able to extract from the soil). In return, the plant roots provide certain nutrients (e.g., sugar molecules) needed by those fungi.

In some cases, the fungal mycelia (i.e., filaments sent out some distance into the surrounding soil) will transfer certain nutrients (e.g., some sugar molecules) from one plant to another plant. The fungi find/locate plant roots via certain signaling chemicals (e.g., strigolactone 5-deoxystrigol) that are exuded by the plant roots.

In some cases, the mycorrhizae will

- Help to protect certain plants from some pathogens
- Transform/render certain toxins present in the soil (e.g., some heavy metals) to be nonavailable for uptake by the host plant's roots

See also *Symbiotic, Fungus, Species, Mycelia, Sugar molecules, Arbuscular mycorrhizae, Pathogen, Glomalin, Signaling molecule.*

Mycotoxins

Toxins produced by fungi. More than 400 different mycotoxins are known to man, but the first ones to be isolated and scientifically characterized (i.e., described) were the *aflatoxins*, in 1961. The second group of mycotoxins to be isolated and characterized was the *ochratoxins*, in 1965. Almost all mycotoxins possess the capacity to harmfully alter the immune systems of animals. Consumption by animals (including humans) of certain mycotoxins (e.g., via eating infected corn/maize, wheat, certain tree nuts, peanuts, cottonseed products) can result in liver toxicity, gastrointestinal lesions, cancer, muscle necrosis, and so on. See also *Toxin, Fungus, Fusarium, Aflatoxin, Vomitoxin, Fusarium moniliforme, Fumonisins, Zearalenone, Ochratoxins, Patulin, Ergotamine, Phomopsins, P53 gene.*

Mycoviruses

Refer to viruses that can infect fungi. Such viral infections have been shown to

- Control/change the virulence (infectiveness) of the fungus, if the fungus was pathogenic
- Increase the thermal tolerance of the fungus

For example, in 2002, Regina Redman and Russell Rodriguez discovered that *Curvularia protuberata* fungi (which live inside *Dichanthelium lanuginosum* grass that grows in hot soils adjacent to magma-heated geysers) impart heat tolerance to both the fungus and the grass it lives in, but *only when those fungi are themselves infected with Curvularia thermal tolerance virus* (*CThTV*). Prior to such infection, neither the fungus nor the grass is able to grow in temperatures exceeding 38°C (100°F). See also *Virus*, *Fungus*, *Pathogenic*.

MyD88 Protein

See *Innate immune system*.

Myelin

Refers to the "insulation" that surrounds nerve fibers (axons) in the spinal cord and brain. See also *Axon*, *Oligodendrocytes*, *Multiple sclerosis*.

Myeloma

A tumor cell line derived from a lymphocyte. It usually produces a single type of immunoglobulin. See also *Hybridoma*, *Lymphocyte*, *Aging*.

Myoblasts

See *Reversine*.

Myoelectric Signals

The nerve signals that are sent by the body in order to control muscle (myotube) movement. See also *Actin*.

Myoepithelial Cells

See *Multipotent progenitors*.

Myoinositol Hexaphosphate

See *Phytate*.

Myosin

See *Actin*.

Myotubes

See *Actin*.

Myristoylation

Transformation of proteins in cells in such a manner that these cells then cause cancer. See also *Cancer*.

Myrosinase

An enzyme, for example, present within cells of the wasabi plant (*Wasabi japonica*) that catalyzes the conversion of that plant's glucosinolates to the isothiocyanates that provide the spicy taste of Wasabi food ingredient. See also *Enzyme*, *Cell*, *Glucosinolates*, *Isothiocyanates*.

N

"Nude" Mouse

Refers to a line of laboratory mice (genotype *Foxn1*^{nu}) that are descended from a mouse in which many years ago occurred a spontaneous mutation which results in (descendants) that:

- Do not develop any thymus-derived T cells.
- Do not develop hair follicles in their epithelium (skin).

Because this line of laboratory mice will therefore "accept" (i.e., immune system will not attack) a wide variety of transplants (e.g., tumors, etc. from other animals), it is a common research test subject for researchers who want to test the impact of a new pharmaceutical compound on—tumor, and so on. Another aspect making it a commonly chosen test subject is the fact that scientists do not have to shave the mouse in order to measure size of the subcutaneous inserted tumors. See also *Genotype, Mutation, Thymus, T cells, Tumor, Epithelium.*

N Glycosylation

See *Glycosylation.*

n-3 Fatty Acids

Also known as "omega-3" fatty acids. Research indicates there are human health benefits (e.g., antithrombotic, anti-inflammatory, reduce/avoid coronary heart disease (CHD)) if the ratio of *n*-6 to *n*-3 fatty acids contained in the diet is higher than 3 but less than 10. Soybean oil has an *n*-6/*n*-3 ratio of approximately 7:1.

Examples of *n*-3 fatty acids include linolenic acid (C18:3*n*-3), eicosapentanoic acid, and docosahexanoic acid. Research indicates that human consumption of *n*-3 fatty acid(s) reduces chronic inflammation and imparts antithrombotic and anti-inflammatory health benefits, plus it lowers the level of triglycerides content in the bloodstream. The *n*-3 fatty acids inhibit an enzyme in the body known as cyclooxygenase (COX), which produces the prostaglandin hormones that can spark chronic inflammation.

At least some *n*-3 fatty acids are essential for membrane synthesis within the human brain and retina, plus some have been shown to be essential for neural development. Recent research indicates that adequate dietary consumption of *n*-3 fatty acids helps preserve telomeres.

During 2000, research was published that indicated a 66% reduction in probability for children to develop "juvenile" (Type I) diabetes, if their mother consumed significant quantities of *n*-3 fatty acids during the pregnancy. See also *Polyunsaturated fatty acids (PUFA), Docosahexaenoic acid (DHA), Eicosapentaenoic acid (EPA), Linolenic acid, Soybean oil, Stearidonic acid, Thrombosis, Triglycerides, Coronary heart disease (CHD), Diabetes, Insulin, Cyclooxygenase, GPR120 receptor, Macrophage, Tamoxifen, Chronic inflammation, Telomeres.*

n-6 Fatty Acids

Also known as "omega-6" fatty acids. Research indicates there are human health benefits (e.g., antithrombotic, reduce/avoid CHD) if the ratio of *n*-6 to *n*-3 fatty acids contained in the diet is higher than 3 but less than 10. Soybean oil has an *n*-6/*n*-3 ratio of approximately 7:1.

Examples of *n*-6 fatty acids include linoleic acid (C18:2*n*-6). Research indicates that consumption of *n*-6 fatty acids has been related to decreased cholesterol levels in the bloodstream and decreased incidence of CHD. See also *Polyunsaturated fatty acids (PUFA), Arachidonic acid, Linoleic acid, Soybean oil, Thrombosis, Coronary heart disease (CHD), Cholesterol.*

NAD (NADH, NADP, NADPH)

Nicotinamide-adenine dinucleotide, also known as diphosphopyridine nucleotide, codehydrogenase 1, coenzyme 1, and coenzymase by its discoverers, Harden and Young. $C_{21}H_{27}O_{14}N_7P_2$. An organic coenzyme (molecule) that functions as a distinct yet integral part of certain enzymes. NAD plays a role in certain enzymes concerned with oxidation–reduction reactions. Meanings: NADH, nicotinamide-adenine dinucleotide, reduced; NADP, nicotinamide-adenine dinucleotide phosphate; and NADPH, nicotinamide-adenine dinucleotide phosphate, reduced. See also *Enzyme, Coenzyme, Oxidation–reduction reaction, Nitric oxide synthase, Sirtuins.*

NaD1

See *Defensins.*

NADA (New Animal Drug Application)

An application to the U.S. Food and Drug Administration (FDA) to begin testing/studies of a new drug for animals (e.g., livestock) that might (eventually) lead to its FDA approval. See also *IND.*

NADH

Nicotine-adenine dinucleotide, reduced. See also *NAD.*

NADP

Nicotine-adenine dinucleotide phosphate. See also *NAD, Nitric oxide synthase.*

NADPH

Nicotinamide-adenine dinucleotide phosphate, reduced. See also *NAD.*

N

Naive T Cells

See *Dendritic cells.*

Naked DNA

See *Naked gene.*

Naked Gene

A bare gene (strand of DNA that codes for a protein) which has been extracted from an organism or otherwise derived (e.g., synthesized from sequence data). During the 1990s, it was discovered that:

- Injecting the Duchenne Muscular Dystrophy "naked gene" into muscle tissue in the bodies of people suffering from Muscular Dystrophy (MD) resulted in temporary production of the relevant protein in that muscle tissue (i.e., temporary MD symptom reduction).
- Injecting the VEGF "naked gene" into relevant tissue in the bodies of people suffering from inadequate local blood supply (e.g., the shortage of blood flow to heart known as myocardial ischemia, lack of blood flow in legs or other extremities, etc.) resulted in (new) growth of blood vessels/endothelium and reduction in symptoms of those inadequate-blood-supply conditions.
- Injecting the "naked gene" for the relevant antigen of certain pathogens into some tissues in the (usual disease host) organism sometimes resulted in those (host organism) tissues taking up the "naked gene" and expressing some of the (pathogen's) antigen(s), such that the (putative host organism's) immune system initiates an immune response (thereby resulting in vaccination against the disease conferred by pathogen). When that happens, such "naked genes" are referred to as "DNA vaccines."

See also *Gene, Deoxyribonucleic acid (DNA), Protein, Organism, Synthesizing (of DNA molecules), Sequencing (of DNA molecules), Duchenne muscular dystrophy gene, Muscular dystrophy (MD), Vascular endothelial growth factor (VEGF), Pathogen, Express, DNA vaccines, Immune response, Cellular immune response, Humoral immunity, Antibody, DNA vector.*

NAM

Acronym for Nested Association Mapping. See *Nested association mapping.*

Nano Robots

See *Nanobots.*

Nanobackpack

A term utilized to refer to a nanometer-scale container that is attached to one spot on the outside of a free-floating cell (e.g., an immune cell within the bloodstream). For example, during 2008 Michael Rubner, Darrell Irvine, and Robert Cohen created and attached to an immune cell a container made of polymerized hyaluronic acid. Inside this "backpack" were magnetic nanoparticles, which enabled the immune cell to be moved (e.g., within the bloodstream) by the scientists simply applying an external magnetic field.

For example, during 2014 Samir Mitragotri and Aaron Anselmo created and attached to monocytes a tiny disc-shaped polymer "backpack" which could contain pharmaceutical(s) that can be released by the backpack when the monocyte reaches the site of (chronic) inflammation within body tissues. These polymer backpacks are coated on one of their sides with an antibody that can bind to receptors on the monocyte's surface. At the site of the chronic inflammation, the other side of the backpack degrades to release the pharmaceutical. See also *Nanometers (nm), Cell, Magnetic particles, Nanoscience, Nanotechnology, Monocytes, Antibody, Receptors, Chronic inflammation.*

Nanobatteries

Refers to electrochemical devices that are small enough in size to be able to provide electrical power to nanotechnology devices (e.g., nanobots). For example, during 2008 Angela Belcher and colleagues created such a device by etching "battery posts" several microns apart into a rubber surface. They deposited on them polymer layers that serve as an electrolyte and onto the surface of the electrolyte a virus that is modified to produce a protein surface coat which attracts molecules of cobalt oxide (that self-assembles into a structure that acts as negative electrode of the nano-sized battery).

In 2009, Angela Belcher changed one gene within the genetically engineered virus M13 to cause it to manufacture for itself a coat of amorphous iron phosphate, thereby forming a positive electrode of a nano-sized battery. Changing another gene in the virus causes that (coated) virus to tightly attach itself to a carbon nanotube, which is utilized to wire all parts of a nano-sized lithium ion battery to each other, thereby making a nanobattery. See also *Nanotechnology, Nanobots, Micron, Polymer, Virus, Protein, Self-assembly (of a large molecular structure), Gene, Genetic engineering, Carbon nanotubes.*

Nanobiology

See *Nanotechnology, Nanocomposites, Nanovaccine, Nanopillars, Nanocapsules, Bioinorganic, Nanocrystals, Nanoelectromechanical system (NEMS), LAAM, Nanobiotechnology.*

Nanobionics

A term coined during 2014 by Michael Strano and Juan Pablo Giraldo to refer to plants into which they have inserted carbon nanotubes or other nanoparticles, which thereby result in those altered plants now functioning as detectors for certain explosives/land mines, as detectors for certain environmental pollutants, and so on. See also *Nanotechnology, Carbon nanotubes, Nanoparticles.*

Nanobiotechnology

Refers to man's utilization of nanotechnology/nanoscience in biological applications. For example, use of man-made nanodiamonds or nanocapsules to carry powerful chemotherapy pharmaceuticals into cancer cells; to maximize the dose delivered to cancerous cells while minimizing adverse impacts on surrounding healthy cells.

Another potential application is to inject nanoshells into a cancer patient's bloodstream. For that application, nanoshells are made of materials/thicknesses selected to absorb specific wavelengths of infrared light (which human tissue is transparent to) and are "surface modified" (via attachment of molecular bridges to surface) so they will accumulate inside tumors. Once the nanoshells have accumulated within the tumor, subsequent shining of intense infrared light of the appropriate wavelength at that tumor causes the nanoshells to heat up and kill the tumor tissue, without harm to adjacent noncancerous tissue.

Another version of that (tumor-destroying) application is to manufacture such nanoshells from magnetic element(s) and then cause them to heat up in the tumor via application of oscillating magnetic fields.

Another potential application is to utilize silica- or iron oxide nanoparticles to swiftly "glue together" soft tissues (especially liver tissue, which can be traumatized by sutures) during surgery. When these nanoparticles (in a water solution) are simply dabbed onto the edges of the tissue, the nanoparticles adsorb onto surface of both pieces of the tissue in a process known as **nanobridging**, and the subsequently pressed-together tissue swiftly adheres together.

See also *Nanotechnology, Nanoscience, Nanovaccine, Nanocrystals, Nanoceria, Nanocapsules, Nanoshells, Cell, Cancer, Carbon nanotubes, Carbon nanohorns, Nanobatteries, Dendrimersomes, Lignin nanotubes, Nanobodies, Nanobionics, Paclitaxel, Albumin, Liposomal nanoparticles.*

Nanobodies

Refers to the smallest possible portion of an antibody which will bind to an antigen/hapten. Nanobodies tend to be approximately one-tenth the size of antibodies. Although nanobodies are not made naturally by humans (i.e., the human body always makes complete/full-size antibodies), human nanobodies (approximately 120 amino acids in length) can be made via genetically engineered cells grown via cell culture.

Camels naturally make nanobodies in their bloodstream. Because such nanobodies can be relatively easily attached to nanoparticles or to large protein molecules, they might be readily targeted to attack certain cancer-related proteins in the future.

Because nanobodies are less sensitive to high temperatures and pH changes than antibodies, they are able to survive highly acidic conditions (e.g., passage through the stomach) without losing their biological activity, so they could potentially be utilized in some future orally administered pharmaceuticals. See also *Antibody, Antigen, Hapten, Combining site, Protein, Amino acid, Genetic engineering, Cell, Cell culture, Nanoparticles, Cancer, Biological activity, Orally administered, Pro-drug therapy, Nanobiotechnology.*

Nanobots

Refers to very small "robots" whose dimensions would be measured in terms of nanometers (nm), and could perform specific tasks, when very specific conditions are encountered by it. For example during 2012, William Shih utilized DNA origami (extended to three dimensions) method to create a long strand of DNA that was "programmed" (via DNA sequences built into the DNA) to self-fold into a very specific 3D shape. When carried by the natural flow of the bloodstream, that "self-assembled nanobot" is subsequently able to detect and attach to a target cell where it then "unfolds" to present the antibody that activates an apoptosis "suicide switch" in the target leukemia or lymphoma cells.

Other nanobots will be powered by a suitable "nanomotor." For example, during 2005, Ben L. Feringa created a nanoscale rotor which is powered by light, by attaching a chiral helical alkene molecule onto a piece of gold.

During 2011, Joseph Wang and Liangfeng Zhang created a nanorocket (i.e., a nanometer-scale-long structure consisting of concentric tubes of platinum, iron, and an outer layer of cold) which is powered on its "flight" through biological fluids by the breakdown of hydrogen peroxide on the nanorocket's inner platinum surface (where oxygen bubbles are formed then expelled out the backside of the nanorocket). The nanorocket is guided through the biological fluid (e.g., bloodstream) via use of directed magnets that act upon the iron contained within the wall of the nanorocket. Antibodies or single strands of DNA can be attached to the outer layer of gold via a thiol group, in order for such nanorockets to capture specific cells (e.g., cancerous cells that the antibodies are specific to) or to hybridize onto specific DNA segments (e.g., in bloodstream) as the nanorockets "fly" through that fluid.

Theoretically at least, such nanobots could also be powered via nanopiezoelectronics, nanobatteries, directed use of Casimir force, and so on. See also *Nanoelectromechanical system (NEMS), Nanoscience, MEMS (nanotechnology), Deoxyribonucleic acid (DNA), DNA origami, Antibody, BioMEMS, Nanometers (nm), Nanotechnology, Self-assembling molecular machines, Chiral compound, Helix, Nanopiezoelectronics, Nanobatteries, Casimir force, Nanomotor, Thiol group, Antibody, Hybridization (molecular genetics), Deoxyribonucleic acid (DNA), Nanorocket, Nanodrills, Apoptosis, Self-assembling molecular machines.*

Nanobridging

See *Nanobiotechnology.*

Nanocantilever

Refers to a testing technology which enables real-time detection of interactions (e.g., "binding"/ligand, etc.) between protein molecules (attached to the surface of a sensitive cantilever of nanometer-scale dimensions) and other molecules (e.g., pharmaceutical candidate compounds, toxins, etc.) passed over those surface-attached protein molecules.

When those interactions occur on the cantilever arm's surface, they cause deformation of the cantilever arm, which can be detected by a change in its electrical resistance or by how it affects relevant wavelength light shined onto the cantilever. See also *Nanometers (nm), Protein, Ligand (in biochemistry), Antibody, Receptors, Protein interaction analysis, Target–ligand interaction screening, Biosensors (electronic), Microfluidics, Nanoelectromechanical system (NEMS), Nanovalve, Nanowire, Bionanotechnology, Nanopiezoelectronics.*

Nanocapsules

Refers to nanometer-scale, hollow, spherically shaped objects that can be utilized to encapsulate small amounts of pharmaceuticals, enzymes or other catalysts, and so on. See also *Nanoshells, Dendrimers, Nanocochleates, Self-assembly (of a large molecular structure), DNA buckyballs, Nanometers (nm), Nanoscience, Nanotechnology, Enzyme, Catalyst, Folic acid, Click chemistry, Dendrimersomes.*

Nanoceria

Refers to cerium oxide nanoparticles. These particles are strong antioxidants that scavenge (grab/neutralize) oxygen free radicals. See also *Nanoparticles, Free radical.*

N

Nanocochleates

See *Phosphatidyl serine.*

Nano-Cocoon

See *Nanoshells.*

Nanocomposites

Nanometer-scale composite structures composed of organic molecules intimately incorporated with inorganic molecules. For example, Abalone shellfish make mother-of-pearl shells via an intimate combination of protein and calcium carbonate (in the form of multiple-sided microscopic "tablets").

Researchers are working on making semiconductor devices (chips) containing peptides and other organic molecules attached to silicon or gallium arsenide. Also working on nanoelectromechanical systems (NEMS), which would have tiny "moving parts" to be able to do "work" at nanometer scale. See also *Nanometers (nm), Nanotechnology, Protein, Biochip, Peptide, Biosensors (electronic), Bioinorganic, Nanoelectromechanical system (NEMS), Nanosolder.*

Nano-Corkscrews

Corkscrew-shaped nanoparticles (e.g., made of gold), which possess optical activity. See also *Nanotechnology, Nanoscience, Nanoparticles, Optical activity.*

Nanocrystal Molecules

Coined by researchers A. Paul Alivisatos and Peter G. Schultz, it is a term used to describe double-stranded DNA molecules that have attached to them several multiatom clusters of gold. As of 1996, these researchers were working to try to create nanometer-scale electrical circuits, semiconductors, and so on. A separate methodology, researched by Chad A. Mirkin et al., utilizes strands of DNA to reversibly assemble gold nanoparticles (nanometer-scale multiatom particles) into supramolecular (many molecule) agglomerations, in which the gold particles are separated from each other by a distance of approximately 60 Å. The aggregation of these DNA-metal nanoparticles causes a visible color change to occur. As of 1996, these researchers were working to try to create simple and rapid tests that would indicate the presence of a virus (e.g., HIV-1 or HIV-2) via a visible color change. Such a test would use two noncomplementary DNA sequences, each of which has attached to it a gold nanoparticle (via a thiol group). The two sequences would be selected for their ability to latch onto a target sequence in the desired virus, but they would be unable to combine with each other, since they are noncomplementary. When double-stranded DNA molecules possessing two "sticky ends" (that are complementary to the sequences attached to virus) are added, the resultant color change indicates virus presence. See also *Double helix, Deoxyribonucleic acid (DNA), Angstrom (Å), Nanometers (nm), Hybridization surfaces, Base pair (bp), Self-assembly, Nanotechnology, Sticky ends, Hybridization (molecular genetics), Sequence (of a DNA molecule), Virus, Biosensors (chemical), Biochip, Microfluidics, Nanocrystals.*

Nanocrystals

A term that is utilized to refer to any crystalline structure possessing dimensions (e.g., overall width) measured in terms of nanometers. A formulation of Rapamune nanocrystals was approved by the U.S. FDA in 1999 for use as the immunosuppressive pharmaceutical Sirolimus™. See also *Nanometers (nm), Quantum dot, Nanoscience, Nanotechnology, Nanocrystal molecules, Nanocomposites, Oriented attachment.*

Nanodiamonds

Refers to diamond-shaped structures (approximately 2 nm in diameter) created by David Ho during 2007. Certain pharmaceutical compounds can be loaded onto the surfaces of these nanodiamonds, so that when the nanodiamonds self-aggregate into cluster structures of up to 100 nm in diameter, they consist of a "container" which can carry the pharmaceutical into specific cells (e.g., cancer tumor cells within the body).

One variant on that is when nanodiamonds are bound to the chemotherapy drug epirubicin and enclosed within a lipid membrane and then coupled to antibodies made to be specific to hard-to-treat tumors.

Only after they reach their "targets" (i.e., the cancer cells) do the nanodiamond clusters break apart and release the pharmaceutical. This maximizes delivery to the cancerous calls and minimizes damage to surrounding normal healthy cells. See also *Nanometers (nm), Nanotechnology, Nanoscience, Cell, Cancer, Tumor, Lipids.*

Nanodisk

Refers to a particular nanometer-scaled (thickness) artificial cell membrane. These have potential use in pharmaceutical discovery research (e.g., where scientists need to know if a particular pharmaceutical candidate is likely to readily cross a cell membrane). For example, during 2011 A.T. Charles Johnson created nanodisks bearing G protein–coupled receptors (GPCRs) which are typically embedded in the outer membrane of cells, where they bind to passing pharmaceutical molecules as the first step in a cascade (sequential series of chemical/biological events) which leads to the pharmaceutical delivering its beneficial impact inside the cell. In addition to providing a more realistic cell membrane model (e.g., in terms of determining if a given pharmaceutical was likely to enter a specific cell type), the nanodisk preserved the biological activity of the GPCRs up to 80 times longer than had previously been possible in such pharmaceutical testing. See also *Nanometers (nm), Nanotechnology, Nanoscience, Cell, Receptors, G-protein-coupled receptors, High-throughput screening (HTS), Biochip, Microarray (testing), Cell array, Target–ligand interaction screening.*

Nanodrills

Refers to certain carbon nanotubes which form into bundles that rotate (spin) rapidly in the presence of a rotating magnetic field. Via precise manipulation of that magnetic field, these "nanodrills" could be utilized to drill into specific cells (e.g., cancer cells) inside the body; to deliver drugs into those cancer cells or to kill them outright. See also *Carbon nanotubes, Cell, Cancer, Nanotechnology, Nanoscience.*

Nanoelectromechanical System (NEMS)

Refers to working (i.e., those with moving "mechanical parts") systems of a scale whose relevant dimensions are measured in terms of nanometers (nm).

For example, in 2000, Carlo Montemagno and colleagues assembled a NEMS in which a tiny metal "propeller" was caused to spin within the domain of the enzyme ATP synthase. The metal propeller was attached (via a biotin–streptavidin "molecular linkage") to the one subunit (designated alpha) of ATP synthase that rotates within the other (hollow) part of ATP synthase molecule—when ATP is "fed" to a freestanding (i.e., not in cell) molecule of ATP synthase. See also *Nanometers (nm), ATP synthase, Enzyme, Adenosine triphosphate (ATP), Biotin, Avidin, Nanocomposites, Nanoscience, Micromachining, Self-assembling molecular machines, Directed self-assembly, Nanopiezoelectronics, Casimir force.*

nanoFET

Refers to a nanometer-scale field effect transistor (FET) created by Charles M. Lieber in 2010. The acronym FET has long referred to the electronics portion of certain miniature biosensors, in which the transistor "gate" (which controls current flow through the transistor) is replaced by a biochemical that serves as a sensing material of the biosensor. A change in the microenvironment (e.g., liquid within tissue, etc.) immediately surrounding the sensing material results in a field effect (impact on the device's electrical field) that drains electrical current off the transistor. Thus, the FET can be utilized to continually monitor/measure that microenvironment for any changes.

The nanoFET consists of a silicon nanowire whose diameter is 15 nm, and whose composition (impurities) is manipulated to result in it functioning as a transistor, and it is small enough to insert into a single cell in order to monitor changes within the cell (e.g., neurons firing, etc.). See also *Field effect transistor, Nanometers (nm), Nanowire, Cell, Biosensors (electronic), Bioelectronics.*

Nanofibers

Refers to any fibers (created by man) possessing diameter(s) of less than 100 nm.

For example, during 2003, Sam Stupp created self-assembling nanofibers that promote (re)growth of nerve cells in laboratory rats. These nanofibers consist of peptide amphiphiles with hydrophilic peptide portions covering the exterior of the fibers, while their hydrophobic portions are in the center of the fibers.

During 2005, Sam Stupp, Kanya Rajangam, and John Lomasney created nanofibers which promote the growth (creation) of blood vessels, a process known as angiogenesis. Stupp and his colleagues synthesized peptide amphiphiles consisting of a specific sequence of eight particular amino acids found in natural blood peptides that bind heparin. When added to a solution containing heparin, fibroblast growth factor (FGF), and vascular endothelial growth factor (VEGF)—like human blood does—these peptide amphiphiles self-assemble into nanofibers. The nanofibers quickly bind heparin, then the bound heparin quickly binds FGF and VEGF, which (when combined in that fashion) promote angiogenesis.

Another category of nanofiber is a "carbon nantube yarn" (i.e., long fiber built of carbon nanotubes oriented longitudinally and twist spun). During 2011, Javad Foroughi and colleagues showed that such a carbon nanotube yarn can be utilized to create a nanomotor (e.g., to power a liquid mixer within a microfluidic chip). See also *Nanometers (nm), Nanoscience, Nanotechnology, Self-assembly (of a large molecular structure), Peptide, Amino acid, Amphiphilic molecules, Carbon nanotubes, Nanomotors, Microfluidic chips, Hydrophilic, Hydrophobic, Cell, Heparin, Angiogenesis, Polypeptide (protein), Sequence (of a protein molecule), Synthesizing (of proteins), Fibroblast growth factor (FGF), Vascular endothelial growth factor (VEGF).*

Nanofluidics

See *Microfluidics.*

Nanogram (ng)

10^{-9} g, or 3.527×10^{-11} oz (avoirdupoir).

Nanohorns

See *Carbon nanohorns.*

Nanolithography

Refers to the practice of using an atomic force microscope tip to apply specific (e.g., DNA) molecules to surfaces such as metals, oxides, and so on. Nanolithography can be used to direct (via DNA interactions) the assembly of tiny structures such as gene chips, catalysts, nanoscale circuits, and so on. See also *Deoxyribonucleic acid (DNA), Atomic force microscopy, Hybridization (molecular genetics), Hybridization surfaces, Gene chips, Catalyst, Directed self-assembly, Bioelectronics, Self-assembling molecular machines, Template.*

Nanomechanical Cantilevers

Refers to nanometer-scale levers that are anchored at one end, like tiny diving boards, for use in detection of certain pathogenic bacteria. Those cantilevers have specific pathogenic-*Salmonella*-bacteria-binding peptides attached to them that bind to applicable *Salmonella* serotypes. When those specific pathogenic *Salmonella* bacteria bind to the peptides on the cantilever, the cantilever arm bends, which generates a signal. This screening system can rapidly distinguish eight different serotypes of pathogenic *Salmonella* from each other. See also *Nanometers (nm), Bacteria, Pathogen, Pathogenic, Peptide, Serotypes, Strain, Salmonella.*

Nanometers (nm)

10^{-9} m. Often used to express wavelengths of light (e.g., in a spectrophotometer), or to express dimensions of nanocomposites, devices (e.g., of miniature "machines" called nanoelectromechanical systems), and so on, in the field of nanotechnology. See also *Spectrophotometer, Nanotechnology, Nanocomposites, Nanoelectromechanical system (NEMS), Microfluidics, Meter.*

Nanomotor

Refers to any nanometer-scale device/phenomenon that provides a useful means to power a "nanomachine" such as a nano-robot (nanobot), a mixer within a microfluidic chip, and so on. One type of nanomotor can be made by torsionally curling "carbon nantube yarn" (i.e., long fiber built of carbon nanotubes oriented longitudinally and twist spun), which subsequently twists/untwists when charge is injected into the electrolyte that nanomotor is immersed in. During 2011, Javad Foroughi and colleagues showed that such a carbon nanotube yarn can be utilized to create a nanomotor (e.g., to power a liquid mixer within a microfluidic chip) that spins at 590 rpm.

N

For example, during 2011, Joseph Wang and Liangfeng Zhang created a nanorocket (i.e., a nanometer-scale long structure consisting of concentric tubes of platinum, iron, and an outer layer of gold) which is powered on its "flight" through biological fluids by the breakdown of hydrogen peroxide on the nanorocket's inner platinum surface (where oxygen bubbles are formed then expelled out the backside of the nanorocket). The nanorocket is guided through the biological fluid (e.g., bloodstream) via use of directed magnets that act upon the iron contained within the wall of the nanorocket. Antibodies or single strands of DNA can be attached to the outer layer of gold via a thiol group, in order for such nanorockets to capture specific cells (e.g., cancerous cells that the antibodies are specific to) or to hybridize onto specific DNA segments (e.g., in bloodstream) as the nanorockets "fly" through that fluid.

For example, during 2014, Zev Bryant and Muneaki Nakamura created nanomotors that are controlled (speed, direction) by light shining on them. The motors consisted of protein molecules coded for by a synthetic gene (DNA sequence) that was itself assembled from DNA segments extracted from pig, slime mold, and oats DNA.

Theoretically at least, such nanomotors could also be powered via nanopiezoelectronics, nanobatteries, light shined onto the nanomotor, directed use of Casimir force, and so on. See also *Nanoelectromechanical system (NEMS)*, *Nanoscience*, *MEMS (nanotechnology)*, *BioMEMS*, *Nanometers (nm)*, *Nanotechnology*, *Self-assembling molecular machines*, *Chiral compound*, *Helix*, *Piezoelectric effect*, *Nanopiezoelectronics*, *Nanobatteries*, *Casimir force*, *Nanofibers*, *Thiol group*, *Antibody*, *Hybridization (molecular genetics)*, *Deoxyribonucleic acid (DNA)*, *Sequence (of DNA)*, *Nanobots*, *Nanorocket*.

Nanoparticles

Term utilized to refer to a variety of nanometer-scale size particles (e.g., nanocrystals, nanoshells, nanostars, nano-corkscrews, nanosponges, quantum dots, lipidoids, spherical nucleic acids, etc.). Depending on the materials used to construct nanoparticles, their uses include:

- Imaging tissues within the body: When injected and subsequently "illuminated" by specific wavelength light, relevant quantum dots emit light in colors which vary depending on the dimensions of the dot, and the particular tissue they happen to reside in.
- Detection of pancreatic cancer cells: During 2005, Ralph Weissleder showed that fluorescent nanoparticles with some molecules of isatoic or 5-chloroisatoic anhydride attached to their surface (in spots) are selectively taken up by only pancreatic cancer cells (in mice that were tested). Since those particular fluorescent nanoparticles show up well in magnetic resonance imaging, this may someday be utilized for early diagnosis of pancreatic cancer in humans.
- Detection of cancer metastasis: Magnetic nanoparticles (100 nm diameter) attached to **antibodies against epithelial cells** can be utilized to detect metastasis of cancer in a human. The magnetized antibodies attach themselves to epithelial cells (a biomarker of metastasis) in a blood sample, enabling the epithelial cells to be detected/counted by doctors.
- Detection of DNA hybridization: When relevant DNA molecules/segments are "labeled" in advance with superparamagnetic particles, those "labeled DNA segments" can be utilized in magnetic DNA microarrays to detect when/which DNA segments are hybridized (by DNA within samples being analyzed).
- Delivery of some pharmaceuticals: For example, when molecules of certain statin drugs (which have an anti-inflammatory impact within blood vessel walls) are delivered to applicable (plaque-lined) blood vessel walls by encasing the statin molecules inside a layer of high-density lipoprotein (HDL, the so-called good cholesterol) molecules, which subsequently attach themselves to HDL receptors on blood vessel walls, thereby delivering those statins where needed (to prevent a second heart attack immediately following first heart attack). For example, when molecules of man-made tumor necrosis factor (TNF) are attached to certain nanoparticles of applicable size (70 nm diameter), TNF can act to disrupt formation of the new vasculature (blood vessels) needed by the tumor for blood supply (e.g., when TNF is delivered in the form of a concentrated pharmaceutical via nanoparticles that preferentially accumulate in tumors). Another potential mode of drug delivery into cancer cells is via drilling of holes in those cells by "nanodrills."
- Delivery of short interfering RNA (siRNA) to specific tissues within the body in order to induce RNA interference (RNAi) in those tissues (e.g., to thereby cure certain diseases). Altering the surface chemistry of the nanoparticles changes the tissue that the siRNA is carried to.
- Prolonging the bloodstream circulation time of some pharmaceuticals in the body. Note that applicable drugs or drug carriers (e.g., certain nanoparticles) bearing a slightly negative charge have a longer circulation time.
- Creation of plastic antibodies via molecular imprinting.
- Killing of the harmful bacteria within some biofilms: When iron oxide particles possessing an average diameter of 8 nm are injected and guided via an externally applied magnetic field to the biofilm that has formed on a medical implant within the body.
- Creation of some liposomal nanoparticles: The term "liposomal nanoparticles" refers to nanoparticles encapsulated within a shell composed of a combination of liposomes and polyethylene glycol. Certain pharmaceutical compounds (e.g., doxorubicin) can thereby be delivered to certain sites within a patient's body (e.g., the immature/growing blood vessels that supply nutrients to certain cancerous tumors) without the pharmaceutical compound dispersing to tissues throughout the body. That is because such pharmaceutical compounds which tend to be rapidly degraded in the bloodstream are protected from degradation/dispersion while in the bloodstream via being enclosed within the liposomes and polyethylene glycol, so that more of the nondegraded/nondispersed pharmaceutical would remain by the time it reached the targeted tissue (i.e., the tumor). For example, during 1995, US regulators approved pharmaceutical-in-liposome Doxil for treatment of Kaposi's sarcoma tumors.
- Creation of some nanoshells.
- Creation of some nanobionic plants.

See also *Nanometers (nm)*, *Nanocrystals*, *Quantum dot*, *Nanocrystal molecules*, *Cancer*, *Metastasis*, *Biomarkers*, *Magnetic particles*, *Deoxyribonucleic acid (DNA)*, *Sequence (of a DNA molecule)*,

Hybridization (molecular genetics), DNA profiling, Microarray (testing), Bio-bar codes, Nanotechnology, Nanocochleates, Nanoshells, Nanostars, Nanobionics, Nanocapsules, Fluorescence, Metamaterials, Van der Waals forces, Casimir force, Tumor necrosis factor (TNF), Pegylation, Folic acid, Click chemistry, Biofilm, Optical activity, Scavenger receptor A, Molecular imprinting, Nanodrills, Short interfering RNA (siRNA), Spherical nucleic acids, Coronary heart disease (CHD), Atherosclerosis, Statins, Plaque, Thrombosis, High-density lipoproteins (HDLPs), Receptors, Modulatory nanotechnologies, Liposomal nanoparticles.

Nanopiezoelectronics

Coined by Zhong Lin Wang during 2006, this term refers to generation of electrical energy (electricity) at the nanometer scale (e.g., to power nano-devices) via mechanical stress to the nanopiezoelectronic device. For example, bending of a zinc oxide nanowire transforms that mechanical energy into electrical energy. See also *Nanotechnology, Nanometers (nm), Piezoelectric effect, Nanowire, Nanobots, Nanoelectromechanical systems (NEMS), Self-assembling molecular machines.*

Nanopillars

Refers to tiny vertical pillars of lipid that naturally form on wing surfaces of dragonflies and cicadas that are bactericidal (i.e., they kill bacteria that land on those pillars by rupturing the bacteria cell wall). The diameter of the pillars is less than 90 nm, though they can be up to several hundred nanometers tall. See also *Nanometers (nm), Nanoscience, Nanotechnology, Bacteria, Lipids.*

Nanoplasmonic Biosensor

See *Surface plasmon resonance (SPR), Surface plasmons, Nanotechnolgy.*

Nanopore

A device that can distinguish between different DNA strands (molecules) that differ—from each other—by a single nucleotide (in the makeup of those molecular strands). Initially conceived by Hagan Bayley, David Deamer, and Mark Akeson in 2001, it can be manufactured in several different ways:

1. Consists of an artificial membrane (lipid bilayer) with a "hole" (nanopore) punctured in that membrane by the protein alpha-hemolysin or by a single-walled carbon nanotube.
2. Consists of a membrane (lipid bilayer) with the "hole" formed by an applicable membrane-spanning protein molecule.
3. Consists of a sheet of graphene or silicon with a hole (between 1 and 100 nm in diameter) bored through it via electron beam equipment.
4. Consists of a sheet of graphene/silicon bearing a much larger hole, then the smaller "pore hole" is made via filling in most of the larger hole with carefully structured DNA origami; leaving the "pore hole" surrounded by DNA/chemical groups which assist in the detection of each DNA sequence as a strand passes through the "pore."

Because a DNA molecule moving through such a "nanopore" temporarily blocks the nanopore (until it dissociates into a single DNA strand and "slides" through), an electrical current/voltage applied to that nanopore varies (in amplitude, modulation, duration, etc.) as the DNA strand "slides through", in a way that provides information (e.g., to scientist) about the nucleotides that makeup that DNA strand. It is expected that nanopores will also be used for DNA sequencing. See also *Nanoscience, Nanometers (nm), Nanotechnology, Plasma membrane, Micelle, Deoxyribonucleic acid (DNA), Nucleotide, Single-nucleotide polymorphism (SNP), Ion channels, Sequencing (of DNA molecules), Carbon nanotubes, DNA origami, Protein, Graphene.*

Nanopore Detection

See *Nanopore, Nanopore sequencing.*

Nanopore Sequencing

Conceived by David Deamer and Daniel Branton during the 1990s, it refers to utilizing a "nanopore" (i.e., <50 nm hole usually within a specific membrane, silicon, or graphene) to sequence DNA molecules.

DNA molecules have a "negative" electrical charge, so when the DNA molecule that a scientist wants to sequence (i.e., determine the identity and sequence of each nucleotide comprising that DNA) is placed in solution on the "positive" side of the membrane, one end of the DNA molecule eventually enters the nanopore. Electrical repulsion (to the solution's positive charge) then causes the DNA molecule to pass through the nanopore slowly.

The nanopore usually consists of an artificial membrane (lipid bilayer) with a "hole" in it caused by the protein alpha-hemolysin or by a single-walled carbon nanotube. It could also be a sheet of graphene or silicon with a hole bored in it. Because a DNA molecule moving through such a nanopore temporarily blocks the nanopore, a separate electrical voltage applied to the nanopore by the scientist will vary (in amplitude, modulation, duration, etc.) as the DNA molecular strand "slides through" in a way that provides information to detection instruments about each of the nucleotides that make up that DNA molecule. See also *Nanopore, Nanoscience, Nanotechnology, Deoxyribonucleic acid (DNA), Nucleotide, Sequencing (of DNA molecules), Plasma membrane, Micelle, Protein, Ion channels, Single-walled carbon nanotubes, Graphene, Nanometers (nm).*

Nanoridges

Nanometer-scale ridge patterns that are naturally created on the surfaces of at least some plants' cells which help those cells to adhere to other cells of that plant's tissues. See also *Nanometers (nm), Cell.*

Nanorocket

Refers to any nanometer-scale device/phenomenon that emits directed gases to provide a useful means to move a "nanomachine" such as a nano-robot (nanobot). For example, during 2011, Joseph Wang and Liangfeng Zhang created a nanorocket (i.e., a nanometer-scale long structure consisting of concentric tubes of platinum, iron, and an outer layer of cold) which is powered on its "flight" through biological fluids by the breakdown of hydrogen peroxide on the nanorocket's inner platinum surface (where oxygen bubbles are formed then expelled out the backside of the nanorocket).

The nanorocket is guided through the biological fluid (e.g., bloodstream) via use of directed magnets that act upon the iron contained within the wall of the nanorocket. Antibodies or single strands of DNA can be attached to the outer layer of gold via a thiol group, in order for such nanorockets to capture specific cells (e.g., cancerous cells that the antibodies are specific to) or to hybridize onto specific DNA segments (e.g., in bloodstream) as the nanorockets "fly" through that fluid.

During 2012, a nanorocket powered by water breakdown was created. See also *Nanoelectromechanical system (NEMS), Nanoscience, MEMS (nanotechnology), BioMEMS, Nanometers (nm), Nanotechnology, Self-assembling molecular machines, Chiral compound, Helix, Nanopiezoelectronics, Nanobatteries, Casimir force, Nanomotor, Thiol group, Antibody, Hybridization (molecular genetics), Deoxyribonucleic acid (DNA), Nanobots, Cell, Cancer.*

Nanorods

A term that is utilized to refer to any man-made tiny structure which is in the shape of a rod (i.e., solid cylinder) possessing dimensions measured in terms of nanometers (nm).

During 2005, Florencio Hernandez made some gold nanorods and suspended those nanorods in a liquid solution of gold salts. When a very small amount of mercury was added to that solution, the gold and the mercury combined into an amalgam which changed the effective length of the nanorods. Because that resulted in an optically detectable change in the surface plasmon resonance of the gold nanorods, this experiment demonstrated that such nanorods could be utilized to make an extremely sensitive (i.e., measure as low as parts per quadrillion) detector of mercury and/or some other biologically relevant elements/compounds.

During 2013, Joseph Tracy discovered that carefully controlled addition of ascorbic acid during solution manufacturing of gold nanorods enables control of the resultant gold nanorods' aspect ratio (i.e., the relative height and width of those "solid cylinders"). The slower the ascorbic acid was added, the shorter—versus their width—the resultant nanorods became. Because the optical properties of gold nanorods depend on their aspect ratio, long "skinny" gold nanorods absorb light at wavelengths of greater than 800 nm (i.e., within the near-infrared spectrum), but shorter "fat" gold nanorods absorb light at wavelengths below 700 nm (red or dark red portion of spectrum). This "tunable" control of the nanorods' wavelength absorbance may enable their future use in destruction of tumors within the body (e.g., get the gold nanorods to accumulate inside tumors, then heat the nanorods enough to kill tumor cells via directing intense radiation of specific wavelength at the tumor). See also *Nanometers (nm), Nanovaccine, Nanoscience, Nanotechnology, Ascorbic acid, Nanowire, Surface plasmon resonance (SPR).*

Nanoscience

A term utilized to refer to the science underlying nanotechnology, nanocrystals, nanocrystal molecules, nanostars, nanopillars, nanosponges, nanocomposites, quantum dots, NEMS, nanovalves, nanodrills, nanostructured materials, and so on.

"Nanoscale" materials (i.e., those whose dimensions are approximately 1–100 nm) generally possess different chemical and physical properties than "bulk" materials. For example, when bulk gold metal is formed into nanoscale rods, the intensity of its fluorescence increases by a factor of approximately 10 million. When bulk silver metal is formed into nanoscale particles, it exhibits extremely potent antimicrobial properties and it emits colors (not fluorescence) when those particles are illuminated with laser light. When they are formed into certain specific nanometer-scale structures, gold, platinum, silver, palladium, and some other transition metals demonstrate magnetic behavior.

Another example is that semiconductor material nanocrystals (i.e., quantum dots) emit larger-than-typical-for-such material amounts of light, when stimulated (i.e., bombarded) with pulses of ultraviolet light. See also *Nanotechnology, Carbon nanotubes, Nanoparticles, Nanocrystals, Quantum dot, Nanocrystal molecules, Nanocomposites, Nanoelectromechanical system (NEMS), Nanovalve, Self-assembly (of a large molecular structure), Nanopore, Optical tweezer, Microfluidics, Template, Self-assembling molecular machines, Scanning tunneling microscope, Atomic force microscope, Nanocochleates, Surface plasmons, Nanocapsules, Bionanotechnology, Nanostructured material, Nanorods, Nanofibers, Metamaterials, Plasmonic metamaterials, Van der Waals forces, Nanopiezoelectronics, Casimir force, Nanodiamonds, Nanbiotechnology, Carbon nanohorns, Nanobatteries, Nanobackpack, Nanocantilever, Optical activity, Nanospheres, Nanosheets, Molecular imprinting, Oriented attachment, Dynamic light scattering, nanoFET, Plasmonic nanohole arrays, Inhibition, Graphene, Click chemistry, Dendrimersomes, Nanorocket, Nanosponges, Nanoceria, Nanostars, Liposomes, Micelle, Nanodrills, DNA origami, Nanosolder, Nanovaccine, Nanopillars, Lignin nanotubes, Nanobionics, Modulatory nanotechnologies.*

Nanosheets

Refers to flat sheets (<3 nm in thickness) comprised of self-assembled peptoids (man-made peptide analogue polymers). Because scientists can attach biologically active ligands to the surface of these nanosheets, and those surface ligands can then selectively bind to (passing) molecules such as specific protein molecules, such nanosheets have the potential to be used in the creation of protein microarrays or other biosensors. See also *Self-assembly (of a large molecular structure), Ligand (in biochemistry), Protein, Protein microarrays, Biosensors (chemical), Nanometers (nm), Target–ligand interaction screening, Nanotechnology, Nanoscience.*

Nanoshells

Refers to nanometer-scale crystalline, DNA or other polymer, or gold structures which form into the shape of hollow balls. For example, nanoshells can be manufactured by surrounding cobalt nanoparticles with sulfur (in a 9:8 ratio). When the proper reaction conditions are subsequently applied to this 9:8 cobalt/sulfur mixture, the Kirkendall effect causes the cobalt atoms within the nanoparticle to diffuse out to (and react with) the sulfur atoms faster than the sulfur atoms diffuse in to the nanoparticle. The end result is a spherical nanoshell comprised of the compound Co_9S_8.

When nanoshells are constructed from polymers, they are called polymersomes. One self-assembling DNA nanoshell, whose surface was studded with folic acid ligands and whose center was filled with the anticancer drug doxorubicin (DOX), was created by Zhen Gu and colleagues in 2014. When the folic acid ligands attach to receptors on the surface of cancer cells, the cell pulls the nanoshell inside, where its anticancer drug payload is released.

Another potential application of certain nanoshells is to inject them into a cancer patient's bloodstream. For that application, nanoshells are made of materials/thicknesses selected to absorb specific wavelengths of near-infrared light (which human tissue is transparent to) and are "surface modified" (via attachment of molecular bridges to surface of the nanoshell) so they will accumulate inside tumors. Once the nanoshells have accumulated within the tumor, subsequent shining of intense near-infrared light of the appropriate wavelength at that tumor causes the nanoshells to heat up and kill the tumor tissue, without harm to adjacent noncancerous tissue.

Another version of that (tumor-destroying) application is to manufacture such nanoshells from magnetic element(s) and then cause them to heat up in the tumor via application of oscillating magnetic fields. This is referred to as magnetic fluid hyperthermia.

The self-assembling capsid (hard outer spherical protein shell) of certain viruses can also be considered to be a "nanoshell" (when empty of contents). During 2004, Gijs Wuite determined with an atomic force microscope that the capsid of one particular spherical virus possesses a Young's modulus (measure of "hardness") that is similar to the Young's modulus of some hard synthetic plastics. Such nanoshells might be used to deliver certain pharmaceuticals to only specific body tissues. See also *Nanoscience, Nanometers (nm), Nanotechnology, Nanospheres, Nanoparticles, Cancer, Folic acid, Ligand (in biochemistry), Receptors, Virus, Capsid, Tobacco mosaic virus (TMV), Self-assembly (of a large molecular structure), Molecular bridge, Polymer, Metamaterials, DNA buckyballs, Dendrimersomes.*

Nanosolder

Refers to certain nanocomposite materials (i.e., comprised of two or more elements intimately mixed together at the nanoscale, such as gold nanorods mixed with peptides) that act to fuse together two pieces of living tissue (e.g., after surgery). See also *Nanocomposites, Nanoparticles, Peptide.*

Nanospheres

Refers to nanometer-scale man-made crystalline structures (e.g., silica–protein composites) which form into the shape of hollow balls. Among their uses are for encapsulation of enzyme molecules to prevent their autolysis (i.e., the reaction in which some enzymes catalyze their own "digestion" (lysis) at room temperature).

This can be minimized by man (e.g., for enzymes utilized in industrial applications) via encapsulation of enzymes inside nanospheres. See also *Nanoshells, Nanoscience, Nanometers (nm), Nanotechnology, Nanoparticles, Protein, Enzyme, Autolysis, Dendrimersomes.*

Nanosponges

Refers to nanoparticles (nanometer-scale size particles) that are capable of adsorbing specific chemicals within a multichemical-containing environment (e.g., adsorb specific toxins present within the bloodstreams of people with certain diseases). For example, during 2014 Liangfang Zhang created a nanosponge to absorb from the bloodstream the toxins produced by the drug-resistant bacteria known as Methicillin-resistant *Staphylococcus aureus*. After these nanosponges have adsorbed the toxins, they can be removed from the bloodstream by the liver. Nanosponges can be designed to adsorb toxins from any almost type of infection (or even a poison) that attacks cellular membranes. See also *Nanometers (nm), Nanoparticles, Cell, Toxin.*

Nanostars

Refers to certain nanoparticles that are made of gold, approximately 25 nm wide, and shaped like a star, with 5–10 points protruding from the surface. The resultant large surface area enables a nanostar to be loaded with a high concentration of relevant drug molecules when the drug is stabilized (adhered) onto the surface of the nanostar via thiol molecular bonds.

Approximately 1000 strands of a single-stranded DNA aptamer named AS1411 can be adhered to each nanostar's surface. Those AS14111 strands bind to nucleolin, a protein that is overexpressed in most cancer cells plus is expressed in the cell surface membrane, as well as within the cancer cell. After binding to a surface nucleolin molecule, the aptamer-laden nanostars are shuttled through the outer cell membrane and carried inside to the cell nucleus. When subsequently exposed to ultrafast pulses of light, the thiol bonds are cleaved by the pulsed light, and the (now) **un**-thiolated DNA aptamers enter the nucleus where they act to kill the cancer cell via restoring its apoptosis pathway. See also *Nanoparticles, Nanometers (nm), Thiol, Aptamer, Protein, Nucleolin, Cell, Cancer, Nucleus, Apoptosis.*

Nanostructured Material

Refers to a (macro) material possessing some properties (e.g., optical, electrical, tensile, etc.) which result from the nanometer-scale structure of the atoms comprising that material. For example, certain cholesteric polymer films can be manufactured, in which the individual polymers comprising the film exist in the form of twisted nanostructures. The pitch that is inherent within those individually twisted "stacked polymer" molecules results in their length over which they twist a full 360° being the same order of magnitude as the wavelengths of visible light. Therefore, they appear to the human eye to be colored.

Because changes in the temperature of such films cause the pitch of the twisted "stacked polymer" molecules to change, and thereby their perceived color to change, these nanostructured cholesteric polymer films can be utilized as temperature-indicating packaging (e.g., on temperature-sensitive food products). See also *Nanometers (nm), Nanoscience, Nanotechnology, Polymer, LAAM, Optical activity, Nanosheets, DNA origami.*

Nanotechnology

From the Latin *nanus* = "dwarf," so it literally means "dwarf technology." The word was originally coined by Norio Taniguchi in 1974, to refer to high precision machining. However, Richard Feynman and K. Eric Drexler later popularized the concept of nanotechnology as a new and developing technology in which man manipulates objects whose dimensions are approximately 1–100 nm. Theoretically, it is possible that in the future a variety of man-made "nano-assemblers" (i.e., tiny [molecular] machines smaller than a grain of sand) would manufacture those things that are produced today in factories. For example, enzyme molecules function essentially as jigs and machine tools to shape large molecules as they are formed in biochemical reactions. The technology also encompasses NANOSTRUCTURED MATERIALS, BIOCHIPS, BIOSENSORS, and manipulating atoms and molecules in order to form (build) bigger, but still vanishingly small functional structures and machines. See also *Enzyme, Genosensors, Nanometers (nm), Carbon nanotubes, Biosensors (electronic), Biochip, Microfluidics, Nanoparticles, Nanocrystals, Nanocrystal molecules, Biosensors (chemical), Quantum dot, Nanocomposites, Nanoelectromechanical system (NEMS), Nanovalve, Nanowire, Self-assembly (of a large molecular structure), Template, Molecular sieves, Nanopore, BioMEMS, Optical tweezer, Directed self-assembly, Self-assembling molecular machines, Dip-pen nanolithography, Nanocochleates, Nanoshells, Surface plasmons, Nanocapsules, DNA,*

N

Nanodiamonds, Buckyballs, Bionanotechnology, Nanostructured material, Nanorods, Nanofibers, Nanoceria, Nanosponges, Metamaterials, Plasmonic metamaterials, Van der Waals forces, Nanopiezoelectronics, Casimir force, Nanobiotechnology, Carbon nanohorns, Nanobatteries, Nanobackpack, Nanocantilever, Optical activity, Nanospheres, Nanosheets, Molecular imprinting, Oriented attachment, Dynamic light scattering, nanoFET, Plasmonic nanohole arrays, Nanostars, Click chemistry, Dendrimers, Dendrimersomes, Nanorocket, Liposomes, Micelle, Nanodrills, DNA origami, Nanobionics, Nanosolder, Nanovaccine, Nanopillars, Lignin nanotubes, Modulatory nanotechnologies.

Nanotube

First created in 1991 by Sumio Iijima, the term refers to a tiny tube whose diameter is measured in nanometers. These have been constructed from a variety of materials such as carbon, tungsten sulfide, titanium dioxide, amino acids, and so on. For example, during 2001–2002, Charles R. Martin et al. were able to manufacture antibody-laced nanotube membranes via:

- First creating alumina films which naturally possess cylindrical pores extending through the film.
- Then "growing" silica nanotubes within those pores using special sol–gel chemistry methods.
- Then coating the interior surfaces of the silica nanotubes with aldehyde silanes.
- And finally, reacting the free amino sites (on antibody fragments—raised against one of the enantiomers in a racemic mixture) with the aldehydes (on the aldehyde silane molecules), thereby attaching antibodies to the interior surfaces of the nanotubes.

Thus constructed, the antibody-laced nanotube membranes were utilized by those scientists to separate out the desired enantiomer from a racemic mixture (racemate).

Another example of the use of nanotubes is the 2004 discovery by Thomas Webster and colleagues that coating a titanium object (e.g., the stem of an artificial hip joint) with nanotubes constructed from "rings" comprising guanine and cytosine results in significantly improved attachment (to the artificial hip) by bone osteoblast cells as the bone grows around the stem following hip replacement surgery. See also *Nanoscience, Nanometers (nm), Nanotechnology, Antibody, Amino acid, Enantiomers, Racemate, Enantiopure, Ultrafiltration, Self-assembly (of a large molecular structure), Guanine, Cytosine, Nanorocket, Nanodrills.*

Nanotube Membranes

See *Nanotube, Carbon nanotubes.*

Nanovaccine

Refers to particularly small gold nanorods (e.g., 21 nm wide and 57 nm long in the case of the respiratory syncytial virus (RSV)—the first one created), which are approximately the same shape and size as the virus that one is planning to vaccinate against. Onto the surface of these special gold nanorods are attached a protein that is present on the virus that one is planning to vaccinate against. In the case of RSV, that surface protein is called the F protein, which the body's natural immune response is directed at.

The RSV F protein molecules strongly bond to the surface of these special nanorods due to the nanorods' unique physical and chemical characteristics. The body's dendritic cells recognize the F proteins and "present" them to lymphocytes, which subsequently initiate the body's immune response. See also *Nanorods, Nanometers (nm), Virus, Protein, Dendritic cells, Lymphocyte, Cellular immune response.*

Nanovalve

Refers to any nanometer-scale device which allows the release of molecules in a manner that can be controlled by man. For example, during 2005 Fraser Stoddart and colleagues were able to create a nanovalve that is "switchable" via chemical energy (i.e., it can be opened when desired via nearby addition of appropriate chemicals that react to yield an electron which causes the nanovalve to open, thereby releasing a molecule). That nanovalve consisted of a rotaxane molecule that is attached DIRECTLY OVER A PORE to the surface of porous glass. When a relevant molecule (e.g., a pharmaceutical) is first inserted into the pore and the rotaxane molecule is subsequently attached above the open end of the pore, this results in a system via which the pharmaceutical molecule can be released at the time desired by man (e.g., when needed to treat a disease, stimulate an organ such as a weakened heart, etc.).

Other nanovalves have been created, that are opened via shining light onto a particular nanovalve, changing the pH (to become basic) in a solution surrounding a particular nanovalve, and so on. See also *Nanotechnology, Nanoscience, Nanoelectromechanical system (NEMS), Nanocapsules, Nanopore, Nanobots, Base (general).*

Nanowhiskers

Term utilized to refer to nanometer-scale cellulose fibers extracted by man from tunicates (sea creatures commonly called "sea squirts"). Those nanowhiskers are utilized in certain wound dressings and could potentially be used as scaffolding for some tissue engineering.

See also *Nanometers (nm), Cellulose, Tissue engineering.*

Nanowire

Term utilized to describe (relatively) long and narrow electrical conductors whose dimensions are measured in nanometers (nm). For example, during 2003, Susan L. Lindquist utilized yeast amyloid proteins (which self-assemble into 60–300 nm long fibers) to create nanowires by subsequently coating those fibers with gold and silver.

Other potential materials with which nanowires can be manufactured include carbon nanotubes.

One of the ways to precisely handle nanowires (e.g., to thereby "build" electrical circuits) is to attach tiny pieces of nickel metal to each end of the nanowire. Because the nickel is ferromagnetic, magnetic field lines of force can then be utilized to precisely move and position the nanowires. See also *Bioelectronics, Nanometers (nm), Carbon nanotubes, Self-assembly (of a large molecular structure), Nanotechnology, Template, Nanorods, Nanopiezoelectronics, Nanobatteries, Metamaterials, nanoFET.*

Napole Gene

See *Redement napole (RN) gene.*

N

Naringen

A glycosylated flavonoid (flavone) which is naturally present in oranges, grapefruit, and other citrus fruits. See also *Flavonoids, Glycosylation, Flavonols.*

Naringenin

See *Naringen.*

NARK Gene

A gene within the DNA of the soybean plant (*Glycine max* (L.) Merrill) which controls the growth of root nodules in which nitrogen fixation takes place. This gene was identified by Peter Gresshoff in 2002, and its acronym stands for Nodule Autoregulation Receptor Kinase.

When *Bradyrhizobium japonicum* bacteria (attracted to the vicinity of soybean plant's roots by the isoflavones those roots exude) are exposed to isoflavones, those bacteria's *nod* genes code for production of specific chemical compounds which then trigger *NARK* gene to cause the soybean plant roots to create/grow nodules which the bacteria subsequently move into and begin to fix nitrogen. See also *Gene, Deoxyribonucleic acid (DNA), Soybean plant, Nodulation, Nitrogen fixation, Bradyrhizobium japonicum, Isoflavones, Nod genes, Receptors.*

NAS

See *National Academy of Sciences.*

Natalizumab

A monoclonal antibody that was approved as a pharmaceutical by the U.S. FDA during 2006, for the treatment of relapsing forms of the disease multiple sclerosis. That pharmaceutical's trade name is TYSABRI™, and it is owned by the Biogen Idec company. See also *Monoclonal antibodies (MAb), Food and Drug Administration (FDA), Multiple sclerosis.*

National Academy of Sciences (NAS)

A private, self-perpetuating society of distinguished scholars in scientific and engineering research, dedicated to the advancement of science and technology and their use for the general welfare. Under the authority of its congressional charter of 1863, the NAS has a working mandate that calls upon it to advise the U.S. Federal Government on scientific and technical matters. See also *Vitamin E.*

National Cancer Institute (NCI)

One of the National Institutes of Health. See also *National Institute of Health (NIH).*

National Heart, Lung, and Blood Institute (NHLBI)

One of the National Institutes of Health. See also *National Institutes of Health (NIH).*

National Institute of Allergy and Infectious Diseases (NIAID)

The main agency of the National Institutes of Health. See also *National Institutes of Health (NIH).*

National Institute of General Medical Sciences (NIGMS)

One of the National Institutes of Health. See also *National Institutes of Health (NIH).*

National Institutes of Health (NIH)

The major U.S. Government sponsor of biotechnology research. It is composed of a group of government institutes that each focus on specific medical areas. See also *Recombinant DNA Advisory Committee (RAC).*

National Plant Protection Organization (NPPO)

Refers to the official service (agency) established by a nation's government to discharge the functions specified by the International Plant Protection Convention (IPPC). Examples of those functions include the enforcement of phytosanitary regulations (e.g., to prevent plant diseases being accidentally brought into a country). See also *International Plant Protection Convention (IPPC), SPS.*

Native Conformation

The normal, biologically active conformation (i.e., the three-dimensional arrangement of its atoms) of a protein molecule. See also *Conformation.*

Native Structure

See *Native conformation.*

Native Trait Recovery

Refers to the creation of crop plants (today) whose genome (DNA) contains a particular trait that had been present within the genome of the wild-type ancestor of that crop plant, but was subsequently lost during that crop plant's domestication process (e.g., 1000 years ago). For example, during 2014, Lijuan Qiu and Rongxia Guan discovered a salt tolerance gene present in the DNA of wild-type soybean plants that are the ancestors of today's domesticated soybean (*Glycine max* (L.) Merrill) varieties. Because today's domesticated soybean varieties do not possess that salt tolerance gene, a soybean breeder wanting to create a modern soybean variety that would grow well in salty soil could utilize a wide cross between a modern soybean variety (germplasm) and one of those salt-tolerant wild-type soybean species (i.e., retrieved from one of the seed banks utilized to store ancestral crop-plant relatives).

In addition to crop breeder use of a WIDE CROSS methodology, such crop "trait restoration" can be accomplished via certain other technologies. See also *Deoxyribonucleic acid (DNA), Gene, Genome, Trait, Soybean plant, Germplasm, Traditional reeding methods, Wide cross, Deletions, Trait recovery.*

nat-miRNAs

Acronym for Natural Antisense microRNAs. See *Antisense RNA, MicroRNAs.*

Naturaceuticals

See *Nutraceuticals.*

N

Natural Killer Cells

These cells of the innate immune system are involved in tumor surveillance. They also kill virus-laden cells. See also *Innate immune system, Innate immune response, Tumor, Virus, Cell.*

NCI

See *National Cancer Institute (NCI).*

N-Cofilin

See *Cell motility.*

ncRNA

Abbreviation for noncoding RNA. See also *Noncoding RNAs.*

ND

Acronym for nanodiamonds. See *Nanodiamonds.*

NDA (to FDA)

New Drug Application (to the U.S. FDA). A detailed application to the U.S. FDA seeking approval of a new drug that has undergone Phase 2 and Phase 3 clinical trials. An NDA is submitted in the form of (thousands of) pages of (clinical and other) data, along with various analyses (e.g., statistical) of that data for efficacy, safety, and so on. See also *CANDA, Food and Drug Administration (FDA), MAA, NDA (to koseisho), Phase I clinical testing.*

NDA (to Koseisho)

New drug application. It is the Japanese equivalent to a U.S. IND (investigational new drug) application; to the Koseisho, the Japanese equivalent of the U.S. FDA. See also *IND, Koseisho, Food and Drug Administration (FDA).*

Near-Infrared Spectroscopy (NIR)

Refers to analytical instruments which shine light (possessing wavelengths between that of visible light and infrared light spectrum) onto samples (e.g., kernels of grain) and measure the reflected or transmitted (near-infrared) light in order to quickly determine the amounts of protein, fat, moisture, lignans, and so on, present in the sample.

In certain samples, the near-infrared light causes cells or (specific molecules) to fluoresce (i.e., as light of very defined wavelength), which can be subsequently utilized for measurement/identification of compounds within the sample.

NIR is also being developed for use in:

- Quantifying (e.g., amounts that are present within the sample) of immunoassays.
- Predicting (e.g., the total amount of digestible energy or actual metabolizable energy in a sample of animal feed) from those quantified amounts.
- Detection of specific molecules (e.g., in DNA sequencing process).

See also *Protein, Fats, Lignans, Immunoassay, Fluorescence, Sequencing (of DNA molecules).*

Near-Infrared Transmission (NIT)

Refers to certain analytical instruments which shine light (possessing wavelengths between that of visible light and infrared spectrum) through samples (e.g., kernels of grain) in order to quickly determine the amounts of protein, fat, moisture, lignans, and so on, present in the sample. See also *Protein, Fats, Lignans, Near-infrared spectroscopy (NIR).*

Necrosis

Refers to cell death caused by physical injury to the cell (e.g., exposure to toxin, exposure to ultraviolet light, lack of oxygen, etc.). See also *Cell, Toxin, Respiration, Tumor necrosis factor.*

Neem Tree

A tropical tree (*Azadirachta indica*) found in India, Somalia, Mauritania, Australia, and other tropical countries that resists insect (e.g., whiteflies, mealybugs, aphids, mites) depredations and certain fungal diseases (e.g., rusts, powdery mildew, etc.) via secretions of liquids that contain Azadirachtin (an insect-repelling chemical). See also *Azadirachtin, Fungus, Crop biologicals.*

Negative Control

Refers to the "turning off" or decrease (i.e., down-regulation) of a given gene's transcription (i.e., coding for production of relevant protein) in an organism due to the binding (to cell's DNA) of negative regulatory elements. See also *Gene, Transcription, Cell, Knockdown, Knockout, Down regulating, Riboswitches, Methylated.*

Negative Supercoiling

Refers to the twisting of a duplex of DNA (deoxyribonucleic acid) in space in the opposite sense to the turns of the strands within the double helix molecule. See also *Deoxyribonucleic acid (DNA), Double helix, Supercoiling.*

Nematodes

Microscopic roundworms, which are the most abundant multicelled creatures on Earth. They are primarily found living in soil.

One nematode named *Caenorhabditis elegans* (*C. elegans*) is commonly used by scientists in genetics experiments; so a large base of knowledge about its genetics has been accumulated by the world's scientific community. For example, of the nearly 300 "disease-causing" genes in the human genome, more than half of them have an analogous gene within the genome of *C. elegans.*

Some nematodes are parasitic. Those nematodes that are parasitic to animals puncture the inner lining of the animal's gut and can impede the absorption of nutrients.

The root-knot nematode (*Meloidogyne incognita*) infests peanut, potato, cotton, and certain other crop plants. As a result of it feeding on plant roots, a knot-like feeding site forms on the host-crop roots that restricts the plant's access to nutrients normally supplied by those roots.

The soybean cyst nematode (*Heterodera glycines*), also referred to as SCN, lives in soil and feeds parasitically on roots of the soybean plant. These nematodes use a spear-like mouthpart, called a stylet, to puncture the plant's root cells so the nematodes can eat their cell contents. That root damage causes the soybean's growth to be stunted, and the plants turn yellow because of a reduction in nodule formation by the nitrogen-fixing *Rhizobium* bacteria (which normally colonize healthy roots of soybean plants). SCN damage can combine with a fungus (*Fusarium solani*) to cause a soybean plant disease known as "sudden death syndrome."

Another phytoparasitic nematode is *Criconemoides* sp., which feeds on the sugarcane plant (*Saccharum* sp.).

One Antarctic nematode (*Panagrolaimus davidi*) is able to survive Antarctic winters by drying out and achieving a state of "suspended animation" (anhydrobiosis) for as long as 39 years. See also *Cell, Caenorhabditis elegans (C. elegans), Genetics, Gene, Genome, Genetic map, Model organism, Soybean cyst nematodes (SCN), CystX, Phytoparasitic.*

NEMS

See *Nanoelectromechanical system (NEMS).*

Neoantigen

An antigen that is expressed by an organism's cell(s) after transformation (e.g., of cell) by an oncogenic virus. See also *Antigen, Organism, Cell, Virus, Oncogenes.*

Neoplasia

New growth. See *Neoplastic growth.*

Neoplasm

Refers to new uncontrolled growth of tissue (e.g., tumor). See also *Neoplastic growth.*

Neoplastic Growth

A new growth of animal or plant tissue resembling (more or less) the tissue from which it arises but having distinct biochemical differences from the parent cell. The neoplastic tissue is a mutant version of the original and appears to serve no physiologic function in the same sense as did the original tissue. It may be benign or malignant (i.e., a cancerous tumor). See also *Tumor, Cancer, Selective apoptotic antineoplastic drug (SAAND), Metastasis, Oncogenes.*

Neovasculogenesis

See *Adult stem cell.*

NER

See *Nucleotide excision repair.*

NERICA

Abbreviation for New Rices for Africa. Refers to new improved (e.g., drought resistant) rice varieties, developed via transgressive segregation, from *Oryza sativa* and *Oryza glaberrima*. See also *Rice, Transgressive segregants.*

Nerve Growth Factor (NGF)

A protein produced by the salivary glands (and also in tumors) that greatly increases growth/reproduction of nerve cells and guides the formation of neural networks. In the brain, NGF is thought to increase the production of the messenger chemical, acetylcholine, by protecting and stimulating those neurons that produce acetylcholine. Because those (acetylcholine-producing) neurons are typically the first to be destroyed in an Alzheimer's disease victim, NGF holds potential to be used to counteract (some of) the effects of the disease. NGF is also necessary for normal development of the hypothalamus, a brain structure that regulates a number of hormones. Human T cells appear to have receptors for NGF, which could explain the "mind–body connection" between a person's emotional well-being and physical health (i.e., NGF may be a go-between for the brain and the immune system). NGF was discovered by Rita Levi-Montalcini in 1954. See also *Growth factor, Epidermal growth factor (EGF), Hypothalamus, Hormone, Protein, Alzheimer's disease.*

Nested Association Mapping

Refers to mapping of units (e.g., individual gene) within a given organism's genome (DNA). See also *Deoxyribonucleic acid (DNA), Genome, Mapping.*

Nested PCR

Refers to a specific polymerase chain reaction (PCR) technique of two consecutive-run PCRs, in which the second PCR amplifies (i.e., makes multiple copies of) a DNA sequence within the product (amplicon) of the first PCR. See also *Polymerase chain reaction (PCR), Polymerase chain reaction (PCR) technique, Sequence (of a DNA molecule), Deoxyribonucleic acid (DNA), Amplicon.*

NETs

Acronym for **neutrophil extracellular traps**. See *Neutrophil extracellular trap.*

Neu5Gc

A gene that is present within the DNA of most animals but not present in human DNA. The Neu5Gc gene controls production/expression of sialic acid (e.g., on surfaces of some cells in the organism). See also *Gene, Deoxyribonucleic acid (DNA), Sialic acid.*

Neuraminidase (NA)

A viral surface (i.e., through the capsid) glycoprotein enzyme that is present in the external surface (capsid) of the influenza virus. See also *Enzyme, Protein, Glycoprotein, Virus, Viral surface proteins.*

Neuron

Cells of the body's nervous system, which transmit nerve impulses (nerve impulses are electrical signals conducted by the flow of ions across the plasma membrane of neuron cells). Neurons are involved in controlling movement (i.e., known as motor control), emotions, and memory.

There are approximately 100 billion neurons in the typical human brain. The nerve impulses within them move at a speed of approximately 400 km per hour (300 miles per hour).

During 1992, Elizabeth Gould proved that adult mammal brains produce new neuron cells, particularly within the hippocampus portion of the brain. See also *Neurotransmitter, Acetylcholine,*

Serotonin, Cell, Parkinson's disease, Dopamine, Plasma membrane, Ion, Dendrites (in brain), Astrocytes.

Neurotransmitter

An organic, low molecular weight compound that is secreted from the (axon) terminal end of a neuron (in response to the arrival of an electrical impulse) into a liquid-filled gap that exists between neurons. The transmitter molecule then diffuses across the small gap and attaches to the next neuron. This attachment causes structural changes in the membrane of the neuron and initiates the conductance of an electrical impulse. In this way, an electrical impulse is transmitted (via this "cascade") along a neuron network of which the neurons themselves do not physically touch. A neurotransmitter serves to transmit a nerve impulse between different neurons.

Examples of neurotransmitters include dopamine, norepinephrine, and so on. A shortage of dopamine in the brain causes the disease known as Parkinson's disease. See also *Molecular weight, Neuron, Serotonin, Acetylcholine, Dopamine, Parkinson's disease, Cascade, Dendrites (in brain), Gated channel.*

Neutraceuticals

See *Nutraceuticals.*

Neutriceuticals

See *Nutraceuticals.*

Neutrophil Extracellular Trap

A sort of mesh that is formed (e.g., in blood vessels) via the pathogen-triggered "programmed death" of immune system neutrophils. Chromatin and some protein molecules from the dead neutrophils adhere to each other, thereby forming the mesh (which entraps pathogens such as bacteria, for other immune system cells to kill). See also *Neutrophils, Pathogen, Immune response, Chromatin, Protein, Bacteria.*

Neutrophils

Phagocytic (ingesting, scavenging) white blood cells that are produced in the bone marrow. The average human has approximately 25 billion neutrophils circulating in their body at any given point in time. Those neutrophils ingest and destroy invading microorganisms and facilitate postinfection tissue repair.

Upon ingestion of pathogens, the neutrophil generates reactive oxygen species such as O_2^- (also known as free radicals). That O_2^- causes an influx of potassium ions into the portion of the neutrophil that contains the pathogen, which thereby releases proteases from existing granules ("storage" sites) in the neutrophil. The proteases then kill the pathogen.

In addition to generating proteases, neutrophils can secrete collagenase and plasminogen activator. They are the immune system's "first line" of defense against invading pathogens, and large reserves are called forth within hours of the start of a "pathogen invasion."

When cells of the body become infected with cytomegalovirus (CMV), those infected cells produce and emit the protein IL-22, which results in neutrophils homing in on those infected cells. At the CMV-infected cells, the neutrophils produce a protein known as TRAIL that directly kills the CMV-infected cells.

See also *Pathogen, Collagenase, Microorganism, Free radical, Ion, Protease, NF-κB, Phagocytosis, Innate immune system, Neutrophil extracellular trap, Cytomegalovirus (CMV).*

New Drug Application

See *NDA (to koseisho), NDA (to FDA), MAA, IND, CANDA.*

N-Extein

See *Intein.*

NF-kappaB

See *NFκB.*

NF-κB

Abbreviation for nuclear factor-κB, NF-kappaB, nuclear factor kappa B, or nuclear factor kappa-light-chain-enhancer of activated B cells; a "family" of nuclear transcription factors that constitute the "master regulator" of the human innate immune system and that help cells regulate the expression of genes which:

- Induce production of neutrophils to fight an infection.
- Induce inflammation (sometimes leading to onset of autoimmune disease, if NFκB proteins are overexpressed).
- Sometimes induce tumorigenesis. However, research indicates that genistein causes down-regulation of this (e.g., thereby helping prevent bladder cancer that would otherwise occur).
- Sometimes prevent apoptosis. Such apoptosis prevention can be used by some cancerous cells to avoid those of the body's defenses that combat cancer via inducing apoptosis. Some peptides activate the NFκB signaling pathway (via binding to Toll-like receptor 5) to prevent apoptosis normally caused by exposure to ionizing radiation.
- Sometimes induce viral replication within cells.
- Help in regulating key mammalian cell processes, including cell proliferation, immune and stress responses.

When activated by applicable signaling molecules (due to infection, radiation, etc.), NF-kB enters the cell's nucleus and binds to DNA to control gene expression. When inactive, it is sequestered in the cytoplasm, away from the DNA.

Too-high levels of NF-kappaB in the body can also result in muscle wastage (atrophy) and/or bone resorption (e.g., some components of bones being reabsorbed by the body, thereby weakening those bones). In those cases, consumption of eicosapentanoic acid helps to reduce bone resorption. See also *Transcription factors, Innate immune system, Chronic inflammation, Nuclear DNA, Deoxyribonucleic acid (DNA), Gene, Cell, Nucleus, Cytoplasm, Gene expression, Transcription, Express, Gene expression cascade, Neutrophils, Autoimmune disease, Tumor, Cancer, Apoptosis, Virus, Adiponectin, Down-regulating, Eicosapentaenoic acid (EPA), Genistein (gen), T cell receptors, Toll-like receptors.*

NFκB Pathway

More properly termed the NFκB-mediated signal transduction pathway, it refers to a pathway within the body that is signaled (turned on) by one of the NFκB (nuclear factor-κB, NF-kappaB, or nuclear factor kappa B) transcription factors. Such a pathway is involved (correctly) in standard immune system responses, cell growth, inflammation, and autoimmune response. When functioning incorrectly, the NF-κB pathway can be involved in diseases such as cancer and diabetes. See also *NFκB pathway, Transcription factors, Immune response, Cell, Autoimmune disease, Cancer, Diabetes.*

NHEJ

Acronym for nonhomologous end-joining. See *Nonhomologous end-joining.*

NIAID

See *National Institute of Allergy and Infectious Diseases.*

Nick

A break in one strand of a double-stranded DNA molecule. One of the phosphodiester bonds between two adjacent nucleotides is ruptured. No bases are removed from the strand; it is just opened at that point. See also *Deoxyribonucleic acid (DNA).*

Nicotine-Adenine Dinucleotide (NAD)

See *NAD.*

Nicotine-Adenine Dinucleotide Phosphate (NADP)

See *NAD.*

Nicotine-Adenine Dinucleotide Phosphate, Reduced (NADPH)

See *NAD.*

Nicotine-Adenine Dinucleotide, Reduced (NADH)

See *NAD.*

NIH

See *National Institutes of Health (NIH).*

NIHRAC

See *Recombinant DNA Advisory Committee (RAC).*

Ninhydrin Reaction

A color reaction given by amino acids and peptides on heating with the chemical ninhydrin. The technique is widely used for the detection and quantitation (measurement) of amino acids and peptides. The concentration of amino acid in a solution (of hydrochloric acid) is proportional to the optical absorbance of the solution after heating it with ninhydrin. α-Amino acids give an intense blue color, and amino acids (such as proline) give a yellow color. One is able to determine concentration of a protein or peptide and also obtain an idea of the type of protein or peptide that is present. See also *Absorbance (A), Amino acid, Peptide.*

NisC Enzyme

See *Nisin.*

Nisin

A powerful antibacterial peptide, first isolated in 1944 from *Lactococcus lactis* bacteria. Because nisin kills the human pathogenic *Salmonella* and *Clostridium* bacteria, it has been utilized by man as a food preservative for cheese and other dairy products for four decades.

The nisin molecule is made via posttranslational modification of the relevant peptide molecule, during which the *L. lactis* bacteria utilizes the NisC enzyme to attach five thioether rings (i.e., molecular substructure) to the peptide.

The nisin molecule acts via binding tightly to Lipid II, which is a precursor molecule utilized by at least some bacteria to build/repair their cytoplasmic membrane (outer cell wall). Because the nisin prevents such cell wall building/repair in those pathogenic bacteria, their cytoplasmic membrane weakens and becomes more permeable (i.e., "leaky"), and they die. See also *Peptide, Cell, Bacteria, Antibiotic, Pathogen, Pathogenic, Salmonella, Clostridium, Enzyme, Posttranslational modification of protein, Lipids, Lipid bilayer, Cytoplasmic membrane.*

Nitrate Bacteria

See *Nitrates, Nitrites, Bacteria, Pink pigmented facultative methylotroph (PPFM).*

Nitrate Reduction

The reduction of nitrate to nitrite or ammonia by an organism. See also *Nitrates, Reduction (in a chemical reaction), Nitrites.*

Nitrates

Refers to nitrogen compounds that exist in a chemical form which plant roots are able to take in (i.e., utilized by the plant to make nitrogen-containing molecules such as proteins). Nitrates are produced from nitrogen:

- Taken out of the atmosphere by nitrogen-fixing bacteria (living among the roots of legume plants such as the soybean, etc.).
- Taken out of nitrites (in soil) by nitrate bacteria.
- Taken out of the atmosphere by blue-green algae.

See also *Protein, Nitrogen fixation, Soybean plant, Nitrites, NUE gene, Pink pigmented facultative methylotroph (PPFM).*

Nitric Oxide

Abbreviated NO, it is a molecule produced in the body of an organism (including plants), which can act as:

- A signaling molecule (e.g., it signals to cause a firefly's tail to begin the chemical reaction of luciferin with luciferase that results in the light emission known as bioluminescence).
- Fostering the formation of new blood vessels (a process known as **angiogenesis**) in an animal's body.

N

- Fostering the body's wound healing process (especially in regeneration of liver cells after liver damage).
- An oxidant (reactive nitrogen species) utilized against pathogens by the immune system.
- An antimicrobial compound which inhibits the growth of certain bacterial pathogens (e.g., *Salmonella* spp.) by rendering them unable to produce two essential amino acids lysine and methionine.
- An instigator of (destructive) free radicals, within the body.
- As an inducer of genes (e.g., in soybean plants) that cause production of certain chemical compounds which protect the organism (e.g., soybean plant) from bacterial diseases.

As a signaling molecule, or "messenger molecule," nitric oxide is utilized by the human body for control of blood pressure (i.e., when the endothelial cells that line blood vessels produce NO that causes neighboring smooth-muscle cells to relax so entire blood vessel dilates, thereby lowering blood pressure and protecting the lining from macrophage adhesion/initiation of plaque deposit). Nitric oxide can bind to the Heme-Nitric Oxide/Oxygen (H-NOX) domain on a particular specific enzyme, activating that enzyme and beginning the chemical cascades that lead to physiological functions such as blood vessel dilation. When needed for such purposes, nitric oxide is released from "storage" in nitrites and nitrosothiols (created earlier via exercise).

Nitric oxide is also utilized by the human body for immune system regulation, and its synthesis in macrophages is required for macrophages to kill pathogens and tumor cells (by oxidizing them after the macrophage has engulfed them).

During the 1980s, John Garthwaite and Solomon H. Snyder showed that nitric oxide is an important messenger molecule utilized in neural signaling (i.e., NO is an important signaling molecule in the human brain). Nitric oxide levels tend to be low in people who have Alzheimer's disease.

Nitric oxide increases the effectiveness of certain other reactive free radicals (e.g., superoxide O_2^-) in killing off any infected cells within a soybean plant.

Nitric oxide also induces certain genes to code for the production of certain chemical compounds which protect the soybean plant and some other plants from bacterial plant diseases.

See also *Signaling molecule, Signaling, Oxidizing agent, Pathogen, Immune response, Human superoxide dismutase (hSOD), Signal transduction, Nitric oxide synthase, Nitrosylation, Soybean plant, Protein, Domain (of a protein), Enzyme, Cascade, Inducers, Gene, Coding sequence, Free radical, Endothelial cells, Endothelium, Macrophage, Adipose, Bacteria, Essential amino acids, Lysine, Tumor, Methionine, Neurotransmitter, Bioluminescence, Angiogenesis, Chemotaxis, Alzheimer's disease.*

Nitric Oxide Synthase

An enzyme that catalyzes the reaction which the body (of animals or plants) utilizes to make nitric oxide from L-arginine (via cleavage, of that molecule). The cofactor for that reaction is nicotine-adenine dinucleotide phosphate (NADP). See also *Endothelial nitric oxide synthase (eNOS), Enzyme, Nitric oxide, Cofactor, NAD (NADH, NADP, NADPH), Arginine (arg), Levorotary* (L) *isomer, Hydrolytic cleavage, Endothelial cells, Endothelium, Macrophage.*

Nitrification

The oxidation of ammonia (e.g., from ammonia-containing substances such as liquid wastes excreted by animals, decomposed animals and plants, etc.) to nitrates by a microorganism. See also *Nitrates, Nitrites, Oxidation (chemical reaction), Endophyte.*

Nitrifying Bacteria

See *Nitrites.*

Nitrilase

An enzyme that catalyzes the degradation (i.e., breaking down) of bromoxynil (an active ingredient in some herbicides).

Nitrilase is naturally produced in the soil bacteria *Klebsiella pneumoniae* subsp. *Ozaenae*. If a gene (called BXN) that codes for the production of nitrilase is inserted via genetic engineering into crop plants, the resultant plant production of nitrilase would enable such plants to survive postemergence applications of bromoxynil-containing herbicides. See also *Enzyme, Bacteria, Bromoxynil, Gene, Coding sequence, Genetic engineering.*

Nitrites

Refers to certain nitrogen compounds that exist in a chemical form which plant roots are unable to take in. After conversion to **nitrates** via internal respiration by nitrate bacteria (in soil), the nitrates can be taken in by plant roots (i.e., utilized by the plant to make nitrogen-containing molecules such as proteins).

Nitrites are made (via internal respiration) by nitrifying bacteria and endophytes (e.g., in soil) from ammonia-containing substances (e.g., liquid wastes excreted by animals, decomposed animals and plants, etc.). See also *Nitrates, Protein, Respiration, Endophytes.*

Nitrogen Cycle

The cycling of various forms of biologically available nitrogen through the plant, animal, and microbial worlds (kingdoms) as well as the atmosphere and geosphere. See also *Nitrates, Nitrites, Nitrification, Denitrification, Nitrogen fixation, Endophytes.*

Nitrogen Fixation

Conversion of atmospheric nitrogen (N_2) into ammonium ion (NH_4^+); a soluble, biologically available form (nitrate) that plants can utilize to synthesize ("manufacture") amino acids/proteins and other nitrogen-containing compounds, so those plants can grow faster and yield more.

First explained during the 1880s by Mikhail Voronin and Hermann Hellriegel, the conversion is carried out by nitrogen-fixing organisms such as:

- *Frankia* bacteria which live symbiotically in the roots of actinorhizal plants (a category of woody plant species that includes cherry trees, beech trees, alder, bayberry, sweet fern, etc.).
- *Rhizobium* bacteria which live symbiotically in the roots of legume plants, for example, alfalfa, soybeans, peanuts, and so on.

- Certain species of *Burkholderia* bacteria which live symbiotically in the roots of legume plants, for example, alfalfa, soybeans, peanuts, and so on.
- *Azospirillum brasilense.*
- *Herbaspirillum seropedicae.*

When not enough nitrogen fixation occurs (when only nonlegume plants are grown), soil is not able to produce maximum crop yields and farmers may need to spread fixed nitrogen onto the field in the form of the fertilizer anhydrous ammonia, ammonium nitrate, or sodium nitrate. See also *Nitrates, Symbiotic, Genistein (gen), Bacteria, Soybean plant, Nitrogenase system, Nitrogen cycle, Isoflavones, Crop biologicals, Hemagglutinin (HA), Nodulation, Rhizobium (bacteria), Bradyrhizobium japonicum, Amino acid, Protein, Ion.*

Nitrogen Metabolism

See *Glutamate dehydrogenase.*

Nitrogen Use Efficiency (NUE) Gene

Refers to a gene that (e.g., when inserted into a plant's DNA or inserted into the DNA of a crop biological that is applied to that plant) increases the efficiency with which that plant utilizes nitrates in its growth processes. See also *Gene, Deoxyribonucleic acid (DNA), Nitrates, Crop biologicals.*

Nitrogenase System

A system of enzymes capable of reducing atmospheric nitrogen to ammonium ion (NH_4^+) in the presence of ATP. That (i.e., ammonium ion) is a soluble form which plants can utilize. See also *Reduction (in a chemical reaction), Enzyme, Nitrogen fixation, Ion.*

Nitrosylation

Refers to the chemical attachment of a nitric oxide molecule to a cysteine amino acid group within a protein molecule. This nitrosylation results in the biological activity of the protein molecule (within the body) being changed (e.g., resulting in greatly improved liver cell regeneration following liver damage). See also *Nitric oxide, Biological activity, Protein, Amino acid, Cell.*

NMR

Acronym for Nuclear Magnetic Resonance. See *Nuclear magnetic resonance.*

NO

See *Nitric oxide.*

Nod Box

See *Nod genes.*

Nod Gene

See *Nodulation.*

Nod Genes

Refers to a category of genes present within the DNA of certain soil-dwelling *Rhizobium* bacteria. When those bacteria are in the presence of specific "signaling molecules" (e.g., isoflavones produced by roots of soybean plant or luteolin produced by roots of alfalfa plant), nod genes code for the production (by those bacteria) of specific chemical compounds which then trigger relevant plant genes (e.g., NARK gene in soybean plant) to cause the plant roots to create/grow nodules (which the bacteria subsequently move into and begin to "fix" nitrogen). See also *Gene, Deoxyribonucleic acid (DNA), Rhizobium (bacteria), Soybean plant, Nodulation, Isoflavones, NARK gene, Nitrogen fixation.*

Nodulation

The process in which certain strains of soil-dwelling bacteria (e.g., *Rhizobium* bacteria) colonize the roots of specific plants (i.e., the legumes) such as soybean (*Glycine max* L.) or alfalfa. As part of that process:

- The *Rhizobium* bacteria are attracted to the vicinity of the plant's roots. For the soybean plant (*Glycine max* L.), that is accomplished by the soybean plant synthesizing the signaling molecules known as isoflavones, which attract *Bradyrhizobium japonicum* bacteria. For the alfalfa plant, that is accomplished by the alfalfa plant synthesizing luteolin molecules, which attract *Sinorhizobium meliloti* bacteria.
- Certain genes (called *nod*) within the relevant *Rhizobium* bacteria are expressed (resulting in the synthesis of specific chemical compounds).
- When the plant roots detect those chemical compounds, certain genes (called *NARK*) within those roots are expressed (resulting in the formation of nodules on those roots).
- The relevant *Rhizobium* bacteria move in and inhabit those plant root nodules, where the bacteria then "fix" nitrogen from the atmosphere, which converts that nitrogen into a chemical form (i.e., nitrates) that is available for use by plants (as fertilizer/plant food).

See also *Rhizobium (bacteria), Chemotaxis, Soybean plant, Bradyrhizobium japonicum, Isoflavones, Genistein (gen), Transcription factors, Gene, Gene expression, Signaling molecule, Nod genes, NARK gene, Nitrogen fixation, Symbiotic, Crop biologicals, Hemagglutinin (HA).*

Nodulin 26-Like Intrinsic Protein

A plant cell aquaporin that can transport arsenic (dissolved in water) into or out of plant cells. See also *Aquaporins.*

Noncoding RNAs

See *Junk DNA, Ribonucleic acid (RNA), Long noncoding RNAs, MicroRNAs, Transcriptome.*

Nonessential Amino Acids

Amino acids of proteins that can be made (biochemically synthesized within the body) by humans and certain other vertebrate

N

animals from simple chemical precursors (in contrast to the essential amino acids). These amino acids are thus not required in the diet (of humans and those other vertebrates). See also *Essential amino acids, Amino acid, Protein.*

Nonheme-Iron Proteins

Proteins containing iron but no porphyrin groups (within which iron atoms are held) in their structure. See also *Heme.*

Nonhomologous End-Joining

Abbreviated NHEJ, it is a DNA repair mechanism (of double-strand breaks in DNA) that often omits one or more nucleotides from the DNA strand ends being joined. NHEJ works especially well for causing gene knockouts because it leads to frameshift. See also *DNA repair, Deoxyribonucleic acid (DNA), Double double-strand breaks (in DNA), Nucleotide, Gene, Knockout, Frameshift.*

Non-O157 STEC

See *Non-O157:H7 shiga-toxin producing E. coli.*

Non-O157:H7 Shiga-Toxin Producing *E. coli*

Refers to the serotypes of *Escherichia coliform* bacteria (additional to serotype O157:H7) which produce shiga toxins. See *Escherichia coliform (E. coli), Bacteria.*

Nonpolar Group

A hydrophobic ("water hating") group on a molecule; usually hydrocarbon (composed of hydrogen and carbon atoms) in nature. These groups are more at home in a nonpolar (oil-like) environment. See also *Polar group, Amphipathic molecules, Amphoteric compound.*

Nonsense Codon

A triplet of nucleotides that does not code for an amino acid. Any one of three triplets (U–A–G, U–A–A, or U–G–A) that cause termination of protein synthesis (in ribosome), and thus the release from ribosome of a (completely translated) protein molecule.

U–A–G is known as amber and U–A–A is known as ochre. See also *Genetic code, Codon, Termination codon (sequence), Translation, Ribosomes, Protein.*

Nonsense Mutation

A mutation that converts a codon that specifies an amino acid into one that does not specify any amino acid. A change in the nucleotide sequence of a codon that may result in the termination of a polypeptide chain (i.e., a stop codon).

Some nonsense mutations are harmful to the organism. For example, a nonsense mutation within the CWC15 gene (in the Jersey breed of cattle) results in spontaneous abortion of pregnant cows. See also *Nonsense codon, Gene, Genetic code, Mutation, Codon, Stop codon.*

Nonstarch Polysaccharides

Term—abbreviated NSP—that is utilized to refer to polysaccharide molecules (in plant seeds) other than starch. These include arabinoxylans, pectins, beta glucans, and alpha galactosides (e.g., raffinose, stachyose, verbascose). See also *Polysaccharides, Stachyose.*

Nontranscribed Spacer

A region between transcription units in a tandem gene cluster. See also *Transcription, Messenger RNA (mRNA), Genetic code, Gene splicing, Gene.*

North American Plant Protection Organization (NAPPO)

One of the international SPS standard-setting organizations that develops plant health standards, guidelines, and recommendations (e.g., to prevent transfer of a disease from one country to another). Subsidiary to the IPPC, it covers the countries of North America. Its secretariat is located in Nepean, Canada. See also *International Plant Protection Convention (IPPC), European Plant Protection Organization (EPPO), Southern Cone Plant Protection Organization (COSAVE), SPS.*

Northern Blotting

A research test/methodology used to transfer RNA fragments from an agarose gel (e.g., following gel electrophoresis) to a filter paper without changing the relative positions of the RNA fragments (e.g., reelectrophoresis separation grid).

Those RNA fragments are then identified via DNA–RNA hybridization (to a known DNA fragment). See also *Ribonucleic acid (RNA), Deoxyribonucleic acid (DNA), Gel electrophoresis, Agarose, Chromatography, Field inversion gel electrophoresis, Hybridization (molecular genetics).*

Northern Corn Rootworm

Latin name *Diabrotica barberi.* See *Corn rootworm.*

NOS Terminator

A termination codon (sequence of DNA) that is frequently utilized in genetic engineering of plants to "terminate" expression of the inserted gene (i.e., to halt synthesis of desired protein in the plant, after the desired protein synthesis has occurred). The NOS acronym stands for *nopaline synthase.*

The NOS terminator was originally extracted from the bacteria species *Agrobacterium tumefaciens.* See also *Termination codon (terminator sequence), Sequence (of a DNA molecule), Deoxyribonucleic acid (DNA), Genetic engineering, Express, Gene, Protein, Synthesizing (of protein molecule), Agrobacterium tumefaciens, Bacteria, Control sequences.*

No-Tillage Crop Production

A methodology of crop production in which the farmer utilizes virtually no mechanical cultivation (i.e., only one pass over the field, with a planter; instead of the conventional four passes per year with mechanical cultivator equipment plus one pass with planter, used for traditional crop production). This reduction in field soil disturbance leaves more carbon in the soil (thereby reducing "greenhouse gases" in the atmosphere), leaves more earthworms (e.g., *Aporrectodea caliginosa, Eisenia foetida,* etc.) per cubic foot or per cubic meter living in the topsoil, and reduces soil compaction (i.e., the reduction of interstitial spaces between individual soil particles), thereby increasing the fertility of such "no till" farm fields.

The plant residue remaining on the field's surface helps to control weeds and reduce soil erosion (by 90%–95% versus traditional mechanical tillage); it also provides sites for insects to shelter and reproduce, leading to a need for increased insect control via methods such as inserting a *Bacillus thuringiensis* (*B.t.*) gene into certain crop plants or utilizing integrated pest management. But, if a farmer needs to apply synthetic chemical pesticides, the plant residue remaining of field's surface helps to cause breakdown (into substances such as carbon dioxide and water) of pesticides. That is because that plant residue helps to retain moisture in the field-surface environment, thereby enhancing growth of the microorganisms that help breakdown pesticides.

Use of No-tillage Crop Production (methodology) helps farmers to reduce the incidence of certain plant diseases such as *white mold disease*. See also *Integrated Pest Management (IPM), Corn, Glomalin, Soybean plant, Bacillus thuringiensis (B.t.), Gene, Genetic engineering, European corn borer (ECB), Helicoverpa zea (H. zea), Corn rootworm, Cold hardening, Microorganism, Low-tillage crop production, Earthworms, White mold disease, Drought tolerance.*

Novel Food (European Union Definition)

Refers to a food or food ingredient that does not have a significant history of human consumption prior to May 15, 1997 within the European Union.

NP

Acronym for **nanoparticles**. See *Nanoparticles.*

NPC

Acronym for **nuclear pore complexes**. See *Nuclear pore complexes.*

NPTII

See *NPTII gene.*

NPTII Gene

A marker gene that codes for (i.e., "causes manufacture of") the enzyme neomycin phosphotransferase II, which can inactivate the antibiotic kanamycin.

The NPTII gene is commonly utilized as a "marker gene" for genetically engineered plants. Neomycin phosphotransferase confers kanamycin resistance to cells expressing it (i.e., cells that contain the NPTII gene in addition to the other gene(s) inserted along with it), so those (engineered) cells will live in a laboratory vessel containing kanamycin. See also *Gene, Marker (genetic marker), Coding sequence, Enzyme, Cell, Genetic engineering.*

NSCLC

An acronym for non small-cell lung cancer. See *Biomarkers.*

NSP

See *Nonstarch polysaccharides.*

NT

An acronym for Nuclear Transfer. See *Nuclear transfer.*

nt

An abbreviation for nucleotide. See *Nucleotide.*

NTR

Acronym for nuclear transport receptor. See *Nuclear pore complexes.*

Nuclear Actin

Also known as G-actin, it is the form of actin which interacts with genes within the nucleus of the cell. See also *Actin, Cell, Nucleus.*

Nuclear DNA

The DNA that is contained within the nucleus of a cell. See also *Deoxyribonucleic acid (DNA), Cell, Genome, Nucleus, Nuclear transfer.*

Nuclear Envelope

Refers to the membrane that surrounds a cell's nucleus. The nuclear envelope contains thousands of nuclear pore complexes (NPCs), which selectively allow passage of water, certain ions, messenger RNA, certain protein molecules, metabolites, and so on. See also *Cell, Nucleus, Nuclear pore complexes, Membranes (of a cell), Messenger RNA (mRNA), Ion, Metabolite.*

Nuclear Factor Kappa B

See *NFκB.*

Nuclear Hormone Receptors

Refers to receptors in a cell's outer membrane that serve to convey the "signal" received (when certain hormones and vitamins latch onto those receptors) all the way into the cell's nucleus. Within the cell's nucleus, they serve as transcription activators/factors that regulate the expression of certain specific genes. See also *Receptors, Nuclear receptors, Cell, Plasma membrane, Nucleus, Hormone, Vitamin, Signaling, Signal transduction, G-proteins, Transcription, Transcription factors, Transcription activators, Gene, Express, Expressivity.*

Nuclear Localization Signal

See *Nuclear proteins.*

Nuclear Magnetic Resonance

Acronym NMR, it is a spectrometry tool/methodology that can be utilized by scientists to determine several fundamental properties of complex/large biomolecules. NMR machines send a very specifically shaped pulse of radiofrequency energy, at the precise resonance frequency (known as Larmor frequency) needed to "pump" (i.e., add) energy to the precessional motion ("rotation") of the atoms' nuclei within the sample (biomolecule) being examined, within a magnetic field.

Those pulses cause the relevant atoms' nuclear magnetic moments to either turn in space 90° or to turn 180°. The former causes the sample's atomic nuclei to send out a signal known as a free induction decay; the latter causes the sample's atomic nuclei to send out detectable spin echos, which are detected by the NMR machine and turned into useful data. See also *Quantitative structure-activity relationship (QSAR), Phenomics.*

Nuclear Matrix Proteins

Protein molecules that are present in cancerous cells but not in normal (nonmutated) cells. See also *Protein, Cell, Mutation, Mutant, Myristoylation, Neoplastic growth, PARP.*

Nuclear Pore Complexes

Refers to the thousands of "gateways" present within the nuclear envelope (membrane that surrounds a cell's nucleus), which selectively allow passage of water, nucleic acids, certain ions, messenger RNA, metabolites, certain protein molecules, and so on.

However, applicable molecules larger than approximately 30 kDa in size can only pass through a NPC when they are first intimately linked to nuclear transport receptor (NTR) protein molecules. Such NTRs, bearing their (>30 kDa) payloads, pass through a given NPC at a rate of up to 1000 per second. To do so, the NTRs interact with specific portions of the nuclear pore proteins (also called nucleoporins) which are present within the NPCs. Those specific portions of the relevant nucleoporin molecules consist of alternating hydrophobic and hydrophilic segments of the protein molecule. Those alternating segments on the nucleoporins anchored within the NPC help to quickly "pull" a payload-carrying NTR through the NPC into the nucleus of the cell (where the NTR then releases its payload). See also *Nuclear envelope, Cell, Nucleus, Ion, Messenger RNA (mRNA), Metabolite, Protein, Sequence (of a protein molecule), Kilodalton (kDa), Hydrophobic, Hydrophilic, Nucleic acids.*

Nuclear Pore Protein

See *Nuclear pore complexes.*

Nuclear Proteins

Refers to a class of protein molecules that are synthesized (i.e., made) in the cytoplasm of eucaryotic cells but are ultimately utilized inside the cell's nucleus. Each nuclear protein molecule gets to the **specific location** where it is needed (within the cell nucleus) via a **nuclear localization signal** (i.e., akin to the postal address written on an envelope) that is present within the amino acid sequence of that protein molecule.

Those nuclear localization signals are recognized (i.e., fit into open "keyholes" of) by the family of receptor molecules (embedded in the nuclear envelope) known as importins/karyopherins. When the applicable nuclear protein molecule inserts its "key" (nuclear localization signal) into the "keyhole" of the relevant importin/karyopherin receptor, that nuclear protein is carried across the nuclear envelope to the specific address inside the cell nucleus where it needs to go. See also *Protein, Eucaryote, Cell, Cytoplasm, Nucleus, Ribosomes, Amino acid, Sequence (of a protein molecule), Nuclear envelope, Receptors.*

Nuclear Receptors

Receptors in a cell's outer membrane that serve to convey a "signal" from outside the cell all the way into the cell's nucleus. They function as ligand-activated transcription factors which thus serve within cells as transcription activators/factors that regulate the expression of certain specific genes (i.e., affecting cell/life processes such as fatty acid metabolism, reproduction, and general development of the organism).

Each nuclear receptor contains two zinc finger proteins, which target hormone response elements (i.e., specific DNA sequences that initiate the activation of applicable gene(s) for that particular receptor). See also *Receptors, Signaling, Signal transduction, Nucleus, G-proteins, Orphan receptors, Retinoid X receptors (RXR), Endocytosis, Vaginosis, CD4 protein, Protein, Organism, Cell, Gene, Express, Gene expression, Zinc finger proteins, Transcription, Transcription factors, Transcription activators, Fatty acid, Metabolism, Ligand (in biochemistry), Polyunsaturated fatty acids (PUFA), Membranes (of a cell), Plasma membrane, Lipid rafts, Farnesoid X receptors (FXR), Liver X receptors (LXR), PPAR, Vitamin.*

Nuclear Transfer

A method of cloning a living organism, in which that organism's entire genetic information is conveyed via transfer of an (adult) cell nucleus into an unfertilized egg (from another animal of the same species) whose nucleus had previously been removed. Discovered by John Gurdon in 1962, this was also the method utilized to produce "Dolly," the first cloned sheep, in 1996.

It is possible to also delete or substitute genes (e.g., brought in from another species) as part of the nuclear transfer process, so nuclear transfer can be utilized to produce transgenic organisms or "knock out" organisms. See also *Clone (an organism), Cell, Nucleus, Genome, Nuclear DNA, Deoxyribonucleic acid (DNA), Gene, Reprogramming, Species, Transgenic (organism), Knock out (gene), Genetic engineering.*

Nuclear Transport Receptors

See *Nuclear pore complexes.*

Nuclease

An enzyme capable of hydrolyzing (cutting) the internucleotide linkages of a nucleic acid (e.g., DNA or RNA). Nucleases present in cells tend to degrade (i.e., hydrolyze, cleave) DNA strands inserted by man, making genetic targeting more difficult. See also *Genetic targeting, Hydrolysis deoxyribonucleic acid (DNA), Ribonucleic acid (RNA), Antisense (DNA sequence), Telomeres, CRISPR/CAS9 gene-editing systems.*

Nucleic Acid Probes

See *DNA probe, Nucleic acids, Polymerase chain reaction (PCR), Rapid microbial detection (RMD).*

Nucleic Acids

A nucleotide polymer. A large, chain-like molecule containing phosphate groups, sugar groups, and purine and pyrimidine bases; two types are ribonucleic acid (RNA) and deoxyribonucleic acid (DNA).

The bases involved are adenine, guanine, cytosine, and thymine (uracil in RNA).

In living cells, nucleic acids are either the specific (genetic) informational molecule (i.e., DNA), or act as agent (i.e., RNA) in causing that information to be expressed (e.g., as a protein). In 1869, Friedrich Miescher discovered nucleic acids in cells. See also *Nucleotide, Cell, Polymer, Informational molecules, Gene, Genetic code, Deoxyribonucleic acid (DNA), Ribonucleic acid (RNA), Express, Extension (in nucleic acids).*

Nucleoid

The compact body that contains the genome in a bacterium. See also *Genome.*

Nucleolar Dominance

An epigenetic process that occurs in some hybrids, in which the ribosomal genes of one parent are silenced. See also *Epigenetic, Epigenetic variation, Hybridization (plant genetics), Gene, Ribosomes, Ribosomal RNA.*

Nucleolus

A round, granular structure situated in the nucleus of eucaryotic cells. It is involved in rRNA (ribosomal RNA) synthesis and ribosome formation. See also *Ribosomes, Nucleus.*

Nucleophilic Group

An electron-rich group with a strong tendency to donate electrons to an electron-deficient nucleus. See also *Polar group, Nonpolar group.*

Nucleoplasm

The protoplasm present within a cell's nucleus. See also *Cell, Nucleus, Protoplasm.*

Nucleoporin

See *Nuclear pore complexes.*

Nucleoproteins

Complexes made up of nucleic acid and protein. These two substances are apparently not linked by strong chemical bonds but are held together by salt linkages and other weak bonds. Most viruses consist entirely of nucleoproteins, although some viruses also contain fatty substances. Nucleoproteins also occur in animal and plant cells and in bacteria. See also *Protein, Nucleic acids, Virus.*

Nucleoside

A "hybrid" molecule consisting of a purine (adenine, guanine) or pyrimidine (thymine, uracil, or cytosine) base covalently linked to a five-membered sugar ring (ribose in the case of RNA and deoxyribose in the case of DNA). See also *Nucleotide, Adenine, Guanine, Guanosine, Uridine, Ribonucleic acid (RNA), Deoxyribonucleic acid (DNA).*

Nucleoside Diphosphate Sugar

A coenzyme-like carrier of a sugar molecule functioning in the enzymatic synthesis of polysaccharides and sugar derivatives. See also *Polysaccharides.*

Nucleosome

Spherical particles composed of a special class of basic proteins (histones) in combination with DNA (146 bp of DNA are wrapped 1.75 times around a "core" of eight histone proteins). The particles are approximately 12.5 nm in diameter and are connected to each other by DNA filaments. Under an electron microscope they appear somewhat like a string of pearls.

Nucleosomes are the basic structural unit of the chromosome and are sometimes called chromosomal packing units. See also *Chromatin, Histones, Protein, Deoxyribonucleic acid (DNA), Base pair (bp), Chromosomes, Nanometers (nm), Epigenetic, Tetrasomes.*

Nucleotide

An ester of a nucleoside and phosphoric acid; it is a subunit of DNA or RNA. Nucleotides are nucleosides that have a phosphate group attached to one or more of the hydroxyl groups of the sugar (ribose or deoxyribose). In short, a nucleotide is a hybrid molecule consisting of a purine or pyrimidine base covalently linked to a five-membered sugar ring which is covalently linked to a phosphate group. While (polymerized) nucleotides are the structural units of a nucleic acid, free nucleotides that are not an integral part of nucleic acids are also found in tissues and play important roles in the cell, for example, ATP and cyclic AMP.

Two nucleotides form each "rung of the ladder" within DNA molecules. See also *ATP, Cyclic AMP, Base (nucleotide), Nucleoside, Nucleic acids, Messenger RNA (mRNA), Ribonucleic acid (RNA), Deoxyribonucleic acid (DNA), Transversion.*

Nucleotide Base

See *Base (nucleotide).*

Nucleotide Excision Repair

See *Double-strand breaks (in DNA), DNA repair.*

Nucleus

From the Latin *nucleus* meaning kernel of a nut. Discovered by Robert Brown in 1833, it is the usually spherical body within each living cell that contains its hereditary biological material (e.g., DNA, genes, chromosomes, etc.) and controls the cell's life functions (e.g., metabolism, growth, and reproduction). The nucleus is a highly differentiated, relatively large organelle lying in the cytoplasm of the cell. The nucleus is surrounded by a (nuclear) membrane which is quite similar to the plasma (cell) membrane, except the nuclear membrane contains holes or pores. It is characterized by its high content of chromatin, which contains most of the cell's DNA. That chromatin is normally (when cell is not in process of dividing) distributed throughout the nucleus in a diffuse manner. See also *Genome, Cell, Gene, Genetic code, RNA, Heredity, Deoxyribonucleic acid (DNA), Chromosomes, Meiosis, Nuclear transfer, Metabolism, Chromatids, Chromatin, Plasma membrane, Organelles, Nuclear receptors, Nuclear proteins.*

N

NUE Gene

Refers to a **nitrogen use efficiency** gene that (e.g., when inserted into a plant's DNA, or inserted into the DNA of a crop biological that is applied to that plant) increases the efficiency with which that plant utilizes nitrates in its growth processes. See also *Gene, Deoxyribonucleic acid (DNA), Nitrates, Crop biologicals.*

Null Allele

Refers to a particular (mutant) version (i.e., allele) of a gene which lacks any activity. For example, for a given gene that normally codes for the production of an enzyme, the null allele would not code for production of any enzyme. See also *Gene, Allele, Mutant, Biological activity, Enzyme, Coding sequence.*

Null Mutation

A mutation that halts a gene's function (i.e., that gene no longer codes for production of a protein or a micro-RNA). See also *Gene, Mutation, Null allele, Protein, Micro-RNAs.*

Nutraceuticals

Coined in 1989 by Stephen DeFelice, this term is used to refer to either a food or portion of food (e.g., a vitamin, essential amino acid, etc.) that possesses medical or health benefits (to the organism that consumes that nutraceutical). For example, saponins (present in beans, spinach, tomatoes, potatoes, alfalfa, clover, etc.) possess some cancer prevention properties. Also sometimes called pharmafoods, functional foods, or designer foods, these are food products that have been designed to contain specific concentrations and/or proportions of certain nutrients (e.g., vitamins, amino acids, etc.) that are critical for good health. See also *Essential amino acids, Amino acid, Vitamin, Food good manufacturing practice (FGMP), Saponins, Essential nutrients, Phytochemicals, Antioxidants, Isoflavones, Genistein (gen), Resveratrol, Phytosterols, Beta carotene, Lycopene, Carotenoids, Lutein, Anthocyanins, Vitamin E, Xanthophylls, Sterols, Sitosterols, Sitostanols, Ellagic acid, Alicin, Proanthocyanidins, Polyphenols, Zeaxanthin, Phyto-manufacturing.*

Nutriceuticals

See *Nutraceuticals.*

Nutricines

See *Nutraceuticals.*

Nutrient Enhanced™

A phrase that is now a trademark of Garst Seed Company; it refers to plants that have been modified to possess novel traits which make those plants more economically valuable for nutritional uses (e.g., higher-than-normal protein content in certain feedgrains). See also *Value-enhanced grains, High-oil corn, Protein, Genetic engineering, High-lysine corn, High-methionine corn, Plant's novel trait (PNT), High-phytase corn and soybeans.*

Nutrigenomics

See *Nutritional genomics.*

Nutritional Epigenetics

See *Epigenetic.*

Nutritional Genomics

Refers to the study of the biological impacts of certain foods or food ingredients on the body due specifically to the different genomes (DNA) of those individual organisms that consume those foods/ingredients. The subgroup consisting of all those individuals whose genome (DNA) causes their body to respond in a specific way to a given food/ingredient is known as a HAPLOTYPE. A haplotype could (theoretically) be as small as one individual (e.g., one man, possessing : genome), because that man's particular response-to-a-food/ingredient could result from one single-nucleotide polymorphism (SNP) that only his genome possesses. See also *Genomics.*

Thus, nutritional genomics applies to the genetically determined biological impact of a given food/ingredient within a specific haplotype. For example, research indicates that consumption of oleic acid by women whose genome possesses the *Her-2/neu* (breast cancer-promoting gene) SNP results in down-regulation of expression of that *Her-2/neu* gene. See also *Genome, Pharmacogenomics, Genomics, Deoxyribonucleic acid (DNA), Gene, Haplotype, Single-nucleotide polymorphisms (SNPs), Coding sequence, Express, Down regulating, Biological activity, Organism, HER-2 gene, Oleic acid.*

O

O Glycosylation

See *Glycosylation (to glycosylate)*.

O'Farrell Gels

Refers to two-dimensional gel electrophoresis, discovered by Patrick O'Farrell in 1975. See *Two-dimensional (2D) gel electrophoresis*.

OAB (Office of Agricultural Biotechnology)

A unit of the U.S. Department of Agriculture that is in charge of a part of the federal regulatory process for biotechnology (e.g., field tests of transgenic plants). See also *Toxic Substances Control Act (TSCA)*, *Recombinant DNA Advisory Committee (RAC)*, *Food and Drug Administration* (FDA), *Transgenic*.

Obeticholic Acid

See *Agonists*, *Farnesoid X receptor (FXR)*.

Obinutuzumab

A monoclonal antibody against CD20 B-cell specific protein (on the surface membrane of malignant B-cells) that was approved in 2013 by the U.S. Food and Drug Administration (FDA) as a pharmaceutical (Gazyva™) for treatment of chronic lymphocyte leukemia. See also *Monoclonal antibodies (MAb)*, *Humanized antibody*, *Protein*, *B cells*, *Food and Drug Administration (FDA)*.

Ochratoxins

A term that refers to a group of related mycotoxins (i.e., toxic metabolites produced by fungi) which are produced by some *Aspergillus* species and some *Penicillium* species of fungi (e.g., *Aspergillus ochraceus* (alutaceus), *Penicillium verrucosum*, *Penicillium viridicatum*).

These particular fungi tend to produce ochratoxins when they grow in damaged grain (e.g., during grain storage), especially when grain temperature is above 4°C (40°F) and grain moisture content is above 18%.

Ochratoxin A (OTA) is a very carcinogenic (cancer-causing) toxin, which also can cause kidney damage when consumed by humans. When dairy cattle consume OTA-containing grain, the OTA soon appears in the milk produced by those cows. See also *Mycotoxins*, *Toxin*, *Fungus*, *Penicillium*, *Carcinogen*.

Octadecanoid/Jasmonate Signal Complex

A chemical signal that is created and emitted by certain plants in response to those plants being wounded (e.g., via chewing) by insects. The octadecanoid/jasmonate signal complex then causes the production and also emission of volatile chemicals such as volicitin, which attract certain types of wasps that are natural enemies of those insects which initially wounded the plants. Thus, the octadecanoid/jasmonate signal complex is crucial part of an (indirect) defense mechanism of such plants. See also *Signaling molecule*, *Signaling*, *European corn borer*, *Integrated Pest Management (IPM)*, *Volicitin*, *Green leaf volatiles*.

OD

See *Optical density*.

ODM

Acronym for oligonucleotide-directed mutagenesis. See *Oligonucleotide*, *Genome editing*, *Mutagen*.

Odorant-Binding Protein

A protein that enhances people's ability to smell odorants in trace quantities much lower than those needed to activate olfactory (i.e., smelling) nerves. The protein accomplishes this by latching onto (odorant) molecules and enhancing their aroma. Hence, it acts as a kind of "helper" entity in bringing about the ability to smell certain odorants present in low concentration. See also *Protein*.

OECD

See *Organization for Economic Cooperation and Development*.

Office International des Epizootics

See *International Office of Epizootics (OIE)*.

OGM

See *GMO*.

OH43

Gene in plants (e.g., corn/maize) that causes production of a seed coat which is more resistant to tearing. Greater tear resistance results in a lower incidence of fungi infestation in seed, which results in less mycotoxin production in seed. See also *Gene*, *Fungus*, *Aflatoxin*, *Mycotoxins*.

OIE

Acronym for Office International des Epizootics. See *International Office of Epizootics (OIE)*.

O

OIF

See *Osteoinductive factor.*

Oils

See *Fatty acid.*

Okazaki Fragments

Refers to a specific combined RNA/DNA segment that is temporarily inserted into the discontinuously copied strand of an organism's genome (DNA molecule) during the genome-copying portion of cell division. That segment consists of approximately 10 RNA ribonucleotides followed by a length of approximately 200–300 DNA deoxyribonucleotides. The ribonucleotides within these Okazaki fragments are subsequently replaced by deoxyribonucleotides (thereby resulting in accurate copying of the cell's original DNA) as part of the final steps of the genome-copying process. See also *Ribonucleic acid (RNA), Deoxyribonucleic acid (DNA), Organism, Cell, Genome, Mitosis, Primosome.*

Olaparib

A poly ADP-ribose polymerase (PARP) inhibitor (i.e., compound that blocks cellular enzymes involved in the repair of damaged DNA) that was approved by the U.S. Food and Drug Administration (FDA) in 2014 as the pharmaceutical Lynparza™ to treat women who have heavily pretreated ovarian cancer resultant from defective BRCA genes. See *PARP, PARP inhibitors, Cell, Enzyme, Deoxyribonucleic acid (DNA), Gene, Food and Drug Administration (FDA), Cancer, BRCA genes.*

Oleate

A term utilized by some, to refer to oleic acid. See *Oleic acid.*

Oleic Acid

A fatty acid that is naturally present in the fat of animals and also in oils extracted from oilseed plants (e.g., soybean, canola, etc.). For example, the soybean oil produced from traditional varieties of soybeans tends to contain 24% oleic acid.

Research indicates that consumption of adequate amounts of oleic acid helps to reduce risk of coronary heart disease and of (over-) expression of the Her-2/neu gene, and thereby conferring via the latter some protection against breast cancer. See also *Monounsaturated fats, Fatty acid, Fats, Canola, Soybean plant, Soybean oil, High-oleic oil soybeans, Gene, Express, Expressivity, Cosuppression, Cancer, Her-2/Neu gene, Down regulating, Coronary heart disease (CHD).*

Oleosomes

The storage bodies for lipids (fats) in the seeds of certain plants. See also *Lipids, Fats, Fatty acid.*

Oligionucleotide

See *Oligonucleotide.*

Oligodendrocytes

One category of neuron cells within the brain (white tissue), which principally produce the "insulation" that surrounds axons. See also *Neuron, Cell, Axon, Myelin, Multiple sclerosis.*

Oligo-Directed Mutagenesis

See *Genome editing.*

Oligofructans

See *Fructan, Fructose oligosaccharides.*

Oligofructose

See *Fructose oligosaccharides.*

Oligomer

A relatively short (the prefix "oligo-" means few, slight) chain molecule (polymer) that is made up of repeating units (e.g., XAXAXAXA or XXAAXXAAXXAA, etc.). Short polymers consisting of only two repeating units are called dimers and those of three repeating units are called trimers. Longer units are called polymers (i.e., many units). As a rule of thumb, oligomers consisting of 11 or more repeating units are called polymers. See also *Polymer.*

Oligonucleotide

Synonymous with oligodeoxyribonucleotide, they are short chains of nucleotides (i.e., single-stranded DNA or RNA) that have been synthesized (i.e., made by man or inside living cells) by chemically linking together a number of specific nucleotides.

When made by man, oligonucleotides (also called, simply "oligos") are used as synthetic (i.e., man-made) genes, DNA probes, and in site-directed mutagenesis. See also *Nucleotide, Gene, Dna probe, Oligomer, Site-directed mutagenesis, Gene machine, Deoxyribonucleic acid (DNA), Ribonucleic acid, Synthesizing (of DNA molecules), Peptide-oligonucleotide conjugates.*

Oligonucleotide Probes

Short chain fragments of DNA that are used in various gene analysis tests (e.g., the single base change in DNA that causes sickle-cell anemia). See also *Oligonucleotide, Deoxyribonucleic acid (DNA), DNA probe, Gene machine.*

Oligonucleotide-Directed Gene Modification

See *Oligonucleotide-mediated mutagenesis, Genome editing.*

Oligonucleotide-Directed Gene Repair

See *Oligonucleotide-mediated mutagenesis, Genome editing.*

Oligonucleotide-Directed Mutagenesis

See *Oligonucleotide-mediated mutagenesis, Genome editing.*

O

Oligonucleotide-Mediated Gene Editing

See *Oligonucleotide-mediated mutagenesis*, *Genome editing*.

Oligonucleotide-Mediated Gene Repair

See *Oligonucleotide-mediated mutagenesis*, *Genome editing*.

Oligonucleotide-Mediated Mutagenesis

Refers to several techniques utilized by scientists to correct or to introduce specific mutations/corrections at a particular site (locus) within the DNA of an organism. In each of those techniques, the mutations are induced via the oligonucleotide (chosen to be specific to the selected DNA locus). The techniques used to accomplish these site-specific corrections or directed mutations (base substitution, addition, or deletion) are called (among other names) targeted gene repair, chimeraplasty, targeted nucleotide exchange, therapeutic nucleic acid repair approach, oligonucleotide-mediated gene editing, oligonucleotide-mediated gene repair, oligonucleotide-directed gene modification, oligodeoxynucleotide-directed gene modification, chimeric oligonucleotide-dependent mismatch repair, triplex-forming oligonucleotide-induced recombination, and so on. See *Mutation*, *Nucleotide*, *Oligonucleotide*, *Deoxyribonucleic acid (DNA)*, *Locus*, *Base substitution*, *Organism*, *DNA repair*, *Mismatch repair*, *Chimeraplasty*, *Gene*, *Gene repair (done by humans)*, *Genome editing*.

Oligopeptide

A relatively short chain molecule that is made up of amino acids linked by peptide bonds. See also *Peptide*, *Polypeptide (protein)*, *Oligomer*, *Amino acid*.

Oligos

Term utilized to refer to man-made "chains" (nucleic acids) consisting of 18–30 nucleotides. They are utilized as synthetic genes, DNA probes, "bio-bar codes" on nanoparticle probes, primers, siRNA, and in site-directed mutagenesis. Oligos can modulate (i.e., increase or decrease) gene expression in cells by directly interacting with the cell's DNA or mRNA. See also *Oligonucleotide*, *Nucleic acids*, *Nucleotide*, *Gene*, *DNA probe*, *Primer (DNA)*, *Deoxyribonucleic acid (DNA)*, *Ribonucleic acid (RNA)*, *Short interfering RNA (siRNA)*, *Site-directed mutagenesis (SDM)*, *Cell*, *Bio-bar codes*, *Express*, *Expressivity*.

Oligosaccharide Microarrays

See *Microarray (testing)*.

Oligosaccharides

Refers generally to relatively short molecular chains made up to 10–100 simple sugar (saccharide) units. These sugar (i.e., carbohydrate) chains are frequently attached to protein molecules. When this happens, the resulting molecule is known as a glycoprotein—that is, a hybrid molecule that is part protein and part sugar. The oligosaccharide portion affects a protein's conformation(s) and biological activity. The oligosaccharide (carbohydrate) portion of a glycoprotein functions as a mediator of cellular uptake of that glycoprotein. Glycosylation thus affects the length of time the molecule resides in the bloodstream before it is taken out of circulation (serum lifetime). It is thought that blood group (e.g., A, B, O, etc.) is based upon an oligosaccharide concept. For example, different oligosaccharide "branches" on a given glycoprotein (e.g., tissue plasminogen activator) could cause that glycoprotein to be perceived by the body's immune system to be another (incorrect) blood type, thus provoking an immune response against it.

One natural source of approximately 200 critically important oligosaccharides is human breast milk. Although humans cannot digest those oligosaccharides, beneficial gut bacteria (known collectively as the infant gut microbiome) thrive on them (which thus benefits the infants). The infant gut microbiome is overwhelmingly dominated by the bacterial species *Bifidobacterium longum biovar infantis* (*B. longum* bv. *infantis*). The *B. longum* bv. *infantis* possess the precise enzymes needed to swiftly digest all the breast milk oligosaccharides, thereby denying those oligosaccharides to "bad" microbes in the gut, and crowding out those "bad" microbes. The now large gut population of *B. longum* bv. *Infantis* also produce short chain fatty acids that favor growth of many beneficial bacterial species, plus those short chain fatty acids help guide the particular cells that line an infant's intestine in how to mount an immune defense and regarding how to best utilize energy.

Oligosaccharides play a critical role in numerous disease processes in adults too (e.g., bacterial and viral infection processes, cancer metastasis processes, inflammation processes, etc.). For example, oligosaccharide molecular "chains" extending from the exterior membrane plasma membrane of cells are utilized by bacteria (and inflammation-triggering immune system cells) to latch onto cells and facilitate entry into cells. See also *Polysaccharides*, *Cell*, *Conformation*, *Monosaccharides*, *Furanose*, *Pentose*, *Pyranose*, *Glycogen*, *Glycoform*, *Fructose oligosaccharides*, *Glycoprotein*, *Sialic acid*, *Mannanoligosaccharides*, *Tissue plasminogen Activator (tPA)*, *Oligomer*, *Serology*, *Humoral immunity*, *Cellular immune response*, *Metastasis*, *Adhesion molecules*, *Hemagglutinin (HA)*, *Transgalacto-oligosaccharides*, *Fatty acids*, *Short chain fatty acids*.

Omega-3 Fatty Acids

More properly called "*n*-3 fatty acids." See *N-3 fatty acids*.

Omega-6 Fatty Acids

More properly called "*n*-6 fatty acids." See *N-6 fatty acids*.

OMM

Acronym for Oligonucleotide-mediated Mutagenesis. See *Oligonucleotide-mediated mutagenesis*.

Oncogenes

Genes within a cell's DNA that code for receptors (proteins on outer surface of cell membrane) for a cellular growth factor (e.g., epidermal growth factor [EGF]). Via that coding for of applicable receptors (or other protein molecules that are part of the signal transduction process of a cell), oncogenes "turn on" the process of cell division (replication) at appropriate time(s) during the life of each cell in an organism.

When oncogenes are mutated (e.g., via exposure to cigarette smoke or ultraviolet light, etc.), those oncogenes can become cancer-causing genes, some of which (e.g., erythroblastosis virus gene) are almost identical to the gene for EGF receptor (i.e., oncogene is a "deformed copy" of that gene). Such mutated oncogenes code for (i.e., cause to be made) proteins (e.g., protein kinases, protein phosphorylating enzymes, etc.) that trigger uncontrolled cell growth. They sometimes may consist of a human chromosome that has viral nucleic acid material incorporated into it and is a permanent part of that chromosome.

See also *Gene, Cell, Deoxyribonucleic acid (DNA), BRCA genes, Her-2 gene, ras gene, Meiosis, Carcinogen, Ribosomes, Protein, Tyrosine kinase, Enzyme, Chromosome, Plasma membrane, Epidermal growth factor (EGF), Signal transduction, Coding sequence, Tumor, Cancer, Proto-oncogenes, Genetic code, Receptors, Mutagen.*

Oncogenic

Refers to a gene that (e.g., when mutated) causes cells to multiply in an uncontrolled fashion (i.e., neoplastic growth). See also *Gene, Mutation, Cell.*

Oncolytics

Refers to a field of science in which certain infective agents (e.g., some bacteria, some viruses) are utilized by scientists to either kill or image (i.e., illuminate/view via bioluminescence or fluorescence) tumors and other metastatic tissues in the body.

This is made possible via the fact that tumors within the body often act as havens where these particular types of bacteria/viruses can safely "hide" from the body's immune system agents (macrophages, monocytes, etc.). See also *Cancer, Bacteria, Virus, Bioluminescence, Fluorescence, Label (fluorescent), Tumor, Metastasis, Macrophage, Monocytes, Cellular immune response.*

Oncomodulin

A growth factor that can stimulate (re)growth of an injured axon (nerve fiber). Discovered in 2006 by Yuqin Yin and Larry Benowitz. See also *Growth factor.*

Oocytes

The cells, produced by an organism's ovaries, that eventually become an ovum ("egg cell") via meiosis. See also *Meiosis, Cell, Organism.*

Oomycetes

See the link. See also *Phytophthora.*

Opague-2

A gene in corn (maize) that (when present in the DNA of a given plant) causes that plant to produce seed that contains higher-than-normal levels of lysine, calcium, magnesium, iron, zinc, and manganese. See also *Lysine (lys), High-methionine corn, Essential amino acids, Corn, Value-enhanced grains, High-lysine corn, Deoxyribonucleic acid (DNA), Gene, MAL (multiple aleurone layer) gene.*

Open Reading Frame (ORF)

Region of a gene (DNA) that contains a series of triplet (bases) coding for amino acids without any termination codons. The ORF sequence is potentially translatable into a protein, but the presence of an open reading frame (sequence) does not guarantee that a protein molecule will be produced (by cell ribosome). See also *Gene, Deoxyribonucleic acid (DNA), Amino acid, Protein, Coding sequence, Genetic code, Translation, Cell, Ribosomes.*

Operator

Also known as the "o locus." The site on the DNA to which a repressor molecule binds to prevent the initiation of transcription. The operator locus is a distinct entity and exists independently of the structural genes and the regulatory gene. It is the structural/biochemical "switch" with which the operon is turned on or off, and it controls the transcription of an entire group of coordinately induced genes. One type of mutation of the operator locus is called operator constitutive mutants. Constitutive mutants continually churn out the protein characteristic for that operon because the operon unit cannot be turned off by the repressor molecule. See also *Operon, Promoter, Regulatory genes, Repression (of gene transcription/translation), Repressor (protein), Structural gene, Structural genomics.*

Operon

A gene unit consisting of one or more genes that specify a polypeptide and an operator unit that regulates the structural gene, that is, the production of messenger RNA (mRNA) and hence, ultimately, of a number of proteins. Generally an operon is defined as a group of functionally related structural genes mapping (i.e., being) close to each other in the chromosome and being controlled by the same (one) operator. If the operator is "turned on," then the DNA of the genes comprising the operon will be transcribed into mRNA, and down the line specific proteins are produced. If, on the other hand, the operator is "turned off" then transcription of the genes does not occur and the production of the operon-specific proteins does not occur. See also *Deoxyribonucleic acid (DNA), Gene, Operator, Transcription, Polypeptide, Protein, Messenger RNA (mRNA), Chromosome.*

Opsonin

A protein present within blood which makes pathogens and other microorganisms more easily engulfed (e.g., by macrophages, etc.). See also *Protein, Pathogen, Phagocyte, Macrophage, Microorganism.*

Opsonization

Refers to (immune response) covering of a pathogen or other microorganism in the bloodstream with opsonin in order to render that pathogen/microorganism more susceptible to being engulfed/destroyed (e.g., by macrophages, etc.). See also *Opsonin, Immune response, Pathogen, Phagocyte, Macrophage.*

Optical Activity

The capacity of a substance to rotate the plane of polarization of plane-polarized light (when examined in an instrument known as a polarimeter). All compounds that are capable of existing in two forms that are nonsuperimposable mirror images of each other exhibit optical activity. Such compounds are called stereoisomers (or enantiomers or chiral molecules) and the two forms arise because compounds having asymmetric carbon atoms to which other atoms are connected may arrange themselves in two different ways.

Historically, such optically active substances required that the light travel through a thickness equivalent to at least several hundred wavelengths of that light in order for the substance to thereby rotate the plane of polarization of that light. However, one metamaterial (gold nano-corkscrews) can rotate the plane of polarization of light after it passes through a thickness equivalent to only one wavelength of the light. See also *Stereoisomers, Enantiomers, Chiral compound, Metamaterials, Nanotechnology, Nanoscience, Nanostructured material.*

Optical Density (OD)

The absorbance of light of a specific wavelength by molecules normally dissolved in a solution. Light absorption depends upon the concentration of the absorbing compound (chemical entity) in the solution, the thickness of the sample being illuminated, and the chemical nature of the absorbing compound. An analytical instrument known as a spectrophotometer is used to (quantitatively) express the amount of a substance (dissolved) in a solution. Mathematically, this is accomplished using the Beer–Lambert Law. See also *Spectrophotometer, Absorbance (A).*

Optical Tweezer

Invented in 1986 by A. Ashkin, this refers to the use of laser or highly focused infrared light beams to "trap" a tiny object (e.g., biomolecules, dendrimers, living cells, supramolecular assemblies, etc.) in three-dimensional space; holding them against gravity, Brownian (molecular) motion, and so on, to be measured or otherwise manipulated by a scientist.

That is accomplished via shining the light beam through a narrow aperture, thereby directing its "radiation pressure" against the tiny object. The radiation pressure derives from the momentum of the photons within the light beam. See also *Nanotechnology, Dendrimers, Supramolecular assembly, Self-assembling molecular machines, Photon.*

Optimum Foods

See *Nutraceuticals, Phytochemicals.*

Optimum pH

The pH (level of acidity) at which maximum growth occurs or maximal enzymatic activity occurs or at which any reaction occurs maximally. See also *Enzyme.*

Optimum Temperature

The temperature at which the maximum growth occurs or maximal enzymatic activity occurs or at which any reaction occurs maximally. See also *Enzyme, Ensiling.*

Optogenetics

Refers to a specific gene therapy process via which only one type of (the many types present in) the brain's neurons is stimulated via introduction of light to applicable deep portions (e.g., dopamine neuron regions, in the case of treating severe tremors resultant from Parkinson's disease) of the brain. That specificity is achieved after a viral vector is utilized to first insert a gene that codes for a light-responsive protein into the relevant brain cells.

Unlike the electrical current that is sometimes now utilized to treat Parkinson's severe tremors, the light-based deep brain stimulation would not have as many adverse side effects, because it discriminates (i.e., only stimulates the particular kind of brain neurons whose functions the doctor is trying to positively impact). See also *Gene, Protein, Virus, Vector, Gene therapy, Parkinson's disease.*

Optrode

A fiberoptic sensor made by coating the tip of a (glass) optic fiber with an antibody that fluoresces when the antibody comes in contact with its corresponding antigen. Alternatively, the fiber tip is sometimes coated with a dye that fluoresces when the dye comes in contact with specific chemicals (e.g., oxygen, glucose, etc.). Functionally, a beam of light is sent down the fiber and strikes ("pumps") the fluorescent complex, which then fluoresces (releases light of a specific wavelength). The light produced by fluorescence travels back up the same optic fiber and is detected by a spectrophotometer upon its return. By application of the Beer–Lambert Law, quantitative detection/measurement of the antigen or chemical *in vivo* in, for example, a patient's bloodstream is possible. See also *Antigen, In vivo, Antibody, Glucose (GLc), Spectrophotometer.*

Oral Cancer

Also sometimes known as "cancer of the mouth," this is a cancer involving the tissues lining the human mouth. Causes include consumption by humans of carcinogens (e.g., tobacco products, certain mycotoxins, etc.). Oral cancerous cells arise from precancerous mouth lesions known as oral leukoplakia.

During 2000, research by Frank Meyskins and William Armstrong indicated that consumption of Bowman–Birk trypsin inhibitor (BB T.I.) derived from soybeans, in a manner that "bathes" mouth tissues in BB T.I. (for extended period of time) inhibits the development of oral leukoplakia. See also *Cancer, Tumor, Mutagen, Mycotoxins, Trypsin inhibitors.*

Oral Leukoplakia

See *Oral cancer.*

Oral Tolerance

See *Glutamic acid decarboxylase (GAD).*

Orally Administered

Refers to the ability of compounds (e.g., pharmaceuticals) to be delivered to the body via the digestive system and still retain their efficacy (biological activity). One method to accomplish that (i.e., protect the compound from being broken down by the digestive system prior to the compound arriving at the site within body where needed) is to encapsulate molecules of the compound within a phosphatidyl serine "nano-coating" (known as a nanocochleate). See also *Biological activity, Phosphatidyl serine, Peyer's patches, Nanobodies.*

ORF

See *Open reading frame (ORF).*

O

Organelles

Membrane-surrounded structures found in eucaryotic cells; they contain enzymes and other components required for specialized cell function (e.g., ribosomes for protein synthesis or lysosomes for enzymatic hydrolysis). Some organelles such as mitochondria and chloroplasts contain DNA and can replicate autonomously (from the rest of the cell). See also *Nucleus*, *Eucaryote*, *Enzyme*, *Ribosomes*, *Lysosome*, *Peroxisome*.

Organism

Refers to any living plant, animal, bacteria, fungus, virus, etc. Also (e.g., in certain international treaties such as the Convention on Biological Diversity), this term includes things (e.g., seeds, spores, eggs) possessing the potential to become plants, animals, fungi, etc. See also *Biology*, *Bacteria*, *Fungus*, *Virus*, *Convention on biological diversity* (*CBD*).

Organismos Geneticamente Modificados

See *GMO*.

Organization for Economic Cooperation and Development (OECD)

An international organization composed of the world's wealthiest (most developed) nations, originally established in 1960 to study trade and related matters. In 1991, the OECD's Group of National Experts on Safety in Biotechnology (GNE) completed a document entitled Report on the Concepts and Principles Underpinning Safety Evaluations of Food Derived from Modern Biotechnology. The "aim of that document was to elaborate the scientific principles to be considered (i.e., by OECD member nations' regulatory agencies) in evaluating the safety of new foods and food components" (e.g., genetically modified soybeans, corn/maize, potatoes, etc.). See also *Biotechnology*, *Soybean plant*, *GNE*, *Canola*, *Mutual recognition agreements* (*MRA*).

Organogenesis

The production of entire organs, usually from basic cells, such as fibroblasts, and structural material such as collagen. See also *Collagen*, *Fibroblasts*.

Oriented Attachment (Nanotechnology)

Refers to a method of controlling how certain molecules assemble/crystallize into nanostructures. For example, a film of oleic acid can be utilized to "guide" the crystallization of lead sulfide into two-dimensional nanocrystals (i.e., flat sheets) instead of the three-dimensional crystals formed in the absence of oleic acid. See also *Nanocrystals*, *Nanoscience*, *Nanotechnology*.

Origin

Point or region where DNA (deoxyribonucleic acid) replication is begun. Often abbreviated "Ori." See also *Deoxyribonucleic acid* (*DNA*), *Replication* (*of virus*), *Replication fork*.

Oropharyngeal Cancer

See *Oral cancer*.

Orphan Drug

The name of the legal status granted by the U.S. Food and Drug Administration's (FDA) Office of Orphan Products Development (to certain pharmaceuticals). This classification provides the sponsors of those pharmaceuticals with special tax and other financial incentives (e.g., market monopoly for a limited time). If companies feel that they possess a cure (drug) for a certain disease, but the number of potential patients is below a certain number and there is potential competition from rival companies, then the high cost of developing and shepherding the drug through the FDA standard pharmaceutical approval process would be such that the company would not be able to regain its development costs and make a profit. Hence, orphan drug status was designed to encourage drug development efforts for otherwise noneconomic pharmaceuticals with less than 200,000 patients a year. See also *Food and Drug Administration* (*FDA*), *Farnesyl transferase inhibitors*.

Orphan Genes

Genes within an organism's genome/DNA that have no apparent function. See also *Gene*, *Organism*, *Genome*, *Deoxyribonucleic acid* (*DNA*), *Functional genomics*.

Orphan Receptors

Refers to nuclear receptors (i.e., embedded in surface of cell's plasma membrane) that are not coupled to G-protein (cell) system complexes. Orphan receptors include receptors for fatty acids (PPARs), bile acids (FXR), xenobiotics/toxins (PXR/SXR), and so on.

Many of the orphan receptors function as lipid sensors that respond to lipid concentrations in cells (e.g., resulting from lipid consumption in diet) and cause changes in gene expression, to protect cells from lipid overload. See also *Bioreceptors*, *Receptors*, *Nuclear receptors*, *Retinoid X receptors* (*RXR*), *Cell*, *Plasma membrane*, *G-proteins*, *Lipids*, *Fatty acids*, *Adhesion molecule*, *Microarray* (*testing*), *Biochips*, *High-throughput screening* (*HTS*), *Target–ligand interaction screening*, *Ligand* (*in biochemistry*), *Bioassay*, *Gene expression analysis*, *Target* (*of a therapeutic agent*).

Ortholog

Refers to a gene (e.g., within the DNA of a small/simple model organism) which:

- Is also present (or at least a very similar sequence gene is) within the DNA of another species (e.g., a more complex species) at an analogous location (e.g., ____ distance from __ end of the other organism's DNA molecule).
- Possesses a similar function to that of the analogous gene in that more complex organism (e.g., human).

Identification and study of such a gene (within a model organism) enables properties/function of its analogous ortholog gene in the more complex organism to be inferred. For example, the *Sir2* gene within the organism *Saccharomyces cerevisiae* is an ortholog of the *SirT1* gene in mammals. See also *Gene*, *Deoxyribonucleic acid* (*DNA*), *Sequence* (*of a DNA molecule*), *Model organism*, *Phylogenetic profiling*, *Functional genomics*, *Sir2 gene*, *SirT1 gene*.

Orthophosphate Cleavage

Enzymatic cleavage of one of the phosphate ester bonds of ATP to yield ADP and a single phosphate molecule known as orthophosphate (designated as P_i). The cleavage of the phosphate bond is energy yielding and is (except in the case of a futile cycle) coupled enzymatically to reactions that utilize the energy to run the cell. An orthophosphate cleavage reaction releases relatively less energy than does a corresponding pyrophosphate cleavage reaction. See also *Adenosine diphosphate* (*ADP*), *Adenosine triphosphate* (*ATP*), *Futile cycle*, *Pyrophosphate cleavage*.

Osmolytes

Refers to certain molecules present within the bloodstreams of some species of fish, amphibians, and reptiles, which act to stabilize/preserve protein molecules at low temperatures. Osmolytes can thereby enable those animals to survive subfreezing temperatures in a sort of semihibernation state. Some osmolytes are utilized by man to similarly preserve pharmaceuticals and human reproductive tissues such as sperm and eggs. See also *Protein*.

Osmosis

From the Greek *osmos* meaning a pushing, it refers to bulk flow of water through a semipermeable (or more accurately, differentially permeable) membrane into another (aqueous) phase containing more of a solute (dissolved compound).

The bulk flow of water has the effect of diluting the solution in the second phase, while concentrating the solution in the first phase. Water will flow from the first phase to the second phase until the salt concentrations of both solutions are equal.

Osmosis is therefore a process in which water passes from regions of low salt concentration to regions of high salt concentration. The process can be viewed as equalizing the number of water and solute molecules on both sides of the membrane. See also *Osmotic pressure*, *Permeable*.

Osmotic Pressure

May be defined as the hydrostatic pressure which must be applied to a solution on one side of a semipermeable membrane (solution B in the example for osmosis) in order to offset the flow of solvent (water) from the other side (solution A in the example for osmosis). It is a measure of the tendency or "strength" of water to flow from a region of low salt concentration (and conversely high water concentration) to regions of high salt concentration (and conversely low water concentration). See also *Osmosis*.

Osmotins

A category of proteins, which are produced by some organisms as a natural defense against pathogenic fungi. See also *Cecrophins*, *Magainins*, *Organism*, *Fungus*, *Pathogenic*.

Osteoarthritis

A disease that affects primarily women older than 45, in which cartilage within the body's joint breaks down. Osteoarthritis encompasses approximately half of all cases of arthritis.

Osteoclasts

Naturally occurring cells within bones that normally breakdown bone material as part of the ongoing bone rebuilding and "remodeling" processes. When signaled by the RANKL protein (sometimes expressed by certain cancerous cells such as prostate cancer cells), osteoclasts then change their behavior in ways that make the bone structures "open" to metastatic tumor entry. See also *Cell*, *Metastasis*.

Osteoinductive Factor (OIF)

A protein that induces the growth of both cartilage-forming cells and bone-forming cells (e.g., after a bone has been broken). When applied in the presence of transforming growth factor-beta, type 2 (another protein), osteoinductive factor first causes connective tissue cells to grow together to form a matrix of cartilage (e.g., across the bone break), then bone cells slowly replace that cartilage. Osteoinductive factor also seems to thwart a type of cell that tears down bone formation, so OIF may someday be used to combat osteoporosis. See also *Growth factor*, *Transforming growth factor-beta* (*TGF-BETA*), *Fibroblasts*, *Fibroblast growth factor* (*FGF*), *Osteoclasts*, *Osteoporosis*.

Osteoporosis

A disease of humans in which the bones gradually weaken and become brittle. A diet containing a large amount of soy isoflavones (i.e., esp. genistein) has been shown to increase bone density, thereby lowering the risk of osteoporosis.

Groups that are especially at risk for osteoporosis include postmenopausal women (particularly of Caucasian or Asian ethnicity), those who have undergone early menopause (i.e., prior to age 45), those who smoked, those who consumed excessive amounts of alcohol, and those who consumed excessive amounts of certain pharmaceuticals (e.g., steroids such as prednisone, thyroid hormone, etc.). See also *Osteoinductive factor* (*OIF*), *Genistein* (*gen*), *Soy protein*, *Isoflavones*, *Steroid*, *Soybean plant*, *High-isoflavone soybeans*, *Haplotype*, *Eicosapentanoic acid* (*EPA*), *NFκB*.

OTA

Acronym for Ochratoxin A. See also *Ochratoxins*.

Outcrossing

The transfer of a given gene or genes (e.g., one synthesized by man and inserted into a plant via genetic engineering) from a domesticated organism (e.g., crop plant) to a wild type (relative of plant). See also *Gene*, *Introgression*, *Synthesizing* (*of DNA molecules*), *Genetic engineering*, *Wild type*.

Overlapping Gene

Refers to a gene whose sequence at least partially overlaps that of another gene (adjacent to the first within the DNA of an organism). See also *Gene*, *Sequence* (*of a DNA molecule*), *Organism*, *Deoxyribonucleic acid* (*DNA*).

Overwinding

Positive supercoiling. Winding which applies further tension in the direction of winding of the two strands about each other in the duplex. See also *Deoxyribonucleic acid (DNA), Supercoiling, Double helix, Duplex.*

Oxalate

A salt or ester of oxalic acid. See also *Calcium oxalate.*

Oxalate Oxidase (OxOx)

An enzyme which catalyzes the breakdown of oxalic acid.

Because certain pathogenic fungi (e.g., *Sclerotina sclerotiorum*) produce oxalic acid to help those fungi to "invade" specific crop plants' tissues, the production of OxOx by those crop plants would make them resistant to those pathogenic fungi.

The crop plant wheat (*Triticum aestivum*) possesses a gene which codes for the production of OxOx. See also *Enzyme, Fungus, Oxalic acid, Pathogen, Gene, Coding sequence.*

Oxalic Acid

A nephrotoxic (i.e., harmful to kidneys) acid which is naturally produced in some organisms in small amounts, such as in starfruit (*Averrhoa carambola*), members of the spinach family and the brassicas (cabbage, broccoli, brussels sprouts), parsley, sorrel, rhubarb. The plant known as jack-in-the-pulpit (*Arisaema triphyllum*) contains crystals of calcium oxalate (a salt of oxalic acid) which deters wild animals from eating it by providing the sensation of stinging needles on their tongue.

Oxalic acid is exuded by some pathogenic fungi (e.g., *Sclerotinia sclerotiorum*) species to help those fungi to invade plant tissues (i.e., as part of the fungal "infection" process). See also *Acid, Fungus, Organism, Pathogen, Oxalate oxidase (OxOx).*

Oxidant

See *Oxidizing agent.*

Oxidation (Chemical Reaction)

Loss of electrons from a compound (or element) in a chemical reaction. When one compound is oxidized, another compound is reduced. That is, the other compound must "pick up" the electrons which the first has lost. See also *Oxidation–reduction reaction, Hydrogenation, Oxidation (of fats/oils/lipids), Reactive oxygen species, Nanoceria.*

Oxidation (of Fats/Oils/Lipids)

A chemical transformation of fat/lipid molecules, in which oxygen (e.g., from air) is combined with those molecules. As a result of that (oxidation chemical reaction), various chemical entities are created (e.g., peroxides, aldehydes, etc.) which possess objectionable flavors/odors and are harmful to animals that consume such (rancid) fats/oils. See also *Fats, Fatty acid, Lipids, Plasma membrane, Oxidation (chemical reaction), Oxidative stress, Hydrolysis, Oxylipins.*

Oxidation (of Fatty Acids)

See *Oxylipins, Carnitine.*

Oxidation–Reduction Reaction

A chemical reaction in which electrons are transferred from a donor to an acceptor molecule or atom. See also *Oxidation* (chemical reaction), *Oxidizing agent, Reactive oxygen species, Reduction (in a chemical reaction), Nanoceria.*

Oxidative Demethylation

The enzymatic removal of methyl submolecule groups from DNA inside living cells (e.g., due to epigenetic gene programming) via DNA-repair dioxygenase enzyme. See also *Methylated, Enzyme, Deoxyribonucleic acid (DNA), Cell, Epigenetic.*

Oxidative Phosphorylation

The enzymatic phosphorylation of ADP to ATP coupled to electron transport from a substrate to molecular oxygen. The synthesis (production) of ATP from the starting materials of ADP and inorganic phosphate (orthophosphate). See also *Adenosine diphosphate (ADP), Adenosine triphosphate (ATP), Orthophosphate cleavage, Enzyme.*

Oxidative Stress

The physiological stress/damage that results from the (chemical reaction-) breakdown of all or part of an organism via oxidation reaction(s). For example, many antibiotics kill bacteria via the action of oxidative stress.

Oxidative stress appears to be present within the brains of all victims of neurodegenerative diseases (e.g., Alzheimer's disease, Parkinson's disease, etc.) in the form of oxylipin molecules.

One common result of such oxidation reactions is the generation (within organism's body) of reactive oxygen species (e.g., "free radicals") that can adversely affect:

- Endothelial function (i.e., the inner lining of blood vessels).
- Platelet aggregation (e.g., inappropriate blood clotting/clumping).
- Atherosclerosis (i.e., buildup of oxidized fatty deposits known as plaque on internal walls of arteries).
- Myocardial function (e.g., heart failure).
- Eye and kidney tissue (especially in diabetics).

A key indicator of oxidative stress is the peroxidation of membrane lipids to form mono- and bifunctional aldehydes (e.g., 4-hydroxy-2-nonenal, also known as HNE). See also *Organism, Reactive oxygen species, Oxidation (chemical reaction), Bacteria, Antibiotics, Hydrogen sulfide (H_2S), Alzheimer's disease, Parkinson's disease, Cell, Antioxidants, Plasma membrane, Lipids, Glutathione, Carotenoids, Endothelial cells, Platelets, Atherosclerosis, Insulin, Coronary heart disease (CHD), Haptoglobin, Cold hardening, Mitogen-activated protein kinase cascade, Sitosterol, Oxylipins, Nanoceria.*

Oxidizing Agent

(oxidant) The acceptor of electrons in an oxidation–reduction reaction. The oxidant is reduced by the end of the chemical reaction. That is, the oxidizing agent is the entity that seeks and accepts electrons. Electron acceptance is, by definition, reduction. See also *Oxidation–reduction reaction, Peroxidase, Reactive oxygen species.*

Oxygen Free Radical

See *Free radical.*

Oxygenase

An enzyme catalyzing a reaction in which oxygen is introduced into an acceptor molecule. See also *Enzyme.*

Oxylipins

A word that was coined in 1998 by Bill Gerwick. It refers to oxygenated fatty acid molecules (i.e., a fatty acid that chemically reacted with an oxygen atom). Some oxylipins (e.g., jasmonic acid) are signaling molecules emitted by plants when those plants are under attack by chewing insect pests. The signal attracts certain predator insects to attack the insects that are chewing on the plants. See also *Fatty acid, Oxidative stress, Jasmonic acid.*

Oxytocin

See *Oxytocin receptor gene.*

Oxytocin Receptor Gene

Gene that codes for the receptors to oxytocin hormone that are present within the surface membranes of cells lining certain body tissues. Via receiving of the oxytocin signal (e.g., by latching on to passing oxytocin molecules and bringing them into their cells), these receptors influence both mother–infant bonding and male–female pair bonding in monogamous species of animals.

The variant of the oxytocin receptor gene present within a given individual person's DNA also influences that particular person's face recognition ability. Oxytocin signaling plays an important role in promoting humans' ability to recognize one another, but approximately one-third of the human population possesses only the gene variant that negatively impacts face recognition ability. See also *Gene, Coding sequence, Receptors, Cell, Signaling, Signaling molecule, Signal transduction.*

P

"Points to Consider" Document

See *Points to consider in the manufacture and testing of monoclonal antibody products for human use.*

P. gossypiella

See *Pectinophora gossypiella.*

P34 Protein

More accurately known as P34/Gly m Bd 30k Allergen, it is the principal cause of the (rare) allergic response of humans to soybean protein. See also *Soybean.*

P34 Protein

One of the primary storage proteins in soybean seeds; it can also cause an allergic response in some people (e.g., approximately 1% of humans) who consume it.

During 2001–2002, Eliot Herman and Rick Helm utilized genetic engineering to induce "gene silencing" of P34 protein production, to create a *reduced-allergen soybean* variety. The resultant biotechnology-derived soybeans contain the same total amount of protein, because production of the other proteins increased exactly enough to offset the (now absent) P34 protein in the resultant seeds. See also *Protein, Storage proteins, Soybean plant, Allergies (food-borne), Reduced-allergen soybeans, Gene, Genetic engineering, Gene silencing, RNA interference (RNAi), Cosuppression, Biotechnology.*

p38 MAP Kinase Pathway

See *Stem cells, Mitogen-activated protein kinase cascade.*

p53 Gene

Discovered in 1978 by David Lane, it is a tumor-suppressor gene, which controls passage (of a given cell) from the *"GI" phase* to the "s" (i.e., DNA synthesis) phase. The p53 protein that is coded for by *p53* gene is a transcription factor (i.e., it "reads" DNA to determine if damaged, then acts to control cell division, while *p53* gene codes for more production of additional p53 protein to repair the DNA damage). Discovered in 1993 by Arnold J. Levine and colleagues, to be responsible for approximately 50% of all human cancer tumors (when the *p53* gene is damaged/mutated).

Normally, the *p53* gene codes for (i.e., causes to be manufactured in cell) the p53 protein, which acts to prevent cells from dividing uncontrollably when the cell's DNA has been damaged (e.g., via exposure to cigarette smoke, ultraviolet light, certain mycotoxins, etc.). If, in spite of the presence of p53 protein, a cell begins to divide uncontrollably following damage to its DNA, the *p53* gene can cause apoptosis, which is also known as "programmed cell death" (to try to prevent tumors).

When even slightly methylated (e.g., a single methyl molecular group attached to a cytosine base in the *p53* gene at a site adjacent to DNA damage), carcinogens such as those within cigarette smoke or aflatoxin are attracted to that site and insert themselves between adjacent base pairs in the DNA, thereby preventing normal DNA repair. See also *Gene, Tumor-suppressor genes, ras gene, Genetic code, Meiosis, Deoxyribonucleic acid (DNA), Carcinogen, Mycotoxins, Ribosomes, Oncogenes, Transcription factors, Cancer, Tumor, p53 protein, Proto-oncogenes, Protein, Apoptosis, Methylated, Cytosine, DNA repair, AMPK.*

p53 Protein

A tumor-suppressor protein, sometimes called the "master transcription factor," or the "guardian of the genome," but whose amino acid sequence alterations (resulting from damage or mutation to the *p53* gene) are believed to be responsible for up to 50% of all human cancer tumors. The p53 protein has four domains, one of which (i.e., the core domain) binds to a specific sequence (recognition site) of the cell's DNA, in order to prevent cell from dividing uncontrollably when the cell's DNA has been damaged (e.g., via exposure to cigarette smoke, ultraviolet light, or other carcinogen), until the damage to that DNA can be repaired (also under the influence of the p53 protein).

As the amount of DNA within a given (damaged) cell increases, the concentration of p53 protein also increases. Because p53 protein is a transcription factor (i.e., "reads" DNA to determine if damaged, then acts to control cell division, while *p53* gene codes for production of more p53), p53 is very efficient at preventing/inhibiting tumors.

However, if the cell's DNA *cannot be repaired*, the p53 protein can cause cell cycle arrest and/or *apoptosis* ("programmed cell death") *to prevent development of (cancerous) tumors.* The p53 protein does this by inducing expression of relevant apoptosis-causing genes (e.g., *noxa*, *puma* (bbc3), etc.). See also *Gene, p53 gene, Tumor-suppressor genes, ras gene, ras protein, Genetic code, Meiosis, Carcinogen, Deoxyribonucleic acid (DNA), DNA repair, Aflatoxin, Ribosomes, Oncogenes, Cancer, Tumor, Proto-oncogenes, Protein, Transcription factors, Domain (of a protein), Apoptosis.*

Paclitaxel

An anticancer compound (pharmaceutical) that was originally isolated from the Pacific yew tree (*Taxus brevifolia*), although it is made synthetically today.

In 1966, Maurice Wall first identified antitumor effects in an extract from *Taxus brevifolia.* In 1992, the U.S. Food and Drug Administration (FDA) approved paclitaxel for use to treat recurrent ovarian cancer. Other anticancer uses were later approved.

When injected into the human body, paclitaxel also inhibits growth of the parasitic microorganism *Toxoplasma gondii* (which can cause loss of sight and neurological disease in humans, if not controlled). See also *Cancer, Taxol, Food and Drug*

P

Administration (FDA), Chemotherapy, Tubulin, Microorganism, Growth (microbial), Plant cell fermentation.

PAF

Acronym for *Platelet Activating Factor.* See *Choline.*

PAGE

See *Polyacrylamide gel electrophoresis (PAGE).*

PAH

Acronym for *polycyclic aromatic hydrocarbons.* See *Activated carbon.*

PAH

Acronym for pulmonary arterial hypertension. See *Pulmonary arterial hypertension.*

Palbociclib

A CDK-4/6 inhibitor that was approved by the U.S. FDA in 2015 as the pharmaceutical Ibrance™ to be used in combination with the anticancer drug Femara™ (letrozole) for treatment of postmenopausal women with advanced, hormone receptor-positive breast cancer who have not previously received endocrine treatment. See also *Food and Drug Administration (FDA), Hormone, Receptor cancer.*

Palindrome

From the Greek *palindromos* meaning *running back again,* it is a DNA molecule sequence that is the same when one strand of the molecule is read left to right and the other strand is read right to left. See also *Deoxyribonucleic acid (DNA), Reading frame.*

PALM

Acronym for *P*hoto-*A*ctivated *L*ight *M*icroscopy.

Palmitate

See *Palmitic acid.*

Palmitic Acid

A saturated fatty acid containing 16 carbon atoms in its molecular "backbone," which tends to increase cholesterol levels in the bloodstream when consumed by humans.

It has been shown that feeding of extruded (whole) high-oleic oil soybeans to dairy cattle did decrease the content of palmitic acid in their milk. See also *Fatty acid, Saturated fatty acids (SAFA), Cholesterol, High-oleic oil soybeans.*

PAMP-Induced Resistance

Refers to the defensive response (e.g., of a plant to some pathogens' attacks) that is triggered via *Pathogen-Associated Molecular Patterns* (i.e., specific molecular segments (pieces) on surface of those pathogens). For example, when it is detected, one particular segment (i.e., 22-amino acid peptide) from bacterial flagella (i.e., the whip-like "tail" on certain bacteria) causes some plants to quickly down-regulate auxin signaling, which increases that plant's resistance to bacterial infection. See also *Pathogen, Bacteria, Innate immune response, Auxins, Flagella, Peptide, Signaling, Down regulating, Pathogenesis-related proteins, PAMPs.*

PAMPs

Acronym for *Pathogen-Associated Molecular Patterns.* These are molecular segments (pieces) on surface of some pathogens which trigger a defensive response to specific pathogen-invaders in certain plants and/or animals. For example, when it is detected, one particular segment (i.e., 22-amino acid peptide) from bacterial flagella (i.e., the whiplike "tail" on certain bacteria) causes some plants to quickly down-regulate auxin signaling, which increases that plant's resistance to bacterial infection.

For example, when the oligosaccharides of chitin (a polymer of *N*-acetyl-D-glucosamine, that is a major component of fungus cell walls) are detected, both plant and mammal cells initiate protective responses. See also *Pathogen, Cell, Bacteria, Innate immune response, Auxins, Flagella, Peptide, Signaling, Down regulating, Pathogenesis-related proteins, Oligosaccharides, Chitin, Fungus.*

Pancreas

An organ (gland) located near the stomach that secretes insulin and glucagon into the bloodstream and digestive fluids into the intestines. See also *DNAse, Insulin, Glucagon, Beta cells, Type I diabetes, Type II diabetes, Diabetes.*

Paneth Cells

Specialized epithelial cells located at the terminal end of intestinal crypts (cave-shaped indentations in the lining of the wall of the small intestine) which secrete several antibacterial/antimicrobial substances in order to protect the lining of the small intestine (e.g., from pathogens within the digesta in the intestine).

For example, Paneth cells secrete (into the mass of digesta inside the small intestine) a lectin known as HIP/PAP, which binds to a specific [carbohydrate] portion of proteoglycans located on the surface of most bacteria, thereby killing those bacteria—especially Gram-positive (pathogenic) bacteria. Paneth cells also secrete lysozyme, alpha-defensins, secretory phospholipase A2, and angiogenin 4. See also *Cell, Epithelium, Lectins, Carbohydrates, Bacteria, Gram-positive (G+), Pathogenic, Lysozyme, Defensins, Angiogenin.*

Panobinostat

A histone deacetylase inhibitor that was approved in 2015 by the U.S. FDA as the pharmaceutical Fairydak™ for the treatment of patients with multiple myeloma that had earlier received at least two "standard therapies," including bortezomib (chemotherapy) and an immunomodulatory agent. Fairydak is supposed to be used in combination with bortezomib, a type of chemotherapy, and dexamethasone, an anti-inflammatory medication. See also *Histone deacetylase inhibitor, Food and Drug Administration (FDA), Bortezomib, Multiple myeloma, Chemotherapy.*

Papovavirus

A class of animal viruses, for example, SV40 and polyoma. See also *Virus.*

PAR

Acronym for *poly(ADP-ribose)*. See *PARP*.

Par Gene

A gene that is present in the DNA of plant and bacterial cells, and which is at least partially responsible for the partitioning of plasmids in those cells. See also *Gene*, *Deoxyribonucleic acid (DNA)*, *Plasmid*.

Paramutation

Refers to an epigenetic phenomenon via which the "instructions" (in the form of expressed RNA) of a given allele (i.e., one specific version of a particular gene) are executed in *subsequent generation(s)* of the organism, when that allele itself has not been inherited by that subsequent generation.

During 2005, researchers discovered that in the plant *Arabidopsis thaliana*, alleles of a gene known as HOTHEAD can be completely *absent* from the plant's DNA in one generation, but then reappear in that plant (offspring's) DNA in the next generation.

During 2008, Karl F. Erhard Jr. and colleagues discovered that *RMR6 gene* (which is required for paramutation in the maize plant) is itself an RNA polymerase (Pol IV). See also *Epigenetic*, *Gene*, *Allele*, *Coding sequence*, *Genetic code*, *Deoxyribonucleic acid (DNA)*, *Ribonucleic acid (RNA)*, *Organism*, *Micro-RNAs*, *Arabidopsis thaliana*, *RNA polymerase*, *Epigenetic*.

Parasite

From the Greek *parasites* meaning *one who eats at the table of another.*

Parkin

A naturally occurring protein that regulates how cells in the human body take up dietary fats and process them (e.g., incorporate them into cell's structures, etc.). Parkin normally roams around inside cells and tags damaged mitochondria as waste. The damaged mitochondria are then degraded by cells' lysosomes, and thereby recycled. Known mutations in the gene that codes for Parkin can prevent tagging, resulting in accumulation of unhealthy mitochondria in the body.

Parkin also acts upon *Mycobacterium tuberculosis* bacteria cells, triggering destruction of those bacteria by alerting immune cells known as macrophages (to engulf those bacteria).

When the Parkin gene is mutated, it can lead to onset of Parkinson's disease. See also *Protein*, *Cell*, *Fats*, *Macrophage*, *Parkinson's disease*, *Mitochondria*, *Gene mutation*, *Bacteria*.

Parkinson's Disease

A disease of the human brain, in which those nerve cells (neurons) associated with emotions and those neurons that are involved in controlling movement (motor control) die due to insoluble aggregates (clumps) of alpha-synuclein in the neurons.

Discovered in 1919 by doctors treating an epidemic of encephalitis lethargica (onset of Parkinson's disease commonly follows that encephalitis, but it can also be induced by certain drugs, etc.). The (natural) cause of Parkinson's disease (i.e., causing a dwindling supply of dopamine in the brain) is unknown although it can be induced by drug misuse. When a human brain is functioning normally, cells within a region of the brain called the *substantia nigra* initiate motor (i.e., muscle) activity by releasing the chemical "messenger" known as dopamine. In the brain of a person suffering from Parkinson's disease, those dopamine-producing cells die off, causing a progressive loss of motor control for that person.

One treatment for some Parkinson's disease symptoms is orally administered L-Dopa, which the body converts to dopamine via a decarboxylation reaction.

One treatment for the severe tremors sometimes present in Parkinson's sufferers is deep brain stimulation. See also *Neurotransmitter*, *Ciliary neurotrophic factor (CNTF)*, *Signaling*, *Glial derived neurotrophic factor (GDNF)*, *Oxidative stress*, *Neuron*, *Alpha-synuclein*, *Carboxyl terminus (of a protein molecule)*, *Molecular tweezers*, *Parkin*, *Optogenetics*.

ParM

A contractile (i.e., periodically contracting) protein that is present in at least some bacteria. Via its contractions during meiosis, ParM is involved in those cells' separation of nuclear DNA (i.e., prior to cell division). See also *Protein*, *Motor proteins*, *Bacteria*, *Meiosis*, *Deoxyribonucleic acid (DNA)*.

PARP

Acronym for *Poly ADP-ribose Polymerase* (an enzyme naturally present in some organisms' cells which is involved in control of DNA repair and/or apoptosis, among other cellular processes). The gene coding for (more) production of PARP enzyme is "switched on" during times of stress to the organism (e.g., during drought in a plant). PARP is cleaved (degraded) by caspases during apoptosis.

During drought conditions, PARP levels in crop plants are often so high (due to overexpression of *PARP* gene) that the plant's yield is decreased versus what it otherwise would have been. Thus, one way that scientists are working to engineer drought-resistant crop plants is to moderate PARP production (e.g., via RNA interference) of PARP enzyme.

PARP enzyme can be commercially produced (e.g., to manufacture certain test devices) via genetically engineered hamster cells grown in cell culture. This enzyme can be utilized by man in order to determine/test if a given substance (e.g., industrial chemical) is carcinogenic to humans.

PARP inhibitor pharmaceuticals have been shown to act against breast and ovarian cancers when used in women with breast and ovarian cancers linked to BRCA mutations. That is because, following DNA damage in human cells, two specific proteins (known as PARP 1 and PARP 2) initially "recruit" other particular protein molecules to repair the resultant DNA damage. However, if that DNA damage is to *BRCA* genes in DNA (i.e., associated with loss of BRCA proteins to cell), such mutations in *BRCA* genes often result in ineffective PARP repair of the cell's damaged DNA, which increases the risk for that cell developing certain cancers (e.g., cancers of the breast cells, ovary cells). Pharmaceutical inhibition of the cell's PARP in such cases thereby prevents the repair of such cell's damaged DNA, leading to apoptosis (cell death) that—appropriately—prevents cell from becoming cancerous.

See also *Enzyme*, *Adenosine diphosphate (ADP)*, *Ribose*, *Polymerase*, *Cell*, *Deoxyribonucleic acid (DNA)*, *DNA repair*, *Gene*, *Coding sequence*, *Express*, *Expressivity*, *RNA interference (RNAi)*, *Apoptosis*, *Abiotic stresses*, *Cell culture*, *Mammalian cell culture*, *CHO cells*, *Carcinogen*, *Cancer*, *Nuclear matrix proteins*, *Genetic*

P

engineering, AMES test, Caspases, Drought tolerance, Drought tolerance trait, BRCA 1 gene, BRCA genes, Olaparib.

PARP Inhibitors

Refers to either RNA interference of (excessive) PARP production in drought-stressed crop plants or to PARP inhibitor pharmaceuticals (e.g., which have been shown to act against some breast and ovarian cancers).

For example, during 2014, the U.S. FDA approved for pharmaceutical use Lynparza (olaparib) as the first poly ADP-ribose polymerase (PARP) inhibitor for patients with germline mutations in BRCA1/2 advanced ovarian cancer who have had three or more lines of chemotherapy. See also *PARP, Cancer, Food and Drug Administration (FDA), BRCA genes, Olaparib.*

Parthenogenesis

From the Greek *parthenos* meaning virgin, and *genesis* meaning origin, it refers to the production of a viable embryo from an egg that has not been fertilized. Examples of parthenogenesis include apomixis, androgenesis, and so on. See also *Apomixis.*

Parthenolide

See *Apoptosis.*

Particle Cannon

See *Biolistic® gene gun, Microparticles.*

Particle Gun

See *Biolistic® gene gun, Microparticles, "Shotgun" method.*

Partition Coefficient

A constant (number) that expresses the ratio in which a given solute will be partitioned (i.e., distributed) between two given immiscible liquids (e.g., oil and water) at equilibrium.

Partitioning Agent

Any one of a number of chemical compounds (e.g., certain hormones, conjugated fatty acids, etc.) which cause a given animal's metabolism to deposit significantly more lean muscle tissue and significantly less fat tissue, within that (growing) animal's body. See also *Bovine somatotropin (BST), Porcine somatotropin (PST), Conjugated linoleic acid (CLA), Carnitine, Metabolism, Fats.*

Passive Immunity

An immune response (to a pathogen) that results from injecting *another* organism's antibodies and/or T-lymphocytes into the organism that is being challenged by the pathogen. See also *Polyclonal antibodies, Humoral immunity, Antibody, Complement, Complement cascade, Immunoglobulin, Pathogen, Antigen, Monoclonal antibodies (MAb), Lymphocyte, T cells.*

Pasteuria

Refers to parasitic-to-SCN (soybean cyst nematode) *Pasteuria* spp. bacteria that can be applied to soybean seeds (e.g., as a crop biological coating) prior to planting, in order to help control the soybean cyst nematodes. The *Pasteuria* bacteria must attach their spores (for reproduction) to juvenile nematodes, so that the *Pasteuria* offspring can consume the SCN when the spores later germinate.

See also *Bacteria, Soybean cyst nematodes (SCN), Crop biologicals.*

Pasteuria nishizawa

See *Pasteuria.*

PAT Gene

A dominant gene *isolated from the Streptomyces viridochromogenes bacterium* which codes for (i.e., causes production of) the enzyme *phosphinothricin acetyl transferase (PAT).* When the *PAT* gene is inserted into a plant's genome, it imparts resistance to glufosinate-ammonium-containing herbicides. Because the glufosinate-ammonium herbicides act via inhibition of glutamine synthetase (an enzyme that catalyzes the synthesis of glutamine), this inhibition of enzyme kills plants (e.g., weeds). That is because glutamine is crucial for plants to synthesize critically needed amino acids. The *PAT* gene is also often used by genetic engineers as a marker gene. See also *Gene, Genome, Genetic engineering, Marker (genetic marker), BAR gene, Dominant allele, Essential amino acids, Herbicide-tolerant crop, GTS, Soybean plant, Canola, Corn, Glutamine, Glutamine synthetase, Phosphinothricin, Phosphinothricin acetyltransferase (PAT).*

Patch Clamping

Developed by Bert Sakmann and Erwin Neher in 1976, this is a research methodology for measuring the current/potential of a single ion channel. The researcher utilizes a microscope to attach tiny electrode tips to the membrane of a cell, and thereby apply a small quantitatively known electrical voltage across that cellular membrane.

Although the initial 1976 technique was slow and laborious (i.e., to "hook up" the cell to electrodes, to measure electrical potential), several companies have since developed easier/faster methodologies to measure the electrical potential across a cell's ion channel(s) as that cell is exposed to one or more pharmaceutical candidate compounds.

Via patch clamping, scientists can monitor extremely fast (e.g., one millionth second) opening or closing of ion channels (e.g., in response to pharmaceutical chemical trigger or to voltage). See also *Cell, Membrane, Ion, Ion channels, Target (of a therapeutic agent), High-throughput screening (HTS).*

Patch-Clamp Recording

See *Whole-cell patch-clamp recording.*

Pathogen

Refers to a virus, bacterium, fungus, parasitic protozoan, or other microorganism that causes infectious disease by invading the body of an organism (e.g., animal, plant, etc.) known as the host. It should be noted that infection is not synonymous with *disease* because infection does not always lead to injury of the host. See also *Virus, Bacteria, Protozoa, Microorganism, Stress proteins, Antigen, Immune response, Phytoalexins, Pathogenesis-related proteins, Quorum sensing, Effectors (fungal).*

Pathogen-Associated Molecular Patterns

See *PAMPs.*

Pathogenesis-Related Proteins

Protective (i.e., disease-fighting) proteins that are produced within certain plants in response to the entry into plant of plant pathogens (e.g., bacteria, fungi, etc., that infect and cause disease in plants).

One pathogenesis-related protein is chitinase, a protein enzyme that degrades (breaks down) the chitin within cell walls of pathogenic fungi.

Production of pathogenesis-related proteins is often initiated by *signaling molecules* (e.g., harpin) produced *by the pathogens.* See also *Protein, Pathogen, Bacteria, Fungus, Chitinase, Chitin, Cell, Enzyme, Signaling, Signaling molecule, Harpin, Hypersensitive response, Septins.*

Pathogenic

Disease causing. See *Pathogen.*

Pathway

A sequential series of chemical reactions, each of which is dependent on previous ones in the pathway (e.g., the third reaction requires chemical product produced by first/second chemical reactions), that—overall—yields a beneficial impact.

For example, metabolism (i.e., the entire set of enzyme-catalyzed chemical reactions which converts food into nutrients that can be used by the body's cells *and* the use of those nutrients by the body's cells to sustain life, grow, etc.) occurs via a very specific METABOLIC PATHWAY. See also *Metabolism, ACC synthase, R genes, Pathway feedback mechanisms, Retinoid X receptors (RXR), Mitogen-activated protein kinase cascade, Bone morphogenetic protein-signaling pathway, Abscisic acid pathway, MicroRNA pathway, Short hairpin RNA, CBF/DREB1 pathway, ras pathway, Differentiation pathways, ABA pathway, Angiogenesis, Systems biology, Anabolic pathway, NFκB pathway.*

Pathway Feedback Mechanisms

Chemically based mechanisms (e.g., series of chemical reactions) that hinder (or increase rate of) a given PATHWAY.

For example, when the body of bacteria needs *catabolism (i.e., energy production)* to be slowed down, it uses the mechanism of *catabolite repression* (to slowdown catabolism via chemical/reaction means). See also *Pathway, Metabolism, Catabolism, Catabolite repression.*

Pattern Biomarkers

Refers to the pattern (group) of genes which are *jointly* "turned-on"/off/up/down as a result of a disease (endpoint or progression), toxicity of a compound (e.g., pharmaceutical ingested), and so on.

In metabonomics, the term PATTERN BIOMARKER refers to the resultant *several compounds* (e.g., metabolites) instead of to genes. See also *Biomarkers, Gene, Express, Expressivity, Gene expression analysis, Metabonomics.*

Pattern Recognition Receptor

A receptor (i.e., protein on cell surface) within certain organisms (e.g., plants) that enable its innate immune defense system to respond to (telltale surface molecular) patterns found on invading pathogens. See also *Receptors, Protein, Cell, Organism, Innate immune response.*

Patulin

A term that refers to a particular mycotoxin (i.e., toxic metabolite produced by fungi) which is produced especially by the fungus *Penicillium expansum.* That fungus tends to grow (postharvest) in apples under certain conditions, causing "soft rot" or "blue mold rot."

Research indicates that patulin can cause adverse impacts on the immune system, can cause mutations, and can have neurotoxic effects. See also *Mycotoxins, Toxin, Fungus, Penicillium, Mutation, Mutagen.*

PBEF

Acronym for *Pre-B cell colony-Enhancing Factor.* See *Visfatin, Pre-B cell colony-enhancing factor.*

PBR

The intellectual property rights that are legally accorded to plant breeders by laws, treaties, and so on. Similar to patent law for inventors. See also *Plant breeder's rights (PBR), Plant's novel trait (PNT), Plant Variety Protection Act (PVP), European Patent Convention, European Patent Office (EPO), U.S. Patent and Trademark Office (USPTO).*

pBR322

An *Escherichia coli (E. coli)* plasmid cloning vector that contains the ampicillin resistance and tetracycline resistance genes. It consists of a circle of double-stranded DNA. See also *Escherichia coliform (E. coli), Plasmid, Vector, Deoxyribonucleic acid (DNA).*

PC

Phosphatidyl choline. See *Lecithin (refined, specific), Lecithin (crude, mixture).*

PCC

See *Protein-conducting channel.*

PCD

Acronym for *Programmed Cell Death.* See *Programmed cell death.*

PCF

Acronym for *Plant Cell Fermentation.* See *Plant cell fermentation.*

P

PCR

See *Polymerase chain reaction (PCR).*

PCSK9 Inhibitors

Refers to humanized monoclonal antibodies created against the cholesterol-regulating protein PCSK9 (proprotein convertase subtilisin/kexin type 9). Because the binding of PCSK9 to LDLP receptors on the surface of liver cells results in a decrease in the total number of those LDLP receptors (which normally remove LDL particles from the bloodstream), PCSK9 inhibitors act to increase the number of liver LDL receptors and thereby reduce bloodstream levels of LDL cholesterol. See also *Monoclonal antibodies (MAb), Humanized monoclonal antibody, Cholesterol, LDL, Protein, Receptors, Low-density lipoproteins (LDLP), LDLP receptors.*

PDCAAS

See *Protein digestibility-corrected amino acid scoring (PDCAAS).*

PDE

See *Phosphodiesterases.*

PDGF

See *Platelet-derived growth factor (PDGF).*

PDWGF

See *Platelet-derived wound growth factor (PDWGF).*

Pectin

From the Greek *pektos* meaning congealed. See *Amylopectin, Soybean plant.*

Pectinophora gossypiella

Also known as the pink bollworm, this is one of three insect species that are called "bollworms" (when they are on cotton plants). The holes that they chew in cotton plant's bolls have been shown to enable the *Aspergillas flavus* fungus to infect those (chewed) cotton plants. See *B.t. kurstaki, Helicoverpa zea (H. zea), Heliothis virescens, Bright greenish-yellow fluorescence (BGYF).*

PEG-SOD (Polyethylene Glycol Superoxide Dismutase)

A modified version of the enzyme human superoxide dismutase (hSOD) in which polyethylene glycol (PEG—a polymer made up of ethylene glycol monomers) is combined with the hSOD molecule. The PEG seems to wrap around or about the enzyme in such a way that the whole complex is able to exist in the blood for longer periods of time than the unmodified hSOD enzyme. This is because the PEG effectively camouflages the hSOD molecule and hence protects it from being inactivated by the body's own immune system defense mechanisms in the bloodstream. This technology is important in that hSOD is used to fight certain diseases by injecting it into the body. However, the SOD must be present in the body for extended periods of time in order to effectively work, and since the injected SOD is a foreign molecule, the body tries to destroy it (and hence its function) as quickly as possible. See also *Human superoxide dismutase (hSOD), Pegylation, Catalase, Enzyme.*

PEGylation

Refers to the process of attaching PEG molecules to the surface of something (e.g., a nanoparticle, pharmaceutical molecule, etc.). This enables the PEGylated nanoparticle/pharmaceutical to "hide" (avoid detection) from the immune system of the organism into which the nanoparticle or pharmaceutical is injected. That "hiding" enables the pharmaceutical to do its work or the nanoparticle to accomplish its task (kill a tumor via heat, etc.) before it gets cleared from the body by the immune system. See also *Nanoparticles, Aptamers, PEG-SOD (polyethylene glycol superoxide dismutase).*

Penicillin G (Benzylpenicillin)

The original penicillin (antibiotic) molecule, discovered by Alexander Fleming in 1928, in a petri dish (experiment) "spoiled" by accidental introduction of a mold. Fleming named the antibiotic after the particular mold (*Penicillium notatum*) that had produced it.

During the 1940s, scientists at the U.S. Department of Agriculture in Peoria, IL (in the United States) discovered how to produce commercial quantities of Penicillin G by utilizing the fungus *Penicillium chrysogenum*, which they found growing on a cantaloupe in Peoria, IL. Penicillin kills bacteria by blocking an enzyme which is crucial to growth and repair of the bacteria's cell wall (peptidoglycan layer), but penicillin does not harm other species, so it is species specific to certain pathogenic bacteria (e.g., streptococcus, meningococcus, and diphtheria bacillus). See also *Antibiotic, Fungus, Bacteria, Enzyme, Species specific, Penicillium, Beta-lactam antibiotics, Bacillus.*

Penicillinases (E.C. 3.5.2.6)

Also known as β-lactamases, these are enzymes that hydrolyze (break down) the β-lactam ring (portion) of the penicillin molecule's structure. Some microorganisms (e.g., pathogenic bacteria) have become able to produce these enzymes as a defense to penicillin and cephalosporin antibiotics (drugs). See also *Enzyme, Hydrolyze, Penicillin G (benzylpenicillin), Pathogenic, Bacteria, Antibiotic, Antibiotic resistance.*

Penicillium

Refers to the genus of fungi (mold) that belongs to the category *Deutromycotina* and often causes (food) spoilage. Some strains (e.g., a strain of *Penicillium bilaii* that grows along the roots of corn, canola, wheat, and legume plants) naturally enhance those crop plants' utilization of phosphate from the soil and applicable fertilizers by 3%–10% and some strains within this genus have been utilized commercially to produce antibiotics. See also *Genus, Fungus, Ochratoxins, Crop biologicals, Antibiotic, Penicillin G (benzylpenicillin).*

Penicillium syringae

A pathogenic fungus that initially adheres to plant cell walls and then forms itself into a sort of "injection syringe" which literally injects approximately 30 proteins into the plant cells through their cell walls. Among other pathogenic functions, some of the injected fungal proteins (e.g., AvrPphB) act to inhibit the plant's mitogen-activated protein kinase cascade defensive response. See also *Fungus, Pathogen, Cell, Pathogenic, Protein, Mitogen-activated protein kinase cascade.*

Pentose

A simple sugar (monosaccharide molecule) whose backbone structure contains five carbon atoms. There exists many different pentoses. Some examples of pentoses are as follows: ribose, arabinose, and xylose, to mention just a few. See also *Monosaccharides.*

Pepsin

A crystallizable proteinase (enzyme) that in an acidic medium digests (breaks down) most proteins to polypeptides. It is secreted by glands in the mucous membrane of the stomach of higher animals. In combination with dilute hydrochloric acid it is the chief active principal (component) of gastric juice. Also used in manufacturing peptones and in digesting gelatin for the recovery (i.e., recycling) of silver from photographic film. See also *Digestion (within organisms), Protein, Peptide, Lactoferrin, Peptone.*

Peptidase

An enzyme that hydrolyzes (cleaves) a protein molecule at a peptide bond (i.e., between amino acids). See also *Peptide bond, Amino acid, Protein, Enzyme, Pepsin, Peptone, Peptide mapping ("fingerprinting").*

Peptide

Two or more amino acids covalently joined by peptide bonds. An oligomer component of a polypeptide. A dipeptide, for example, consists of two (di)amino acids joined together by a peptide bond or linkage. By analogy, this structure would correspond to two joined links of a chain. See also *Polypeptide (protein), Oligomer, Amino acid, Peptide bond, Peptide-oligonucleotide conjugates.*

Peptide Bond

A covalent bond (linkage) between the α-amino group of one amino acid and the α-carboxyl group of another amino acid. This is the linkage or bond which holds the amino acids (chain links) together in a polypeptide chain. It is the all-important bond that holds the amino acid monomers together to form the polymer known as a polypeptide. See also *Peptide, Polypeptide (protein), Oligomer, Peptidyl transferase, Carboxyl terminus (of a protein molecule).*

Peptide Mapping (Fingerprinting)

Refers to the characteristic pattern of peptides (i.e., pieces that make up a protein molecule) resulting from partial hydrolysis (cleavage, digestion) of a protein.

The pattern (fingerprint) is obtained by separating the peptides via two-dimensional chromatography, in which the peptides are first subjected to chromatography using one solution which separates many, but not all peptides. The chromatogram is then turned 90°, and is again chromatographed using a second solution, which then separates all of the peptides—thereby producing the final "fingerprint" of the protein. See also *Chromatography, Peptide, Protein, Hydrolysis, Mass spectrometer.*

Peptide Nanotube

See *Self-assembly (of a large molecular structure).*

Peptide Nucleic Acid

Polymeric molecules possessing a molecular "backbone" that is structurally similar to that of a DNA molecule. A peptide nucleic acid (PNA) molecule can be made that will hybridize with (bind tightly to) any piece of DNA (e.g., a gene, etc.) whose molecular sequence is known. Thus, PNA's can be utilized when attached to an appropriate surface (e.g., microarray, biochip, etc.) in the manufacture of testing devices to detect the presence of such pieces of DNA (e.g., gene) in something being sampled (e.g., a given foodstuff, etc.).

The PNA molecular "backbone" is essentially a polypeptide molecular structure, bearing attached purine and pyrimidine bases which can pair up with their counterpart bases on the specific DNA piece (e.g., gene, etc.) via hybridization process. See also *Polymer, Nucleic acids, Deoxyribonucleic acid (DNA), Hybridization (molecular genetics), Microarray (testing), Biochips, Hybridization surfaces, Sequence (of a DNA molecule), Peptide, Peptide bond, Polypeptide (protein), Base (nucleotide), Base pair (bp), Gene.*

Peptide-Oligonucleotide Conjugates

Abbreviated POCs, these are chimeric molecules consisting of a peptide or protein linked to an oligonucleotide. They can thereby possess some of the valuable properties of both. For example, the peptide's cellular uptake/cell-membrane-crossing ability and the oligonucleotide's ability to control gene expression (after it enters the targeted cell). The most commonly used peptides within POCs are cell-penetrating peptides, which have the ability to carry the linked oligonucleotide through the cell membrane into the cell's cytosol. See also *Peptide, Protein, Oligonucleotide, Conjugate, Chimera, Chimeric proteins, Cell, Plasma membrane, Gene, Gene expression, Gene expression cascade.*

Peptidoglycan

Refers to a polymer ("molecular chain") which comprises equal amounts of peptides and polysaccharide units. See also *Peptide, Glycan, Polymer, Polysaccharides.*

Peptido-Mimetic

See *Biomimetic materials, Peptide.*

Peptidyl Transferase

An enzyme within the ribosome, which catalyzes the formation of peptide bonds (within protein molecule) as protein molecules are being synthesized (i.e., "manufactured") during the process of translation. See also *Enzyme, Ribosomes, Peptide bond, Protein, Gene, Translation.*

Peptoids

See *Nanosheets.*

P

Peptone

A protein that has been partially hydrolyzed (i.e., cleaved) by the peptidase pepsin. See also *Protein, Hydrolytic cleavage, Peptidase, Pepsin, Peptide mapping ("fingerprinting")*.

Perforin

A 70 kDa (kilodalton) protein that is instrumental in the lysis of infected cells. A series of reactions occur on the surface of a cell which results in the polymerization of certain monomers to form transmembrane (i.e., through the membrane) pores 100 Å (Angstroms) wide, which allows ions to rush into the cell (due to osmotic pressure) and thus burst (lyse) that cell, so the (formerly) internal pathogens can be attacked by the body's immune system. Perforin is a protein that is akin to the C9 component of the complement. See also *Osmotic pressure, Complement, Complement cascade, Kd, Cytotoxic T cells, Cecrophins, Magainins, Osmotins.*

Periodicity

The number of base pairs per turn of the DNA double helix. See also *Deoxyribonucleic acid (DNA).*

Periodontium

Tissue that anchors teeth in the jaw. Regrowth of periodontal tissue can be stimulated by a combination of platelet-derived growth factor (PDGF) and insulin-like growth factor-1. See also *Platelet-derived growth factor (PDGF), Insulin-like growth factor-1 (IGF-1).*

Peritoneal Cavity/Membrane

The smooth, transparent, serous membrane that lines the cavity of the abdomen of a mammal.

Permeable

From the Latin *permeabilis* meaning *that which can be penetrated.*

Peroxidase

An enzyme that catalyzes the *oxidation of a substrate with hydrogen peroxide* (as the electron acceptor, so the hydrogen peroxide is *reduced*).

Peroxidase is naturally produced in soybeans by approximately half of all commercial soybean varieties. Peroxidase very effectively inhibits (stops) growth of any *Aspergillus flavus* fungi that might be present (e.g., in the soil). Peroxidase can be used to replace more toxic and environmentally problematic chemicals in certain industrial processes. Among other applications, peroxidase can replace formaldehyde use in paints, varnishes, glues, and computer chip manufacturing. See also *Enzyme, Oxidizing agent (oxidant), Oxidation, Substrate (chemical), Oxidation–reduction reaction.*

Peroxisome

Special organelles within the interior of cells that carry out some of the chemical reactions (i.e., oxidative reactions) involved in the cell's metabolism of lipids. See also *PPAR, Organelles, Cell, Lipids, Metabolism.*

Persistence

The tendency of a compound (e.g., an insecticide) to resist degradation by biological means (e.g., metabolism by microorganisms) after that compound has been introduced into the environment (e.g., sprayed onto a field) or by physical means (e.g., degradation caused by exposure to sunlight, moisture, etc.).

The term persistence can also be utilized to refer to the ability of an organism to live/remain in a particular (new) environment after it has been placed in that environment. See also *Metabolism, Organism, Microorganism, Biodegradable.*

Personalized Medicine

Refers to the use of an individual's genetic information (e.g., by doctors) to customize the prevention, detection, or treatment of a disease. See also *Deoxribonucleic acid (DNA), Gene, Genomics, HER-2 gene, Pharmacogenomics, Pharmacogenetics, Immune profiling.*

Perturb

See *Copy number variant.*

Pest Free Area

See *Introduction, Pest risk analysis (PRA).*

Pest Risk Analysis (PRA)

A process delineated by the International Plant Protection Convention, consisting of the following:

- *Analyzing risk*: Identification of potential pests and/or pathways via which a pest (e.g., weed, insect, disease, etc.) might enter a "pest free area," plus determination of whether a pest is a "quarantine pest" and evaluation of its potential "introduction" to the pest free area.
- *Assessing pest risk*: Determination of whether a given pest is a "quarantine pest" and evaluation of the potential for that pest to be "introduced" into a "pest free area."
- *Managing pest risk*: The decision-making process, and measures instituted, to reduce the risk of a "quarantine pest" being "introduced" into a "pest free area."

See also *international Plant Protection Convention (IPPC), Quarantine pest, Introduction, Establishment potential.*

Petiole

The part of the plant that connects a leaf to the plant's stem. Growth (elongation) of the stem is promoted by the plant hormone ethylene.

Peyer's Patches

A specific set of lymphoid organs found in the intestinal wall of many mammals. These patches filter out antigens that enter the intestine in food or come from bacteria growing in the intestine, and "present" those intact antigens to adjacent lymphoid tissues via special *M cells* of the Peyer's patches. This activates the lymphocytes in the patches, which then migrate out of the node and into the blood where they float in the tissue spaces just inside the intestinal lining. There they secrete antibodies (primarily IgA), which are then transported into the lumen (contents) of the gut and subsequently attack (bind) the antigens. See also *Humoral immunity, Antigen, Lymphocyte, Antibody, Immunoglobulin, M cells, Edible vaccines.*

Pfiesteria piscicida

A single-celled microscopic alga which has a predator/prey relationship with fish in its ecosystem. During a large portion of its life cycle, *Pfiesteria piscicida* exists in a nontoxic cyst form at the bottom of a river. When those (cysts) detect certain substances (e.g., excreta) emitted by live fish, the *P. piscicida* transform into an amoeboid or dinoflagellate form, which secretes a water-soluble neurotoxin into the water (which incapacitates nearby fish). The *P. piscicida* next attach themselves to those fish and excrete a lipid-soluble toxin which destroys the epidermal layer of the fish's skin, allowing the *P. piscicida* to begin "eating" the fish's tissue. Human exposure to the neurotoxin apparently causes short-term memory loss. See also *Ecology, Cell, Toxin, Lipids.*

PGHS

Abbreviation for Prostaglandin H Synthase, an enzyme that is colloquially referred to as *cyclooxygenase* or COX. PGHS exists in two different forms (isozymes): PGHS-1 (COX-1) and PGHS-2 (COX-2). See also *Enzyme, Cyclooxygenase, Isozymes.*

P-Glycoprotein

Refers to a specific glycosylated protein (e.g., which occurs in high numbers on interior cell surfaces of blood vessels within the brain, on surfaces of cells of certain cancer tumors, etc.) that act as a "pump" to remove specific chemicals (e.g., naturally occurring toxins, some chemotherapy drugs, etc.) from within those brain cells or tumor cells. See also *Protein, Glycosylation, Glycoprotein, Cell, Blood–brain barrier (BBB), Cancer, Tumor, Multidrug resistance, Efflux pump.*

P-gp

Abbreviation for *P-glycoprotein.* See also *P-glycoprotein, Efflux pump.*

PGx

Abbreviation for pharmacogenetics. See *Pharmacogenetics.*

PHA

See *Polyhydroxyalkanoic acid (PHA).*

Phage

Abbreviation for *bacteriophage*, which is another name for a specific type of virus. It is a virus that attacks bacteria (and "consumes" those bacteria via the making of more copies of that virus) is known as a bacteriophage. Phage is derived from the Greek word meaning "to eat."

Bacteriophages are frequently used as vectors for carrying (foreign) DNA into cells by genetic engineers. See also *Bacteriophage, Vector, Genetic engineering, Transfection, Deoxyribonucleic acid (DNA), Phage display.*

Phage Display

A methodology in which capsid proteins (i.e., on surface of bacteriophages) are attached to selected peptides. Because determination of each bacteriophage's DNA (gene) sequence thus determines the peptide (sequence) that is "displayed" on its surface, large "libraries" of phage-displayed peptides can be created (e.g., to be utilized by scientists in the screening of candidate-compounds, etc., in search for new pharmaceuticals). For example, the pharmaceutical ADALIMUMAB (Humira™) was developed using phage display technology. See also *Protein, Capsid, Bacteriophage, Genetic engineering, Transfection, Vector, Peptide, Library, Adalimumab, Gene, Deoxyribonucleic acid (DNA), Biopanning.*

Phagocyte

A cell such as a leukocyte that engulfs and digests cells, cell debris, microorganisms, and other foreign bodies in the bloodstream and tissues (phagocytosis). The ingested material is then degraded via enzymes. A whole class of cells is known to be phagocytic. See also *Macrophage, Microphage, Monocytes, T cells, Polymorphonuclear leukocytes (PMN), Cellular immune response, Polymorphonuclear granulocytes, Lysosome.*

Phagocytosis

A process via which certain immune system cells (e.g., macrophages, polymorphonuclear leukocytes (PMN), etc.) extend toward and engulf (by surrounding/enveloping) specific cells (e.g., pathogens that are invading that organism). Subsequent to engulfing, such immune system cells later fuse with lysosomes, resulting in the digestion/destruction of the engulfed (pathogen). See also *Cellular immune response, Phagocyte, Macrophage, Lysosome, Pathogen.*

Phagophores

See *Lysosome, Autophagy.*

Pharmacoenvirogenetics

A word coined during 2000 by Tim Studt to describe the fact that *environmental factors* interact with a given individual organism's (human/animal/plant) genetic makeup (i.e., genome) to determine the individual's (body's) response to a given pharmaceutical/nutraceutical/crop biological (and/or progression of a disease). Those environmental factors include the following:

- Foods eaten/nutrients absorbed by the individual organism.
- The stress the individual is exposed to.
- Air and water pollution the individual is exposed to.
- Temperature and humidity the individual is exposed to.
- Geographical elevation the individual is exposed to.
- Bacteria the individual is exposed to. For example, when *Rhizobium japonicum* bacteria grow in the soil near the roots of a soybean plant (*Glycine max* L.), that causes certain specific genes in the soybean plant to be expressed (i.e., "turned on") so that soybean plant's roots *become*

P

more hospitable as a "home" for those R. japonicum bacteria to live symbiotically (in nodules on the roots) with the soybean plant.

See also *Nutraceuticals*, *Crop biologicals*, *Pharmacokinetics*, *Genetics*, *Pharmacology*, *Pharmacogenetics*, *Absorption*, *Metabolism*, *SNP*, *Allele*, *Haplotype*, *Haptoglobin*, *Rhizobium* (*bacteria*), *Nodulation*, *Symbiotic*, *Central dogma* (*new*), *Acclimatization*.

Pharmacogenetics

A branch of pharmacokinetics that deals with the reactions between drugs, free radicals, nutraceuticals, or extracted/synthetic food ingredients, and *specific individuals* due to the genetics of those individuals.

The subgroup of all those individuals whose DNA causes their bodies to respond in a specific way to a given drug or synthetic food ingredient is known as a HAPLOTYPE. For example, one haplotype (subgroup) of *pediatric leukemia* patients suffers severe and life-threatening reactions to some commonly used leukemia treatment drugs, due to the variation (i.e., SNP) in the *thiopurine S-methyl transferase gene* (allele) in their genome.

Another example is the 2010 discovery by John Bartlett that a SNP known as CEP17, found on chromosome 17 in the DNA of one haplotype of women, is a "highly significant indicator" that breast cancer tumors in those women will respond to the class of chemotherapy drugs known as anthracyclines (i.e., antitumor antibiotics which interfere with enzymes involved in DNA replication).

Another example is that consumption of the tyrosine kinase inhibitor drug *gefinitib* (Iressa) works to control tumors of non-small-cell lung cancer in people who possess a mutation (i.e., the SNP which codes for the epidermal growth factor receptor) within their lung cancer tumors.

Another example is that consumption of the pharmaceuticals tolbutamide, warfarin, or phenytoin can be riskier for people who possess a mutation (i.e., an SNP which codes for less or no expression of CYP2C9) within their liver tissue. That is because the CYP269 enzyme causes rapid metabolism of tolbutamide, warfarin, phenytoin (and some other pharmaceuticals), so the "typical dose" could result in higher-than-expected bloodstream levels of those pharmaceuticals in people possessing that particular SNP.

Another example is that consumption of plant sterols helps to lower blood cholesterol levels of people in some haplotypes but is ineffective in people of other haplotypes.

Another example is that consumption of sodium-containing food ingredients tends to cause dangerous increase in blood pressure (hypertension) among the African-American people living in the United States, more often than among other ethnic groups living in the United States.

See also *Pharmacokinetics*, *Pharmacogenomics*, *Genetics*, *Pharmacology*, *Absorption*, *Metabolism*, *Haplotype*, *Deoxyribonucleic acid* (*DNA*), *Gene*, *Mutation*, *SNP*, *Allele*, *Coding sequence*, *Express*, *Expressivity*, *Enzyme*, *Cytochrome P450* (*CYP*), *Cancer*, *Haptoglobin*, *Transversion*, *Fluorescence in situ hybridization* (*FISH*), *Tyrosine kinase inhibitors* (*TKI*), *Receptors*, *Epidermal growth factor receptor* (*EGF*), *Chromosomes*, *Chemotherapy*, *Replication* (*of DNA*), *Antibiotic*, *Copy number variation*, *Immune profiling*, *Nutraceuticals*, *Streols*.

Pharmacogenomics

A branch of pharmacokinetics that deals with the biological impacts of pharmaceuticals, nutraceuticals, or extracted/synthetic food ingredients, and the *specific differences in response/reaction* of living structures (e.g., tissues, organs, etc.) *due to different genomes (DNA) of those individual organisms that consume those pharmaceuticals or food ingredients*.

The subgroup consisting of all those individuals whose genome (DNA) causes their body to respond in a specific way to a given pharmaceutical, free radical, or synthetic food ingredient is known as a HAPLOTYPE. A haplotype could (theoretically) be as small as one individual (e.g., one woman, possessing __ genome), because that woman's particular specific response to a pharmaceutical could result from one single-nucleotide polymorphism (SNP) that only her genome possesses.

Thus, pharmacogenomics is the *pharmacokinetics (of a given pharmaceutical or food ingredient) within a specific haplotype*. For example, some ethnic minorities, genders, and some individuals have far different biological reactions/responses to certain pharmaceuticals (e.g., the painkiller morphine works better in women, aspirin "thins" men's blood better than women's blood, the painkiller ibuprofen works better in men, the AstraZeneca tyrosine kinase inhibitor drug *gefinitib* (Iressa™) works to control tumors of non-small-cell lung cancer in patients whose lung cancer tumors possess the *SNP which codes for the epidermal growth factor receptor*, the blood-thinning drug *warfarin* can cause life-threatening internal bleeding in patients possessing certain SNPs (i.e., approximately one in every 250 people), the diuretic drug thiazide works to control hypertension in 60% of U.S. African Americans but only 8% of U.S. Caucasian people, etc.) and food ingredients (e.g., monosodium glutamate, lactase, ethanol, etc.) impact some members of some ethnic minorities more than they do the majority of humans. That is due to the fact that different gene(s) within their genomes (DNA) cause synthesis of certain different proteins (generally enzymes), which thereby cause the tissues/bodies of those individuals/ethnic minorities to react differently to specific pharmaceuticals or food ingredients in terms of the following:

- *Absorption*: Transport of the drug (pharmaceutical) or food ingredient into the bloodstream (e.g., from the intestinal tract, in the case of food ingredients or orally administered drugs).
- *Distribution*: Initial physical disposition/behavior of the substance in the body after the substance enters the body tissues. For example, does the substance preferentially concentrate in the fat cells (adipose tissue) of the body, or in other specific tissues?
- *Metabolism*: Breakdown of the substance (if breakdown does occur) into other chemical compounds and the ultimate disposition in body of those compounds (or the original substance, if breakdown does not occur).
- *Elimination*: The speed and thoroughness with which the substance is excreted or is otherwise removed from the body.

Another example is the 2010 discovery by John Bartlett that a SNP known as CEP17, found on chromosome 17 in the DNA of one haplotype of women, is a "highly significant indicator" that breast cancer tumors in those women will respond to the class of chemotherapy drugs known as anthracyclines (i.e., antitumor antibiotics which interfere with enzymes involved in DNA replication). See also *Pharmacokinetics*, *Genomics*, *Pharmacology*, *ADME tests*, *Absorption*, *Metabolism*, *Genome*, *Deoxyribonucleic acid* (*DNA*), *Digestion* (*within organisms*), *Phase I clinical testing*, *Haplotype*, *Consensus sequence*, *Pharmacogenetics*, *Gene*, *Allele*, *Single-nucleotide polymorphisms* (*SNPs*), *Protein*, *Enzyme*, *Haptoglobin*,

Adipose, Fluorescence in situ hybridization (FISH), Tyrosine kinase inhibitors (TKI), Receptors, Epidermal growth factor (EGF), Chromosomes, Cancer, Tumor, Antibiotic, Chemotherapy, Replication (of DNA), Copy number variation, Immune profiling, Nutraceuticals.

Pharmacokinetics (Pharmacodynamics)

A branch of pharmacology dealing with the reactions between drugs or synthetic food ingredients and living structures (e.g., tissues, organs). The study of the:

- *Absorption*: Transport of the drug (pharmaceutical) or food ingredient into the bloodstream (e.g., from the intestinal tract, in the case of food ingredients).
- *Distribution*: Initial physical disposition/behavior of the substance in the body after the substance enters the body. For example, does the substance preferentially concentrate in the fat cells of the body? And so on.
- *Metabolism*: Breakdown of the substance (if breakdown does occur) into other compounds and ultimate disposition of those compounds (or the original substance, if breakdown does not occur). For example, some pharmaceuticals break down into smaller compound(s), one of which then acts upon the relevant body cells (e.g., to relieve pain, lower blood pressure, etc.).
- *Elimination*: The speed and thoroughness with which the substance is excreted or otherwise removed from the body. Note that applicable drugs or drug carriers (e.g., certain nanoparticles) bearing a slightly negative charge have a long circulation time.

In short, pharmacokinetics deals with what happens to a substance that is introduced into a living system. For example, how quickly it is broken down, to what intermediates and metabolites it is broken down, and what the pathway of this breakdown is. See also *Pharmacology, ADME tests, Absorption, Metabolism, Intermediary metabolism, Digestion (within organisms), Phase I clinical testing, Pharmacogenomics, Nanoparticles, Pharmacogenetics, Pharmacoenvirogenetics, Pathway.*

Pharmacology

The study of chemicals (e.g., pharmaceuticals) and their effects on living organisms. See also *Pharmacokinetics, Pharmacogenomics, Pharmacogenetics.*

Pharmacophore

The portion of a molecule (e.g., a pharmaceutical) that is responsible for its biological activity (i.e., therapeutic action on recipient's tissue, etc.). See also *Biological activity, Active site, Catalytic site, Miniproteins, Pharmacophore searching.*

Pharmacophore Searching

One type of *in silico* screening in which scientists utilize the (known) chemical/molecular attributes and physical aspects of selected/known molecules—entered into a molecular modeling software program—to identify compounds which would be likely to interact with selected target(s) in a beneficial (e.g., pharmaceutically useful) manner. See also *In silico screening, Pharmacophore, Docking (in computational biology), Rational drug design, Computer-assisted drug design (CADD), Target (of a therapeutic agent), Ligand (in biochemistry), High-throughput screening (HTS), Target–ligand interaction screening, Quantitative structure–activity relationship (QSAR), Receptor mapping.*

Pharming

Refers to the production of pharmaceuticals (or intermediate chemicals utilized to manufacture pharmaceuticals) in agronomic plants (which have been genetically engineered). See also *Genetic engineering, Plantibodies™, Plantigens, Edible vaccines.*

Phase I Clinical Testing

The first in a series of human tests of new (proposed) pharmaceuticals, mandated by the United States' FDA. The primary purpose of the Phase I clinical test is to detect if the new pharmaceutical is toxic or otherwise harmful to the 20–100 normal, healthy humans usually selected for the testing. Phase I clinical testing also seeks to determine how the pharmaceutical is absorbed, distributed, how long it is active in the body, how it is metabolized by the body, and how/when excreted. The conclusion of Phase I testing leads to Phase II and Phase III testing.

During the 1990s, the FDA began to require the inclusion of ethnic minorities and women (in addition to men) as subjects in these tests, to enable pharmacogenomics (i.e., the testing to determine if a given pharmaceutical causes nontypical response in the bodies of members of these subgroups). See also *Food and Drug Administration (FDA), Kefauver rule, Koseisho, Bundesgesundheitsamt (BGA), Committee for Proprietary Medicinal Products (CPMP), IND, IND exemption, Pharmacogenomics, Haplotype, Phase II clinical tests, Fluorescence in situ hybridization (FISH).*

Phase I Detoxification Enzymes

See *Inducible enzymes.*

Phase II Clinical Tests

The second in a series of human tests of new pharmaceuticals, mandated by the United States' FDA. The primary purpose of the Phase II clinical tests (on 100–500 people) is to determine the pharmaceutical's *efficacy* (i.e., does it work?). Also to determine side effects, effective dose strength, and dosing schedule.

Successful conclusion of Phase II tests allows Phase III clinical tests to begin. See also *Phase I clinical testing, Food and Drug Administration (FDA), Kefauver rule, Koseisho, Bundesgesundheitsamt (BGA), Committee for Proprietary Medicinal Products (CPMP), IND, IND exemption, Fluorescence in situ hybridization (FISH).*

Phase II Detoxification Enzymes

See *Inducible enzymes.*

Phase III Clinical Tests

The third in a series of human tests of new (proposed) pharmaceutical compounds, mandated by the US FDA. The primary purpose of Phase III clinical tests (on several hundred or several thousands of

P

people) is to verify efficacy and proper dosage of the new pharmaceutical. Also to further assess safety and side effects.

In phase III, the outcomes for the humans taking the new drug are typically compared head to head with outcomes for those receiving a placebo or the current standard-of-care therapy (if one exists for that disease). Generally, the FDA requires that two "adequate and well-controlled" trials confirm that a new drug is safe and effective before it approves it for sale.

Extremely rigorous statistical tests are performed on the resultant data to ensure that the new drug's demonstrated benefit is genuine, instead of the result of chance.

See also *Phase I clinical testing, Phase II clinical tests, Food and Drug Administration (FDA), Kefauver rule, Koseisho, Bundesgesundheitsamt (BGA), Committee for Proprietary Medicinal Products (CPMP).*

Phaseolus vulgaris

Term utilized to refer to the broad "family" of edible beans which includes kidney beans, pinto beans, snap beans, and so on. See *Amylase inhibitors.*

PHB

See *Polyhydroxylbutylate.*

Phenolic Hormones

A category of compounds found in the human body that are synthesized (i.e., "manufactured") by the body from certain phenolic dietary substances (phytochemicals) such as isoflavones.

Research indicates that phenolic hormones act to prevent a number of cancers such as those of the prostate, breast, large bowel, and so on. See also *Hormone, Phytochemicals, Isoflavones, Cancer, Selective estrogen effect.*

Phenomics

Utilized to refer to the relationship between genomics and phenotype/traits, it includes study of how the genetic makeup of an organism determines its appearance, function, and performance.

Plant breeders are able to *quantitatively assess* the appearance and function of growing plants (i.e., as those plant breeders are modifying those plants' genetic makeups) via use of the following:

- Magnifying *infrared cameras* to determine temperature profiles of cells within specific tissues.
- *Spectroscopes* and/or *fluorescence microscopy* to quantify differing rates of photosynthesis in different cells within specific tissues.
- *Lidar* to quantitatively measure growth rates of different cells within specific tissues.
- *MRI* (nuclear magnetic resonance imaging) to assess plant root physiology.

Phenomics thus helps guide some plant breeders and plant genetic engineers in their quest to create plants possessing more desirable phenotypes (e.g., drought tolerant, salt tolerant, aluminum tolerant, etc.) faster/easier. See also *Functional genomics, Phenotype, Genetic engineering, Trait, Gene function analysis, Nuclear magnetic resonance, Fluorescence mapping, Drought tolerance, Salt tolerance, Aluminum tolerance.*

Phenotype

Coined in 1909 by Wilhelm Johannsen, this term refers to the outward appearance (structure) or other visible characteristics of an organism (which of course, is determined by the DNA of its genotype, plus some impacts of gene interactions with the environment and/or some other genes within the organism's own DNA during development). This also includes (and/or determines) how that organism's body responds to a given physical agent (e.g., a pharmaceutical, a toxin, sunlight, etc.). For example, genetically fair-skinned people tend to get sunburned easier/faster than other people do. See also *Genotype, Deoxyribonucleic acid (DNA), Morphology, Gene, Haplotype, Gene expression profiling, Epigenetics, Epistatic genetic interactions, Epigenetic variation.*

Phenotype Plasticity

Refers to changes in the appearances of some plants and animals in response to the environment they grow up within. For example:

- The teosinte plant (*Zea diploperennis*) which is the "wild ancestor" of today's domesticated corn/maize (*Zea mays* L.) looks much more like today's domesticated corn/maize plant if it is grown inside a special greenhouse that mimics the environment of late Pleistocene and early Holocene era (e.g., the temperature was 3.5°–5.4° celsius cooler than today, and atmospheric carbon dioxide content was approximately 260 parts per million—versus today's carbon dioxide level of approximately 405 ppm).
- A domesticated male pig looks very different (i.e., classic "wild boar") when it grows up "wild" (i.e., fending for food on its own in the forest) than when grown in modern commercial pig production conditions (i.e., provided complete nutritious feed rations daily by man, housed in a barn, etc.).

See also *Phenotype, Teosinte, Wild type, Corn.*

Phenylalanine (phe)

An essential amino acid, which is utilized by both plants and animals to build protein molecules. Phenylalanine is also a precursor (building block molecule) to make approximately 8000 other chemical compounds needed by plants for their growth and/or metabolism.

L-Phenylalanine is one of the raw materials used to manufacture NutraSweet® (NutraSweet Co.) synthetic sweetener. See also *Amino acid, Levorotary* (L) *isomer, Essential amino acids, Stereoisomers, Protein, Metabolism.*

Pheromones

From the Greek words "pherein" (to carry) and "hormon" (to excite), they are sex hormones emitted by insects and animals, and spread through the air by the wind and diffusion for the purposes of attracting the opposite sex. Some pheromones have been produced artificially and used in lure traps to attract and catch male insects so as to prevent their mating with females (i.e., a biological pesticide). Pheromone-based traps for Japanese beetles are commonplace in infested areas (e.g., when utilizing Integrated Pest Management). See also *Hormone, Integrated pest management (IPM).*

Philadelphia Chromosome

Refers to a particular human chromosome that is (visibly) distorted by the *mutated gene* that results in the disease known as chronic myelogenous leukemia (abbreviated *CML*, also known as chronic myeloid leukemia). That is because that gene codes for extensive production of the tyrosine kinase known as Bcr-Abl, an enzyme which causes neoplastic (aberrant) cell growth and cell division. As a result, people with CML disease tend to have 10–25 times more white blood cells than normal.

The pharmaceutical known as Gleevec™ induces *apoptosis—"programmed" (self-destruct) cell death*—in the cells that have the Philadelphia chromosome, thus leading to cessation of CML. See also *Chromosomes, Karyotype, Karyotyper, Gene, Coding sequence, Mutation, Cancer, Cell, White blood cells, Gleevec™, Apoptosis.*

Philadelphia Translocation

Refers to a particular human chromosome change that is visible as a result of the distortion caused by the mutated gene that results in the disease known as chronic myelogenous leukemia (abbreviated CML, also known as chronic myeloid leukemia). See *Philadelphia chromosome.*

Phloem

The "plant food"-conducting tissues within plants.

Phomopsins

A "family" of mycotoxins produced by the fungus *Diaporthe toxica* (formerly referred to as *Phomopsis leptostromiformis*). Ingestion of phomopsins (e.g., by livestock consuming infested lupin plant seed) can lead to toxicity resulting from disruption of microtubule functions in the body's cells. See also *Mycotoxins, Fungus, Microtubules.*

Phosphate Transporter Genes

Gene(s) within the genomes of at least some plants, which code for proteins that enable/increase the ability of those plants to extract and utilize *phosphate (form of phosphorous)* from the soil. Since all plants require phosphorous for proper growth and functioning, yet most plants are not inherently very adept at extracting and utilizing soil phosphate, adding (more) phosphate transporter genes to a given (crop) plant is likely to increase that plant's growth and yield (e.g., of seeds). See also *Gene, Genetic engineering.*

Phosphate-Group Energy

The decrease in free energy as one mole of a phosphorylated compound at 1.0 M concentration undergoes hydrolysis to equilibrium at pH 7.0 and 25°C (77°F). The energy that is available to do biochemical work. The energy arises from the breakage (cleavage) of a *phosphate to phosphate* bond. See also *Free energy, Hydrolysis, Fats, Mole, Phospholipids.*

Phosphatidyl Choline

See *Lecithin.*

Phosphatidyl Serine

A lipid that is naturally produced in soybeans, certain green leafy vegetables, and some meats. Research indicates that human consumption of large enough amounts of phosphatidyl serine can help to improve brain function.

During 2003, the U.S. FDA approved a qualified (label) health claim that associates consumption of phosphatidyl serine with reduced risk of cognitive dysfunction in people and reduced risk of dementia in elderly people.

Because phosphatidyl serine in healthy living cells is only present within the "interior layer" of the cell's plasma membrane, one way that *a cell undergoing apoptosis* is marked for degradation/cleanup (i.e., done by macrophages) is the presence of phosphatidyl serine on the exterior surface of the cell's plasma membrane.

Some formerly injected pharmaceuticals (e.g., Amphotericin B) can be rendered to be orally administrable via encapsulation inside a "coating" of phosphatidyl serine (i.e., known as a nanocochleate). See also *Lipids, Soybean plant, Food and Drug Administration (FDA), Cell, Plasma membrane, Apoptosis, Macrophage, Orally administered, Nanocochleates.*

Phosphatidylinositol

See *Inositol, Effectors (fungal).*

Phosphatidylserine

See *Phosphatidyl serine.*

Phosphinothricin

Another name for the herbicide active ingredient *glufosinate.* See also *Glufosinate, Phosphinothricin acetyltransferase (PAT), PAT gene, BAR gene.*

Phosphinothricin Acetyltransferase (PAT)

An enzyme which degrades (breaks down) phosphinothricin (also known as glufosinate), which is an active ingredient in some herbicides.

PAT is naturally produced in some strains of soil bacteria (e.g., *Streptomyces viridochromogenes*). If a gene (called the "*PAT* gene") that codes for the production of phosphinothricin acetyltransferase is inserted via genetic engineering into a crop plant's genome, that would enable such plants to survive postemergence applications of phosphinothricin-containing herbicides. See also *Enzyme, Phosphinothricin, Glufosinate, Bacteria, Gene, PAT gene, Genetic engineering, Genome, BAR gene, Marker (genetic marker).*

Phosphinotricine

See *Phosphinothricin.*

Phosphodiesterases

A category of enzymes that inhibit apoptosis. Abbreviation for this term (category) is *PDE.* See also *Enzyme, Apoptosis.*

Phospholipids

The principal class of lipids that are present in cell membranes; phospholipids are diglycerides (i.e., two fatty acids attached to a glycerol

P

"*molecular backbone*") to which is also attached a phosphate group. As a component of cell membranes, they serve to establish a barrier that limits ingress and egress of water-soluble compounds.

Another function of phospholipids is to provide raw materials for the body to manufacture leukotrienes, prostaglandins, and other signaling molecules.

The principal sites in plants of lipid and fatty acid biosynthesis (i.e., "manufacturing") are the endoplasmic reticulum, chloroplasts, and the mitochondria. See also *Lipids, Plasma membrane, Cell, Fats, Fatty acid, Phosphate-group energy, Leukotrienes, Prostaglandins, Signaling, Signaling molecule, Endoplasmic reticulum (ER), Chloroplasts, Mitochondria, Choline.*

Phosphorylation

The introduction of a phosphate group into a molecule. Formation of a phosphate derivative of a biomolecule, usually by enzymatic transfer of a phosphate group from ATP.

In some forms of cancer, certain very specific phosphorylations are produced (e.g., in posttranslational modification (PTM) of protein molecules). See also *Adenosine triphosphate (ATP), Kinases, Enzyme, Cancer, Posttranslational modification of protein, Epigenetic.*

Phosphorylation Potential

Abbreviated ΔG_p, it is the actual free-energy change of ATP hydrolysis under a given set of conditions. See also *Phosphorylation, Free energy, Hydrolysis, Adenosine triphosphate (ATP).*

Photolyases

A class of enzymes that are present in plants, frogs, fish, and snakes, which harness the energy of ultraviolet and/or near-ultraviolet light to repair damaged DNA within cells. See also *Enzyme, DNA repair.*

Photon

A single (smallest possible) unit of light energy. See also *Photosynthesis, Photosynthetic phosphorylation.*

Photoperiod

The optimum length or period of illumination required for the growth and maturation of a plant. The photoperiod is distinct from photosynthesis. See also *Phytochrome, Central dogma (new).*

Photophore

See *Bioluminescence.*

Photophosphorylation

See *Cyclic photophosphorylation.*

Photorhabdus luminescens

A soil-dwelling bacterium which produces certain toxins (effective against a variety of insect pests), antibiotics, antifungal compounds, lipases, proteases, and bioluminescent (light-producing) compounds.

P. luminescens naturally colonizes the gut of the *Heterorhabditis* nematode which attacks certain insect pests (e.g., tobacco hornworm, mealworm, cockroaches, etc.). When that nematode enters those insects, the *P. luminescens* is released inside the insect, which it subsequently kills via the toxins secreted by *P. luminescens.*

P. luminescens synthesizes (i.e., "manufactures") a protein that is high in content of the amino acids methionine and lysine, and that protein constitutes approximately 50% of the *total protein content* of *P. luminescens.* See also *Bacteria, Antibiotic, Toxin.*

Photosynthesis

The synthesis (production) of bioorganic compounds (molecules) using light energy as the power source. The synthesis of carbohydrates (e.g., hexose) occurs via a complicated, multistep process involving reactions that occur both in the light (light reactions) and in the dark (dark reactions). In eucaryotic cells the photosynthetic machinery necessary to capture light energy and subsequently utilize it is contained within structures called chloroplasts, which contain the molecule that initially captures light energy, called chlorophyll. Chlorophyll appears green in color. Green plants synthesize carbohydrates from carbon dioxide and water, which are used as a hydrogen source. The synthesis reaction, which is light driven, utilizes a catalyst comprising manganese, calcium, and oxygen to split water molecules, and which liberates oxygen in the process. Other organisms use this oxygen to sustain life.

From initial carbohydrates, plants subsequently also synthesize (i.e., manufacture) other compounds (e.g., fatty acids, amino acids, etc.).

Plants are not the only users of photosynthesis technology. Other organisms such as green sulfur bacteria and purple bacteria also carry out photosynthesis, but they use other compounds besides water as a hydrogen source. See also *C3 photosynthesis, C4 photosynthesis, Carbohydrates, Chloroplasts, Organism, Eucaryote, Hexose, Cyclic photophosphorylation, Carotenoids, Golden rice, Fatty acids, Amino acids, Bacteria, Alternative splicing, Ion channels.*

Photosynthetic Phosphorylation

Also called *photophosphorylation*, it is the formation of ATP from the starting compounds ADP and inorganic phosphate (P_i). The formation is coupled to light-dependent electron flow in photosynthetic organisms. See also *Photon, Photosynthesis, Adenosine triphosphate (ATP), Adenosine diphosphate (ADP), Cyclic photophosphorylation.*

Phototropism

See *Cortical microtubules.*

Phylogenetic Constraint

The limitations inherent in an organism as a result of what its ancestors were. For example, a horse will never fly and an ape will never speak, because the ancestors of neither possessed those capabilities. See also *Genotype, Phenotype, Genome, Morphology.*

Phylogenetic Profiling

A research methodology that is utilized to try to predict the *function of a protein molecule* within a larger, complex organism (e.g., man)

P

from the function of a similar/related protein molecule in a smaller/simple organism (e.g., a "model organism") which is easier to study. See also *Protein, Organism, Model organism, Ortholog.*

Physical Map (of Genome)

A diagram showing the linear order of genes or genetic markers on the genome, with units indicating the actual distance between the genes or markers. See also *Genetic map, Gene, Genome, Position effect.*

Physiology

The branch of biology dealing with the study of the functioning of living things. The materials of physiology include all life: animals, plants, microorganisms, and viruses.

Phytase

A digestive enzyme which is present in the digestive systems of many plant-eating animals to enable breakdown of phytate (also known as "phytic acid"). Phytase is sometimes present within the plant material consumed by animals. For example, phytase is naturally produced in the seed coat of wheat. See also *Enzyme, Digestion (within organisms), Phytate, High-phytase corn/soybeans, Low-phytate corn, Low-phytate soybeans.*

Phytate

The salt form of phytic acid, it is a chemical complex (large molecule) substance (inositol hexaphosphate plus a metal) that is the dominant (i.e., 60%–80%) chemical form of phosphorus present within cereal grains, oilseeds, and their by-products. Monogastric animals (e.g., swine, poultry) cannot digest and utilize the phosphorous within phytate, because they lack the enzyme known as phytase in their digestive system, so that phosphorus (phytate) is excreted into the environment. When phytase enzyme is present in the ration of a monogastric animal, at a high enough level, the monogastric animal is then able to digest the phytate (thereby "releasing" most of that phosphorus for absorption by the body of the animal).

However, the (cleaved-off, "free" inositol) that was "liberated" from (six phosphate atoms per molecule of phytate) than can quickly chelate (i.e., "combine" with) other minerals in the feed ration (e.g., iron, calcium, zinc, etc.) and the combination is excreted in the animal's wastes.

Thus, *low-phytate* crop varieties (i.e., containing inherently smaller amounts of inositol) are less likely to chelate important dietary minerals such as iron (which can exacerbate malnutrition in typically iron-poor diets such as in developing countries where adequate iron content/iron fortification of human diets is not common).

In adult humans (e.g., those past childbearing age), the chelating ("combining"-with) property of the phytate-source inositol causes it to act as a beneficial antioxidant in the human body, which can help to protect against certain cancers (e.g., prostate cancer). See also *Phytase, Low-phytate corn, Low-phytate soybeans, Enzyme, Digestion (within organisms), High-phytase corn and soybeans, Prostate, Cancer, Antioxidants, Chelation, Iron deficiency anemia (IDA).*

Phytic Acid

Also known as inositol hexaphosphate. See *Phytate.*

Phytoalexins

Term utilized to refer to chemical compounds (e.g., antimicrobials, etc.) that are produced by certain plants in response to the presence of infectious agents (e.g., fungus, bacteria) or their metabolic products.

From the Greek words *phyton* = "plant" and *alexein* = "to defend," phytoalexins possess antimicrobial (i.e., fungus-killing, bacteria-killing) properties, so they can help plants to protect themselves against those microorganisms. See also *Phytotoxin, Metabolism, Isoflavones, Allelopathy, Stress proteins, Pharmacoenvirogenetics, Antibiotic, Phytochemicals, Enzyme, Fungus, Bacteria, Isoflavones, Pathogenic, Microbe, Microbicide, Salicylic acid (SA), Pathogenesis-related proteins, Systemic acquired resistance (SAR).*

Phytochemicals

A term used to refer to certain biologically active chemical compounds that occur in fruits, vegetables, grains, herbs, flowers, bark, and so on. Phytochemicals act to repel or control insects, prevent plant diseases, control fungi and adjacent weeds. Phytochemicals also sometimes confer beneficial health effects to the animals (e.g., humans) that consume the plant (portions) containing those applicable phytochemicals.

For example, vitamin C in citrus fruits, beta carotene in carrots and other orange vegetables, *d*-limonene in orange peels, tannins in green tea, capsaicin in chili peppers, *n*-3 (omega-3) fatty acids in soybean oil and fish oil, *genistein, saponins, vitamin E, and phytosterols in soybeans,* and so on.

Beta carotene has been found to aid eyesight and may help prevent lung cancer. *d*-Limonene has been found to protect rats against breast cancer. Tannins appear to help prevent stomach cancer. Quercitin appears to help prevent prostate cancer. Capsaicin can reduce arthritis pain. *N*-3 (omega-3) fatty acids help to lower triglyceride levels in the blood. Genistein appears to block growth of breast cancer tumors, prostate cancer tumors, and to prevent the loss of bone density that leads to the disease osteoporosis. Tocotrienols act as antioxidants and also inhibit synthesis of cholesterol (in humans). See also *Cancer, Dextrorotary* (D) *isomer, Fatty acid, Linolenic acid, Linoleic acid, Genistein (Gen), Biological activity, Molecular pharming, Flavonoids, Resveratrol, Nutraceuticals, Cholesterol, N-3 fatty acids, Phytotoxin, Allelopathy, Antibiotic, Phytoalexins, Antioxidants, Abrin, Ricin, Pfiesteria piscicida, Phytosterols, Lignans, Polyphenols, Glycyrrhizic acid, Saponins, Fructose oligosaccharides, Lycopene, Lutein, Anthocyanin, Soybean plant, Soybean oil, Vitamin E, Xanthophylls, Sitosterols, Carotenoids, Sterols, Alicin, Ellagic acid, Proanthocyanidins, Caffeine, Quercitin, Rosemarinic acid, Zeaxanthin, Catechins, Phosphatidyl serine, Sulforaphane.*

Phytochrome

A protein plant pigment that serves to direct the course of plant growth and development and differentiation in a plant. The response is independent of photosynthesis, for example, in the photoperiod (i.e., length of light period) response. See also *Photoperiod, Protein, Photosynthesis, Plant hormone.*

Phytoene

See also *Golden rice, Lycopene, Carotenoids.*

P

Phytoestrogens

Compounds possessing molecular structures that are somewhat similar to that of estrogen, which are naturally found in all plants on Earth. As a result every vegetable, fruit, cereal, and legume contains at least one type of "phytoestrogen."

For example, flavones and flavonols are beneficial phytoestrogens (mostly red- and yellow-colored pigments) found in colored vegetables and fruits (e.g., in red grapes, yellow grapefruit, oranges, etc.). See also *Phytochemicals, Flavonoids, Flavonols, Lignans, Selective estrogen effect, Isoflavones, Estrogen.*

Phytohormone

See *Plant hormone.*

Phytol

The hydrophobic side chain of a chlorophyll molecule. Phytol is also present in tocopherol and fatty acid phytyl ester molecules. See also *Hydrophobic, Chloroplasts, Fatty acid.*

Phyto-manufacturing

Refers to the production of valuable substances (e.g., polyhydroxybutylate biodegradable plastic, industrial process enzymes, etc.) in plants (e.g., genetically engineered plants). See also *Polyhydroxylbutylate (PHB), Biopolymer, Polyhydroxyalkanoic acid (PHA), Extremozymes, Nutraceuticals.*

Phytonutrients

See *Phytochemicals.*

Phytoparasitic

Refers to organisms that act as parasites to plants (e.g., to crop plants).

Phytopharmaceuticals

See *Phytochemicals, Nutraceuticals, Phyto-manufacturing.*

Phytophthora

From the Greek meaning "plant destroyer," it is a genus of fungus-like pathogenic organism (oomycetes) whose genus includes many types which attack numerous plant species.

For example, *Phytophthora infestans* caused the Irish potato famine in the mid-1840s by destroying Ireland's potato crop. See also *Fungus, Phytophthora megasperma* f. sp. *glycinea, Phytophthora sojae, Phytophthora root rot, Pathogenic, RNA interference (RNAi).*

Phytophthora megasperma f. sp. *glycinea*

A strain of *Phytophthora* that can infect the soybean plant [*Glycine max* (L.) Merrill] under certain conditions, and thereby cause that soybean plant's stem and root to degrade (so-called rot). See also *Phytophthora, Phytophthora root rot, Fungus, Pathogenic, Soybean plant, Strain, Isoflavones.*

Phytophthora Root Rot

A plant disease that is caused by a certain phytophthora pathogen (*Phytophthora sojae*). Some soybean varieties are genetically resistant to as many as 21 races/strains of *P. sojae*. See also *Phytophthora, Fungus, RPS1c gene, RPS1k gene, Genotype, Strain, Pathogenic, Soybean plant, RPS6 gene, Isoflavones.*

Phytophthora sojae

See *Phytophthora root rot.*

Phytoplankton

Algae that are floating or freely suspended in the water.

Phytoremediation

Refers to the use of specific plants to remove contaminants or pollutants from either soils (e.g., polluted fields) or water resources (e.g., polluted lakes). For example, the Brazil water hyacinth (*Eichhornia crassipes*) naturally accumulates in its tissues toxic metals such as lead, arsenic, cadmium, mercury, nickel, copper, and so on, so has been utilized as a "biofilter" (e.g., in India).

Insertion of the *Escherichia coliform* bacteria gene known as *gsh 11* into the plant known as Indian mustard causes that plant to accumulate 40%–90% higher amounts of cadmium (from cadmium-tainted soil) in its tissues than before, so such genetically engineered plants could be utilized to extract cadmium from polluted sites. See also *Bioremediation, Biorecovery, Escherichia coliform, Bacteria, Gene, Genetic engineering, Endophyte.*

Phytosanitary

Refers to the topic of plant health (e.g., prevent plant disease, etc.).

Phytosterols

A group of phytochemicals (i.e., solid alcohols consisting of ring-structured molecules) that are present in seeds produced by certain plants (e.g., the soybean plant *Glycine max* L.).

Evidence shows that human consumption of certain phytosterols can help to prevent certain types of cancers (e.g., cancers of the colon, prostate, breast), and can help lower total serum cholesterol and low-density lipoproteins (LDLP) levels, thereby reducing the risk of coronary heart disease (CHD). Evidence also indicates that those phytosterols (e.g., campesterol, stigmasterol, beta-sitosterol) interfere with absorption of dietary cholesterol by the intestines, and decrease the body's recovery and reuse of cholesterol-containing bile salts, which causes more cholesterol to be excreted from the body than previously.

In 2000, the researcher Joseph Judd fed phytosterols extracted from soybeans (*Glycine max* L.) to human volunteers that were consuming a "low-fat" diet. Their total blood serum cholesterol and LDLP levels decreased by more than 10% in a short time.

During 2003, the U.S. FDA approved a (label) health claim that associates consumption of phytosterols (added into a broad range of food products) with reduced CHD in humans. See also *Phytochemicals, Sterols, Sitostanol, Soybean plant, Low-density lipoproteins (LDLP), Cholesterol, Campesterol, Stigmasterol, Beta-sitosterol, Sitosterol, Coronary heart disease (CHD), Food and Drug Administration (FDA).*

Phyto-sterols

See *Phytosterols*.

Phytotoxin

Any toxic compound produced by a plant. See also *Allelopathy*, *Antibiotic*, *Phytochemicals*, *Phytoalexins*, *Toxin*, *Abrin*, *Ricin*, *Pfiesteria piscicida*, *Solanine*, *Glucosamines*, *Psoralene*, *Glucosinolates*, *Gossypol*, *Alkaloids*.

pI

See *Isoelectric Point*.

Picogram (pg)

10^{-12} g or 3.527×10^{-14} oz (avoirdupoir). See also *Microgram*.

Picorna

A "family" of the smallest known viruses. The viruses of this family are a cause of the common cold and Hepatitis A in humans, one form of hoof and mouth disease in animals, and at least one disease in corn (maize).

In 1994, Dr. Asim Dasgupta discovered a cellular molecule within ordinary baker's yeast that prevents picorna virus reproduction. This advance could lead to the creation of a treatment, in the future, to cure one or more of the above-mentioned diseases after infection has begun. See also *Virus*, *Cladistics*, *Clades*.

Piezo Motor

See *Piezoelectric effect*.

Piezoelectric Effect

Discovered in 1880, this refers to the facts that:

- Applying a mechanical stress (e.g., deformative) to applicable piezoelectric material (e.g., certain ceramics, quartz, etc.) causes production of an electrical charge.
  or
- Applying an electrical charge produces a mechanical strain (or *motion* if the piezoelectric material is designed in a "motor" format).

Such "piezo motors"/actuators can be manufactured at a very small scale, and utilized to:

- Position the tiny stylus very precisely (i.e., "nanoposition" it) in atomic force microscopy.
- Power some applications of nanotechnology (i.e., act as a "nanomotor").

See also *Atomic force microscopy*, *Nanotechnology*, *Nanometers (nm)*, *Nanowire*, *Nanopiezoelectronics*, *Nanobots*, *Nanoelectromechanical systems (NEMS)*, *Self-assembling molecular machines*.

PIM-1 Protein

A protein molecule that promotes cell survival and growth.

When this protein is added by man to adult stem cells that have been removed via biopsy from (an aged) heart, those adult stem cells are rejuvenated because telomerase enzyme activity is enhanced (resulting in elongation of the cells' telomeres) and stem cell proliferation is increased (when the PIM-1-modified stem cells are reinserted into the heart muscle).

See also *Protein*, *Cell*, *Adult stem cell*, *Differentiation*, *Telomerase*.

Pink Bollworm

See *Pectinophora gossypiella*.

Pink Pigmented Facultative Methylotroph (PPFM)

A type of bacteria that is naturally present on virtually all plants and which lives symbiotically with the host plants. PPFM consume methanol (one of the natural waste products from plant metabolism), and in return PPFM produce:

- Cytokinin, which aids the cell division (growth) process in plants.
- Crucial plant nutrients, including nitrates.

PPFM also produces a chemical substance similar to vitamin B-12. In 1996, Joe Polacco discovered that impregnation of aged seeds with PPFM improved the germination (sprouting) rate of those aged seeds. See also *Bacteria*, *Symbiotic*, *Cytokinins*, *Nitrates*, *Nitrate bacteria*, *Mitosis*, *Cell differentiation*, *Vitamin*.

Pink Stalk Borer

Sesamia calamistis, an insect that attacks corn (maize) plants. See also *Corn*.

piRNAs

Abbreviation for *Piwi-interacting RNAs*, because they are RNAs that interact with members of the Piwi subclass of Argonaute proteins. These are RNA segments of approximately 24–31 nucleotides in length. All of their gene-regulation functions are not yet known, but it is known that:

- They are involved in the regulation of transposons.
- Blocking the production of piRNAs usually renders an animal to be infertile.

See also *Ribonucleic acid (RNA)*, *Nucleotide*, *Gene*, *RNA interference (RNAi)*, *Transposon*, *Protein*.

Pituitary Gland

One of the endocrine glands, it lies beneath the hypothalamus (at the base of the brain). Along with the other endocrine glands, the pituitary helps to control long-term bodily processes. This control is accomplished via interdependent secretion of hormones along with the other glands comprising the total endocrine system. For example, the pituitary helps to control the body's growth from birth until the end of puberty, by secreting growth hormone (GH). Secretion of GH by the pituitary is itself governed by the hormone known as growth hormone-releasing factor, received by the pituitary gland from the hypothalamus. The pituitary gland also helps

P

to control reproduction (e.g., development and growth of ovaries, timing of ovulation, maturation of oocytes, etc.) by secreting two gonadotropic (reproductive) hormones named luteinizing hormone (LH) and follicle-stimulating hormone (FSH). Secretion of LH and FSH by the pituitary is itself governed by the hormones gonadotropin-releasing hormone (GnRH, received by pituitary from the hypothalamus) and estrogen/progesterone (received by pituitary from the ovaries). See also *Endocrine glands, Endocrine hormones, Hormone, Endocrinology, Hypothalamus, Follicle-stimulating hormone (FSH), Estrogen, Growth hormone-releasing factor (GRF or GHRF), Growth hormone (GH).*

Piwi-Interacting RNAs

See *piRNAs.*

PK/PD

Acronym for pharmacokinetics/pharmacodynamics. See *Pharmacokinetics.*

Planar Patch Clamping

See *Patch clamping.*

Plant Breeder's Rights (PBR)

The intellectual property rights that are legally accorded to plant breeders for newly created plant varieties by various laws, international treaties, and so on. Similar to patent law for inventors. See also *Plant's novel trait (PNT), Plant Variety Protection Act (PVP), Plant Protection Act, European Patent Convention, European Patent Office (EPO), U.S. Patent and Trademark Office (USPTO), Union for Protection of New Varieties of Plants (UPOV), Community Plant Variety Office, Variety (e.g., of crop plants).*

Plant Cell Culture

Refers to the growing of cells from plants (or plant roots) in a man-made vessel. See also *Plant cell fermentation, Cell, Cell culture, Tissue culture, Culture medium, Amino acids, Vitamin, Paclitaxel.*

Plant Cell Fermentation

Refers to a cell/tissue culture production technology in which plant calluses (an undifferentiated cluster of plant cells formed after plant injury or analogous stimulation by scientists) are propagated/kept alive in a water-based system containing needed amino acids, sugars, vitamins, trace elements, and other nutrients while they produce a desired chemical substance, such as paclitaxel. See also *Cell, Cell culture, Tissue culture, Culture medium, Amino acids, Vitamin, Paclitaxel.*

Plant Hormone

An organic compound that is synthesized in minute quantities by certain plants. It influences and regulates plant physiological processes. Also called a *phytochrome.* The five general types of hormones that together influence cell division, enlargement, differentiation, and dormancy are the auxins, ethylene, gibberellins, cytokinins, and abscisic acid. See also *Hormone, Gibberellins, Auxins, Phytochrome, GPA1, Ethylene, Lysophosphatidylethanolamine, Abscisic acid, Stress hormones.*

Plant Protection Act

A law passed by the US Congress in 1930 that enabled intellectual property protection via patents for new plants (developed by scientists) which are propagated asexually (e.g., via grafting). See also *U.S. Patent and Trademark Office (USPTO), European Patent Convention, European Patent Office (EPO), Plant's novel trait (PNT), Plant breeder's rights (PBR), Community Plant Variety Office, Plant Variety Protection Act (PVP).*

Plant Sterols

See *Phytosterols.*

Plant Variety Protection Act (PVP)

A law passed by the US Congress in 1970 that enables intellectual property protection (analogous to copyright protection) for new varieties of seed plants and seeds in America. See also *U.S. Patent and Trademark Office (USPTO), European Patent Convention, European Patent Office (EPO), Plant's novel trait (PNT), Plant breeder's rights (PBR), Plant Protection Act, Union for Protection of New Varieties of Plants (UPOV), Community Plant Variety Office, Variety (e.g., of crop plants).*

Plant's Novel Trait (PNT)

The new (novel) trait added to a plant (e.g., crop plant such as cotton, corn/maize, soybean, etc.). Example novel traits are *herbicide tolerance* (via inserted *CP4 EPSPS* gene, *PAT* gene, etc.), *insect resistance* (via inserted *B.t.* gene, *Photorhabdus luminescens* gene, etc.), *resistance to aluminum toxicity* (via inserted *CSb* gene, etc.), among others. See also *Trait, Corn, Soybean plant, CP4 EPSPS, Gene, PAT gene, B.t., Bacillus thuringiensis (B.t.), Event, Citrate synthase (CSb) gene, Genetic engineering, Agrobacterium tumefaciens, Photorhabdus luminescens.*

Plantibodies™

A trademark owned by EPIcyte Pharmaceutical, Inc.

It refers to antibodies (e.g., akin to mammalian ones) produced in plants which are genetically engineered to produce those (specific) antibodies. That process (i.e., genetically engineering plants to cause them to produce *plantibodies*) was invented during the 1990s by Andrew Hiatt and Mich Hein.

Although plants do not always glycosylate (i.e., attach oligosaccharide units to protein molecules such as these antibodies) in the same manner as animal cells, an *antibody against HSV-2 pathogen* expressed in genetically engineered soybean plants has proven comparable to that same antibody expressed in genetically engineered animal cells. See also *Antibody, Genetic engineering, Glycosylation, Oligosaccharides, Express, Soybean plant, Pathogen, Molecular Pharming™, Pharming.*

Plantigens

Antigens (e.g., of pathogenic bacteria) produced in plants which are genetically engineered to produce those (specific) antigens. That process (i.e., genetically engineering plants to cause them to

P

produce specific antigens) can be utilized to produce *edible vaccines* for the *pathogenic bacteria possessing those antigens*. Then, people could be "vaccinated" against disease merely by eating the genetically engineered plant (e.g., banana). See also *Antigen, Pathogenic, Bacteria, Vaccine, Genetic engineering, Edible vaccines, Pharming.*

Plaque

Refers to deposits of (oxidized) cholesterol intermixed with smooth-muscle cells and some macrophages, lining the inside of certain blood vessels. These deposits can result in the disease *atherosclerosis* and/or adversely increasing blood platelet aggregation (e.g., clotting/thrombosis). See also *Vitamin E, Atherosclerosis, Thrombosis, Arteriosclerosis, Cholesterol, Adipose, Macrophage, Epithelium, Endothelin, APO A-1 milano, Nanoparticles.*

Plasma

A pale, amber-colored fluid constituting the fluid portion of the blood in which are suspended the cellular elements. Plasma contains 8%–9% solids. Of these, 85% are proteins consisting of three major groups, which are as follows: fibrinogen, albumin, and globulin. The other components are the lipids, which include the neutral fats, fatty acids, lecithin, and cholesterol. Also present are sodium, chloride and bicarbonate, potassium, calcium, lycopene, and magnesium. A most essential function of plasma is the maintenance of blood pressure and the exchange (with tissues) of nutrients for waste. See also *Absorption, Homeostasis, Lycopene.*

Plasma Cell

An antibody-producing B lymphocyte. See *B lymphocytes, Antibody.*

Plasma Membrane

From the Greek word *plasm* = "something formed." Also sometimes known as a *plasmalemma*, it is a thin structure that completely surrounds the cell as a sort of "skin." This membrane may be seen with the aid of an electron microscope. The entire membrane appears to be about 100 Angstroms (Å; 0.1 mm) thick and is composed of two dark lines each about 30 Å thick which are, however, separated by a lighter area. This trilaminar "sandwich" structure is referred to as the unit membrane.

The plasma membrane is composed of lipoidal (fat-like) material in which proteins and protein complexes and whole functional systems are embedded within specific regions known as *lipid rafts* (flat "islands" within the plasma membrane) or *caveolae* (cave-like structures within the plasma membrane). For example, the folate receptor (i.e., the cellular receptor for the B vitamin folic acid) is embedded in caveolae.

In the plasma membrane are incorporated such energy-dependent transport systems as Na^+ and K^+ transporting ATPase and amino acid transport systems. Besides the cell, membranes surround such systems as the endoplasmic reticulum, vacuoles, lysosomes, Golgi bodies, mitochondria, chloroplasts, and the nucleus, to mention just a few. The plasma membrane and membranes in general function in part as a permeability barrier to the free movement of substances between the inside and exterior of the cell or organelles that they surround. See also *Cell, Protein, Cecrophins (lytic proteins), Magainins, Membranes (of a cell), Golgi bodies, Endoplasmic reticulum (ER), Lysosomes, Transmembrane proteins, Receptors, Lipids, Lipid bilayer, Lipid rafts, Caveolae, Membrane transport, Tryanslocon, ATPase, Amino acid, Vitamin, Intrinsic protein, Phosphatidyl serine, Aquaporins, Efflux pump, Effectors (fungal), Peptide-oligonucleotide conjugates.*

Plasma Protein Binding

See *ADME tests, ADME/Tox.*

Plasmalemma

See *Plasma membrane.*

Plasmid

An independent, stable, self-replicating piece of DNA in bacterial cells that is not a part of the normal cell genome and that never becomes integrated into the host chromosome. This is in contrast to a similar genetic element known as an episome plasmid which may exist independently of the chromosome or may become integrated into the host chromosome. Plasmids are sometimes called extrachromosomal DNA to signify that they are DNA that isn't part of the chromosome. Plasmids can be readily extracted from bacteria by scientists, and they can both add some extra DNA (e.g., to cause production of a valuable pharmaceutical) and reintroduce that genetically enlarged plasmid back into living bacteria, where the reintroduced plasmid has the ability to multiply inside the bacteria and be inherited by all future generations that arise via cell division.

Plasmids are known to confer resistance to antibiotics and may be transferred by cell-to-cell contact (by conjugation via the sex pilus) or by viral-mediated transduction. Plasmids are commonly used in recombinant DNA experiments as acceptors of foreign DNA. Known forms of plasmids include both linear and circular molecules. See also *Episome, Vector, Copy number, Multicopy plasmids, Deoxyribonucleic acid (DNA), Cell, Bacteria, Genome, Chromosome, Antibiotic, Ti plasmid, MreB, ParM, PAR gene.*

Plasmocyte

Another name for a blast cell. See *Blast cell.*

Plasmonic Metamaterials

See *Surface plasmons, Metamaterials.*

Plasmonic Nanohole Arrays

Refers to arrays consisting of apertures (i.e., small holes) approximately 200–350 nm in size, through thin metallic films. When a particular pathogen (e.g., a virus) within a sample solution (e.g., in blood) binds to this array's surface near an aperture, surface plasmon resonance causes directly proportional changes in the refractive index of reflected light striking the metallic film. By shining a highly focused beam of light (e.g., laser, polarized light, etc.) on the metallic surface and measuring the readily detectable change (i.e., shift) in the resonance frequency of light transmitted through the nanoholes, the *mass change* (resulting from the pathogen binding to the metallic film) *and thus the presence/identity and concentration of the pathogen within the solution* can be determined. See also *Surface plasmons, Nanometers (nm), Nanoscience, Nanotechnology, Surface plasmon resonance (SPR), Pathogen, Ligand (in biochemistry),*

P

Protein interaction analysis, Target–ligand interaction screening, Microarray (testing).

Plastic Antibodies

See *Molecular imprinting.*

Plastid

From the Greek word *plastis* = "builder," it is an independent, stable, self-replicating organelle containing a piece of DNA, inside a plant cell's cytoplasm. Plastids, which include chloroplasts (where photosynthesis occurs), chromoplasts (colored organelles in flower petals and fruits), amyloplasts (starch-storage organelles in plant roots), nonphotosynthetic etioplasts, and so on, are not a part of the reproduction cell genome (i.e., in nucleus). Because there can exist up to 10,000 plastids in a given plant cell, the insertion of a gene (e.g., via genetic engineering) into plastids can result in a higher yield (of the specific protein coded for by that gene) than is achieved via insertion of the gene into the cell's nuclear DNA.

Plastids are where fatty acids are synthesized (manufactured) in plant cells via FAD2 desaturase enzymes that float freely within those plastids. Prior to 2000, it had proven almost impossible for scientists to genetically engineer the plastids of plants other than the tobacco plant. However during 2000, the plastids in the tomato plant (*Lycopersicon esculentum*) were successfully genetically engineered by man. See also *Deoxyribonucleic acid (DNA), Cell, Cytoplasm, Nuclear DNA, Copy number, Genome, Promoter, Gene, Genetic engineering, Fats, Fatty acid, Desaturase, Chloroplasts, Photosynthesis, Tomato, Plastidome, Starch, Fatty acid.*

Plastidome

The complete set of plastid compounds contained within a cell. See also *Plastid, Cell.*

Platelet Activating Factor (PAF)

See *Choline.*

Platelet-Derived Growth Factor (PDGF)

An angiogenic growth factor produced by the blood's platelet cells which attracts the growth of capillaries into the vicinity of a fresh wound. This action releases still other growth factors, and starts the process of building a fibrin network, to support the subsequent (blood) clot. PDGF is a competence factor (i.e., a growth factor that is required to make a cell able or competent to react to other growth factors). PDGF is normally contained within the platelet cells, so does not circulate in the blood in a form enabling it to be freely available to its "target cells." This "containment" of PDGF in platelets ensures site-specific delivery of the PDGF directly to a wound site, so stimulus (i.e., of capillary growth) is localized to the actual wound site. After PDGF has caused the formation of the initial clot at a wound site, PDGF attracts connective tissue cells into the vicinity of the wound (to start the tissue-repair process). PDGF also acts as a mitogen (substance causing cell to divide and thus multiply) for connective tissue cells, granulocytes, and monocytes (each of which is involved in the wound's healing process). See also *Angiogenic growth factors, Fibrin, Fibronectin, Platelets, Mitogen, Granulocytes, Monocytes, Cyclooxygenase.*

Platelet-Derived Wound Growth Factor (PDWGF)

See *Platelet-derived growth factor (PDGF).*

Platelet-Derived Wound Healing Factor (PDWHF)

See *Platelet-derived growth factor (PDGF).*

Platelets

Disk-shaped blood cells that stick to the (microscopically "jagged") edges of wounds. The aggregation of platelets at the wound site leads to blood clotting, forming a temporary wound covering. During this blood clotting process, the platelets release PDGF which attracts fibroblasts to the wound area (for subsequent healing process).

Platelets are created when tendrils from megakaryocytes (i.e., large bone marrow cells) enter a given blood vessel by growing to pierce through its wall. Blood flowing past that point breaks off some pieces of the megakaryocyte cell, and those pieces subsequently become platelets. See also *Fibrin, Fibronectin, Platelet-derived growth factor (PDGF), Fibroblasts, Cyclooxygenase, Choline, Oxidative stress, Serotonin, Megakaryocyte-stimulating factor (MSF).*

Plectonemic Coiling

Refers to the *inter*-twining (within cell) of a double helix (DNA) molecule in such a manner, that correction of that inter-twining requires "unwinding" of the double helix molecule (DNA). See also *Deoxyribonucleic acid (DNA), Double helix, Cell, Unwinding protein.*

Pleiotropic

Adjective used to describe a gene that affects more than one trait (apparently unrelated) characteristic of the phenotype (appearance of an organism). For example, biologist David Ho in 1993 discovered a single gene in the barley (*Hordeum vulgare*) plant that controls the traits of the plant's height, drought resistance, strength, and time to maturity. See also *Gene, Genetic code, Deoxyribonucleic acid (DNA), Informational molecules, Phenotype.*

Pluripotent Stem Cells

Refers to those stem cells which can differentiate into the numerous different types of tissues comprising the body of an organism.

Human pluripotent stem cells refer to those from which each of the human body's 210 different types of tissues could arise, via differentiation. See also *Stem cells, Induced pluripotent stem cells, Stem cell growth factor (SCF), Differentiation, Human embryonic stem cells, Erythropoiesis, Mesenchymal stem cell (MSC).*

PMP

Acronym for *Plant-made Pharmaceuticals*. See *Pharming.*

PNA

Acronym for *peptide nucleic acid.* See *Peptide nucleic acid.*

PNT

See *Plant's novel trait (PNT)*.

POC

See *Peptide-oligonucleotide conjugates.*

Point Mutation

A mutation consisting of a change of only one nucleotide or one base pair in a DNA molecule.

At "hot spots" (i.e., certain specific locations on the DNA within some organisms), numerous point mutations can occur.

In the case of SNPs, the *same* point mutation occurs at the same location (on the DNA within some organisms) across a population of individuals of that organism. See also *Mutation, Heredity, Mutant, Mutagen, Deoxyribonucleic acid (DNA), Nucleotide, Substitution, Base substitution, Hot spots, Base excision sequence scanning (BESS), Organism, Site-directed mutagenesis (SDM), Single-nucleotide polymorphisms (SNPs), Traditional breeding methods, Zinc finger proteins.*

Points to Consider in the Manufacture and Testing of Monoclonal Antibody Products for Human Use

The FDA's governing rules for IND (investigational new drug) submission for monoclonal antibody (MAb)-based pharmaceuticals. See also *IND.*

Pol III

Abbreviation for *DNA Polymerase III.* See *DNA polymerase.*

Polar Group

A hydrophilic ("water loving") portion of a molecule; it may carry an electrical charge. A group that "likes" to be in the presence of water molecules or other polar compounds. See also *Nonpolar group, Polarity (chemical), Polar molecule (dipole), Amphipathic molecules, Amphoteric compound, Lipid bilayer.*

Polar Molecule (Dipole)

A molecule in which the centers of positive and negative (electrical) charge do not coincide, so that one end of the molecule carries a positive (or partial positive) charge and the other end a negative (or partial negative) charge. See also *Polarity (chemical), Polar group, Ion-exchange chromatography, Nonpolar group.*

Polar Mutation

A mutation in one gene which, because transcription occurs only in one direction, reduces the expression of subsequent genes in the same transcription unit further down the line. See also *Transcription, Translation, Express, Nucleic acids.*

Polarimeter

An instrument used for measuring the degree of rotation of plane-polarized light by an optically active compound/solution. See also *Stereoisomers, Optical activity, Levorotary* (L) *isomer, Dextrorotary* (D) *isomer, Polarized light, Racemate.*

Polarity (Chemical)

The degree to which an atom or molecule bears an electrical charge or a partial electrical charge. In general, the more polar (i.e., separation or partial separation of charge) a molecule is, the more hydrophilic ("water loving") it is. Polarity results from an uneven distribution of electrons between the atoms comprising a molecule. See also *Polar group, Hydrophilic, Polar molecule (dipole).*

Polarity (Genetic)

Having to do with the one way or unidirectionality of gene transcription in an operon unit. That is, the region near the operator is always transcribed before the more distant regions. By analogy, transcription begins at the left end of an operon unit and proceeds (reads, transcribes) toward the right end of the operon unit. The distinction between the 5′ and the 3′ ends of nucleic acids. See also *Polar mutation, Transcription.*

Polarized Light

Refers to one of the two forms of polarized light (left-handed circular polarization or right-handed circular polarization). See also *Polarimeter, Racemate.*

Pollen Tube

See *Style.*

Polyacrylamide Gel

A "sieving" gel that is used in electrophoresis. See also *Polyacrylamide gel electrophoresis (PAGE).*

Polyacrylamide Gel Electrophoresis (PAGE)

A form of chromatography in which molecules are separated over a period of time on the basis of size and charge. The stationary phase (called the polyacrylamide gel) consists of interlinked polymerized acrylamide monomers. The gel looks and feels a lot like Jello™ gelatin. On a molecular basis it consists of an intertwined and cross-linked mesh of polyacrylamide "strings"; tiny "holes" in the gel (much like there are in a plastic mesh grocery bag). The size of those holes is approximately the size of the molecules which are to be separated. Since some of the molecules will be larger and some smaller, portion(s) of them will be able to pass through the gel matrix more easily than other, which is part of the basis for PAGE separation.

The other part of the basis for PAGE separation is electrical charge of the molecule. Functionally, the gel serves to hold and separate the molecules, after the scientist places a small amount of the solution containing the (unknown) molecules into wells (i.e., grooves on surface of gel to hold the sample liquid) and the entire gel is subjected to an electric field. Over the course of time (minutes to hours), the sample's fractions of molecules bearing different charge/mass separate by traveling different distances within the gel. See also *Bioluminescence, Chromatography, Two-dimensional (2D) gel electrophoresis, Field inversion gel electrophoresis (FIGE), Electrophoresis, Molecular-weight size marker.*

P

Polyadenylation

The addition of a sequence of polyadenylic acid to the 3′ end of a eucaryotic mRNA after its transcription (posttranscriptional). See also *Messenger RNA (mRNA), Transcription.*

Polycation Conjugate

Refers to a man-made macromolecule which possesses several positive charges (cations), that is attached (conjugated) to a specific protein which binds to a cell receptor (thereby gets "admitted" into the cell's interior). Some polycation conjugates can be utilized (e.g., in gene therapy) to deliver gene(s) into cells. See also *Cation, Polymer, Vector, Conjugate, Protein, Macromolecules, Cell, Receptors, Gene, Gene therapy.*

Polycistronic

Coding regions representing more than one gene in mRNA (i.e., they code for two or more polypeptide chains). Many mRNA molecules in procaryotes are polycistronic. See also *Ribosomes, Procaryotes.*

Polyclonal Antibodies (Used in Humans)

A mixture of antibody molecules that are specific for given antigen(s), which have been purified from an immunized (to that given antigen) animal's blood. Such antibodies are polyclonal in that they are the products of many different populations of antibody-producing cells (within the animal's body). Hence they differ somewhat in their precise specificity and affinity for the antigen.

Decades ago, antibodies (then called antitoxin) that were purified from an immunized animal's blood (e.g., a horse) were injected into humans suffering from certain diseases (e.g., diphtheria, caused by *Corynebacterium diphtheriae*). In these cases the pathogen had caused disease by secreting large amounts of toxin into the victim's bloodstream. The antitoxin combined quantitatively (e.g., 1:1, 2:1, 1:2, 1:3, 3:1, etc.) with, and neutralized the toxin (for those few diseases for which it was applicable). Vaccines are now generally used instead, because of the adverse immune response caused by the horse's blood (antigens).

During 2003, polyclonal antibodies from (immunized) goat blood serum were utilized by Angus Dalgleish to treat some multiple sclerosis patients, with good results reported. See also *Antibody, Passive immunity, Monoclonal antibodies (MAb), Antigen, Pathogen, Toxin, Multiple sclerosis.*

Polyclonal Response (of Immune System to a Given Pathogen)

Because a given pathogen generally has several antigenic sites on its surface, the B lymphocytes (activated by helper T cells in response to a pathogen invading the body) synthesize several (subtly different) antibodies against that pathogen. And since the antibodies are made by different cells the response is known as poly (many)clonal. See also *Pathogen, Antigen, Antibody, Hapten, Epitope, Helper T cells (T4 cells), Lymphocyte, B lymphocytes, Lymphokines.*

Polyethylene-Glycol Superoxide Dismutase (PEG-SOD)

See *PEG-SOD (polyethylene glycol superoxide dismutase), Human superoxide dismutase (hSOD).*

Polygalacturonase (PG)

An enzyme (e.g., present in tomatoes) that starts the breakdown (softening) of the fruit tissue. Recent advances make it possible to significantly delay the softening (i.e., spoilage) process by reducing the production of polygalacturonase through genetic engineering of the plant. In 1986, William Hiatt of the American company Calgene discovered the gene for polygalacturonase, which led to that company commercializing a tomato variety that had been genetically engineered to reduce production of polygalacturonase in that variety's tomatoes (in 1994). See also *EPSP synthase, Genetic engineering, Antisense (DNA sequence), Enzyme, Gene, ACC synthase.*

Polygenic

A trait or end product (e.g., in a grain crop) that requires simultaneous expression of more than one gene. For example, the level of protein produced in soybeans is controlled by five genes. See also *Polyhydroxylbutylate (PHB), Protein, Soybean plant, Gene, Trait, Soybean oil, Bce4, Arabidopsis thaliana, Plastid.*

Polyhydroxyalkanoates

See *Polyhydroxyalkanoic acid (PHA).*

Polyhydroxyalkanoic Acid (PHA)

A "family" of chemically related "energy storage" substances (i.e., polyesters) that is naturally produced by certain bacteria (90 strains known). When PHA is removed from the bacteria and purified, this substance has physical properties quite similar to thermoplastics like polystyrene. PHA can quickly be broken down by soil microorganisms, so PHA is a biodegradable plastic.

During the 1990s, Daniel Solaiman and coworkers at the U.S. Department of Agriculture developed some bacteria strains (e.g., *Bacillus thermoleovorans*) that can produce PHA utilizing vegetable oils (e.g., soybean oil) as a major part of their "diet" (energy source). The precise chemical composition (and physical characteristics) of the PHA thereby produced varies according to the particular vegetable oil that is used as the *energy source* for those bacteria. For example, PHA thus produced utilizing soybean oil is very amorphous (formable).

In 1994, researchers transferred genes for the production of one PHA into the weed plant *Arabidopsis thaliana* and the crop plant rapeseed (canola). In 1997, researchers transferred *phaB* and *phaC* genes into the crop plant cotton (*Gossypium hirsutum*), which caused those transformed plants to express (i.e., produce) PHA inside the fibers (seed hair cells) in amount of 0.34% of the fiber weight. That PHA (inside those cotton fibers) resulted in a fabric (i.e., cotton-PHA "blend") possessing better insulation properties than traditional cotton fabric. See also *Polyhydroxylbutylate (PHB), Starch, Bacteria, Biopolymer, Arabidopsis thaliana, Canola, Gene, Transformation, Express, Biodegradable, Microorganism, Soybean oil.*

Polyhydroxylbutylate (PHB)

One of the PHAs, polyhydroxybutylate is an "energy storage" substance that is naturally produced by certain bacteria, yeasts, and plants. It was initially discovered within the bacterium *Alcaligenes eutrophus*.

When removed from the bacteria and purified, this substance has physical properties quite similar to thermoplastics like polystyrene. PHB can quickly be broken down by soil microorganisms, so PHB is a biodegradable plastic. Three separate enzymes are utilized by the organism in order to make the PHB molecule. In 1994, researchers succeeded in transferring genes for PHB production into the weed plant *Arabidopsis thaliana* and the crop plant rapeseed (canola).

Later (1997), researchers transferred *phaB* and *phaC* genes into the crop plant cotton (*Gossypium hirsutum*), which caused those transformed plants to express (produce) PHA inside the fibers (seed hair cells) in amount of 0.34% of the fiber weight. That PHA (inside those cotton fibers) resulted in a fabric (i.e., cotton-PHA "blend") possessing better insulation properties than traditional cotton fabric. See also *Starch, Bacteria, Biopolymer, Enzyme, Polygenic, Microorganism, Polyhydroxyalkanoic acid (PHA), Canola, Arabidopsis thaliana, Gene, Express, Biodegradable.*

Polymer

A molecule possessing a regular, repeating, covalently bonded arrangement of smaller units called monomers. By analogy, a chain (polymer) that is composed of links (monomer) hooked together. See also *Oligomer, Protein, Nucleic acids, Actin, Tandems.*

Polymerase

Refers to an enzyme that catalyzes the assembly of nucleotides into RNA (RNA polymerase) and of deoxynucleotides into DNA (DNA polymerase). See also *DNA polymerase, RNA polymerase, Reverse Transcriptases, DNA, RNA, TAQ.*

Polymerase Chain Reaction (PCR)

A reaction that uses the enzyme DNA polymerase to catalyze the formation of more DNA strands from an original one by the execution of repeated cycles of DNA synthesis. Functionally, this is accomplished by heating and melting double-stranded (hydrogen bonded) DNA into single-stranded (nonhydrogen bonded) DNA and producing an oligonucleotide primer complementary to each DNA strand. The primers bind to the DNA and mark it in such a way that the addition of DNA polymerase and deoxynucleoside triphosphates causes a new strand of DNA to form which is complementary to the target section of DNA. The process described previously is repeated (trait, product, etc.) again and again to produce millions of *copies* (*amplicons*) of the desired strand of DNA. PCR and its registered trademarks are the property of F. Hoffmann-La Roche & Co. AG, Basel, Switzerland. See also *Polymerase chain reaction (PCR) technique, Nested PCR, Deoxyribonucleic acid (DNA), DNA probe, Probe, Q-beta replicase technique, Cocloning (of molecules), Positive and negative selection (PNS), Amplicon, Nested PCR, Primer (DNA), Capillary electrophoresis.*

Polymerase Chain Reaction (PCR) Technique

Developed in 1984 and 1985 by Kary B. Mullis, Randall K. Saiki, Stephen J. Scharf, Fred A. Faloona, Glenn Horn, Henry A. Erlich, and Norman Arnheim, the PCR technique is an *in vitro* method that greatly amplifies (makes millions of copies of) DNA sequences that otherwise could not be detected or studied. It can be utilized to amplify a given DNA sequence that constitutes less than one part per million of initial sample (e.g., a 100-base-pair target DNA sequence within the genome of one of the higher organisms, which can contain up to 500 million base pairs). The procedure alleviates the necessity of *in vivo* replication of a target DNA sequence or of replication of one-of-a-kind tiny DNA samples (e.g., from a crime scene). See also *In vitro, In vivo, Polymerase chain reaction (PCR), Amplicon, Nested PCR, Deoxyribonucleic acid (DNA), Base pair (bp), Genome, Sequence (of a DNA molecule), TAQ, DNA polymerase, Primer (DNA).*

Polymersomes

See *Nanoshells.*

Polymery

From the Greek word *polys* meaning many. This term refers to how *more than one* gene (i.e., many genes) must sometimes be expressed simultaneously in order to produce a given effect, in an organism.

For example, the level of protein produced in soybeans is controlled by five genes. See also *Gene, Express, Polygenic.*

Polymorphism (Chemical)

The property of a chemical substance crystallizing (or simply existing) in two or more forms having different structures. For example, diamond and graphite and graphene are three different structures (manifestations) of the element carbon. Deoxyribonucleic acid (DNA) is a polymorphic compound because the polymer can take on different forms. See also *A-DNA, B-DNA, Z-DNA, Deoxyribonucleic acid (DNA), DNA profiling, Polymorphism (genetic), Graphene.*

Polymorphism (Genetic)

A name applied to a condition in which a species of plant or animal is represented by several distinct, nonintegrating forms or types unrelated to age or sex. The differences are often in coloration, though any characteristic of the organism may be involved (e.g., nuclei shape for polymorphonuclear leukocytes). See also *Polymorphonuclear leukocytes (PMN), Polymorphonuclear granulocytes, Single-nucleotide polymorphisms (SNPs), Short tandem repeats, Copy number polymorphisms, Polymorphism (chemical).*

Polymorphonuclear Granulocytes

Neutrophils, eosinophils, and basophils are collectively known as polymorphonuclear granulocytes. This is due to the fact that collectively their nuclei are segmented into lobes, and they have granule-like inclusions within their cytoplasm. See also *Granulocytes, Basophils, Eosinophils, Neutrophils, Cytoplasm.*

Polymorphonuclear Leukocytes (PMN)

Formerly named *microphages*, they are phagocytic (i.e., foreign particle-ingesting) white blood cells that have a lobed nucleus. For example, during an attack of the common cold (when virus first invades mucous membranes of the human nose), the body responds by making interleukin-8 (IL-8); a glycoprotein that attracts large quantities of PMN to the mucous membranes of the nose (to try to combat the infection).

Another example is when PMN migrate in to a female pig's uterus within 6 hours after semen is introduced via breeding. PMN remove excess sperm and bacteria, resulting in a *"friendly" environment* for embryos to develop in the uterus. See also *Cellular immune response, Leukotrienes, Leukocytes, Phagocytosis, Polymorphism*

P

(genetic), Virus, Bacteria, Glycoprotein, Interleukin-8 (IL-8), Cell, Nucleus, Plasma membrane.

Polypeptide (Protein)

A molecular chain of amino acids linked by peptide bonds. Synonymous with protein. Via the synthesis (of this "chain") performed by ribosomes, each polypeptide (protein) in nature is the ultimate expression product of a gene. All of the amino acids commonly found in proteins have an asymmetric carbon atom, except the amino acid glycine. Thus, the polypeptide is potentially chiral in nature. See also *Protein, Amino acid, Gene, Peptide, Stereoisomers, Chiral compound, Express, Ribosomes, Polyribosome (polysome), Messenger RNA (mRNA).*

Polyphenol Oxidase

The enzyme present within apple fruit that is responsible for browning (e.g., when apple is sliced open its interior exposed to oxygen). While useful for wild trees (to liberate the seeds from the fruit, for dispersal/planting), such rapid browning is undesirable for commercial apple producers. Thus, one company has decreased such browning by silencing the gene that codes for polyphenol oxidase in fruit of an apple tree. See also *Enzyme, Gene, Gene silencing, Coding sequence.*

Polyphenols

Refers to a group of phytochemicals naturally found in coffee, pomegranates (*Punica granatum* L.), certain types of grapes, red wines, green tea (*Camellia sinensis*), certain nuts, carob, cocoa, red onions, olives, blueberries, and so on, that act as antioxidants when consumed by humans. Polyphenols (especially those in red wines) and the pterostilbene polyphenol produced in blueberries also inhibit the production of endothelin-1, a peptide that causes constriction of blood vessels in humans; so they help lower human blood pressure.

Some polyphenols are naturally produced within the beans of the cocoa (cacao) tree (*Theobroma cacao*), and thus are present in chocolate made from those beans. Some polyphenols are naturally produced within the fruit of the carob (*Ceratonia siliqua*); sometimes called St. John's Bread.

Polyphenols naturally produced in apples have been shown to inhibit certain bacteria in the human mouth from producing the particular glucans that lead to a buildup of plaque on teeth; prevention of such plaque buildup may help prevent cavities from forming in teeth. Research indicates that certain polyphenols naturally produced in grapeseed inhibit bowel cancer cells.

During 2005, Dr. Thomas Smith and colleagues discovered that two polyphenols from green tea have an indirect impact on insulin production in the human body. Those polyphenols, known as EGCG and ECG, regulate downward the production of the enzyme GDH (glutamine dehydrogenase). GDH was shown in 1998 to help regulate insulin production. See also *Phytochemicals, Flavonoids, Catechins, Atherosclerosis, Endothelin, Tannins, Antioxidants, Oxidative stress, Phenolic hormones, Nutraceuticals, Bacteria, Glucans, Insulin, Resveratrol.*

Polyploid

From the Greek *polys* meaning *many*, and *ploos* meaning *fold*. This term refers to cells or organisms containing *more than two* sets of chromosomes.

Arising initially from a number of different natural causes, scientists have observed (or sometimes themselves caused) the following:

- Triploid (i.e., three sets of chromosomes). For example, the wheat plant is triploid due to a natural crossing of three Middle East grasses (*Triticum monococcum, Aegilops speltoids*, and *Triticum tauscii*) centuries ago.
- Tetraploid (i.e., four sets of chromosomes).
- Pentaploid (i.e., five sets of chromosomes).
- Hexaploid (i.e., six sets of chromosomes).
- Heptaploid (i.e., seven sets of chromosomes).
- Octoploid (i.e., eight sets of chromosomes).

See also *Cell, Organism, Chromosomes, Wheat, Triploid, Tetraploid, Colchicine.*

Polyribosome (Polysome)

A complex of a messenger RNA (mRNA) molecule on which ribosomes (ribosomal RNA; rRNA) are anchored. A number of ribosomes bound to only a single mRNA molecule. One mRNA molecule hence functions as a template for a number of polypeptide chains at one time. See also *Ribosomes, rRNA (ribosomal RNA), Messenger RNA (mRNA).*

Polysaccharides

From the Greek words *polys* meaning *many*, and *sakcharon* meaning *sugar*, this term refers to linear and/or branched (structure) macromolecules (i.e., large molecules) composed of many monosaccharide units (monomers such as glucose, cellulose, etc.) linked by glycosidic bonds. See also *Glycoside, Monosaccharides, Amylose, Amylopectin, Carbon nanohorns.*

Polysome

See *Ribosomes, Polyribosome.*

Polyunsaturated Fatty Acids (PUFA)

Unsaturated fatty acids, possessing more than one molecular double bond in their molecular "backbone" (i.e., they contain *at least two less than the maximum possible number of hydrogen atoms*).

Enzymes (e.g., Δ12 desaturase) present in some oilseed plants (e.g., soybean, canola, corn/maize, etc.) convert some monounsaturated fatty acids (e.g., oleic acid) to some polyunsaturated fatty acids (PUFA) (e.g., linoleic acid), within their developing seeds. For example, soybean oil contains (historical average) 60% PUFA.

Extensive research shows that PUFA impart a variety of health benefits to humans that consume them. In general, those health benefits include anti-inflammatory, antihypertensive (i.e., prevention of high blood pressure), reduction in cancer risk, reduction in the blood cholesterol levels, reduction in the risk of CHD, plus aiding in the development of retina and brain tissues.

For example, the *n-3* ("omega-3") PUFAs possess antithrombotic effects and also reduce blood concentrations of triglycerides. High dietary levels (in human diet) of the *n-6* ("omega-6") PUFAs have been related to a decreased risk of CHD.

Research indicates that some of the beneficial effects of PUFAs occur via PUFA interactions with several types of nuclear receptors (present in cells of some human tissues), which results in (*PUFA-*) *modulation* of certain gene(s) expression in those cells. PUFAs can also be converted (via enzymatic oxidation inside an organism) to prostaglandins, leukotrienes, or jasmonates. See also *Unsaturated fatty acid, Essential fatty acids, Thrombosis, Triglycerides, Coronary heart disease (CHD), Cancer, N-3 fatty acids, Soybean oil, N-6 fatty acids, Enzyme, Docosahexanoic acid (DHA), Highly unsaturated fatty acids (HUFA), Eicosapentanoic acid (EPA), Conjugated linoleic acid (CLA), Cell, Gene, Receptors, Nuclear receptors, Deoxyribonucleic acid (DNA), Express, Gene expression, Transcription, Transcription factors, Soybean plant, Oleic acid, Linoleic acid, Linolenic acid, Prostaglandins, Leukotrienes, Jasmonates, Oxylipins.*

Porcine Somatotropin (PST)

A hormone, produced in the pituitary gland of pigs, that increases a swine's muscle tissue production efficiency. Injecting this hormone causes a faster growing, leaner pig. See also *Hormone.*

Porin

A transmembrane (i.e., through the cell's membrane) protein which forms pore(s) through the membrane. Porins are present in the outer membranes of bacteria and mitochondria. See also *Protein, Cell, Membranes (of a cell), Plasma membrane, Transmembrane proteins, Bacteria, Mitochondria.*

Porphyrins

Complex nitrogenous compounds containing four substituted pyrroles covalently joined into a ring structure. When complexed with a central metal atom it is called a metalloporphyrin.

Position Effect

A change in the expression of a gene that is brought about by its translocation to a new site in the genome. For example, a previously active gene may become inactive if placed on a new site in the genome. See also *Genome, Translation, Genetic map, Map distance, Promoter.*

Positional Cloning

A technique used by researchers to zero in on the gene(s) responsible for a given trait or disease. A genetic map of the organism's genome is used to make an educated guess as to the precise location of the gene of interest (e.g., near marker _or marker _, etc.). Then those guessed genes are cloned, inserted into living organisms or cells, and tested to see if the guessed gene causes expression of the protein of interest (e.g., a protein that causes the disease that the researcher is attempting to cure). See also *Clone (a molecule), Gene, Gene amplification, Gene delivery, DNA probe, Gene machine, Genetic engineering, Genetic map, Genetic marker, Genome, Map distance, Functional genomics, Position effect, Express.*

Positive and Negative Selection (PNS)

A separation technique; a technique to speed up the task of selecting, from thousands of laboratory specimens, the few cells with precisely the desired genetic changes induced (via genetic engineering). The thousands of genetically altered cells are brought about (produced) by genetic engineering experiments. Many genetic alterations are accomplished by injecting or flooding (specimen) cells with fragments of new genetic material (genes). A few cells are produced that have precisely the desired genetic changes among a large number of cells that do not have the desired changes. Sort of like a "needle in a haystack." By analogy, the few cells possessing the desired trait represent the needles while the multitude of cells not possessing the trait represent the hay. In order to isolate the few desired cells, the needles must be separated from the hay. PNS gets rid of the nondesired cells and leaves only the cells possessing the desired genetic change. This is accomplished in the following way. The pieces of newly injected genetic material are composed not only of the desired sequence of DNA, but also another piece of DNA (known as a marker) which renders only those cells possessing the desired (genetic) change resistant to certain antibiotic drugs (such as neomycin) and certain antiviral drugs (e.g., Ganciclovir™). When all of the engineered cells are exposed to the drug (which normally kills all of the cells) only those cells possessing the desired genetic change (and the concomitant piece of DNA providing drug resistance) survive and hence are "selected." The other cells not having the drug resistance are selected against, and die. See also *Genetic engineering, Gene, Marker (genetic marker), Q-beta replicase technique, Polymerase chain reaction (PCR) technique.*

Positive Control

Refers to activation (start/increase) of the transcription of a gene due to the binding (e.g., of a transcription factor, etc.) to a regulatory element. See also *Gene, Transcription, Transcription activators, Transcriptional activator, Regulatory element, Riboswitches, Methylated.*

Positive Supercoiling

Occurs in double-stranded cyclic DNA molecules having no breaks at all in either strand. If the double helix (of DNA) is wound further in the same direction as the winding of the two strands of the double helix molecule, then the circular duplex itself takes on superhelical turns.

By analogy, supercoiling or superhelicity may be described as follows. A piece of rope can be composed of two or three smaller strands of rope wound around each other to yield the finished rope. This is equivalent to the normal double-stranded DNA. If the ends of the rope are then joined or tied together and the resultant circle of rope is again wound in the same direction as the winding that produced the rope in the first place, supercoils will be formed and the rope will become a much thicker (supercoiled) but shorter piece of rope. See also *Deoxyribonucleic acid (DNA), Double helix, DNA gyrase, Supercoiling.*

Postentry Measures

Refers to a country's mandatory restrictions on the transport and/or use of imported agricultural commodities (e.g., to prevent accidental introduction of weed seeds within commodity shipments into that country). Examples of postentry measures include *mandatory covering with tarps* of trucks or railcars transporting the commodity from the port to the processing plant (e.g., flour mill), mandatory frequent mowing of ditches along the roadways between port and processing plant to prevent any (spilled) seeds from growing tall enough to reproduce, and so on. See also *International Plant Protection Convention (IPPC), Introduction, Pest risk analysis (PRA), Quarantine pest, Treatment system.*

P

P

Posttranscriptional Gene Silencing (PTGS)

Refers to an automatic natural response (e.g., in certain plants) to the high buildup (i.e., within such plant cells) of identical mRNA molecules. Because such a high buildup typically occurs as a result of viral infection (of plant), the plant's natural defense system systematically breaks down those mRNA molecules (to fight the viral infection).

This (i.e., triggering the plant to "attack" an unwanted mRNA) can be employed by genetic engineers to "silence" a given gene (i.e., by destruction of *that gene's* mRNA), via the (cosuppression of) plant's natural PTGS response. See also *Gene transcription, Messenger RNA (mRNA), Gene silencing, Knockout, Genetic engineering, Virus, Cosuppression, RNA interference (RNAi).*

Posttranscriptional Processing (Modification) of RNAs

The enzyme-catalyzed processing or structural modifications that RNAs such as mRNAs, rRNAs, and tRNAs must undergo before they are functionally finished products. For example, in eucaryotes a block of poly A containing at least 200 AMP residues is enzymatically attached to the 3′ end of mRNA in the nucleus of the cell. The mRNAs with the "tail" are then transferred to the cytoplasm and the tail enzymatically removed to form the functional mRNAs. It is believed that the poly A tail aids in the transfer of the complex and/or targets the complex to the cytoplasm.

For example, pseudouridylation, in which the RNA's base nucleoside uridine (U) has its chemical structure altered to form a molecule known as pseudouridine (ψ). See also *Posttranslational modification of protein, mRNA, rRNA, tRNA, Primary transcript, RNA editing.*

Posttranslational Modification of Protein

Refers to enzymatic processing of a polypeptide chain (i.e., protein molecule) after its translation from its mRNA transcript, for example:

- *Glycosylation*: Addition of carbohydrate moieties to the protein molecule. For example, glycosylation of asparagine, serine, or threonine portions of certain protein molecules is critical for enabling those protein molecules to function properly regarding (the cell they are in) being discerned by the body's immune system to be indigenous or foreign.
- *Phosphorylation*: Addition of a phosphate molecular group to the protein molecule. For example, phosphorylation of serine, threonine, or tyrosine portions of a protein molecule is critical for enabling that protein molecule to be able to function in signaling (e.g., thereby triggering cell growth, cell death, etc.). Sometimes, apoptosis is inhibited by phosphorylating and inactivating several target proteins.
- *Eliminylation*: Removal of a phosphate molecular group (leaving a carbon=carbon double bond) from the protein molecule.
- *Sulfation*: Addition of a sulfate molecular group to the protein molecule.
- *Methylation*: Addition of a methyl molecular group to the protein molecule (e.g., histone). For example, methylation serves as a predominant form of PTM of Histone H3, and has also been linked to transcriptional regulation and to epigenetic silencing via heterochromatin assembly.
- *Acetylation*: Addition of an acetyl molecular group to the protein molecule (e.g., histone). For example, acetylation of core histones in applicable cells regulates gene expression in those cells.
- *Deacetylation*: Removal of an acetyl molecular group from the protein molecule (e.g., histone).
- *Ribosylation*: Addition of a ribose molecular group to the protein molecule.
- *Deamidation*: The loss of their side-chain molecular groups by some of the glutamine and asparagine portions of a given protein molecule. For example, such deamidation of some asparagine portions of certain apoptosis-blocking proteins causes loss of that molecule's apoptosis-blocking ability.
- *Prenylation*: Attachment of certain hydrophobic molecular groups (e.g., lipid groups) to the protein molecule; for example, which enable that protein to be embedded in a cell's (lipid-based) plasma membrane. Sometimes referred to as isoprenylation or lipidation.
- *Ubiquitination*: Addition of a ubiquitin moiety to the protein molecule, thereby "tagging" (marking) them for subsequent destruction (via proteolytic cleavage).
- *Cleavage*: The removal of a portion of the polypeptide chain in order to produce a functional protein molecule in the correct environment.

PTMs—including phosphorylation, acetylation, and ubiquitination—are specific modifications that can alter the activity of an individual protein target. The cumulative effect of these small modifications is the regulation of large signaling pathways and networks within cells.

See also *Polypeptide (protein), Moiety, Cell, Messenger RNA (mRNA), Enzyme, Cell, Ribosomes, Carbohydrates, Protein, Glycosylation, Glycoprotein, Autoimmune disease, Phosphorylation, Signaling, Apoptosis, Ribose, Intein, Histones, Nisin, Epigenetic, Lipids, Lipid bilayer, Plasma membrane, Receptors, Heterochromatin, Ubiquitin.*

Posttranslational Modification of Protein

Enzymatic processing of a polypeptide chain (i.e., protein molecule) after its translation from its mRNA transcript:

- *Glycosylation*: Addition of carbohydrate moieties to the protein molecule. For example, glycosylation of asparagine, serine, or threonine portions of certain protein molecules is critical for enabling those molecules to function properly, that is, the cell they are in must be discerned by the body's immune system to be indigenous or foreign.
- *Phosphorylation*: Addition of a phosphate molecular group to the protein molecule. For example, phosphorylation of serine, threonine, or tyrosine portions of a protein molecule is critical for enabling that molecule to be able to function in signaling (e.g., thereby triggering cell growth, cell growth, etc.).
- *Sulfation*: Addition of a sulfate molecular group to the protein.
- *Acetylation*: Addition of an acetyl molecular group to the protein molecule.
- *Ribosylation*: Addition of a ribose molecular group to the protein molecule.
- *Ubiquitination*: Addition of a ubiquitin molecular group to the protein molecule.

- *Deamidation*: The loss of their side chain molecular groups by some of the glutamine and asparagine portions of a given protein molecule. For example, such deamidation of some asparagine portions of certain apoptosis-blocking proteins causes loss of that molecule's apoptosis-blocking ability.
- Prenylation (also sometimes called isoprenylation or lipidation) is the addition of certain hydrophobic molecular groups to a protein. The prenyl molecular groups (3-methyl-but-2-en-1-yl) facilitate attachment of prenylated proteins to cell membranes.
- *Cleavage*: The removal of a portion of the polypeptide molecular chain in order to produce a functional protein in the correct environment.

See also *Polypeptide (protein), Moiety, Cell, Messenger RNA (mRNA), Enzyme, Ribosomes, Carbohydrates, Protein, Glycosylation, Glycoprotein, Autoimmune disease, Phosphorylation, Signaling, Apoptosis, Ribose, Intein, Histones, Ubiquitin, Annotation (bioinformatics), Micromodification, Plasma membrane, Membranes (of a cell), Membrane transporter protein.*

Potato Late Blight

A fungal disease of the potato plant (*Solanum tuberosum*) that is caused by the fungus *Phytophthora infestans*.

During the 1840s, this plant disease struck the potato crops of Ireland and Europe leading to the starvation of more than one million people (principally in Ireland, because that nation was very dependent on potatoes for food). See also *Fungus*.

PPA

See *Plant Protection Act*.

PPAR

Acronym for *peroxisome proliferators activated receptor*. They constitute a "family" of nuclear receptors (i.e., receptor molecules located on cell nucleus) which influence a cell's metabolism of lipids and glucose. PPARs are grouped into two "subfamilies," *PPAR alpha* and *PPAR gamma*.

PPAR agonist pharmaceuticals such as GlaxoSmithKline's Avandia™ (resiglitazone), and Takeda Pharmaceuticals' Actos™ (pioglitazone) can help control hyperglycemia and dyslipidemia associated with Type II diabetes. See also *Cell, Nucleus, Receptors, Nuclear receptors, Gene, Express, Peroxisome, Metabolism, Lipids, Glucose (GLc), Polyunsaturated fatty acids (PUFA), Agonists, Type II diabetes, Epigenetic, Histone.*

PPAR Alpha

See *PPAR*.

PPAR Gamma

See *PPAR*.

PPB

See *ADME tests, ADME/Tox.*

PPFM

See *Pink pigmented facultative methylotroph.*

PPO

Acronym for *Protoporphyrinogen Oxidase*. See *Acuron*™ gene.

PR Proteins

See *Pathogenesis-related proteins.*

Pre-B Cell Colony-Enhancing Factor

Abbreviated *PBEF*, it is a hormone (also known as visfatin) which was discovered to act as a growth factor for immature B cells of the immune system, during 1994 by B. Samal and colleagues. See also *Hormone, Visfatin, B cells, Growth factor.*

Prebiotics

Chemical compounds or microorganisms (e.g., yeasts)—administered alone or in combination (e.g., in the feed rations of animals)—that (generally) act to stimulate growth of beneficial types of bacteria within the digestive system of animals (e.g., livestock). Those compounds can include some organic acids (e.g., propionic acid, malic acid, etc.).

For example, adding certain strains of yeast (culture) and malate (malic acid) to cattle feed rations has been shown to stimulate *Selenomonas ruminantium* bacteria (growth) in the rumen (i.e., the "first stomach" in cattle). *S. ruminantium* tend to constitute 22%–51% of the total bacteria in a typical rumen and are important for optimal digestion (e.g., of the grass eaten by that animal).

Inulin, and several fructose oligosaccharides, and so on, act as prebiotics in the human digestive system (e.g., by stimulating growth of *Bifidus*) species of bacteria in the digestive system.

For animal feed rations—in addition to *fructose* oligosaccharides, transgalacto-oligosaccharides may be added, to also act as prebiotics. See also *Probiotics, Yeast, Bacteria, Bifidus, Inulin, Fructose oligosaccharides, Transgalacto-oligosaccharides, Strain.*

Precision Farming

Refers to application of variable amounts per acre (hectare) of certain farming inputs (e.g., seeds, fertilizers, irrigation water, etc.) and the selection of which crop seed varieties/hybrids to plant based on the differing soil types, plant nutrient levels in soil, soil pH (which affects bioavailability of the nutrients), that field's crop disease history, that field's crop–pest pressure (predictable from pest population/reproduction, etc. during previous growing season), microorganism populations in soil, field hydrology, and microclimate(s) found within each individual farm field or each homogenous subportion of the field (such homogenous subportions are known as *ERU maps* or *Yield Environments* or *Management Zones*). As a result of greater input-use efficiency for seed/fertilizer/water, the farmer's costs per kilogram of crop produced are lower and the risk of nutrient run-off (e.g., after a major rainfall) into adjacent waterways is decreased. See also *Microorganism, Crop biologicals, ERU maps.*

Predictive Breeding

See *Phenomics.*

P

Pre-mRNA

See *Primary transcript.*

Prenylation

Also sometimes called isoprenylation or lipidation, it is the addition of certain (lipid) hydrophobic molecular groups to a protein molecule. The prenyl molecular groups (3-methyl-but-2-en-1-yl) facilitate attachment of prenylated proteins to cell membranes. See *Protein, Posttranslational modification of protein, Farnesyl transferase inhibitors, Membranes (of a cell), Plasma membrane, Transmembrane proteins, Membrane transporter protein, Lipids, Lipoprotein.*

Pribnow Box

The consensus sequence T-A-T-A-A-T-G centered about 10 base pairs before the starting point of bacterial genes. It is a part of the promoter and is especially important in binding RNA polymerase. See also *RNA polymerase, TATA homology, Homeobox, Promoter, Base pair (bp).*

Primary Structure

Refers to the sequence of amino acids in a protein *"molecular" chain,* or to the linear sequence of nucleotides in a polynucleotide (RNA or DNA) *molecular chain.* See also *Polypeptide (protein), Amino acid, Protein, Structural biology, Structural gene, Structural genomics, Nucleotide, Proteomics, Deoxyribonucleic acid (DNA), Ribonucleic acid (RNA).*

Primary Transcript

Refers to the mRNA that is produced via the transcription process before any posttranscriptional modifications to that RNA molecule occur. In eucaryotic organisms, the primary transcript is known as *pre-mRNA.* See also *Ribonucleic acid (RNA), Messenger RNA (mRNA), Transcript, Transcription, Organism, Eucaryotes, Posttranscriptional processing (modification) of RNAs.*

Primer (DNA)

A short sequence deoxyribonucleic acid (DNA) that is paired with one strand of the template DNA, in the polymerase chain reaction (PCR) technique. In PCR testing (e.g., a paternity test), the primer is selected to be complementary to the analytically relevant sequence of DNA. It is the growing end of the DNA chain and it simply provides a free 3′-OH end at which the enzyme DNA polymerase adds on deoxyribonucleotide units (monomers). Which deoxyribonucleotide is added is dictated by base pairing to the template DNA chain. Without a DNA primer sequence a new DNA chain cannot form since DNA polymerase is not able to initiate DNA chains. See also *Deoxyribonucleic acid (DNA), Sequence (of a DNA molecule), Template, Complementary (molecular genetics), Double helix, Polymerase, Polymerase chain reaction (PCR), Polymerase chain reaction (PCR) technique, Nested PCR.*

Primordium

The initial small group of cells which subsequently become an organ in the body. See also *Cell.*

Primosome

An agglomeration consisting of DNA helicase, primase, and so on, which "unwinds" the DNA molecule (within cell) prior to replication. It also helps start the synthesis of Okazaki fragments (of DNA) during discontinuous replication of DNA. See also *Deoxyribonucleic acid (DNA), DNA helicase, Cell, Unwinding protein, Okazaki fragments.*

Prion

Proteinaceous structures (molecules) found in the plasma membrane (surface) of cells, in the brains and various other tissues of all vertebrate animals. In addition to a role in signaling, one of the functions of (normal) prions (PrP^C) is to help *"capture" and deactivate* oxygen free radicals (i.e., oxygen atoms bearing an extra electron, thus high in energy; for example, which are sometimes generated in a biological system such as within the body of an organism).

In 1982, Stanley Prusiner discovered that misshapen (mutated) versions (PrP^{Sc}) can cause the neurodegenerative disease Bovine Spongiform Encephalopathy in cattle, and the neurodegenerative diseases Creutzfeld–Jakob disease, kuru, Gerstmann–Straussler–Scheinker syndrome, and Fatal familial insomnia in humans. Stanley Prusiner named these molecules *prions for "proteinaceous infected particle,"* because unlike infectious pathogenic bacteria or viruses, prions do not contain DNA.

When misshapen (isoform of) prions are introduced into an animal's central nervous system, they can cause (normal helical shape) prions to adopt mis-shaped form (i.e., akin to the pleated folds in an accordion bellows); resulting is massive neurodegeneration (and death).

The dye named Congo Red, and IDX (a derivative of the chemotherapeutic doxorubicin) have shown some ability to slow prion-caused neurodegeneration. See also *Protein, Cell, Plasma membrane, Mutant, Bacteria, Deoxyribonucleic acid (DNA), Protein structure, BSE, Proto-oncogenes, Stress proteins, Monoclonal antibodies, Free radical, Antioxidants, Human superoxide dismutase (hSOD), Signaling, Organism.*

Proanthocyanidins

Refers to phytochemical components (i.e., condensed tannins) within North American cranberries (*Vaccinium macrocarpon*), red grapes, and blueberries (genus *Vaccinium*) that impart health benefits to humans who consume those cranberries/red grapes/blueberries.

For example, when humans consume these fruits, the proanthocyanidin molecules "tie up" the free radicals which otherwise can cause oxidative stress to the human body. Some research also indicates these molecules help protect against some cancers.

For example, when humans consume cranberries, these chemical compounds prevent *Escherichia coli* bacteria from adhering to the cells lining the human urinary tract (thereby helping to prevent some urinary tract infections). See also *Phytochemicals, Free radical, Oxidative stress, Antioxidants, Anthocyanidins, Nutraceuticals, Cell, Escherichia coliform (E. coli), Cancer.*

Probe

A relatively small molecule that can be used to sense the presence and condition of a specific protein, DNA fragment, RNA fragment, or nucleic acid by a unique interaction with that macromolecule.

For example, the phrase PROBE DNA or DNA PROBE refers to a known-sequence piece of DNA that is first labeled (e.g., via autoradiography or fluorescence *in situ* hybridization) then utilized to detect the presence of a particular gene (or shorter piece of DNA) within a biological sample. When a florescent-labeled DNA probe hybridizes to the desired gene (or shorter piece of DNA) in sample, its "fluorescent tag" begins to fluoresce at a different (known) wavelength/color, thereby indicating positively the presence in sample of that particular gene (or shorter piece of DNA). See also *DNA probe, Hybridization (molecular genetics), Bacterial artificial chromosomes (BAC), Yeast artificial chromosomes (YAC), Human artificial chromosomes (HAC), Marker assisted selection, Southern blot analysis, Autoradiography, Fluorescence, Label (fluorescent), Fluorescence in situ hybridization (FISH), Bio-bar codes.*

Probe DNA

See *Probe.*

Probiotics

Refers to specific species/strains of microorganisms that are very beneficial to the digestive system of humans or animals, or that restrain harmful bacteria/fungi (e.g., by "crowding them out" within the digestive system).

For example, while growing within the digestive system numerous strains of *Bifidus* bacteria produce organic acids (e.g., propionic, acetic, lactic), which make the host animal's digestive system more acidic. Because most pathogens (i.e., disease-causing microorganisms) grow best at a neutral pH (i.e., neither acidic nor base/caustic), the growth rates of pathogens are thereby inhibited. That enables the *Bifidus* bacteria to "crowd out" enteric pathogens, since *Bifidus* bacteria grow fast in the acidic environment created by those organic acids.

For example, the bacterial strain *Bifidobacterium animalis* ssp. *animalis* has been shown to increase the vaso-active intestinal peptide levels within the human digestive system, which improves barrier integrity function (e.g., prevent leaky gut syndrome, irritable bowel syndrome, etc.).

Some people would include within the definition of probiotics compounds that (generally) act to stimulate growth of beneficial types of bacteria within the digestive system of animals (e.g., livestock). For example, organic acids (e.g., propionic acid, acetic acid, lactic acid, citric acid, etc.) act to inhibit the growth/multiplication of pathogens (i.e., disease-causing microorganisms) in the digestive system of monogastric (i.e., single-stomach) animals such as poultry and swine. Those acids are able to pass through the outer cell membrane (i.e., plasma membrane) of pathogenic bacteria and fungi. Once inside those pathogens' cells, the acids dissociate, and acidify the cell interior (which disrupts the cell's protein synthesis, growth, and replication of the pathogen). However such compounds are more appropriately known as prebiotics. See also *Microorganism, Bacteria, Species, Bifidus, Gut leakage, Cell, Acid, Propionic acid, Citric acid, Pathogen, Fungus, Peptide, Protein, Prebiotics, Chronic inflammation.*

Procambium

Refers to a primary meristem (i.e., undifferentiated plant cells from which plant "organs" are derived) that becomes the plant's primary vascular (i.e., liquid-conveying) tissue. See also *Cell, Differentiation.*

Procaryotes

From the Latin *pro* meaning before and the Greek *karyon* meaning nut. Simple organisms that lack a distinct nuclear membrane and other organelles (which would look a bit like nuts within a cell, under a microscope).

Many structural systems are different between procaryotes and eucaryotes including the DNA arrangement, composition of membranes, the respiratory chain, the photosynthetic apparatus, ribosome size, the presence or lack of cytoplasmic streaming, the cell wall, flagella, the mode of sexual reproduction, and the presence or lack of vacuoles. Some representative procaryotes are the bacteria and blue-green algae. See also *Cell, Eucaryote, Bacteria, Nucleus.*

Process Validation (for Production of a Pharmaceutical)

Defined by America's FDA as "Establishing documented evidence which provides a high degree of assurance that a specific process will consistently produce a (pharmaceutical) product meeting predetermined specifications and quality characteristics." See also *Food and Drug Administration (FDA), Good manufacturing practices (GMP), Good laboratory practices (GLP), cGMP.*

Pro-drug Therapy

Refers to a regime in which pharmaceutical(s):

- Are first administered (e.g., injected intravenously) in the form of biologically *inactive* compound(s).
- Accumulate in the targeted tissue (e.g., tumor).
- Are then caused to change into biologically active chemical(s) at that targeted tissue/location via a "trigger" (e.g., orally administered nanobody/zymogen).

For example, the blood-thinning (platelet inhibiting) pharmaceutical Plavix (clopidogrel) is inactive until it gets metabolized in the human liver by cytochrome P450 enzymes. See also *Zyme systems, Biological activity, Absorption, Tumor, Orally administered, Zymogens, Nanobodies, Platelets, Metabolism, Cytochrome P450 (CYP).*

Proenzyme

See *Zymogen.*

Progesterone

A female sex hormone, secreted by the ovaries, that supports pregnancy (e.g., it prepares the lining of the uterus for attachment of the fertilized egg cell) and subsequent lactation (i.e., milk production). See also *Hormone, Pituitary gland, Estrogen, Stem cells.*

Programmed Cell Death

See *p53 gene, Apoptosis, Hypersensitive response.*

Prokaryotes

See *Procaryotes.*

P

Promoter

The region on DNA to which RNA polymerase binds and initiates transcription (of RNA). The promoter "promotes" the transcription (expression) of that gene, but the promoter's impact on the timing/degree of gene expression is itself regulated by the molecules that bind to the promoter. For example, the "binding" of RNA polymerase causes transcription of RNA to begin, and the "binding" to promoter of other STATs (i.e., signal transducers and activators of transcription) can regulate the degree to which a given gene is expressed.

A promoter is a region of DNA (deoxyribonucleic acid) which lies "upstream" of the transcriptional initiation site of a gene. The promoter controls where (e.g., which portion of a plant, which organ within an animal, etc.) and when (e.g., which stage in the lifetime of an organism) that the gene is expressed. For example, the promoter named "Bce4" is "seed-specific" [i.e., it only "promotes" the expression of a given gene's product (e.g., protein, fatty acid, amino acids, etc.) within a plant's seed]. See also *Polymerase, Gene, Express, RNA polymerase, Control sequences, Gene expression, Bce4, Plastid, Deoxyribonucleic acid (DNA), Polygenic, Transcription, Cauliflower mosaic virus 35S promoter, Signal transducers and activators of transcription (STATs), Methylated, Long noncoding RNAs.*

Proof-Reading

Any mechanism for correcting errors in nucleic acid (DNA) synthesis that involves scrutiny of individual (chemical) units after they have been added to the DNA (molecular) chain. This function is carried out by DNA polymerase, a 3′–5′ exonuclease, among others. Proof-reading dramatically increases the fidelity (accuracy) of the DNA base-pairing mechanism. See also *Replication (of DNA), DNA polymerase, Sequencing (of DNA molecules), Mismatch repair, Gene repair (natural), Base pair (bp), Exonuclease.*

Pro-phage

Refers to when a scientist inserts the entire DNA of a bacteriophage into the chromosome of a bacterium so that it is replicated along with the chromosome of the bacterium. See also *Deoxyribonucleic acid (DNA), Bacteriophage, Chromosome, Bacteria.*

Programmed Cell Death

See *p53 gene, Apoptosis, Hypersensitive response.*

Prophase

From the Greek words *pro* meaning *before* and *phasis* meaning *appearance*, it is the first of the four phases of eucaryotic mitosis (i.e., cell replication via division) during which the now-doubled chromosomes condense and become visible (under a microscope), and the membrane surrounding the cell nucleus dissolves. See also *Mitosis, Eucaryotes, Cell, Chromosomes, Membranes (of a cell), Nucleus, S-phase.*

Propionic Acid

See *Probiotics, Bifidus.*

Prostaglandin Endoperoxide Synthase

An enzyme that can exist in several different forms within the human body, to catalyze the production of prostaglandins. See also *Enzyme, Cyclooxygenase, Arachidonic acid, Isozymes, Prostaglandins, Highly unsaturated fatty acids (HUFA).*

Prostaglandins

A group of cyclic (i.e., circle-shaped molecule) fatty acids that act as hormones in the body (i.e., promote inflammation during infections, help promote maintenance of the tissues of the stomach/kidney/intestines, etc.). Their primary mode of action is through certain G-protein-coupled receptors.

Originally isolated from sheep and human prostates, prostaglandins are synthesized (i.e., "manufactured") by most cells in the body via chemical reactions catalyzed by the enzyme cyclooxygenase/prostaglandin endoperoxide synthase, usually from arachidonic acid (also docosahexanoic acid). Consumption of significant amounts of aspirin inhibits prostaglandin production.

When prostaglandins are present in excess within the joints of people who have arthritis, prostaglandins can cause swelling of tissue and pain. Excess presence of prostaglandins has also been linked to an increased risk of colon cancer. See also *Eicosanoids, Prostaglandin endoperoxide synthase, Cyclooxygenase, Arachidonic acid, Fatty acid, G-protein-coupled receptors, G-proteins, Hormone, Enzyme, Highly unsaturated fatty acids (HUFA), Docosahexanoic acid (DHA), Cancer, Glycyrrhizic acid, Polyunsaturated fatty acids (PUFA).*

Prostate

The gland in the body of males that produces the liquid which carries sperm into the females (during mating).

In older human males, the prostate will often become enlarged (e.g., by "antagonism" when estrogen molecules circulating in the blood contact its surface). Via the *selective estrogen effect*, isoflavones (e.g., from soybeans) consumed by such males can displace and replace those estrogen molecules from the surface of the prostate (thereby preventing enlargement). See also *Estrogen, Isoflavones, Selective estrogen effect, RNAse 1.*

Prostate Cancer

See *Metastasis, Signaling.*

Prostate-Specific Antigen (PSA)

An antigen whose concentration increases significantly 5–10 years prior to the (clinical) diagnosis of prostate cancer. This means that PSA level measurements can be utilized in (biomarker) diagnosis of prostate cancer before symptoms appear. However, a series of tests are required in order to accurately gauge the probability of cancer because PSA levels can also be elevated when a man develops a noncancerous enlarged prostate. See also *Antigen, Tumor, Tumor-associated antigens, Gene, Cancer, Prostate, RNAse 1, RNAse 1 gene, Biomarkers.*

Prostatitis

Refers to noncancerous enlargement of the prostate, which tends to occur in men as they get older. See the links *Prostate, Prostate-specific antigen (PSA).*

Prosthetic Group

A heat-stable metal ion or an organic group (other than an amino acid) that is covalently bonded to the apoenzyme protein. It is required for enzyme function. The term is now largely obsolete. See also *Ion, Amino acid, Protein, Enzyme, Apoenzyme, Coenzyme.*

Protease

An enzyme that catalyzes the hydrolytic cleavage (breakdown) of proteins. By analogy, the enzyme breaks the link (peptide bond) holding a protein molecular chain together. Proteases represent a whole class of protein-degrading enzymes.

There are approximately 560 human proteases. See also *Hydrolytic cleavage, Enzyme, Peptide bond, Trypsin, Chymotrypsin, Lactoferrin, Neutrophils, Protein.*

Protease Inhibitor

Refers to a compound that slows or stops the action (cleavage of protein molecules) of a specific protease. If that particular protein is utilized by a pathogen to cause disease, such a protease inhibitor can prevent or halt the disease. For example, during 2014 the U.S. FDA approved simeprevir (a hepatitis C virus NS3/4A protease inhibitor) as the pharmaceutical Olysio, for oral treatment of genotype 1 chronic hepatitis C infection in adult patients. See also *Protease, Protein, Pathogen, Food and Drug Administration (FDA), Virus.*

Protease Nexin I (PN-I)

A specific protein that acts as an inhibitor of certain proteases. See also *Protease, Protein, Protease nexin II (PN-II).*

Protease Nexin II (PN-II)

A protein that is thought to regulate important activities in the body and brain by inhibiting specific enzymes and interacting with certain body cells. PN-II is formed from a precursor molecule known as beta-amyloid, via metabolic processing of the beta-amyloid. Recent research indicates that incorrect metabolic processing of beta-amyloid by the body results in amyloid plaques in the brain. The amyloid plaques are generally found in victims of Alzheimer's disease, and directly correlate (in number) with the degree of dementia. See also *Protease nexin I (PN-I), Regulatory enzyme, Protein, Enzyme, Inhibition, Metabolism.*

Proteasome Inhibitors

Refers to any compounds which halt or slow the action of proteasomes in living cells.

During 2003, the U.S. FDA approved one proteasome-inhibiting pharmaceutical known as Velcade™ (bortezomib) for the treatment of mantle cell lymphoma blood cancer. See also *Proteasomes, Food and Drug Administration (FDA), MCL, Cancer.*

Proteasomes

Refers to enzymatic/catalytic bodies that are present within all mammal cells, which activate certain transcription factors, are involved in causing the cell to "present" antigens (i.e., from pathogens that invaded that cell) on the cell's surface, and various other cellular functions. For example, the 26S proteasome degrades (i.e., breaks-down) all ubiquinated (i.e., ubiquitin-"tagged") proteins in that cell. See also *Enzyme, Protein, Cell, Transcription factors, Antigen, Pathogen, Ubiquitin.*

Protein

Coined in 1838 by Jons Berzelius. From the Greek word *proteios*, which means "the first" or "the most important" or "of the first rank." Any of a class of high molecular weight polymer compounds composed of a variety of α-amino acids joined by peptide linkages. Via the synthesis (of this "chain") performed by ribosomes, each protein is the ultimate expression product of a gene. More than one protein can be expressed from a given gene (the particular protein expressed is determined by factors such as the cell's temperature or other environmental variable, presence of STATs—some of which themselves are proteins, presence of certain bacteria, etc.).

During their synthesis (after emerging from cell's ribosome), proteins may also be phosphorylated (i.e., a "phosphate group" is added to the protein molecule), glycosylated (i.e., one or more oligosaccharides is added onto the protein molecule), acetylated (i.e., one or more "acetyl groups" is added to the protein molecule), farnesylated (i.e., a "farnesyl group" is added to the protein molecule), ubiquinated (i.e., a ubiquitin "tag" is added to the protein molecule), sulfated (i.e., a "sulfate group" is added to the protein molecule), or otherwise chemically modified.

Proteins are the "workhorses" of living systems and include enzymes, antibodies, receptors, peptide hormones, and so on. Proteins in living organisms respond to changing environmental and other conditions by changing their location within cells, by getting cut into (specific) pieces, by changing which (other) molecules they will bind (adhere) to, and so on. All of the amino acids commonly found in (each and every one of the) proteins have an asymmetric carbon atom, except the amino acid glycine. Thus, the protein is potentially chiral in nature. See also *Amino acid, Gene, Peptide, Absolute configuration, Stereoisomers, Chiral compound, Express, Oligomer, Protein folding, Messenger RNA (mRNA), Ribosomes, Polyribosome (polysome), Organism, Cell, Signal transducers and activators of transcription (STATs), Central dogma (NEW), Phosphorylation, Ubiquitin, Glycosylation (to glycosylate), Farnesyl transferase, Docking proteins, Peptide-oligonucleotide conjugates.*

Protein Arrays

See *Protein microarrays.*

Protein Biochips

See *Protein microarrays.*

Protein Bioreceptors

See *Receptors.*

Protein C

An anticlotting (glyco) protein that prevents postoperative arterial clot formation when administered intravenously. May be synergistic (in its anticlotting effect) with tissue plasminogen activator (tPA). See also *Thrombomodulin, Tissue plasminogen activator (tPA), Protein, Glycoprotein.*

P

Protein Chaperones

See *Chaperones.*

Protein Chips

See *Protein microarrays.*

Protein Digestibility-Corrected Amino Acid Scoring (PDCAAS)

A method of expressing the quality of a given (food) protein source, in terms of its digestible protein (amino acid constituents') ability to support growth in young growing humans (i.e., if that protein supplies all needed essential amino acids in their proportions required by humans—that protein scores 1.00). For example, two complete ("ideal") protein sources are soy protein (concentrate) with a PDCAAS of 0.99, and soy protein (isolate) with a PDCAAS of 1.0.

PDCAAS has been recommended by the U.S. FDA, and the Food and Agricultural Organization of the United Nations/World Health Organization (FAO/WHO). See also *Protein, Amino acid, Essential amino acids, Ideal protein concept, Soy protein, Food and Drug Administration (FDA), Digestion (within organisms), Deamination.*

Protein Docking

See *Docking proteins.*

Protein Engineering

The selective, deliberate (re)designing and synthesis of proteins. This is done in order to cause the resultant proteins to carry out desired (new) functions. Protein engineering is accomplished by changing or interchanging individual amino acids in a normal protein. This may be done via chemical synthesis or recombinant DNA technology (i.e., genetic engineering). "Protein engineers" (actually genetic engineers) use recombinant DNA technology to alter a particular nucleoside or triplet (codon) in the DNA (genes) of a cell. In this way it is hoped that the resulting DNA codes for the different (new) amino acid in the desired location in the protein produced by that cell. See also *Protein, Polypeptide (protein), Gene, Codon, Genetic engineering, Amino acid, Essential amino acids, Synthesizing (of proteins).*

Protein Expression

Consists of (the combination/total of) both translation *and* PTM of a given protein molecule. See also *Protein, Express, Translation, Posttranslational modification of protein.*

Protein Folding

The complex interactions of a polypeptide molecular chain with its environment and itself and other protein entities, which cause the polypeptide molecule to fold up into a highly organized, tightly packed, three-dimensional structure. Proven to occur spontaneously, by Christian B. Anfinsen during the 1960s; for protein molecules outside of living cells.

This ability of polypeptide chains to fold into a great variety of topologies, combined with the large number of sequences (in the molecular chain) that can be derived from the 20 common amino acids in proteins, confers on protein molecules their great powers of recognition and selectivity. How a protein folds up determines its chemical function.

During the 1990s, it was discovered that inside living cells, "chaperone" molecules are needed for proper protein folding to occur. These chaperones are protein molecules (e.g., certain heat-shock proteins) that form a loosely bound complex to suppress incorrect protein folding as the protein molecule is emerging from the cell's ribosome, so protein folding is both complete and correct as soon as the newly formed protein molecule is released from the cell's ribosome.

Some diseases (e.g., Alzheimer's disease) can be caused by large amounts of mis-folded proteins. See also *Amino acid, Protein, Polypeptide (protein), Ribosomes, Chaperones, Prion, Absolute configuration, Conformation, Enzyme, Protein structure, Alzheimer's disease, Rapid protein folding assay, Unfoldases.*

Protein Inclusion Bodies

See *Refractile bodies (RB).*

Protein Interaction Analysis

Refers to a number of different analyses/technologies utilized to determine if a given (e.g., "unknown") protein molecule interacts with a protein molecule whose function is already known (e.g., from previous research, its use as a pharmaceutical, etc.). Through that analysis (e.g., inferring the "new" protein's function by its interactions with the "known" protein), useful information about the "new"/unknown protein can be gathered.

Technologies utilized include two-hybrid systems (e.g., yeast two-hybrid system), surface plasmon resonance, nuclear magnetic resonance, mass spectroscopy, tandem affinity purification tagging, light resonant wave technology, acoustic wave technology, and so on. See also *Proteomics, Two-hybrid systems, Proteome chip, Biochips, Gene expression analysis, Protein, Genomics, Functional genomics, Protein microarrays, Fluorescence resonance energy transfer (FRET), Surface plasmon resonance (SPR), Nuclear magnetic resonance, Mass spectrometer, Tandem affinity purification tagging, Quantum dot, Bioluminescence resonance energy transfer.*

Protein Interaction Profile Sequencing

Refers to specific technologies that can quantitatively capture both the pattern of protein-RNA binding sites (e.g., in the nucleus of plant cells) and the larger structure of nuclear DNA or other DNA-sequence-resultant transcripts. See also *Sequence (of a DNA molecule), Deoxyribonucleic acid (DNA), Sequencing (of DNA molecules), Protein, Ribonucleic acid (RNA), Cell, Nucleus, Nuclear DNA, Transcription, Transcript.*

Protein Kinases

Enzymes capable of phosphorylating (covalently bonding a phosphate group to) certain amino acid residues in specific proteins. Protein kinases play crucial roles in the regulation of signaling within, and between cells. See also *Kinases, Phosphorylation, Tyrosine kinase, Enzyme, Amino acid, Protein, Protein signaling, Cell, Tyrosine kinase inhibitors (TKI).*

Protein Ladder

See *Molecular-weight size marker.*

Protein Microarrays

Refers to a piece of glass, plastic, silicon, or a nanosheet… onto which has been attached a number of *capture agents* (e.g., antibodies, aptamers, enzymes, antigens, receptors, ligands, or other molecules of other chemical compounds that bind/interact with proteins in a specific manner) at specific/known locations on the microarray.

These microarrays (sometimes called "biochips," protein biochips, protein arrays, etc.) can then be utilized to test (e.g., a single sample) for a wide variety of attributes or effects (on, or by the protein molecules in the sample that is exposed to that microarray). See also *Protein, High-throughput screening (HTS), Target–ligand interaction screening, Receptors, Protein interaction analysis, Protein structure, Proteomics, Antibody, Aptamers, Proteome chip, Microarray (testing), Biochip, Quantum dot, Enzymes, Capture agent, Functional protein microarrays, Nanosheets.*

Protein Pump

See *Efflux pump.*

Protein Quality

See *Amino acid profile, PDCAAS.*

Protein Sequencer

See *Sequencing (of protein molecules), Gene machine, Sequencing (of DNA molecules).*

Protein Signaling

The "communication" by protein molecules (e.g., to cells) that governs their transport and localization (i.e., destination that they go to, in cell). Discovered and delineated by Guenter Blobel during the 1970s, protein signaling (e.g., via a short sequence of amino acids attached to end of newly synthesized protein molecules) results in proteins traveling to the appropriate cell compartments (e.g., organelles) and/or out of the cell (i.e., secretion). See also *Protein, Signaling, Signaling molecule, Cell, Amino acid, Signal transduction, G-proteins, Ribosomes, Protein kinases.*

Protein Solubilization

Refers to the process of dis-aggregating protein molecules (e.g., in preparation for putting those proteins through two-dimensional gel electrophoresis, etc.). This dis-aggregation process involves breaking the intermolecular protein interactions/attractions resulting from van der Waals forces, disulfide bonds, ionic and other weak interactions, hydrophobic interactions, and so on. See also *Protein, Two-dimensional (2D) gel electrophoresis, van der Waals forces, Weak interactions, Hydrophobic, Disulfide bond.*

Protein Splicing

See *Splicing (of protein molecule).*

Protein Structure

A polypeptide chain may take on a certain structure in and of itself because of the amino acid monomers it contains and their location within the chain. The chain may furthermore interact with other polypeptide chains to form larger proteins known as oligomeric proteins. In what follows, the levels of protein structure normally encountered will be highlighted:

- *Primary structure*: Refers to the backbone of the polypeptide chain and to the sequence of the amino acids of which it is comprised.
- *Secondary structure*: Refers to the shape (recurring arrangement in space in one dimension) of the individual polypeptide chain. In some cases, because of its primary structure, the chain may take on an extended or longitudinally coiled conformation.
- *Tertiary structure*: Refers to how the polypeptide chain (the primary structure) is bent and folded in three-dimensional space in order to form the normal tightly folded and compact structure.
- *Quaternary structures*: Refers to how, in larger proteins made up of two or more individual polypeptide chains, the individual polypeptide chains are arranged relative to each other. These large multipolypeptide proteins are called oligomeric proteins and the individual chains are called subunits. An example of such a protein is hemoglobin.

See also *Conformation, Protein folding, Polypeptide (protein), Tertiary structure, Proteomics, Native conformation, Chaperones, Unfoldases.*

Protein Tyrosine Kinase Inhibitor

Any compound that inhibits the action of the enzyme *tyrosine kinase*. Examples include genistein and the pharmaceuticals Gleevec™ (imatinib mesylate), Iressa™ (gefinitib), and Tarcera. See also *Enzyme, Inhibition, Tyrosine kinase, Genistein (gen), Gleevec™, Kinases.*

Protein Tyrosine Kinases

Refers to a "family" of kinase enzymes in the human body, which assist/facilitate the transfer of "phosphoryl groups" (from one molecule to another molecule that is "targeted" by that kinase.

Protein tyrosine kinases are critical components in the signaling pathways involved in tumorigenesis (tumor creation) and angiogenesis (i.e., creation of new blood vessels to "feed" the growing tumor). See also *Tyrosine kinase, Enzyme, Phosphorylation, Protein, Kinases, Signaling, Signaling molecule, Pathway, Tumor, Angiogenesis, Kinome.*

Protein-Based Lithography

See *Bioelectronics, Protein.*

Protein-Conducting Channel

Refers to transmembrane (i.e., through a plasma membrane) holes through which can pass newly synthesized protein molecules, under appropriate conditions. It is also thought that *membrane proteins* (e.g. receptors) enter the relevant membrane (e.g., plasma membrane) via protein-conducting channels (whereupon much of the membrane protein molecule remains embedded in that membrane). See also *Protein, Cell, Plasma membrane, Receptors.*

P

Protein–Protein Interactions

See *Protein, Protein interaction analysis, Protein microarrays, Structure–activity models, Two-hybrid systems.*

Proteolytic

Refers to the breakdown of protein molecules into smaller pieces. See also *Protein, Proteolytic enzymes.*

Proteolytic Enzymes

Enzymes which catalyze the hydrolysis (breakdown) of proteins or peptides. Proteins (enzymes) that destroy the structure (by peptide bond cleavage) and hence the function of other proteins. These other proteins may or may not themselves be enzymes. See also *Enzyme, Protein, Protease, Ubiquitin.*

Proteome Chip

A microarray ("biochip") developed by Michael Snyder et al. during 2001 which:

1. Has a large number of *known sequence protein molecules* (e.g., all proteins present in a given organism) attached to its surface at known locations (i.e., specific "addresses" on the microarray).
2. Utilizes specific *bioactive agents such as certain lipids or biotinylated calmodulin (i.e., calmodulin molecules to which a molecule of biotin is "attached")* in order to determine which of the protein molecules in #1 interacts with (relevant bioactive agents). Because calmodulin is a well-known and very well-characterized calcium-binding protein (i.e., bioactive agent) involved in (known) cellular processes, the binding of calmodulin to specific *protein molecules attached to the microarray/biochip* provides critical information about the (cellular, protein–protein, etc.) functions and interactions of those protein molecules *in the organism.*
3. Reveals a large amount of data concerning *protein–protein interactions* (e.g., via subsequent application to the microarray of dye-labeled streptavidin to identify the protein molecules VIA THEIR ADDRESSES on the biochip) and *protein–lipid interactions.*

all of which are needed, in order to determine the organism's proteome. See also *Biochips, Protein microarray, Protein interaction analysis, Target–ligand interaction screening, Microarray (testing), Proteome, Biotin, Organism, Avidin, Metabonomics.*

Proteomes

See *Proteomics.*

Proteomics

The scientific study of an organism's proteins and their role in an organism's structure, growth, health, disease (and/or the organism's resistance to disease, etc.). Those roles are predominantly due to each protein molecule's tertiary structure/conformation, but some are also due to some proteins' interaction with the organism's DNA (e.g., transcription factors).

Some methods utilized to determine which impact results from which protein, are as follows:

- *Chemical genetics,* to compare two same-species organisms (one of which has protein—or a portion of protein—at least partially inactivated by a specific chemical).
- *Gene expression analysis,* to determine the protein(s) produced when a given gene is "switched on"; by measuring fluorescence of individual messenger RNA (mRNA) molecules (specific to which particular gene is "switched on" at the time), when that mRNA hybridizes (with DNA pieces corresponding to genes analyzed, that were attached to hybridization surface on biochip).
- Gene expression analysis, to determine impact when a given gene is "knocked out"/"turned off."
- *Protein interaction analysis,* to determine if a newly discovered protein molecule interacts with a protein molecule whose function is already known (e.g., from previous research or use as a pharmaceutical). If the newly discovered protein molecule interacts with one whose function is already known, it generally has the same or similar function (in living cells) as the previously known protein molecule. Thus, the function of a newly discovered human protein can sometimes be inferred from a protein molecule discovered earlier in a microorganism (e.g., via Expressed Sequence Tags, model organism, Raman optical activity spectroscopy, etc.).
- *In silico biology (modeling),* to compare computer-predicted events (e.g., the constituent peptides resulting from protein digestion) with actual or *in vitro* outcomes.
- *Chromatin immunoprecipitation* to determine all points on an organism's DNA that a particular protein (e.g., transcription factor) interacts with.

See also *Protein, Primary structure, Conformation, Native conformation, Tertiary structure, Deoxyribonucleic acid (DNA), Gene, Genetic map, Genomics, Electrophoresis, Two-dimensional (2D) gel electrophoresis, Sequencing (of protein molecules), Genetic code, Cell, Sequence (of a protein molecule), Structural genomics, Functional genomics, Combinatorial chemistry, Bioinformatics, High-throughput screening, Biochips, Chemical genetics, Gene expression analysis, Fluorescence, Messenger RNA (mRNA), Organism, Microorganism, Hybridization (molecular biology), Hybridization surfaces, Express, Expressed sequence tags (EST), Organism, Protein interaction analysis, In silico biology, In vitro, Metabonomics, Phylogenetic profiling, Raman optical activity spectroscopy, Knockout, Knockin, Model organism, MUDPIT, Transcription factors, Chromatin Immunoprecipitation, Chromatin immunoprecipitation method, Unfoldases.*

Proto-Oncogenes

Certain genes within a cell's DNA that code-for receptors (proteins on outer surface of cell membrane) for a cellular growth factor (e.g., epidermal growth factor). Via that coding-for of applicable receptors (or other protein molecules that are part of the signal transduction process of a cell), oncogenes "turn on" the process of cell division (replication) at appropriate time(s) during the life of each cell in an organism.

Proto-oncogenes can become cancer producing. Proto-oncogenes are activated to oncogenes via different mechanisms, including point mutation, chromosome translocation, insertional mutation, and amplification. See also *Gene, Cell, Deoxyribonucleic acid (DNA), Receptors, Protein, Growth factor, Epidermal growth factor, Signal transduction, Oncogenes, Amplification, Mutation, Point mutation, Cancer.*

Protoplasm

Coined by J. E. Parkinje in 1840, it is a general term referring to the entire contents of a living cell; living substance. See also *Cell, Nucleoplasm.*

Protoplast

From the Greek word *protoplastos* meaning *formed first*, it refers to the cell structure consisting of the cell membrane and all of the intracellular components, but devoid of a cell wall. This (removal of cell's outer wall) can be done to plant cells via treatment with cell-wall-degrading enzymes or electroporation. Under specific conditions (e.g., electroporation), certain DNA sequences (genes) prepared by man, can enter protoplasts. The cell then incorporates some or all of that DNA into its genetic complement (genome), and produces whatever product the newly introduced gene codes for. In the case of plant protoplasts, whole plants can be regenerated from the (genetically engineered) protoplasts, resulting in plants that produce whatever product(s) the introduced gene(s) codes for. See also *Cell, Enzyme, Electroporation, Gene, Genetic engineering, Deoxyribonucleic acid (DNA), Coding sequence, Protein, Soybean plant, Corn, Canola.*

Protoplast Fusion

Refers to the practice of fusing two living cells together by first making each cell into a protoplast and then fusing together the two in order to result in a (combined) cell which possesses traits from both of the original cells. See also *Protoplast, Cell, Trait.*

Protoxin

A chemical compound that only becomes a toxin after it is altered in some way. For example, the B.t. protoxins (e.g., Cry9C, Cry1A (b), Cry1A (c), etc.) only become toxic after they are chemically altered by the alkaline environment inside the gut of certain insects. See also *Bacillus thuringiensis (B.t.), B.t. kurstaki, CRY proteins, CRY1A (b) protein, CRY1A (c) protein, CRY9C protein, B.T. israelensis, B.T. tenebrionis, Target (of a herbicide or insecticide).*

Protozoa

A microscopic, single-celled animal form. A unicellular organism without a true cell wall, that obtains its food phagotropically. See also *Phagocyte.*

Provitamin A

See *Beta carotene, Golden rice.*

PrP^{C}

Abbreviation for *prion protein cellular.* See *Prion.*

PrP^{Sc}

Abbreviation for *prion protein scrapie*, the mis-shapen (infectious) form of prion. See *Prion.*

PRR

Acronym for pattern recognition receptor. See *Phytophthora root rot.*

PS

See *Phosphatidyl serine.*

PSA

See *Prostate-specific antigen (PSA).*

P-Selectin

Formerly known as GMP-140 and PADGEM, it is a selectin molecule that is synthesized by endothelial cells before (adjacent) tissues are infected. Thus "stored in advance," the endothelial cells can present P-selectin molecules on the internal surface of the endothelium within minutes after an infection (of adjacent tissue) begins. This presentation of P-selectin molecules attracts leukocytes to the site of the infection, and draws them out of the bloodstream (the leukocytes "squeeze" between adjacent endothelial cells). See also *Selectins, Lectins, ELAM-1, Adhesion molecule, Leukocytes, Endothelium.*

Pseudogene

A segment of a DNA molecule that acts like a gene (i.e., it codes for a protein molecule product), but its protein product is generally not biologically active. See also *Deoxyribonucleic acid (DNA), Gene, Coding sequence, Protein, Biological activity.*

Pseudomonas aeruginosa

See *Citrate synthase (CSb) gene.*

Pseudomonas fluorescens

A normally harmless soil microorganism (bacteria) that colonizes the roots of certain plants. At least one company has incorporated the gene for a protein that is toxic to insects (taken from *Bacillus thuringiensis*) into a *Pseudomonas fluorescens.* This was done in order to confer insect resistance to the plants the roots of which the genetically engineered *Pseudomonas fluorescens* has colonized. See also *Bacillus thuringiensis (B.t.), Bacteria, Wheat take-all disease, Genetic engineering, Endophyte, Commensal.*

Pseudomonas syringae

A pathogenic (i.e., plant disease causing) bacteria which gains entry into plant cells by facilitating the formation of ice crystals on the surface of leaves at a temperature slightly higher than 0°C (32°F). This causes the leaf surface cells to burst open, resulting in an infection of the leaves by the *Pseudomonas syringae.* See also *Bacteria, Pathogenic.*

Pseudopodia

See *Actin.*

P

Pseudouridylation

See *Posttranscriptional processing (modification) of RNAs.*

Psoralen

See *Psoralene.*

Psoralene

A toxic chemical (furanocoumarin) to ward off insects, that is naturally produced by (wild type) plants related to the domesticated celery plant. Also present in small amounts in celery, parsley, parsnips, and dill. See also *Toxin, Phytotoxin, Wild type, Food and Drug Administration (FDA), Traditional breeding methods.*

PST

See *Porcine somatotropin.*

Psychrophile

An organism that requires a cold environment such as 0°C (32°F) for growth. See also *Mesophile, Thermophile, Psychrophilic enzymes.*

Psychrophilic Enzymes

Enzymes found within certain organisms that are adapted to function in cold environments. See also *Psychrophile, Enzyme.*

PTEN Activity

Acronym for *phosphatase and tensin homolog* activity. Refers to activity of a pathway within the body, which helps to regulate insulin signaling and insulin sensitivity in adipose tissue.

That pathway's normal product is a specific lipid, but in some advanced cancers the PTEN activity is greatly altered. Thus, some pharmaceutical research programs screen cancer drug candidate compounds against PTEN activity. See also *PTEN gene, Insulin, Signaling, Diabetes, Adipose, Pathway, Lipids, Cancer, High-throughput screening (HTS), Homology.*

PTEN Gene

Refers to a tumor-suppressor gene (in human DNA) that also serves a function in the regulation of insulin signaling and insulin sensitivity. The presence of genistein has been shown to induce the *PTEN* gene (thereby resulting in apoptosis of breast cancer cells). See also *Gene, Deoxyribonucleic acid (DNA), Tumor, Cancer, Tumor-suppressor genes, PTEN activity, Signaling, Signaling molecule, Insulin, Genistein (gen), Inducible promoter.*

Pterostilbenes

See *Polyphenols.*

PTK

Acronym for *protein tyrosine kinase*. See *Protein tyrosine kinase.*

PTM

Acronym for *Posttranslational Modification (of protein molecules).* See *Posttranslational modification of protein.*

PTPN22 Gene

Refers to one gene that (in humans) is one of the factors causing rheumatoid arthritis. See also *Arthritis, Rheumatoid arthritis, Gene.*

PUFA

See *Polyunsaturated fatty acids (PUFA).*

Pulmonary Arterial Hypertension

Abbreviated PAH, it is a progressive and life-threatening form of pulmonary hypertension (high blood pressure within lungs) in which the blood pressure in the pulmonary arteries is significantly increased due to constriction of those particular blood vessels and which can lead to heart failure and death.

Pure Culture

A culture containing only one species of microorganism. See also *Culture, Culture medium.*

Purification Tag

See *Affinity tag.*

Purine

A basic nitrogenous heterocyclic compound found in nucleotides and nucleic acids; it contains fused pyrimidine and imidazole rings. Adenine and guanine are examples. See also *Nucleotides, Nucleic acids.*

PVP

See *Plant Variety Protection Act (PVP).*

PVPA

See *Plant Variety Protection Act (PVP).*

PVR

Plant Variety Rights. See also *Plant Variety Protection Act.*

PWGF

See *Platelet-derived wound growth factor, Growth factor.*

Pyralis

An insect that is also known as the European corn borer (*Ostrinia nubilalis*). See also *European corn borer (ECB).*

Pyranose

The six-membered ring forms of sugars are called pyranoses. This is because they are derivatives of the heterocyclic compound pyran. See also *Sugar molecules*.

Pyrexia

Fever; elevation of the body temperature above normal (e.g., to combat disease).

During 2005, research conducted by Clodagh O'Shea indicated that heating certain cancer cells to a temperature of 102.9°F (39.4°C) made those cancer cells easier to kill via one type of therapy. See also *Pyrogen*, *Cancer*, *T cells*.

Pyrimidine

A heterocyclic organic compound containing nitrogen atoms at (molecular ring) positions 1 and 3. Naturally occurring derivatives are components of nucleic acids and coenzymes, uracil, thymine, and cytosine. See also *Nucleic acids*, *Coenzymes*, *Uracil*, *Thymine*, *Cytosine*, *Toxicogenomics*.

Pyrogen

A substance (typically of bacterial origin) that is capable of producing pyrexia (i.e., fever) in mammals. See also *Pyrexia*, *Bacteria*.

Pyrophosphate Cleavage

The enzymatic removal of two phosphate groups (designated as PP_i) from ATP in one piece leaving AMP as another product. This cleavage releases more energy, which can be used in certain reactions that require more of a "push" to get them going. See also *ATP*, *Orthophosphate cleavage*.

Pyrrolizidine Alkaloids

A class of toxic chemical compounds which are produced naturally by certain plants, as a defense mechanism (against predators).

One of the pyrrolizidine alkaloids, *monocrotaline* is consumed (preferentially) by the larvae (caterpillars) of the moth *Utetheisa ornatrix*. That moth subsequently utilizes the monocrotaline content of its body as a defense mechanism itself, against spiders that would otherwise eat that moth. See also *Alkaloids*, *Toxin*.

Pyruvate Dehydrogenase

See *Dehydrogenases*, *Ac-CoA*.

Q

Q-Beta Replicase

A viral RNA polymerase secreted by a bacteriophage that infects *Escherichia coli* bacteria. Q-beta replicase can copy a naturally occurring RNA (molecule) sequence (e.g., from bacteria, viruses, fungi, or tumor cells) at a geometric (i.e., very fast) rate. See also *Polymerase, Bacteriophage, Ribonucleic acid (RNA), Q-beta replicase technique.*

Q-Beta Replicase Technique

An RNA assay (test) that "amplifies RNA probes" that a researcher is seeking. For instance, by using the Q-beta replicase technique to assay for the presence of RNA that is specific to the AIDS virus, it is possible to detect an AIDS infection in a patient's blood sample long before that infection has progressed to the point where antibodies would appear in the blood. See also *Q-beta replicase, RNA probes, Ribonucleic acid (RNA), Positive and negative selection (PNS), Assay, Immunoassay, Antibody, Polymerase chain reaction (PCR) technique, Cocloning, Western blot test.*

QCM

Acronym for Quartz Crystal Microbalances. See *Quartz crystal microbalances.*

QD

Acronym for "quantum dot." See *Quantum dot.*

qPCR

Acronym for "Quantitative Polymerase Chain Reaction."

Uses include "gene expression analysis" (i.e., quantitatively determine the amounts of each protein being expressed by a cell), genotyping, DNA quantification, and so on. See also *Polymerase chain reaction (PCR), Real-time PCR, Cell, Gene expression profiling, Protein, Genotype, Deoxyribonucleic acid (DNA).*

qRT-PCR

Acronym for *quantitative real-time reverse transcription polymerase chain reaction.* See the links. See also *RT-PCR, Real-time PCR (testing), Polymerase chain reaction (PCR) technique.*

QS

Acronym for *quorum sensing.* See *Quorum sensing.*

QSAR

See *Quantitative structure–activity relationship (QSAR).*

QSPR

See *Quantitative structure–property relationship (QSPR).*

QTL

See *Quantitative trait loci (QTL).*

Quadrupole Ion Trap

See *ion trap.*

Qualitative Trait

A measurable trait in an organism, that is *digital* (i.e., varies in the form of only a few discrete possibilities), so it does not manifest itself as a continuously varying property.

For example, the cattle trait of *hide color* only manifests itself in the form of a few discrete colors (e.g., black, white, red, etc.), so that is a qualitative trait.

By contrast,

- The human trait of *height* (for an adult) can vary from approximately 1 m to approximately 2 m.
- The soybean plant trait of *yield per hectare* can vary from 0 metric tons per hectare (e.g., during a severe drought year) to as much as 9 metric tons per hectare.

so those are quantitative traits. See also *Trait, Soybean plant, Quantitative trait.*

Quantitative Real-Time Reverse Transcription PCR

See *qRT-PCR.*

Quantitative Structure–Activity Relationship (QSAR)

Invented in 1963 by Corwin Hasch, it is a computer modeling technique that enables researchers (e.g., drug development chemists) to predict the likely activity (e.g., effect on tissue) of a new compound before that compound is actually created. QSAR is based on data from decades of research investigating the impact on "activity" of the chemical structures of thousands of thoroughly studied molecules.

For example, the biological activity (i.e., bacteria-killing effectiveness) of most antibiotics correlates with their tendency to dimerize (i.e., link two molecules into a single molecular unit).

During the late 1990s, Stephen Fesik and Phil Hajduk created *SAR by NMR* which is a means for researchers in pharmaceutical companies to utilize nuclear magnetic resonance (NMR) to build the structure–activity model (e.g., of a "candidate pharmaceutical" molecule) for interactions with its "target molecule" (e.g., cell receptors).

In SAR by NMR, NMR is utilized to detect even weak binding of "ligands (fragments of the pharmaceutical candidate molecule)" to receptor and then the ligands that successfully bind to target are assembled together into an optimized-to-target pharmaceutical molecule. See also *Biological activity, Pharmacophore, Antibiotic, Cell, Receptors, Pharmacokinetics, Pharmacology, Analogue, Rational drug design, In silico screening, Polymer, Structure–activity models, Nuclear magnetic resonance, Target (of a therapeutic agent), ligand (in biochemistry), Target–ligand interaction screening.*

Quantitative Structure–Property Relationship (QSPR)

A computer modeling technique which enables scientists to predict the likely properties of a new chemical compound before that chemical compound is actually created. See also *Quantitative structure–activity relationship (QSAR), Analogue, Rational drug design.*

Quantitative Trait

A measurable trait in an organism, that is "analog (i.e., varies in a continuous fashion)," so it does not manifest itself as a few discrete categories. For example,

- The human trait of "height" (for an adult) can vary from approximately 1 m to approximately 2 m.
- The soybean plant trait of *yield per hectare* can vary from 0 metric tons per hectare (e.g., during a severe drought year) to as much as 9 metric tons per hectare.

so those are quantitative traits. In stark contrast, the cattle trait of *hide color* only manifests itself in the form of a few discrete colors (e.g., black, white, red, etc.), so that is a qualitative trait. See also *Trait, Organism, Soybean plant, Qualitative trait.*

Quantitative Trait Loci (QTL)

Individual specific DNA sequences that are related to known quantitative traits (e.g., litter size in animals, height in humans, annual egg production in birds, yield per hectare in crop plants). QTL can be utilized by crop plant breeders as one of the inputs to guide their breeding program choices. See also *Marker (DNA sequence), Trait, Quantitative trait, Linkage, Deoxyribonucleic acid (DNA), Linkage group, Linkage MAP, Gene, Sequence (of a DNA molecule), Marker-assisted selection, Corn, High-oil corn, Restriction fragment length polymorphism (RFLP) technique, Random amplified polymorphic DNA (RAPD) technique, AFLP, Simple sequence repeat (SSR), DNA marker technique.*

Quantum Dot

A nanocrystal ("molecular structure" that is between 1 and 100 nm in size, so it is midway between molecular and solid states). Quantum dots have been constructed of semiconductor materials (e.g., cadmium selenide, zinc sulfide), crystallites (grown via molecular beam epitaxy), and so on. When excited via illumination by certain light, or via application of an applied electrical voltage, these semiconductor crystals emit light in the visible, UV (ultraviolet), and IR (infrared) wavelengths of the spectrum depending on their chemical composition, dimensions of the quantum dot(s), and the specific light source utilized to illuminate them.

Quantum dots possessing specific color (emission) combinations can be "attached" to

- Receptors or other proteins via *molecular bridges.*
- Specific types of cells (e.g., cancerous cells) via coating them with peptides or other relevant molecules.

For example, by encapsulating clusters of selected quantum dots within polymer beads, which are subsequently attached to a molecular functional group (ligand) that preferentially attaches to *specific types* of cells (e.g., the cancer cells desired to be "color tagged"). When later the tissue is then illuminated by relevant wavelength light, the "tagged" cells glow with the selected colors.

Quantum dots can be utilized to illuminate different living tissues (or different structures within a given cell) with different colors, inside an organism. The color emitted is impacted by the specific tissue each quantum dot is within.

Quantum dots could conceivably be constructed to act as receptors (e.g., on "biochips") for specific ligands (e.g., a blood component that is only present in a diseased patient), in a way that would signal the presence of disease when (blood) sample was passed over the quantum dot. That signal might be electronic, emission of specific wavelength light, and so on. See also *Nanometers (nm), Nanotechnology, Receptors, MEMS (nanotechnology), Biochip, Bioelectronics, Microarray (testing), Molecular bridge, Ligand (in biochemistry), Cell, Protein, Receptors, Label (fluorescent), Protein interaction analysis, Nanoparticles, Peptide, Metamaterials.*

Quantum Tags

See *Quantum dot.*

Quantum Wire

A strip or "wire" of electrically (super-) conducting material that is 10 nanometers (nm) or less in its thickness or width. DNA molecules can be created that have an affinity (preferentially attracted to) the superconducting form of carbon nanotubes (known as "armchair" form). Once those DNA molecules have attached themselves to the "armchair form" of nanotubes (within the mixture of different carbon nanotubes resulting from the carbon-nanotube-manufacturing process), chromatography can then be utilized to purify the "armchair form" nanotubes from the mixture in order to subsequently form them into superconducting quantum wires. See also *Nanometers (nm), Nanotechnology, Deoxyribonucleic acid (DNA), Carbon nanotubes, Chromatography, MEMS (Nanotechnology), Bioelectronics.*

Quarantine Pest

A pest (e.g., weed, insect, disease, etc.) of potential economic importance to the area (e.g., "pest free area") which is thereby endangered (e.g., a weed which would harm local crops, etc.) and *not yet present in that area*, or present but not widely distributed and being *officially controlled.* See also *International Plant Protection Convention (IPPC), Pest risk analysis (PRA).*

Quartz Crystal Microbalances

Abbreviated QCM. Refers to biosensors consisting of small quartz crystals (to which is attached a source of appropriate electric current), with sensitive measurement devices utilized to detect when the "attachment" of *specific molecules (e.g., viruses, DNA sequences, antigens)* to the quartz (or to layers of certain materials previously deposited on the quartz surface) causes the *specific oscillation frequency* of that quartz crystal to change in a way that enables (electronic) identification of the specific molecule(s) that attached themselves to the QCM. See also *Biosensors (Electronic), Virus, Sequence (of a DNA molecule), Antigen.*

Quaternary Structure

The three-dimensional structure of an oligomeric protein, particularly the manner in which the subunit chains fit together. See also *Protein, Oligomer, Configuration, Native conformation.*

Quelling

Refers to the impact (on gene expression) of RNA interference. See also *RNA interference (RNAi), Express, Expressivity, Gene.*

Quencher Dye

See *Molecular beacon.*

Quercetin

A "family" of phytochemicals (flavonoids) that is naturally produced in apples, pears, raspberries, red grapes, cherries, citrus fruits, tomatoes, onions, cranberries, and some other plants. The white clover plant increases the concentration of quercitin in its tissues, to become drought resistant, in response to drought stress. Quercitin is an antioxidant. Research indicates that human consumption of quercitin helps to prevent prostate cancer and some other cancers.

Research indicates that human consumption of quercitin leads to suppression of the body's tendency to release histamine. See also *Flavonoids, Chalcone isomerase, Phytochemicals, Nutraceuticals, Cancer, Biological activity, Histamine.*

Quick-Stop

The term used to describe how DNA mutants of *Escherichia coli* cease replication immediately when the temperature is increased to 42°C (108°F). See also *Escherichia coliform (E. coli).*

Quorum Sensing

Refers to the signaling mechanism utilized by certain microorganisms (e.g., in a biofilm, a population of enteric pathogens within the digestive system of an animal, etc.) in which those microorganisms emit/receive chemical signals (known as autoinducers) until they collectively determine that "enough" of that microorganism are present, to initiate a *collective* action. Such collective actions can include

- "Turning on" one or more pathways for production of specific product(s) from certain substrate(s). For example, certain pathogenic bacteria (e.g., *Vibrio cholerae*) will often live benignly within the digestive system of an animal until "enough" of them are present; as determined via quorum sensing (e.g., utilizing an acyl homoserine lactone signaling molecule). At that point in time, those bacteria collectively turn on a pathway for virulence/production of their particular enterotoxin.
- DIFFERENTIATING into specialized subtypes of cells, which perform different needed functions (for the biofilm/colony to survive).
- INFECTING another (host) organism; if pathogenic bacteria or yeasts.
- SPORULATING (creation of spores for survival and/or reproduction).
- BIOLUMINESCING (creation of light).

Recent research has shown that sulforaphane can prevent quorum sensing by/of certain pathogenic bacteria (thus preventing virulence in body of certain diseases).

See also *Signaling, Microorganism, Bacteria, Pathogen, Biolfilm, Signaling molecule, Enterotoxin, Differentiation, Pathway, Substrate (chemical), Cholera toxin, Bioluminescence, Yeast, Indole-3-acetic acid, Sulforaphane.*

R

rAAV

Acronym for *Recombinant adeno-associated virus*–based genome editing. It is a genetic engineering technique in which a scientist uses rAAV vectors to do precise insertion, deletion, or substitution of DNA sequences at a predetermined location in the genome (DNA) of mammalian cells. See also *Recombinant DNA (rDNA), Deoxyribonucleic acid (DNA), Sequence (of a DNA molecule), Virus, Adenovirus, Genome, Genetic engineering.*

RAC

See *Recombinant DNA Advisory Committee (RAC).*

Racemate

An equimolar (i.e., equal number of molecules) mixture of the D and L stereoisomers of an optically active compound. A solution of dextrorotary (D) isomer (enantiomer) will rotate the plane in which the light was polarized a specific number of degrees to the right (*dextro*) while a solution containing the same number of levorotary (L) isomer molecules will rotate the plane in which the light was polarized the same number of degrees (as in the D isomer case) to the left (*levo*). The difference between D and L enantiomers is that the rotations of the plane of plane-polarized light are equal in magnitude but opposite in sign. Hence, a 50:50 mixture of both enantiomers (known as a racemic mixture) shows no optical activity. That is, a solution containing a 50:50 mixture of enantiomers will not rotate the plane of plane polarized light when it is passed through the solution. See also *Enantiomers, Stereoisomers, Levorotary (L) isomer, Dextrorotary (D) Isomer, Polarimeter, Polarized light.*

Racemic (Mixture)

See *Racemate.*

Radioactive Isotope

An isotope with an unstable (atomic) nucleus that spontaneously emits radiation. The radiation emitted includes alpha particles, nucleons, electrons, and gamma rays. See *Isotope.*

Radioimmunoassay

Invented by Rosalyn Yarlow and Solomon Berson in 1959, it is a very sensitive method of quantitating a specific antigen using a specific radiolabeled antibody. Functionally, the antibody is made radioactive by the covalent incorporation of radioactive iodine. The radioimmuno probe thus prepared is exposed to its antigen (which may be a protein, or a receptor, etc.) in excess (the exact amount will have to be determined). The radiolabeled probe then binds to the antigen and the unbound, free probe is washed away. The radioactivity is then determined (counted) and by comparison to a standard plot, which has been constructed previously, the amount of antigen (binding) is determined. See also *Antibody, Assay, Hormone, Radioimmunotechnique.*

Radioimmunotechnique

A method of using a radiolabeled antibody to quantitate a known antigen. See also *Radioimmunoassay, Antigen, Antibody.*

Radioimmunotherapy

Refers to the use of a monoclonal antibody (MAb) immunoconjugate to deliver a radiation source (e.g., molecules of beta particle-emitting isotopes like iodine-131 or yttrium-90) to the MAb-targeted cancer cells within the body. It is safer than using an external radiation beam aimed into the body, which can cause significant damage to adjacent healthy tissues that such an external beam must pass through in order to reach the cancerous cells. See also *Monoclonal antibodies (MAb), Cancer, Immunoconjugate.*

Radiolabeled

From the Latin *radiare* = "to emit beams." See *Label (radioactive).*

Rafts

Term used to refer to *lipid rafts*. See *Lipid rafts, Plasma membrane.*

Rag1 Gene

A gene (present within the DNA of some soybean varieties) that imparts to the soybean plant some resistance to predation by the soybean aphid (*Aphis glycines*). The "R" stands for *resistance* and "AG" stands for Aphis glycines. See also *Gene, Soybean aphid.*

Rag2 Gene

A gene (present within the DNA of some soybean varieties) that imparts to the soybean plant some resistance to predation by the soybean aphid (*Aphis glycines*). The "R" stands for *resistance* and "AG" stands for *Aphis glycines*. See also *Gene, Soybean aphid.*

Raman Optical Activity Spectroscopy

A chiral-optical spectroscopy tool which is used to investigate the molecular behavior in solution of certain biomolecules such as viruses, nucleic acids, proteins, and carbohydrates.

In ROA spectroscopy, selected wavelength light is shined onto the biomolecules of interest. Those biomolecules absorb some of the light's energy, and become "excited" (i.e., their molecular vibrations increase), then they emit (reflected/scattered) light. That light can be analyzed in order to determine detailed information about the structure and behavior of the biomolecules being analyzed.

For example, an ROA spectrum (i.e., plot of the difference in intensities versus wavelengths reflected from the biomolecules) also provides information about the conformations and tertiary structures of biomolecules.

ROA spectroscopy can also be utilized for structural classification of biomolecules (e.g., in the field of proteomics). See also *Protein, Virus, Nucleic acids, Carbohydrates, Conformation, Native conformation, Tertiary structure, Proteomics.*

Random Amplified Polymorphic DNA (RAPD) Technique

A genetic mapping methodology that utilizes as its basis the fact that specific DNA sequences (polymorphic DNA) are "repeated" (i.e., appear in sequence) with gene of interest. Thus, the polymorphic DNA sequences are linked to that specific gene. Their linked presence serves to facilitate genetic mapping (i.e., "location" of specific gene(s) on an organism's genome). See also *Genetic map, Sequence (of a DNA molecule), Restriction fragment length polymorphism (RFLP) technique, Linkage, Deoxyribonucleic acid (DNA), Physical map (of genome), Linkage group, Marker (genetic marker), Linkage map, Trait, Genome, Gene, Quantitative trait loci (QTL).*

Ranibizumab

A monoclonal antibody (fragment) approved by the U.S. Food and Drug Administration (FDA) in 2006 for use as the pharmaceutical Lucentis™ treatment to inhibit the growth of abnormal blood vessels in/front of the eye's retina/macula. Such abnormal blood vessels are a cause of approximately 10% of the cases of the disease known as age-related macular degeneration.

The antibody fragment binds to vascular endothelial growth factor (VEGF), thereby inhibiting VEGF's usual causation of new blood vessel growth.

In 2010, research was completed indicating that ranibizumab (LUCENTIS®) also is effective in treating diabetic macular edema, in concert with standard laser treatment. See also *Monoclonal antibodies (MAb), Antibody, Food and Drug Administration (FDA), Age-related macular degeneration (AMD), Minimized proteins, Vascular endothelial growth factor (VEGF).*

Rapamycin

A pharmaceutical that minimizes the entry of viruses into cells and also slows cell growth. Currently utilized in some chemotherapies. Research indicates that it may someday be used in gene therapy because it slows down differentiation of certain stem cells. See also *Cell, Virus, Chemotherapy, Cancer, Gene therapy, Stem cells, Differentiation.*

RAPD

See *Random amplified polymorphic DNA (RAPD) technique.*

Rapid Microbial Detection (RMD)

A broad term used to describe the various testing products/technologies that can be utilized to quickly detect the presence of microorganisms (e.g., pathogenic bacteria in a food processing plant). These testing products are based on immunoassay, DNA probe, electrical conductance and/or impedance, bioluminescence, and enzyme-induced reactions (e.g., which produce fluorescence or a color change to indicate the presence of specific microorganism). See *Bioluminescence, Microbe, Bacteria, Pathogen, Immunoassay, Enzyme, Probe, DNA probe, Electrophoresis, Hazard analysis and critical points (HACCP).*

Rapid Protein Folding Assay

Refers to an assay/methodology developed by Geoff Waldo in 2001, which initially utilized a fusion protein (made via fusing a gene which codes for Green Fluorescent Protein [GFP] to a gene which codes for the protein being analyzed) in order to indicate when proper/correct folding of that particular protein had occurred (within a cell). Proper/correct folding of the (fusion) protein resulted in green fluorescence.

Due to limitations inherent in the initial version of the Rapid Protein Folding Assay (RFPA) such as the quite large GFP portion hindering the movement of fusion protein within the cell, during 2005 Geoff Waldo developed a new version of RFPA. That new RFPA incorporates only a *portion* of GFP within the fusion protein (thereby making the resultant fusion protein small enough to both move around the cell and fold without hindrance). See also *Assay, Cell, Protein, Fusion protein, Gene, Gene fusion, Deoxyribonucleic acid (DNA), Coding sequence, Fluorescence, Label (fluorescent), Green fluorescent protein, Conformation, Protein folding.*

ras Gene

Discovered in 1978 by Edward Scolnick, who named it *ras* for "rat sarcoma" (i.e., the particular diseased tissue in which he found it). The *ras* gene is also present in the human genome, and it is an oncogene that (when mutated) is believed to be responsible for up to 90% of all human pancreatic cancer, 50% of human colon cancers, 30% of lung cancers, and 30% of leukemias. The *ras* gene codes for the production (i.e., "manufacture") of *ras proteins* via the *ras* pathway, which help to signal each cell to divide and grow at appropriate time(s); for example, when free EGF "attaches" to relevant cell receptor on plasma membrane.

When the *ras* gene has been damaged or mutated (e.g., via exposure to cigarette smoke or ultraviolet light, etc.), it codes for (i.e., causes to be manufactured in the cell's ribosome) a mutated version of the *ras protein* that can cause the cell to become cancerous (i.e., divide and grow uncontrollably, independent of any presence of growth factors). See also *Gene, Oncogenes, Mutation, p53 Gene, Genetic code, Meiosis, Deoxyribonucleic acid (DNA), Carcinogen, Ribosomes, Cancer, Tumor, ras protein, Farnesyl transferase, Proto-oncogenes, Protein, Epidermal growth factor (EGF), EGF receptor, Reovirus.*

ras Pathway

See *ras Gene, Pathway, Reovirus.*

ras Protein

A transmembrane (i.e., through the cell membrane) protein that is coded for by the *ras* gene. The ras protein end that is outside the cell membrane acts as a receptor for applicable growth factors (e.g., fibroblast growth factor), and conveys that signal (i.e., to divide/grow) into the cell when that chemical signal (i.e., the growth factor) touches "receptor end" of the ras protein. *Ras* proteins are involved in almost every cell function (e.g., metabolism, activation of genes, growth signals, inflammation signals, apoptosis, etc.).

When the *ras* gene has been damaged or mutated (e.g., via exposure to cigarette smoke or ultraviolet light), that gene causes excess ras proteins to be manufactured, which causes oversignaling of the cell to divide and grow (i.e., cell becomes cancerous). See also *Gene, Transmembrane proteins, ras Gene, Fibroblast growth factor (FGF), Oncogenes, Genetic code, Protein, p53 Protein, Meiosis, Carcinogen, Ribosomes, Deoxyribonucleic acid (DNA), Cancer, Tumor, Proto-oncogenes, Receptors, EGF receptor, CD4 protein, Signaling, Signal transduction, Mitogen-activated protein kinase cascade, Metabolism, Apoptosis.*

rasiRNA

Abbreviation for *repeat-associated small interfering RNA*. See *Small interfering RNA.*

Rational Drug Design

The "engineering" (building) of chemically synthesized drugs based on knowledge of receptor modeling and drug/target interaction(s) with the aid of supercomputers/interactive graphics, etc.); the educated, creative design of the three-dimensional structure of a drug atom by atom, that is, "from the ground up." This approach represents a major advance over the prior practice of first synthesizing large numbers of compounds (or finding them in nature), followed by thousands of tedious screenings to test for efficacy against a given disease (target). The approach of rational drug design has, however, not yet been perfected and optimized due, in part, to gaps in our knowledge of drug/receptor interaction (called "docking") and to gaps in our knowledge in general. See also *Receptors, Receptor mapping (RM), Analogue, Molecular diversity, Target (of a therapeutic agent), In silico biology, Free energy, Homology modeling, Docking (in computational biology), In silico screening, X-ray crystallography, Pharmacophore searching.*

RB

See *Refractile bodies.*

RBS1 Gene

A gene which confers to any soybean plant (possessing that gene in its DNA) resistance to the adverse effects of the soilborne fungus *Phialophora gregata*, which can cause the plant disease *brown stem rot (BSR)* in soybean plants. See also *Gene, Deoxyribonucleic acid (DNA), Brown stem rot (BSR), Fungus, Pathogenic, Soybean plant.*

RBS3 Gene

A gene which confers to any soybean plant (possessing that gene in its DNA) resistance to the adverse effects of the soilborne fungus *Phialophora gregata*, which can cause the plant disease known as BSR in soybean plants. See also *Gene, Deoxyribonucleic acid (DNA), Brown stem rot (BSR), Fungus, Pathogenic, Soybean plant.*

RdDM

Acronym for *RNA-directed DNA methylation*. See *Epigenetic marks.*

rDNA

See *Recombinant DNA.*

Reactive Nitrogen Species

See *Nitric oxide.*

Reactive Oxygen Species

Abbreviated ROS, this phrase refers to several different chemicals (e.g., hydrogen peroxide, hydrogen sulfide, etc., which are sometimes produced within the body) that oxidize most compounds that they come in contact with. During 2012, Enrique Amaya showed that:

- Elevated ROS levels are needed for tissue regrowth and regeneration to occur (e.g., after wounding or amputation).
- ROS production is essential to activate Wnt signaling, which helps regulate tissue regeneration in organisms.

See *Free radical, Oxidation (chemical reaction), Oxidative stress, Hydrogen sulfide* (H_2S), *Tumor necrosis factor (TNF).*

Reading Frame

The particular nucleotide sequence that starts at a specific point and is then partitioned into codons. The reading frame may be shifted by removing or adding a nucleotide(s). This would cause a new sequence of codons to be read. For example, the sequence CATGGT is normally read as the two codons: CAT and GGT. If another adenosine nucleotide (A) were inserted between the initial C and A, producing the sequence CAATGGT, then the reading frame would have been shifted in such a way that the two new (different) codons would be CAA and TGG, which would code for something completely different. See also *Codon, Genetic code, Frameshift, Deoxyribonucleic acid (DNA), Mutation, Read-through, Start codon.*

Read-Through

Refers to transcription/translation which occurs and continues beyond the normal stopping point (e.g., at end of transcription unit, *for transcription*) due to the absence of the transcription/termination signal of the gene being transcribed/translated. See also *Transcription, Translation, Transcription unit, Control sequences, Reading frame, Termination codon, Alternative splicing.*

Real-Time PCR

Refers to the use of qPCR to attempt quantitative (determination of a given DNA sequence within a sample) via coupling of a "molecular beacon" (with a "quencher molecule" attached to it) to the polymerase chain reaction (PCR) probe.

Thus, at the same time the PCR reaction (cycling) is producing copies of the relevant DNA sequence, the *molecular beacon* (i.e., fluorescent marker) is "un-quenched" so it fluoresces in direct proportion to the amount of DNA present (which can theoretically be back-calculated to infer the original amount of that particular DNA present in sample prior to initiation of PCR cycling).

Common uses of real-time PCR are gene expression analyses, diagnosing of infectious diseases, and detecting food pathogens.

See also *Polymerase chain reaction (PCR), Probe, Polymerase chain reaction (PCR) technique, qPCR, Molecular beacon, Microarray (testing), Sequence (of a DNA molecule), Fluorescence, Deoxyribonucleic acid (DNA), Gene expression analysis.*

Reassociation (of DNA)

The pairing of complementary single strands (of the molecule) to form a double helix (structure). See also *Double helix.*

RecA

The product of the RecA locus (in a gene of) *Escherichia coli* and some other bacteria. It is a protein with dual activities, acting as a protease and also able to exchange single strands of DNA (deoxyribonucleic acid) molecules. The protease activity controls the SOS response. The nucleic acid handling facility (i.e., ability to exchange single strands of DNA) is involved in recombination/repair pathways. See also *SOS response, Locus, Protein, Ribosomes, Escherichia coliform (E. coli).*

Receptor-Binding Mapping

Refers to quantitative genomic receptor-fitting assessment of all receptors in the body, regarding which of them bind (in lock-and-key manner) a given entity (e.g., hormone, vitamin, antibody, etc.). For example, during 2012, some researchers utilized receptor-binding mapping to link vitamin D deficiency to an increased risk for cancer and autoimmune diseases (e.g., rheumatoid arthritis, multiple sclerosis, lupus, etc.). See also *Receptors, Receptor fitting (RF), Genomics, Hormone, Vitamin, Antibody, Cancer, Autoimmune disease, Rheumatoid arthritis, Multiple sclerosis, Lupus.*

Receptor Engineering

Refers to the replacement of one *entire category of receptor molecules (e.g., in the plasma membrane of applicable cells of an organism)* with another receptor molecule. For example during 2014, Sang-Youl Park and his colleagues replaced the abscisic acid receptor (ABA receptor) protein molecule known as PYRABACTIN RESISTANCE 1 (PYR1) in a plant with a variant protein molecule that exhibits sensitivity to the fungicide mandipropamid i.e., when that plant was subsequently sprayed with mandipropamid, these new receptor molecules activated the plant's abscisic acid pathway (i.e., decreased the size of all stomatal pores in that plant's leaves, which decreases the amount of water vapor leaving the plant). This resulted in a plant that—if a drought occurred during the growing season—a farmer could swiftly render to be DROUGHT TOLERANT by simply spraying mandipropamid onto it. See also *Receptors, Drought tolerance, Cell, Plasma membrane, Protein, Organism, Abscisic acid, Pathway.*

Receptor Fitting (RF)

A research method used to determine the macromolecular structure that a chemical compound (e.g., an inhibitor) must have in order to fit (in a lock-and-key fashion) into a receptor. For example, a pain inhibitor compound blocking a pain receptor on the surface of a cell. See also *CD4 protein, T cell receptors, Receptors, Receptor-binding mapping, Receptor mapping (RM), Interleukin-1 receptor antagonist (IL-1ra), Rational drug design.*

Receptor Mapping (RM)

A method used to guess at (determine) the three-dimensional structure of a receptor-binding site extrapolating from the known structure of the molecule binding to it. This approach can be carried out because of the complementary shape of the receptor and the binding molecule. Functionally, the researcher projects the (guessed-at) properties of the receptor ligands into a mathematical model in which the profile of the receptor is predicted by complementariness (to known chemical molecular structures). The receptor mapping process requires repetitive refinement of the mathematical model to fit properties continually being discovered via the use/interaction of chemical reagents bearing the known molecular structures. See also *CD4 protein, T cell receptors, Receptors, Receptor fitting (RF).*

Receptor Tyrosine Kinase

Refers to a "family" of cell surface receptors which respond (i.e., signal transduction) when epidermal growth factor (EGF), certain other growth factors, or structurally related ligands dock at those receptors.

In cancerous tissues, some receptor tyrosine kinases (RTKs) help initiate tumor proliferation/spread and related angiogenesis (i.e., creation of new blood vessels to "feed" the growing tumor). For example, the tyrosine kinase inhibitor pharmaceutical GLEEVEC™ can be utilized to treat gastrointestinal stromal tumors, where it targets the *receptor tyrosine kinase* known as *KIT.* See also *Receptors, Cell, Signal transduction, Growth factor, Epidermal growth factor (EGF), Ligand (in biochemistry), Endocytosis, Cancer, Tumor, Angiogenesis, Gleevec™, Mitogen-activated protein kinase cascade.*

Receptor-Mediated Endocytosis

See *Endocytosis.*

Receptor-Mediated Transcytosis

See also *Blood–brain barrier (BBB), Transferrin receptor.*

Receptors

Functional proteinaceous structures typically found in the plasma membrane (surface) of cells that tightly bind specific molecules (organic, proteins or viruses). Some (relatively rare) receptors are located inside the cell's plasma membrane (e.g., free-floating receptor for Retin-A). Both (membrane, internal) types of receptors are a functional part of information transmission (i.e., signaling) to the cell. A general overview is that once bound, both the receptor and its "bound entity" as a complex is internalized by the cell via a process called endocytosis, in which the cell membrane in the vicinity of the bound complex invaginates. This process forms a membrane "bubble" on the inside of the cell, which then pinches off to form an endocytic vesicle. The receptor then is released from its bound entity by cleavage in the cell's lysosomes. It is recycled (returned) to the surface of the cell (e.g., low-density lipoprotein receptors). In some cases the receptor, along with its bound molecule, may be degraded by the powerful hydrolytic enzymes found in the cell's lysosomes (e.g., insulin receptors, EGF receptors, and nerve growth factor receptors). Endocytosis (internalization of receptors and bound ligand such as a hormone) removes hormones from the circulation and makes the cell temporarily less responsive to them because of the decrease in the number of receptors on the surface of the cell. Hence the cell is able to respond (to new signal). A receptor may be thought of as a butler who allows guests (in this case, molecules that bind specifically to the receptor) to enter the house (cell) and who accompanies them as they enter.

Another mode of "reception" occurs when, following binding, a transmembrane protein (e.g., one of the G proteins) activates the portion of the transmembrane (i.e., through the cell membrane) protein lying inside the cell. That "activation" causes an effector inside cell to produce a "signal" chemical inside the cell which causes the cell's nucleus (via gene expression) to react to the original external chemical signal (that bound itself to the receptor portion of the transmembrane protein). See also *CD4 protein, T cell receptors, Receptor fitting (RF), Receptor mapping (RM), Lysosomes, Agonists, Interleukin-1 receptor antagonist (IL-1ra), CD95 protein, Transferrin, Vaginosis, Signal transduction, Endocytosis, G proteins, Cell, Signaling, Protein, Nuclear receptors, Gene, Gene expression, Liver X receptors (LXR), Retinoid X receptors (RXR), Farnesoid X receptors (FXR), Human immunodeficiency virus type 1 (HIV-1), Nuclear pore complexes, Estrogen receptors, Receptor engineering, Resveratrol, Human immunodeficiency virus type 2 (HIV-2), Scavenger receptor A, Nuclear proteins, Importins, Karyopherins, Antiepidermal growth factor receptor monoclonal antibodies.*

Recessive (Gene)

See *Recessive allele.*

Recessive Allele

Discovered by Gregor Mendel in the 1860s, this refers to an allelic gene whose existence is obscured in the phenotype of a heterozygote by the dominant allele. In a heterozygote the recessive allele does not produce a polypeptide; it is "switched off." In this case, the dominant allele is the one producing the polypeptide chain (via cell's ribosome). See also *Genetics, Allele, Dominant allele, Homozygous, Heterozygote, Polypeptide (protein), Cell, Ribosomes.*

Recognition Site

See *p53 protein, Meganuclease.*

Recombinant Adeno-Associated Virus (rAAV)-Based Genome Editing

Refers to a genetic engineering technique in which a scientist uses rAAV vectors to do precise insertion, deletion, or substitution of DNA sequences at a predetermined location in the genome (DNA) of mammalian cells. See also *Recombinant DNA (rDNA), Deoxyribonucleic acid (DNA), Sequence (of a DNA molecule), Virus, Adenovirus, Genome, Genetic engineering.*

Recombinant DNA (rDNA)

DNA formed by the joining of genes (genetic material) into a new combination. See also *Recombination, Genetic engineering, Editing.*

Recombinant DNA Advisory Committee (RAC)

The former standing U.S. national committee set up in 1974 by the U.S. National Institutes of Health (NIH) to advise the NIH director on matters regarding policy and safety issues of recombinant DNA research and development. Over time, it had evolved to become part of the American government's regulatory process for recombinant DNA research and product approval. The RAC was terminated by the director of the NIH in 1996 because the "human health and environmental safety concerns expressed at the inception (of genetic engineering/biotechnology) had not materialized." See also *Interim Office of the Gene Technology Regulator (IOGTR), Gene Technology Office, Genetic engineering, ZKBS (Central Committee on Biological Safety), National Institutes of Health (NIH), Recombinant DNA (rDNA), Biotechnology, Recombination, Indian Department of Biotechnology, Commission of Biomolecular Engineering, Gene technology regulator (GTR), Genetic Manipulation Advisory Committee (GMAC).*

Recombinase

A category of enzymes that acts to "cut open" the strand of DNA within a cell (e.g., to "splice-out" or "splice in") a given gene. Recombinases normally circulate throughout the cell containing them and initiate repair of cell's (damaged) DNA under certain circumstances.

During 2000, Nam-Hai Chua and Jian-ru Zuo showed that activation of the *gene for recombinase* (via β-estradiol transcription factor) could be done to cause expression of recombinase in a manner that "spliced out" (removed) *antibiotic-resistance* "marker genes" from genetically engineered plants. See also *Enzyme, Deoxyribonucleic acid (DNA), Gene, Cell, Gene splicing, Homologous recombination, Genetic engineering, Transcription factors, Antibiotic resistance, Marker genes (genetic marker), DNA glycosylase.*

Recombinase Polymerase Amplification

An isothermal enzymatic (molecule) amplification technology capable of single molecule detection in 10–15 min. It utilizes recombinase enzymes, which are capable of pairing oligonucleotide primers with homologous sequence in duplex DNA. The synthesis of new DNA is directed to defined points in a sample of DNA, and an amplification reaction is initiated. See also *Deoxyribonucleic acid (DNA), Duplex, Enzyme, Recombinase, Oligonucleotide, Primer (DNA), Homologous (chromosomes or genes).*

Recombination

The joining of genes, sets of genes, or parts of genes, into new combinations, either biologically or through laboratory manipulation (e.g., genetic engineering). See also *Genetic engineering, Gene, Recombinant DNA (rDNA), Editing.*

Recruitment

See *Long noncoding RNAs, Transcription, Transcription activators.*

Red Biotechnology

Term utilized in some countries to refer to *medical* applications of genetic engineering. See also *Genetic engineering.*

Red Blood Cells

See *Erythrocytes.*

R

Redement Napole (*RN*) Gene

A swine gene that causes animals (possessing at least one negative allele of this gene) to produce meat which is more acidic than average, and thus that meat has a lower "water-holding" capacity. The *RN* gene was first identified in the Hampshire breed of swine in France. The Hampshire breed has been known to produce meat that is more acidic than average since the 1960s. See also *Gene*, *Allele*, *Acid*.

REDOX

Abbreviation for *oxidation–reduction* reaction. See also *Oxidation–reduction reaction*.

Reduced-Allergen Soybeans

Refers to a biotechnology-derived soybean variety developed by Eliot Herman and Rick Helm in 2002, in which production of the allergenic P34 storage protein (within the seeds) is prevented via gene silencing. See also *Allergies (foodborne)*, *Soybean plant*, *Biotechnology*, *Gene*, *Gene silencing*, *Protein*, *Storage proteins*, *P34 protein*, *RNA interference (RNAi)*.

Reduction (Biological)

The decomposition of complex compounds and cellular structures by heterotrophic organisms. In a given ecological system, this heterotrophic decomposition serves the valuable function of recycling organic materials. This occurs because the heterotrophs absorb some of the decomposition products (for nourishment) and leave the balance of the (decomposed) substances for consumption (recycling) by other organisms. For example, bacteria break down fallen leaves on the floor of a forest, thus releasing some nutrients to be utilized by plants. See also *Heterotroph*.

Reduction (in a Chemical Reaction)

The gain of (negatively charged) electrons by a chemical substance. When one substance is reduced by another, the other compound is oxidized (loses electrons) and is called the reducing agent. See also *Oxidation–reduction reaction*, *Oxidizing agent*, *Template*.

Redundancy

A term used to describe the fact that some amino acids have more than one codon (that codes for production of that amino acid). There are approximately 64 possible codons available to code for 20 amino acids. Therefore, some amino acids will be specified by more than one codon. These (extra) codons are redundant. See also *Codon*, *Genetic code*, *Ribosomes*.

Redundant Codons

See *Silent mutation*.

Refractile Bodies (RB)

Dense, insoluble (i.e., not easily dissolved) protein bodies (i.e., clumps) that are produced within the cells of certain microorganisms. The refractile bodies function as a sort of natural storage device for the microorganism. They are called refractile bodies because their greater density (than the rest of the microorganism's body mass) causes light to be refracted (bent) when it is passed through them. This bending of light causes the appearance of very bright and dark areas around the refractile body and makes them visible under a microscope.

Relatively rare in natural occurrence, refractile bodies can be induced (i.e., caused to occur) in procaryotes (e.g., bacteria) when the procaryotes are genetically engineered to produce eucaryotic (e.g., mammal) proteins. The proteins are stored in refractile bodies. For example, the *E. coli* bacterium can be genetically engineered to produce bovine somatotropin (BST, a cow hormone) which is stored within refractile bodies in the bacterium. After some time of growth when a significant amount of BST has been synthesized the *E. coli* cells are disrupted (i.e., broken open), and the refractile bodies are removed by centrifugation and washed. They are then dissolved in appropriate solutions to release the protein molecules. This step denatures (unfolds, inactivates) the BST molecules and they are refolded to their native conformation (i.e., restored to the natural conformation found within the cow) in order to regain their natural activity. The protein is then formulated in such a way as to be commercially viable as a biopharmaceutical.

Refractile bodies are also known as inclusion bodies, protein inclusion bodies, and refractile inclusions. One point of interest is that the prerequisite for the generation of a mammalian protein by (in) a living foreign system such as *E. coli* is that the system used to generate the protein (1) must not have an immune system capable of destroying the foreign protein it is making or (2) the foreign protein made must be camouflaged or protected from any defense mechanisms possessed by the synthesizing organism. See also *Protein*, *Genetic engineering*, *Genetic code*, *Procaryotes*, *Eucaryote*, *Escherichia coliform (E. coli)*, *Bovine somatotropin (BST)*, *Ultracentrifuge*, *Conformation*, *Native conformation*, *Protein folding*.

Regeneration

The production of new organisms (e.g., plants) or new limbs/organs (e.g., in animals) from undifferentiated totipotent cells. For example, the salamander newt (*Notophthalmus viridescens*) can thereby grow a new limb if one gets amputated.

For example, after genetic engineers insert a new gene into a plant cell, an entire plant can then be grown by the scientists from that single cell (microcallus). See also *Organism*, *Cell*, *Differentiation*, *Cell differentiation*, *Totipotency*, *Totipotent stem cells*, *Cell differentiation proteins*, *Gene*, *Genetic engineering*.

Regional Plant Protection Organization (RPPO)

See *International Plant Protection Convention (IPPC)*, *SPS*, *National Plant Protection Organization (NPPO)*.

Regulatory DNA

See *Regulatory sequence*.

Regulatory Element

See *Regulatory sequence*.

Regulatory Enzyme

A highly specialized enzyme having a regulatory (controlling) function through its capacity to undergo a change in its catalytic activity.

There exist two major types of regulatory enzymes: (1) covalently modulated enzymes and (2) allosteric enzymes.

Covalently modulated enzymes are enzymes that can be interconverted between active and inactive (or less active) forms by the covalent attachment (or removal) of a modulating metabolite by other enzymes. Hence the activity of one enzyme can, under certain conditions, be regulated by other enzymes. Glycogen phosphorylase, an oligomeric protein with four major subunits (tetramer), is a classic example of a covalently modulated enzyme. The enzyme occurs in two forms: (1) phosphorylase a, the more active form and (2) phosphorylase b, the less active form. In order for the enzyme to possess maximal catalytic activity (i.e., be phosphorylase a) certain serine residue on all four subunits must have a phosphate covalently attached. If, due to other regulatory signals it has received, the enzyme phosphorylase phosphatase hydrolytically cleaves and removes the phosphate group from the four subunits, the tetramer dissociates into the inactive (or much less active) dimer, phosphorylase b. Another enzyme, phosphorylase kinase, is able to rephosphorylate the four specific serine residues of the four subunits at the expense of ATP and regenerate the active phosphorylase a tetramer.

Allosteric enzymes are enzymes that possess a special site on their surfaces that is distinct from the enzyme's catalytic site and to which specific metabolites (called effectors or modulators) are reversibly and noncovalently bound. The allosteric-binding site is as specific for a particular metabolite as is the catalytic site, but it cannot catalyze a reaction, only bind the effector. The binding of the effector causes a conformation change in the enzyme such that its catalytic activity is impaired or stopped. Allosteric enzymes are normally the first enzymes in, or are near the beginning of, a multienzyme system. The very last product produced by the multienzyme system (the end product) may act as a specific inhibitor of the allosteric enzyme by binding to that enzyme's allosteric site. The binding consequently causes a conformation change to occur in the enzyme, which inactivates it. A classic example of an allosteric enzyme in a multienzyme sequence is the enzyme L-threonine dehydratase, which is the initial enzyme in the enzyme sequence that catalyzes the conversion of L-threonine to L-isoleucine. This reaction occurs in five enzyme-catalyzed steps. The end product, L-isoleucine, strongly inhibits L-threonine dehydratase, the first enzyme in the five-enzyme sequence. No other intermediate in the sequence is able to inhibit the enzyme. This kind of repression is called feedback or end-product inhibition. It should be noted that allosteric control may be negative (as in the example earlier) or positive. In positive control the effector binds to an allosteric site and stimulates the activity of the enzyme. Furthermore, some allosteric enzymes respond to two or more specific modulators with each modulator having its own specific binding site on the enzyme. An allosteric enzyme that has only one specific modulator is called monovalent whereas an enzyme responding to two or more specific modulators is called polyvalent. Combinations of the previous possibilities could lead to very fine tuning of the enzymes involved in the synthesis and/or degradation of metabolites. Note that in the two examples earlier, the common denominator is the structural change that occurs upon execution of the mechanism. See also *Metabolite*, *Repressible enzyme*.

Regulatory Genes

Genes whose primary function is to control the state of synthesis of the products of other genes. See also *Gene*, *Micro-RNAs*, *Regulatory sequence*.

Regulatory Sequence

A DNA sequence involved in regulating the expression of a gene, e.g., a promoter, enhancer, or operator region (in the organism's DNA). See also *Operator*, *Promoter*, *Down promoter mutations*, *Micro-RNAs*, *Down regulating*, *Transcription factors*, *Transcriptional activator*, *Transcriptional repressor*, *Copy number variation*, *Enhancer*.

Regulatory T Cells

Refers to one class of T cells (thymus-derived lymphocytes) which was formerly called *suppressor T cells*, and which constitutes less than 10% of the total T cells in the body of the average human.

Discovered by Tomio Tada in 1971, these T cells suppress B cell activity (i.e., after the body's immune system has fought off an infection).

Research has shown that regulatory T cells can reduce inflammation and regulate the body's other immune system cells involved in some autoimmune diseases (e.g., resulting when those other immune system cells are overly active, are active against the body's own cells, etc.). Absence of regulatory T cells has been shown to lead to some autoimmune diseases (e.g., immune dysregulation, polyendocrinopathy, enteropathy, X-linked syndrome—also known as IPEX).

Recent research indicates that some cancer tumors can "recruit" regulatory T cells to leave the lymph nodes in order to come and protect those tumors from being destroyed by the body's immune system. See also *Cell*, *T cells*, *Suppressor T cells*, *Cellular immune response*, *Lymphocyte*, *B lymphocytes*, *Autoimmune disease*, *Tumor*, *Cancer*.

Remediation

The cleanup or containment (if chemicals are moving) of a hazardous waste disposal site to the satisfaction of the applicable regulatory agency (e.g., the Environmental Protection Agency [EPA]). Such cleanup can sometimes be accomplished via use of microorganisms that have been adapted (naturally or via genetic engineering) to consume those chemical wastes that are present in the disposal site. See also *Acclimatization*.

Renaturation

The return to the natural structure of a protein or nucleic acid from a denatured (more random coil) state. For example, a protein may be denatured (lose its native [natural] structure) by exposure to surfactants such as SDS or to changes in the pH of the medium, etc. If the surfactant is slowly removed or the pH is slowly readjusted to the optimum for the protein, it will refold (snap) back into its original (native) form. See also *Native configuration*, *denaturation*, *SDS*.

Renin

A proteolytic enzyme that is secreted by the juxtaglomerular cells of the kidney. Its release is stimulated by decreased arterial pressure and renal blood flow resulting from decreased extracellular fluid volume. It catalyzes the formation of angiotensin I from hypertensinogen. Angiotensin I is then converted to angiotensin II by another enzyme located in the endothelial cells of the lungs. Angiotensin II then causes the increase in the force of the heartbeat and constricts the arterioles. This scenario causes a rise in the blood pressure and is thus a cause of hypertension (high blood pressure). See also *Homeostasis*, *Renin inhibitors*, *Atrial peptides*.

R

Renin Inhibitors

Those chemicals that act to block the hypertensive (i.e., high blood pressure-inducing) effect of the enzyme, renin. See also *Homeostasis, Renin inhibitors, Atrial peptides.*

Rennin

See *Chymosin.*

Reovirus

A virus containing double-stranded RNA; first characterized by man in 1951. Reovirus is isolated from the respiratory and intestinal tracts of humans and other mammals. The prefix "reo-" is an acronym for respiratory enteric orphan.

Reovirus enters cells by latching onto sialic acid molecules on their surface, but reovirus is able to replicate only in cells possessing an *activated ras pathway.* Since those tend to be cancerous cells, reovirus actually commonly kills cancer cells in humans. See also *Virus, Retroviruses, Ribonucleic acid (RNA), Cell, Sialic acid, ras Gene, Pathway, Cancer.*

Repeat Unit

See *Microsatellite DNA.*

Reperfusion

The restoration of blood flow to an occluded (i.e., blocked) blood vessel. May be done biochemically (e.g., via injection of tissue plasminogen activator) or via surgery. Often, the restoration of blood (and also thus oxygen) supply results in:

- Activation of the lectin pathway of complement activation (thereby resulting in damage via complement-caused over-inflammation of the formerly occluded tissues). Research indicates that timely administration of a therapeutic (i.e., created by man) antibody-against-*Mannan-Binding lectin-Associated Serine Protease-2* can help minimize tissue damage from complement-caused overinflammation.
- Generation of oxygen free radicals (thereby resulting in tissue damage from the highly energized free radicals). Research indicates that timely administration of human superoxide dismutase can help minimize tissue damage from oxygen free radicals (by "capturing" and inactivating them).

See also *Tissue plasminogen activator (tPA), Complement (component of the immune system), Free radical, Antibody, Human superoxide dismutase (hSOD), Lazaroids.*

Replication (of DNA)

Reproduction of a DNA molecule (inside a cell). This process can be viewed as occurring in stages, in which the first stage consists of a helicase enzyme "unwinding" the double helix of the DNA molecule at a replication origin, forming a replication fork. At the replication fork, the two separated (DNA) strands serve as templates for new DNA synthesis.

That new DNA synthesis is accomplished on each strand via enzymes known as DNA polymerase, which travel along each (single) strand making a second complementary strand by catalyzing the addition of DNA bases (to the new, growing strands).

The end result is two new double helices (DNA molecules), each of which has one chain from the original DNA molecule and one chain that was newly synthesized by the DNA polymerase enzymes. See also *Deoxyribonucleic acid (DNA), DNA polymerase, Helicase, Enzyme, Replication fork, Duplex, Double helix, Base pair (bp), Mismatch repair, Leading strand.*

Replication (of Virus)

Reproduction of the original virus. This process can be viewed as occurring in stages, in which the first stage consists of the adsorption of the virus to the host cell, followed by penetration of the virus (or its nucleic acid) into the cell, the taking over of the cell's biomachinery and harnessing of it to replicate viral nucleic acid along with the synthesis of other virus constituents, the correct assembly of the nucleic acids and other constituents into a functional virus, followed finally by release of the virus from the confines of the cell. See also *Virus, Cell, Nucleic acids.*

Replication Fork

The point at which strands of parental duplex DNA are separated in a Y shape. This region represents a growing point in DNA replication. See also *Replication (of DNA), Deoxyribonucleic acid (DNA), Duplex.*

Replicon

Refers to the nonreplicating viral RNA particles (e.g., derived by scientists from the polio virus) utilized to induce apoptosis (i.e., "programmed cell death") in brain tumors. Replicons are able to cross the blood–brain barrier, they preferentially infect cancer/tumor cells, and they then produce proteins which cause tumor cells to die via apoptosis. See also *Virus, Replication (of virus), Ribonucleic acid (RNA), Apoptosis, Cell, Blood–brain barrier (BBB), Cancer, Tumor.*

Reporter Gene

A specific gene that is inserted into the DNA of a cell so that cell will "report" (to researchers) when

- Signal transduction has occurred in that cell, or
- A (linked) gene was successfully expressed.

The gene that codes for production of the enzyme luciferase (which catalyzes bioluminescence—light production) is one of the most commonly used *reporter genes.*

For example, when researchers are testing numerous candidate drugs for their ability to stop cells from (over-) producing a hormone or growth factor, the researchers need to *quickly* know when one of the candidate drugs has had the desired effect on the cell of interest. By prior insertion into that cell of a gene (e.g., which causes bioluminescence or a certain chemical to be produced by the cell when signal transduction has taken place), that cell "reports" (when a candidate drug has had the desired effect on the cell) by producing the bioluminescence or chemical (coded for by the reporter gene) which can be rapidly detected by the researcher (e.g., via light sensors or biosensors placed adjacent to the cell).

Another example is use of (inserted) *lux gene* as a reporter gene. The *lux* gene, which codes for the bioluminescent *lux protein* (a luminophore) can be inserted into the DNA of certain bacteria species that can be genetically engineered to biodegrade diesel fuel spilled in soil. Then, when those engineered bacteria encounter diesel fuel and begin "eating" it (i.e., breaking it down), those engineered bacteria will glow (bioluminesce) to "report" that they are biodegrading the spilled diesel fuel.

Another example is the use of such a fluorescent reporter gene (e.g., for *green fluorescent protein*) in "sentinel bacteria" sprayed onto battlefields after a war is ended. Those engineered bacteria produce fluorescent pigments in the presence of explosive chemicals (e.g., TNT), thereby marking landmines and unexploded ordnance for safe removal. See also *Gene, Genetic engineering, Genetic code, Coding sequence, Protein, Cell, Bioluminescence, Cell culture, Signal transduction, Linkage, hormone, Growth factor, Green fluorescent protein, Biosensors (electronic), Luminophore, Lux gene, Lux protein, Deoxyribonucleic acid (DNA), Gus gene, Bacteria, Bioremediation, High-throughput screening (HTS).*

Reporter Molecules

Refers to molecular tags (e.g., molecules chemically attached to protein) that serve to indicate when a particular biological interaction occurs (e.g., a protein–protein interaction). For example, the protein-interaction analysis assessment known as BRET (bioluminescence resonance energy transfer) utilizes a bright red fluorescent dye molecule attached to one protein (e.g., a receptor or a signaling protein) and a bioluminescent enzyme from deep-sea shrimp (*Oplophorus*) attached to the other protein (e.g., a biopharmaceutical candidate). When those two proteins interact (i.e., the biopharmaceutical exerts the desired effect on receptor or on signaling protein), the bioluminescent enzyme excites the red dye so the dye fluoresces in a manner that scientists can detect in BRET test. See also *Protein, Enzyme, Bioluminescence, Signaling protein, Protein interaction analysis.*

Repressible Enzyme

An enzyme whose synthesis (rate of production) is inhibited (repressed) when the product that it (or it in a multienzyme sequence) synthesizes is present in high concentrations. It is a way of shutting down the synthesis of an enzyme whose product is not required because so much of it is readily available to the cell. When that enzyme product is no longer available (e.g., because the cell has consumed that product) more of the enzyme is synthesized (to catalyze production of more product). See also *Repression (of an enzyme), Regulatory enzyme, Enzyme.*

Repression (of an Enzyme)

The prevention of synthesis of certain enzymes when their reaction products are present. See also *Repressible enzyme.*

Repression (of Gene Transcription/Translation)

The inhibition of transcription (or translation) by the binding of a repressor protein to a specific site on the DNA (or RNA) molecule. The repressor molecule is the product of either:

- A repressor gene.

or

- Demethylation of a relevant histone.

or

- Sumoylation of relevant histones.

See also *Repressor (protein), Gene, Transcription, Translation, Deoxyribonucleic acid (DNA), Histones, Methylated, Small ubiquitin-related modifier.*

Repressor (Protein)

Discovered in 1967 by Walter Gilbert, et al., this term refers to the product of a regulatory gene; it is a protein that combines both with an inducer (or corepressor) and with an operator region (e.g., of DNA). See also *Inducers, Corepressor, Operator, Repression (of gene transcription/translation).*

Reprogramming

A term utilized to refer to the massive change in the status of a given cell resulting from:

- Complete replacement of its nucleus (e.g., the total replacement of a germ cell's genetic material during *cloning via nuclear transfer process*). In that process, the recipient cell (an unfertilized egg) is thereby "reprogrammed" to become a genetic clone of the organism from which the nucleus was extracted.
- Reversion of an already differentiated cell to its undifferentiated state. For example during 2005, Kevin Eggan and Douglas A. Melton caused a fully differentiated (i.e., adult) human skin cell to be "reprogrammed" into an embryonic stem cell (i.e., exhibited totipotency) via a cell fusion process.

See also *Cell, Nucleus, Nuclear DNA, Deoxyribonucleic acid (DNA), Gene, Genome, Organism, Genetic engineering, Germ cell, Clone (an organism), Differentiation, Cell differentiation, Stem cells, Totipotency, Embryonic stem cells, Cell fusion, Germ cell, Reversine.*

Research Foundation for Microbiological Diseases (Includes Institute of Physical and Chemical Research)

Also known as RIKEN. A Japanese institution that performs research on infectious diseases, among other research. See also *National Institute of Allergy and Infectious Diseases (NIAID), Koseisho.*

Residue (of Chemical within a Foodstuff)

See *Maximum residue level (MRL).*

Residue (Portion of a Protein Molecule)

See *Minimized proteins.*

Resistin

Refers to a gene that is:

- In some obese animals linked to insulin resistance (i.e., the animals can thereby get Type II diabetes).

R

- In some humans who are being treated with highly active antiretroviral therapy linked to the side effects of insulin resistance, hyperlipidemia, and elevated blood triglyceride levels.

See also *Gene, Insulin, Type II diabetes.*

Resolvins

A class of lipids that shutdown inflammation within the body after it is no longer needed (e.g., following an infection or injury) via preventing neutrophils from leaving the bloodstream and encouraging macrophages to consume proinflammatory cell debris. See also *Lipids, Chronic inflammation, Cell, Neutrophils, Macrophage.*

Respiration

Oxidative process in living cells in which oxygen or an inorganic compound serves as the terminal (final, ultimate) electron acceptor. Aerobic organisms obtain most of their energy from the oxidation of organic fuels. This process is known as respiration. See also *Oxidation–reduction reaction, Reduction (in a chemical reaction), Oxidation, Oxidizing agent.*

Restriction Endoglycosidases

A class of enzymes, each of which cleaves (i.e., cuts) oligosaccharides (e.g., the side chains on glycoprotein molecules) at a specific location within the chain. They are an important tool in carbohydrate engineering, enabling the carbohydrate engineer to sequence (i.e., determine the structure of) existing oligosaccharides, to create different oligosaccharides, and to create different glycoproteins via removal/addition/change of the oligosaccharide chains on glycoprotein molecules. See also *Oligosaccharides, Glycoprotein, Carbohydrate engineering, Glycosidases, Endoglycosidase, Exoglycosidase, Glycoform, Glycobiology, Glycosylation.*

Restriction Endonucleases

A class of enzymes that cleave (i.e., cut) DNA at a specific and unique internal location along its length. These enzymes are naturally produced by bacteria that use them as a defense mechanism against viral infection. The enzymes chop up the viral nucleic acids and hence their function is destroyed.

Discovered in 1970 by Werner Arber, Hamilton Smith, and Daniel Nathans, restriction endonucleases are an important tool in genetic engineering, enabling the biotechnologist to splice new genes into the location(s) of a molecule of DNA where a restriction endonuclease has created a gap (via cleavage of the DNA). See also *Vector, Enzyme, Polymerase, Gene, Genetic engineering, Gene splicing, Electrophoresis, EcoRI.*

Restriction Enzymes

See *Restriction endonucleases.*

Restriction Fragment Length Polymorphism (RFLP) Technique

A "genetic mapping" technique which analyzes the specific sequence of bases (i.e., nucleotides) in a piece of DNA (from an organism). Since the specific sequence of bases in their DNA molecules is different for each species, strain, variety, and individual (due to DNA polymorphism), RFLP can be utilized to "map" those DNA molecules (e.g., for plant breeding purposes, for criminal investigation purposes, etc.). See also *Genetic map, Sequence (of a DNA molecule), Random amplified polymorphic DNA (RAPD) Technique, Deoxyribonucleic acid (DNA), Genome, Physical map (of genome), Linkage, Linkage group, Marker (genetic marker), Linkage map, Trait, Base pair (bp), DNA profiling, Polymorphism (chemical), Nucleic acids, Genetic code, Informational molecules.*

Restriction Map

A pictorial representation of the specific restriction sites (i.e., nucleotide sequences that are cleaved by given restriction endonucleases) in a DNA molecule (e.g., plasmid or chromosome). See also *Restriction site, Restriction endonucleases, DNA.*

Restriction Site

A nucleotide sequence (of base pairs) in a DNA molecule that is "recognized" and cleaved by a given restriction endonuclease. See also *Nucleotide, Sequence (of a DNA molecule), Base pair (bp), DNA, Restriction endonucleases, Restriction map.*

Resveratrol

Also known as 3,5,4 trihydroxy stilbene, it is a naturally occurring (e.g., in red grapes) antifungal agent (e.g., against grape fungus). Resveratrol is responsible for the fact that consumption of red wine by humans helps those humans' *blood fat (triglycerides)* levels and blood cholesterol levels to be lowered, thereby reducing risk of cardiovascular disease. Resveratrol consumption also reduces inflammation and the risk of some types of cancers because the resveratrol molecule acts as such an agonist with the estrogen receptor (without stimulating estrogenic cell proliferation) to beneficially control the body's inflammation response.

Resveratrol is a polyphenol that is produced by certain plants in response to "wounding" (e.g., by fungal growth on plant) or other stress. Plants that produce resveratrol include red grapes, mulberries, blueberries, cranberries, soybeans, peanuts, and pistachios. Resveratrol inhibits cell mutations, stimulates at least one enzyme that can inactivate certain carcinogens, and (when consumed by humans) lowers blood cholesterol and blood fat levels.

Research indicates that consumption of resveratrol by humans can reduce the risk of blood clots and stroke. When administered to certain tumors, resveratrol renders those tumors more susceptible to being killed by radiation treatments. See also *Polyphenols, Phytochemicals, Soybean plant, Fungus, Carcinogen, Cell, Cancer, Tumor, Mutation, Triglycerides, Cholesterol, Enzyme, Inducible enzymes, Agonists, Estrogen, Selective estrogen effect, Receptors, Estrogen receptors, Atherosclerosis, Coronary heart disease (CHD), Chronic inflammation, SIRT1 gene, Sirtuins.*

Retinoid X Receptors (RXR)

Refers to one "subfamily" among the so-called *orphan receptors*, which sense (via "docking" at RXRs) the presence of retinoids within cell, and thereby regulate the expression of certain genes (e.g., initiating/controlling certain retinoid-dependent regulatory pathways). For example, the retinoid *9-cis retinoic acid* (a derivative of vitamin A) can dock at RXRs to initiate one or more crucial regulatory pathways.

RXRs can also be activated via docking by several dietary lipids, including docosohexanoic acid.

Retinoid X receptors (after docking) function as transcription activators/factors, thus controlling/preventing cellular differentiation and proliferation (i.e., can act to prevent neoplastic growth or cancer). See also *Receptors, Nuclear receptors, Orphan receptors, Retinoids, Cell, Nucleus, Signaling, Pathway, Express, Transcription, Transcription activators, Transcription factors, Gene, Differentiation, Docosahexanoic acid (DHA), Neoplastic growth, Cancer.*

Retinoids

A group of biologically active compounds that are chemical derivatives of vitamin A. Among other effects on living cells, some of the retinoid compounds act to deprive cancerous cells of their ability to proliferate endlessly, so these (formerly cancerous) cells then progress to a natural death (after exposure to an applicable retinoid). See also *Cell, Apoptosis, Vitamin, Biological activity, Cancer, Neoplastic growth, Retinoid X receptors (RXR).*

Retroelements

See *Transposon.*

Retroviral Vectors

Certain retroviruses that are used by genetic engineers to carry new genes into cells. These molecules become part of that cell's protoplasm. See also *Retroviruses, Genetic engineering, Vector, Gene, Protoplasm.*

Retroviruses

(From the Latin word *retrovir,* which means "backward man") Oncogenic (i.e., cancer-producing), single-stranded, diploid RNA (ribonucleic acid) viruses that contain (+) RNA in their virions and propagate through a double-helical DNA intermediate. They are known as retroviruses because their genetic information flows from RNA to DNA (reverse of normal). That is, the viruses contain an enzyme that allows the production of DNA using RNA as a template. Retroviruses can only infect cells in which DNA is replicating, such as tumor cells (since they are constantly replicating) or cells comprising the lining of the stomach (since that lining must replace itself every few days). See also *Oncogenes, Diploid, Ribonucleic acid (RNA), Reverse transcriptases, Central dogma.*

Reverse Breeding

Refers to the creation of crop plants (today) whose genome (DNA) contains a particular trait that had been present within the genome of the wild-type ancestor of that crop plant, but was subsequently lost during that crop plant's domestication process (e.g., 1000 years ago). For example, during 2014, Lijuan Qiu and Rongxia Guan discovered a salt tolerance gene present in the DNA of wild-type soybean plants that are the ancestors of today's domesticated soybean (*Glycine max* (L.) Merrill) varieties. Because today's domesticated soybean varieties do not possess that salt tolerance gene, a soybean breeder wanting to create a modern soybean variety that would grow well in salty soil could utilize a wide cross between a modern soybean variety (germplasm) and one of those salt-tolerant wild-type soybean species (i.e., retrieved from one of the seed banks utilized to store ancestral crop-plant relatives).

In addition to crop breeder use of a WIDE CROSS methodology, such crop "trait restoration" can be accomplished via certain other technologies. See also *Deoxyribonucleic acid (DNA), Gene, Genome, Trait, Soybean plant, Germplasm, Traditional reeding methods, Wide cross, Deletions, Trait restoration, Rewilding.*

Reverse Micelle (RM)

Also known as reversed micelle or inverted micelle. A spheroidal structure formed by the association of a number of amphipathic (i.e., bearing both polar and nonpolar domains) surfactant molecules dissolved in organic, nonpolar solvents such as benzene, hexane, isooctane, and oils such as corn and sesame. The structure of an RM is the reverse of that of a micelle. Reverse micelles may be characterized by a structure in which the polar groups of the surfactant and any water present are centrally located with the surfactant hydrocarbon chains pointing outward into the surrounding hydrocarbon medium. Reverse micelles may be used to solubilize polar molecules (i.e., water, enzymes) in organic nonpolar solvents and oils. See also *Amphipathic molecules, Micelle, Surfactant.*

Reverse Phase Chromatography (RPC)

A method of separating a mixture of proteins or nucleic acids or other molecules by specific interactions of the molecules with a hydrophobic (i.e., "water hating") immobilized phase (i.e., stationary substrate) which interacts with hydrophobic regions of the protein (or nucleic acid) molecules to achieve (preferential) separation of the mixture. See also *Chromatography.*

Reverse Transcriptases

Also known as RNA-directed DNA polymerases, reverse transcriptases were discovered by Howard Martin Temin and David Baltimore in 1970. They are a class of enzymes first discovered to be present in RNA tumor virus, which allows the synthesis of DNA (complementary to the RNA) using the RNA present in the virus as a template. This is the reverse of what normally happens and hence the name.

Reverse transcriptases closely resemble the DNA-directed DNA polymerases (DNA polymerases) in that they require the same materials and conditions as the DNA polymerases (e.g., for RT-PCR). See also *Enzyme, Virus, Ribonucleic acid (RNA), Central dogma (new), Polymerase, RT-PCR.*

Reversed Micelle

See *Reverse micelle* (RM).

Reversine

A protein that—when adult myoblast cells are exposed to it—causes those (differentiated) myoblast cells to be reprogrammed into multipotent adult stem cells which can subsequently differentiate into bone or fat (adipose) cells. See also *Protein, Cell, Differentiation, Cell differentiation, Reprogramming, Multipotent, Multipotent adult stem cell, Adipose.*

Rewilding

Refers to the creation of crop plants (today) whose genome (DNA) contains a particular trait that had been present within the genome of

R

the wild-type ancestor of that crop plant, but was subsequently lost during that crop plant's domestication process (e.g., 1000 years ago). For example, during 2014, Lijuan Qiu and Rongxia Guan discovered a salt tolerance gene present in the DNA of wild-type soybean plants that are the ancestors of today's domesticated soybean (*Glycine max* (L.) Merrill) varieties. Because today's domesticated soybean varieties do not possess that salt tolerance gene, a soybean breeder wanting to create a modern soybean variety that would grow well in salty soil could utilize a wide cross between a modern soybean variety (germplasm) and one of those salt-tolerant wild-type soybean species (i.e., retrieved from one of the seed banks utilized to store ancestral crop-plant relatives).

In addition to crop breeder use of a WIDE CROSS methodology, such crop "trait restoration" can be accomplished via certain other technologies. See also *Deoxyribonucleic acid (DNA), Gene, Genome, Trait, Soybean plant, Germplasm, Traditional reeding methods, Wide cross, Deletions, Native trait restoration.*

RFLP (Restriction Fragment Length Polymorphism)

Refers to a DNA-testing technology/methodology which is based on the detection of variation in the *length of restriction fragments* when the sample's DNA is first (digested) with restriction endonucleases, then separated via Southern blot analysis or via electrophoresis. See also *Polymorphism (chemical), Deoxyribonucleic acid (DNA), Restriction endonucleases, Southern blot analysis, Electrophoresis, Restriction fragment length polymorphism (RFLP) technique.*

RFP

Acronym for *Red Fluorescent Protein.* See *Visible fluorescent proteins.*

R Genes

This refers to genes within some plants that confer resistance (to certain plant diseases) through common signaling pathways involved in ("surveillance" and activation of) natural plant defense responses.

rh

Used to denote compounds (human molecules) made through the use of recombinant DNA technology. Recombinant (r) human (h). See also *rhTNF, Recombinant DNA (rDNA), Recombination, Genetic engineering.*

Rheumatoid Arthritis

A chronic autoimmune disease which causes joint inflammation, swelling, and stiffness, resulting in eventual destruction of the interiors of affected joints. It affects 0.5%–1% of adults in the developed nations of the world and is thought to be resultant from a combination of genetic and environmental factors.

One example of a genetic factor is that people whose DNA contains the *MHC2TA* gene variant have a greater probability (est. 20%) of developing rheumatoid arthritis. Environmental factors that could increase the probability of developing rheumatoid arthritis include a deficiency of vitamin D.

Angiogenesis also contributes to rheumatoid arthritis via development of new (inappropriate) blood vessels in the joints.

Monoclonal antibody-based pharmaceuticals which help prevent/minimize tissue damage from rheumatoid arthritis include adalimumab, infliximab, and rituximab.

Fusion protein-based pharmaceuticals which help to prevent/minimize tissue damage from rheumatoid arthritis include etanercept. See also *Autoimmune disease, Tumor necrosis factor (TNF), B lymphocytes, Deoxyribonucleic acid (DNA), Gene, Allele, Monoclonal antibodies (MAb), Adalimumab, Infliximab, Rituximab, Fusion protein, Etanercept, PTPN22 gene, Receptor-binding mapping, Vitamin, Angiogenesis.*

Rhizobium (Bacteria)

Refers to one or more species of bacteria that lives symbiotically in the roots of legume plants (e.g., alfalfa, soybeans, peanuts, etc.) and converts atmospheric nitrogen (N) into ammonium ion (NH_4^+); a soluble, biologically available form (nitrate) that plants can utilize to synthesize ("manufacture") amino acids and other nitrogen-containing compounds, so those plants can grow faster and yield more.

First explained during the 1880s by Mikhail Voronin and Hermann Hellriegel.

When not enough nitrogen fixation occurs (when only non-legume plants are grown), soil is not able to produce maximum crop yields and farmers may need to spread fixed nitrogen onto the field in the form of the fertilizer *anhydrous ammonia, ammonium nitrate,* or *sodium nitrate.* See also *Nitrates, Symbiotic, Genistein (gen), Bacteria, Soybean plant, Flavonoids, Nitrogenase system, Nitrogen cycle, Isoflavones, Crop biologicals, Hemagglutinin (HA), Nodulation, Rhizobium (bacteria), Bradyrhizobium japonicum, Amino acid, Ion.*

Rhizoremediation

See *Phytoremediation, Rhizobium (bacteria).*

Rho Factor

A protein involved in (chemically) assisting *E. coli* RNA polymerase in the termination of transcription at certain (rho dependent) sites on the DNA molecule. See also *Transcription, Polymerase, Escherichia coliform (E. coli).*

rhTNF

Acronym for *Recombinant Human TNF.* See also *Tumor necrosis factor (TNF).*

RIA

See *Radioimmunoassay.*

Ribonuclease 1 Gene

A human gene which plays a major role in regulation of cell proliferation. Research indicates that a mutation in the *RNASE 1* gene predisposes certain men to get early onset prostate cancer. See also *Gene, Rnase 1 gene, Rnase 1, Cell, Prostate, Mutation, Cancer.*

Ribonucleic Acid (RNA)

A long-chain, usually single-stranded nucleic acid consisting of repeating nucleotide units containing four kinds of heterocyclic, organic bases: adenine, cytosine, guanine, and uracil. These bases are conjugated to the pentose sugar ribose and held in sequence by phosphodiester (chemical) bonds.

The primary function of RNA is protein synthesis within a cell. However, RNA is involved in various ways in the processes of expression and repression of hereditary information. The four main functionally distinct varieties of RNA molecules are as follows: (1) messenger RNA (mRNA) which is involved in the transmission of DNA information, (2) ribosomal RNA (rRNA) which makes up the physical machinery of the synthetic process, (3) transfer RNA (tRNA) which also constitutes another functional part of the machinery of protein synthesis, and (4) long noncoding RNA (lncRNA) which is involved both in epigenetic regulation of the genome and in the host innate immune response to viral infection.

Research indicates RNA can also sometimes be directly involved in protein synthesis and in the activity of certain enzymes. See also *Heredity, Genetic code, Protein, Ribosomes, Ribosomal RNA, Informational molecules, Information RNA (iRNA), Messenger RNA (mRNA), Transfer RNA (tRNA), Nanotechnology, Enzyme, Paramutation, Large intervening noncoding RNA, Long noncoding RNA (lncRNA), Innate immune response.*

Ribose

D-Ribose, a five-carbon-atom monosaccharide (i.e., a sugar). It is important to life because it and the closely allied compound deoxyribose form a part of the molecules that constitute the backbone of nucleic acids. See also *Nucleic acids, Monosaccharides.*

Ribosomal Adaptor

See *Transfer RNA (tRNA).*

Ribosomal RNA

See *rRNA (Ribosomal RNA).*

Ribosomes

The molecular "machines" within cells that coordinate the interplay of tRNAs, mRNA, and proteins in the complex process of protein synthesis (manufacture). RNA constitutes nearly two-thirds of the mass of these large (mega-Dalton) molecular assemblies, which are technically *ribozymes* (i.e., an enzyme in which the catalysis is performed by RNA).

The formation of a ribosome (in the endoplasmic reticulum of a cell) from individual RNA and protein molecules (i.e., into aminoacyl tRNA synthetases [AARSs], a family of 20 enzymes whose function is to utilize the nucleotide codes contained in genes to synthesize the corresponding 20 amino acids within the ribosome). Because AARSs are essential for the translation of genetic information into working proteins, they are found in all life forms on the planet.

Ribosome formation is largely a self-assembly process, because all of the information needed for the correct assembly of this structure is contained in the primary structure of its (molecular) components. The assembly process is ordered and proceeds in stages. Many ribosomes (in a given cell) can simultaneously translate an mRNA molecule. The structure, consisting of a group of ribosomes bound to an mRNA molecule that is actively synthesizing protein, is called a polyribosome or a polysome. The ribosomes in this (polysome) unit operate independently of each other, each synthesizing a complete polypeptide (protein) "molecular chain." See also *Protein, Polypeptide (protein), Protein signaling, Protein folding, Polycistronic, Protein structure, Primary structure, Translation, Transcription, Transcription unit, Messenger RNA (mRNA), Cell, Endoplasmic reticulum (ER), Transfer RNA (tRNA), rRNA (ribosomal RNA), Dalton, Self-assembly (of a large molecular structure), Ribozymes, Ribonucleic acid (RNA).*

Riboswitches

Refers to certain noncoding segments within messenger RNA molecules, which act to regulate gene expression (i.e., can *decrease/stop* or *increase* it) when specific molecules (e.g., metabolites/ligands such as glycine) bind to those riboswitches. For example, during 2004, Ronald R. Breaker and coworkers discovered one riboswitch in bacteria that activates the genes involved in the glycine cleavage pathway, in that bacteria.

Many other riboswitches bind to certain metabolites and halt gene expression when the concentration of those metabolite(s) in the cell increases. See also *Cell, Messenger RNA (mRNA), Transcription, Coding sequence, Gene, Gene expression, Positive control, Negative control, Down regulating, Ligand (in biochemistry), Glycine (gly), Metabolite.*

Ribozymes

Discovered by Thomas Cech and Sidney Altman, they are RNA molecules which act as enzymes; that is, possess catalytic activity and can specifically cleave (cut) other RNA molecules. The ribozyme (RNA) molecule and the other RNA molecule come together, whereupon the ribozyme molecule cuts the other RNA molecule at a specific defined (three-base) site. Because the ribozyme molecule acts as an enzyme in this reaction, the ribozyme molecule is not consumed or destroyed, but goes on to similarly "cut" other RNA molecules.

During 2000, Thomas Steitz and Peter Moore, et al. proved that *ribosomes* (i.e., the cell's internal protein synthesis "machinery") *are functionally ribozymes.*

During 2002, Scott Baskerville and David P. Bartel created a ribozyme via rational design that catalyzes the ligation (i.e., joining) of RNA molecules to protein molecules. Such ribozymes could potentially be utilized to thereby attach "affinity tags" to certain protein molecules. See also *Ribonucleic acid (RNA), Catalytic RNA, Base (nucleotide), Enzyme, Cell, Ribosomes, Catalyst, Affinity tag, Rational drug design.*

Rice

The domesticated form of crop, from *Oryza sativa* and/or *Oryza glaberrima*. Within *Oryza sativa* are the two subspecies (groups) known as *Indica* and *Japonica*. See also *Nerica, Rice blast.*

Rice Blast

A disease which can afflict the domesticated rice (*Oryza sativa*) plant. Caused by the filamentous fungi *Magnaporthe oryzae, Magnaporthe grisea*, or *Pyricularia grisea*, rice blast disease can result in damage to the plant's leaves, stems, and grain to the point that the plant *appears* to have been "blasted" with projectiles.

To infect the rice plant, the fungus (peg) grows a large appendage known as an appressorium, which it forces through the wall of rice plant cell(s). See also *Fungus.*

Ricin

A lethal-to-cells lectin that is naturally produced in castor beans (seeds of the plant *Ricinus communis*). The ricin molecule cleaves

R

purine from ribosomal RNA molecules, which halts protein synthesis and thereby kills.

In 1994, Robert J. Ferl and Paul C. Sehnke genetically engineered a tobacco plant to produce ricin. Attached to a pharmaceutical "guided missile" or "magic bullet" such as a monoclonal antibody or the CD4 protein, ricin is potentially useful for treatment against some tumors and has been investigated as a possible treatment against acquired immune deficiency syndrome (AIDS). See also *Lectins, Ribosomal RNA, Purine, Immunotoxin, Monoclonal antibodies (MAb), Cell, CD4 protein, Genetic engineering, Fusion protein, Fusion toxin, Soluble CD4, Phytochemicals, Magic bullet.*

RIKEN

Abbreviation for *Rikagaku Kenkyusho*, a network of Japan's national laboratories (research organizations). It carries out research in a large number of fields, including physics, chemistry, medical science, biology, and engineering. See also *Research Foundation for Microbiological Diseases.*

RISC

Acronym for *RNA-induced silencing complex.* See *RNA interference (RNAi), RNA silencing.*

RIT

Abbreviation for the term radioimmunotherapy. See *Radioimmunotherapy.*

Rituximab

A monoclonal antibody against CD20 B-cell specific protein (on the surface membrane of malignant B-cells) that is utilized as a pharmaceutical (Rituxan™) against non-Hodgkin's lymphoma and against certain types of rheumatoid arthritis.

Rituximab (i.e., its monoclonal antibodies) act against certain B lymphocytes in the body, thereby preventing them from "maturing" (via affinity maturation) into harmful B cells (e.g., those which secrete antibodies that attack some of the body's own tissues).

During 2010, the U.S. Food and Drug Administration approved use of rituximab to treat certain patients with chronic lymphocytic leukemia, a slowly progressing blood and bone marrow cancer that arises from B cells. See also *Monoclonal antibodies (MAb), Antibody, B lymphocytes, Rheumatoid arthritis, U.S. Food and Drug Administration (FDA), Cancer.*

RMD

See *Rapid microbial detection.*

RN Gene

See *Redement napole (RN) gene.*

RNA

See *Ribonucleic acid (RNA).*

RNA Chaperone

See *Chaperones.*

RNA Editing

Refers to various chemical and enzymatic processes that occur following transcription, which must change the "message" (information) that is encoded in the RNAs such as mRNAs, rRNAs, and tRNAs before they are functionally finished products.

For example, in eucaryotes a block of poly A containing at least 200 AMP residues is enzymatically attached to the 3′ end of mRNA in the nucleus of the cell. The mRNAs with the "tail" are then transferred to the cytoplasm and the tail enzymatically removed to form the functional mRNAs. It is believed that the poly A tail aids in the transfer of the complex and/or targets the complex to the cytoplasm. See also *Cell, Nucleus, Ribonucleic acid (RNA), Transcription, Posttranscriptional processing (modification) of RNAs, Cytoplasm, mRNA, rRNA, tRNA, Primary transcript, Enzyme.*

RNA Interference (RNAi)

Coined by Andrew Fire and Craig Mello when they discovered it in 1998, this term refers to what happens when applicable short strands of (complementary) double-stranded RNA (dsRNA) are introduced into living cells. That introduction can be done either by physical insertion of the dsRNA (e.g., via spherical nucleic acids, etc.), cellular uptake from bloodstream (e.g., when manmade dsRNA is placed inside lipidoid nanoparticles, etc.) by genetic engineering of the organism so the organism's cell(s) themselves produce that (new) dsRNA, by pest organisms consuming such genetically engineered crop plants, etc. For example, genetic engineers can utilize *T7 RNA polymerase* to cause the production of such dsRNA within living cells.

Viral infection (i.e., insertion of viral dsRNA) and also microRNAs can also cause RNA interference. For example, the fungus-like pathogenic organisms (oomycetes) known as *Phythophtora* attack numerous plant species via insertion of effectors (a class of essential virulence proteins produced by a broad range of pathogens) that disable the immune systems of the plants via RNA interference.

If those dsRNAs are relatively long, they are cleaved (cut) by enzymes known as *dimeric RNase III ribonucleases* (also called *dicer enzymes*) into segments approximately 19–21 bp (base pairs) in length; called *siRNAs* (*short interfering RNAs* or *small interfering RNAs*). That siRNA (i.e., specific to a selected gene's mRNA) causes specific cellular "cutting enzymes" in RISC (RNA-induced silencing complex) to adhere to the transcribed-from-gene mRNA which the *dsRNA was chosen to be specific to.* Those "cutting enzymes" cut-up and mark for destruction the transcribed-from-gene mRNA, thereby negating the effects of that gene. That effect is known as *gene silencing*, and it can persist even in the (first generation) offspring of that affected organism.

Thus, RNAi is one methodology which can be utilized by scientists to cause gene silencing/knockout. During 2002, Thomas A. Rosenquist and Gregory J. Hannon created "knockdown" mice via genetic engineering so those mice (continually) produced the dsRNA which silenced a selected gene. Later, the first-generation offspring of those *knockdown mice* also "silenced" that selected gene in their bodies.

To control a pest known as the soybean cyst nematode (microscopic roundworm), scientists can design and insert a gene into a soybean plant that, when expressed in the plant:

- That gene's resultant RNA is chopped into pieces by dicer molecule, as usual.

- One or more of the small RNA pieces enter into the soybean cyst nematode when it chews on the soybean plant, where they silence gene(s) responsible for SCN feeding. The resultant nematode starvation results in control of the nematode.

Under some conditions, RNAi can also inhibit gene transcription (i.e., causing formation of silent heterochromatin in some organisms). See also *Short interfering RNA (siRNA)*, *Ribonucleic acid (RNA)*, *dsRNA*, *Cell*, *Micro-RNAs*, *Gene*, *Messenger RNA (mRNA)*, *Transcription*, *Gene silencing*, *Chromatin remodeling*, *Knockout*, *Knockdown*, *Enzyme*, *Base pair (bp)*, *Posttranscriptional gene silencing (PTGS)*, *Reduced-allergen soybeans*, *Epigenetic*, *Cellular pathway mapping*, *Spherical nucleic acids*, *Lipidoids*, *Soybean plant*, *Soybean cyst nematodes (SCN)*, *Effectors (fungal)*, *DNA-directed RNA interference*.

RNA Ladder

See *Molecular-weight size marker*.

RNA Origami

Refers to one method of creating manmade molecular-scale structures or devices that are comprised of nucleic acids (i.e., RNA strands in this case). Via preplanned design the creator encodes applicable DNA sequences into a gene (DNA strand), making that gene code for specific RNA strands (subsequently created via the gene's transcription) that self-assemble into the desired nanoscale devices or structures. See also *Deoxyribonucleic acid (DNA)*, *Nucleic acids*, *Ribonucleic acid (RNA)*, *Sequence (of a DNA molecule)*, *Self-assembly (of a large molecular structure)*, *Self-assembling molecular machines*, *Nanoscience*, *Nanotechnology*.

RNA Polymerase

Discovered by Severo Ochoa in 1955, it is an enzyme that catalyzes the synthesis of a complementary mRNA (messenger RNA) molecule from a DNA (deoxyribonucleic acid) template in the presence of a mixture of the four ribonucleotides (ATP, UTP, GTP, and CTP). Also called transcriptase. See also *Transcription*, *Central dogma (old)*, *Polymerase*, *DNA polymerase*, *Promoter*, *Long noncoding RNAs*, *Alternative splicing*, *Template*.

RNA Probes

See *DNA probe*.

RNA Processing

See *Alternative splicing*.

RNA Repeats

See *Tetranucleotide repeats*.

RNA Silencing

See *RNA interference (RNAi)*.

RNA Silencing

See *Alternative splicing*.

RNA Silencing

Refers to gene silencing effected via RNA interference. See *RNA interference (RNAi)*.

RNA Transcriptase

See *RNA polymerase*.

RNA Vectors

An RNA (ribonucleic acid) vehicle for transferring genetic information from one cell to another. See also *Vector*, *Retroviral vectors*.

RNA-Directed DNA Methylation

Abbreviated as RdDM, it is the cell's use of short-interfering RNAs (siRNA, i.e., RNA molecules that are 24 nucleotides long and that guide the addition of methyl groups to matching DNA strands in a gene) to accomplish DNA methylation, thereby rendering the gene inactive (silencing it). See also *Ribonucleic acid (RNA)*, *Cell*, *Short interfering RNA (siRNA)*, *Nucleotides*, *Deoxyribonucleic acid (DNA)*, *Gene*, *Gene silencing*, *DNA methylation*.

RNAi

See *RNA interference (RNAi)*, *RNA silencing*, *Corn rootworm*.

RNAi

Acronym for *RNA interference*. See *RNA interference (RNAi)*.

RNA-Induced Silencing Complex

See *RNA interference (RNAi)*.

RNAP

Acronym for *RNA Polymerase*. See *RNA Polymerase*.

RNase

A category of enzymes, which catalyze the destruction of nucleic acids within cells. See also *Enzyme*, *Barnase*.

RNase 1

An enzyme which is coded for by a *RNase 1* gene (located in "chromosome 1" of the human genome) that has been shown to be associated with an inherited form of prostate cancer in some human families.

The *RNase 1* enzyme protects certain cells from viral infections (by triggering apoptosis—cell death), so absence of RNase 1 (e.g., via mutation of its gene, the inactivation of RNase 1, etc.) tends to predispose such individuals (due to that SNP) to prostate cancer. See also *Enzyme*, *Gene*, *Cell*, *RNase*, *Coding sequence*, *Chromosome*, *RNASE 1 gene*, *Transcription*, *Linkage*, *Cancer*, *Prostate*, *Apoptosis*, *Mutation*, *Single-nucleotide polymorphisms (SNPs)*.

RNASE 1 Gene

A human gene which plays a major role in the regulation of cell proliferation. Research indicates that a mutation in the *RNASE 1*

R

gene predisposes men (possessing that SNP) to get early onset prostate cancer. See also *Gene, Ribonuclease 1 gene, Cell, Prostate, Mutation, Cancer, Rnase 1, Single-nucleotide polymorphisms (SNPs).*

Root-Knot Nematode

Refers to plant-parasitic nematodes from the genus *Meloidogyne*, which live in the soil. By chewing on roots, they cause crop plant losses around the world in areas with hot climates or short, mild winters. Once they are established in the soil of a farm field, root-knot nematodes can survive winter seasons in a fallow field and infect crop plants grown in that field during the next crop growing season. As part of their reproduction cycle, these nematodes need to infect a living plant root, but they can only complete their reproduction on certain suitable species/strains of plants.

Farmers can minimize a field's root-knot nematode populations by planting *trap crops* (i.e., host plant species/strains that the nematodes cannot reproduce in, but that "trick" the nematodes into starting their life cycle) instead of letting the field lie fallow between nematode-susceptible crops. For example, California tomato growers can reduce their tomato losses to root-knot nematodes by planting a strain of wheat known as Lassik in the field between tomato crops. See also *Nematode, Strain.*

Rootworm

See *Corn rootworm.*

ROS

Acronym for *Reactive Oxygen Species.* See also *Reactive oxygen species, Free radical.*

Rosemarinic Acid

A phenolic compound (naturally found in some plants) that acts as an antioxidant in the body's tissues when consumed by humans. For example, rosemarinic acid is naturally produced in Rosemary (*Rosemarinus officinalis*) and also in the edible herbs *Origanum vulgare* and *Salvia officinalis.* See also *Phytochemicals, Antioxidants, Oxidative stress, Nutraceuticals.*

Rotation

See *Crop rotation.*

Roving Gene

See *Jumping genes, Transposition, Transposase, Gene, Genome, Deoxyribonucleic acid (DNA).*

RPA

Acronym for *Recombinase Polymerase Amplification.* See *Recombinase polymerase amplification.*

RPFA

Acronym for *Rapid Protein Folding Assay.* See *Rapid protein folding assay.*

Rps1c Gene

A gene that confers to any soybean plant (possessing that gene in its DNA) resistance to several strains/races of phytophthora root rot (PRR) disease. See also *Gene, Deoxyribonucleic acid (DNA), Soybean plant, Phytophthora root rot.*

Rps1k Gene

A gene that confers to any soybean plant (possessing that gene in its DNA) resistance to as many as 21 strains/races of PRR disease. See also *Gene, Deoxyribonucleic acid (DNA), Phytophthora root rot, Soybean plant.*

Rps6 Gene

A gene that confers to any soybean plant (possessing that gene in its DNA) resistance to some strains/races of PRR disease. See also *Gene, Deoxyribonucleic acid (DNA), Phytophthora root rot, Soybean plant.*

Rps8 Gene

A gene that confers to any soybean plant (possessing that gene in its DNA) resistance to as many as 50 strains/races of PRR disease. See also *Gene, Deoxyribonucleic acid (DNA), Phytophthora root rot, Soybean plant.*

rRNA (Ribosomal RNA)

The nucleic acid component of ribosomes, making up approximately two-thirds of the mass of the bacteria *E. coli* ribosome, and approximately one-half of the mass of mammalian ribosomes. Ribosomal RNA accounts for nearly 80% of the RNA content of the bacterial cell. See also *Nucleic acids, Ribosomes, Escherichia coliform (E. coli), Ribonucleic acid (RNA).*

RSV F Protein

See *Nanovaccine.*

RTK

Acronym for *Receptor Tyrosine Kinase.* See *Receptor tyrosine kinase.*

RT-PCR

Acronym for Reverse Transcriptase Polymerase Chain Reaction, a PCR technique which starts with cellular RNA (transcript), then utilizes reverse transcriptase to create its counterpart DNA, which is then amplified/copied by PCR technique. See also *Ribonucleic acid (RNA), Transcription, Transcriptome, Reverse transcriptases, DNA polymerase, Polymerase chain reaction (PCR), Polymerase chain reaction (PCR) technique, Capillary electrophoresis, Deoxyribonucleic acid (DNA), Gene expression analysis.*

Rubisco

See *C3 photosynthesis, C4 photosynthesis.*

Rubitecan

A pharmaceutical that either shrinks or halts the growth of pancreatic cancer tumors in humans.

The pharmacophore (i.e., *active portion* of molecule) in rubitecan was derived from a Chinese flowering tree (*Camptotheca acuminata*); thus that "family" of drugs is known as *camptothecins*. Camptothecins inhibit a critical enzyme that is required for cell division to occur (thus it inhibits rapidly growing tumors). See also *Cancer*, *Pancreas*, *Tumor*, *Pharmacophore*, *Enzyme*.

Rumen (of Cattle)

The "first stomach" of cattle (and other bovines). See *Prebiotics*, *Maillard reaction*.

Rumenic Acid

See *Conjugated linoleic acid* (*CLA*).

Rusts

Various fungal diseases (*Puccinia* spp.) which attack small grains plants such as wheat, corn/maize, sorghum, oats, barley, and rye. Its visual appearance is like that of rust on the surfaces of those plants. See also *Fungus*, *Wheat*, *Corn*.

RXR

See *Retinoid X Receptors* (*RXR*).

S

S

S1 Nuclease

An enzyme that specifically degrades (destroys) single-stranded sequences of DNA. See also *Restriction endonucleases, Enzyme, Deoxyribonucleic acid (DNA).*

S1P

See *Sphingosine-1-phosphate.*

SAAND

Acronym for *Selective Apoptotic Antineoplastic Drug*. See *Selective apoptotic antineoplastic drug (SAAND).*

SAGB

Senior Advisory Group on Biotechnology. See *Senior Advisory Group on Biotechnology (SAGB).*

SAGE

Acronym for Serial Analysis of Gene Expression. See *Serial analysis of gene expression (SAGE).*

Saint John's Wort

An herb, that when consumed by humans, has the effect of boosting the activity of CYP3A4 liver enzyme. Consumption of Saint John's Wort together with a number of different modern pharmaceuticals could be risky for people because that heightened enzyme activity could cause too rapid metabolism of pharmaceuticals such as statins, certain chemotherapy drugs, birth control pills (and some other pharmaceuticals); so the "typical dose" could result in lower-than-expected bloodstream levels. See also *Serotonin, CYP3A4, Cytochrome P4503A4, Enzyme, Commission e monographs.*

Salicylic Acid (SA)

SA is a plant hormone that is a signaling molecule in the Hypersensitive Response or Systemic Acquired Resistance (SAR) when SAR is triggered in plants (e.g., via spray application of COBRA® herbicide to soybean plants, via spray application of harpin protein to various plants, via chewing by insects on the leaves of tomato plants, and/or the entry into plant of certain pathogenic bacteria/fungi, etc.).

Salicylic acid also helps many plants to modulate the microbiome (community of bacteria) within their root system, so that microbiome helps to protect the plant from certain pathogens, assists roots in the uptake of nutrients, and so on.

Since at least the fifth century BC, mankind has consumed salicylic acid (e.g., in the form of dried willow or myrtle leaves then; in the form of aspirin now) to treat pain and inflammation. The name salicylic acid arose from *Salix alba*, the Latin name for the white willow tree.

Research indicates that human aspirin consumption correlates with halted growth of vestibular schwannomas (also known as acoustic neuromas); a potentially lethal intracranial tumor that typically causes hearing loss and tinnitus.

Research indicates that adequate human aspirin consumption correlates with a 30% decrease in colorectal cancer risk, except in people who possess rare single-nucleotide polymorphisms (SNPs) at sites on their chromosomes 12 and 15. The people possessing those rare variants actually increased their risk of colorectal cancer via daily aspirin consumption. See also *Hypersensitive response, Systemic acquired resistance (SAR), Signaling molecule, Soybean plant, Harpin, Fungus, Pathogen, Protein, Pathogenesis-related proteins, Jasmonic acid, Azelaic acid, Alternative splicing, Tumor, Single-nucleotide polymorphisms (SNPs), Synthetic biology, Microbiome.*

Salinity Tolerance

See *Salt tolerance.*

Salmonella

A genus of bacteria, consisting of more than 2400 serovars (strains/types) that are classified within two species (*Salmonella enterica* and *Salmonella bongori*). All of these serovars are potentially pathogenic (disease causing) to humans. For example, some variants of *Salmonella typhimurium* can cause typhoid fever. The nontyphoid strains of *Salmonella* generally cause enterocolitis; although that enterocolitis can lead to/become more serious systemic infections.

Salmonella enteritidis, *Salmonella Heidelberg*, and *S. typhimurium* are increasingly causing outbreaks of foodborne illnesses (e.g., when foods are not washed or cooked thoroughly enough prior to consumption by humans). See also *Bacteria, Pathogen, Pathogenic, Strain, Salmonella enteritidis, Salmonella typhimurium, Salmonella heidelberg, Commensal.*

Salmonella enterica

A pathogenic strain of *Salmonella* bacteria which can cause gastroenteritis, nontyphoidal septicemia, or (serovar Typhi) the disease known as typhoid fever in humans. *S. enterica* can cause human macrophages to undergo apoptosis (programmed cell death), thereby enabling *S. enterica* to resist the human cellular immune response. See also *Bacteria, Pathogen, Pathogenic, Strain, Salmonella, Macrophage, Cell, Cellular immune response.*

Salmonella enterica serovar Enteritidis

A pathogenic strain of *Salmonella* bacteria which can cause gastroenteritis. *S. enterica* serovar Enteritidis can cause human macrophages to undergo apoptosis (programmed cell death), thereby enabling *S. enterica* to resist the human cellular immune response. See also *Bacteria, Pathogen, Pathogenic, Strain, Salmonella, Macrophage, Cell, Cellular immune response.*

Salmonella enteritidis (Se)

A pathogenic strain of *Salmonella* bacteria, which can cause fatal infections in poultry and humans (e.g., when undercooked eggs are eaten by humans). See also *Bacteria, Pathogen, Pathogenic, Strain, Salmonella.*

Salmonella Heidelberg

A pathogenic and antibiotic-resistant strain of *Salmonella* bacteria, which can cause disease in humans (e.g., when contaminated food is not washed and cooked enough prior to consumption). See also *Bacteria, Pathogen, Pathogenic, Strain, Commensal.*

Salmonella typhimurium

A pathogenic strain of *Salmonella* bacteria, which can cause disease in humans (e.g., when contaminated food is not washed and cooked enough prior to consumption). See also *Bacteria, Pathogen, Pathogenic, Strain, Commensal.*

Salt Tolerance

Refers to the trait (of a plant) which enables a plant to grow/survive in soil that contains a high level of salt. For example, during 2001, Eduardo Blumwald and Hong-Xia Zhang inserted an *AtNHX1* gene from *Arabidopsis thaliana* into a tomato plant (*Lycopersicon esculentum*) and thereby made that tomato plant resistant to salt concentrations up to 200 mM (i.e., far higher than it could previously survive).

That (*Arabidopsis*-origin) gene enables the tomato plant to extract salt from the soil and then sequester and store the salt in vacuoles (i.e., small compartments) within its leaf cells.

For example, during 2014, Lijuan Qiu and Rongxia Guan discovered a salt tolerance gene present in the DNA of wild-type soybean plants that are the ancestors of today's domesticated soybean (*Glycine max* (L.) Merrill) varieties. Because today's domesticated soybean varieties do not possess that salt tolerance gene, a soybean breeder wanting to create a modern soybean variety that would grow well in salty soil could utilize a wide cross between a modern soybean variety (germplasm) and one of those salt-tolerant wild-type soybean species (i.e., retrieved from one of the seed banks utilized to store ancestral crop-plant relatives).

See also *Arabidopsis thaliana, Vacuoles, Tomato, Antiporter, Phenomics, Wild type, Germplasm, Traditional breeding methods, Native trait restoration, Soybean plant, Wide cross, Deletions.*

Salting Out

A technique used for forcing (dissolved) proteins out of a solution by increasing the concentration of salt in the solution. The Na^+ and Cl^- ions derived from the salt compete for and "tie up" water molecules that are solubilizing the protein molecules thereby rendering them insoluble or more insoluble. See also *Protein.*

SAM

See *Sam-K gene.*

Sam-K Gene

A gene that is naturally present within the *E. coli bacteriophage T3.*

If the *sam-k* gene is inserted via genetic engineering into a (fruit crop) plant's genome, that causes greatly *reduced production* of the chemical compound *S*-adenosylmethionine (SAM) in that plant's fruit.

Because the SAM is normally converted (chemically) into L-aminocyclopropane-1-carboxylic acid (ACC) in the fruits of traditional varieties of (fruit crop) plants, such *sam-k* gene-containing plants produce fruits which ripen/soften far *slower* than fruit from traditional varieties of those plants, which can reduce spoilage/loss in the harvest and transport of such fruit. That is because ACC is required for fruits to produce ethylene, the plant hormone which triggers (over-) ripening/softening of fruit. See also *Gene, Bacteriophage, Escherichia coliform (E. coli), Genetic engineering, Genome, ACC, ACC synthase.*

Sanger Sequencing

See *Sequencing (of DNA molecules).*

Sanitary and Phytosanitary (SPS) Agreement

The agreement to GATT/World Trade Organization (WTO) via which WTO member nations agreed to base their technical barriers (regarding some imports, designed for the protection of human health or the control of animal and plant pests/diseases)—only on an assessment of *actual risks* posed by the particular import in question, and to only utilize scientific methods in assessing those risks. See also *Sanitary and phytosanitary (SPS) measures, World Trade Organization (WTO), SPS.*

Sanitary and Phytosanitary (SPS) Measures

Technical barriers (i.e., against some imports) that are designed for the protection of human health or the control of animal and plant pests/diseases.

In the Sanitary and Phytosanitary (SPS) Agreement to GATT/WTO, the WTO member nations agreed to base their SPS measures only on an assessment of *actual risks* posed by the particular import in question and to only utilize scientific methods in assessing those risks. See also *Sanitary and phytosanitary (SPS) agreement, World Trade Organization (WTO), SPS.*

Saponification

Alkaline hydrolysis of triacyl glycerols to yield fatty acid salts. The molecules thus produced are known as surfactants (surface active agents) commonly called soap. The process of soapmaking. See also *Hydrolysis.*

Saponins

A group of phytochemicals (i.e., sugars linked to a triterpene or a steroid molecular subunit) which are produced by certain plants (e.g., the soybean plant, spinach plant, tomatoes, potatoes, ginseng plant, etc.). Evidence suggests that human consumption of saponins (e.g., produced in soybeans) can help to lower a person's blood content of low-density lipoproteins (LDLPs) and can help to prevent certain types of cancer. See also *Phytochemicals, Sugar molecules, Soybean plant, Low-density lipoproteins (LDLP), Cancer, Steroid.*

Saponnins

See *Saponins*.

SAR

Acronym for *Systemic Acquired Resistance*. See also *Systemic acquired resistance (SAR)*.

SAR by NMR

See *Quantitative structure–activity relationship (QSAR)*.

Satellite DNA

Many tandem repeats (identical or related) of a short basic repeating unit (in the DNA molecule). See also *Deoxyribonucleic acid (DNA)*.

Saturated Fatty Acids (SAFA)

Fatty acids containing fully saturated alkyl chains (on their molecules). This means that the carbon atoms comprising the chains are held together by one carbon-to-carbon bond and not two or three. High levels of dietary SAFA have been related to increased blood cholesterol levels, which tends to lead to coronary heart disease (CHD) in humans. The sole exception is *stearic acid* (also known as stearate), which research has shown has no impact on the blood cholesterol levels of humans that consume it.

Beef fat typically contains approximately 54% saturated fatty acids. Sheep fat typically contains approximately 58% saturated fatty acids. Pork fat typically contains approximately 45% saturated fatty acids. Chicken fat typically contains approximately 32% saturated fatty acids. In general, fats possessing the highest levels of saturated fatty acids tend to be solid at room temperature; and those fats possessing the highest levels of unsaturated fatty acids tend to be liquid at room temperature. That rule of thumb was the original "dividing line" between the terms "fats" and "oils," respectively. See also *Fatty acid*, *Dehydrogenation*, *Cholesterol*, *Monounsaturated fats*, *Saponification*, *LPAAT protein*, *Unsaturated fatty acid*, *Polyunsaturated fatty acids (PUFA)*, *Coronary heart disease (CHD)*, *Palmitic acid*, *Stearate (stearic acid)*, *High-stearate soybeans*, *High-stearate canola*.

Saxitoxins

Paralytic poisons that are produced by certain shellfish. See also *Ricin*.

SBH

Acronym for *sequencing by hybridization*. See *Hybridization (molecular genetics)*, *Sequence (of a DNA molecule)*, *Sequencing (of DNA molecules)*, *Comparative sequencing*, *Biochips*.

SBO

Abbreviation for *soybean oil*. See *Soybean oil*.

Scab

The common (colloquial) name for Fusarium Head Blight, a disease of wheat (*Triticum aestivum*) that is caused by *Fusarium graminearum* fungus. See also *Fusarium graminearum*.

Scaffolding (Utilized in Tissue Engineering)

Refers to physical structures (e.g., manufactured from a biocompatible plastic) that are utilized to provide a support to which applicable cells can attach and grow (e.g., bone cells induced to span a break/gap that is too large to achieve healing without the scaffolding inserted across the bone break, bone cells being grown in a bioreactor from stem cells, etc.). Scaffolding can be manufactured via simple extrusion of the plastic, electrospinning (resulting in creation of microscopic fibers which then form into a randomly entangled "mat" for cells to adhere to), foaming/leaching (resulting in creation of a surface akin to Swiss cheese for cells to adhere to), freeform short-fiber fabrication (resulting in a "mat" of fibers staked in a criss-cross pattern). The electrospun "mat" was found to be the most effective at encouraging one type of adult stem cell (human bone marrow stromal cell) to develop into the branched and elongated form that is typical of mature bone cells. See also *Tissue engineering*, *Biocompatible*, *Adult stem cell*, *Nanowhiskers*, *Bioreactor*.

Scale-Up

The transition step in moving a (chemical) process from experimental (e.g., "test tube," small, bench) scale to a larger scale producing more or much more product than the bench scale (e.g., production of tons/year in a chemical plant). A process may require a number of scale-ups, which each scale-up producing more product than the last one.

Scanning Tunneling Electron Microscopy

See *Electron microscopy (EM)*.

Scavenger Receptor A

A receptor molecule on the surface of macrophages that takes up (via adherence) many nanoparticles after they are inserted into the human bloodstream. See also *Receptors*, *Macrophage*, *Nanoparticles*.

$scCO_2$

Abbreviation for supercritical carbon dioxide. See *Supercritical carbon dioxide*.

SCD-1

Acronym for *stearoyl-CoA-desaturase 1*. See *Desaturase*.

SCFA

Acronym for short chain fatty acids. See *Short chain fatty acids*.

scF_v

Acronym for *single chain variable fraction* (of an antibody). See *Antibody*.

Sclerotinia spp.

Refers to the several different species of the *Sclerotinia* fungus (i.e., *Sclerotinia sclerotiorum*, *Sclerotinia minor*, and *Sclerotinia trifoliorum*) that parasitically attack approximately 400 different plant species, resulting (when conditions are conducive to fungal growth) in the disease known as White Mold. See also *Fungus*, *Crop biologicals*.

SCNT

Acronym for *somatic cell nuclear transfer*. See also *Clone (an organism)*.

SCP

See *Single-cell protein (SCP)*.

sd1 Gene

See *Gibberellins*.

SDA

Acronym for *stearidonic acid*. See *Stearidonic acid*.

SDM

Acronym for *site-directed mutagenesis*. See *Site-directed mutagenesis (SDM)*.

SDS

Acronym for *sudden death syndrome* (a soybean disease).

SDS

Acronym for *sodium dodecyl sulfate*. Also known as sodium lauryl sulfate. A surfactant commonly used in biochemical and biotechnological applications for the solubilization of membrane components and hard-to-solubilize (dissolve) molecules. For example, it is often utilized at high concentration in water solution (e.g., along with potassium acetate) to dissolve plant DNA samples (e.g., when a scientist wants to sequence that sample of plant DNA). The SDS/PA in water solution helps the scientist to separate out contaminants that are commonly present in samples from plant tissues (i.e., polysaccharides, proteins, etc.) because DNA molecules are much more soluble in SDS/PA solution than are those contaminant molecules. Above a critical concentration (CMC), SDS forms micelles in water which are thought to be responsible for its solubilizing action. SDS is also used in such items as shampoo. See also *Critical micelle concentration*, *Micelle*, *Reverse micelle (RM)*, *Protein*, *Membrane (of a cell)*, *Surfactant*, *Deoxyribonucleic acid (DNA)*, *Polysaccharides*, *Sequencing (of DNA molecules)*, *Hexadecyltrimethylammonium bromide (CTAB)*.

Secondary Transporters

Refers to transport protein molecules that utilize an ion gradient (i.e., higher concentration of ions in fluid on one side of membrane than the other side) to carry (i.e., transport) compounds across the plasma membrane of cells. See also *Transport proteins*, *Cell*, *Plasma membrane*, *Ion*, *Ion channels*.

Seed Amendments

See *Crop biologicals*.

Seed Bank

A phrase and concept invented during the 1920s by Nikolai Vavilov. It refers to an institution that stores multiple lots of carefully cataloged seeds of numerous crop plant varieties (including wild type) in low-temperature vaults. On a rigid schedule (e.g., every 5 years, etc.) it also periodically thaws out applicable seeds and plants them in one growing season in order to regenerate fresh seeds (i.e., certain crops' seed only remains viable for a limited number of years in cold storage).

For the seeds of certain plant species (e.g., apple, black currant, etc.) which won't remain viable at all in cold storage, the species are "stored alive" in the form of programs to (eternally) grow and replant those plants on the protected grounds of the seed bank.

Seed banks are very valuable, because they preserve genes which might be needed (for introgression) by future crop plant breeders to impart disease resistance, drought tolerance, aluminum resistance, or higher yield to some future crop/variety. See also *Species*, *Wild type*, *Gene*, *Trait*, *Introgression*, *Drought tolerance*, *Drought tolerance trait*, *Aluminum resistance*.

Seed Region

See *Artificial interfering RNA (aiRNA)*.

Seed Treatments

Refers to the application of either relevant biological organisms or chemicals to seeds (of crop plants) to kill; suppress; or repel plant pathogens, insects, and other pests that attack crop seedlings or plants. See also *Pathogen*, *Crop biologicals*.

Seedless Fruits

See *Mutation*, *Triploid*.

Seed-Specific Promoter

See *Promoter*.

Segregant

Refers to a hybrid (e.g., crop plant) created via mating of two *genetically unlike* parents. See also *Hybrid*, *Transgressive segregation*.

Segregation

Refers to the separation of allele pairs (within chromosomes) during cell meiosis (preparatory to cell's division), resulting in only *one* copy of each allele. Plant breeders sometimes utilize this segregation to remove a "marker allele" inserted earlier by genetic engineers or to increase genetic diversity (because every single allele is available—versus only one allele of each pair, prior to segregation) within their plant breeding programs. See also *Gene*, *Allele*, *Chromosomes*, *Cell*, *Meiosis*, *Marker (genetic marker)*, *Genetics*, *Transgressive segregation*, *Transgressive segregants*, *Epistasis*.

SELDI

See *Mass spectrometer*.

S

Selectable Marker Genes

See *Marker (genetic marker)*.

Selectins

Also called LEC-CAMs (leukocyte-cell adhesion molecules). A class of molecular structurally related lectins that mediate (i.e., control, cause, etc.) the contacts between a variety of cells (e.g., leukocytes and endothelial cells) and function as cellular adhesion receptors. See also *Receptors, Lectins, Adhesion molecule, Leukocytes, Endothelial cells, Endothelium, Signal transduction.*

Selective Apoptotic Antineoplastic Drug (SAAND)

A category of pharmaceuticals that acts to prevent neoplastic growth (i.e., cancer) by allowing normal cell apoptosis to occur again (e.g., by blocking an enzyme that is hindering normal apoptosis) in abnormal precancerous cells and cancerous cells.

Examples of SAANDs include *sulindac*, which blocks phosphodiesterases (enzymes). See also *Neoplastic growth, Cancer, Tumor, Apoptosis, Cell, Enzyme, Phosphodiesterases.*

Selective Estrogen Effect

A term that is used to describe how certain phytochemicals (e.g., flavones, flavonols, isoflavones, etc.) and pharmaceuticals (e.g., Evista/raloxifene, tamoxifen, etc.) possessing molecular structures that are similar to estrogen (a hormone) impart some *beneficial* effect on the human body when consumed by humans, without any of the *adverse* impacts of estrogen (e.g., promotion of the growth of certain tumors by estrogen). See also *Phytochemicals, Flavonols, Isoflavones, Flavonoids, Estrogen, Phytoestrogens, Prostate, Genistein (gen).*

Selective Estrogen Receptor Modulators

Abbreviated SERM. This term refers to chemical compounds (e.g., isoflavones, the pharmaceuticals Evista/raloxifene and tamoxifen, etc.) which impart some *beneficial* effect on the human body when consumed by humans, without any of the *adverse* impacts of estrogen (e.g., promotion of the growth of certain tumors by estrogen). See also *Selective estrogen effect, Estrogen, Isoflavones, Phytochemicals.*

Selenocysteine

See *Amino acid, Archaea.*

Self-Assembled Monolayer

See *Self-assembly (of a large molecular structure), van der Waals forces, Oriented attachment.*

Self-Assembling Molecular Machines

Refers to nanometer (nm)-sized devices, which can be caused to *self*-assemble (from carefully preplanned man-made components) via affinity/hybridization to each other of molecular (e.g., thiol-, DNA, RNA, etc.) segments (attached to the relevant man-made components).

Theoretically at least, these devices could be powered via nanopiezoelectronics, directed use of the Casimir force, and so on. See also *Self-assembly (of a large molecular structure), Nanobots, Atomic force microscopy, Dip-pen nanolithography, Nanotechnology, Directed self-assembly, Nanometers (nm), Optical tweezer, Nanoelectromechanical system (NEMS), Biomotors, Deoxyribonucleic acid (DNA), Ribonucleic acid (RNA), Hybridization (molecular genetics), Thiol group, Nanopiezoelectronics, Nanobatteries, Casimir force, DNA origami, RNA origami.*

Self-Assembly (of a Large Molecular Structure)

The essentially automatic ordering and assembly of certain molecules into a large structure. Examples of such large molecular structures (often called supramolecular structures or supramolecular assemblies) include nanofibers, nanowires, nanocrystals, nanobots, micelles, reverse micelles, ribosomes, nanotubes, Tobacco Mosaic Virus (TMV), and peptide hydrogels.

The first discovery of a self-assembling active biological structure occurred in 1955, when Heinz Frankel-Conrat and Robley Williams showed that TMV will reassemble into functioning, infectious virus particles (after TMV has been dissociated into its components via immersion in concentrated acetic acid).

During 2000, Samuel Stupp designed two-part molecules known as *peptide amphiphiles*, which *assemble themselves* (e.g., when inserted into the space between broken bones inside humans) *into rigid nanofibers possessing specific peptides on their exteriors, that encourage the growth of hydroxyapatite crystals* (a constituent of bone).

In the future, it is hoped that man will be able to "direct" the self-assembly of molecular structures which will:

- Serve as "cages" to carefully protect and deliver sensitive/unstable pharmaceuticals to targeted tissues within the body. For example, during 2006, Jeremy K. M. Sanders discovered that certain amino acid naphthalenedimide derivative molecules will self-assemble into helical nanotubes which might serve as such "cages."
- Serve as "crucibles" (i.e., reaction vessels) for small-scale chemical reactions to occur within.
- Serve as computer logic or memory devices (i.e., bioelectronics) connected to each other by nanowires.
- Serve as antibiotics. For example, during the 1990s, M. Reza Ghadiri created "peptide nanotubes" made via self-assembly of certain peptides into tubes (cylinders) of nanometer dimensions. These peptide nanotubes are "membrane active" (i.e., insert one end of themselves into the outer membrane of a cell) and cause the cell (e.g., pathogenic bacteria) contents to "leak out," which kills the bacteria.

See also *Nanowire, Micelle, Critical micelle concentration, Reverse micelle (RM), Ribosomes, Nanotube, Nanoshells, Nanobots, DNA buckyballs, Tobacco mosaic virus (TMV), Nanocrystals, Nanocrystal molecules, Nanoscience, Nanotechnology, Nanometers (nm), Bioelectronics, Antibiotic, Pathogen, Bacteria, Hairpin loop, Self-assembling molecular machines, Amino acid, Peptide, Tissue engineering, Amphiphilic molecules, Dip-pen nanolithography, SP-1, Van der waals forces, Carbon nanohorns, Nanobatteries, Nanosheets, Oriented attachment, Dendrimersomes, DNA origami, RNA origami.*

Self-Pollination

The process by which a given plant's pollen fertilizes that same plant's ovule(s).

S

For example, pollen of a soybean plant is transferred to its ovule before the flowers even open on the soybean plant. See also *Fertilization, Stigma, Soybean plant.*

Semisynthetic Catalytic Antibody

An antibody that is produced (e.g., via monoclonal antibody techniques) in response to a carefully selected antigen (i.e., one of the molecules involved in the chemical reaction that you are trying to catalyze). Such an antibody is then made to be catalytic by "attaching" a (molecular) group that is known to catalyze the desired chemical reaction. This attaching is done either via chemical modification of the antibody or via genetic engineering of the cell (DNA) that produces that antibody. See also *Catalyst, Antibody, Catalytic antibody, Site-directed mutagenesis (SDM), Monoclonal antibodies (MAb), Antigen, Genetic engineering, Abzymes.*

Senescence

Refers to the:

1. Stage of (an annual) plant's life after its seed/fruit has ripened, but before the plant dies. During this time period, the plant is primarily respiring (using oxygen) and some metabolites' content increasing. Leaves will sometimes wilt and begin to decompose.

   or
2. State of an organism's cell in which proliferation is halted and gene expression is changed from normal, that is triggered by certain stresses. One of its effects is to suppress tumors. The altered gene expression results in senescent cells hyperactively secreting proinflammatory cytokines, chemokines, growth factors, and proteases. That hyperactive secretion is known as the Senescence-associated Secretory Phenotype.

A senescence-like state can be induced by nitrogen deprivation of plants (i.e., when the plant gets less nitrogen than it needs). See also *Metabolism, Metabolite, Organism, Respiration, Cell, Gene, Gene expression, Tumor, Cytokines, Chemokines, Growth factors, Proteases, Sphingolipids, Nitrates.*

Senior Advisory Group on Biotechnology (SAGB)

An association of approximately 35 of the largest European companies that are engaged in at least some form of genetic engineering research or production. Similar to America's Biotechnology Industry Organization, the SAGB works with governments and the public to promote safe and rational advancement of genetic engineering and biotechnology. It was formed in 1989 and is based in Brussels, Belgium. See also *Biotechnology, Genetic engineering, Recombinant DNA (rDNA), Japan Bio-Industry Association, International Food Biotechnology Council (IFBC), Biotechnology Industry Organization (BIO).*

Sense

Normal (forward) orientation of DNA sequence (gene) in genome. See also *Gene silencing, Antisense (DNA sequence).*

Sepsis

Also known as systemic inflammatory response syndrome or (in its final stage) "septic shock," this life-threatening condition is characterized by an infection plus multiple-organ dysfunction. It occurs when the body's immune system overresponds to an infection (e.g., by gram-negative bacteria), most often in which release of bacterial endotoxin (lipopolysaccharide) occurs.

For unknown reason(s), sepsis usually occurs when the applicable immune system cells (e.g., macrophages, etc.) overproduce numerous inflammatory and signaling agents (e.g., cytokines), which result in the body's blood vessels becoming slack and permeable (thereby allowing plasma from the blood to "leak" out into surrounding tissues). That reduces blood pressure, and the nonplasma blood components begin to form clots in the body's smallest blood vessels (which results in inadequate oxygen supply reaching the body's organs). With time, the kidneys and other organs fail, the heart's electrical activity becomes erratic, and death often results. Even if they survive, patients often have blindness or their limbs must be amputated.

In some cases of pneumonia, the infecting bacteria activate an F-box protein known as *Fbxo3* to form a molecular complex that degrades another protein called *Fbxl2* that is needed to suppress the body's inflammatory response. If that occurs, the result is a harmfully overactive inflammatory response that can cause further damage of the lung tissue, multiple-organ failure, and septic shock (sepsis). See also *Gram-negative (G–), Bacteria, Cytokines, Endotoxin, Macrophage.*

Septic Shock

See *Sepsis.*

Septins

Structural protein molecules produced within certain animals, yeasts, and also in the fungus *Magnaporthe oryzae*. Discovered in 1970 by Leland H. Hartwell, septins act as guanosine-5′-triphosphate-binding proteins and are used to build "scaffolding" applicable to cell structural support during cell division and to compartmentalize parts of the cell (e.g., the "rings" that divide a cell into two daughter cells during the cell replication process). Septins build "cages" around certain bacterial pathogens (e.g., *Shigella*) that have entered the cell, thereby targeting those pathogens for destruction by autophagy and to prevent them from invading other cells.

See *GTPases, Protein, Cell, Bacteria, Pathogen, Pathogenesis-related proteins.*

Sequence (of a DNA Molecule)

The specific nucleic acids (and the order in which they occur) that comprise a given segment of a DNA molecule. See also *Deoxyribonucleic acid (DNA), DNA looping, Genetic code, Gene, Chromosomes, Nucleic acids, Control sequences, Sequencing (of DNA molecules), Structural genomics, Complementary (molecular genetics), Whole-genome sequencing.*

Sequence (of a Protein Molecule)

The specific amino acids (and the order in which they are coupled together) that comprise a given segment of a protein molecule. See also *Protein, Amino acid, Structural gene, Leader sequence (protein molecule*), *Genomics*, *Structural genomics*, *Chemical genetics*, *Sequencing* (*of protein molecules*).

Sequence Map

A pictorial representation of the sequence of amino acids in a protein molecule, the sequence of nucleic acids in a DNA molecule, or the sequence of oligosaccharide components in a glycoprotein/carbohydrate molecule. See also *Sequencing* (*of DNA molecules*), *Sequencing* (*of protein molecules*), *Sequencing* (*of oligosaccharides*), *Sequence* (*of a DNA molecule*), *Sequence* (*of a protein molecule*), *Restriction map*, *Binning*.

Sequencing (of DNA Molecules)

The process used to obtain the sequential arrangement of nucleotides in the DNA molecule's backbone.

Most of the sequencing processes involve the cleavage into fragments (followed by separation of those fragments, which can then be sequenced individually) of DNA molecules by one or more of several methods:

1. A chemical cleavage method followed by polyacrylamide gel electrophoresis (PAGE) or capillary electrophoresis.
2. A method consisting of controlled interruption of enzymatic replication methods followed by PAGE.
3. A dideoxyl method utilizing fluorescent "tag" atoms attached to the DNA fragments, followed by use of spectrophotometry to identify the respective DNA fragments by their differing "tags" (which fluoresce at different wavelengths). This (fluorescent tag) variant of the dideoxy method can be automated to "decipher" large DNA molecules (i.e., genomes). Such automated machines are sometimes called "gene machines."

One DNA sequencing process that does *not* involve the cleavage into fragments of DNA molecules is *nanopore sequencing*.

Sequencing of DNA was first done in the mid-1970s by Frederick Sanger. See also *Polyacrylamide gel electrophoresis* (*PAGE*), *Gene machine*, *Capillary electrophoresis*, *Deoxyribonucleic acid* (*DNA*), *Sequence* (*of a DNA molecule*), *Nanopore sequencing*, *Base excision sequence scanning* (*BESS*), *Shotgun sequencing*, *Nanopore*, *Near-infrared spectroscopy* (*NIR*), *Comparative sequencing*, *Biochips*, *Whole-genome sequencing*, *Binning*.

Sequencing (of Oligosaccharides)

See *Restriction endoglycosidases*, *Sequence map*.

Sequencing (of Protein Molecules)

The process used to obtain the sequential arrangement of amino acids in a protein molecule. See also *Protein*, *Amino acid*, *Sequence* (*of a protein molecule*).

Sequon

A (potential) site on a protein molecule's "backbone" where a sugar molecule (or a chain of sugar molecules, i.e., an oligosaccharide) may be attached. See also *Protein*, *Sugar molecules*, *Glycoprotein*, *Glycogen*, *Glycosylation*, *Protein engineering*, *Oligosaccharides*.

SER

Acronym for *smooth endoplasmic reticulum*. See also *Endoplasmic reticulum* (*ER*).

Serial Analysis of Gene Expression (SAGE)

Refers to a methodology of gene expression analysis that is based upon identification of the amount of mRNA transcribed (from each relevant gene) via a "tag" (i.e., a *specific* short mRNA fragment found in the *3′ region* of each mRNA transcript).
Developed during the 1990s, SAGE involves:

- Collecting a single sample of tissue (e.g., *specific type* of cells from a tumor).
- Separating (digesting) short oligonucleotides from each mRNA (messenger RNA) present in collected cell/tissue.
- Ligating/sequencing those oligonucleotides and comparing each (sequence) to *known sequences* garnered from the Human Genome Project. The ratios determined of the expressed proteins thereby identified result in a *quantitative determination of gene expression in that specific population of cells* (e.g., tumor cells in this example).

See also *Gene*, *Express*, *Gene expression analysis*, *Messenger RNA* (*mRNA*), *Oligonucleotide*, *Transcription*, *Transcription unit*, *Sequence* (*of a DNA molecule*), *Ligation*.

Serine (ser)

A nonessential amino acid; a biosynthetic precursor of several metabolites, including cysteine, glycine, and choline.

In 1999, Solomon H. Snyder, Herman Wolosker, and Seth Blackshaw conducted research that showed some mammals synthesize ("manufacture") D-serine within their brains and it functions as a neurotransmitter there. See also *Essential amino acids*, *Metabolite*, *Cysteine* (*cys*), *Glycine* (*gly*), *Choline*, *Neurotransmitter*.

SERM

Acronym for *Selective Estrogen Receptor Modulators*. See *Selective estrogen receptor modulators*.

Seroconversion

The development of antibodies (specific to that disease-causing microorganism) in response to vaccination or natural exposure to a disease-causing microorganism. See also *Serology*, *Antibody*, *Immunoglobulin*, *Humoral immunity*, *Pathogen*, *Polyclonal antibodies*, *Passive immunity*.

Serologist

See *Serology*.

Serology

A subdiscipline of immunology, concerned with the properties and reactions of blood sera. It includes the diverse techniques used for the "test tube" measurement of antibody–antigen reactions since 1929, including blood typing (e.g., for transfusions). The different human blood types (A, B, O) were first identified by Karl Landsteiner in 1901. See also *Major histocompatibility complex (MHC), Oligosaccharides, Serum lifetime.*

Seronegative

Refers to negative results of a serology test. See *Serology, Humoral immunity, Antibody.*

Serotonin

An important neurochemical (5-hydroxytryptamine) whose effects upon the human brain include mood elevation. Production of serotonin is increased by ingestion of the amino acid tryptophan (a chemical precursor to serotonin), typically via eggs, fish, meat, or milk. Serotonin aids blood clotting, liver cell regeneration (e.g., after liver is injured), and increases the number of insulin-producing islet cells in the pancreas of pregnant women.

Elevation of brain levels of serotonin can also be caused by consumption of the herb known as *Saint John's Wort* (*Hypericum perforatum*), a West African plant alkaloid known as ibogaine, or by consumption of certain pharmaceuticals such as the antidepressants Prozac™ (trademarked product of Eli Lilly & Company), Zoloft™ (sertraline, trademarked product of Pfizer, Inc.), or Paxil™ (paroxetine, trademarked product of GlaxoSmithkline PLC).

During pregnancy, a mother provides serotonin to the infant (i.e., across the placenta) before the infant has the capacity to make its own serotonin. After pregnancy, serotonin produced in human mammary glands helps control milk production.

In 1997, Marianne Regard and Theodor Landis discovered that humans afflicted with hemorrhagic lesions in the brain (cause of abnormal serotonin activation/production) often became "passionate culinary afficionados."

Research indicates that serotonin (i.e., carried in the bloodstream by platelets) acts as a hormone to assist the carefully structured regeneration of the liver after major liver tissue loss (e.g., via surgery, etc.).

In 2008, Stephen M. Rogers discovered that an increase in serotonin causes the desert locust (*Schistocerca gregaria*) to swarm. See also *Tryptophan (trp), Essential amino acids, Blood–brain barrier (BBB), Neurotransmitter, Alkaloids, Platelets, Insulin, Beta cells, Islets of langerhans, Blood clotting, Commission E monographs.*

Serotonin Receptors

Refers to cell surface receptors (i.e., protein molecules that "latch on" to the neurochemical serotonin) specific to serotonin that are present throughout most organs of the body. At least 14 different subtypes are known, and various pharmaceuticals can activate different subtypes. If the subtype known as 5-HT2B gets activated, it can result in harmful cardiac side effects. See also *Serotonin, Receptors, Cell.*

Serotypes

A variety (substrain) of a microorganism that is distinguished from others in (strain) via its serological effects (within immune system of the host organism it inhabits).

For example, *serovar Typhi* of the bacteria *S. enterica* can cause the human disease known as *typhoid fever.* See also *Bacteria, Strain, E. coli 0157:H7, Serology, Human immunodeficiency virus type 1* (*HIV-1*), *Human immunodeficiency virus type 2* (*HIV-2*).

Serovar

See *Serotypes.*

Serum

Blood plasma that has had its clotting factor removed. See also *Factor VIII, Factor IX, Plasma.*

Serum Half Life

See *Serum lifetime.*

Serum Immune Response

See *Humoral immunity.*

Serum Lifetime

The average length of time that a molecule circulates in an organism's bloodstream before it is cleared from the bloodstream. See also *Immune response, Antigen.*

Sessile

(Micro)organisms that are attached to a (support) substrate directly by their base and not attached via an intervening peduncle (i.e., stalk). Can also refer to fruit or leaves that are attached directly to the main stem or branch of a plant. See also *Vagile.*

Sex Chromosomes

Those chromosomes whose content is different in the two sexes of a given species. They are usually labeled X and Y (or W and Z); one sex has XX (or WW), the other sex has XY (or WZ). XX (WW) is female and XY (WZ) is male. See also *Chromosome.*

Sexual Conjugation

An infrequent occurrence in which two adjacent bacteria stretch out portions of their (cell) membranes to touch one another, fuse, and then pass transposons, jumping genes, or plasmids to each other. See also *Asexual, Bacteria, Cell, Conjugation, Plasmid, Transposon, Jumping genes.*

SFE

Acronym for *supercritical fluid extraction.* See *Supercritical fluid.*

sgRNA

Acronym for *short single guide RNA* or *single guide RNA* (utilized in CRISPR/CAS9). In the CRISPR/Cas9 gene-editing system, carefully designed man-made sgRNA molecule is utilized to guide the Cas9 nuclease (a DNA-cutting enzyme) to any specific desired site (e.g., a particular gene) on a given DNA molecule. The CRISPR/Cas9 gene-editing system can thereby be utilized to insert specific new gene(s) in precisely determined DNA location (e.g., to create a genetically engineered crop plant), to cure certain animal disorders/diseases caused by a single genetic mutation, and so on.

The sgRNA is a chimera molecule consisting of three RNA-segment regions: a 20–25 bp long base-pairing region for specific DNA binding, a 42 bp long dCas9 handle hairpin for Cas9 protein binding, and a 40 bp long transcription terminator hairpin RNA. See also *Ribonucleic acid (RNA), Deoxyribonucleic acid (DNA), Gene, Messenger RNA (mRNA), RNA interference (RNAi), Short hairpin RNA, Gene silencing, CRISPR/Cas9 gene-editing systems, Genetic engineering, Mutation, Protein, Enzyme.*

Shigellosis

Refers to a food- and waterborne gastrointestinal illness which annually kills more than one million people, primarily in developing countries. It is caused by release of Shiga toxin by either *Shigella* bacteria or by some *Escherichia coli* strains of bacteria. Recent research indicates that administration of manganese enables cells' lysosomes to destroy the Shiga toxin. See also *Lysosome, Cell, Bacteria.*

Short Chain Fatty Acids

Abbreviated SCFA, it refers to molecules which otherwise meet the formal definition of *fatty acids*, but whose "molecular chain" length is shorter than that of fatty acids. Some SCFAs cause epigenetic changes to some cellular DNA. Some SCFAs promote the growth of beneficial bacteria within the infant gut microbiome.

For example, some fruits are broken down in the human digestive system to yield the SCFA known as butyrate within the human colon. Butyrate can act via an epigenetic mechanism (i.e., histone modification) to cause apoptosis of cancerous cells in the colon, thereby reducing the incidence of human colon cancer. See also *Fatty acids, Essential fatty acids, Oligosaccharides, Epigenetic, Histone modification, Apoptosis, Beta-glucan.*

Short Hairpin RNA

Abbreviated *shRNA*. It refers to specific segments of dsRNA (i.e., that bend into a "hairpin-like" molecular shape after they self-assemble) that can either be chemically synthesized by man to cause RNA interference or else are formed inside cells when the applicable DNA-directed RNA interference methodology is utilized.

An example of the latter is genetic targeting in which a particular viral vector is utilized to deliver the applicable shRNA into cells. That viral vector actually makes the shRNA after it enters the cell and then the shRNA undergoes additional processing within the cell's microRNA pathway (which results in siRNA). See also *Ribonucleic acid (RNA), RNA interference (RNAi), Gene, Knockdown, Cell, Short interfering RNA (siRNA), DNA-directed RNA interference, dsRNA, Genetic targeting, Vector, Pathway, MicroRNAs.*

Short Interfering RNA (siRNA)

Refers to specific short sequences of double-stranded RNA (dsRNA) of 21–24 base pairs (bp) in length, which trigger degradation of messenger RNA (mRNA) possessing the same sequence (as those siRNAs) within a cell, as part of the cellular process known as *RNA interference (RNAi).*

Because that degradation of mRNA thereby shuts down (quells) production of the corresponding protein, siRNA (via RNAi) constitutes a pathway that cells utilize to regulate/silence gene expression. Relevant promoters within the DNA are silenced via DNA methylation and/or chromatin remodeling. The siRNA can be utilized by man to cause gene silencing/knockout.

In plants and nematodes, such RNA interference-induced gene silencing spreads (e.g., from the site of dsRNA entry into organism) throughout the organism, apparently via mediation/transport by the transmembrane (i.e., through the plasma membrane) protein known as SID-1.

In animals, such RNA interference tends to be localized at/near site of dsRNA entry into organism. For example, relevant siRNA has been utilized to quell the (over)production of vascular endothelial growth factor in laboratory animals, thereby helping prevent some age-related macular degeneration damage. Relevant siRNA has been utilized to quell (over) production of apolipoprotein B in laboratory animals, thereby lowering serum cholesterol and bloodstream levels of LDLP. See also *Ribonucleic acid (RNA), Cell, Organism, dsRNA, Messenger RNA (mRNA), Protein, Transmembrane proteins, Pathway, Deoxyribonucleic acid (DNA), Gene, Gene expression, RNA interference (RNAi), Gene silencing, Chromatin remodeling, DNA methylation, Knockout, Base pair (bp), Promoter, Posttranscriptional gene silencing (PTGS), AMD, Angiogenesis, Nematodes, Plasma membrane, Epigenetic, Liposomes, Apolipoprotein B, Cholesterol, Low-density lipoproteins (LDLP), Age-related macular degeneration (AMD), Short hairpin RNA, Artificial interfering RNA (aiRNA).*

Short Single Guide RNA

See *sgRNA.*

Short Tandem Repeats

Abbreviated as "STRs," these are genetic polymorphisms that consist of short sequences of DNA (e.g., a length of 2–5 bp) that are repeated numerous times within an organism's DNA. The differing numbers of copies of the repeat element that can occur in each individual's DNA (within a population of individuals) constitutes the genetic polymorphism (DNA polymorphism). See also *Deoxyribonucleic acid (DNA), Base pair (bp), Polymorphism (genetic), Amyotrophic lateral sclerosis.*

Shotgun Cloning Method

A technique for obtaining the desired gene that involves "chopping up" the entire genetic complement of a cell using restriction enzymes, then attaching each (resultant) DNA fragment to a vector and transferring it into a bacterium, and finally screening those (engineered) bacteria to locate the bacteria that are producing the desired product (e.g., a protein). See also *Genetic engineering, Genome, Restriction endonucleases, Vector.*

Shotgun Sequencing

Sometimes called *Whole-genome Shotgun Sequencing*, it was invented by J. Craig Venter and Hamilton O. Smith during the mid-1990s. Shotgun sequencing is a technology for rapid sequencing of (eucaryotic and procaryotic) DNA, in which an organism's genome (DNA) is first fragmented ("broken up"), and then randomly selected pieces of the DNA are individually sequenced.

Those individual pieces' sequences must subsequently be "bridged" (i.e., "assembled" in an overlapping end-by-end pattern) in order to assemble a complete map (e.g., of an organism's chromosome or genome). See also *Sequencing (of DNA molecules), Deoxyribonucleic acid (DNA), Sequence (of a DNA molecule), Genome, DNA "bridges," Chromosome, Genetic map, Chromosome walking, Eucaryote, Procaryotes, Whole-genome sequencing, Binning.*

shRNA

Acronym for *short hairpin RNA*. See *Short hairpin RNA.*

Shuttle Vector

A vector capable of replicating in two unrelated species. See also *Vector, Replication (of virus).*

Sialic Acid

A sugar (branched-molecule carbohydrate) which the body attaches to surfaces of certain glycoprotein and glycolipid molecules which it manufactures; to enable those glycoproteins/glycolipids to thereby evade/avoid "clearance mechanisms" (e.g., parts of body's immune system), and thus circulate (e.g., in bloodstream) longer.

In the human body, the *Neu5Gc* gene causes sialic acid molecules to be produced on the outer surface of some types of cells (typically attached to the end of cell-surface glycan molecules).

Some sialic acid (branched) molecular structures on the surface of animal cells are also utilized as binding sites by some pathogens (e.g., certain viruses and bacteria) to infect those animals. For example, the avian H5N1 influenza ("bird flu") virus preferentially attaches to *sialic acid-α2,3-galactose* structures located on the surface of *bird* cellular receptor (protein) molecules. But, the human H5N1 influenza virus preferentially attaches to *sialic acid-α2,6-galactose* structures located on the surface of *human* cellular receptor molecules.

Sialic acid is absent from plant-produced glycoproteins. See also *Sugar molecules, Oligosaccharides, Carbohydrates, Protein, Glycoprotein, Glycans, Glycosylation, Antibody, Lipids, Humoral immunity, Cellular immune response, Gene, Neu5Gc, Deoxyribonucleic acid (DNA), Cell, Pathogen, Virus, Reovirus, Receptors, Species specific.*

Sickle Cell Disease

A human disease that results from a mutation in the gene that codes for the protein known as hemoglobin (i.e., the component of red blood cells (erythrocytes) that binds to oxygen molecules when the blood carries red blood cells through the lungs, where the blood picks up oxygen via hemoglobin binding to the oxygen). The hemoglobin releases that oxygen when the blood carries the red blood cells to tissues of the body that need oxygen.

After releasing its oxygen, the mutant hemoglobin polymerizes (i.e., improperly binds itself to other hemoglobin molecules), which results in the red blood cell warping and stiffening. Such warped (i.e., sickle-shaped) red blood cells tend to clog blood vessels instead of flowing through them properly, thereby failing to deliver needed oxygen to tissues. See also *Mutation, Gene, Protein, Hemoglobin, Erythrocytes, Polymer.*

SID-1 Protein

See *Short interfering RNA (siRNA).*

Siderophore

Derived from the Greek words for "iron carrier," this refers to a chemical compound that functions as an iron chelator within bacteria, mitochondria of mammalian cells, fungi, grasses, and so on. Applicable organisms synthesize (i.e., manufacture) and release siderophores in order to procure soluble iron from insoluble mineral forms/phases the iron is typically found in their environment via formation of soluble Fe^{3+} chelated molecular complexes.

For example, both bacterial cells (e.g., those that live inside the digestive systems of animals) and mammals' cell mitochondria (i.e., membrane-encased organelles within cells that generate the majority of the cell's energy) each have their own siderophore mechanism that delivers iron from the cell's/mitochondria's environment for utilization by those respective entities. One siderophore of the *E. coli* bacteria that live within the human digestive system is enterobactin. See also *Chelation, Chelating agent, Cell, Bacteria, Mitochondria, Fungi.*

Signal Amplification

See *Synthetic biology.*

Signal Sequence

A sequence of approximately 15–30 amino acids within a given protein molecule (i.e., located at the N terminus of the protein), which enables that particular protein molecule to pass out of the cell membrane.

As the protein molecule is being "led out" of the cell membrane, the signal sequence is simultaneously removed from the protein molecule. See also *Cell, Amino acid, Polypeptide (protein), Protein, Sequence (of a protein molecule), Membranes (of a cell), Plasma membrane, Leader sequence (protein molecule).*

Signal Transducers and Activators of Transcription (STATs)

Molecules that cause *signal transduction* to occur (i.e., when a hormone or other chemical "binds" to it), or molecules that cause *transcription* to occur [i.e., when transcription factor(s) "bind" to it].

STATs can be attached to solid surfaces (e.g., in a bioassay or biosensor) for use in such research applications as *high-throughput screening.* See also *Signal transduction, Hormone, Transcription factors, Biochips, Biosensors (electronic), Bioassay, High-throughput screening (HTS), Microarray (testing), Target (of a herbicide or insecticide), Cascade, Activator (of gene), Transcription activators.*

Signal Transduction

Coined in 1972, this is a phrase meaning the "reception" and "conversion" of a "chemical message" (e.g., hormone) by a cell. For example, G-proteins (which are embedded in the surface membrane of certain cells, but extend through to outside and inside of the membrane) accomplish signal transduction. When a hormone, drug, neurotransmitter, or other signal chemical binds (i.e., "docks") to the receptor (on the exterior of the cell's plasma membrane), the receptor activates the G-protein, which causes an effector inside cell to produce a "signal" chemical inside cell, which causes the cell to react to the original external chemical signal received. See also *Cell, Plasma membrane, Transmembrane proteins, Receptors, EGF receptor, ras gene, Nuclear receptors, Signaling, G-proteins, Mast cells, CD95 protein, Hormone, Substance P, Lecithin, Cascade, Kinases, Lipids.*

Signaling

The "communication" that occurs between and within cells of an organism, for example, via hormones, nitric acid, and so on. Such signaling "tells" certain cells to grow, change, or produce specific proteins at specific times. The term is also utilized to refer to the quorum sensing process of bacteria. See also *Receptors, Protein, Nuclear receptors, G-proteins, Signal transduction, Transduction (gene), CD95 protein, Hormone, Parkinson's disease, Harpin, Abscisic acid, Substance P, Lecithin, Nitric oxide, Signal transducers and activators of transcription (STATs), Protein signaling, Cascade, Gene expression analysis, Choline, Kinases, Quorum sensing, Cell motility, Bone morphogenetic protein-signaling pathway, Anergy, Angiogenesis, Extracellular matrix, Stem cells, Quorum sensing, Metastasis, Osteoclasts, Signal amplification.*

Signaling Molecule

A molecule utilized to *signal (communicate)* with cells in same organism or to deliver a *signal* to other organisms (e.g., a signal by the soybean plant to attract beneficial *Rhizobium* bacteria to colonize the roots of that soybean plant).

For example, the young offspring of fleas can remain immature (larvae) for up to 2 years' time in the absence of a food source, until carbon dioxide molecules and heat from a nearby mammal (potential host/food source) *signal them to mature into adults* in order to prey on the mammal.

For example, the larvae of North American Tree Frogs (*Rana temporaria*) are *signaled* by chemicals which are released into a pond's water when the first such frog larva is killed by a (predatory) dragonfly nymph (i.e., when those dragonflies first arrive each year at a given pond, to prey on the frog larvae). That chemical "signal" causes all of the North American Tree Frog larvae in that pond to subsequently grow *tails that are twice as large as were grown by them prior to that chemical signal*, to facilitate their escape from the dragonfly nymphs. See also *Signaling, Nitric oxide, G-proteins, Hormone, Substance P, Ethylene, Angiogenesis, Leukotrienes, Isoflavones, Soybean plant, Rhizobium (bacteria), Harpin, Octadecanoid/Jasmonate signal complex, Salicyclic acid (SA), Azelaic acid, MicroRNAs, Signal amplification.*

Signaling Protein

See *Signaling molecule.*

Silencing

Refers to loss of gene expression impact via either:

- The cell's natural gene regulation (e.g., occurs with some genes in an organism as the organism matures (e.g., from an embryo to a seedling/juvenile) via formation of heterochromatin, and so on.
- Natural epigenetic regulation.
- Infection of plant cells by a geminivirus (which commandeers the cell's nucleus to instead transcribe the genes of that particular DNA virus).
- Genetic engineering done by man (e.g., silencing of a fruit plant's gene for polygalacturonase which causes fruit to ripen, of the FLC gene in *Arabidopsis thaliana* plant which results in it subsequently flowering, of the gene for allergenic P34 protein in soybeans, etc.) via a variety of methods (e.g., via RNA interference (RNAi), chemical genetics, effect of certain viruses, via "zinc finger proteins," via sense or antisense genes, via epigenetic silencing, etc.).
- A mutation alteration in gene's DNA sequence.
- An alteration of DNA sequence in a given gene's regulatory sequence (e.g., a promoter or operator region).
- RNA interference (RNAi) interaction(s) between the gene's messenger RNA (mRNA) transcript and other mRNA present within the cell (i.e., because scientist caused the production of those other mRNAs in order to thereby silence that particular gene's expression impact).

See *Gene, Cell, Nucleus, Gene expression, Deoxyribonucleic acid (DNA), Sequence (of a DNA molecule), Regulatory sequence, Promoter, Operator, RNA interference (RNAi), Mutation, Epigenetic, Chromatin remodeling, Gene silencing, Virus, Geminivirus.*

Silent Mutation

A mutation (single letter change within a DNA codon) in a gene that causes no detectable change in the chemical or biological characteristics of that gene's product (e.g., a protein). However, when such a mutation occurs with a redundant codon, it can slow the rate of gene expression (i.e., protein production) to one-tenth of its normal rate or less. See also *Express, Gene, Protein.*

Silk

A natural, protein polymer with a predominance of alanine and glycine amino acids. Silk is produced by silkworms that have fed on mulberry tree leaves. The body of a silkworm can retain proteins (i.e., raw material for silk) amounting to as much as 20% of its body weight. It is thought that silk may be altered, via genetic engineering of silkworms, to produce fibers of very high strength. See also *Genetic engineering, Protein engineering, Amino acid.*

SIM

Acronym for *S*tructure *I*llumination *M*icroscopy.

Simeprevir

A hepatitis C virus NS3/4A protease inhibitor that the U.S. Food and Drug Administration (FDA) approved in 2014 as the pharmaceutical Olysio, for oral treatment of genotype 1 chronic hepatitis C infection in adult patients. See also *Protease inhibitor, Food and Drug Administration (FDA).*

Simple Protein

A protein that yields only amino acids on hydrolysis (i.e., cleavage of the protein molecule into fragments) and does not have other molecular constituents such as lipids or polysaccharide attachments. See also *Protein, Amino acid, Glycoprotein, Lipids, Polysaccharides.*

S

Simple Sequence Repeat (SSR) DNA Marker Technique

A "genetic mapping" technique which utilizes the fact that microsatellite sequences "repeat" (i.e., appear repeatedly in sequence within the DNA molecule) in a manner enabling them to be used as "markers." See also *Genetic map, Sequence (of a DNA molecule), Random amplified polymorphic DNA (RAPD) technique, Restriction fragment length polymorphism (RFLP) technique, Deoxyribonucleic acid (DNA), Physical map (of genome), Linkage, Linkage group, Marker (genetic marker), Linkage map, Trait, Microsatellite DNA, Quantitative trait loci (QTL).*

Simple Sequence Repeat (SSR) Genetic Markers

See *Simple sequence repeat (SSR) DNA marker technique.*

Single Guide RNA

See *sgRNA.*

Single-Cell Protein (SCP)

Protein that is derived from single-celled organisms with a high protein content. Yeast is an example. Generally used in regard to those organisms that are edible by domesticated animals or humans. Single-Domain Antibodies (dAbs) VH "heavy chains" (portion of antibody molecules) produced by genetically engineered *E. coli* cells that act to bind antigens in a manner similar to antibodies or monoclonal antibodies (MAbs). Similar to MAbs, dAbs can be produced in large quantities, to be used as human or animal therapeutics (e.g., to combat diseases). See also *Antibody, Monoclonal antibodies (MAb), Antigen, Escherichia coli.*

Single-Nucleotide Polymorphisms (SNPs)

Variations (in individual nucleotides) that occur within DNA at the rate of approximately 1 in every 1300 base pairs in most organisms (approx. 1 in every 1200 base pairs in humans' DNA). SNPs usually occur in the same genomic location (e.g., on the organism's DNA) in different individuals. These variations account for:

- Diversity within a given species (e.g., black cattle and white cattle, different human eye colors, different strains/serotypes within a given bacteria species, etc.).
- Some genetic diseases [e.g., the disease Cystic Fibrosis is due to one SNP, the disease Sickle Cell Anemia is due to one SNP, the disease known as Familial Dysautonomia is due to one SNP, the disease known as (Duchenne) Muscular Dystrophy is due to one SNP, Huntington's Disease is due to one SNP, the disease known as neurofibromatosis is due to one SNP, Tay-Sachs disease is due to one SNP, etc.].
- The body's response to certain pharmaceuticals and food ingredients. For example, fewer than 50% of humans can metabolize the isoflavone daidzein to the metabolite known as equol (due to a particular SNP). For example, the diuretic drug thiazide works to control hypertension in 60% of U.S. African Americans, but only 8% of U.S. Caucasian people (due to one SNP). Certain pharmaceuticals do not have the desired effect in some groups of humans possessing certain specific "grouped SNPs" known as haplotypes). Because those "GROUPINGS OF SNPs" are linked (i.e., tend to "travel together" AS A GROUP within the genetics of a given population), they can collectively confer a given "multiple SNP trait" to an identifiable subpopulation of individuals. For example, the pharmaceuticals acetaminophen, aspirin, and Valium remain in the bodies of women (who constitute a haplotype for that pharmacogenomic trait) longer than in men. For example, adequate human aspirin consumption correlates with a 30% decrease in colorectal cancer risk, except in people who possess rare SNPs at sites on their chromosomes 12 and 15. The people possessing those rare variants actually increased their risk of colorectal cancer via daily aspirin consumption.

Methods utilized to identify SNPs include:

- Examination of the DNA of populations of individuals with and without a given (genetically related) disease.
- Examination of the DNA of populations of individuals with and without a given trait.

Four technologies/methodologies for the detection of SNPs (within a sample of DNA taken from such individuals) are as follows:

- Capillary electrophoresis.
- Denaturing high performance liquid chromatography.
- FP-TDI assay (i.e., template-directed dye-terminator incorporation with fluorescence polarization detection), which was developed by Pui-Yan Kwok in 1999.
- SNP chip.

"SNP mapping" (or haplotype mapping) is a "genetic mapping" technique that utilizes the fact that individual nucleotides (within DNA molecule) can exist in different forms (for a particular "site"/location on that DNA molecule), which enables such SNPs to be utilized as "markers." One example would be to track a given SNP versus occurrence of genetically related disease in a given human population. See also *Point mutation, Deoxyribonucleic acid (DNA), Sequence (of a DNA molecule), Capillary electrophoresis, Nucleotide, Polymorphism (genetic), Genetics, Genetic map, Physical map (of genome), Genome, Trait, Marker (genetic marker), Quantitative trait loci (QTL), Diversity (within a species), Base pair (bp), Transversion, Cystic fibrosis transmembrane regulator protein (CFTR), Muscular dystrophy (MD), Huntington's disease, Microarray (testing), Pharmacogenetics, Pharmacogenomics, Haplotype, Linkage map, SNP map, Haplotype map, Toxicogenetics, Organism, Hemoglobin, Chromatography, Alpha-synuclein, Fluorescence polarization (FP), SNP chip, Salicylic acid (SA), Leptin, Metabolism, Daidzein, Isoflavones, Metabolite, GWA.*

Single-Stranded DNA

Refers to molecules of "unwound" DNA (i.e., half of the double helix DNA).

During 2004, Ming Zheng and coworkers discovered that single-stranded DNA (ssDNA) will under certain circumstances (i.e., sonication in water) "coat" the exteriors of carbon nanotubes, thereby rendering those carbon nanotubes to be soluble. See also *Deoxyribonucleic acid (DNA), Double helix, Aptamers, Carbon nanotubes.*

Single-Walled Carbon Nanotubes

See *Carbon nanotubes.*

Sir2 Gene

See *Sirtuins.*

siRNA

Acronym for *short interfering RNA*. See *Short interfering RNA (siRNA).*

SirT1 Gene

The gene (in mammalian cells) which codes for production of SirT1 proteins (NAD-dependent histone deacetylase (HDAC) enzymes). Certain environmental conditions (e.g., greatly reduced caloric intake or presence of the polyphenol known as resveratrol) cause increased *SirT1* gene activity (i.e., increased production of SirT1 protein), which often significantly increases lifespan of the organism. See also *Gene, Coding sequence, Cell, Protein, Sirtuins, Enzyme, Polyphenols, Resveratrol.*

Sirtuins

Also known by the designation *Sir2*, which stands for *Silent Information Regulator 2*, these are a "family" of proteins (NAD^+-dependent HDAC enzymes) coded for by the *Sir2* gene (e.g., in *Saccharomyces cerevisiae*) or the *SirT1* gene (in mammals).

Sirtuins are usually overproduced (i.e., *Sir2* gene is activated) in *S. cerevisiae* yeast by very low-calorie environment (i.e., near "starvation" of the yeast). During 2003, researchers discovered that the phytochemical resveratrol (3,5,4-trihydroxy stilbene) will also activate sirtuins. When those sirtuins are thus activated (e.g., in *S. cerevisiae* yeast), the lifespan of that yeast is greatly extended.

The Sir2 proteins deacetylate (i.e., remove *acetyl* molecular groups from) the lysine molecular portions of histones. Because histones are components of chromatin (which makes up chromosomes), the deacetylation causes the chromatin/chromosome structure to become compressed in a manner which silences certain segments of the DNA in the chromosome. See also *Protein, Enzyme, Deoxyribonucleic acid (DNA), Coding sequence, Gene, Transcriptional activator, Positive control, Resveratrol, NAD (NADH, NADP, NADPH), Histones, Chromatin, Gene silencing, Sir2 gene, SirT1 gene.*

Site-Directed Mutagenesis (SDM)

A technique that can be used to make a protein that differs slightly in its structure from the protein that is normally produced (by an organism or cell). A single mutation (in the cell's DNA) is caused by hybridizing the region in a codon to be mutated with a short, synthetic oligonucleotide. This causes the codon to code for a *different* specific amino acid in the protein gene product.

Site-directed mutagenesis holds the potential to enable man to create modified (engineered) proteins that have desirable properties not currently available in the proteins produced by existing organisms. For example, during the 1990s, Georges Fuller and Charles Gerday utilized SDM (starting with a *Bacillus* bacteria from Antarctica that naturally produces subtilisin) to create an enzyme (for subtilisin production) which possessed 20 times the catalytic activity of other subtilisin-production enzymes. See also *Mutant, Mutation, Point mutation, Organism, Cell, Protein, Enzyme, Catalyst, Catalysis, Gene, Informational molecules, Heredity, Genetic code, Genetic map, Amino acid, Deoxyribonucleic acid (DNA), Codon, Oligonucleotide, Protein engineering, Bacteria, Bacillus.*

Site-Directed Nucleases

Refers to natural DNA repair and replication enzymes found in organisms. See also *Deoxyribonucleic acid (DNA), Enzyme, Nuclease, Zinc finger nuclease, DNA repair, Point mutation, Gene repair (done by man), Gene repair (natural), Editing, CAS9, Cas proteins.*

Site-Specific Mutagenesis

See *Site-directed mutagenesis (SDM).*

Sitostanol

A chemical (ester) that is derived from sitosterol (a sterol that is present in pine trees, and fibers (e.g., the hull or seed coat) of corn/maize (*Zea mays*) or soybeans (*Glycine max* L.). When sitostanol is consumed by humans in sufficient quantities, it causes their total serum cholesterol and their LDLP levels to be lowered by approximately 10% via inhibition (i.e., the sitostanol is preferentially absorbed by the gastrointestinal system instead of cholesterol). During 2000, the U.S. FDA approved a (label) health claim that associates consumption of sitostanols with reduced blood cholesterol content and with reduced CHD. See also *Absorption, Digestion (within organisms), Soybean plant, Low-density lipoproteins (LDLP), Serum lifetime, Cholesterol, Sterols, Phytosterols, Sitosterol, Coronary heart disease (CHD).*

Sitosterol

A phytosterol that is naturally produced in fibers within soybean (*Glycine max* L.) hulls, pumpkin seeds, pine trees, fibers of corn/maize (*Zea mays*) seed coats, and so on. Sitosterol can exist in several different molecular forms (e.g., known as alpha α, beta β, etc.).

A human diet containing large amounts of sitosterol and/or certain other phytosterols (e.g., campesterol, stigmasterol, etc.) has been shown to lower total serum (blood) cholesterol and LDLP levels, and thereby lower the risk of CHD. Evidence indicates that certain phytosterols (including sitosterol) interfere with absorption of cholesterol by the intestines and decrease the body's recovery and reuse of cholesterol-containing bile salts, which causes more cholesterol to be excreted from the body than previously.

During 2000, the U.S. FDA approved a (label) health claim that associates consumption of sitosterols with reduced blood cholesterol content and with reduced CHD. Recent research indicates that β sitosterol increases the activity of superoxide dismutase and other natural antioxidant enzymes within the body, thereby reducing the oxidative stress. See also *Phytosterols, Soybean plant, Corn, Sterols, Sitostanol, Campesterol, Stigmasterol, Coronary heart disease (CHD), Beta-sitosterol, Cholesterol, Food and Drug Administration (FDA), Superoxide dismutase (SOD), Human superoxide dismutase (hSOD), Oxidative stress.*

Size Exclusion Chromatography (SEC)

See *Gel permeation chromatograph (GPC).*

SK

See *Substance K.*

S

SLAC1

See *Ion channels.*

SLE

Acronym for Lupus Erythematosus (SLE). See *Lupus.*

Sliding Clamps

Term utilized to refer to certain protein molecules (e.g., human PCNA protein, etc.) which act to regulate DNA repair by "recruiting" DNA-repair enzymes (e.g., DNA ligase) to docking sites which the sliding clamp protein creates when it slides to the site (on cell's DNA molecule) where a break in the DNA has occurred. Such breaks in cellular DNA can occur due to ultraviolet radiation, certain mutagenic chemicals, free radicals in cells, and so on. See also *Protein, DNA repair, Deoxyribonucleic acid (DNA), DNA ligase, Cell, Mutagen, Free radical.*

Slime

An extracellular (i.e., outside of the cell) material that is produced by some (micro)organisms, characterized by a slimy consistency. The slime is of varied chemical composition. However, usual components are polysaccharides (polysugars) and specific protein molecules. See also *Cell, Biofilm, Polysaccharides, Protein.*

Slit1 Protein

See *Astrocytes.*

Small Interfering RNA

See *Short interfering RNA (siRNA).*

Small RNA

Refers to a class of double-stranded RNA molecules that are 20–25 nucleotides in length, which function to help regulate growth and the expression of certain genes. See also *Ribonucleic acid (RNA), Nucleotide, Gene, Express, Short interfering RNA (siRNA), Micro RNAs, Transcriptome.*

Small Ubiquitin-Related Modifier

Abbreviated SUMO, it is a "partner protein" which readily fuses with certain other protein molecules and causes:

- Enhanced expression of those other protein molecules.
- Enhanced solubility of those other protein molecules.
- Correct folding of those other protein molecules.

In some cases, the SUMO protein binds to histones (i.e., certain proteins complexed with DNA in chromosomes) of chromosomes/genes across an organism's genome, thereby repressing the transcription of many of the genes across that genome in order to protect it (e.g., from cancer). Such a SUMO-binding process is called *sumoylation.*

When sumoylation occurs in applicable plant tissues (i.e., those containing receptors for abscisic acid), it negatively regulates (or even halts) the seed germination and/or root growth which normally results from abscisic acid signaling.

Research indicates that SUMO might help to prevent the aggregations of alpha-synuclein in neurons that occur in Parkinson's disease.

See also *Protein, Fusion protein, Express, Expressivity, Protein folding, Protein structure, Conformation, Ubiquitin, Histones, Deoxyribonucleic acid (DNA), Gene, Chromosomes, Organism, Genome, Repression (of gene transcription or translation), Cancer, Parkinson's disease, Abscisic acid, Receptors, Abscision, Signaling.*

Smoothened

One of the proteins within the hedgehog signaling pathway, which is required for hedgehog signaling to occur. See also *Protein, Hedgehog proteins, Hedgehog signaling pathway.*

Smut

See *Telethia controversia koon* smut.

SNP

See *Single-nucleotide polymorphisms (SNPs).*

SNP Chip

Refers to a piece of glass, plastic, or silicon onto which has been placed a large number of *strands of DNA that are complementary to one or more known SNPs* (single-nucleotide polymorphisms). Such "SNP chips" (sometimes known as microarrays) can then be utilized to test a single biological sample for the presence of given SNP(s).

For example, human blood samples could potentially be tested for the presence of:

- The SNP responsible for the disease Cystic Fibrosis.
- The SNP responsible for the disease known as Sickle Cell Anemia.
- The SNP responsible for the disease known as (Duchenne) Muscular Dystrophy.
- The SNP responsible for the disease known as Tay-Sachs disease, and so on.

For example, human blood samples could potentially be tested for the presence of:

- The SNP responsible for the body's response *or failure to respond to* certain pharmaceuticals (e.g., the diuretic thiazide).
- The SNP(s) responsible for the body's response *or lack of response to* certain toxins.

See also *Single-nucleotide polymorphisms (SNP), Point mutation, Deoxyribonucleic acid (DNA), Hybridization (molecular biology), Complementary (molecular genetics), Microarray (testing), DNA chip, Biochip, Cystic fibrosis transmembrane regulator protein (CFTR), Muscular dystrophy (MD), High-throughput screening (HTS), Pharmacogenetics, Pharmacogenomics, SNP map, Haplotype, Haplotype map, Toxin, Toxicogenetics, GWA.*

SNP Map

A group of known/detailed SNPs, superimposed onto the genome map of an organism (e.g., to facilitate genetic/population studies, such as of genetically related disease susceptibility). See also *Single-nucleotide polymorphisms (SNPs), Organism, Genome, Genomic sciences, Mapping (of genome), Map distance, Marker (DNA sequence), YSTR DNA.*

SNP Markers

See *Single-nucleotide polymorphisms (SNPs).*

snRNA

Acronym for *small nuclear RNA*. See *Ribonucleic acid (RNA).*

snRNP

Acronym for *small nuclear ribonucleoproteins*. See *Alternative splicing, Intron, Spliceosomes.*

SOCS3 Protein

See *Chronic inflammation.*

Sodium Dodecyl Sulfate

See *SDS.*

Sodium Lauryl Sulfate

See *SDS.*

Sofosbuvir

A nucleoside analogue polymerase inhibitor that was approved by the U.S. FDA in 2013 as the pharmaceutical Sovaldi™ for the treatment of hepatitis C disease (in combination with the oral antiviral ribavirin and the pharmaceutical interferon). Because the NS5B polymerase enzyme is required for replication of the Hepatitis C virus, the fact that sofosbuvir inhibits that enzyme helps to stop the hepatitis C. See also *Nucleoside, Analogue, Enzyme, Polymerase, Food and Drug Administration (FDA), Interferons.*

Soft Laser Desorption

See *ICM.*

Solanine

A glycoside neurotoxin (glycoalkyloid) that is naturally present at low levels within potatoes. As a result of that, solanine is present at detectable levels in the bloodstream of humans that consume potatoes.

When consumed by humans, solanine acts as a plasma cholinesterase inhibitor.

The United States' FDA prohibits the sale in U.S. of potatoes which contain more (than a very low level of solanine); for example, the naturally present level in potatoes can unfortunately increase in harvested potatoes that are exposed to direct sunlight. See also *Toxin, Phytotoxin, Chaconine, Glycoside, Wild type, Food and Drug Administration (FDA), Traditional breeding methods, Plasma, Cholinesterase, Inhibition.*

Solid Support

See *Substrate (structural).*

Solid-Phase Synthesis

See *Synthesizing (of proteins), Synthesizing (of DNA molecules).*

Soluble CD4

A synthetic version of the CD4 protein that (when in solution in the bloodstream) interferes with the ability of adhesion molecules on HIV (i.e., AIDS) viruses to infect the relevant human immune system cells with the acquired immune deficiency syndrome (AIDS) virus. See also *CD4 protein, Adhesion molecule, Selectins, Lectins, Protein.*

Soluble Fiber

See *Water soluble fiber.*

Somaclonal Variation

The genetic variation (i.e., new traits) that results from the growing of entire new plants from plant cells or tissues (e.g., maintained in culture). Frequently encountered when plants are regenerated (grown) from plant cells that have been altered via genetic engineering. However, somaclonal variation (i.e., new genetic traits) can occur even in standard tissue culture when plants are regenerated from cells that were part of the same original plant.

Somaclonal variation can result from either epigenetic or genetic changes. See also *Cell culture, Somatic variants, Clone (an organism), Agrobacterium tumefaciens, Biolistic® gene gun, "Explosion" method, Shotgun method.*

Somatacrin

See *Growth hormone-releasing factor (GRF or GHRF).*

Somatic Cell Nuclear Transfer

See *Nuclear transfer.*

Somatic Cells

All eucaryote body cells except the gametes and the cells from which they develop. See also *Cell, Gamete, Oocytes.*

Somatic Variants

Regenerated plants (i.e., clones) that were derived (produced) from cells that originally came from the same plant—that are not genetically identical. Such plants (clones) are called "sports" or somatic variants because they vary (genetically) from the "parent" plant. Sometimes, such somatic variants are developed by man to become a new plant variety (e.g., the nectarine is an example of this). See also *Somaclonal variation, Cell culture, Clone (an organism), Genotype.*

Somatomedins

A family of peptides that mediates the action of growth hormone on skeletal tissue and stimulates bone formation. See also

Human growth hormone (HGH), Peptide, Bone morphogenetic proteins (BMP).

Somatostatin

A 14 amino acid peptide that inhibits the release of growth hormone. See also *Human growth hormone (HGH), Growth hormone-releasing factor (GRF or GHRF), Peptide.*

Somatotropin

Category of hormone that is produced naturally in the bodies of all mammals, including man. See also *Hormone, Growth hormone, Bovine somatotropin (BST), Porcine somatotropin (PST).*

Somites

Refers to differentiated tissue segments formed in the mesoderm of vertebrates' embryos that develop into the ribs, vertebrae, and muscle in the adult vertebrates. See *Embryology, Differentiation.*

Sonic Hedgehog Protein (Shh)

See *Hedgehog proteins.*

SorLA

See *Docosahexaenoic acid (DHA).*

SOS Protein

See *SOS response (in Escherichia coli bacteria).*

SOS Repair System

First postulated by Miroslav Radman in 1970, it refers to a *secondary/alternative* DNA-repair system utilized by living cells to repair the cell's DNA when damaged (e.g., by radiation) and that DNA damage prevents usage of the cell's primary DNA-repair system. See also *Cell, Deoxyribonucleic acid (DNA), DNA repair, SOS response (in Escherichia coli bacteria), Gene repair (natural).*

SOS Response (in *Escherichia coli* Bacteria)

The "switching on" of genetic repair machinery in this bacteria when its DNA has been damaged (e.g., by ionizing radiation, ultraviolet light, etc.). See also *Deoxyribonucleic acid (DNA), DNA repair, Escherichia coliform (E. coli), Gene repair (natural).*

SOS1 Gene

See *Antiporter.*

Southern Blot Analysis

Invented in 1975 by Edwin Mellor Southern, it is a test that is performed on biological samples such as restriction endonuclease-digested (i.e., fragmented) plant DNA (e.g., to ascertain if genetic-engineering-"inserted" DNA is present in particular plant cells). Gel electrophoresis is used to separate (the DNA fragments) according to the size of those fragments and then those are transferred to a filter (blot).

Radiolabeled DNA probes or RNA probes are added, and the ones which are complementary to each of the (separated, on blot) fragments will hybridize to those respective DNA fragments. The location (i.e., on the blot) and "radioactive label" of those hybridized probes can then be utilized to determine the nature of the DNA that was in those plant cells. See also *Deoxyribonucleic acid (DNA), Ribonucleic acid (RNA), Genetic engineering, Restriction endonucleases, Electrophoresis, Two-dimensional (2D) gel electrophoresis, Polyacryamide gel electrophoresis (PAGE), Radiolabeled, DNA probe, Complementary (molecular genetics), Hybridization (molecular genetics), Radioimmunoassay.*

Soybean Cyst Nematodes (SCN)

Microscopic roundworms (Heterodera glycines) living in the soil, which feed parasitically on roots of the soybean plant. The nematodes use a spear-like mouth part, called a stylet, to puncture the plant's root cells so the nematodes can eat their cell contents. That root damage causes the soybean's growth to be stunted, and the plant turn yellow because of a reduction in nodule formation by the nitrogen-fixing Rhizobium bacteria (which normally colonize roots of soybean plants). SCN can combine with a fungus (Fusarium solani) to cause a soybean plant disease known as "sudden death syndrome." See also *Soybean Plant.*

Soybean Oil

An edible oil that is produced within its beans (seeds) by the soybean plant (botanical name Glycine max (L.) Merrill). When removed from soybeans via crushing and refining processes, soybean oil is (historical average) composed of 60.8% polyunsaturated fatty acids (PUFA), 24.5% monounsaturated fatty acids, and 15.1% saturated fatty acids. However, soybean varieties have recently been created that possess as little as 7% saturated fatty acids. See also *Ac-CoA, Fatty Acid, High-Oleic Oil Soybeans, Hydrogenation, Oleic Acid, Polyunsaturated Fatty Acids (PUFA), Saturated Fatty Acids (SAFA), Soybean Plant, Essential Fatty Acids, Low-linolenic Oil Soybeans, Conjugated Linoleic Acid (CLA), Lecithin, Linoleic Acid, Linolenic Acid, Monounsaturated Fatty Acids (MUFA).*

Soybean Plant

Botanical name Glycine max (L.) Merrill. A green, bushy legume that is the world's single largest provider of protein and edible oil for mankind's use. This summer annual plant varies in height from less than a foot (0.3 meter) to more than three feet (one meter) tall. The seeds (soybeans) are borne in pods and historically have contained 13%–26% oil and 38%–45% protein (on a moisture-free basis). Its leaves contain some carotenoids.

The soybean plant has approximately 80,000 genes. It is a self-pollinating plant (i.e., male and female reproductive structures on the same plant—so is monoecious). See also *Soy Protein.*

Soy Protein

An edible protein (after heat processing) that is produced within its beans (seeds) by the soybean plant (botanical name Glycine max (L.) Merrill). When removed from soybeans via crushing, extrusion, or other process(es) involving adequate heat treatment, soy protein is (historical average) composed of 2.5% cysteine, 3.4% histidine,

5.2% isoleucine, 8.2% leucine, 6.8% lysine, 1.1% methionine, 5.6% phenylalanine, 4.2% threonine, 1.3% tryptophan, 4.2% tyrosine, 5.4% valine, 4% alanine, 7.7% arginine, 6.9% aspartic acid, 19% glutamic acid, 3.7% glycine, 0.1% 4-hydroxyproline, 5.3% proline, and 5.4% serine.

Soy protein (concentrate) is a complete (i.e., "ideal") protein (i.e., it provides all essential amino acids) for humans. It is a good dietary source of calcium, with an absorption rate equivalent to milk.

In its initial form (i.e., following crushing/extrusion from soybeans as described earlier), soy protein is known as soybean meal and contains a bit less than half protein by weight. If the soy is washed with water (following crushing/extrusion) to remove soluble polysaccharides (e.g., carbohydrates known as stachyose, raffinose, etc.), the resultant soy protein is known as soy protein concentrate and contains approximately 60% protein by weight.

If the soy is washed with water-and-alkali solution, followed by isoelectric precipitation of the soluble protein, the result is "isolated soy protein" (ISP) and is often known as soy protein isolate or soy isolate.

See also *Soybean Plant, Protein Digestibility-corrected Amino Acid Scoring (PDCAAS).*

Species

From the Latin *species* meaning *kind* or *form*, this refers to a single type (taxonomic group) of organism as determined by the distinguishing characteristics used for the particular group of life forms (e.g., the horse is one species among the mammals).

While the horse is easily distinguished from other *obviously* nonsimilar mammals, such as humans (e.g., due to the horse's four legs versus the human's two legs and two arms), it is less easy to distinguish a horse from a more closely related animal such as a donkey or a zebra.

The so-called *boundary between different species* is determined by human assessment/categorization (e.g., whether *systematics* or *cladistics* are utilized by those doing the species categorization and definitions), and sometimes *changes* when more information becomes known at a later date (e.g., if new 2D electrophoresis tests reveal certain ones to be genetically related, or not). See also *Strain, Systematics, Cladistics, Conserved, Diversity (within a species), Electrophoresis, Two-dimensional (2D) gel electrophoresis, Organism.*

Species Specific

Refers to a compound (e.g., a protein) or a disease (e.g., a viral infection) or some other effect that only acts in/on one specific species of organism.

For example, the antibiotic penicillin kills bacteria by blocking an enzyme which is critical for growth and repair of the bacterial cell wall (i.e., peptidoglycan layer), but penicillin does not harm other species (e.g., man).

For example, consumption of grapes or raisins (i.e., dried grapes) can kill dogs via causing kidney failure, but grape consumption does not harm humans. Consumption of erythritol (a naturally occurring sugar alcohol that is used to make some artificial sweeteners) kills fruit flies (*Drosophila melanogaster*) but does not harm humans. Consumption of xylitol artificial sweetener can kill dogs but does not harm humans.

For example, the pharmaceutical acetaminophen is highly toxic to snakes but is only a mild painkiller for man. The pharmaceutical diclofenac is highly toxic to vultures but is merely a painkiller for man.

For example, bovine somatotropin is a protein hormone that increases growth rate of young cattle and also increases the efficiency of mature cows in converting their feed into milk. Bovine somatotropin has no effect in humans and (if it were eaten by humans) is simply digested like any other food protein. Research indicates that most growth hormones are species specific.

For example, the first statin discovered by man had virtually no pharmaceutical effect (i.e., lowering of blood cholesterol level) in rats, but it did lower cholesterol levels in chickens and humans. See also *Species, Hormone, Penicillin G (benzylpenicillin), Statins, Sialic acid.*

Specific Activity

An enzyme unit defined as the number of moles of substrate converted to product by an enzyme preparation per unit time under specified conditions of pH, substrate concentration, temperature, and so on. Specific enzyme activity units may be expressed as follows: moles of product produced/minute/mg of protein used (or mole of enzyme used if the preparation is pure). See also *Mole, Enzyme, Substrate (chemical).*

Spectrophotometer

An instrument that measures the concentration of a compound that has been dissolved in a solvent (such as water, alcohol, etc.). The instrument shines a light through the solution, measures the fraction of the light that is absorbed by the solution, and calculates the concentration from that absorbance value. See also *Optical density (OD), Absorbance (A).*

SPFMV

Acronym for *sweet potato feathery mottle virus.* See also *Virus.*

S-Phase

The prelude to start of eucaryotic mitosis (i.e., cell replication via division); it is marked by production of identical copies of the cell's existing chromosomes. See also *Mitosis, Eucaryotes, Cell, Chromosomes.*

Spherical Nucleic Acids

Refers to nanoparticles that consist of inert (e.g., gold particle) cores surrounded by a dense shell of highly oriented short strands of DNA or RNA. When loaded with applicable RNA, these enter cells and are a way of triggering RNA interference (e.g., RNAi-based therapy for treatment of glioblastoma) or gene silencing (for treatment of psoriasis). See also *Nanoparticles, Deoxyribonucleic acid (DNA), Ribonucleic acid (RNA). RNA interference (RNAi), Gene silencing.*

Sphingadienes

A specific category of lipid molecules present in soybeans, which induce apoptosis in at least some human colon cancer cells. See also *Lipids, Soybean plant, Cancer, Cell, Sphingolipids.*

Sphingolipids

A class of lipid molecules found in all eukaryotes and several bacteria. The many functions of sphingolipids range from structural roles

S

(e.g., lipid bilayer) to signal transduction mediators which impact the regulation of some cells' growth, differentiation, motility, senescence, apoptosis, proliferation, and inflammation. See also *Lipids, Eukaryote, Bacteria, Cell, Differentiation, Cell differentiation, Lipid bilayer, Motility, Cell motility, Apoptosis, Senescence.*

Sphingosine-1-Phosphate

Abbreviated as S1P, it is a lipid within the body that determines the distribution of lymphocytes (i.e., between being in lymph nodes and being in the bloodstream). See also *Lipids, Lymphocyte, Sphingolipids.*

Spinosad™

A pesticide which is active against certain insects and mites, whose active ingredients (i.e., spinosyn A and spinosyn D) are naturally produced by the soil bacterium *Saccharopolyspora.*

Research shows that Spinosad™ is effective in controlling the lesser grain borer and other stored-grain insects. Spinosad and its registered trademark are owned by Dow Chemical Company. See also *Spinosyns, Bacteria, Weevils.*

Spinosyns

A "family" of pesticidal compounds, active against certain insects and mites, which are naturally produced by some species of bacteria (e.g., *Saccharopolyspora spinosa*). See also *Spinosad™, Bacteria, Species.*

Spiral Polypeptides

Refers to specific polypeptides whose amino acids (i.e., the "links" in the polypeptide molecular "chain") each have attached to them a side chain bearing a small positive electrical charge. Such polypeptide molecules can adopt helical shapes because those side chains are long enough that the positive charges do not interfere with the "winding" (into helical shape) of the polypeptide chain. However, the positively charged side chains do readily bind to negatively charged DNA, forming complexes that can be utilized (e.g., as part of gene therapy) because they are internalized into a living cell's endosomes. The helical structures rupture the endosomal membranes, letting the DNA escape into the cell, where it begins to direct the cell. See also *Polypeptides, Amino acid, Gene therapy, Cell, Endosomes.*

Splice Forms

See *Splice variants.*

Splice Variants

Refers to the several different proteins which can be expressed from a single given gene via all possible gene transcripts (i.e., different mRNAs resulting from alternative splicing). See also *Gene, Protein, Express, Expression, Alternative splicing, Transcriptome, Mutation.*

Spliceosomes

The cellular entity (protein–RNA hybrid complex) which processes primary RNA to remove introns and to ligate (i.e., attach together) exons, resulting in the mRNA transcript that the cell uses for translation.

Spliceosomes consist of a molecular complex made up of both RNA (ribonucleic acid) and snRNPs (small nuclear ribonucleoproteins). See also *Transcription, Protein, Translation, Cell, Intron, Ribonucleic acid (RNA), Exon, Ligation, Transcript, Messenger RNA (mRNA), Alternative splicing, Splice variants, Editing.*

Splicing

The removal of introns and joining of exons in RNA (e.g., genes). Thus, introns are spliced out, while exons are spliced together. See also *Exon, Intron, Spliceosomes, Genetic engineering, Ribonucleic acid (RNA), Central dogma (new), Alternative splicing, Differential splicing.*

Splicing (of Protein Molecule)

The removal of an intein (i.e., an intervening protein *domain in "center"* of a protein molecule) either spontaneously or by man's manipulation, followed by joining together of the two exteins (i.e., *end segments* of the protein molecule). See also *Protein, Intein, Extein, Domain, Sequence (of a protein molecule), Excision (of protein molecule), Domain (of a protein).*

Splicing Junctions

The sequences (in RNA molecules) of nucleotides immediately surrounding the exon–intron boundaries. See also *Exon, Intron, Splicing, Alternative splicing, Differential splicing, Splice variants, Ribonucleic acid (RNA), Nucleotide, Sequence (of a DNA molecule).*

SPM

See *Atomic force microscopy.*

Spontaneous Assembly

See *Self-assembly.*

Sports

Refers to new individual plants which arise by spontaneous mutation. See also *Mutation, Somaclonal variation.*

SPR

Acronym for *Surface Plasmon Resonance*. See *Surface plasmon resonance (SPR).*

SPS

Acronym for the SPS Standards Agreement of the WTO, a multinational trading agreement that "sets the rules" that govern international trade. Sanitary (i.e., human and animal) and phytosanitary (i.e., plant) standards are important in preventing the transfer of diseases from one nation to another via international trade. SPS standards are designed to protect animal, plant, and human life/health (within WTO member countries) from:

- Entry of pests (e.g., insects, weeds, etc.)
- Entry of disease-carrying organisms (e.g., European Corn Borer).

- Entry of disease-causing organisms (e.g., *Aspergillus flavus*).
- Toxins, contaminants, or disease-causing organisms in foods, beverages, or feedstuffs.

WTO member nations are required to base their SPS standards as much as possible on *existing* (e.g., Codex Alimentarius, IPPC, and OIE) *international sanitary/phytosanitary standards* and practices. See also *Sanitary and phytosanitary (SPS) agreement, Sanitary and phytosanitary (SPS) measures, International Plant Protection Convention (IPPC), International Office of Epizootics (OIE), Codex Alimentarius Commission, Maximum residue level (MRL), World Trade Organization (WTO), European corn borer (ECB), Aspergillus flavus.*

Squalamine

A potent antimicrobial agent (steroid, antibiotic) that was discovered by Michael Zasloff in the tissues of the dogfish shark in 1992. It has been found to be active against a broad spectrum of bacteria, viruses, protozoa, and fungi. It also inhibits the cell growth of certain rapidly growing blood vessels (e.g., in some cancers, macular degeneration, etc.). Squalamine was chemically synthesized by man in 1993. See also *Magainins, Steroid, Fungus, Bacteria, Bacteriocins, Protozoa, Antibiotic, Cancer, Macular degeneration.*

Squalene

A sterol that is produced in some plants. See also *Sterols.*

SRB (Sulfate-Reducing Bacterium)

Any organism that metabolically reduces sulfate to H_2S (hydrogen sulfide). This includes a variety of microorganisms. See also *Reduction (in a chemical reaction), Metabolism, Microorganism, Ferrobacteria.*

sRNA

Acronym for small RNA. See *Small RNA.*

ssDNA

Acronym for *single-stranded DNA.* See also *Deoxyribonucleic acid (DNA), Single-stranded DNA, Aptamers.*

SSR

See *Simple sequence repeat (SSR) DNA marker technique.*

ssRNA

Acronym for *single-stranded RNA.* See also *Ribonucleic acid (RNA), Innate immune response.*

STa

See *Enterotoxin.*

Stable (Mutation/Insertion)

See *Mutation.*

Stacchyose

See *Stachyose.*

Stachyose

A carbohydrate (oligosaccharide) that is naturally produced in soybeans (and some other plants). Stachyose is relatively insoluble in water and much less available for digestion by monogastric animals (e.g., swine, poultry) than the other carbohydrate components within soybeans. See also *Carbohydrates (saccharides), Low stachyose soybeans, Oligosaccharides, Soybean plant.*

Stacked Genes

Refers to the insertion of two or more (synthetic) genes into the genome of an organism. One example of that would be a plant into which has been inserted a gene from *Bacillus thuringiensis (B.t.)* and a gene for resistance to a specific herbicide. See also *Gene, Biotechnology, Genetic engineering, Bacillus thuringiensis (B.t.), B.t. kurstaki, Genetically engineered microbial pesticides (GEMP), EPSP synthase, PAT gene, BAR gene.*

Staggered Cuts

Scissions (cuts) made in duplex DNA when the two strands of DNA that make up the duplex DNA are cleaved at different points near each other by restriction endonucleases. What is produced is a single-stranded structure (in which the single strands are a number of nucleotide bases long) with a double-stranded core section. This core section is much longer than the single-stranded region. See also *Deoxyribonucleic acid (DNA), Restriction endonucleases, Sticky ends.*

Stanol Ester

See *Sitostanol.*

Stanol Fatty Acid Esters

See *Sitostanol, Fatty acid.*

STAP Cells

Acronym for Stimulus-Triggered Acquisition of Pluripotency cells. See *Stimulus-triggered acquisition of pluripotency (STAP) cells.*

Starch

From the Old English *strechen* meaning *to stiffen*, it is a polymer of glucose molecules (i.e., a polysaccharide) used by plants to store energy. Plants produce starch in two different molecular forms, amylopectin and amylose. For example, the starch content in traditional corn (maize) kernels averages 72%–76% amylopectin and 24%–28% amylose. The starch in traditional potatoes averages 80% amylopectin and 20% amylose.

Starch is broken down by enzymes (amylases) to yield glucose, which can be used as an energy source. The analogous polymer that is used by mammalian systems is called glycogen or, in old terminology, "animal starch." See also *Glucose (GLc), Enzyme, Amylase, Alpha amylase (α-amylase), Corn, Amylose, Amylopectin, Plastid.*

S

Start Codon

Refers to the specific set of three nucleotides within each mRNA molecule at which the ribosome begins the translation process, thereby establishing the reading frame. See also *Reading frame, Nucleotide, Messenger RNA (mRNA), Ribosomes, Translation.*

Startpoint

Refers to the position on a DNA molecule corresponding to the first base incorporated into mRNA. See also *Deoxyribonucleic acid (DNA), Transcription, Messenger RNA (mRNA), Exon, Ribonucleic acid (RNA), Kozak sequence.*

STAT4

See *CD8+ T cells.*

Statins

Refers to a class of biologically active compounds which reduce blood cholesterol levels by blocking the enzyme HMG-CoA reductase, and reduce some kinds of inflammation, in humans. First discovered in a mold (*Penicillium citrinum* growing on oranges) by Akira Endo in 1973.

Synthetic (i.e., man-made) statins commercially available today include pravastatin, simvastatin, atorvastatin, and so on.

Research published in 2002 by Dr. Leon Simons et al. reported that when such statin pharmaceuticals are taken while the humans are also consuming sterol-containing foods (e.g., sterol-fortified margarine), the beneficial impact on their LDLP levels was more than the impact of either one alone (i.e., the beneficial impacts are *additive*). See also *Cholesterol, Coronary heart disease (CHD), C-reactive protein (CRP), Mold, Species specific, Sterols, Low-density lipoproteins (LDLP), Chronic inflammation, Nanoparticles.*

STATs

See *Signal transducers and activators of transcription (STATs).*

Stearate (Stearic Acid)

A saturated fatty acid containing 18 carbon atoms in its molecular "backbone," which is essentially neutral in effect on CHD in humans (i.e., doesn't appreciably increase LDLPs in the bloodstream). Because of the heart disease neutrality, stearate-containing oils (e.g., high-stearate soybean oil) are an acceptable cooking oil choice, with the resistance to oxidation/breakdown of a saturated fatty acid, but no bloodstream-cholesterol increasing effect. In the mid-1990s, the American Cocoa Research Institute/Chocolate Manufacturers Association filed a petition with America's FDA to differentiate stearate (on food product labels) from the other saturated long-chain fatty acids used as food ingredients.

In order to make milk, dairy cows require more stearic acid than a conventional digestive system alone could provide from the cow's (mainly carbohydrate) diet. Therefore, cows utilize microorganisms living in their rumen (i.e., a special sort of prestomach) to convert carbohydrate (grass) to stearic acid. Thus, high-performance dairy cows might benefit from a diet that contained high-stearate soybeans, if their milk output is limited by dietary stearate availability. See also *Fatty acid, Low-density lipoproteins (LDLP), Saturated fatty acids, Food and Drug Administration (FDA), High-stearate soybeans, Fats, Enoyl-acyl protein reductase, High-stearate canola.*

Stearic Acid

See *Stearate.*

Stearidonate

Another name for stearidonic acid. See *Stearidonic acid.*

Stearidonic Acid

A fatty acid which is naturally produced in the seeds of some plants (e.g., *Nasa carunculata, Nasa hornii, Nasa* cf. *magnifica, Borago officinalis* L., etc.). When consumed by humans, stearidonic acid is readily converted into the *n*-3 fatty acids *eicosapentanoic acid* and *docosahexanoic acid.* See also *Fatty acid, N-3 fatty acids, Eicosapentanoic acid (EPA), Docosahexanoic acid (DHA).*

Stearidonic Acid-Containing Soybeans

Refers to soybeans from soybean plants which have been genetically engineered to produce soybeans bearing oil that contains more than 18% stearidonic acid, instead of the typical 0% stearidonic acid content of soybean oil produced from conventional varieties of soybeans. See also *Stearidonic acid, Soybean plant, Soybean oil, Genetic engineering.*

Stearoyl-ACP Desaturase

A "family" of enzymes that is naturally produced in oilseed plants. They play the central role in determining the ratio of saturated to unsaturated fatty acids (in the vegetable oils produced from such plants). See also *Fats, Fatty acid, Enzyme, Genetic engineering, Genetic code, Laurate, High-stearate soybeans, High-stearate canola.*

Steckling

Refers to the plant structure that is formed during first year (autumn) by the sugar beet (*Beta vulgaris* ssp. *vulgaris*) plant. Only after vernalization (i.e., enduring a cold winter) is the sugar beet plant able to produce flowers (from the stecklings). See also *Vernalization.*

STED

Acronym for *ST*imulated *E*mission *D*epletion microscopy.

Stem Cell Engineering

See *PIM-1 protein.*

Stem Cell Growth Factor (SCF)

A growth factor (glycoprotein hormone) that acts upon certain stem cells in a wide variety of ways to increase growth, proliferation, and maturity (into red blood cells or white blood cells). See also *Stem cells, Growth factor, Hormone, Glycoprotein, Differentiation, Totipotent stem cells, Colony stimulating factors (CSFs), Adult stem cell.*

Stem Cell One

The single stem cell in the bone marrow of a fetus from which every immune system cell in the adult is subsequently derived. The primordial stem cell is stimulated to develop into the mature immune system's differentiated, specialized cells by interleukin-7. See also *Stem cells, Totipotent stem cells, Interleukin-7 (IL-7), Embryonic stem cells, Differentiation.*

Stem Cells

Certain cells—present in the bodies of mammals even prior to birth, although *also present in adult mammals*—that can grow/differentiate into different cells/tissues of the (adult organism) body. For example, bone marrow (stem) cells, some of which eventually mature into red blood cells or white blood cells. The stem cells that remain in the bone marrow maintain their own numbers by self-renewal divisions, yielding more (adult stem cell) cells to start the maturation process. This maturation process is stimulated and controlled by stem cell growth factor, granulocyte colony stimulating factor, and by granulocyte-macrophage colony stimulating factor.

As they age, muscle stem cells tend to eventually lose much of their ability to thereby rejuvenate (e.g., damaged) muscle tissue via activation of the p38 MAP kinase pathway (which over time impedes the self-renewal proliferation of the stem cells and encourages them to instead differentiate/become nonstem, muscle-progenitor cells. During 2014, research led by Helen Blau indicated that if the p38 MAP kinase pathway could be blocked (e.g., by an applicable pharmaceutical), old muscle stem cells could regain the ability to self-renew proliferate and disseminate into damaged/aged muscle tissue to repair it.

During 2000, research by Richard Childs showed that stem cells (i.e., collected from a sibling's bloodstream and) transplanted into a patient suffering from kidney cancer could induce generation of a *"new" immune system* which could help stop/reverse the kidney cancer.

During 2003, research led by Songtao Shi showed that living stem cells (i.e., easily collected) remain in a child's baby teeth (i.e., the temporary teeth that begin falling out when child is approximately 6 years old) after they fall out of the child's mouth.

During 2010, research led by Rama Khokha showed that when progesterone concentration in a woman's bloodstream (and thus in her breast tissue) peaks during the second half of the menstrual cycle, it causes a cross-signaling between stem cells and neighboring breast tissue cells that results in an increase in the numbers of normal breast stem cells. See also *Cell, Adult stem cell, Multipotent adult stem cells, Ectodermal adult stem cells, Endodermal adult stem cells, Mesodermal adult stem cells, Hematopoietic stem cells, Red blood cells, White blood cells, Basophils, Stem cell one, Stem cell growth factor (SCF), Totipotent stem cells, Totipotency, Embryonic stem cells, Differentiation, Differentiation pathways, MAPK system, Immune response, Cancer, Monocytes, Induced pluripotent stem cells, Large intervening noncoding RNA, Signaling, Stimulus-triggered acquisition of pluripotency (STAP) cells.*

Stereoisomers

Molecules that have the same structural formula but different spatial arrangements of dissimilar groups (of atoms) bonded to a common atom (in the molecule). Many of the physical and chemical properties of stereoisomers are the same, but there are differences in the crystal structures, in the direction in which they rotate polarized light (which has been passed through a solution of the stereoisomer), and in their use in an enzyme-catalyzed (biological) reaction. See also *Racemate, Polarimeter, Dextrorotary (D) isomer, Epimers, Isomer, Levorotary (L) isomer, Isomerase, Diastereoisomers, Enantiomers, Nanotube.*

Steric Hindrance

This term refers to the compression that a molecule group (chemical entity) suffers by being too close to its nonbonded neighbors. If an enzyme and a substrate try to come together in order to react, but the substrate has on it a bulky group that disallows close contact between the two (because the group bumps into the enzyme), then the reaction will not occur because of steric hindrance. Seen in another way, two chemical groups bump into each other and cannot get by each other because they are held in place by the bonds binding them to other atoms. Hindrance of movement or activity occurs because chemical groups bump into each other and cannot occupy the same space. See also *Repression (of an enzyme), Enzyme, Inhibition, Corepressor, Structure–activity models.*

Sterile (Environment)

One that is free of any living organisms or spores. For example, a hypodermic needle that has been sterilized (e.g., by heating it) and is free of living microorganisms is said to be sterile. See also *Microorganism.*

Sterile (Organism)

One that is unable to reproduce. For example, a bull which has been castrated is rendered sterile. See also *Triploid, Barnase.*

Sterilization

See *Sterile (environment), Sterile (organism).*

Steroid

A category of chemical compounds composed of a series of four carbon rings joined together to form a (molecular) structural unit called cyclopentanoperhydrophenanthrene. Any of a group of naturally occurring, fat-soluble substances, essential to life, usually classed as lipids. Steroids of importance to the body are the sterols, which are bile acids (produced by the liver, characterized by the presence of a carboxyl group in the molecule's side chain), and the hormones of the sex glands and the adrenal cortex. In addition, the plant kingdom possesses a wide variety of steroid glycosides. See also *Glycoside, Lipids, Hormone, Cholesterol, Sterols, Saponins, Cortisol, Brassinosteroids.*

Sterols

Solid alcohols consisting of ring-structured molecules (i.e., a "ring" made of atoms). Evidence suggests that human consumption of certain phytosterols (i.e., sterols produced in plant seeds) can help to prevent certain types of cancers and can help lower levels of total blood serum cholesterol and LDLP, thereby reducing risk of CHD.

Evidence indicates that those phytosterols interfere with absorption of cholesterol by the intestines, and they decrease the body's

S

recovery and reuse of cholesterol-containing bile salts, which causes more cholesterol to be excreted from the body. Elevated levels in the body of those phytosterols activate LXRs (Liver X Receptors), which thereby act as *cholesterol sensors*; transactivating a "family" of genes that collectively control the catabolism (i.e., breakdown to yield energy), transport, and elimination of cholesterol, thereby lowering the body's blood cholesterol levels.

During 2000, researcher Joseph Judd fed phytosterols extracted from soybeans (*Glycine max* L.) to human volunteers who were already consuming a "low fat" diet. Their total blood serum cholesterol and LDLP levels decreased by more than 10%, in a short time.

During 2001, the United States' FDA approved a (label) health claim that associates the consumption of plant sterols with reduced blood cholesterol content and with reduced CHD.

Some of the sterols known to impart health benefits when consumed by humans include β-sitosterol (beta-sitsterol) and squalene.

Research published in 2002 by Dr. Leon Simons et al. reported that when such sterols are consumed while the humans are also taking relevant statin pharmaceuticals, the beneficial impact on their LDLP levels was more than the impact of either one alone (i.e., the beneficial impacts are *additive*). See also *Phytosterols, Steroid, Cholesterol, Receptors, Liver X receptors* (*LXR*), *Bile acids, Bile, Sitostanol, Soybean plant, Campesterol, Stigmasterol, Beta-sitosterol, Coronary heart disease* (*CHD*), *Low-density lipoproteins* (*LDLP*), *Food and Drug Administration* (*FDA*), *Protein, Transactivating protein, Catabolism, Transactivation, Statins.*

Sticky Ends

Complementary single strands of DNA (deoxyribonucleic acid) that protrude from opposite ends of a DNA duplex or from ends of different DNA duplex molecules. They can be generated by staggered cuts in DNA (via Type II restriction endonucleases). They are called "sticky" because the exposed single strands can bind (stick) to complementary single strands on another DNA molecule. A hybrid piece of DNA is hence produced (by that binding). See also *Staggered cuts, Hybridization* (*molecular genetics*), *Duplex, Anneal, Deoxyribonucleic acid* (*DNA*), *Blunt-end ligation, Restriction endonucleases, Complementary* (*molecular genetics*).

Stigma

From the Latin *stigma* meaning *a mark* or *a spot*. It refers to the "female" (i.e., receptive) portion of the style (i.e., the thin column of tissue arising from top of ovary) within a plant's flower. From the stigma located at the tip of that column, a *long slender pollen tube grows* (from pollen grains that land and attach to stigma) *down to fertilize the* (*embryo* within ovary).

In many land plants' flowers, the pollen must come from a *different* plant of that species in order for the pollen to successfully fertilize (embryos in that plant ovary). However, some plants such as the soybean plant (*Glycine max* (L.) Merrill) are self-pollinated. See also *Style, Fertilization, Self-pollination, Species, Soybean plant.*

Stigmasterol

A phytosterol that is produced within the seeds of the soybean plant (*Glycine max* L.), among others. Evidence indicates that human consumption of stigmasterol helps to reduce levels of total serum cholesterol and LDLP, thereby lowering risk of CHD.

Evidence indicates that certain phytosterols (including stigmasterol) interfere with absorption of cholesterol by the intestines and decrease the body's recovery and reuse of cholesterol-containing bile salts, which causes more cholesterol to be excreted from the body. See also *Phytosterols, Phytochemicals, Sterols, Soybean plant, Cholesterol, Campesterol, Beta-sitosterol, Coronary heart disease* (*CHD*).

Stimuli-Responsive Polymers

See *Tissue culture.*

Stimulus-Triggered Acquisition of Pluripotency (STAP) Cells

Term coined in 2014 by Haruko Obokata and colleagues, when they published research asserting they had caused certain cells to revert back to pluripotent state by immersing those cells in mild acid or hypoxic (i.e., low oxygen) environment.

As they age, many tissues tend to eventually lose much of their ability to rejuvenate damaged (e.g., muscle) tissue via the self-renewal proliferation of their embedded stem cells that then move to damaged areas to repair it by differentiating into (e.g., muscle-progenitor cells in this example). The 2014 research by Obokata et al. asserted that aged embedded stem cells in certain tissues can be restored to such STAP (self-renewal/proliferation) cell status via exposure to a slightly acidic or hypoxic environment. See also *Cell, Stem cells, Pluripotent stem cells, Differentiation.*

STM

Acronym for *Scanning Tunneling Microscope.* See *Scanning tunneling microscope.*

Stomatal Pores

From the Greek word *stoma* meaning *mouth*, these are openings in the epidermis (surface/"skin") of plant leaves that allow gases and water vapor to enter and exit the leaves.

The ion channel known as SLAC1 is located in the outer cell membrane of the cells constituting stomatal pores. SLAC1 is utilized by plants to control the opening and closing (e.g., in response to drought conditions) of the stomatal pores. Because those stomatal pores must be open *enough* to allow carbon dioxide to enter the plant leaves (i.e., it is needed for photosynthesis) and for oxygen to enter/exit the leaves—but must not allow *too much* water vapor to exit the leaves (especially during drought conditions)—the survival of a plant depends on the precise regulation of stomatal pore openings in response to environmental stimuli that it achieves via SLAC1. See also *Ion channels, GPA1, Abscisic acid.*

Stop Codon

See *Termination codon.*

Storage Proteins

Refers to proteins whose molecules are utilized as a "store" (for a time) of amino acids to be consumed later. For example, the storage protein known as casein (i.e., in mammal mother's milk) is such a "store" source of amino acids for baby mammals. See also *Protein, Amino acids, P34 protein.*

STORM

Acronym for sub-diffraction-limit imaging by *ST*ochastic *O*ptical *R*econstruction *M*icroscopy. It is a special type of microscopy which "builds" an optical image out of the sample's individual photon emissions of fluorescent molecules ("switched on" by the scientist utilizing various means). Each point of light thereby detected is "built" by STORM apparatus into an image with an effective resolution of 20 nanometers (nm), which is far smaller than can be achieved via a conventional light microscope. STORM can be used to "see" *intra*cellular details (i.e., within a cell). See also *Cell, Fluorescence, Nanometers (nm).*

STR

Acronym for short tandem repeats. See *Short tandem repeats.*

STR Markers

See *YSTR DNA*

Strain (When Referring to a Prion)

One molecular shape type of a given prion, which typically is the cause of one particular neurological disease. See also *Prion, BSE,* PrP^c, PrP^{Sc}.

Strain (When Referring to an Organism)

A group of organisms of the same species that possesses distinctive genetic characteristics which set it apart from others within the same species, but which differences are not "severe" enough for it to be considered a different breed or variety (of that species). The basic taxonomic unit of microbiology. The word strain can also be used to designate a population of cells derived from a single cell. See also *Species, Cell, Clone (an organism).*

Streptavidin

A bacteria-origin protein which possesses natural anticancer properties (e.g., it causes human promyelocytic leukemia cells to die when pure streptavidin enters those cells).

Streptavidin has a specific and high affinity for biotin (i.e., it "sticks" tightly to the biotin molecule). This can be utilized by researchers to:

- "Label" certain large molecules of interest, by attaching biotin molecules to them via a chemical reaction (this is known as *biotinylation*).
- Similarly attach a fluorophore, enzyme, colored bead/quantum dot, and so on, to molecules of streptavidin.
- Apply the specific high affinity of streptavidin–biotin for research/diagnostics within microarrays, affinity chromatography, and other separation methodologies, and so on.

See also *Protein, Cell, Cancer, Biotin, Molecular bridge, Fluorophore, Enzyme, Quantum dot, Microarray (testing), Affinity chromatography, Carbon nanotubes, Bacteria, Carbohydrate microarrays.*

Streptococcus

Refers to bacteria of the genus *Streptococcus.* Among the diseases that can be caused by some strains of *Streptococcus* is *necrotizing fasciitis* (so-called flesh eating bacteria disease), which is caused by *Group A Streptococcus.* See also *Bacteria, Genus, Streptococcus mutans.*

Streptococcus mutans

The strain of *Streptococcus* bacteria that grows on the surface of teeth and can contribute to causing tooth "decay." See also *Strain, Bacteria, Streptococcus.*

Stress Erythropoiesis

See *Erythropoiesis, Macrophage.*

Stress Hormones

Refers to hormones (or phytohormones) which signal an organism (e.g., plant) to respond in a defensive way to a particular environmental stress. For example, in response to drought stress, plants can initiate multiple hormone response pathways involving production of strigolactone, abscisic acid, ethylene, and cytokinins(s) to reduce growth of new shoots and result in additional moisture-conserving plant responses. See also *Hormone, Phytohormone, Plant hormone, Abscisic acid, Pathway, Ethylene.*

Stress Proteins

Discovered by Italian biologist Ferruchio Ritossa in the 1960s, these molecules are also called heat-shock proteins (HSP). They are proteins made by many organisms' (plant, bacteria, and mammal) cells when those cells are stressed by environmental conditions such as certain chemicals, pathogens, or heat. Among other functions, HSPs act to chaperone irreparably misfolded/abnormal proteins out of cells.

HSPs also function as molecular chaperones to protect protein molecules from folding prematurely/incorrectly (within the cells they are produced in).

When corn/maize (*Zea mays* L.) is stressed during its growing season by high nighttime temperatures, that plant switches from its normal production of ("immune system" defense) chitinase to production of heat-shock (i.e., stress) proteins, instead.

Stress proteins are also produced by tuberculosis and leprosy bacteria after these bacteria have invaded (i.e., infected) cells in the human body, in an attempt by those bacteria to mimic the stress proteins that (mammal) cells would normally manufacture to repair damage done to the (mammal) cells. This mimicry makes it more difficult for the immune system to recognize and attack those pathogenic bacteria (and/or repair misshaped protein molecules in the body's cells).

Similarly, production of stress proteins helps some types of cancer cells to avoid being attacked by the immune system. Because consumption of genistein by humans causes a reduction in the production of stress proteins, genistein may thereby help the human immune system to destroy cancerous cells.

In 1996, Richard I. Morimoto discovered that two stress proteins known as HSP 90 and HSP 70 help to ensure that certain crucial proteins in cells are folded into the configuration/conformation needed by that cell. See also *Antigen, Immune response, Pathogen, Protein, Protein folding, HSP 90 conformation, Chaperones, Protein structure, Heat shock proteins, Absolute configuration, Prion, Chitinase, Aflatoxin, Genistein, Cancer, Lipoxygenase (LOX), Phytoalexins, Hormesis, tRNA synthetase.*

S

Stress Response

See *Stress proteins*, *Stress hormones*.

Stress Response Proteins

See *Stress proteins*.

Stress Responsive Proteins

See *Stress proteins*.

Strigolactone

See *Stress hormones*.

Stroma

See *Stromal cells*.

Stromal Cells

See *Multipotent*.

Stromelysin (MMP-3)

A collagenase (enzyme) that "clears a path" through living tissue, ahead of tumor cells, thereby enabling a cancer to spread within the body. See also *Collagenase*, *Enzyme*, *Cancer*, *Tumor*.

Structural Biology

Refers to the study of *molecular physical structures* and their impact on life processes (e.g., whether or not a given protein molecule can bind to another, how "tightly" an antibody or an enzyme bind to the molecules they each act upon, etc.).

For example, *enzyme molecules possessing structures that make those molecules more rigid* tend to have greater resistance to denaturation caused by high temperatures. Molecular structure features imparting such rigidity include hydrogen bonds, salt bridges (i.e., between two portions of the molecule), and hydrophobic interactions at particular locations on the molecule which prevent the molecule from unfolding at elevated temperatures. See also *Nuclear magnetic resonance*, *Gene*, *Structural gene*, *Homology modeling*, *Protein*, *Enzyme*, *Active site*, *Catalytic site*, *Antibody*, *Avidity (of an antibody)*, *Primary structure*, *Protein folding*, *Conformation*, *Native conformation*, *Tertiary structure*, *Intrinsically unstructured proteins*, *Steric hindrance*, *Substrate (chemical)*, *Structure–activity models*, *Cytoskeleton*, *Domain (of a protein)*, *Denaturation*, *Enzyme denaturation*, *Hydrophobic*, *Van der waals forces*.

Structural Gene

A gene that codes for any RNA (ribonucleic acid) or protein product other than a regulator molecule. It determines the primary sequences (i.e., the amino acid sequences) of a polypeptide (protein). See also *Gene*, *Express*, *Polypeptide (protein)*, *Amino acid*, *Primary structure*, *Ribonucleic acid (RNA)*.

Structural Genomics

Study of or discovery of where (gene) sequences are located within the genome, and what (DNA) subunits comprise those sequences. See also *Gene*, *Sequence (of a DNA molecule)*, *Deoxyribonucleic acid (DNA)*, *Sequencing (of DNA molecules)*, *Genome*, *Genomics*, *Primary structure*, *Structural biology*.

Structural Proteomics

See *Proteomics*, *Structural biology*, *Structure–activity models*.

Structure–Activity Models

Refers to models (e.g., of protein molecules) from which their biological activity (e.g., impact on a cell's metabolism, etc.), ligand binding sites, steric hindrance, avidity, and so on, can be calculated or inferred. See also *Protein*, *Conformation*, *Native conformation*, *Tertiary structure*, *Biological activity*, *Cell*, *metabolism*, *Protein folding*, *Disulfide bond*, *Ligand (in biochemistry)*, *Steric hindrance*, *Avidity (of an antibody)*, *Structural biology*, *Quantitative structure–activity relationship (QSAR)*.

STS Sulfonylurea (Herbicide)-Tolerant Soybeans

These are soybeans that have been bred (via insertion of *ALS* gene by traditional breeding methods) to resist the (weed killing) effects of sulfonylurea-based herbicides. The *ALS* gene was discovered by Scott Sebastian in 1986. See also *Gene*, *Genetic engineering*, *HTC*, *ALS*, *ALS gene*, *BAR gene*, *PAT gene*, *EPSP synthase*, *Glyphosate oxidase*, *Herbicide-tolerant crop*.

Stx

Shiga-like toxins. See also *Toxin*, *Toxigenic E. coli*, *Enterohemorrhagic E. coli*, *Escherichia coliform 0157:H7 (E. coli 0157:H7)*.

Style

From the Greek word *stylos* meaning *column*, it is the thin hollow column of tissue arising from (top of ovary) within a plant's flower. At the tip of that column is the stigma (pollen-receptive spot) via which a *long slender pollen tube grows* (from pollen grains that land and attach to stigma) *down to fertilize the (embryo* within ovary).

In many land plants' flowers, the pollen must come from a *different* plant of that species in order for the pollen to successfully fertilize (embryos in that plant ovary). However, some plants such as the soybean plant (*Glycine max* (L.) Merrill) are self-pollinated. See also *Fertilization*, *Self-pollination*, *Species*, *Soybean plant*, *Stigma*.

Subculture Interval

Refers to the elapsed chronological time between separate/subsequent times that cell(s) have been transplanted (i.e., transferred) from one ongoing cell culture (e.g., in a fermentation vat) to a new one (where growth of the culture is continued). See also *Cell*, *Cell culture*, *Mammalian cell culture*, *Subculture number*.

S

Subculture Number

Refers to the number of separate/subsequent times that cell(s) have been transplanted (i.e., transferred) from one ongoing cell culture (e.g., in a fermentation vat) to a new one (where growth of the culture is continued). See also *Cell, Cell culture, Mammalian cell culture*.

Subspecies

See *Strain*.

Substance K

See *Tachykinins*.

Substance P

A neuropeptide (i.e., peptide produced by cells of the nervous system) which is involved in activation of the immune system, pain sensation, and (when in excess) some psychiatric disorders. In the case of chronic, intractable pain (hypersensitivity), approximately one percent of the nerve cells in the human spine process substance P (thereby "transmitting" its pain message via signal transduction). In 1997, Patrick Mantyh showed that killing those (one percent) cells relieved chronic pain hypersensitivity without impairing sense of touch or normal (beneficial) pain sensation, in humans. See also *Tachykinins, Protein, Polypeptide (protein), Signal transduction, Signaling, Peptide, Neurotransmitter.*

Substantial Equivalence

Refers to a foundation concept in the safety evaluation of biotechnology-derived crops by national government regulatory agencies. If the composition (nutrients, antinutritional factors, etc.) of a biotechnology-derived crop is found to be equivalent to that of nonbiotech varieties of the same crop, and those nonbiotech crop varieties are considered safe, then further safety assessment of the transgenic crop can focus on only the intended modification (e.g., the expression of a transgenic protein that is novel in that crop). See *Canola, Biotechnology, Organization for Economic Cooperation and Development (OECD), Antinutritional factors.*

Substantially Equivalent

See *Substantial equivalence.*

Substitution

Refers to a point mutation within a DNA molecule, in which the original base pair at that locus is substituted for a different base pair. See also *Deoxyribonucleic acid (DNA), Point mutation, Locus, Base pair (bp), Base substitution.*

Substrate (Chemical)

The substance acted upon, for example, by an enzyme. For example, the enzyme amylase catalyzes the breakdown of starch molecules into glucose polysaccharide molecules; starch is the substrate (of the enzyme amylase). See also *Enzyme, Amylase, Catalyst, Substrate (structural), Luciferin.*

Substrate (in Chromatography)

The (usually solid or gel) substance that attracts and noncovalently binds (interacts) with one or more of the molecules in a solution that is passed over that substrate (e.g., in a chromatography column). This preferential binding (interaction with the substrate) enables one or more of the solution's molecular ingredients to be separated from the other(s). See also *Chromatography, Monolithic chromatography substrates.*

Substrate (Structural)

The substance (support) to which the agent of interest (e.g., molecule, cell, etc.) is attached. For example, some catalyst molecules are chemically attached to nonreactive solids to preserve the catalyst from being flushed away when the chemical substrate (the molecule to be converted by the catalyst) is washed by the catalyst immobilized on the structural substrate. See also *Substrate (chemical), Catalyst, Hybridization surfaces, Extracellular matrix, Biofilm.*

Sudden Death Syndrome

A plant disease caused by the soilborne *Fusarium virguliforme* fungus that sometimes afflicts soybean plants. See also *Soybean plant, Soybean cyst nematodes (SCN).*

Sugar Molecules

See *Oligosaccharides, Polysaccharides, Monosaccharides, Carbohydrates, Aldose, Glycobiology, Pyranose, Glucose (GLc), Furanose, Glycoprotein, Sialic acid.*

Suicide Genes

See *Gene, P53 gene, Apoptosis.*

Sulfate-Reducing Bacterium

See *SRB (Sulfate-reducing bacterium).*

Sulfenic Acid

See *Alicin.*

Sulfolobus solfataricus

An archaeon (i.e., single-celled organism) that naturally lives in and near certain hot volcanic springs (known as *sulfurous cauldrons*) which have significant amounts of sulfur in their waters. *Sulfolobus solfataricus* can metabolize sulfur into sulfuric acid. See also *Archaea, Metabolism, Extremophilic bacteria, Thermoacidophilic bacteria.*

Sulforaphane

An isothiocyanate-breakdown compound that is naturally produced (from glucoraphanin) within cruciferous plants such as broccoli, cabbage, and kale. Also is present within horseradish.

Consumption of sulforaphane results in higher levels of (induced) Phase II detoxification enzymes in the digestive system. Research indicates that human consumption of significant amounts

S

of sulforaphane helps to lower the risk of several cancers (e.g., prostate, colon, breast, lung, and skin cancers). Sulforaphane inhibits HDACs, which are enzymes that can interfere with the normal function of genes that suppress tumors.

Consumption of sulforaphane also helps to prevent inflammatory processes associated with certain respiratory diseases such as asthma, inhalation of certain air pollutants (e.g., prelude to emphysema), chronic obstructive pulmonary disease, and so on. Research has shown that sulforaphane can kill *Helicobacter pylori* bacteria. Recent research has shown that sulforaphane can prevent *quorum sensing* by/of certain pathogenic bacteria (thus preventing virulence in body of certain diseases).

Research published by Duxin Sun and Max Wicha in 2010 indicated that when injected into mouse cell culture, the chemical known as sulforaphane (extracted from cruciferous plants such as broccoli, cabbage, kale, etc.) induces apoptosis in mouse *breast cancer stem cells* (i.e., the progenitor cells from which all resultant mouse breast cancer cells in that cell culture would be "descended"). See also *Nutraceuticals, Phytochemicals, Gene, Tumor, Cancer, H. pylori, Quorum sensing, Enzyme, Inducible enzymes, Phase II detoxification enzymes, Apoptosis, Histone deacetylases.*

Sulfosate

An active ingredient in some herbicides, it kills plants (e.g., weeds) by inhibiting the crucial plant enzyme EPSP Synthase.

Chemically, sulfosate is a trimethylsulfonium salt of the same organic acid as glyphosate, so sulfosate can be applied over crops (e.g., soybeans) that have been genetically engineered to be tolerant to glyphosate-based herbicides. See also *Enzyme, EPSP synthase, CP4 EPSPS, Glyphosate, Acid, Soybean plant, Herbicide-tolerant crop, Genetic engineering.*

Sulphoraphane

See *Sulforaphane.*

SUMO

Acronym for *Small Ubiquitin-Related Modifier.* See *Small ubiquitin-related modifier.*

Sumoylation

See *SUMO.*

Superantigens

Certain types of antigens that activate a large proportion of an organism's immune system T cells. These superantigens, which thus overactivate the organism's immune system, are thought to be responsible for some autoimmune diseases (in which T cells attack and destroy the organism's own, healthy tissues). See *Antigen, T cells, Autoimmune disease.*

Supercoiling

Also known as superhelicity, it refers to the coiling of a closed duplex DNA (deoxyribonucleic acid molecule) in space so that it crosses over its own axis. In other words, it would be a DNA molecule with "extra" twists.

See *Deoxyribonucleic acid (DNA), Helix, Duplex, Double helix, Positive supercoiling, Overwinding, Negative supercoiling.*

Supercritical Carbon Dioxide

A solvent that, when combined with water and an appropriate surfactant (e.g., fluoroethers), forms a solvent system that can effectively dissolve large biological molecules without causing those molecules to lose biological activity. Carbon dioxide is a gas at normal (atmospheric) pressure and ambient temperature, but in its *supercritical state*—temperature above 31.3°C (88°F) and pressure greater than 72.9 atmospheres—carbon dioxide becomes a dense (sort of) liquid. Some coffee processors have used supercritical carbon dioxide as a solvent to remove caffeine from coffee.

In 1995, Keith Johnston added the surfactant ammonium carboxylate perfluoropolyether to a supercritical carbon dioxide system containing water and proved that the large biological molecule bovine serum albumin dissolved inside the micelles that form via water droplet surrounded by fluoroether molecules. Subsequent to that, Eric Beckman proved that the protease *subtilisin Carlsberg* can be extracted from crude (impure) cell broth because that protease preferentially dissolves in a supercritical carbon dioxide/water system containing fluoroether amphiphiles as surfactants. See also *Biological activity, Surfactant, Micelle, Reverse micelle (RM), Broth, Protease, Supercritical fluid, Albumin, Amphiphilic molecules.*

Supercritical Fluid

Refers to a material that has been heated to a temperature above its (normal atmospheric pressure) boiling point, but which is kept in a state that resembles a liquid via the application of high pressure. Less commonly—refers to a liquid that has been cooled to a temperature below its normal freezing point, but which is kept in a liquid state by various means.

For example, water will remain "liquid" up to a temperature of 375°C (617°F) if it is placed under enough pressure. Ammonia will remain "liquid" up to a temperature of 133°C (271°F) if it is placed under enough pressure, despite the fact that ammonia normally becomes a gas (at std. atmospheric pressure) whenever the temperature is higher than −33.35°C (−30°F).

One predatory mite (*Alaskozetes antarcticus*) living in Antarctica is able to survive subfreezing temperatures by preventing ice crystals from forming (i.e., supercritical water) inside its body, even when the environmental temperature is below the freezing point (i.e., supercritical). Most supercritical fluids have unique physical properties (e.g., they are often better solvents than their true liquid forms). Some supercritical fluids (e.g., supercritical carbon dioxide) can be used to extract biological molecules (e.g., chlorophyll) from mixtures (e.g., ground up plant leaves). After the biological molecule has dissolved out of the mixture, the biological molecule is recovered by releasing pressure so the carbon dioxide returns to gaseous form and drifts away. See also *Supercritical carbon dioxide.*

Superoxide Dismutase (SOD)

See *Human superoxide dismutase (hSOD).*

Superparamagnetic Nanoparticles

See *Nanoparticles.*

S

Super-Resolution Optical Imaging

See *STORM, STED, SIM, PALM.*

Suppressor Gene

A gene that can reverse the effect of a specific type of mutation in other genes, such as a premature termination sequence. See also *Gene, Transwitch®.*

Suppressor Mutation

A mutation that totally or partially restores a function that was lost by a primary mutation. It is located at a site in the gene different from the site of the primary mutation. See also *Gene, Mutation.*

Suppressor T Cells

Those T cells (thymus-derived lymphocytes) that are triggered (after other types of T cells and other immune system cells have successfully fought off an infection) to gradually slow down and halt the body's immune response (to the now-conquered pathogen). Discovered by Tomio Tada in 1971, suppressor T cells suppress B cell activity. Failure to halt the immune response in time could lead to harm to the body by its own immune system. The B and T lymphocytes are indistinguishable in size and general morphology. Only the existence or nonexistence of certain proteins (e.g., CD4, CD25) on their cell surfaces distinguishes the two classes of lymphocytes.

Today, most scientists refer to (former) suppressor T cells as *Regulatory T Cells*. See also *Cellular immune response, Pathogen, B lymphocytes, T cells, Autoimmune disease, Regulatory T cells.*

Supramolecular Assembly

Phrase utilized to refer to a very large molecular structure. See *Self-assembly (of a large molecular structure), Optical tweezer, Nanowire.*

Surface Plasmon Polaritons

See *Surface plasmons.*

Surface Plasmon Resonance (SPR)

A testing technology which enables real-time detection of interactions (e.g., "binding"/ligand, etc.) between protein molecules (attached to a gold surface on a sensitive glass "sensor chip") and other molecules (e.g., pharmaceutical candidate compounds, toxins, etc.) passed over those attached protein molecules.

When certain metal surfaces (e.g., gold, silver, etc.) are struck by relevant wavelength light, *electromagnetic charge oscillations known as surface plasmons* are generated. By shining a highly focused beam of light (e.g., laser, or polarized, etc.) onto the bottom of the gold surface (i.e., through the transparent glass "chip"), and measuring changes in refractive index of the chip/gold/reflected light, the *mass change* (e.g., caused by pharmaceutical molecule binding to an attached protein such as an antibody or receptor-target) *can be determined* from the SPR it induces on gold surface. See also *Surface plasmons, Protein, Ligand (in biochemistry), Antibody, Receptors, Protein interaction analysis, Target–ligand interaction screening, Nanorods, Plasmonic nanohole arrays.*

Surface Plasmons

Refers to propagating excitations that occur when relevant wavelength light couples with collective oscillations of electrons on the surface of a metal—a sort of nanometer-scale "shock wave" (electromagnetic excitation) which occurs when light of applicable wavelength is shined onto certain metal surfaces (i.e., the plasmons are caused by the light waves striking the "free" electrons at surface of those metal atoms).

When such light is shined through tiny holes in those metals, such as through a nanometer-scale metal mesh, MORE LIGHT EMERGES FROM THE FAR SIDE OF THE NANO-MESH THAN WOULD BE PREDICTED VIA CLASSICAL OPTICS. That is because—in addition to the light which actually *passes through* the mesh openings—some of the light which strikes the "edge of the openings"

- Is converted to surface plasmons.
- Which transit the thickness of the metal mesh in the form of surface plasmons *traveling through the free electrons that rim each mesh opening* while also creating large electric fields around the openings.
- Then are reconverted into light, on the far side of the metal mesh.

Materials possessing this effect as a result of the tiny holes are known as *plasmonic metamaterials*. See also *Nanometers (nm), Nanoscience, Nanotechnology, Surface plasmon resonance (SPR), Nanorods, Plasmonic nanohole arrays.*

Surfactant

Acronym for surface active agent. Amphipathic molecules (i.e., molecules that contain both a polar and nonpolar domain) which, due to their unique properties, position themselves at interfacial regions (surfaces) such as an oil/water interface. When surfactants are dissolved above a certain critical concentration in either water or nonpolar solvents they may form micelles or reverse micelles, respectively. Surfactants are commonly used to solubilize cell membrane components and other hard to solubilize molecules. See also *Amphipathic molecules, Amphiphilic molecules, Micelle, Reverse micelle (RM), SDS, Adjuvant (to a herbicide).*

Sustainable Development

Defined in the 1987 United Nations report OUR COMMON FUTURE to be development (e.g., economic development) that meets the needs of the present without compromising the ability of future generations to meet their own needs. See also *Conservation tillage, Glomalin, No-tillage crop production, Low-tillage crop production, Earthworms.*

Switch Proteins

Refers to certain protein molecules which signal a plant when environmental conditions are so dry (or cold, etc.) that the plant need to protect itself (via extreme measures) to survive. See also *Trehalose, Protein, Signaling, Transcription factors, CBF1, Sequence (of a DNA molecule), Regulatory sequence.*

S

Switching (e.g., On/Off) of Genes

See *Gene*, *Genetic code*, *Coding sequence*, *Deoxyribonucleic acid (DNA)*, *Sequence (of a DNA Molecule)*, *Regulatory sequence*, *Transcription factors*, *CBF1*, *Cold hardening*, *Cessation cassette*.

SWNT

Acronym for *single-walled carbon nanotube*. See *Carbon nanotubes*.

Syk Protein

See *Mast cells*.

Symbionts

See *Symbiotic*.

Symbiotic

Refers to the mutually beneficial living together of organisms, in an intimate association or union. For example, *lichens* are a life form consisting of algae and a fungus growing together as a unit on a solid surface (e.g., a tree trunk or a rock). Each helps the other to survive and grow.

For example, *mycorrhizae* consist of plant roots and a fungus growing together as a unit within soil. Each helps the other to survive and grow.

During 2002, Regina Redman and Russell Rodriguez discovered that *Curvularia protuberata* fungi (which live inside *Dichantelium lanuginosum* grass that grows in hot soils adjacent to magma-heated geysers) impart heat tolerance to the grass they live in when those fungi are themselves infected with *Curvularia* thermal tolerance virus (CThTV). Redman, Rodriguez, and Joan Henson were later able to show that when these CThTV-infected fungi were inserted into tomato and watermelon seedlings, those plants/roots were also able to withstand far higher temperatures than before. See also *Algae*, *Fungus*, *Rhizobium (bacteria)*, *Mycorrhizae*, *Pink-pigmented facultative methylotroph (PPFM)*, *Pharmacoenvirogenetics*, *Antibiosis*, *Virus*.

Synapse

From the Greek, meaning *point of contact*. They are contacts between nerves within the brain. See *Dendrites (in brain)*.

Synthase

See *ACC synthase*, *ATP synthase*, *EPSP synthase*, *Enzyme*, *CP4 EPSPS*, *Citrate synthase (CSb) gene*, *Glutamine synthetase*, *ALS gene*, *Low-phytate soybeans*, *Thermotolerant wheat*.

Synthase Inhibitors

See *ATP synthase*.

Synthesizing (of DNA Molecules)

The building (i.e., polymerization manufacture) of a known sequence of nucleotides into a chain called an oligonucleotide (of which genes are made) or DNA (deoxyribonucleic acid). Invented by Har Goribind Khorana and his colleagues at the University of Wisconsin–Madison in 1968, this process enables scientists to create genes or gene fragments for use in research.

In 1973, Robert Bruce Merrifield developed a means to partially automate the oligonucleotide assembly process. This led to automated machines that can now rapidly manufacture a gene fragment, gene, or DNA probe. See also *Gene machine*, *Nucleotide*, *Oligomer*, *Oligonucleotide*, *Synthesizing (of proteins)*, *Deoxyribonucleic acid (DNA)*, *DNA probe*, *Synthesizing (of oligosaccharides)*.

Synthesizing (of Oligosaccharides)

Chemical synthesis (i.e., manufacture) of a known oligosaccharide (structure). For example, a synthesis of a defined-sequence oligosaccharide (molecular) "branch" at a specific site on a glycoprotein in order to "cover up" an antigenic site on that glycoprotein molecule (e.g., so the glycoprotein can be used as a pharmaceutical). See also *Oligosaccharides*, *Glycoprotein*, *Antigen*, *Antigenic determinant*, *Restriction endoglycosidases*.

Synthesizing (of Proteins)

Chemical synthesis (manufacture) of a known protein molecule. Devised based upon the solid phase synthesis methodology developed by Robert Bruce Merrifield in 1963, the desired proteins are assembled by repetitive coupling of the constituent amino acids to a growing polypeptide backbone which itself is attached to a polymeric support (substrate). This procedure has been automated, so it is now possible to make proteins via automated synthesizers. See also *Protein*, *Polypeptide (protein)*, *Amino acid*, *Substrate (structural)*, *Combinatorial chemistry*, *Synthesizing (of DNA molecules)*, *Copy number (protein molecules)*.

Synthetase

See *Synthase*.

Synthetic Biology

Refers to the study and modification of biology/life via utilizing engineering and molecular biology to create synthetic versions of "modular parts" of it:

1. To observe how those parts "work." For example, during 2003 Hamilton O. Smith and J. Craig Venter (re)created the phiX174 bacteriophage by synthesizing its 5386 base pairs from original chemicals. They discovered that it functioned just like other phiX174 bacteriophages.

   or
2. To add new function(s) to existing living organisms. For example during 2011, scientists in Colombia *created a modular* (made from two different genetically modified *E. coli* bacteria) *"detect, amplify and alert" system*, to function as a defense aid for coffee plants against pathogenic fungi. The modular system's first bacteria *detects* chitin (a polysaccharide compound present in fungal cell walls) and as a result it synthesizes chitinase plus a signaling molecule that triggers the second bacteria to produce a second signaling molecule known as salicylic acid (i.e., to *amplify* the initial chitin-presence signal), and thereby *alerts* the coffee plant via that salicylic acid stimulating within the plant a hypersensitive response against fungal infection. The salicylic acid also acts to control the

(synthetic biology) modular system's *E. coli* population, because too high salicylic acid concentration is toxic to those two bacteria.

3. To model, design, and build synthetic gene "circuits" and other biomolecular components in order to "rewire" cells/organisms for (useful to man) purposes.

For example, during 2014 George Church and colleagues modified *E. coli* bacteria so that only those bacteria producing a particular bio-based industrial chemical (selected by George Church) proliferated, because their DNA had been modified so that antibiotic-resistant genes were activated, but only in the presence of that selected industrial chemical, in subsequent culture medium that contained relevant antibiotic. Only the most productive *E. coli* cells generate enough of the industrial chemical to be totally resistant to the antibiotic and survive to the next round of (directed) evolution. When Dr. Church and his colleagues subsequently exposed the *E. coli* bacteria to an antibiotic that killed all but the most efficient of those (industrial) chemical-producing bacteria, the bacterial production of that industrial chemical increased 30-fold. When it is thus utilized to improve an organism's output (e.g., of an industrial chemical) against a self-sought (by the living organism across multiple generations) goal, this particular application of synthetic biology is known as *directed evolution*.

During 1990, Peter G. Schultz had developed cells whose DNA coded for amino acids *additional to* the 20 naturally occurring amino acids. During the late 1990s, Drew Endy developed numerous distinct segments of DNA which *both* code for specific desired proteins *and also work well together* (when both of those segments are inserted into a given cell to each do a specific task). And such "parts"/working together in living systems have been extensively modeled *in silico*.

The recipient organism (of the above-delineated changes) is referred to as a *chassis,* and the particular (modular) introduced/modifying DNA is referred to as a *bio brick*. See also *Biology, Organism, Bacteriophage, Protein, Synthesizing (of proteins), Enzyme, Base pair (bp), Cell, Deoxyribonucleic acid (DNA), Synthesizing (of DNA molecules), Sequence (of a DNA molecule), DNA shuffling, Coding sequence, Bacteria, Antibiotic, Escherichia coliform (E. coli), Amino acid, In silico biology, Farnesene, Chitin, Chitinase, Signaling molecule, Salicylic acid (SA), Hypersensitive response, Receptor engineering, Directed evolution.*

Systematic Activated Resistance

See *Systemic acquired resistance (SAR).*

Systematics

An extension of taxonomy, it is the scientific classification of living organisms. See also *Organism.*

Systemic Acquired Resistance (SAR)

Discovered in 1992 (applicable to harpin-induced SAR) and in 1996 by J.A. Ryals, U.H. Neuenschwander, M.G. Willits, A. Molina, H.-Y. Steiner, and M.D. Hunt, SAR is a sort of "immune (cascade) response" by a plant against infection (e.g., by bacteria, fungus, virus, etc.) or attack by chewing insect pests. One example of this is the production of stress proteins or pathogenesis-related proteins when certain plants are attacked by certain pathogens. Via such SAR response triggered by low-level fungal or viral infection, many plants successfully resist fungal/bacterial/viral attacks.

In 1998, the U.S. Environmental Protection Agency (EPA) approved one herbicide (COBRA™ owned by Valent Corp.), whose active ingredient is the chemical LACTOFIN, to be applied to soybean plants "at or near bloom stage" in order to trigger SAR against *white mold disease*. In 2000, the U.S. EPA approved harpin protein to be applied to some crops in order to trigger SAR against certain plant diseases. The Syngenta product known as ACTIGARD™ (Acibenzolar-*S*-methyl) and the Marrone Bio Innovations product known as REGALIA RX™ (extract of *Reynoutria sachalinensis*) also trigger some crop plants' SAR. See also *Pathogenesis-related proteins, Phytoalexins, R genes, Isoflavones, Soybean plant, Tomatidine, Fungus, Immune response, Virus, Pathogen, Stress proteins, Oxylipins, Salicylic acid (SA), Jasmonic acid, Harpin, Cascade, White mold disease, Crop biologicals, Systems biology.*

Systemic Inflammatory Response Syndrome

See *Sepsis.*

Systemic Lupus Erythematosus (SLE)

See *Lupus.*

Systems Biology

Refers to the use by scientists of engineering models that *predict how a living organism's systems* (e.g., the various gene-governed interconnected chemical pathways containing 21+ enzymes that determine a plant's lignin content and composition) *each act and interact* (e.g., with the plant's environment, with each other, etc.). Those engineering models help to guide future research and develop (e.g., to create strong fast-growing trees containing less lignin, to subsequently be utilized to make more energy-efficient biofuels, etc.).

Some of the factors that can impact on a living organism's systems include protein–protein interactions, shared genes, shared environmental factors, common treatments (e.g., certain herbicides, certain fungicides, etc.).

See also *Organism, Lignins, Gene, Gene function analysis, Gene expression profiling, Gene expression analysis, Gene expression cascade, Gene editing, CRISPR/CAS9 gene-editing systems, Genetic engineering, Linkage map, Protein, Protein-protein interactions, Protein interaction analysis, Enzyme, Pathway, Pathway feedback mechanisms, Metabolism, Metabolic engineering, Genomics, Phenomics, Systemic acquired resistance (SAR).*

Systeomics

A term coined during 2002 by the California Separation Science Society. Defined as the integration of genomics, proteomics, and metabonomics. See also *Genomics, Proteomics, Metabonomics.*

Syx Protein

See *Endothelial cells.*

T

"Treatment" IND Regulations

Food and Drug Administration (FDA) regulations promulgated in 1987, to provide a more rapid formal pharmaceutical approval mechanism than the usual IND (Investigational New Drug) regulatory approval process. Its purpose is to enable drug developers to provide promising experimental drugs to patients suffering from immediately life-threatening diseases or certain serious conditions (e.g., acquired immune deficiency syndrome, or AIDS) before complete data on that drug's efficacy or toxicity are available. See also *IND*, *Food and Drug Administration (FDA)*, *Delaney clause*, *Koseisho*, *Committee for Proprietary Medicinal Products (CPMP)*.

T Cell Growth Factor (TCGF)

Also known as Interleukin-2. See *Interleukin-2 (IL-2)*.

T Cell Modulating Peptide (TCMP)

A short protein chain that is thought to restrain certain types of T cells from attacking an (arthritis) afflicted patient's tissues (mainly cartilage). Arthritis is caused by the arthritis sufferer's own immune system attacking the body's cartilage tissues. See also *Cytotoxic T cells*, *Helper T cells (T4 cells)*, *Lymphocyte*, *Suppressor T cells*, *T cell receptors*, *Autoimmune disease*, *Tumor necrosis factor (TNF)*.

T Cell Receptors

Antibody-like transmembrane (i.e., across the cell's surface membrane) proteins located on the surface of T cells. These trigger the (cellular) immune response that is mounted by T cells when these receptors bind to antigens (foreign pieces of antigenic protein) which have been "presented" to these receptors by an MHC protein which itself is located on the surface of phagocytic (i.e., scavenging, pathogen-ingesting) B lymphocyte.

Antibodies in the blood recognize native antigen macromolecules (i.e., large molecules), whereas T cell receptors recognize fragments derived from those antigen macromolecules (upon presentation at the surface of B lymphocytes following ingestion and digestion by the B lymphocytes).

Activation of the T cell receptors leads to activation of the transcription factor NFκB. The NFκB then causes T cell proliferation and activation as part of the body's immune response. Activated T cells also direct B cells in their response (producing antibodies against the antigen).

When beta interferon later binds to T cell receptors, that binding slows down the body's immune response (e.g., normally after the immune response has defeated an infection). See also *Receptors*, *Antibody*, *Antigen*, *Major histocompatibility complex (MHC)*, *Protein*, *T cells*, *Cellular immune response*, *Phagocyte*, *B lymphocytes*, *Cytotoxic T cells*, *Helper T cells*, *Suppressor T cells*, *Transcription factors*, *NFκB*, *Beta interferon*.

T Cells

A class of (thymus- or tonsil-derived) lymphocytes which include helper T cells (also known as T helper cells or T_H cells), suppressor T cells, and cytotoxic T cells (also known as killer cells or CTL for cytotoxic T lymphocyte). These cells mediate (i.e., control/direct) the cellular response of the human immune system in very complex ways (e.g., synthesis of leukotrienes). T cells are often "switched on" by pyrexia (fever), and T cells are involved in the activation of B cells.

Follicular helper T cells (abbreviated TFH) are a subset of *helper T cells* that are specialized to B cell responses. See also *Cellular immune response*, *Cytotoxic T cells*, *Helper T cells (T4 cells)*, *CD8⁺ T cells*, *Lymphocyte*, *Suppressor T cells*, *T cell receptors*, *T cell modulating peptide (TCMP)*, *Allergies (foodborne)*, *Dendritic cells*, *Leukotrienes*, *Pyrexia*.

T Lymphocytes

See *T cells*, *Lymphocyte*, *Lymphokines*, *Thymus*.

T2D

Acronym for *Type II diabetes*. See *Type II diabetes*.

T3

See *SAM-K gene*.

T4 Cells

See *Helper T cells (T4 cells)*.

T6P

See *Trehalose 6-phosphate*.

Tachykinins

A class of neuropeptides (i.e., peptides produced by cells of the nervous system; neurons) that includes neurokinin A, neurokinin B, eledoisin, physalaemin, kassinin, substance P, and substance K. Some of these neuropeptides (e.g., substance P) are picked up by mast cells, lymphocytes, and/or monocytes and cause those three types of immune system cells to release certain lymphokines (e.g., tumor necrosis factor (TNF), interleukin-1, etc.), thus activating the immune system. See also *Mast cells*, *Lymphocyte*, *Monocytes*, *Tumor necrosis factor (TNF)*, *Interleukin-1 (IL-1)*.

TAG

See *Triacylglycerols*.

T

Tagged (Molecules or Cells)

Also sometimes referred to as *labeling (molecules or cells)*. See *Affinity tag, Affinity, Affinity chromatography, Expressed sequence tags (EST), Bacterial expressed sequence tags (BEST), Label (fluorescent), Label (radioactive), Probe, Molecular beacon, Quantum dot, Cell surface engineering, Microarray (testing), DNA microarray, Bio-bar codes, Nanoparticles.*

TAL Code Technology

See *TALEs.*

TAL Effectors

See *TALEs.*

TALENs

Acronym for *Transcription Activator-Like Effector Nucleases*, which are artificial restriction enzymes generated by fusing the TALE DNA-binding domain (i.e., proteins originally discovered to be secreted by *Xanthomonas* bacteria) to a nonspecific DNA cleavage domain (nuclease).

TALENs can be utilized by scientists to do an almost infinite variety of customized, precise DNA manipulations (e.g., insertions, deletions, and rearrangements) in almost any organism. For example, during 2012, Bruce Whitelaw created pigs that modeled human atherosclerosis disease, by utilizing TALENs to disrupt genes that code for low-density lipoprotein (LDL) receptors in those animals' livers. Without these LDL receptors to remove cholesterol-containing LDLs from the blood, LDLs buildup and lead to atherosclerosis. See also *TALEs, Nuclease, Deoxyribonucleic acid (DNA), Gene, Deletions, Insertional mutagenesis, Enzyme, Restriction enzymes, Protein, Fusion protein, Deoxyribonucleic acid (DNA), Gene, Gene editing, Knockout, Atherosclerosis, Receptors.*

TALEs

Acronym for *Transcription Activator-Like Effectors*, which are protein molecules that selectively bind to very specific DNA sequences. The DNA-binding site is determined by a specific amino acid sequence within each TALE, and that amino acid sequence can be tailored by the maker of each TALE, so the TALE binds to only one spot on a cell's DNA. When fused to an applicable *nuclease* (i.e., enzyme that cuts DNA at specific location), the fused molecule (now called a *Transcription Activator-Like Effector Nucleases* or TALENs) can be utilized by man to change the sequence of a selected gene (thereby changing its function) or knock out that gene.

Sometimes called TAL Code technology, TALEs was discovered by Ulla Bonas, Jens Boch, Thomas Lahaye, and Sebastian Schornack. See also *Protein, Deoxyribonucleic acid (DNA), Gene, Sequence (of a DNA molecule), Amino acid, Sequence (of a protein molecule), Cell, Fusion protein, Enzyme, Nuclease, Knockout.*

Tamoxifen

A pharmaceutical compound which acts as an antagonist to the estrogen receptor within breast tissue cells. By acting as an antagonist (i.e., Tamoxifen binds to the estrogen receptor molecules, thereby preventing estrogen from binding to them), Tamoxifen helps to prevent estrogen-dependent breast cancer from returning (in a patient being treated for breast cancer).

During 2011, Jose Russo published research showing that consumption of omega-3 fatty acids (*n*-3 fatty acids) along with the Tamoxifen administration in such cancer patients resulted in reduced expression of genes applicable to growing and spreading of the tumor(s).

See also *Antagonists, Receptors, Cell, N-3 fatty acids, Melatonin.*

Tandem Affinity Purification Tagging

Abbreviated "TAP Tagging," it refers to one particular method of *protein interaction analysis*. In "TAP tagging," the scientist first creates a *fusion protein* by fusing a "TAP tag" (short segment of known amino acids) onto the thoroughly known target protein. Next, the scientist introduces that fusion protein into (e.g., living cells in which *the proteins that will hopefully interact with the target protein*) are present.

When the "TAP tagged" fusion protein is recovered after some time, the *cell protein(s) which ligand interacted with that fusion protein* are determined by:

- Utilizing an *antibody which is specific to the TAP tag* to capture the tagged fusion protein along with the ligand(s) that are bound to that fusion protein.
- Utilization of a protease buffer to cleave off those bound ligand(s).
- Utilization of a calcium-containing solution with calmodulin-coated beads to remove any remaining protease and impurities.
- Use of mass spectrometry or two-dimensional gel electrophoresis to determine precisely *what those ligands* (to the known fusion protein) *are.*

See also *Protein interaction analysis, Protein, Amino acid, Cell, Polypeptide (protein), Fusion protein, Ligand (in biochemistry), Protease, Mass spectrometer, Two-dimensional (2D) Gel Electrophoresis, Antibody, Affinity chromatography.*

Tandems

Refers to certain electrically conductive light-harvesting polymer molecules that (when coupled to an applicable dye molecule) nonradiatively passes energy (which it procured via light photons absorbed by the polymer) to that dye molecule. That transfer of energy causes the dye molecule's emissions (resultant from its own light absorption) to shift further into the red portion of the light spectrum. That results in said tandem-augmented dye molecule able to be better utilized (e.g., to make more effective cell/molecule labels) in RT-PCR, DNA sequencing, flow cytometry, cell sorters, or microscopy.

See also *Polymer, Photon, RT-PCR, Sequencing (of DNA molecules), Flow cytometry, Cell sorting, Fluorescence-activated cell sorter (FACS), Confocal microscopy.*

Tannins

Refers to a large category of polyphenolic chemical compounds that are produced in many plant species. Some of the tannins are beneficial to human health when consumed by humans (e.g., the proanthocyanidins within cranberries, cocoa, chocolate, certain tannins within grapeseed, etc.). Most of the tannins have antimicrobial properties. See also *Polyphenols, Roanthocyanidins, Microbicide.*

TAP Tagging

See *Tandem affinity purification tagging.*

Taq DNA Polymerase

A 94 kDa DNA polymerase, which was originally isolated from the thermophilic archaean *Thermus aquaticus*. Commonly utilized to catalyze PCR reactions due to its heat resistance (needed for thermal cycles utilized in the PCR technique). See also *DNA polymerase, Polymerase, Kilodalton (kDa), Deoxyribonucleic acid (DNA), Bacteria, Thermophilic bacteria, PCR, Polymerase chain reaction (PCR) technique, Archaea.*

Taq Polymerase

See *Taq DNA polymerase.*

Target (of a Herbicide or Insecticide)

The molecule (e.g., receptor, enzyme, etc.) within a weed plant or within a pest insect that a given herbicide or insecticide is "aimed" at (e.g., when scientists are conducting research aimed at creating that herbicide or insecticide). For example, glyphosate-containing herbicides act on the (target) crucial plant enzyme EPSP synthase.

For example, insect-resistant transgenic plants containing "B.t. gene(s)" act on (target) receptors inside the digestive system of specific insect species via the *B.t. protoxin*. See also *Receptors, Enzyme, Glyphosate, EPSP synthase, Transgenic (organism), Protoxin, Herbicide-tolerant crop, Pat gene, Glutamine, Glutamine synthetase, Corn, Biological activity, Target–ligand interaction screening.*

Target (of a Therapeutic Agent)

The molecule (e.g., receptor) or moiety that a given drug or therapeutic regimen (e.g., gene delivery) is "aimed" at (i.e., when scientists are working to create/discover that drug or regimen).

Targets can be normally occurring constituents of the body (e.g., receptors, enzymes, factors, hormones, ion channels, nuclear receptors, DNA, etc.), or nonnormal constituents of the body (e.g., tumors, antigens on tumor surfaces, etc.), or (external, invading) pathogenic agents (e.g., microorganisms, viruses, parasites, etc.). See also *Enzyme, Factor, Hormone, ION channels, Nuclear receptors, Deoxyribonucleic acid (DNA), Tumor, Microorganism, Biological activity, Pathogen, Pathogenic, Virus, Pharmacophore, Gene delivery, Receptors, Moiety, Combinatorial chemistry, Combinatorial biology, Signaling, Signal transduction, G-proteins, Tumor necrosis factor (TNF), High-throughput screening (HTS), Multiplexed assay, Target–ligand interaction screening, Fluorescence mapping, Label (radioactive), Validation (of target), Biochips, Quantitative structure-activity relationship (QSAR), Whole-cell patch-clamp recording, Pharmacophore searching, Click chemistry.*

Target Validation

See *Validation (of target).*

Targeted Gene Repair

See *Oligonucleotide-mediated mutagenesis, Genome editing.*

Targeted Nucleotide Exchange

See *Oligonucleotide-mediated mutagenesis, Genome editing.*

Target–Ligand Interaction Screening

A methodology of high-throughput screening (HTS) that is utilized to screen a large number of candidates (e.g., pharmaceutical compounds) based upon their interaction (e.g., chemical "binding") to a preselected "target" (e.g., receptor molecule present within a cell membrane, molecule placed on a biochip or other bioassay to facilitate HTS, molecule present on the surface of a nematode utilized in HTS, etc.). See also *High-throughput screening (HTS), Target (of a therapeutic agent), Target (of a herbicide or insecticide), Combinatorial chemistry, Combinatorial biology, Ligand (in biochemistry), Receptors, Signal transduction, Nuclear receptors, Biochip, Signal transducers and activators of transcription (STATs), Caenorhabditis elegans (C. elegans), Surface plasmon resonance (SPR), Two-hybrid systems, Quantitative structure-activity relationship (QSAR), Pharmacophore searching, Nanosheets.*

TAT

The name of a protein which helps the HIV ("AIDS virus") to cross the human cell plasma membrane, thereby enabling infection of those cells by HIV (human immunodeficiency virus).

TAT is the main activator of HIV gene expression in cells; it is a protein which complexes with TAR (a 60-nucleotide sequence found in all viral messenger ribonucleic acid) to mediate synthesis of proteins (in an infected cell) necessary for HIV to reproduce. See also *Tata homology, Human immunodeficiency virus type 1 (HIV-1), Human immunodeficiency virus type 2 (HIV-2), Gene, Express, Nucleotide, Messenger RNA (mRNA), Virus, Protein, Plasma membrane, Cell.*

TATA Box

See *Tata homology.*

TATA Homology

An adenine-thymidine-rich (gene) sequence present 20–30 nucleotides "upstream" of the transcription start site on most eucaryotic protein coding genes. Because it is the binding site for RNA polymerase (RNAP), it is required for correct gene expression.

Recent research indicates that blocking this portion of the (gene) sequence may inhibit ability of the AIDS virus to reproduce. See also *Gene, Genetic code, RNA polymerase, Nucleotide, Adenine, Deoxyribonucleic acid (DNA), Sequence (of a DNA molecule), Tat, Express, Transcription, Startpoint, Eucaryote, Coding sequence, Homology, Pribnow box, Promoter, Sequence (of a protein molecule).*

Tau Protein

Refers to a protein present within brain neurons that serves as a structure (e.g., somewhat like a railroad track) that is utilized to clear from the neurons the normal daily accumulation of broken down and toxic proteins and/or cellular "garbage" resultant from metabolism. Tau also stabilizes certain proteins that are responsible for general transport within neuron cells. See also *Alzheimer's disease, Protein, Cell, Neuron, Metabolism.*

Taxol

A phytochemical that is naturally produced in some plants and functions to protect those plants from the plant pathogen known as *water mold.*

Coined during the 1960s by Monroe E. Wall when it was originally isolated from the Pacific yew tree (genus *Taxus*) this word is now a trademark of the Bristol-Myers Squibb Co. Taxol now refers to the antitumor pharmaceutical sold by Bristol-Myers Squibb Company. The active compound from Pacific yew tree is now known as paclitaxel.

Both Taxol and paclitaxel act by binding and stabilizing microtubules in cells (thereby halting/preventing the uncontrolled cell growth/proliferation that is cancer). They are used to treat breast cancer, ovarian cancer, some forms of lung cancer, and so on. See also *Chemotherapy, Paclitaxel, Cancer, Cell, Microtubules, Tubulin.*

TBT

Acronym for the *Technical Barriers to Trade (TBT) Agreement* to WTO. See also *Technical Barriers to Trade (TBT) Agreement, World Trade Organization (WTO).*

TCGF

See *T cell growth factor (TCGF).*

TCK Smut

See *Telethia controversia koon smut.*

T-DNA

See *Ti plasmid.*

Technical Barriers to Trade (TBT) Agreement

The agreement to GATT/WTO via which WTO member nations agreed to base their import (restrictive) regulations and standards (e.g., mandatory packaging, package marking, testing, certification, labeling requirements, etc.)—known as TBT measures—only on scientific assessments of actual risks (i.e., for those TBT measures intended to protect human health, animal and plant health, or the environment) and to require *only* those TBT measures that do not create unnecessary obstacles to international trade. See also *World Trade Organization (WTO), SPS, Sanitary and phytosanitary (SPS) agreement, Sanitary and phytosanitary (SPS) measures, Technical barriers to trade (TBT measures).*

Technical Barriers to Trade (TBT) Measures

These are (restrictive) import regulations, standards (e.g., mandatory packaging, package marking, testing, certification and labeling requirements, etc.). Some of them are designed to protect human health, animal and plant health, and/or the environment. In the Technical Barriers to Trade (TBT) Agreement to GATT/WTO, the WTO member nations agreed to base their TBT measures *only* on requirements that do not create unnecessary obstacles to international trade. See also *Technical barriers to trade (TBT) agreement, SPS, World Trade Organization (WTO).*

Technology Protection System

See *Cessation cassette.*

Telethia Controversia Koon Smut

A fungal disease that sometimes afflicts wheat (*Triticum aestivum*) plants. See also *Fungus, Wheat.*

Telomerase

An enzyme that enables the "repair" of telomeres (thereby stabilizing their length and limiting the "shortening" of the telomeres which can occur during cell division). The telomerase enzyme is only present and "working on a full-time basis" in cancerous cells (thereby enabling the "immortality" of cancerous cells).

In certain rare people possessing a gene "active" telomerase, their telomere "lifetime" is extended by a limited amount of lengthening of the telomeres. During 2012, Sadia Mohsin extracted adult stem cells from elderly people's hearts (who were suffering from heart disease) and modified those extracted stem cells in the laboratory using PIM-1 (a protein that promotes cell survival and growth) which enhanced activity of their telomerase. That resulted in elongated telomere length and renewed the ability of those elderly people to regenerate heart tissue.

Human telomerase contains an RNA component and a catalytic-protein component (i.e., a member of the reverse transcriptase "family" of enzymes). See also *Reverse transcriptases, Cancer, Neoplastic growth, Zygote, Telomeres, Enzyme, Oncogenes, Hybridoma, Monoclonal antibodies (MAb), Aging, Ribonucleic acid (RNA), Protein, PIM-1 protein, Adult stem cell.*

Telomeres

Molecular assemblies consisting of protein and DNA sequences (that do not code for proteins, but *do* code for some RNA), which are located at the (end) tips of chromosomes.

Telomeres protect the ends of chromosomes, prevent chromosomes from fusing to each other, and serve to limit the maximum number of times that a given cell divides (in the cell's "lifetime"). Telomeres' DNA consists of the sequence GGGGTT repeated many times (which can code for nodding TERRA RNA).

With the exception of certain types of cells (e.g., zygotes, cancerous cells, "immortal" hybridoma cells), portions of each telomere "break off" each time that the cell containing that chromosome divides. This "shortening" process serves to limit the lifetime (i.e., number of replications) of those (noncancerous, nonzygote, nonhybridoma, etc.) cells, so shorter telomeres are indicative of a shorter lifespan for the organism too. Related to that, shorter telomeres are also indicative of a greater chance for the organism getting lung disease, bone marrow failure, liver disease, or skin disease. Recent research indicates that adequate dietary consumption of *n*-3 ("omega-3") fatty acids can slow/decrease this telomere-shortening process.

In certain rare people possessing a gene for a more active telomerase, that "lifetime" is extended by a limited amount of lengthening of the telomeres. During 2012, Sadia Mohsin extracted adult stem cells from elderly people's hearts (who were suffering from heart disease) and modified those extracted stem cells in the laboratory using PIM-1 (a protein that promotes cell survival and growth) which enhanced activity of their telomerase. That resulted in elongated telomere length and renewed the ability of those elderly people to regenerate heart tissue. See also *Deoxyribonucleic acid (DNA), Coding sequence, Protein, Chromosomes, Sequence (of a DNA molecule), Telomerase, Mitosis, Mitogen, Cancer, Gamete, Cell, Aging, Protein, PIM-1 protein, Retinoids, Hybridoma, Zygote, Ribonucleic acid (RNA), Adult stem cell, N-3 fatty acids.*

Telophase

From the Greek word *telos* meaning *end*, it refers to the final (i.e., fourth) of the four phases of eucaryotic meiosis or mitosis (i.e., cell replication via division) during which a new membrane is made to envelope each *set of the newly divided chromosomes* (which now each constitute a new cell nucleus).

See also *Mitosis, Eucaryotes, Cell, Chromosomes, Membranes (of a cell), Nucleus, Meiosis.*

Temperate Phage

Refers to a phage (bacteriophage), which attacks bacteria but tends to not totally "consume"/destroy those bacteria (via the usual making of many more copies of that phage). Showing how unusual such *temperate* behavior is among phages, the word phage is derived from the Greek word meaning "to eat."

See also *Phage, Bacteria.*

Temperature-Sensitive Protein

Refers to certain protein molecules which are functional (e.g., carry out certain enzymatic/chemical reactions) at one range of temperatures (e.g., from 0°C to 90°C) but cannot carry out that function at higher temperatures (e.g., >91°C).

For example, one function of the specific proteins known as *kinases* is to transfer phosphoryl molecular groups to certain molecules "targeted" by those kinases. See also *Protein, Enzyme, Functional group, Protein kinases, Tyrosine kinase, Receptor tyrosine kinase, Functional protein microarrays, Protein–protein interactions, Phosphorylation.*

Template

In general terms, it is a mold or pattern that can be copied or its shape reproduced. When used with reference to molecular dimensions, it is a macromolecular mold or pattern for the synthesis of another macromolecule.

For example, during 2003 Angela Belcher, Daniel Solis, and Chuanbin Mao genetically engineered a pencil-shaped bacteriophage known as *M13* so that it expressed and incorporated into its capsid a *peptide which causes and controls nucleation/condensation onto it* of specific nanometer-size (conductor) particles. After those (conductor) nanoparticles thereby become *deposited in very specific order* onto the bacteriophage pencil-shaped "template," exposure to very high temperature removes the bacteriophage, leaving a solid NANOWIRE.

For example, during 2003, Susan Lindquist:

- Triggered the self-assembly of *Saccharomyces cerevisiae* (yeast) amyloid protein to thereby create 10 nm wide fibers.
- Placed those fibers onto specially designed electrodes and then reacted colloidal gold particles with cysteine molecular groups "protruding" from those fibers.
- Subsequently filled in the spaces between the gold particles bound to fiber via a reductive deposition procedure which deposited both gold and silver atoms.

This resulted in a NANOWIRE possessing an average diameter of 100 nm.

For example, during 2002, William A. Drucker and Chang-Hyun Jang were able to utilize the enzyme acetylcholinesterase as a "template" to create a precisely structured strip 70 nm wide deposited onto a prepared gold surface. After first coating the gold surface with a carboxylic acid (film), they were able to adhere onto it a film of acetylcholinesterase; then by scratching away a strip via utilization of an atomic force microscope tip (stylus), they were able to lay down a 70 nm wide strip (onto the gold) of thiocholine cleaved from acetylcholine-containing solution. See also *Macromolecules, Enzyme, Nanometers (nm), Nanowire, Bacteriophage, Genetic engineering, Peptide, Acetylcholinesterase, Acetylcholine, Deoxyribonucleic acid (DNA), RNA polymerase, Structural gene, Informational molecules, Heredity, Gene, Genetic code, Genetic Map, Biosensors (chemical), Genosensors, Ribonucleic acid (RNA), Gene repair (done by man), Codon, Exon, Chimeraplasty, Nanotechnology, Nanoparticles, Bioelectronics, Primer (DNA), Atomic force microscope, Yeast, Self-assembly (of a large molecular structure), Cysteine (cys), Reduction (in a chemical reaction).*

Teosinte

Refers to a "family" of specific wild plants (*Zea diploperennis*) native to southern Mexico, Guatemala, Honduras, and Nicaragua, which are related to (domesticated) corn/maize (*Zea mays* L.). See also *Corn, Wild type.*

Termination Codon

Also known as *terminator sequence.* One of three triplet sequences (U-A-G, U-A-A, or U-G-A) found in DNA molecules (genes) that cause termination of protein synthesis; they are also called *stop codons* or *nonsense codons.* Those sequences cause the termination of the ribosome's synthesis of (growing) polypeptide molecular chain and its release from the ribosome in free form. See also *Protein, Polypeptide (protein), Coding sequence, Codon, Deoxyribonucleic acid (DNA), Genetic code, Nonsense codon, Sequencing (of DNA molecules), Control sequences.*

Terminator

See *Termination codon.*

Terminator Cassette

See *Cessation cassette.*

Terminator Region

See *Termination codon.*

Terminator Sequence

See *Termination codon.*

Terpenes

A category of chemical compounds (cyclic hydrocarbon molecules) which are produced by plants, especially conifers, oranges, and certain blue-green algae (e.g., *Oscillatoria peronata*).

For example, the drug Taxol (paclitaxel) is a diterpene (i.e., molecule consisting of two terpene units) extracted from the Pacific yew tree (genus *Taxus*).

For example, the blue-green algae *Oscillatoria perornata* produce the terpene *2-methylisoborneol.* When catfish (*Ictalurus furcatus*) consume such algae, it accumulates in those catfish and imparts an undesirable (musty) flavor to their meat. See also *Taxol, Paclitaxel, Saponins, Allelopathy.*

Terpenoids

A category of chemicals possessing complicated molecular structures. Made by plants to:

- Assist in the development and growth of the plant.
- Assist in attracting certain pollinators (e.g., bees) to the flower(s) of the plant.
- Repel some predators from the plant.
- Attract to the plant's vicinity (e.g., after certain pest insects begin chewing on plant) some *predators of those pest insects*.

See also *Terpenes*.

TERRA RNA

See *Ribonucleic acid (RNA)*, *Telomeres*.

Tertiary Structure

The three-dimensional folding of the polypeptide (i.e., protein) molecular chains that characterizes a protein molecule in its native state. See also *Protein structure*, *Protein*, *Polypeptide (protein)*, *Conformation*, *Protein folding*, *Native conformation*, *Proteomics*, *Transcriptome*.

Testosterone

An androgen (steroid hormone) that is biochemically synthesized (made) from androstenedione, which is itself synthesized from progesterone. Testosterone is responsible for the development of male secondary sex characteristics in humans such as greater strength, larger body size, facial hair and a deeper voice, and so on. See also *Steroid*, *Estrogen*.

Tetracycline

An antibiotic which disrupts protein expression in (and thus the life of) procaryotic organisms (e.g., bacteria). See also *Antibiotic*, *Protein*, *Protein expression*, *Procaryotes*, *Organism*, *Bacteria*.

Tetrahydrofolic Acid

The reduced, active coenzyme form of the vitamin folic acid; involved in C_1 transfers. Tetrahydrofolate (also known as FH_4) serves as an intermediate carrier (molecule) of methyl, hydroxy-methyl, or formyl groups (all containing one carbon atom) in a relatively large number of enzymatic reactions in which such one-carbon groups are transferred from one metabolite to another. See also *Coenzyme*.

Tetranucleotide Repeat

See *Click chemistry*.

Tetraploid

Refers to organisms that possess four sets of chromosomes, instead of the normal two sets of chromosomes. Conversion of a diploid (i.e., two sets of chromosomes) organism to tetraploid can be done by man. For example, by soaking seeds in certain chemicals such as colchicine, scientists can cause the resultant (plant) to become tetraploid. This (tetraploidism) is utilized to create (mutated) plant varieties with new traits. See also *Chromosomes*, *Colchicine*, *Polyploid*, *Diploid*, *Mutation breeding*.

Tetrasomes

Refers to protein complexes within nucleosomes that are formed from histone proteins. During formation, a nucleosome is formed in two steps. In the first step, a tetrasome is formed from four histone protein molecules; while the second nucleosome-formation step involves the subsequent addition of four histones to create a complete nucleosome (DNA wrapped around the protein). See also *Protein*, *Histones*, *Nucleosome*, *Deoxyribonucleic acid (DNA)*.

Tetraspanin Proteins

A category of protein molecules that are present in both mammals and in some plants.

In plants, tetraspanin proteins are involved in the regulation of the plant's growth (especially of vascular systems).

In mammals, tetraspanin proteins are involved in the development of at least some vascular systems, in some cell's differentiation processes, and in certain ongoing cellular functions such as cell motility and cell adhesion. See also *Protein*, *Protein signaling*, *Cell*, *Differentiation*, *Cell differentiation*, *Cell motility*, *Adhesion protein*.

TFH

Abbreviation for *follicular helper T cells*. See *T cells*.

TfR

Abbreviation for transferrin receptor. See *Transferrin receptor*.

TG

See *Triglycerides*.

TGA

The government regulatory agency charged with approving all pharmaceutical products sold within Australia. See also *Food and Drug Administration (FDA)*, *Koseisho*, *Committee for Proprietary Medicinal Products (CPMP)*, *European Medicines Evaluation Agency (EMEA)*, *Medicines Control Agency (MCA)*, *Committee on Safety in Medicines*, *Bundesgesundheitsamt (BGA)*, *Gene Technology Office*.

TGF

See *Transforming growth factor-alpha (TGF-alpha)*, *Transforming growth factor-beta (TGF-beta)*.

Thale Cress

Common name for *Arabidopsis thaliana*. See *Arabidopsis thaliana*.

Theranostics

A term created from the words "therapeutic" and "diagnostics"; it refers to a class of pharmaceuticals and their directly related diagnostic accouterments. For example, a diagnostic test (for presence of *HER-2* gene in breast tissue) is administered to relevant breast cancer patients prior to those patients receiving the pharmaceutical trastuzumab (Herceptin™) which is a monoclonal antibody specific to the protein receptor expressed from *HER-2* gene. Only if the test is positive for presence of the *HER-2* gene, is the trastuzumab pharmaceutical administered to the patient (thereby essentially ensuring the pharmaceutical is only administered to those in which it will be effective). See also *Gene, HER-2 Gene, Cancer, Trastuzumab, Monoclonal antibodies (MAb), Protein, Receptor, HER-2 Receptor.*

Therapeutic Nucleic Acid Repair Approach

See *Oligonucleotide-mediated mutagenesis, Genome editing.*

Thermal Hysteresis Proteins

Also referred to as "antifreeze proteins" or AFPs, these are a class of proteins/glycoproteins (possessed by some organisms) which lowers the freezing point of the organisms' blood and inhibit the formation of ice crystals inside the cells of that organism when those cells are exposed to temperatures colder than 32°F (0°C).

For example, certain glucanase (enzyme) molecules in the plant known as rye (*Secale cereale*) act to inhibit the formation of ice crystals inside rye, thereby imparting frost tolerance to that plant. See also *Protein, Organism, Cell, Glycoprotein.*

Thermoacidophilic Bacteria

Refers to those particular extremophilic bacteria that grow well in conditions of high temperatures and high acidity. See also *Bacteria, Extremophilic bacteria, Thermophilic bacteria, Sulfolobus solfataricus.*

Thermoduric

An organism that can survive high temperatures but does not necessarily grow at such temperatures. See also *Thermophile, Mesophile, Extremophilic bacteria, Psychrophile, Endophyte.*

Thermolabile

Refers to a compound/molecule which is unstable at elevated temperatures (e.g., it disintegrates).

This is in contrast to a compound/molecule which is merely temperature *sensitive* (i.e., ceases to function at elevated temperatures).

This is in contrast to a protein molecule or DNA molecule which denatures (i.e., loses biological activity) at high temperature. See also *Protein, Temperature-sensitive protein, Denaturation, Deoxyribonucleic acid (DNA), Denatured DNA, Biological activity.*

Thermophile

An organism whose optimum temperature for growth is close to, or exceeds, the boiling point of water (100°C, 212°F). See also *Extremophilic bacteria, Thermophilic bacteria, Thermoduric, Mesophile, Psychrophile, Eucaryote.*

Thermophilic Bacteria

Literally "heat loving" bacteria. They are a category of thermophiles generally found near geothermal vents beneath bodies of water. See also *Thermophile, Thermoacidophilic bacteria, Thermoduric, Extremophilic bacteria, Mesophile, Psychrophile.*

Thermotolerant Wheat

Refers to a transgenic wheat that tolerates warmer-than-ideal temperatures during wheat's critical grain filling stage (i.e., the stage of the wheat plant's life during which the wheat kernels achieve their final/maximum weight and thus yield per acre). The optimum temperature for the wheat plant (*Triticum aestivum*) during the grain filling stage is 59°F–64°F (15°C–18°C). Approximately 3%–4% of the yield potential of the wheat plant could be lost for each 1°C in temperature above that optimum temperature level during the grain-filling stage.

During the grain-filling stage, the wheat kernels accumulate 75%–85% starch and approximately 10%–15% protein. The starch is synthesized from (photosynthesis-created) sucrose via the enzyme known as soluble starch synthase (SSS).

During 2014, Harold Trick and Allan Fritz increased a wheat plant's tolerance to warmer-than-ideal temperatures during grain-filling stage, by inserting a single rice plant *SSS* gene that imparts greater thermotolerance into the wheat plant's DNA. See also *Wheat, Transgenic, Starch, Sucrose, Photosynthesis, Enzyme, Synthase, Deoxyribonucleic acid (DNA), Gene.*

Thioesterase

A "family" of enzymes that is naturally produced within some plants, such as the California bay tree (*Umbellularia californica*). Thioesterase catalyzes those plants' production of the fatty acid *laurate*. See also *Fats, Fatty acid, Lauroyl-ACP thioesterase, Enzyme, Laurate, Canola, High-laurate canola.*

Thiol Group

Refers to a specific chemical entity (on a molecule). See *Cysteine (cys), Cystine.*

Thioredoxin

See *Allergies (foodborne).*

Threonine (thr)

A crystalline, α-amino acid considered essential for normal growth of animals. It is biosynthesized (i.e., made) from aspartic acid and is a precursor of isoleucine in microorganisms. See also *Essential amino acids.*

Thrombin

The key to thrombus (blood clot) formation. Thrombin is a proteolytic enzyme that cleaves fibrinogen into (molecular) pieces, which then spontaneously assemble themselves into fibrin, which forms a clot. See also *Thrombus, Thrombosis, Thrombomodulin, Thrombolytic agents, Fibrin, Fibrinolytic agents, Cascade.*

Thrombolytic Agents

Blood-borne compounds (such as tissue plasminogen activator) that work to disintegrate (break up or lyse) blood clots. See also *Fibrin, Fibrinolytic agents, Tissue plasminogen activator (tPA).*

T

Thrombomodulin

A cell surface protein found on endothelial cells that plays a key role in modulating the final step in the coagulation process. After thrombin binds to thrombomodulin, thrombin loses its ability to cleave fibrinogen to form fibrin. In addition, once thrombin binds to thrombomodulin, thrombin's activation of protein C is increased 200-fold and this activated protein C then degrades factors Va and VIIIa which are both required for the production of thrombin from prothrombin. Hence, thrombomodulin modulates the activity of the enzyme thrombin causing a cessation of full-blown clotting activity. See also *Thrombin, Protein, Protein C, Thrombosis, Pathway, Pathway feedback mechanisms.*

Thrombosis

The intravascular (i.e., inside of blood vessel) formation of a blood clot. See also *Thrombin, Thrombus, Thrombolytic agents, Triglycerides, Fibrin, Fibrinolytic agents, Tissue plasminogen activator (tPA), Plaque, Von Willebrand factor.*

Thrombus

The blood clot itself. The mass of blood coagulated *in situ* in the heart or other blood vessel. For example, such a clot causes a heart attack when the coagulation occurs in the vessels feeding the heart. See also *Thrombin, Thrombosis, Thrombolytic agents, Fibrin, Triglycerides, Fibrinolytic agents.*

Thymine (thy)

A pyrimidine component of nucleic acid first isolated from the thymus. Its hydrogen-bonding counterpart in RNA is uracil. See also *Nucleic acids, Pyrimidine, Base (nucleotide), Thymus, Ribonucleic acid (RNA).*

Thymoleptics

A class of drugs that primarily exerts their effect on the brain influencing "feeling" and behavior.

Thymus

An immune system gland that enables cells of the immune system of mammals to mature. In humans, it lies behind the breast bone and extends upwards as far as the thyroid gland. The thymus is the place in the body where T lymphocytes (created in thymus or tonsils) are "taught" to distinguish "foreign" (e.g., pathogen's) antigens from "self" cell antigens, to avoid immune responses in which the body's immune system attacks organs and other cells within the body (resulting in autoimmune disease). Any T lymphocytes that remain "autoreactive" (i.e., would tend to attack "self" cells, such as organs in the body) are destroyed by the thymus via a cytotoxic mechanism.

An example of an autoimmune disease is multiple sclerosis (MS), where the body's acetylcholine receptors are attacked by the body's immune system. Since acetylcholine is crucial in the transmission of nerve impulses to the body's muscles, such destruction of acetylcholine receptors results in loss of control of the body's muscles. See also *T lymphocytes, Cytotoxic, Receptors, T cells, Immune response, Pathogen, Antigen, Neurotransmitter, Acetylcholine, Autoimmune disease.*

Thyroid Gland

A gland that is found on both sides of the trachea ("windpipe") in humans. This gland secretes the iodine-containing hormone thyroxine, which increases the rate of the body's metabolism. Other thyroid hormones act to increase the number of cells within bones (resulting in growth/strengthening of bones). See also *Thyroid stimulating hormone (TSH), Graves' disease, Cell.*

Thyroid Stimulating Hormone (TSH)

A hormone that causes the thyroid gland to secrete additional amounts of thyroxine. See also *Thyroid gland, Grave's disease.*

Ti Plasmid

Abbreviation for *tumor-inducing plasmid* or *tumor induction plasmid*. It is the plasmid of *Agrobacterium tumefaciens* bacteria that naturally has a part of its DNA (T DNA) transferred to a plant when *Agrobacterium tumefaciens* infects that plant (e.g., via a wound in the plant). Discovered in 1974 by Marc Van Montagu.

After it has been transferred into the plant, that Ti plasmid DNA segment (now known as T-DNA or *transferred DNA*) inserts itself into the plant's DNA, where it causes cells to grow into tumor-like structures known as galls.

The *Ti plasmid* can be modified so that it can be utilized (by genetic engineers) to insert genes from other organisms into plants. See also *Plasmid, Bacteria, Agrobacterium tumefaciens, Cell, Deoxyribonucleic acid (DNA), Gene, Genetic engineering.*

Tight Junction Proteins

A specific class of proteins that are present between the cells that line the interior of the intestines, and thereby prevent leakage of the contents (e.g., microorganisms, partially digested food, etc.) of the intestines into the body's tissues. When tight junction proteins are depleted (e.g., one consequence of chronic kidney disease), the leakage of intestine contents into the body can result in chronic inflammation. See also *Protein, Chronic inflammation.*

Tiling Arrays

Refers to any arrays (e.g., microarrays) utilized in biotechnology applications which test/interact with the sample in a manner which is even spaced across all possible informational sites.

For example, some microarrays are available to test for the following, in a manner which is evenly spaced (e.g., every __nucleotides) across the DNA of a given organism's genome:

- Sites of DNA methylation.
- Transcription factor-binding sites.
- Transcription.
- Chromatin modification sites.

See also *Microarray (testing), Whole-genome association, Deoxyribonucleic acid (DNA), Nucleotide, Methylated, Transcription, Transcription factors, Genome, Chromatin, Chromatin remodeling, Protein microarrays, Carbohydrate microarrays, Oligosaccharide microarrays, Gene expression analysis.*

t-IND

Treatment Investigational New Drug Application to America's Food and Drug Administration (FDA). See also *"Treatment" IND regulations.*

t-IND Treatment

Investigational New Drug Application to America's Food and Drug Administration (FDA). See also *"Treatment" IND regulations*.

TIR1

See *Auxins*.

TIRF Microscopy

Refers to microscopy systems/microscopes which utilize *total internal reflection fluorescence* to help visualize (e.g., a thin layer of biological tissue, a layer adsorbed out of solution, a layer of tissue culture cells adhered onto quartz/glass wall of container, etc.). A coherent light source (e.g., a laser) is aimed into the sample at a steep incident angle so that the *difference in refractive index* between the cells/tissue and the quartz/glass causes the light to be totally reflected. This is analogous to a person seeing his reflection when attempting to look in through a window into a darkened room.

That total reflection within a cell/tissue sample results in the creation of an extremely thin (~200 nm) electromagnetic field, known as an *evanescent wave*. The evanescent wave results in excitation (and thus fluorescence) of only the atoms *within the 200 nm thick evanescent wave*, thereby allowing clear viewing of only the desired portion of the cells/tissue (without interference by adjacent tissue/materials or solution).

For example, instead of using chemical stains to try to "color" various tissues/cells differently to make them easier to see/differentiate under a conventional light microscope, scientists can instead:

- Utilize *fluorescent proteins* such as Green Fluorescent Protein (GFP) or Kusabira Orange (which will differentially "adhere" to differing tissues/cells).
- Genetically engineer (e.g., living cells) to express one or more of the fluorescent proteins in one or more cellular structures (or within tissues).

With the cell/tissue sample thus *fluorescently labeled*, the scientist can then utilize one of the fluorescent microscopes available (e.g., confocal microscope equipped with a relevant laser source to "light up" the fluorescent proteins) to view the fluorescence-illuminated/colored cell/tissue in three dimensions.

The "lighting up" (i.e., the fluorescence) of those proteins results from an *evanescent wave* (i.e., a sort of "shock wave" that arises when light passes from a denser to a less dense medium under conditions of total internal reflection). See also *Fluorescence, Cell, Cell culture, Protein, Green fluorescent protein, Kusabira orange, Genetic engineering, Transfection, Express, Gram stain, Confocal microscopy, Nanometers (nm).*

Tissue Array

See *Live cell array*.

Tissue Culture

The growth and maintenance (by researchers) of cells from higher organisms *in vitro*, that is, in a sterile (e.g., test tube, petri dish, etc.) environment which contains the nutrients and substrate/structure necessary for cell growth. Animal cells generally require a three-dimensional substrate for the cells to attach to, while they are growing. Such substrates are best made from *stimuli-responsive polymers*. Via changes in temperature, pH, magnetic or electrical fields, the surface properties of stimuli-responsive polymers (also sometimes called "smart materials") are changed in a manner that releases the cultured cells without harming them (e.g., so they can be inserted into body of organism, to continue to grow there).

One plant use of tissue culture is to produce *disease-free* offspring from certain (valuable, high quality) crop plants.

Another use of tissue culture methods is for "embryo rescue" to enable "wide crosses" between two different species of plants. In that procedure, pollen from one plant species (e.g., a wild plant possessing disease resistance) is induced to fertilize a plant from another species (e.g., a domesticated crop). The resultant fertilized plant embryo, which would not grow on its own, is "rescued" via tissue culture methods. Following maturation, that wide cross (i.e., a hybrid plant from two species that normally would not cross) produces fertile seeds on its own without any need for further intervention by man. See also *Cell, Organism, Culture medium, Species, Hybridization (plant genetics), In vitro, Polymer, Micropropagation.*

Tissue Engineering

Refers to the technologies utilized to induce:

- (Injected) liver, cartilage, mesenchymal stem cells, and so on, cells to grow (within recipient organism's body) and form entire (integral) tissues.
- (Extant) cells within the body to grow and form desired (integral) tissues via precise injection of relevant compounds (e.g., certain growth factors, growth hormones, hydrogel, scaffolding, etc.).
- Certain cells (e.g., stem cells) extracted from the body, to grow and/or differentiate while temporarily outside the body (e.g., on a hydrogel within a vessel designed to provide oxygen and needed nutrients) and form desired integral tissues. Among many other cues to guide these cells regarding the specific type of cell they become, the physical stiffness of the extracellular matrix (ECM) touching them is a major factor.

During the 1980s, Robert S. Langer invented degradable "scaffolds" made of lactic acid–glycolic acid copolymer fibers. Others have made such scaffolds using hydrogels, and so on. When these "scaffolds" are "seeded" with cells (e.g., propagated in vats or transplanted from another organism), the cells grow together to form tissue. This (scaffold) technology can also be utilized to grow skin tissue, blood vessels, corneas, nerves, and so on.

During 2000, Samuel Stupp created two-part molecules known as *peptide amphiphiles*, which self-assemble into nanofibers which encourage bone growth (e.g., when inserted into the space between human broken bone fragments).

During 2010, Thomas Petersen was able to create a scaffold composed of ECM by utilizing detergent to remove the non-ECM cellular constituents of a rat lung. When that scaffold was subsequently "seeded" with a mixture of lung endothelial and epithelial cells, and cultured, this ENGINEERED LUNG TISSUE produced micro blood vessels, alveoli, and appropriate airways populated with applicable cell/tissue types. See also *Tissue culture, Cell, Organism, Growth factor, Growth hormone (GH), Self-assembly (of a large molecular structure), Extracellular matrix, Hydrogels, Endothelial cells, Epithelium, HSE, Nanowhiskers, Mesenchymal stem cells.*

T

Tissue Plasminogen Activator (tPA)

A glycoprotein that possesses thrombolytic (i.e., blood clot-dissolving) activity. It is used as a drug to dissolve clots and acts by first binding to fibrin (clots). It then activates (i.e., proteolytically cleaves) plasminogen (molecules) to yield plasmin, a bloodborne enzyme that itself cleaves molecular bonds in the fibrin clot. The plasmin molecules diffuse through the fibrin clot and cause the clot to dissolve rapidly. With the dissolution of the clot, blood flow to the formerly blocked blood vessel (e.g., the heart) is restored. See also *Thrombus*, *Thrombin*, *Thrombolytic agents*, *Glycoprotein*, *Fibrin*, *Fibrinolytic agents*.

Tissue Scaffold

See *Extracellular matrix*, *Tissue engineering*.

TKI

See *Tyrosine kinase inhibitors*.

TLR

Acronym for *toll-like receptors*. See *Innate immune response*.

TME (N)

Abbreviation for "true metabolizable energy (corrected for nitrogen)"; a measure of the amount of energy that a given animal (e.g., chicken) can extract from a given feed ration. See also *Metabolism*, *Chemometrics*, *Calorie*.

TMEn

See *TME (N)*.

TNBC

Acronym for triple-negative breast cancer. See *Chronic inflammation*.

TNF

See *Tumor necrosis factor (TNF)*.

TNF Blockers

Refers to the class of pharmaceuticals that block the harmful action of excess TNF. See *Tumor necrosis factor (TNF)*.

TNFR

Acronym for Tumor Necrosis Factor Receptor. See *Tumor necrosis factor (TNF)*, *Receptors*, *Etanercept*, *Fusion protein*.

Tobacco Budworm

See *Heliothis virescens (H. virescens)*.

Tobacco Hornworm

Caterpillars (pupae) of the Lepidopteran insect *Manduca sexta*.

Tobacco Hornworm is susceptible to Cry1A(b) protein (e.g., they are killed if they eat plants genetically engineered to contain Cry1A(b) protein). See also *CRY1A(b) Protein*.

Tobacco Mosaic Virus (TMV)

One of the smallest viruses, consisting of some 2200 chains of identical polypeptides and a molecule of RNA. All of the genetic/heredity information of the Tobacco Mosaic Virus is contained in its RNA.

The first discovery of a self-assembling, active biological structure occurred in 1955, when Heinz Frankel-Conrat and Robley Williams showed that TMV, will reassemble into functioning, infectious virus particles (after the TMV has been dissociated into its components via immersion in concentrated acetic acid). The TMV virus infects the leaves of tomato and tobacco plants causing disease.

Tobacco plants can be genetically engineered to resist TMV infection. A tomato plant, genetically engineered to resist TMV infection, has been commercially available since 1992. See also *Genetic engineering*, *Capsid*, *Virus*, *RNA*, *Polypeptide (protein)*, *Gene*, *Informational molecules*, *Heredity*, *Self-assembly (of a large molecular structure)*.

Tocopherols

A "family" of different molecular forms of vitamin E, each of which has a saturated phytyl "tail" attached to (the "backbone" of the molecule). Among this family, alpha-tocopherol is the active form of vitamin E in humans.

Commercial tocopherols are extracted from soybeans, inside which they protect biological membranes. Some tocopherols are also naturally present in canola and sunflower seeds. See also *Vitamin*, *Soybean plant*, *Vitamin E*, *Plasma membrane*.

Tocotrienols

A "family" of different molecular forms of vitamin E, each of which has an unsaturated isoprenoid "side chain" attached to (the "backbone" of the molecule).

Tocotrienols are naturally present in oil palm (*Elaeis guineensis*) and in cereal grains (e.g., oats, barley, rye, and rice bran). See also *Vitamin*, *Isoprene*, *Vitamin E*.

Toll-Like Receptors

See *Innate immune response*.

Tomatidine

A naturally produced hexacyclic (six molecular ring structure) steroidal chemical compound that is found in green tomatoes (*Lycopersicon esculentum*). It is responsible for initiating the tomato plant's defenses (e.g., SAR response) against attack by bacteria, viruses, fungi, and insect pests. Tomatidine disappears from tomato tissues as the fruit ripens.

When consumed by animals and digested, tomatidine has been shown to inhibit muscle atrophy ("wasting away") and to increase the growth of muscle tissue. See also *Systemic acquired resistance (SAR)*, *Tomato*, *Immune response*.

Tomato

A green bushy plant, botanical name *Lycopersicon esculentum*. The wild type is native to South America, but the (domesticated) tomato is grown worldwide today.

Its fruit, known as tomatoes, are a natural source of the antioxidant carotenoid *lycopene*, a phytochemical whose consumption has been linked to a reduction in coronary heart disease and some cancers (e.g., prostate cancer). See also *Lycopene*, *Phytochemicals*, *Antioxidants*, *Cancer*, *Carotenoids*, *Coronary heart disease* (*CHD*), *Wild type*.

Tomato Fruitworm

An insect pest that is also known as soybean podworm (when found on soybean plants) and as the corn earworm (when it is on corn/maize plants). See *Helicoverpa zea* (*H. zea*).

Topo-Isomerase

Refers to a particular isomerase enzyme which either causes, or reduces, supercoiling in DNA molecules. See also *Enzyme*, *Isomerase*, *Deoxyribonucleic acid* (*DNA*), *Supercoiling*.

Topotaxis

See *Tropism*.

TOS

See *Transgalacto-oligosaccharides*.

Total Internal Reflecton Fluorescence

See *TIRF microscopy*.

Totipotency

The ability to grow/differentiate into all of the types of cells/tissues constituting an (adult) organism's body. See also *Stem cell one*, *Cell*, *Zygote*, *Cell differentiation*, *Cell-differentiation proteins*, *Totipotent stem cells*.

Totipotent Stem Cells

Bone marrow cells that (when signaled) mature into both red blood cells and white blood cells. Receptors on the surface of totipotent stem cells "grasp" passing blood cell growth factors (e.g., Interleukin-7, Stem Cell Growth Factor, etc.), bringing them inside these stem cells and thus causing the maturation and differentiation into red and white blood cells. These receptors are called FLK-Z receptors. See also *Stem cell one*, *Stem cells*, *White blood cells*, *Growth factor*, *Receptors*, *Cell-differentiation proteins*, *Cell differentiation*, *Cell*.

Toxic Substances Control Act (TSCA)

A 1976 American federal law under which the U.S. Environmental Protection Agency has regulated the release of genetically engineered organisms (e.g., bacteria or plants) that produce natural insecticides. This is based on legal analogy to synthetic chemical insecticides, which are clearly regulated under TSCA. See also *OAB* (*Office of Agricultural Biotechnology*), *Federal Insecticide Fungicide and Rodenticide Act* (*FIFRA*), *Genetically engineered microbial pesticides* (*GEMP*), *Wheat take-all disease*, *Bacillus thuringiensis* (*B.t.*).

Toxicogenomics

A branch of toxicology that deals with the reactions between toxins and the *specific differences in response* of different organisms *due to their different genomes/DNA* (of the different individuals that consume the same toxin). For example, some rare humans can tolerate eating certain poisonous mushrooms (which sicken or kill all other humans that consume those particular mushroom species).

Some humans can tolerate intimate exposure to urushiol oil toxin in poison oak (*Rhus diversiloba*), which harms other humans (e.g., causes oozing, itching rash, etc.).

During 2001, scientists identified a human genetic variation which makes some cancer patients approximately seven times more likely to have a toxic reaction to the common chemotherapy drug *irinotecan* used to treat colorectal cancer.

Some rare humans are unable to degrade (breakdown) pyrimidine-containing pharmaceuticals, due to those humans' lack of a gene coding for production in their body of a specific pyrimidine-degrading enzyme (abbreviated DHPDH). The goal of toxicogenomics in that case is to *avoid* administering pyrimidine-containing pharmaceuticals (e.g., the anticancer drug *5-fluorouracil*) to those particular humans. One way to accomplish that goal would be via genetic testing, to determine which individuals lack the gene for DHPDH before any pyrimidine-containing pharmaceuticals are administered.

During 2001, Fred Gould, David Heckel, and Linda Gahan showed that a rare, recessive gene (allele) known as *BtR-4* could confer (to tobacco budworms possessing two copies of that particular gene) resistance to at least some of the "cry" proteins (which kill all other tobacco budworms that consume those "cry proteins").

The subgroup of *all those individuals whose DNA* (*genome*) *causes their bodies to resist the effects of a given toxin*, or be unable to degrade other toxin(s), is known as a HAPLOTYPE. A haplotype could (theoretically) be as small as one individual, because the particular resistance to toxin could result from one single-nucleotide polymorphism. See also *Gene*, *Genomics*, *Pharmacogenomics*, *Toxin*, *Mutagen*, *Pharmacogenetics*, *Genome*, *Deoxyribonucleic acid* (*DNA*), *Haplotype*, *Single-nucleotide polymorphisms* (*SNPs*), *Recessive allele*, *Cry proteins*, *Cell array*, *Tobacco budworm*, *Live cell array*, *Enzyme*, *Chemotherapy*, *ADME/TOX*, *Cellular pathway mapping*.

Toxigenic *E. coli*

See *Enterohemorrhagic E. coli*, *Escherichia coliform 0157:H7* (*E. coli 0157:H7*).

Toxin

From the Latin word *toxicum* meaning *poison*, it refers to a substance (e.g., produced in some cases by fungi, weeds, ants, disease-causing microorganisms, etc.) which is poisonous to certain other living organisms. See also *Antitoxin*, *Abrin*, *Ricin*, *Colicins*, *Bacteriocins*, *Escherichia coliform 0157:H7* (*E. coli 0157:H7*), *Enterohemorrhagic E. coli*, *Pfiesteria piscicida*, *Domoic acid*, *Phytotoxin*, *Photorhabdus luminescens*, *Enterotoxin*, *Glucosinolates*, *Alkaloids*, *Aflatoxin*, *Mycotoxins*, *Fungus*.

TPS

See *Technology protection system*.

Tracer (Radioactive Isotopic Method)

A metabolite that is labeled by incorporation of an isotopic atom into its structure. The metabolic fate of the labeled metabolite can then be traced in intact organisms. That is, one is able to ascertain where (in what kind of structure) the metabolite ends up as well as the transformation products (intermediate molecules) which were involved in its formation. Certain atoms of a given metabolite are labeled. This is done by substituting radioactive isotopes for the atom in question. Because an atom is replaced by an isotope, the metabolite as a whole is chemically and biologically indistinguishable from its normal analogue. The presence of the isotope allows the metabolite and its transformation products to be detected and measured. Without this technique, many aspects of metabolism could not have been studied. These include the process of photosynthesis, metabolic turnover rates, and the biosynthesis of proteins and nucleic acids. See also *Reassociation (of DNA), Radioactive isotope, Radioimmunoassay.*

Traditional Breeding Methods

A phrase utilized by some people to refer to some or most techniques/technologies utilized by crop plant breeders prior to some arbitrarily chosen date (after which some people feel that "genetic engineering" arrived abruptly).

For example, in 1992 Tim Croughan discovered a single rice (*Oryza sativa*) plant that had survived (what should have been a lethal dose of) an imidazolinone-based herbicide, due to a (mutated) gene in its DNA that made it resistant to imidazolinones. That plant was then propagated via straightforward breeding to yield seeds still sown today.

Many years ago, some other crops similarly were given new traits (e.g., herbicide tolerance, compositional improvements, etc.) via MUTATION BREEDING (i.e., soaking seeds or pollen in mutation-causing chemicals such as colchicine, or bombardment of seeds by ionizing radiation to cause random genetic mutations, followed by grow out and selection of the particular mutation desired such as herbicide tolerance—as described above).

Other crops were given new traits via crossing them with related wild plants, which occasionally resulted in extremely high levels of natural toxicants in those plants/seeds (e.g., solanine, psoralene, etc.).

Still others were given new traits via wide-crossing them with other domesticated species (e.g., the tangelo is a hybrid of the grapefruit and the tangerine).

The United States' Food and Drug Administration (FDA) regulates all new crop plants similarly (e.g., also requires testing of plants produced via "traditional breeding methods" for the potential presence of introduced or increased natural toxicants). See also *Genetic engineering, Herbicide-tolerant crop, Genetics, Mutation, Mutation breeding, High-oleic sunflowers, Trait, Canola, Soybean plant, Corn, Solanine, Psoralene, Food and Drug Administration (FDA), Barley, Hybridization (plant genetics), Marker (DNA sequence), Marker assisted selection, Point mutation, Somaclonal variation, Somatic variants, Wide cross, Embryo rescue, Tissue culture, Colchicine.*

Traditional Breeding Techniques

See *Traditional breeding methods.*

TRAIL

See *Neutrophils.*

Trait

A characteristic of an organism, which manifests itself in the phenotype (physically). Many traits are the result of the expression of a single gene, but some are polygenic (result from simultaneous expression of more than one gene). For example, the level of protein content in soybeans is controlled by five genes. See also *Phenotype, Genotype, Express, Gene, Polygenic, Protein, Soybean plant, Callipyge.*

Trait Restoration

Refers to the creation of crop plants (today) whose genome (DNA) contains a particular trait that had been present within the genome of the wild-type ancestor of that crop plant, but was subsequently lost during that crop plant's domestication process (e.g., 1000 years ago). For example, during 2014, Lijuan Qiu and Rongxia Guan discovered a salt tolerance gene present in the DNA of wild-type soybean plants that are the ancestors of today's domesticated soybean (*Glycine max* (L.) Merrill) varieties. Because today's domesticated soybean varieties do not possess that salt tolerance gene, a soybean breeder wanting to create a modern soybean variety that would grow well in salty soil could utilize a wide cross between a modern soybean variety (germplasm) and one of those salt-tolerant wild-type soybean species (i.e., retrieved from one of the seed banks utilized to store ancestral crop-plant relatives).

In addition to crop breeder use of a WIDE CROSS methodology, such crop "trait restoration" can be accomplished via certain other technologies. See also *Deoxyribonucleic acid (DNA), Gene, Genome, Trait, Soybean plant, Germplasm, Traditional reeding methods, Wide cross, Deletions.*

Trans Fatty Acids

One of the two isomeric forms that fatty acids can exist in. Trans fatty acids are naturally present in some meat and dairy products (which constitute approximately 5% of the average American diet). See also *Fatty acid, Isomer, Stereoisomers, Hydrogenation.*

Trans-Acting

Refers to a diffusible chemical/substance which can simultaneously impact at least some of the organelles within a living cell. See also *Cell, Organelles.*

Trans-Acting Protein

A *trans*-acting protein has the exceptional property of acting (having an effect) only on the molecule of DNA (deoxyribonucleic acid) from which it was expressed. See also *Protein, Deoxyribonucleic acid (DNA), Express, cis-acting protein, ChIP.*

Transactivating Protein

Refers to a specific protein which "switches on" a cascade of genes/gene regulation. See also *Protein, Transactivation, Gene, Cascade, Gene expression cascade, Sterols, Transcription activators, Viral transactivating protein, Liver X receptors (LXR).*

Transactivation

Refers to the activation (i.e., start/increase) of transcription via the "binding" of a transcription factor to a given DNA regulatory sequence. See also *Transcription, Transcription factors, Regulatory sequence, Sterols, Liver X receptors (LXR).*

T

Transaminase

A large group of enzymes that catalyze the transfer of the amino group from any one of at least 12 amino acids to a keto acid to form another amino acid. Also known as aminotransferases. See also *Enzyme, Amino acid.*

Transamination

The reaction of the enzymatic removal and transfer of an amino group from one specific compound to another. See also *Transaminase, Amino acid.*

Transcript

Term used to refer to the various segment(s) of messenger RNA (mRNA) that result from transcription of a gene. See also *Gene, Transcription, Primary transcript, Messenger RNA (mRNA), Transcriptome, Central dogma (NEW).*

Transcriptase

See *RNA polymerase.*

Transcription

The enzyme-catalyzed process whereby the genetic information contained in one strand of DNA (deoxyribonucleic acid) is used as a template to specify and produce a complementary mRNA strand. Transcription may be thought of as a rewriting of the information contained in DNA into RNA. The language, however, is the same—both are nucleic acid based. This is in contrast to translation, in which the information is translated from one language (RNA, nucleic acid-based) into another language (protein, amino acid-based). See also *Gene expression, Translation, Messenger RNA (mRNA), Genetic code, Deoxyribonucleic acid (DNA), Transcription factors, Transcription unit, Anticoding strand, Activator (of gene), Editing, Startpoint, Bursting, Long noncoding RNAs.*

Transcription Activator-Like Effector Nucleases

Acronym TALENs, they are artificial restriction enzymes generated by fusing the TALE DNA-binding domain (i.e., proteins originally discovered to be secreted by *Xanthomonas* bacteria) to a DNA cleavage domain (nuclease). During 2012, Bruce Whitelaw created pigs that modeled human atherosclerosis disease, by utilizing TALENs to disrupt genes that code for LDL receptors in those animals' livers. Without these LDL receptors to remove cholesterol-containing LDLs from the blood, LDLs buildup and lead to atherosclerosis. See also *Talens, Nuclease, Enzyme, Restriction enzymes, Protein, Fusion protein, Deoxyribonucleic acid (DNA), Gene, Knockout, Atherosclerosis, Receptors.*

Transcription Activator-Like Effectors

Acronym TALEs, they are protein molecules that selectively bind to very specific DNA sequences. The DNA-binding site is determined by a specific amino acid sequence within each TALE, and that amino acid sequence can be tailored by the maker of each TALE, so the TALE binds to only one spot on a cell's DNA. When fused to an applicable *nuclease* (i.e., enzyme that cuts DNA at specific location), the fused molecule (now called a) *Transcription Activator-Like Effector Nucleases* or TALEN) can be utilized by man to change the sequence of a selected gene (thereby changing its function) or knock out that gene. See also *TALEs, Protein, Deoxyribonucleic acid (DNA), Gene, Sequence (of a DNA molecule), Amino acid, Sequence (of a protein molecule), Cell, Fusion protein, Enzyme, Nuclease, Knockout.*

Transcription Activators

Refers to transcription factors (proteins and/or other molecules) which interact with regulatory sequences within DNA (in cell). By binding directly to those regulatory sequences (usually at multiple sites on the sequence), and "recruitment" of modifying molecules (chromatin remodeling elements) to also come to the site(s) on the DNA, transcription activators cause transcription (of a given gene) to begin or to increase.

Classes of transcription activators include the following:

- *Nuclear receptors*: These receptors (in cell's outer membrane) convey a "signal" from outside the cell all the way into the DNA within the cell's nucleus. For example, when the steroid hormone cortisol (i.e., a chemical "signal") binds to the glucocorticoid receptor (GR) in cells, the GR (protein molecule) enters the cell's nucleus and binds to the *glucocorticoid response element* (i.e., a specific regulatory sequence) in that cell's DNA. That then causes a *second* GR molecule to bind to that same glucocorticoid response element. The binding of the two (i.e., a GR dimer) to the glucocorticoid response element "activates" (i.e., starts) transcription of the gene (in the DNA molecule) immediately adjacent to the glucocorticoid response element.
- *Catabolite activator proteins (CAP)*: CAPs activate transcription by binding to the DNA near Class I CAP promoter sites or Class II CAP promoter sites on the DNA molecule and RNAP, which results in an amalgamated molecular structure known as *RNAP-promoter complex.*

See also *Protein, Genetic code, Coding sequence, Cell, Nucleus, Regulatory sequence, Transcription, Transcription factors, Deoxyribonucleic acid (DNA), Gene, Activator (of gene), Signal transducers and activators of transcription (STATs), Nuclear receptors, CAP, RNA Polymerase, Signaling, G-proteins, CD4 Protein, Long noncoding RNAs, Hormone, Retinoid X receptors, Cortisol, Positive control, Bursting.*

Transcription Factor-Binding Site

Refers to a sequence (segment) of DNA within an organism's genome (DNA) which is "recognized" and bound (i.e., "adhered to") by a transcription factor, thereby activating (or repressing, for *repressor*) transcription. See also *Sequence (of a DNA molecule), Organism, Deoxyribonucleic acid (DNA), Transcription factors, Transcription.*

Transcription Factors

Proteins and/or other chemical compounds that interact with each other, and with regulatory sequences within DNA (when immediately adjacent to the DNA in a cell), to either facilitate (i.e., "turn on") or inhibit (i.e., "turn off") the activity (i.e., coding for proteins) of that DNA's genes. Transcription factors hold potential to:

T

- Cure diseases (e.g., by blocking the deleterious effects of certain disease-causing genes).
- To assist farmers in crop protection (e.g., by *switching on* the genes that cause crop plants to initiate "cold hardening" or certain types of insect resistance mechanisms).
- To improve human health (e.g., PUFA modulation of genes, modulation of genes by some vitamins, etc.).

Some transcription factors are an integral component in certain *gene expression cascades*. For example, a gene expression cascade is initiated by the first gene causing expression of a transcription factor, which then *itself* interacts with cell's DNA to either cause or speedup yet *another* gene expression. The protein resulting from that second gene expression is yet *another* transcription factor which triggers another (i.e., third) gene expression, and so on. See also *Protein, Genetic code, Coding sequence, Deoxyribonucleic acid (DNA), Cell, Inhibition, Gene, P53 gene, Transcription, P53 Protein, CBF1, Cold hardening, Regulatory sequence, Express, Gene expression, Gene expression cascade, Down regulating, Vitamin, Polyunsaturated fatty acids (PUFA), Recombinase, Activator (of gene), Zinc finger proteins, Transcription activators, Transcription factor-binding site, Oleic acid, Fatty acid-binding proteins, NFκB, FT protein, Large intervening noncoding RNA, Bursting.*

Transcription Unit

A *group of genes* that code for functionally related RNA molecules or protein molecules. This group of genes is expressed (transcribed) together (i.e., as a unit, thus the name). See also *Express, Gene, Transcription, Translation, Genetic code, Coding sequence, Deoxyribonucleic acid (DNA), Ribonucleic acid (RNA), Ribosomes.*

Transcriptional Activator

A regulatory sequence which binds (i.e., adheres) to a DNA transcription control sequence, and thereby activates (i.e., begins/increases) the transcription of a gene. See also *Regulatory sequence, Gene, Deoxyribonucleic acid (DNA), Transcription, Transcription factor, Control sequence, Gene expression, Transactivation, Bursting.*

Transcriptional Profiling

See *Transcription, Gene expression profiling, Metabolite profiling, Gene expression analysis.*

Transcriptional Repressor

A regulatory sequence (segment of DNA) which "binds" (i.e., adheres to) a *DNA transcription control sequence*, and thereby represses (decreases/halts) the transcription of a gene. See also *Regulatory sequence, Gene, Deoxyribonucleic acid (DNA), Transcription, Transcription factor, Control sequence, Gene expression, Positive control, Down regulating, Long noncoding RNAs, Transactivation, Bursting.*

Transcriptome

Refers to the entire (complete, possible) set of all gene *transcripts* (i.e., RNA segments resulting from gene transcription process) in a given organism. Also to knowledge of their roles in that organism's structure, growth, health, disease (and/or that organism's resistance to disease), and so on. This set of transcripts produced by the organism's genome includes mRNAs, noncoding RNAs, miRNAs, and other small RNAs. Their roles are predominantly resultant from the impact of each protein molecule (i.e., resulting from the mRNA segments being *translated* in cells' ribosomes), which is itself due to the protein molecule's composition *and its tertiary conformation* (which determines the protein's impact in the organism's tissues, metabolism, etc.).

More than one protein can result from each gene in an organism's genome, due to:

- Interactions *between* genes.
- Interactions between genes and their (protein) products.
- Interactions between genes and some environmental factors.

Mechanistically, this results in different proteins being produced (during translation process) via:

- *Alternative splicing* of the mRNA transcript. For example, a single intronic base substitution which is present within the *IKAP* gene (i.e., the allele responsible) for the disease known as *Familial Dysautonomia* affects the splicing of the IKAP transcript (i.e., the mRNA segment that determines which specific protein is subsequently "manufactured" by the ribosomes). Up to eight different proteins can be produced from each human gene, via alternative splicing.
- Varying translation start or stop site (on the gene).
- *Frameshifting* (i.e., different set of triplet codons in the mRNA/transcript is translated by the ribosome).

See also *Gene, Transcript, Messenger RNA (mRNA), Small RNA, Micro RNAs, Coding sequence, Translation, Codon, Protein, Genome, Genetic code, Central dogma (new), Organism, Conformation, Metabolism, Tertiary structure, Intron, Base, Alternative splicing, Frameshift, Splice variants, Spliceosomes, Long noncoding RNAs.*

Transducing Phage

See *Transduction (gene).*

Transduction (Gene)

The transfer of bacterial genes (DNA) from one bacterium to another by means of a (temperature-disabled or defective) bacterial virus (bacteriophage). There exist two kinds of transduction: specialized and general.

In the case of *specialized transduction*, a restricted group of host genes becomes integrated into the virus genome. These "guest" genes usually replace some of the virus genes and are subsequently transferred to a second bacterium. In the case of *generalized transduction*, host genes become a part of the mature virus particle in place of or in addition to the virus DNA. However, in this case the genes can come from virtually any portion of the host genome and this material does not become directly integrated into the virus genome. In the case of plants, the vector can be *Agrobacterium tumefaciens.* See also *Bacteriophage, Vector, Genetic code, Agrobacterium tumefaciens, Retroviral vectors, Gene delivery, Transfection.*

Transduction (Signal)

See *Signal transduction.*

Transfection

This term has several different meanings, depending on the context in which it is used:

- A word utilized most generally to refer to insertion of DNA segments (e.g., genes) into cells (e.g., via electroporation, endocytosis, etc.). For example, insertion of a gene that codes for GFP into a cell in such a manner that it causes the cell (or *class of cells*) to fluoresce under certain conditions/illumination.
- A word utilized since 1998 to refer to insertion of certain double-stranded RNA (dsRNA) segments into cells (via electroporation, certain RNAPs, certain viral infections, etc.); to cause RNA interference (RNAi)/knockout/silencing.
- A word utilized to refer to insertion of (already "diced") siRNA segments (abbreviated d-siRNA); to cause RNA interference/knockout/silencing.
- A word utilized to refer to insertion of (*complementary-to-constitutive mRNA*) antisense oligonucleotides, to cause cosuppression/knockdown.
- A word utilized to refer to insertion of a *DNA-biologically active protein* (e.g., transcription factors, STATs, helical peptides, etc.) or other DNA/RNA-biologically active small molecule into cells in order to impact gene expression.
- A special case of transformation in which an appropriate recipient strain of bacteria is exposed to (free) DNA isolated from a transducing phage with the "take-up" of that DNA by some of the bacteria and consequent production and release of complete virus particles. The process involves the direct transfer of genetic material from donor to recipient.

See also *Marker* (*Genetic marker*), *Transformation*, *Electroporation*, *Gene*, *Virus*, *Cell*, *Bacteria*, *Deoxyribonucleic acid* (*DNA*), *Transduction* (*gene*), *Green fluorescent protein* (*GFP*), *Coding sequence*, *Fluorescence*, *Protein*, *Ribonucleic acid* (*RNA*), *RNA interference* (*RNAi*), *dsRNA*, *Messenger RNA* (*mRNA*), *Knockout*, *Gene silencing*, *Antisense* (*DNA sequence*), *Knockdown*, *Down regulating*, *Complementary* (*molecular genetics*), *Transcription factors*, *Signal transducers and activators of transcription* (*STATs*), *Biological activity*, *Reporter gene*, *siRNA*, D-siRNA.

Transfer RNA (tRNA)

Discovered in 1957 by Mahlon Bush Hoagland, they are a class of relatively small RNA (ribonucleic acid) molecules of molecular weight 23,000 to about 30,000. tRNA molecules act as carriers of specific amino acids during the process of protein synthesis. Each of the 20 amino acids found in proteins has at least one specific corresponding tRNA. Attachment of an amino acid to a tRNA molecule forms an active (amino-acyl) tRNA, which then functions as a *ribosomal adaptor* (tRNA adaptor).

The tRNA binds covalently with "its" specific amino acid and "leads" it to the ribosome for incorporation into the growing polypeptide (protein molecule) chain. See also *Ribonucleic acid* (*RNA*), *Molecular weight*, *Amino acid*, *Messenger RNA* (*mRNA*), *Polypeptide* (*protein*).

Transferases

Enzymes that catalyze the transfer of functional groups to certain molecules (from other molecules). See also *Transaminase*, *Enzyme*, *Hedgehog proteins*, *Functional group*, *Glycosyltransferases*.

Transferred DNA

See *Ti plasmid*.

Transferrin

The category of protein molecule(s) responsible for transporting iron (molecules) to tissues throughout the body via the circulatory system. See also *Protein*, *Transferrin receptor*, *HEME*, *Blood–brain barrier* (*BBB*), *Lactoferrin*.

Transferrin Receptor

The receptor molecule (located on the surface of cells throughout the body) that is responsible for binding to transferrin molecules, then bringing those iron-rich transferrin molecules into the cell via a process called receptor-mediated transcytosis, where the iron is subsequently released to be used by the cell. This process can also be utilized to get certain drugs through the blood–brain barrier to treat disease in the brain. See also *Transferrin*, *Receptors*, *HEME*, *Blood–Brain Barrier* (*BBB*).

Transformation

The process in which free DNA is transferred directly into a competent recipient cell. The direct transfer of genetic material from donor to recipient. The acquisition (e.g., by bacteria cells) of new genetic markers (new traits coded for by the new DNA) via the process of transformation.

See also *Deoxyribonucleic acid* (*DNA*), *Transfection*, *Marker* (*genetic marker*).

Transforming Growth Factor-Alpha (TGF-Alpha)

An angiogenic growth factor produced by tumor cells. It is able to induce specific malignant characteristics in normal cells (such as fibroblasts), thereby "transforming" those cells. TGF-alpha appears to possess a variety of potentially useful pharmaceutical properties, such as powerful stimulation of scar tissue formation following wounding of a tissue, as indicated by preliminary research. See also *Transforming growth factor-beta* (*TGF-beta*), *Growth factor*, *Nerve growth factor* (*NGF*), *Tumor*, *Fibroblasts*, *Angiogenic growth factors*.

Transforming Growth Factor-Beta (TGF-Beta)

An angiogenic growth factor produced by platelets and by tumor cells, it is able to induce specific malignant characteristics in normal cells (such as fibroblasts), thereby "transforming" those cells from epithelial phenotype to mesenchymal phenotype (thereby enabling cell motility), leading to metastasis.

TGF-beta stimulates blood vessel growth, even though it inhibits the division of endothelial cells. TGF-beta is a strong "attracting agent" for macrophages (i.e., TGF-beta is chemotactic) and appears to be responsible for the high concentrations of macrophages that are often found in tumors. TGF-beta has shown immunosuppressive activity (i.e., it suppresses the immune system). For example, transforming growth factor-beta works together with osteoinductive factor to promote bone formation by first causing connective tissue cells to grow together to form a matrix of cartilage (e.g., across a bone break) then bone cells slowly replace that cartilage. Research indicates that when injected (i.e., in the form of a pharmaceutical),

TGF-beta can inhibit ovarian cancer. An excess of TGF-beta tends to reduce the amount of muscle repair activity in the body. See also *Platelets, Metastasis, Transforming growth factor-alpha (TGF-alpha), Growth factor, Osteoinductive factor (OIF), Epithelium, Immunosuppressive, Nerve growth factor (NGF), Tumor, Fibroblasts, Angiogenic growth factors, Mitogen, Endothelial cells, Chemotaxis, Macrophage, Cell motility.*

Transgalacto-Oligosaccharides

A "family" of oligosaccharides (produced via enzymatic conversion of lactose, using β-glucosidase enzyme); some of which help to foster the growth of beneficial *bifidobacteria* in the lower colon of monogastric animals (e.g., humans, swine, etc.). See also *Oligosaccharides, Prebiotics, Bacteria, Bifidobacteria, Bifidus, Enzyme.*

Transgene

A "package" of genetic material (i.e., DNA) that is inserted into the genome of a cell via gene splicing techniques. May include promoter(s), leader sequence, termination codon, and so on. See also *Deoxyribonucleic acid (DNA), Gene splicing, Genome, Leader sequence, Promoter, Genetic code, Termination codon (sequence), Genetic engineering, Cassette.*

Transgenic

An organism whose gamete cells (sperm/egg) contain genetic material originally derived from an organism *other* than the parents or in addition to the parental genetic material. See also *Genetic engineering, Gamete, Nuclear transfer.*

Transgressive Segregants

Refers to offspring (e.g., created within a formal crop seed company breeding program) which possess significantly *different* traits/phenotypes than their parents. It results when allele pairs get separated from each other during meiosis (and subsequently sorted into different cells). See also *Gene, Allele, Segregation, Trait, Phenotype, Cell, Meiosis, Epistasis, Nerica.*

Transgressive Segregation

A plant breeding (propagation) technique, in which *genetically very different* members of the *same species* (e.g., derived via segregation) are mated with each other. The offspring of that mating can be more healthy, productive (e.g., fast growing), and uniform than their parents, a phenomenon known as "hybrid vigor." See also *Genetics, Species, F1 Hybrids, Hybridization (plant genetics), Segregation, Segregant.*

Transgressive Variation

Refers to the more pronounced development of a trait (or traits) in *subsequent generations* than were present in either of the initial parent organisms, due to the phenomenon of transgressive segregants. See also *Trait, Transgressive segregants.*

Transient Receptor Potential Channels

See *TRP channels.*

Transit Peptide

A peptide that, when fused to a protein, acts to transport that protein between compartments within eucaryotic cells. Once inside the "destination compartment," the transit peptide is cleaved off the protein and that protein is then free (to do its designed task). See also *Peptide, Protein, Eucaryote, Cell, Fusion protein, Gated transport, Vesicular transport, Chloroplast transit peptide (CTP).*

Transition

Refers to the replacement (i.e., in DNA or RNA molecule) of one purine by another purine or one pyrimidine by another pyrimidine. See also *Purine, Pyrimidine, Deoxyribonuncleic acid (DNA), Ribonucleic acid (RNA), Base substitution.*

Transition State (in a Chemical Reaction)

That point in the chemical reaction at which the reactants (i.e., chemical entities about to react with each other) have been "brought to the brink." It is a point in the chemical reaction process in which an "activated condition" is reached. From this point the probability of the reaction going to completion and producing a product is very high. The transition state separates (energetically) products from reactants. It is viewed as being at the top of the energy barrier separating reactants and products. The reacting species in the transition state can, because of their location at the "top" of the energy barrier, "fall" to either products or reactants. See also *Catalyst, Endergonic reaction, Activation energy, Free energy, Catalytic antibody, Semisynthetic catalytic antibody, Exergonic reaction.*

Translation

The process via which protein molecules are synthesized (made) whereby the genetic information present in an mRNA molecule directs the order of incorporation of specific amino acids, and hence the growth of the polypeptide chain during protein synthesis. One can think of translation as the process of translating one language into another. In this particular case, the nucleic acid-based language represented by mRNA is translated into the amino acid-based language of proteins. See also *Coding sequence, Codon, Ribosomes, Messenger RNA (mRNA), Amino acid, Ribosomes, Spliceosomes, Protein, Gene, Genetic code, Alternative splicing.*

Translational Repression

See *Micro-RNAs.*

Translocation

Genetic mutation in which a section of a chromosome "breaks off" and moves to a new (abnormal) position in that (or a different) chromosome. See also *Gene, Chromosomes, Genetic code, Coding sequence, Transposition, Deoxyribonucleic acid (DNA), Mutation, Introgression, Jumping genes, Hot spots.*

Translocation (of Protein Molecules)

The movement of a protein molecule:

- From one location/compartment within a cell to another location.
- Across a cellular membrane (e.g., a plasma membrane).

See also *Protein, Cell, Membranes (of a cell), Membrane transport, Membrane transporter protein, Leader sequence (protein molecule), Gated transport, Chaperones, Plasma membrane.*

Translocon

A transmembrane (i.e., through the membrane) protein molecule present within a cell's plasma membrane, that is able to:

- Secrete a newly synthesized protein molecule (i.e., transits through the cell's plasma membrane).

Or

- Insert a newly synthesized protein molecule *into the cell membrane* where it remains, to be subsequently modified (e.g., glycosylated, phosphorylated, etc.) or to function in place (e.g., as a cell receptor) without further chemical modification.

See also *Protein, Transmembrane proteins, Cell, Plasma membrane, Membrane transport, Synthesizing (of proteins), Receptors, Glycosylation, Phosphorylation, Posttranslational modification of protein.*

Transmembrane Proteins

Refers to those protein molecules that extend from one side of a cell membrane to the other side of that membrane.

For example, G-proteins are transmembrane proteins that act to accomplish signal transduction (i.e., convey "signal" from outside the cell to one or more internal cell parts). EGF receptors bind to EGF molecules (e.g., passing by in the blood), then both enter the cell (through the cell membrane) together, where the EGF stimulates growth/division of that cell. See also *Protein, Cell, Plasma membrane, Receptors, Membranes (of a cell), Translocon, Membrane transport, ABC transporters, EGF receptor, G-proteins, Cecrophins (lytic proteins), Magainins, Signal transduction, Signaling, Epidermal growth factor (EGF), Gated transport, Porin, SID-1 protein.*

Transport Proteins

Refers to protein molecules which are utilized to carry (i.e., transport) compounds within the body of an organism. For example, the transport protein known as hemoglobin is used by the human body to transport oxygen from the lungs to (all of) the cells of the body.

For example, fatty acid-binding proteins are used by cells to transport specific fatty acids from the cell's plasma membrane to the needed destination within the cell's interior. See also *Protein, Organism, Cell, Hemoglobin, Fatty acid, Fatty acid-binding proteins, Plasma membrane, Nuclear pore complexes.*

Transporters

See *Transport proteins.*

Transposable Element

A short sequence (segment of molecule) of DNA (deoxyribonucleic acid) that is able to replicate and insert one copy (of itself) at a new location on the genome (i.e., DNA molecule within same organism).

For example, the mouse pigment gene known as *agouti* can be rendered defective when a certain transposable element inserts itself in that gene's nearby regulatory DNA sequence, resulting in yellow or mottled mouse fur which is less effective in camouflaging the mouse from predators.

Approximately 85% of the genome of corn/maize (*Zea mays* L.) consists of transposable elements. See also *Transposon, Deoxyribonucleic acid (DNA), Sequence (of a DNA molecule), Organism, DNA methylation, Genome, Corn.*

Transposase

An enzyme that is required for transposition to occur (i.e., this enzyme assists movement of a transposon from one location to another within a cell's DNA). It is coded for by the transposon known as the P element. See also *Transposition, Transposon, Enzyme, Genetic code, Coding sequence, Deoxyribonucleic acid (DNA).*

Transposition

Movement of a gene or set of genes from one site in the genome to another without a reciprocal exchange (of DNA). See also *Gene, Jumping genes, Genome, Transposon, Transposase, Hot spots, Deoxyribonucleic acid (DNA).*

Transposon

A DNA (deoxyribonucleic acid) sequence (segment of molecule) able to replicate and insert one copy (of itself) at a new location in the genome (i.e., a transposition of location). Discovered in 1950 by geneticist Barbara McClintock in corn (maize) plants (*Zea mays* L.); and in bacteria a decade later by Joshua Lederberg. Transposons can either carry genes along one organism's genome or even into another organism's genome (e.g., via sexual conjugation, in bacteria). By such sexual conjugation, transposons can carry genes that confer new phenotypic properties (e.g., resistance to certain antibiotics, for a given bacterial cell). See also *Deoxyribonucleic acid (DNA), Replication (of virus), Genome, Transposition, Transposase, Sequence (of a DNA molecule), Corn, Jumping genes, Gene, Sexual conjugation, Phenotype, Conjugation, DNA methylation.*

Transversion

The substitution of a purine for a pyramidine or of a pyramidine for a purine (at a specific site, within a given nucleotide in a molecule of DNA). That substitution generally results from a mutation in an organism's DNA. See also *Nucleotide, Deoxyribonucleic acid (DNA), Single-nucleotide polymorphisms (SNPs), Mutation, Base substitution.*

TRANSWITCH®

A "sense" technology used to "turn off " (suppress) a gene (e.g., the one that causes tomato to ripen) that causes an unwanted effect (e.g., premature softening of tomato). TRANSWITCH® and its registered trademark are owned by DNA Plant Technology Corp. See also *Gene silencing, Suppressor gene, Sense.*

Trap Crop

See *Root-knot nematode.*

Trastuzumab

A ("humanized") monoclonal antibody against mutated-HER2 receptor that was approved by the U.S. FDA during 1998 as the

T

pharmaceutical Herceptin to be utilized in conjunction with chemotherapy against metastatic (mutated HER2) breast cancer. See also *Monoclonal antibodies (MAb), Cancer, HER-2 receptor, Metastasis, Fluorescence in situ hybridization (FISH), Food and Drug Administration (FDA), Humanized antibody, Gene, HER-2 gene, ADO-Trastuzumab emtansine.*

Treatment Investigational New Drug

See *"Treatment" IND regulations.*

Treatment System

Also sometimes called *Treatment Process*. Refers to the measure(s) utilized (e.g., by an agricultural-commodity-importing country) to prevent the *introduction* of a "quarantine pest" into a "pest free area."

For example, some countries which are free of relevant insect pests may require that certain agricultural commodities (i.e., containing that particular live pest) be fumigated with specific pesticide(s) before those commodity shipments are allowed to enter that country. See also *International Plant Protection Convention (IPPC), Quarantine pest, Introduction.*

Trehalose

A disaccharide (simple sugar) that is naturally synthesized (i.e., "manufactured") by many plants and animals in response to the stresses of freezing, heating, or drying. That is because trehalose protects certain proteins (needed for life) and prevents loss of crucial volatile (i.e., easily evaporated) compounds from organisms during those stressful (e.g., dry, frozen, or hot) conditions. Trehalose also provides a source of quick energy after the stressful conditions have passed. That is why dried baker's yeast (which contains up to 20% trehalose by weight) can be stored in its dry state for many years, yet quickly leavens bread dough within minutes of being rehydrated (i.e., rewetted).

Trehalose accomplishes this protection by forming a nonhygroscopic "glass" on the surfaces of cells and large molecules. It immobilizes and stabilizes large molecules (e.g., proteins) but still allows water to diffuse out so complete drying can occur. Thus, trehalose holds potential as a food additive to keep proteins (e.g., eggs) fresh in the dried form. In 1991, the United Kingdom approved trehalose for use in food. Trehalose hydrolyzes (e.g., during digestion) into two molecules of glucose.

Due to its trehalose content, the Resurrection Plant (also known as *Rose of Jericho*) can be "resurrected" (after years in a state of total dessication) by rehydrating it.

When larvae of the midge *Polypedilum vanderplanki* detect the water disappearing from the small rock pools they live in, those larvae synthesize (i.e., manufacture) trehalose and utilize it to protect their body tissue as it dehydrates (e.g., during a drought period). When the pools later fill with water again, the larvae rehydrate and continue their development into adult midges. See also *Disaccharides, Protein, Glucose (GLc), Hydrolysis, Conformation, "Switch" Proteins, Tertiary structure, Protein folding.*

Trehalose 6-Phosphate

Abbreviated T6P, it is a phosphate-containing that exerts control over the yield of the corn (maize) plant because it drives the allocation of that plant's sucrose, to different parts of the plant during growth and development. Via altering the amounts of T6P in key cells that deliver sucrose to the developing seeds within the cobs, additional sucrose is transported into the growing corn kernels. This increases the seed numbers per cob and the corn plant's yield of grain. See also *Trehalose, Sucrose, Cell.*

Tremorgenic Indole Alkaloids

A "family" of toxic alkaloids (chemical compounds) that are naturally produced (e.g., within some plants) by certain fungi (i.e., which sometimes grow in those plants).

For example, the alkaloid known as *Penitrem D* is produced by certain fungi which grow in some grass species. It causes tremors, weakness, lack of coordination, and convulsions in animals that consume those fungus-infested grasses. See also *Alkaloids, Toxin, Fungus, Endophyte.*

TR-FRET

Acronym for *time-resolved fluorescence resonance energy transfer.* See *Fluorescence resonance energy transfer (FRET).*

Triacyglycerides

See *Triglycerides.*

Triacylglycerols

See *Triglycerides.*

Trichoderma harzianum

A microorganism that possesses (natural) fungicide activity. See also *Bacillus thuringiensis (B.t.), Wheat take-all disease, Fungus, Fungicide, Crop biologicals.*

Trichomes

Refers to certain outgrowths of some plants (e.g., where excess soil-source salt is stored, after it has been taken in by halophyte plants). See *Halophytes.*

Trichosanthin

An enzyme extracted from a specific Chinese plant. It has been discovered to "cut apart" the ribosomes in some cells that are infected with the HIV (i.e., AIDS) virus, thus potentially stopping the virus and preventing infection of additional cells. See also *Ribosomes, Acquired immune deficiency syndrome (AIDS), Enzyme, Protein, Human immunodeficiency virus type 1 (HIV-1), Human immunodeficiency virus type 2 (HIV-2).*

Triglycerides

The primary constituent of fats or oils; triglycerides are molecules that consist of three fatty acids attached to a glycerol "molecular backbone." More accurately called *triacylglycerols*, although long-term historical usage of "triglycerides" has made the latter term more common (though not totally accurate).

Similarly, the term "diglyceride" is often used to refer to those molecules which consist of two fatty acids attached to a glycerol "molecular backbone." "Diglycerides" (more accurately called *diacylglycerols*) can result from the splitting off (i.e., hydrolysis) of one fatty acid from a triacylglycerol ("triglyceride") molecule (e.g., during fat breakdown/oxidation), or from the combination of two fatty acids with glycerol (e.g., during synthesis of fats).

The "triglyceride level" in human bloodstream refers to the blood's content of noncholesterol total fats. Research during the 1990s provided evidence that high blood levels of triglycerides in humans (e.g., immediately after meals) can contribute to thrombosis.

In the case of plant vegetable oils (i.e., found in seeds), the triacylglycerols get broken down (hydrolyzed) during and following germination of the seeds, in order to supply the energy and carbon needed by the growing plant. See also *Fats, Thrombosis, Fatty acid, Saturated fatty acids (SAFA), LPAAT protein, Unsaturated fatty acid, Hydrolysis, Oxidation (of fats/oils/lipids), Adipocytes, Fructose oligosaccharides, Bifidus, Polyunsaturated fatty acids (PUFA), Diacylglycerols, Medium chain triacylglycerides, Hydrolysis.*

Trinucleotide Repeat

Refers to when a particular sequence of three nucleotides is repeated within an organism's DNA.

Some human diseases (e.g., polyglutamine disease) result when a person's DNA contains an *expanded* trinucleotide repeat (i.e., one in which it is repeated *too many* times within the DNA). See also *Nucleotide, Deoxyribonucleic acid (DNA), Organism.*

Triplex-Forming Oligonucleotides Induced Recombination

See *Oligonucleotide-mediated mutagenesis, Genome editing.*

Triploid

Refers to organisms that possess three sets of chromosomes, instead of the normal two sets of chromosomes. Conversion of a diploid (i.e., two sets of chromosomes) organism to triploid can be done by man (e.g., certain fish, "seedless," grapes, etc.). For example, fish are ordinarily diploid. By exposing fish eggs to certain specific combinations of temperature and pressure, immediately after fertilization of those eggs, scientists can cause the resultant fish to become triploid. Triploid fish are unable to reproduce. This sterility is desired by man, in order to prevent certain fish (e.g., those that have been genetically engineered) from mating with wild fish.

Such induced (triploid) sterility also prevents the (genetically engineered) fish from wasting energy on the act of reproduction, so they grow faster and larger. That transfer (of energy use from reproduction to growth) also holds true for "seedless" grapes, watermelons, and so on. See also *Diploid, Chromosomes, Wheat, Anuploid, Polyploid.*

tRNA

Abbreviation for *transfer RNA*. See *Transfer RNA (tRNA).*

tRNA Synthetase

Abbreviated TyrRS, it is an enzyme that links the amino acid tyrosine to the DNA (gene) that codes for production of tyrosine within the body. When the organism is under stress, tyrosine can move to the cells' nuclei where it takes on a protective, stress-response role. See also *Enzyme, Synthase, Cell, Amino acid, Tyrosine (TYR), Deoxyribonucleic acid (DNA), Gene, Nucleus, Coding sequence, Stress proteins, Aminoacyl-transfer RNA synthetases.*

Tropism

Orientation movement of a sessile organism in response to a stimulus. Movement of curvature due to an external stimulus that determines the direction of movement. Also known as topotaxis. See also *Sessile, Chemotaxis.*

TRP Channels

Abbreviation for transient receptor potential channels. Refers to a "family" of approximately 30 ion channels that are present within the cell surface membrane (plasma membrane) of cells located throughout the human body. Among other functions, TRP channels enable humans to sense temperature, certain types of pain, and so on. Most TRP channels are *nonselective ion channels (i.e., they admit sodium ions, calcium ions, etc.).* When certain TRP channels (e.g., TRPV1) open on nerve cells, the resultant ion flow initiates a series of events within the cell that result in a pain signal getting sent to the brain. See also *Ion channels, Ion, Cell, Plasma membrane.*

Trypsin

A proteolytic (protein *molecular chain*-cutting) enzyme that is produced by the pancreas, to facilitate digestion within certain animals.

Trypsin cleaves polypeptide (protein) molecular chains on the carboxyl (group) side of arginine and lysine units (residues), and it is often utilized by man to break-apart protein molecules (e.g., to enable scientists to study that protein's constituent peptides). See also *Arginine (arg), Lysine (lys), Protein, Peptide, Polypeptide (protein), Proteolytic enzymes, Proteases, Chymotrypsin, Trypsin inhibitors, Digestion (within organisms), Cowpea trypsin inhibitor (CpTI).*

Trypsin Inhibitors

Compounds present in certain plants (e.g., squash, soybeans, etc.) that inhibit the activity (i.e., protein cleavage, which aids digestion) of proteases (i.e., protein-cleaving enzymes such as trypsin or chymotrypsin) in the digestive systems of monogastric (i.e., single-stomach) animals (which include swine, poultry, and humans). Trypsin inhibitors (TIs) present in some varieties of squash include the *Ecballium elaterium trypsin inhibitor (EETI).*

TIs present in traditional varieties of soybeans (botanical name *Glycine max* (L.) Merrill) include:

- The Kunitz TI, which was first isolated and crystallized by M. Kunitz in 1945. It combines tightly with molecules of trypsin on a 1:1 basis, and thereby reduces the rate of protein cleavage affected by the trypsin enzyme, which inhibits the animal's digestion of protein(s).
- The Bowman–Birk trypsin inhibitor (BB T.I.), which was first described by D. E. Bowman in 1944. It combines with molecules of trypsin and chymotripsin, and thereby reduces the rate of protein cleavage affected by the trypsin and chymotrypsin enzymes, which inhibits the animal's digestion of protein(s).
- *Note*: During 2000, research by Frank Meyskins and William Armstrong indicated that consumption of BB T.I. in a manner that "bathes" mouth tissues in it (for extended

period of time) inhibits the development of the precancerous mouth lesions that can become oral cancer.

During 2007, research by B. Gran and colleagues indicated that consumption of BB T.I. suppresses autoimmune encephalomyelitis, by inhibiting the proteases which result from inflammation (thus consumption of BB T.I. could potentially help decrease the inflammatory demyelination that occurs in the disease MS).

- Certain free fatty acids and their acyl CoA esters, which reduce the rate of protein cleavage affected by the trypsin enzyme, which inhibits the animal's digestion of protein(s).

Heating of soybeans to a temperature of 212°F (100°C) for 15 minutes causes these TIs to be rendered inactive in soybeans, so the animal's digestion is unimpeded when it is fed soy that has been thus heated. See also *Trypsin, Chymotrypsin, Soybean plant, Protein, Proteases, Enzyme, Proteolytic enzymes, Digestion (within organisms), Polypeptide (protein), Biological activity, Acyl COA, Cowpea trypsin inhibitor (CpTI), EETI, Oral cancer, Autoimmune disease, Multiple sclerosis.*

Tryptophan (trp)

Discovered in 1900 by Frederick Hopkins, tryptophan is an essential amino acid. It is a precursor of the important biochemical molecules: indoleacetic acid, serotonin, and nicotinic acid. L-Tryptophan is used as a common feed additive for livestock to ensure that their diet includes an adequate amount of this essential amino acid.

In the field of biophysics, it has been shown that the presence of tryptophan within (the "peptide chain" of) a long protein molecule facilitates electron transfer between two metal atoms sited at distant locations in that protein molecule (i.e., metalloprotein). See also *Essential amino acids, Stereoisomers, Serotonin, Amino acid, Protein, Peptide, Metalloproteins, Metalloenzyme, ET.*

TSH

See *Thyroid stimulating hormone (TSH).*

Tuberculosis

See *Mycobacterium tuberculosis.*

Tubulin

Also known as αβ-*Tubulin*, it is a cell protein that polymerizes (i.e., links together into a long chain) to form microtubules in eucaryotic cells. Such microtubules fulfill a number of cellular functions and are required for intracellular component organization and for cell mitosis (i.e., the *cell division/reproduction process* in which a cell becomes two identical daughter cells).

When the drugs *paclitaxel* or Taxol™ are administered to body (e.g., in chemotherapy), they bind tubulin, which halts cell division and causes apoptosis in the affected cells (e.g., tumor cells) by binding *Bcl-2* (a protein that prevents apoptosis in cells).

Tubulin analogues present in bacterial cells include *FtsZ*, which is the principal cytoskeleton component of the *Z-ring* that first constricts and then divides the cells into two different cells during mitosis.

Tubulin imparts structural strength to the roots of many plants. See also *Cell, Protein, Eucaryote, Polymer, Microtubules, Mitosis, Paclitaxel, Taxol, Cancer, Chemotherapy, Apoptosis, Analogue, Cytoskeleton, Motor proteins, Gallic acid.*

Tumor

A mass of abnormal tissue that resembles normal tissues in structure, but which fulfills no useful function (to the organism) and grows at the expense of the body. Tumors may be malignant or benign. Malignant tumors (which infiltrate adjacent healthy tissues) can result from oncogenes and/or carcinogens. They can eventually kill their host if unchecked.

Epidermal growth factor encourages rapid cell growth in more than 50% of human tumors. See also *Cancer, Angiogenesis, Oncogenes, Proto-oncogenes, Cell, Carcinogen, Tyrosine kinase, Tyrosine kinase inhibitors (TKI), ATP synthase, Epidermal growth factor (EGF), Regulatory T cells, Oncolytics.*

Tumor Necrosis Factor (TNF)

Literally, *tumor death factor.* More precisely called *tumor necrosis factor-α*, it is an adipokine (i.e., protein synthesized by adipose cells which helps regulate the immune system) that has shown potential to combat (kill) malignant (cancer) tumors. TNF also can act to disrupt formation of the new vasculature (blood vessels) needed by the tumor for blood supply (e.g., when TNF is delivered in the form of a concentrated pharmaceutical via nanoparticles that preferentially accumulate in tumors).

During certain infections such as tuberculosis (TB), synthesis of TNF by the immune system at first rapidly causes macrophages to scavenge and kill TB bacteria. However, if the immune system produces TNF in excess, that excess TNF forces the macrophages to die and expel their TB bacteria captives alive. Both the antibacterial activity inside the macrophages, and the (later) death of the macrophages, are resultant from the production of reactive oxygen species by mitochondria in the macrophages.

TNF was discovered to be 10,000 times more toxic in humans than in rodents, where it had been tested for toxicity prior to human clinical tests. This example illustrates one potential pitfall of nontarget animal testing in that sometimes animal testing does not accurately reflect or foretell what will happen in humans.

Another drawback to using TNF-α as a drug to combat human tumors is the fact that it is one of the substances released by the body in excess (in the disease rheumatoid arthritis) that destroys tissue in the joints (TNF causes both inflammation and joint damage).

When released as part of the AIDS (disease), TNF causes cachexia, which is a "wasting away" of the body due to the body's reduced ability to process nutrients received via digestion.

The pharmaceuticals known as Remicade™ (infliximab) and Humira™ (adalimumab) are monoclonal antibodies approved by the U.S. FDA as treatments to inhibit the structural damage (to body joints) of the autoimmune disease rheumatoid arthritis.

The pharmaceutical known as Enbrel™ (etanercept) is a fusion protein approved by the U.S. FDA as a treatment to inhibit the structural damage (to body joints) of the autoimmune disease rheumatoid arthritis.

Each of those three pharmaceuticals specifically block TNF-α. See also *Adipokines, Adipose, Lymphokines, Necrosis, Tumor, Tumor-infiltrating lymphocytes (TIL cells), Protein, Autoimmune disease, T cell modulating peptide (TCMP), Digestion (within organisms), Toxicogenomics, Rheumatoid arthritis, Monoclonal antibodies (MAb), Adalimumab, Fusion protein, Food and Drug Administration (FDA), Nanoparticles, Macrophage, Mitochondria, Reactive oxygen species.*

Tumor Necrosis Factor-α

See *Tumor necrosis factor (TNF)*.

Tumor-Associated Antigens

Discovered by Thierry Boon in 1991, these are distinctive protein molecules that are produced in the surface membrane of tumor cells. These protein molecules are used by the body's cytotoxic T cells to recognize (and destroy) tumor cells, so such proteins hold promise for use in vaccines.

See also *Major histocompatibility complex (MHC), Macrophage, Tumor, T cell receptors, Antigen, T cells, Protein, Cell, Cytotoxic T cells, Human leukocyte antigens (HLA)*.

Tumor-Infiltrating Lymphocytes (TIL Cells)

The white blood cells of a cancer patient which have been:

1. Taken from that patient's tumor (where those white blood cells had been attempting to combat the cancer, albeit unsuccessfully).
2. Stimulated with doses of interleukin-2 (to make the lymphocytes more effective against the cancer).
3. Multiplied *in vitro* (i.e., outside of the patient's body) to make them more numerous (and thus more likely to successfully combat the cancer).

When these "souped up" lymphocytes (white blood cells) are reintroduced into that same patient's body, the lymphocytes (now called TIL cells because they have been "souped up") attack the cancer tumor (malignant growth) more vigorously than before. See also *Tumor, White blood cells, Lymphocyte, Lymphokines, T cells, Cytotoxic T cells*.

Tumor-Suppressor Genes

Also called anticancer genes. Genes within a cell's DNA that code for (i.e., cause to be manufactured in cell's ribosomes) proteins that hold the cell's growth in check. If these genes are damaged (e.g., by radiation, by a carcinogen, or by chance accident in normal cell division), they no longer hold cell growth in check and the cell becomes malignant (if the cell's DNA also contains a gene called an oncogene). Oncogenes must be present for the cell to become malignant, but oncogenes cannot cause a cell to become malignant until a tumor-suppressor gene is damaged. As with all genes, tumor-suppressor genes are inherited in two copies (alleles, one from each parent) and either copy can code for the proteins necessary for cell growth control. However, an organism that is born with one defective copy of a tumor-suppressor gene (or in whom one copy is damaged early in life) is especially prone to cancer (malignancy). See also *Gene, p53 gene, Genetic code, Meiosis, Deoxyribonucleic acid (DNA), DNA methylation, Carcinogen, Ribosomes, Oncogenes, Cancer, Tumor, Proto-oncogenes, Protein, Epigenetic therapy*.

Tumor-Suppressor Proteins

Proteins that are coded for (i.e., caused to be manufactured in the cell's ribosomes) by tumor-suppressor genes (e.g., the *p53* gene). Such proteins (e.g., the p53 protein) then act upon the cell's DNA in order to prevent uncontrolled cell growth and division (i.e., cancer).

In the case of human cervical cells, when they are infected by the human papilloma virus (HPV), the *E6* and *E7* genes of the HPV "turn off" tumor-suppressor genes in the cervical cells. As a result, approximately 70% of human cervical cancers are caused by HPV. See also *Tumor-suppressor genes, Gene, p53 gene, Protein, Genetic code, Meiosis, Deoxyribonucleic acid (DNA), Ribosomes, Oncogenes, Cancer, Tumor, Cell, Proto-oncogenes*.

Turnover Number

The number of molecules of a product produced per minute by a single-enzyme molecule when that enzyme is working at its maximum rate. That is, the number of substrate molecules converted into a product by one enzyme molecule per minute when that enzyme is "going (catalyzing) as fast as it can." See also *Enzyme, Transferases, Protease, Protein kinases, Proteolytic enzymes, Transaminase*.

Two-Dimensional (2D) Gel Electrophoresis

Abbreviated as DIGE, it is a technology/methodology discovered by Patrick O'Farrell in 1975, to separate the various proteins within a given biological sample, prior to their analysis. The proteins are moved by applying an electrical field in two distinct directions.

The sample is moved through two different gels (i.e., two different dimensions). The initial gel has a pH gradient that separates the different proteins based on their respective isoelectric points (i.e., separated on the basis of the protein molecule's charge).

The second gel (dimension) the sample is moved through is a gel that separates the protein molecules based on their individual molecular weights. That gel acts as a "molecular sieve" (i.e., smaller proteins move faster—and farther—than larger proteins do through this gel; in a fixed amount of time).

A fixed-time gel run (i.e., with appropriate gel and the appropriate electrical fields applied to the gel) leaves a scientist with as many as 1000 "spots" (of individual protein molecules) on the gel. Each "spot" is a collection of the molecules of one protein—from the original sample (mixture). To identify the protein(s) in the "spots," the scientist can:

- Stain them (e.g., with ethidium bromide, etc.), then illuminate them with a special (UV) light and assesses the entire gel with an electronic image scanner (or he assesses it visually).
- Cut out each spot from the gel, then analyze them (e.g., via use of MALDI-TOF mass spectrometer) to determine the identity of the protein in each spot.

From the pattern (coupled with intensity) of the "spots," two such gels could be utilized to:

- Confirm if two sample organisms were the same species/strain/variety.
- Determine the differences (in gene expression) between samples of diseased versus healthy tissues.

See also *Protein, Gel, Electrophoresis, Agarose, Polyacrylamide gel electrophoresis (PAGE), Page, Isoelectric point, Gene expression analysis, Proteomics, Molecular weight, Species, MALDI-TOF-MS, Isoelectric focusing (IEF), Protein solubilization, Molecular-weight size marker*.

T

Two-Hybrid Systems

Refers to yeast or bacterial "systems" (test systems built by scientists) which are utilized to detect specific protein–protein interactions (e.g., identification of the gene which codes for the *protein which is found to specifically interact with a known protein* when the known protein is exposed to a sample containing numerous "unknown" proteins).

Two-hybrid systems take advantage of the fact that certain transcriptional factors possess *two* distinctly separate functional domains. One of those two domains must interact with a second domain (e.g., in fusion protein containing portion of known protein) in order to cause transcription (e.g., of the *GAL4* gene in the yeast two-hybrid system).

The oldest two-hybrid system is that utilizing the *GAL4* transcriptional activation system of *Saccharomyces cerevisiae* yeast. *GAL4* is required to initiate expression of proteins that are central to galactose metabolism in that yeast. See also *Protein*, *Fusion protein*, *Gene*, *Coding sequence*, *Domain* (*of a protein*), *Gene expression*, *Target–ligand interaction screening*, *Ligand* (*in biochemistry*), *Protein microarrays*, *Transcription*, *Transcriptional activator*, *Yeast two-hybrid system*.

Type I Diabetes

The form of diabetes disease that usually strikes young people (thus, it was formerly known as juvenile or insulin-dependent diabetes). This disease is characterized by the body's immune system (antibodies) destroying the insulin-producing cells (Beta cells) of the pancreas.

If not treated in time (i.e., via insulin injections), the person can die suddenly. Even when treated, the person is at increased risk of blindness, atherosclerosis, coronary heart disease, heart attack, stroke, and kidney disease. See also *Diabetes*, *Beta Cells*, *Pancreas*, *Insulin*, *Insulin-dependent diabetes mellitis* (*IDDM*), *Calpain-10*, *Atherosclerosis*, *Coronary heart disease* (*CHD*), *Type II diabetes*, *Antibody*.

Type II Diabetes

The form of diabetes disease that usually strikes people who are older than 40 years old. Also known as ADULT-ONSET DIABETES or NON-INSULIN-DEPENDENT DIABETES, this disease is characterized by the body's tissues becoming insensitive to insulin.

Effects on the body include increased likelihood of blindness, atherosclerosis, coronary heart disease, heart attack, stroke, and kidney disease. See also *Diabetes*, *Inositol*, *Insulin-dependent diabetes mellitis* (*IDDM*), *Insulin*, *Calpain-10*, *Atherosclerosis*, *Coronary heart disease* (*CHD*), *Type I diabetes*, *PPAR*, *Resistin*.

Type II Restriction Endonucleases

See *Sticky ends*.

Type II Restriction Enzymes

See *Type II restriction endonucleases*.

Type Specimen

The actual physical specimen (e.g., a stuffed lizard or a dried insect) that a scientist (who describes and names a previously unknown species) must place in a museum (or other recognized repository) in order to have the right to name that newly discovered species. This "officially deposited specimen" is required for three purposes:

1. So that comparisons can later be made if there is ever a doubt whether another "new" species is simply a member of this same species (and thus already named).
2. So that taxonomists (who determine and keep the official scientific names by which scientists must refer to each of the world's organisms) can name each of the newly discovered species in accordance with the complex rules of the International Codes for Nomenclature; examples of such names in this glossary are *Arabidopsis thaliana*, *Escherichia coli*, and *Agrobacterium tumefaciens*.
3. So that patent claims for genetically engineered organisms can later be enforced.

See also *Species*, *Strain*, *Cladistics*, *Chakrabarty decision*, *American type culture collection* (*ATCC*), *Consultative Group on International Agricultural Research* (*CGIAR*).

Tyrosine (tyr)

A phenolic α-amino acid. It is a precursor of the hormones epinephrine, norepinephrine, thyroxine, and triiodothyronine. It is also a precursor of the molecule known as melanin (which is the pigment of a suntan). See also *Amino acid*, *Hormone*.

Tyrosine Kinase

A category of kinases (enzymes that facilitate the transfer of molecular phosphoryl groups from a molecule to another molecule which is targeted by that particular kinase) present throughout the human body. See also *Enzyme*, *Phosphorylation*, *Protein*, *Protein kinases*, *Receptor tyrosine kinase*, *Tyrosine kinase inhibitors* (*TKI*), *Protein tyrosine kinase inhibitor*, *BCR-ABl protein*, *Gleevec*™.

Tyrosine Kinase Inhibitors (TKI)

Refers to various compounds that inhibit the activity of tyrosine kinase enzyme (inside the body). Examples of TKI include genistein and the pharmaceuticals Iressa™ (gefinitib), Gleevec (imatinib mesylate), ibrutinib, and Tarcera™.

Because the activity of tyrosine kinase helps cancerous (tumor) cells to metastasize (spread/grow), consumption by humans of relevant TKI acts to help prevent (spreading of) certain cancers. See also *Enzyme*, *Kinases*, *Tyrosine kinase*, *Protein*, *Protein tyrosine kinase inhibitor*, *Biological activity*, *Cancer*, *Cell*, *Tumor*, *Genistein*, *Isoflavones*, *Gleevec*™, *Ibrutinib*, *Bcr-Abl protein*.

U

U.S. Patent and Trademark Office (USPTO)

The Washington, DC-based American Government agency that is responsible for common patent protection matters for all of America's 50 states and its territorial possessions. The USPTO allows the patenting of new and unique microbes, plants, and animals, as well as the new and unique methods to produce such biotechnology advances. See also *European Patent Office (EPO), Chakrabarty decision, Microbe, Genetic engineering, Plant's Novel Trait (PNT), Plant Breeder's Rights (PBR), Biotechnology, American Type Culture Collection (ATCC).*

Ubiquinone

Also called *coenzyme Q*, ubiquinone is present in all cells of eucaryotic organisms. Ubiquinone "escorts" nutrients across cell walls, where it plays an assisting role in the formation of adenosine triphosphate (ATP), in the cell's mitochondria.

Significant dietary sources of ubiquinone include soybeans, fish, meat, nuts, and so on. See also *Enzyme, Coenzyme, Eucaryote, Organism, Adenosine triphosphate (ATP), Mitochondria, Soybean plant, Soy protein.*

Ubiquitin

A small protein molecule present in all eucaryotic cells (i.e., it is ubiquitous) which plays an important role in "tagging" other protein molecules (e.g., old ones which are degraded/mis-folded or no longer needed) so they are destined (marked) for destruction (via proteolytic cleavage). Such proteins are then broken down in cell's proteasomes and their constituents recycled because they are damaged or no longer needed by the body. The ubiquitin molecules are then available for reuse.

Ubiquitin also similarly "tags" damaged chloroplasts in plants. A chloroplast can be damaged (i.e., its capacity to transfer electrons exceeded) by drought, extreme temperatures, or by too high irradiance (sun too bright for that plant), resulting in the generation of free radicals that could subsequently damage DNA, lipids, proteins, and so on. Thus, rapid removal of such chloroplasts is required to protect the plant cells.

Ubiquitin can also sometimes be covalently attached to DNA or its associated histones, thereby resulting in an epigenetic change to thus impacted cells. That tagging/marking for destruction occurs via fusion of the ubiquitin to the relevant protein molecules, and those "tagged" protein molecules are said to have been *ubiquitinated* or *ubiquitylated*. The overall ubiquitination process was discovered/delineated during the 1970s and 1980s by Irwin Rose, Aaron Ciechanover, and Avram Hershko.

Ubiquitination is also critical in cell division processes, DNA repair processes, apoptosis, transport of certain proteins from a cell's surface to the interior of cell, and protein synthesis processes.

In some specific instances, the fusion of ubiquitin to certain protein molecules in cells causes that "partner protein" to be expressed in larger amounts than previously. See also *Eucaryote, Protein, Protein folding, Conformation, Proteolytic enzymes, Proteasomes, Denaturation, Cell, Gene, Express, Ribosomes, Apoptosis, Auxins, Histones, Epigenetic, Posttranslational modification of protein, Free radical, Chloroplasts.*

Ubiquitinated

See *Ubiquitin, Auxins, Epigenetic.*

Ubiquitination

See *Ubiquitin.*

Ubiquitin-Proteasome Pathway

See *Ubiquitin.*

Ubiquitylation

See *Ubiquitin, Auxins.*

Ultracentrifuge

A high-speed centrifuge that can attain revolving speeds up to 85,000 rpm and centrifugal fields up to 500,000 times gravity. The machine is used to sediment (i.e., cause to settle out) and hence separate macromolecules (i.e., large molecules) and macromolecular structures in a mixture/solution. In general, a centrifuge is a machine that whirls test tubes around rapidly, like a merry-go-round, to force the heavier suspended materials (in the solutions in the test tubes) to the bottoms of those test tubes before the lighter material.

Ultrafiltration

A (mixture) separation methodology that uses the ability of synthetic semipermeable membranes (possessing appropriate physical and chemical natures) to discriminate between molecules in the mixture, primarily on the basis of the molecules' size/shape. Invented and developed by Dr. Roy J. Taylor in the 1950s and 1960s, ultrafiltration is typically utilized for the separation of relatively high-molecular-weight solutes (e.g. proteins, gums, polymers, and other complex organic molecules) and colloidally dispersed substances (e.g., minerals, microorganisms, etc.) from their solvents (e.g., water). See also *Dialysis, Membrane transport, Microorganism, Molecular weight, Protein, Polymer, Hollow fiber separation.*

Unfoldases

A category of enzymes that are sometimes involved in the "unfolding" (degradation) of a protein molecule's tertiary structure.

U

See also *Enzyme*, *Protein*, *Protein structure*, *Tertiary structure*, *Conformation*, *Native conformation*, *Protein folding*, *Denaturation*.

Union for Protection of New Varieties of Plants (UPOV)

A group of the world's countries that have jointly agreed to mutually protect the intellectual property (of owners, breeders) that is inherent in new plant varieties developed by man. These *intellectual property protections* are often collectively referred to as "Breeder's Rights."

Established in 1961, the secretariat for this union (UPOV) is in Geneva, Switzerland. See also *Plant Variety Protection Act (PVP)*, *U.S. Patent and Trademark Office (USPTO)*, *Plant's Novel Trait (PNT)*, *Plant Breeder's Rights (PBR)*, *European Patent Convention*, *European Patent Office (EPO)*, *Mutual Recognition Agreements (MRAs)*, *Community Plant Variety Office*.

Units (U)

A measure (quantitation) of biological activity of a substance, as defined by various standardized assays (tests). See also *Assay*, *Bioassay*.

Unsaturated Fatty Acid

A fatty acid containing one or more double bonds (between individual atoms of the molecule). See also *Fatty acid*, *Desaturase*, *Monounsaturated fats*, *Polyunsaturated fatty acids (PUFA)*.

Unwinding Protein

A protein which binds (i.e., adheres to) single-stranded DNA and thereby aids/stabilizes the unwinding of the DNA molecule's normal (helical/wound-up) structure. See also *Protein*, *Double helix*, *Deoxyribonucleic acid (DNA)*, *Plectonemic coiling*, *Helicase*, *Gene repair (natural)*, *Mismatch repair*, *Primosome*.

Up Regulating

See *Transcriptional activator*, *Positive control*.

UPOV

See *Union for Protection of New Varieties of Plants (UPOV)*.

Up-regulation

See *Positive control*, *Transcriptional activator*.

Uracil

A pyrimidine base, important as a component of ribonucleic acid (RNA). Its hydrogen-bonding counterpart in DNA is thymine. See also *Pyrimidine*, *Ribonucleic acid (RNA)*, *Base (Nucleotide)*, *Deoxyribonucleic acid (DNA)*.

Uridine

A nucleoside form of uracil. See also *Uracil*, *Nucleoside*.

Urokinase

A thrombolytic (i.e., clot-dissolving) enzyme used as a biopharmaceutical. See also *Thrombolytic agents*, *Tissue plasminogen activator (tPA)*, *Fibrinolytic agents*.

Urushiol

An oily liquid allergen found in poison ivy and poison oak plants.

USPTO

See the link *U.S. Patent and Trademark Office (USPTO)*.

V

Vaccine

Any substance, bearing antigens on its surface, that causes activation of an animal's immune system without causing actual disease. The animals' immune system components (e.g., antibodies) are then prepared to quickly vanquish those particular pathogens when they later enter the body. See also *DNA vaccines*, *"Naked" gene*, *Edible vaccines*, *Peyer's patches*, *Antigen*, *Cellular immune response*, *Antibody*, *Humoral immunity*, *Carrier protein*, *Cell surface engineering*, *Lutein*.

Vaccinia

A nonpathogenic virus that is believed to be a (modified) form of the virus that causes cowpox. Vaccinia readily accepts genes (inserted into its genome via genetic engineering) from pathogenic viruses so it can be used to make vaccines that do not possess the risk inherent in attenuated-virus vaccines (i.e., that the attenuated virus "revives" and causes disease). Such genetically engineered *vaccinia* codes for (presents) the proteins of the pathogenic virus on its surface, which activates the immune system (e.g., of vaccinated animal) to produce antibodies against that pathogenic virus. See also *Vaccine*, *Pathogenic*, *Virus*, *Gene*, *Gene delivery*, *Genetic engineering*, *Attenuated (pathogens)*, *Antibody*, *Macrophage*, *Complement cascade*, *Cellular immune response*, *Phagocyte*.

Vacuoles

A membrane-bound sac within a cell, within which water, food, waste, or salt, and so on, is temporarily stored. Also pigments, in certain plant cells. See also *Plasma membrane*, *Cell*, *Anthocyanidins*.

VAD

Acronym for *Vitamin A Deficiency*. See also *Golden rice*, *Vitamin*, *Beta carotene*, *Carotenoids*.

Vagile

Wandering or roaming (e.g., a microorganism that is not attached to a solid support tends to "wander" through its environment as it gets pushed about by currents of air or liquid). See also *Sessile*, *Vagility*.

Vagility

The ability of organisms to disseminate (e.g., spread throughout a given habitat). See also *Vagile*.

Vaginosis

The process whereby a cell internalizes an entity (such as a virus or a protein) that has bound to the cell's outer membrane. Once that "bound entity" is inside the cell, the cell membrane fuses together again. See also *Nuclear receptors*, *Receptors*, *Endocytosis*, *Transferrin*, *Virus*, *Blood–brain barrier (BBB)*.

Validation

See *Validation (of target)*, *Process validation*.

Validation (of Target)

The process of verifying that a selected "target" (e.g., receptor, ion channel, DNA segment, etc.) *of pharmaceuticals* functions as expected in a given disease process.

For protein "targets" (e.g., receptors), validation is done by cloning and expressing the gene that codes for that protein target (to evaluate its *biological impact* on the disease, *which cells* it is expressed in, etc.). Another tool utilized is short interfering RNA (siRNA).

Validation of a target often involves linking together the results of several different studies (e.g., clinical studies with patients who are suffering from the specific disease but were treated with different agents, etc.). See also *Target (of a therapeutic agent)*, *Biomarkers*, *Receptors*, *Nuclear receptors*, *Ion channels*, *Deoxyribonucleic acid (DNA)*, *Coding sequence*, *High-throughput screening (HTS)*, *Target–ligand interaction screening*, *Short interfering RNA (siRNA)*.

Valine (val)

An amino acid considered essential for normal growth of animals. It is biosynthesized (made) from pyruvic acid. See also *Amino acid*, *Essential amino acids*, *ALS gene*.

Value-Added Grains

See *Value-enhanced grains*.

Value-Enhanced Grains

Those grains that possess novel traits that are economically valuable (e.g., higher-than-normal protein content, better *quality* protein, higher-than-normal oil content, etc.). For example, high-oil corn (maize) possesses a kernel oil content of 5.8% of greater versus oil content of 3.5% or less for traditional No. 2 yellow corn.

Glutamate dehydrogenase (GDH) corn (maize) possesses a kernel protein content that tends to be approximately 10% greater than the protein content of traditional corn (maize) varieties.

High-amylose corn possesses a kernel amylose content of 50% or more of the total kernel starch, and so on. See also *High-oil corn*, *Protein*, *Amylose*, *High-amylose corn*, *Opague-2*, *Floury-2*, *Genetic engineering*, *Low-phytate corn*, *Low-phytate soybeans*, *Trait*, *High-lysine corn*, *High-methionine corn*, *High-phytase corn and soybeans*, *High-oleic oil soybeans*, *High-stearate soybeans*, *Glutamate dehydrogenase*, *High-sucrose soybeans*, *High-laurate canola*, *High-lactoferrin rice*.

V

Van der Waals Forces

The relatively weak forces of attraction between molecules (resultant from fluctuations in electron distribution) that contribute to intermolecular bonding (i.e., binding together two or more adjacent molecules). Historically, it was thought that van der Waals forces were always weaker than the hydrogen bond forces responsible for intramolecular bonding. However, in 1995, Dr. Alfred French discovered that van der Waals forces are primarily responsible for holding together a mass of cellulose molecules, with hydrogen bonding playing a lesser role.

During 2000, Kellar Autumn discovered that van der Waals forces (acting between *foot skin hairs* and the surface climbed) are responsible for enabling the Tokay gecko (*Gecko gecko*) to climb vertical surfaces and also to hang upside down. These forces work (i.e., to "adhere" a gecko's foot) even underwater or in a vacuum.

When certain nanoparticles (e.g., consisting of metallic core surrounded by a self-assembled monolayer of thiol-term. ligands) are placed on a very flat gold surface, those nanoparticles automatically arrange themselves into a two-dimensional crystal-like structure which maximizes the van der Waals forces attractions to all neighboring such nanoparticles. See also *Cellulose, Cellulase, Molecular weight, Weak interactions, Immobilization, Hydrogen bonding, Nanoparticles, Self-assembly (of a large molecular structure), Thiol group.*

Variable Number of Tandem Repeats (VNTRs)

Refers to certain short segments of DNA consisting of repeated sequences such as CACACACA that tend to occur in noncoding DNA and are sometimes referred to as variable number of tandem repeats (VNTRs). Also called microsatellite DNA. See also *Deoxyribonucleic acid (DNA), Sequence (of a DNA molecule), Microsatellite DNA.*

Variable Region

See *Antibody.*

Variant Calls

Refers to the base pair data specific to one particular sample (e.g., one patient's particular base pairs versus analogous ones within a reference genome) resultant from a DNA sequencing process. See also *Sequencing (of DNA molecules), Sequence (of a DNA molecule), Base (nucleotide), Base pair (bp), Base calls, Algorithm (bioinformatics).*

Variant Histones

See *Histones, Epigenetic marks.*

Variety (e.g., of a Crop Plant)

Also known as *cultivars (from "cultivated varieties")*, it refers to a distinct (and genetically stable) version of the crop, created by man. Crop varieties have been created with shortened growing seasons, increased resistance to diseases or pests, larger seeds, larger fruits, increased nutritional content, shelf life, and better adaptation to diverse ecological conditions under which crops are grown. Selective breeding, genomic recombination, new allele creation, mutation, and gene insertion/deletion/duplication/silencing have all been utilized to create new plant varieties. See also *Gene, Allele, Silencing, Introgression.*

Vascular Endothelial Growth Factor (VEGF)

A human growth factor (GF) that causes angiogenesis (i.e., growth/proliferation of blood vessels/endothelium and also endothelial cells). Discovered in 1989 by Napoleone Ferrara. See also *Growth factor, Endothelium, Endothelial cells, Angiogenesis, Aptamers, Nanofibers, Ranibizumab.*

VDR

Acronym for *vitamin D receptor.* See also *Vitamin.*

Vector

The agent used (by researchers) to carry new genes into cells. Plasmids currently are the biological vectors of choice; though viruses and other biological vectors such as *Agrobacterium tumefaciens* bacteria or BACs are increasingly being used for this purpose.

Nonbiological vectors include the metal microparticles (coated with genes) which are "shot" into cells by the Biolistic® gene gun. See also *Plasmid, Gene, Cell, Retroviral vectors, Protoplasm, Agrobacterium tumefaciens, Bacteria, Biolistic® gene gun, Microparticles, Baculovirus expression vectors (BEVs), BAC.*

VEGF

Acronym for vascular endothelial growth factor. See *Vascular endothelial growth factor (VEGF).*

Velvetbean Caterpillar

Refers to the Lepidopteran insect pest *Anticarsia gemmatalis* Hübner.

Vernalization

Refers to the process via which plants "sense" the cold of winter and activate the plant's innate defense responses to cold stress (i.e., thereby waiting for spring to begin their flowering).

For example, in the mouse-ear cress plant (*Arabidopsis thaliana*), a gene known as *FLC* represses flowering. However, exposure of that plant to prolonged cold will epigenetically silence that gene (thereby enabling flowering to begin when spring's warm weather arrives).

For example, in the wheat plant (*Triticum aestivum*), a gene known as *VRN2* represses flowering. However, exposure of that plant to prolonged cold will down-regulate that gene (thereby enabling flowering to begin when spring's warm weather arrives).

During 2003, Kan Wang and coworkers were able to initiate earlier vernalization in soybean, corn (maize), and rice plants via insertion of a *VRN2-analogous* tobacco plant gene. One result of that was to confer greater resistance to cold temperatures. See also *Wheat, Gene, Down regulating, Cold hardening, Gene silencing, Epigenetic, Steckling.*

Vernolic Acid

A Δ12-epoxy fatty acid (i.e., a fatty acid molecule which contains an epoxy ring molecular substructure), which is naturally produced in small amounts in the seeds of the wild plants *Crepis palaestina*, *Vernonia galamensis*, and *Euphorbia lagascae*.

Vernolic acid is utilized by man as a plasticizer in some plastics, adhesives, and paints. Also, in some composite materials (e.g., reinforced plastic products). See also *Fatty acid*.

Vero

Abbreviation for specific cell lines (utilized in cell culture) which are "descended" from kidney cells that were taken from an African green monkey. See also *Cell*, *Cell culture*.

Vertical Gene Transfer

See *Outcrossing*.

Very Low-Density Lipoproteins (VLDL)

VLDLs and LDLPs are the specific lipoproteins that are most likely to *deposit cholesterol* on artery walls inside the human body, which increases risk of coronary heart disease. See also *Low-density lipoproteins (LDLP)*, *Lipoprotein*, *Apolipoproteins*, *Cholesterol*.

Vesicle

A small vacuole. See *Vesicular transport*, *Vacuoles*, *Dynein*.

Vesicular Transport (of a Protein)

One of three means for a protein molecule to pass between compartments within eucaryotic cells. The compartment "wall" (membrane) possesses a "sensor" (receptor) that detects the presence of correct protein (e.g., after that protein has been synthesized in the cell's ribosomes) and then bulges outward along with that protein molecule. The membrane bulge-containing protein then "breaks off" and carries (transports) the protein to its destination in another compartment in the cell. See also *Protein*, *Eucaryote*, *Cell*, *Ribosomes*, *Microtubules*, *Signalling*, *Vaginosis*, *Endocytosis*, *Gated transport*.

VFP

Acronym for *Visible Fluorescent Proteins*. See *Visible fluorescent proteins*.

Viral Surface Proteins

Refers to external proteins extending from the viral surface (capsid) that are responsible for the virus' initial interactions with living cells. Examples include hemagglutinin and neuraminidase (i.e., the *H* and the *N* within the designation of a given viral strain such as the "H1N1 flu strain"), the GP120 protein, and so on. See also *Virus*, *Protein*, *Capsid*, *Hemagglutinin*, *Neuraminidase (NA)*, *Cell*, *Strain*.

Viral Transactivating Protein

The specific protein used by a lytic virus to "switch on" the cascade of gene regulation by which that virus "takes over" a healthy cell and subverts its molecular processes (machinery) to produce virus components. This (transactivating) protein is key to the whole lytic cycle of the virus and therefore a potential target for therapeutic intervention. See also *Lytic infection*, *Transactivating protein*, *Virus*, *Protein*, *Cell*, *Gene cascade*, *Transactivating protein*.

Virion

An intact virus particle (e.g., prior to it entering a cell to "infect" the cell).

A virion consists of a protein coat (known as a *capsid*) around a central core composed of genetic material (DNA or RNA). See also *Virus*, *Cell*, *Capsid*, *Protein*, *Ribonucleic acid (RNA)*, *Deoxyribonucleic acid (DNA)*.

Viroid

Refers to certain plant pathogens (i.e., plant-disease-causing virus-like entities) capable of reproducing on their own (unlike a true virus). See also *Pathogen*, *Virus*.

Virotherapy

Refers to the use of viruses (i.e., modified to prevent any harm to patent) in the treatment of a disease (e.g., cancer).

For example:

- During the 1950s, researchers showed that an adenovirus was able to attack and kill cervical cancer cells.
- During 2005, Stephen Russell and Irvin S. Y. Chen were able to "infect" cancer cells in mice with modified viruses that killed the cancer cells but did not harm the adjacent healthy tissues.
- During 2014, Stephen Russell and Mark Federspiel were able to infect the cancer cells of numerous tumors (multiple myeloma) in a human patient with a measles virus that killed those cancer cells but spared adjacent normal tissues.

See also *Magic bullet*, *Virus*, *Cancer*, *Adenovirus*.

Virtual HTS

See *In silico screening*, *High-throughput screening (HTS)*.

Virus

A simple, noncellular particle (entity) that can reproduce only inside living cells (of other organisms), which was first proved to exist in 1892 by Dimitry Ivanovsky. The simple structure of viruses is their most important characteristic. Most viruses consist only of a genetic material—either DNA (deoxyribonucleic acid) or RNA (ribonucleic acid)—and a protein coating. This (combination) material is categorized as a nucleoprotein.

Some viruses also have membranous envelopes (coatings). They are able to "inject" their genetic material into (host) cells because the internal pressure inside viruses is greater than inside the host cells.

Viruses are "alive" in that they can reproduce themselves—although only by taking over a cell's "synthetic genetic machinery"—but they have none of the other characteristic of living organisms. After invading the cell(s) of an organism, viruses reprogram that cell to make it more conducive to virus reproduction. For example, the adenovirus reprograms the cell after invading it, to grow larger in size and to be able take in more glucose (e.g., from nearby bloodstream). The reprogramming also causes the cell to increase its usage of the glucose taken in to create energy and facilitate the virus to begin replicating inside the cell.

Viruses cause a large variety of significant diseases in plants and animals, including humans. Viruses can (e.g., as one part of the disease) cause *gene silencing.*

In 1970, it was discovered that some RNA viruses (known as retroviruses) are able to utilize viral reverse transcriptase to cause (infected) cells to make DNA copies of those particular viruses. That "viral DNA" is then able to incorporate itself into the (infected) cell in a manner which transforms that cell into a cancer cell. For example, infections by hepatitis B virus, hepatitis C virus, and some human papilloma viruses can lead to cancer.

Viruses present a philosophical problem to those who would speak of living and nonliving systems because in and of itself a virus is not "alive" as we know life, but rather represents "life potential" or "symbiotic life." See also *Vaccinia, Nucleoproteins, Retroviruses, Tobacco mosaic virus (TMV), Viral transactivating protein, Gene delivery, Adenovirus, Glucose, Replicon, NFκB, Cancer, Reverse transcriptases, Neoantigen, Kinesin, Viral surface proteins, dsRNA.*

Viscosity

A measure of a liquid's resistance to flow, as expressed in units called poise (P; g/cm/s). The degree of "thickness" or "syrupiness" of a liquid.

Visfatin

A protein discovered during 2004 by Iichiro Shimomura, which is secreted by visceral adipose cells (i.e., *vis*ceral *fat* tissue surrounding the body's organs).

In some aspects, visfatin has some of the same effects as insulin (e.g., stimulate glucose uptake by the body, which lowers blood sugar levels). Visfatin binds to insulin receptors in the body, without interfering with the same binding by insulin.

Visfatin is sometimes called *PBEF.* See also *Protein, Adipokines, Adipose, Glucose (GLc), Insulin, Receptors, PBEF.*

Visible Fluorescent Proteins

Refers to a number of proteins, many of them naturally present in some species of organisms, which visibly fluoresce when illuminated with light of relevant wavelength. Examples include green fluorescent protein, cyan fluorescent protein, yellow fluorescent protein, red fluorescent protein, kusabira orange, and so on. See also *Green fluorescent protein, Kusabira orange.*

Vitafoods

See *Nutraceuticals.*

Vitamers

See *Vitamin.*

Vitamin

The modern term that is descended from the original phrase "vital amine" (or "vitamine"), which was coined by Casimir Funk in 1912.

Most vitamins are actually "families" of chemically related isomers (i.e., vitamers) which cause same or similar metabolic impact (benefit) in most animals (including humans) that consume those vitamins. Some compounds are vitamins for certain species of animals but are not for certain other species.

In general, a vitamin is an organic compound required in tiny amounts (for the optimal growth, proper biological functioning, and maintenance of health of an organism). For example, vitamin K is utilized by plants in the photosynthesis process. For example, vitamin D enhances an animal's absorption and utilization of calcium (e.g., to build bones while growing).

Vitamins are commonly classified into two categories, the fat soluble and the water soluble. Vitamins A, D, E, and K are fat soluble whereas vitamin C (ascorbic acid) and members of the vitamin B complex group are water soluble. In general, the vitamins play catalytic and regulatory roles in the body's metabolism. Among the water-soluble vitamins, the B vitamins apparently function as coenzymes (nonprotein parts of enzymes). Vitamin C's coenzyme role, if any, has not been established. Part of the importance of vitamin C to the body may arise from its strong antioxidant action. The functions of the fat-soluble vitamins are less well understood. Some of them, too, may contribute to enzyme activity, and others are essential to the functioning of cellular membranes (on surface of cells).

Some vitamins act as transcription factors. Vitamin A is able to regulate the expression of certain genes in the embryos of mammals via one of its metabolites, retinoic acid. Those embryo cells contain nuclear receptors (which bring the retinoic acid "signal" from outside into the cell's nucleus) on their cell membrane surface. The retinoic acid then (via the nuclear receptors) regulates the expression of the genes that cause embryonic cell differentiation into complex body structures, such as legs and arms, of the growing embryo.

Vitamin D exerts its effects on genes via the vitamin D receptor, a nuclear receptor that binds to specific locations on the cell's DNA to thereby regulate applicable gene expression. During 2012, some researchers utilized receptor-binding mapping to link vitamin D deficiency to an increased risk for certain cancers (e.g., premenopausal breast cancer) and autoimmune diseases (e.g., rheumatoid arthritis, multiple sclerosis, lupus, etc.). See also *Enzyme, Catalyst, Coenzyme, Metabolism, Metabolite, Gene, Express, Beta carotene, Embryology, Retinoids, Protein, Cell, Receptors, Signaling, Choline, Signaling molecules, Signal transduction, Nuclear receptors, Nuclear hormone receptors, Lycopene, Lutein, Fats, Transcription factors, Species, Avidin, Vitamin E, Biotin, Tocopherols, Tocotrienols, Antioxidants, Inositol, Folic acid, Alternative splicing, Crohn's disease, Receptor-binding mapping.*

Vitamin E

Refers to a group of related, naturally occurring compounds consisting of TOCOPHEROL and TOCOTRIENOL "families." It is a fat-soluble vitamin with antioxidant properties (i.e., helps prevent lipids in body from breaking down). Vitamin E is especially effective at preventing oxidation of low-density lipoproteins (so-called bad cholesterol), whose oxidation products (e.g., beta hydroxycholesterol) can be deposited onto the interior walls of blood vessels (e.g., arteries) in the form of *plaque* which can result in the disease atherosclerosis and/or adversely increasing blood platelet aggregation (e.g., clotting).

Vitamin E occurs naturally in soybeans, cereal grains, and so on, so it can be considered a phytochemical. In 2000, the Institute of Medicine of the United States' National Academy of Sciences issued a report that called for an increase in the amount of vitamin E consumed each day, to improve citizens' health. See also *Vitamin, Oxidative stress, Antioxidants, Phytochemicals, Oxidation, Lipids, Cholesterol, Low-density lipoproteins (LDLP), Atherosclerosis, Plaque, Platelets, Phytochemicals, National Academy of Sciences (NAS), Tocopherols, Tocotrienols, Soybean plant.*

VNTRs

Acronym for Variable Number of Tandem Repeats. See *Variable number of tandem repeats (VNTRs).*

Volicitin

A chemical compound produced by Beet Armyworm Caterpillars (*Spodoptera exigua*) after they have consumed some linoleic acid (in plants they chew on, such as corn/maize). The body cells of Beet Armyworm caterpillars conjugate (i.e., chemically join together) the linoleic acid molecules onto glutamine molecules. The conjugated molecule, consisting of one linoleic acid (molecule) joined to one glutamine (molecule), is known as volicitin.

When Beet Armyworm caterpillars subsequently chew on corn/maize plants, some volicitin is inadvertently inserted by those caterpillars into the tissue of the corn (maize). That volicitin causes the corn (maize) plant to emit certain compounds known as *green leaf volatiles*, which

- Attract type(s) of wasps which are natural enemies of the Beet Armyworm, leading those wasps to attack those Beet Armyworm caterpillars (which are feeding on the maize/corn).
- Activate natural innate chemical defense systems within nearby corn (maize) plants.

See also *Linoleic acid, Corn, Glutamine, Cell, Green leaf volatiles, Octadecanoid/Jasmonate signal complex.*

Voltage-Gated Ion Channel

A transmembrane (i.e., extending from one side of a cell membrane to the other side) ion channel. Passage (e.g., of ions, atoms) through a voltage-gated ion channel is controlled (i.e., this channel is opened, or closed) by membrane potential (i.e., *electrical charge difference* between one side of the cell membrane and the other). See also *Ion channels, Ion, Cell, Membranes (of a cell), Plasma membrane.*

Volume Rendering

Refers to technologies utilized (e.g., in a confocal microscope, ultrasound imaging machines, data mining, etc.) to convert numerous "slices" (2D) of a given 3D volume (e.g., the inside of a living cell, a living organ within a body, etc.) into a compiled-together 3D (three-dimensional) image. See also *Data mining, Confocal microscopy, Cell.*

Vomitoxin

See *Fusarium, Mycotoxins.*

von Willebrand Factor

Abbreviated VWF, it is a very long molecule (polymer) that is made and secreted into the bloodstream by the endothelial cells of the human body. It exerts two different effects on the blood:

- Promotion of thrombosis, when needed.
- Following cleavage of the very long VWF molecule by a metalloproteinase, the now-shorter molecules promote homeostasis (e.g., by controlling any arterial bleeding).

See also *Thrombosis, Endothelial cells, Homeostasis, Metalloproteins, Enzyme, Polymer.*

VRN2 Gene

See *Vernalization.*

W

Walking

See *Chromosome walking.*

Water Activity (A_w)

A measure of the "free" *unbound* water (e.g., in a processed food product) that is available to sustain the growth of microorganisms (spoilage) and/or to sustain undesired chemical reactions (e.g., "staling" of baked food products).

Most bacteria are unable to grow in foods possessing a water activity below 0.90. Most yeasts and molds that cause spoilage cannot grow in foods possessing a water activity below 0.80. Sugars can be added to certain foods in order to increase A_w, as they "bind up" the (formerly) free water present. See also *Microorganism, Hydrophilic, Bacteria, Yeast, Penicillium.*

Water-Soluble Fiber

Food fiber (e.g., oat fiber, barley fiber, carbon fiber, soybean fiber, guar gum fiber) that dissolves in water. It absorbs low-density lipoproteins (LDLPs) in the intestine, before the fiber passes from the body; plus it inhibits absorption of LDLP by the body's intestinal walls due to increasing the viscosity of the intestine's contents. Those two effects thus lower the amount of "bad" cholesterol (i.e., LDLP can lead to hardening/blockage of arteries) in the body and thereby lower the risk of coronary heart disease (CHD).

Additional to those two effects, water-soluble fiber also absorbs/binds bile acid and causes it to be excreted along with that water-soluble fiber. That helps to lower cholesterol levels in the body (bloodstream) because the liver synthesizes ("manufactures") more bile acids (to replace those absorbed and removed by the fiber) from cholesterol.

Water-soluble fiber from oat bran is a polysaccharide in a form known as beta-glucan (β), composed entirely of glucose (molecular) units. Wheat and barley also contain some beta-glucan. U.S. FDA regulations also include gums, pectins, mucilages, and certain hemicelluloses in the category of water-soluble fiber. Soybean flour/meal is also a source of some water-soluble fiber.

In 1997, the U.S. FDA approved a (label) health claim that associates consumption of food products containing a minimum of 0.75 g of soluble beta-glucan oat fiber per serving with reduced blood cholesterol content and with risk of reduced CHD. In 1998, the U.S. FDA approved a (label) health claim that associates soluble fiber from psyllium husks with reduced risk of CHD.

See also *High-density lipoproteins (HDLPs), Low-density lipoproteins (LDLP), Polysaccharides, Glucose (GLc), Food and Drug Administration (FDA), Atherosclerosis, Coronary heart disease (CHD), Soybean meal, Soybean plant, High-mannogalactan soybeans, Amylose, High-amylose wheat, Cholesterol, Plaque.*

Water Use Efficiency

See *Drought resistance.*

Waxy Corn

Refers to corn (maize) hybrids that produce kernels in which the starch that is contained within those kernels is at least 99% amylopectin, versus the average of 72%–76% amylopectin in traditional corn starch. See also *Corn, Starch, Amylopectin.*

Waxy Wheat

Refers to varieties of wheat (*Triticum aestivum*) that produce a higher amylopectin content and thus a lower amylose content in the starch within their seeds than traditional varieties of wheat. For example, bread flour made from *waxy wheat* would contain 0%–3% amylose, versus 24%–27% amylose in bread flour made from traditional varieties of wheat. Because bread made from such waxy (i.e., lower amylose) wheat becomes firm at a much slower rate than bread made from traditional wheat varieties, bread made from waxy wheats would require less shortening (added to the flour) to keep that bread soft. See also *Wheat, Starch, Amylose, Amylopectin.*

WCR

Acronym for western corn rootworm. See also *Corn rootworm.*

Weak Interactions

The forces between atoms that are less strong than the forces involved in a covalent (chemical) bond (between two atoms). Weak interactions include ionic (chemical) bonds, hydrogen bonds, and van der Waals forces. See also *van der Waals forces, Hydrogen bonding.*

Weevils

A term that is utilized for a number of insects that consume grains (i.e., grown and used by man). Many of the weevils consume and proliferate in stored grains (e.g., the Indian meal moth *Plodia interpunctella*) and stored grain products (e.g., flour).

One example of a weevil is the insect known as the pea weevil, which lay their eggs on pea pods or dried peas. When the larvae hatch, they burrow into the pod and eat the peas inside.

The insect *Theocolax elegans* attacks the larvae of maize weevils (*Sitophilus granarius, Tribolium castaneum*), rice weevils (*Sitophilus oryzae*), and the lesser grain borer. Thus, it could potentially be added to grain storage bins (silos) as part of an integrated pest management program.

W

Research indicates that the "green" pesticide Spinosad is effective in controlling the lesser grain borer and other weevils. Spinosad and its registered trademark are owned by Dow Chemical Company. See also *Integrated pest management (IPM), Biotin, Avidin, Alpha amylase inhibitor-1, Spinosad, Spinosyns.*

Western Blot Test

A test that is performed on biological samples such as blood (after centrifugation to remove red blood cells from the blood) to *detect proteins* (e.g., AIDS-related antibodies or other biomarkers) individually. The western blotting procedure was first published in 1979; that published paper explained how to transfer protein molecules from a polyacrylamide gel electrophoresis (PAGE) to a sheet of nitrocellulose in order to obtain a replica of the original PAGE gel pattern of (segregated) protein molecules. Because those protein molecules are thereby then outside of the gel matrix, they can subsequently be detected via a specific-to-a-given-protein antibody (e.g., to confirm the presence within an initially mixed-protein sample of a particular protein).

For example, gel electrophoresis is used to separate the AIDS antigen proteins of killed (known) AIDS viruses. Next, the protein bands (resulting from the gel electrophoresis) are exposed to the blood being tested and (AIDS) antibodies stick to specific individual antigens (bands), which are then identified (as being present in the tested blood) via special dyes. See also *Biomarkers, Protein, Acquired immune deficiency syndrome (AIDS), Antibody, Antigen, Electrophoresis, Polyacrylamide gel electrophoresis (PAGE), Basophilic, Buffy coat (cells).*

Western Corn Rootworm

Latin name *Diabrotica virgifera virgifera* LeConte. See *Corn rootworm.*

WGA

Acronym for *whole-genome amplification.* See *Whole-genome amplification.*

WGS

Acronym for whole-genome sequencing. See *Whole-genome sequencing.*

WGSS

Acronym for *whole-genome shotgun sequencing.* See *Whole-genome shotgun sequencing.*

Wheat

Refers to a family of related small grains that are descended from the natural crossing of three Middle East grasses (*Triticum monococcum, Aegilops speltoides,* and *Triticum tauschii*) centuries ago. As a result, wheat's genome is triploid (i.e., it incorporates three complete sets of deoxyribonucleic acid [DNA]) and contains approximately 17 billion base pairs.

Wheat is historically an annual plant that can attain a height of 4 ft (1.2 m), although variations (e.g., shorter) have been bred.

The Latin name for traditional (bread) wheat is *Triticum aestivum* and for durum (pasta) wheat is *Triticum durum desf.*

Historically, wheat kernels have contained 15% or less protein. Most of the rest of the kernel is composed of starch (amylose and amylopectin). See also *Genome, Deoxyribonucleic acid (DNA), Base pair (bp), Hybridization (plant genetics), Triploid, Thermotolerant wheat, Wheat take-all disease, Wheat scab, Karnal bunt, Wheat head blight, Gluten, Glutenin, Protein, Starch, Amylose, High-amylose wheat, Amylopectin, Telethia controversia koon smut, Oxalate oxidase, Polyploid.*

Wheat Head Blight

See *Fusarium.*

Wheat Scab

See *Fusarium.*

Wheat Take-All Disease

A disease caused by *Gaeumannomyces graminis* var. *tritici* fungus that attacks wheat (*Triticum aestivum*) plant roots and causes dry rot and premature death of the plant.

Certain strains of *Brassica* plants and *Pseudomonas* bacteria produce compounds that can act as natural antifungal agents against the *Gaeumannomyces graminis* var. *tritici* fungus. For example, *Pseudomonas* strain AN5 produces gluconic acid that inhibits the wheat take-all fungus. See also *Fungus, Bacteria, Genetically engineered microbial pesticides (GEMP), Brassica, Allelopathy.*

Whiskers™

A trademarked method for inserting DNA (genes) into plants cells, so that those plant cells will then incorporate that new DNA and express the protein(s) coded-for by that DNA.

Developed by ICI Seeds Inc. (Garst Seed Company) in 1993, Whiskers™ is an alternative to other methods of inserting DNA into plant cells (e.g., the Biolistic® Gene Gun, *Agrobacterium tumefaciens,* the "shotgun" method); it consists of needlelike crystals ("whiskers") of silicon carbide.

The crystals are placed into a container along with the plant cells, then mixed at high speed, which causes the crystals to pierce the plant cell walls with microscopic "holes" (passages). Then the new DNA (gene) is added, which causes the DNA to flow into the plant cells. The plant cells then incorporate the new gene(s) and thus they have been genetically engineered. See also *Biolistic® gene gun, Agrobacterium tumefaciens, "Shotgun" method, Genetic engineering, Gene, Bioseeds, Coding sequence, Protein, Cell, Deoxyribonucleic acid (DNA).*

White Adipocytes

Refer to certain adipocyte (body fat) cells that function mainly as highly flexible energy "storehouses" that are "filled up" in times of calorie abundance for an organism. This "white" fat is stored in the form of lipid droplets, which are mobilized when energy is needed. Under certain conditions (organism in cold environment), white adipocytes can interconvert to become brown adipocytes via the mTORC1 molecular pathway, which is regulated by the Grb10 protein. Chronic excessive amounts of white adipose tissue in the body are associated with chronic conditions/diseases such as obesity and type 2 diabetes. See also *Brown adipocytes, Type II diabetes.*

White Fat Cells

See *White adipocytes.*

White Biotechnology

Term utilized in some countries to refer to *industrial* applications of genetic engineering. One example would be production of laccase in genetically engineered organisms. See also *Genetic engineering, Laccase, Ionic liquids.*

White Blood Cells

See *Leukocytes, Dendritic cells.*

White Corpuscles

See *Leukocytes.*

White Mold Disease

The common name that is used to refer to a plant disease that is caused under certain conditions (e.g., moist, humid) by the *Sclerotinia sclerotiorum* fungus.

In 1998, the United States Environmental Protection Agency approved one herbicide (COBRA™ owned by Valent Corporation), whose active ingredient is the chemical *Lactofin*, to be applied to soybean plants "at or near bloom stage" in order to trigger *systemic acquired resistance* (a sort of "immune response") in those soybean plants against white mold disease. Use of no-tillage crop production (methodology) for some crops helps to reduce the incidence of white mold disease. See also *Fungus, Systemic acquired resistance* (SAR), *No-tillage crop production, Soybean plant.*

Whole-Genome Amplification

Whole-genome amplification refers to a methodology that is required for sequencing an entire genome (i.e., the entire DNA of an organism), that of first "amplifying (i.e., making copies of) every sequence" within the selected organism's DNA.

Typically, the genome (i.e., DNA molecule) is first fragmented into numerous small, overlapping segments. Next, relevant chemicals are added that convert those overlapping segments into a "library" of DNA segments. Then, conventional polymerase chain reaction technique is utilized to amplify (i.e., make copies of) the entire library. See also *Deoxyribonucleic acid (DNA), Genome, sequence (of a DNA molecule), Genetic code, Sequencing (of DNA molecules), Organism, Polymerase chain reaction (PCR) technique, Library.*

Whole-Cell Patch-Clamp Recording

Developed by Bert Sakmann and Erwin Neher in 1976, this is a method for measuring the current/potential of a single ion channel.

Although the initial 1976 technique was slow and laborious (i.e., to "hook up" the cell to electrodes, to measure electrical potential), several companies have since developed easier/faster methodologies to measure the electrical potential across a cell's ion channel(s) as that cell is exposed to one or more pharmaceutical candidate compounds. See also *Cell, Ion, Ion channels, Target (of a therapeutic agent), High-throughput screening (HTS).*

Whole-Genome Association

Refers to the use of gene expression analysis across tissue (e.g., blood) samples taken from a very large number of organisms (e.g., people), in order to find out which genes (SNPs) are associated with a given disease or with a given phenomenon (e.g., short term memory).

For example, during 2005 whole-genome association was utilized to find that variations (polymorphisms) of two human genes account for approximately 75% of the risk for people to develop the disease known as macular degeneration. See also *Gene expression analysis, Gene, Single-nucleotide polymorphisms (SNPs), Organism, Tiling arrays, Charm.*

Whole-Genome Sequencing

Refers to determining the DNA sequences of an individual organism's entire genome.

See also *Deoxyribonucleic acid (DNA), Genome, Sequencing (of DNA molecules), Sequence (of a DNA molecule), Shotgun sequencing.*

Whole-Genome Shotgun Sequencing

Abbreviated WGSS, it was invented by J. Craig Venter and Hamilton O. Smith during the mid-1990s. WGSS is a technology for rapid sequencing of (eucaryotic and procaryotic) DNA, in which an organism's genome (DNA) is first fragmented ("broken up"), and then randomly selected pieces of the DNA are individually sequenced. Those individual pieces' sequences must subsequently be "bridged" (i.e., "assembled" in an overlapping end-by-end pattern) in order to assemble a complete map (e.g., of an organism's chromosome or genome).

WGSS sequencing data can be utilized to identify genetic markers in plants to facilitate phylogenetic studies of the relationship between various species within a genus. See also *Sequencing (of DNA molecules), Deoxyribonucleic acid (DNA), Sequence (of a DNA molecule), Gene, Genome, DNA "bridges," Chromosomes, Genetic marker, Genetic map, Chromosome walking, Eucaryote, Procaryotes, Whole-genome sequencing, Binning, Phylogenetic profiling.*

Wide Cross

Developed during the 1930s, this refers to certain plant breeding technologies/techniques that are utilized to cross two plant species that would not normally cross in nature. For example, during 2014, Lijuan Qiu and Rongxia Guan discovered a salt tolerance gene present in the DNA of wild type soybean plants that are the ancestors of today's domesticated soybean (*Glycine max* (L.) Merrill) varieties. Because today's domesticated soybean varieties do not possess that salt tolerance gene, a soybean breeder wanting to create a modern soybean variety that would grow well in salty soil could utilize a wide cross between a modern soybean variety (germplasm) and one of those salt-tolerant wild type soybean species. See also *Traditional breeding methods, Deoxyribonucleic acid (DNA), Gene, Salt tolerance, Wild type, Soybean plant, Germplasm, Tissue culture, Species.*

Wide Spectrum

See *Gram stain.*

W

Wild Type

The traditional/historical form of an organism as it is ordinarily encountered in nature, in contrast to domesticated strains, natural mutant, or laboratory mutant individuals (organisms). One example of a measurable difference between the two types is that wild strains of animals respond to the presence of EMF fields (e.g., weak magnetic fields such as those generated near power transmission cables), and laboratory strains of the same animals do not.

The term "wild type" is also sometimes utilized to refer to the "natural" version (allele) of a given gene. See also *Strain*, *Gene*, *Allele*, *Mutant*, *Phenotype*, *Genotype*, *Psoralene*, *Solanine*.

Wnt Signaling

See *Reactive oxygen species*, *Signaling*.

Wobble

The ability of the third base in a tRNA (transfer RNA) anticodon to hydrogen bond with any of two or three bases at the 3′ end of a codon. This wobble (nonspecificity) allows a single tRNA species to recognize several different codons. See also *Transfer RNA* (*tRNA*), *Codon*, *Base pair* (*bp*), *Redundancy*.

World Organization for Animal Health

See *International Office of Epizootics* (*OIE*).

WUE

Acronym for water use efficiency. See *Drought resistance*.

X

X Chromosome

A sex chromosome that usually occurs paired in each female cell and single (i.e., unpaired) in each male cell in those species in which the male typically has two unlike sex chromosomes (e.g., humans). See also *Chromosomes*, *Imprinting*, *Epigenetic*, *DNA methylation*.

X-Chromosome Inactivation

Refers to the genomic imprinting process (i.e., the epigenetic process in which an offspring's X-chromosome gene expression is ensured to be a *mixture* of the genes contributed by *each* of its parents).

See also *X chromosome*, *Long noncoding RNAs*, *Genomic imprinting*, *Epigenetic*, *Gene*, *Express*.

X Receptors

See *Liver X receptors (LXR)*, *Farnesoid X receptors (FXR)*, *Retinoid X receptors (RXR)*.

X-Ray Crystallography

Invented in 1915 by William and Lawrence Bragg, based on Max von Laue's 1912 discovery that crystals diffract x-rays, it is the use of diffraction patterns produced by x-ray scattering from crystals (of a given material's molecules) to determine the 3D atomic structure of the molecules.

First used for biological molecules in 1958 by Max Perutz and John Cowdery Kendrew (to determine a myoglobin molecule structure), x-ray crystallography data are also utilized today for efforts such as rational drug design. See also *Protein*, *Configuration*, *Conformation*, *Tertiary structure*, *Protein folding*, *Rational drug design*.

Xanthine Oxidase

An enzyme responsible for production of free radicals in the body. See also *Enzyme*, *Human superoxide dismutase (hSOD)*.

Xanthophylls

A "family" of carotenoids (i.e., plant-produced pigments that act as protective antioxidants in photosynthetic plants and in the bodies of animals that consume those carotenoids). Among other plants, xanthophylls are produced by yellow carrots.

Consumption of xanthophylls by humans and animals assists development of healthy eye tissue.

Research indicates that consumption of xanthophylls by humans helps prevent lung cancer and some other cancers. See also *Carotenoids*, *Antioxidants*, *Oxidative stress*, *Cancer*.

Xenobiotic Compounds

Those compounds (e.g., veterinary drugs, agrochemical herbicides) that are designed to be used in an ecosystem composed of more than one species. For example, herbicides intended to kill weeds but leave commercial crops undamaged, or veterinary drugs that are intended to kill parasitic worms but leave the host livestock unharmed. See also *Species*.

Xenogeneic Organs

From the Greek word *xenos*, meaning "stranger." Xenogeneic literally means "strange genes." Refers to genetically engineered (e.g., "humanized") organs that have been grown within an animal of another species. For example, several companies are working to engineer and grow—inside swine—a number of organs to be transplanted into humans that need those organs (e.g., due to loss of their own organs via disease or accident). If successful, this would free human organ transplant recipients from having to continually use immunosuppressive drugs in order to keep their body from "rejecting" the new organ. See also *Immunosuppressive*, *Graft-versus-host disease (GVHD)*, *Cyclosporin*, *Major histocompatibility complex (MHC)*, *Genetic engineering*.

Xenogenesis

The (theoretical) production of offspring that is genetically different from and genotypically unrelated to either of the parents or offspring. See also *Genotype*, *Transgenic*, *Heredity*, *Genetics*, *Meiosis*, *Genetic code*.

Xenogenetic Organs

See *Xenogeneic organs*.

Xenogenic Organs

See *Xenogeneic organs*.

Xenograft

See *Xenotransplant*.

X

Xenotransplant

From the Greek word *xenos*, meaning "stranger." Xenotransplant is the implantation of an organ or limb from one species to another organism in a different species. When performed in animals, "rejection" of the transplant by the recipient's immune system is a common response. See also *Graft-versus-host disease (GVHD)*, *Xenogeneic organs.*

Xenotropic Virus

A virus that can grow/reproduce within one or more species *other* than its normal host species. See also *Virus*, *Zoonoses*, *Species.*

Xylanase

A term that is used to refer to a category of enzymes that catalyze the chemical reaction in which xylan (one of the components of cellulosic plant cell walls) is "broken" into molecular pieces.

When appropriate xylanases are added to a wood pulp solution, they help break down the xylan (in tree cell walls) and thereby facilitate more efficient bleaching of the pulp in preparation for paper manufacturing. See also *Enzyme*, *Catalyst.*

Y

Y Chromosome

A sex chromosome that is characteristic of male zygotes (and cells) in species in which the male typically has two unlike sex chromosomes. See also *Chromosomes.*

Y2H

Acronym for *yeast two-hybrid system*. See *Yeast two-hybrid system.*

YAC

See *Yeast artificial chromosomes (YAC).*

Yeast

A fungus of the family Saccharomycetaceae that is used by man especially in the making of alcoholic liquors and as a leavening agent in bread making.

Some strains of yeast cells are also commonly used in bioprocesses, because they are relatively simple to genetically engineer (via recombinant DNA) and relatively easy to propagate (via fermentation) to yield desired products (e.g., proteins). See also *Fungus, Strain, Prebiotics, Fermentation, Genetic engineering, Yeast artificial chromosomes (YAC), Recombinant DNA (rDNA), Indole-3-acetic acid.*

Yeast Artificial Chromosomes

Pieces of DNA (usually human DNA) that have been cloned (made) inside living yeast cells. While most bacterial vectors cannot carry DNA pieces that are larger than 50 base pairs, yeast artificial chromosomes (YACs) can typically carry DNA pieces that are as large as several hundred base pairs. See also *Yeast, Chromosomes, Human artificial chromosomes (HAC), Bacterial artificial chromosomes (BAC), Arabidopsis thaliana, Deoxyribonucleic acid (DNA), Clone (a molecule), Vector, Base pair (bp), Mega-yeast artificial chromosomes (mega YAC).*

Yeast Episomal Plasmid

A cloning vehicle used for introduction of constructions (i.e., genes and pieces of genetic material) into certain yeast strains at high copy number. Yeast episomal plasmid (YEP) can replicate in both *Escherichia coli* and certain yeast strains. See also *Plasmid, Cassette, Clone (an organism), Gene, Genetic engineering, Escherichia coliform (E. coli), Copy number.*

Yeast Two-Hybrid System

The oldest of the two-hybrid systems, utilized to analyze protein–protein interactions.

Discovered in 1989 by Stanley Fields, it utilizes the fact that the *gene activation system* for *Saccharomyces cerevisiae* yeast's *GAL4* gene is activated by a transcriptional factor possessing *two* distinctly separate functional domains. One of those two domains must interact with a second domain (i.e., in a genetically engineered yeast-created fusion protein containing a portion of the "known" protein) in order to cause transcription of the *GAL4* gene. *GAL4* is required to initiate expression of proteins that are central to galactose metabolism in that yeast. See also *Two-hybrid systems, Protein, Protein interaction analysis, Gene, Gene expression, Genetic engineering, Transcription, Transcriptional activator, Coding sequence, Domain (of a protein), Fusion protein, Galactose (gal), Metabolism.*

YFP

Acronym for *yellow fluorescent protein.* See *Visible fluorescent proteins.*

Yield Environment

Refers to precision agriculture software products that utilize a combination of public soil databases (including soil depth, texture, soil organic matter, and water-holding capacity), company proprietary (analytics, high-resolution field elevation data, field topography, and watersheds/hydrogeology), and electrical conductivity soil testing (basis: topsoil depth, pH, salt concentrations, and available water-holding capacity) to divide farm fields into contiguous *yield environments* (subportions of field, also sometimes known as management zones or ERU maps) where crops planted there will respond positively to farmer management decisions regarding crop inputs such as amounts and timing of fertilizer applied to field, amounts and timing of irrigation water applied to field, and number of crop seeds planted per hectare.

For example, such precision agriculture software might thereby recommend far less irrigation water be applied by the farmer to a *yield environment* located in a low-elevation, heavy-soil-type area of farm field that is naturally wet throughout the growing season due to the field's hydrogeology, plus a higher number of crop seeds planted per hectare (of a wet-environment-tolerant crop variety). The precision agriculture software would recommend different inputs/rates for a *yield environment* located in a sandy soil, higher-elevation area of the farm field (e.g., more irrigation water applied, lower number of crop seeds planted per hectare, of a drought-tolerant crop variety). Such properly integrated precision management of *yield environments* would maximize the field's crop yield while minimizing its consumption of inputs such as fertilizer, irrigation water, and crop seeds. See also *Drought-tolerance trait, ERU maps, Management zones, pH.*

Y

YSTR DNA

Refers to *Y chromosome short tandem repeat* DNA, which is DNA found only in males (since they are the only humans with Y chromosome).

Utilizing certain specific *STR markers* (i.e., based on detectable repeated DNA sequences in Y chromosome), YSTR DNA can be utilized for some genetic studies (e.g., to track parentage) and for some forensic efforts (e.g., crime evidence determination).

SNPs within such YSTR DNA can be utilized to increase the "completeness" of some SNP maps. See also *Deoxyribonucleic acid (DNA)*, *Y chromosome*, *Sequence (of a DNA molecule)*, *Marker (DNA sequence)*, *Single-nucleotide polymorphisms (SNPs)*, *SNP map.*

Z

Z-DNA

A left-handed helix (molecular structure) of DNA, in contrast to A-DNA and B-DNA that are right-handed helix structures. The difference is in the direction of the double-helix twist (Z-DNA is an alternative form of B-DNA in which the double helix unwinds to the left instead of the right). Z-DNA has the most base pairs per turn (in the helix) and so has the least twisted structure; it is very "skinny" and its name is taken from the zigzag path that the sugar-phosphate "backbone" follows along the helix. This is quite different from the smoothly curving path of the backbone of *B-DNA*. The Z-form of DNA has been found in polymers that have an alternating purine–pyrimidine sequence.

One possible biological importance of Z-DNA is that it is much more stable at lower salt concentrations, and there is a possibility that the Z-DNA form (of DNA within cells) is the cause of certain diseases (e.g., certain cancers).

During 2000, Jonathan Chaires/Waldemar Priebe/John Trent showed that *WP 900* (i.e., the enantiomer of daunorubicin, a natural chemical compound that inhibits cancer) binds tightly (and selectively) to a Z-DNA polymer. See also *Cell, Deoxyribonucleic acid (DNA), B-DNA, Helix, Double helix, A-DNA, Purine, Base pair (bp), Pyrimidine, Enantiomers, Cancer.*

Z-Ring

See *Tubulin.*

Zearalenone

One of the mycotoxins (i.e., toxins produced by a fungus), it causes reproductive difficulties in swine (e.g., reduced sperm production, halting of estrus) when consumed by animals (e.g., in contaminated grain such as corn/maize). It causes a variety of adverse health impacts when consumed by some other species of animals.

Zearalenone is produced by certain strains of *Fusarium* fungi (e.g., *Fusarium culmorum, Fusarium graminearum, Fusarium crookwellense, Fusarium oxysporum*) when climate (e.g., moisture and temperature) conditions during the grain growing season, combined with entry points (e.g., holes chewed into the grain plants by insects), facilitate growth of those *Fusarium* strains in grain. See also *Toxin, Mycotoxins, Fungus, Strain, Fusarium, Lactonase, Fusarium graminearum.*

Zeaxanthin

A carotenoid (i.e., "light harvesting" compound utilized in photosynthesis) that is naturally produced in Brussels sprouts, summer squash, maize, avocado, green beans, and dark green leafy vegetables. Zeaxanthin is also naturally present within the retina of the human eye.

Zeaxanthin is a phytochemical/nutraceutical whose consumption by humans has been shown to reduce the risk of *age-related macular degeneration*, a leading cause of blindness in older people. See also *Carotenoids, Photosynthesis, Phytochemicals, Nutraceuticals.*

Zebra Fish

The fish *Danio rerio*. See also *Model organism.*

ZFN

Acronym for *zinc finger nuclease*. See *Zinc finger nuclease.*

ZFP

Acronym for *zinc finger protein*. See *Zinc finger proteins.*

Zinc Finger Nuclease Gene Insertion

See *Zinc finger nuclease.*

Zinc Finger Nuclease Point Mutation

See *Zinc finger nuclease.*

Zinc Finger Nuclease

Abbreviated ZFN, it refers to a category of nuclease molecules bearing at least one "finger-shaped" molecular appurtenance that acts to either repress or activate transcription (i.e., of the gene the "finger" touches within the DNA molecule). For example, nuclear receptors contain two zinc fingers that target those receptors to *hormone response elements* (i.e., specific DNA sequences that initiate the activation of applicable gene(s) for that receptor).

Manmade ZFNs were invented in the mid-1990s by Carlos F. Barbas III and are artificial constructs made of two types of protein: a "zinc-finger" structure that can be designed to bind to a specific short DNA sequence and a nuclease enzyme that will cut DNA at that binding site in a way that cells can't repair easily.

Thus, ZFNs can be specifically constructed by man to cause DNA-site-specific modifications, which serve to

- Introduce a new genetic trait (gene) to the organism
- Introduce a point mutation (e.g., within an existing gene)
- "Knock out"/silence an existing (e.g., wild type) trait, by disrupting that particular existing gene

ZFNs could also potentially be utilized in *functional genomics* (i.e., to study the specific "function" of a given gene).

ZFNs possess two domains:

- A "DNA recognition domain" that locates and binds itself to a specific DNA sequence
- A "functional domain" that initiates (or inhibits) transcription

ZFNs are also sometimes involved in DNA repair. See also *Nuclease, Nuclear receptors, Protein, Gene, Transcription, Repression (of gene transcription), Promoter, Deoxyribonucleic acid (DNA), Functional genomics, Domain (of a protein), Point mutation, Sequence (of a DNA molecule), Initiation factors, Inhibition, Transcription factors, Editing, DNA repair, Trait, Organism, Knockout, Knockout (gene), Silencing, Wild type.*

Zinc Finger Proteins

Protein molecules (transcription factors) bearing at least one "finger-shaped" molecular appurtenance that acts to either repress or activate transcription (i.e., of the gene the "finger" touches within the DNA molecule). For example, nuclear receptors contain two zinc fingers that target those receptors to *hormone response elements* (i.e., specific DNA sequences that initiate the activation of applicable gene(s) for that receptor).

Zinc finger DNA-binding proteins possess two domains:

- A "DNA recognition domain" that locates and binds itself to a specific DNA sequence
- A "functional domain" that initiates (or inhibits) transcription

Zinc finger proteins are also sometimes involved in DNA repair.

During 2000, Yen Choo and Aaron Klug developed a system of man-made "zinc fingers" (i.e., zinc-coordinated protein-based compounds bearing "fingers" that can be manufactured to specifically target virtually any desired 18-base-pair sequence within a DNA molecule). Using this technology system (e.g., fused with a nuclease), scientists are able to insert a new segment of DNA at almost any selected spot in DNA within cells, thereby

- Creating a new gene (e.g., and thus a commercially valuable trait) in an organism
- Creating a point mutation within a known gene (e.g., to facilitate research)
- Knocking out a particular gene
- "Fixing" a DNA-sequence defect within a particular gene (e.g., thus enabling researchers to determine if that specific defect is the cause of a genetic disease)

Such zinc finger proteins could also potentially be utilized in *functional genomics* (i.e., to study the specific "function" of a given gene). See also *Cell, Nucleus, Nuclear receptors, Protein, Gene, Trait, Transcription, Repression (of gene transcription), Promoter, Deoxyribonucleic acid (DNA), Functional genomics, Domain (of a protein), Sequence (of a DNA molecule), Initiation factors, Inhibition, Transcription factors, Editing, DNA repair, Point mutation, Gene repair (done by man), Gene repair (natural), Editing, Knockout, Site-directed nucleases.*

ZKBS (Central Committee on Biological Safety)

The advisory body on safety in gene-splicing labs and plants for the German Government's Ministry of Health. It is the German counterpart of the American Government's Recombinant DNA Advisory Committee, Australia's Genetic Manipulation Advisory Committee, Brazil's National Biosafety Commission, and the Kenya Biosafety Council.

The ZKBS is composed of 10 experts from the biology and ecology sectors, trade union representatives, plus representatives from the industrial sector and environmental pressure groups. The ZKBS advises the Ministry of Health and the individual German States (Länder), which regulate all recombinant DNA (i.e., gene-splicing) activities in Germany. See also *Genetic Manipulation Advisory Committee (GMAC), CTNBio, Kenya Biosafety Council, Gene Technology Office, Recombinant DNA Advisory Committee (RAC), Genetic engineering, Recombinant DNA (rDNA), Recombination, Biotechnology, Indian Department of Biotechnology, Commission of Biomolecular Engineering.*

Zn Finger Nuclease

Abbreviation for zinc finger nuclease. See *Zinc finger nuclease.*

Zoonoses

Diseases that are communicable from animals to humans.

Examples would include the diseases anthrax (caused by *Bacillus anthracis*) and tularemia (caused by *Francisella tularensis*). See also *International Office of Epizootics (OIE).*

Zoonotic

See *Zoonoses.*

Zygote

A fertilized egg formed as a result of the union of the male (sperm) and female (egg) sex cells. The zygote gives rise to the placenta (lining of the uterus) in addition to growing into (adult organism) body. See also *X chromosome, Y chromosome, Telomeres, Gamete, Organism, Cell, Cell differentiation.*

Zyme Systems

Chemical reactions characterized by the presence of an inactive precursor of an enzyme. The enzyme is activated via another enzyme that normally removes an extra piece of peptide chain at a physiologically appropriate time and place. See also *Zymogens, Fibrin, Prodrug therapy, Digestion (within organisms), Nanobodies.*

Zymogens

The enzymatically inactive precursors of certain proteolytic enzymes. The enzymes are inactive because they contain an extra piece of peptide chain. When this peptide is hydrolyzed (clipped away) by another proteolytic enzyme, the zymogen is converted into the normal, active enzyme. Zymogens are sometimes referred to as *proenzymes.*

The reason for the existence of zymogens may be to protect the cell, its machinery, and/or the place of manufacture within the cell from the potentially harmful or lethal effects of an active, proteolytic enzyme. In other words, the strategy is to activate the enzyme only when and especially where it is needed. See also *Enzyme, Proteolytic enzymes, Fibrin, Zyme systems, Lipoprotein-associated coagulation (clot) inhibitor (LACI), Prodrug therapy, Nanobodies.*

An environmentally friendly book printed and bound in England by www.printondemand-worldwide.com

PEFC Certified

This product is from sustainably managed forests and controlled sources

www.pefc.org

PEFC/16-33-415

This book is made of chain-of-custody materials; FSC materials for the cover and PEFC materials for the text pages.